Volvo 740 & 760 Owners Workshop Manual

Matthew Minter

Models covered
Volvo 740 & 760 models with petrol engines, including Turbo & special/limited editions
1986 cc, 2316 cc & 2849 cc

Does not cover Diesel engine models, or 2.3 litre 16-valve DOHC engine

(1258-2X4)

ABCDE
FGHIJ
KLMNO
P

2

Haynes Publishing
Sparkford Nr Yeovil
Somerset BA22 7JJ England

Haynes North America, Inc
861 Lawrence Drive
Newbury Park
California 91320 USA

Acknowledgements

Thanks are due to Champion Spark Plug, who supplied the illustrations showing spark plug conditions, to Holt Lloyd Limited who supplied the illustrations showing bodywork repair, and to Duckhams Oils, who provided lubrication data. Certain other illustrations are the copyright of AB Volvo, and are used with their permission. Thanks are also due to Sykes-Pickavant Limited, who provided some of the workshop tools, and to all those people at Sparkford who helped in the production of this manual.

© Haynes Publishing 1995

A book in the **Haynes Owners Workshop Manual Series**

Printed by J. H. Haynes & Co. Ltd., Sparkford, Nr. Yeovil, Somerset BA22 7JJ, England

All rights reserved. No part of this book may be reproduced or transmitted in any form or by any means, electronic or mechanical, including photocopying, recording or by any information storage or retrieval system, without permission in writing from the copyright holder.

ISBN 1 85960 070 0

British Library Cataloguing in Publication Data
A catalogue record for this book is available from the British Library.

We take great pride in the accuracy of information given in this manual, but vehicle manufacturers make alterations and design changes during the production run of a particular vehicle of which they do not inform us. No liability can be accepted by the authors or publishers for loss, damage or injury caused by any errors in, or omissions from, the information given.

Restoring and Preserving our Motoring Heritage

Few people can have had the luck to realise their dreams to quite the same extent and in such a remarkable fashion as John Haynes, Founder and Chairman of the Haynes Publishing Group.

Since 1965 his unique approach to workshop manual publishing has proved so successful that millions of Haynes Manuals are now sold every year throughout the world, covering literally thousands of different makes and models of cars, vans and motorcycles.

A continuing passion for cars and motoring led to the founding in 1985 of a Charitable Trust dedicated to the restoration and preservation of our motoring heritage. To inaugurate the new Museum, John Haynes donated virtually his entire private collection of 52 cars.

Now with an unrivalled international collection of over 210 veteran, vintage and classic cars and motorcycles, the Haynes Motor Museum in Somerset is well on the way to becoming one of the most interesting Motor Museums in the world.

A 70 seat video cinema, a cafe and an extensive motoring bookshop, together with a specially constructed one kilometre motor circuit, make a visit to the Haynes Motor Museum a truly unforgettable experience.

Every vehicle in the museum is preserved in as near as possible mint condition and each car is run every six months on the motor circuit.

Enjoy the picnic area set amongst the rolling Somerset hills. Peer through the William Morris workshop windows at cars being restored, and browse through the extensive displays of fascinating motoring memorabilia.

From the 1903 Oldsmobile through such classics as an MG Midget to the mighty 'E' type Jaguar, Lamborghini, Ferrari Berlinetta Boxer, and Graham Hill's Lola Cosworth, there is something for everyone, young and old alike, at this Somerset Museum.

Haynes Motor Museum

Situated mid-way between London and Penzance, the Haynes Motor Museum is located just off the A303 at Sparkford, Somerset (home of the Haynes Manual) and is open to the public 7 days a week all year round, except Christmas Day and Boxing Day.

Telephone 01963 440804.

Contents

	Page
Acknowledgements	2
About this manual	5
Introduction to the Volvo 740 and 760	5
General dimensions, weights and capacities *(also see Chapter 13, page 326)*	6
Jacking, towing and wheel changing	7
Buying spare parts and vehicle identification numbers	9
General repair procedures	11
Tools and working facilities	12
Safety first!	14
Routine maintenance *(see also Chapter 13, page 326)*	15
Recommended lubricants and fluids	21
Conversion factors	22
Fault diagnosis	23
Chapter 1 Engine *(also see Chapter 13, page 326)*	27
Chapter 2 Cooling, heating and air conditioning systems *(also see Chapter 13, page 326)*	78
Chapter 3 Fuel and exhaust systems *(also see Chapter 13, page 326)*	96
Chapter 4 Ignition system *(also see Chapter 13, page 326)*	142
Chapter 5 Clutch	158
Chapter 6 Manual gearbox, overdrive and automatic transmission *(also see Chapter 13, page 326)*	164
Chapter 7 Propeller shaft *(also see Chapter 13, page 326)*	200
Chapter 8 Rear axle *(also see Chapter 13, page 326)*	205
Chapter 9 Braking system *(also see Chapter 13, page 326)*	209
Chapter 10 Steering and suspension *(also see Chapter 13, page 326)*	229
Chapter 11 Bodywork and fittings *(also see Chapter 13, page 326)*	247
Chapter 12 Electrical system *(also see Chapter 13, page 326)*	276
Chapter 13 Supplement: Revisions and information on later models	326
Index	391

Spark plug condition and bodywork repair colour section between pages 32 and 33

Volvo 760 Turbo Saloon

Volvo 740 Turbo Estate

About this manual

Its aim

The aim of this manual is to help you get the best value from your vehicle. It can do so in several ways. It can help you decide what work must be done (even should you choose to get it done by a garage), provide information on routine maintenance and servicing, and give a logical course of action and diagnosis when random faults occur. However, it is hoped that you will use the manual by tackling the work yourself. On simpler jobs it may even be quicker than booking the car into a garage and going there twice, to leave and collect it. Perhaps most important, a lot of money can be saved by avoiding the costs a garage must charge to cover its labour and overheads.

The manual has drawings and descriptions to show the function of the various components so that their layout can be understood. Then the tasks are described and photographed in a step-by-step sequence so that even a novice can do the work.

Its arrangement

The manual is divided into thirteen Chapters, each covering a logical sub-division of the vehicle. The Chapters are each divided into Sections, numbered with single figures, eg 5; and the Sections into paragraphs (or sub-sections), with decimal numbers following on from the Section they are in, eg 5.1, 5.2, 5.3 etc.

It is freely illustrated, especially in those parts where there is a detailed sequence of operations to be carried out. There are two forms of illustration: figures and photographs. The figures are numbered in sequence with decimal numbers, according to their position in the Chapter – eg Fig. 6.4 is the fourth drawing/illustration in Chapter 6. Photographs carry the same number (either individually or in related groups) as the Section or paragraph to which they relate.

There is an alphabetical index at the back of the manual as well as a contents list at the front. Each Chapter is also preceded by its own individual contents list.

References to the 'left' or 'right' of the vehicle are in the sense of a person in the driver's seat facing forwards.

Unless otherwise stated, nuts and bolts are removed by turning anti-clockwise, and tightened by turning clockwise.

Vehicle manufacturers continually make changes to specifications and recommendations, and these, when notified, are incorporated into our manuals at the earliest opportunity.

We take great pride in the accuracy of information given in this manual, but vehicle manufacturers make alterations and design changes during the production run of a particular vehicle of which they do not inform us. No liability can be accepted by the authors or publishers for loss, damage or injury caused by any errors in, or omissions from, the information given.

Project vehicles

The vehicles used in the preparation of this manual, and which appear in many of the photographic sequences, were: a 760 Turbo Saloon, a 760 GLE Saloon, and a 740 GL Estate.

Introduction to the Volvo 740 and 760

The Volvo 760 Saloon was introduced to the UK market in July 1982, followed by the 740 Saloon in October 1984. Estate versions became available for the 1986 model year. The models represent the top of the Volvo range; besides the solidity and attention to safety characteristic of the marque, they are luxuriously equipped.

Engines available in the 760 range are a 2.8 litre V6 and a turbocharged 2.3 litre four-cylinder, both with fuel injection. The turbocharged engine is also available in the 740 range, other options being the same engine without the turbo, with either fuel injection or a carburettor.

Both manual and automatic transmissions are available throughout the range. The manual gearbox may be 5-speed, or 4-speed plus overdrive. The automatic transmission fitted to UK models may be 4-speed, or 3-speed plus overdrive. Drive is taken to the rear wheels via a traditional live rear axle. A limited slip differential is available as an optional extra.

Braking is by discs all round, the handbrake acting on separate drums on the rear wheels. Anti-lock braking (ABS) is available on later models, sometimes in conjunction with electronic traction control (ETC). Together these two systems offer the best possible handling and safety characteristics on loose or slippery surfaces. Steering is power-assisted on all models.

Changes to the range in later model years include the introduction of a 2-litre fuel injected engine to the 740 range, and provision of independent rear suspension on later 760 Saloons. These and other changes are covered in Chapter 13 of the manual.

The home mechanic will find the vehicles well-constructed and pleasant to work on, provided that tools and facilities appropriate to the vehicle are available.

General dimensions, weights and capacities

For modifications, and information applicable to later models, see Supplement at end of manual

Dimensions
Wheelbase .... 2.770 m (9ft 1.1 in)
Overall length .... 4.785 m (15 ft 8.4 in)
Overall width .... 1.760 m (5 ft 9.3 in)
Track .... 1.460 m (4 ft 9.5 in)
Overall height .... 1.430 m (4 ft 8.3 in)

Weights
Kerb weight (depending on equipment):
    740 .... 1270 to 1460 kg (2800 to 3219 lb)
    760 .... 1330 to 1500 kg (2932 to 3307 lb)
Gross vehicle weight, gross train weight etc .... See type designation plate (under bonnet)

Capacities (approx)
Engine oil (drain and refill, including filter change):
    B23/B230 .... 3.85 litres (6.8 pints) – plus 0.6 litre (1 pint) for Turbo oil cooler if drained
    B28 .... 6.5 litres (11.4 pints)
Cooling system:
    B23/B230 .... 9.5 litres (16.7 pints)
    B28 .... 10.0 litres (17.6 pints)
Fuel tank .... 60 or 82 litres (13.2 or 18.0 gallons) depending on model and year
Manual gearbox:
    M46 (4-speed plus overdrive) .... 2.3 litres (4.0 pints)
    M47 (5-speed) .... 1.3 litres (2.3 pints)
Automatic transmission (drain and refill):
    AW 71 .... 3.9 litres (6.9 pints)
    ZF 4HP 22 .... 2.0 litres (3.5 pints)
Rear axle:
    1030 .... 1.3 litres (2.3 pints)
    1031 .... 1.6 litres (2.8 pints)

Jacking, towing and wheel changing

Jacking

Use the jack supplied with the vehicle only for wheel changing during roadside emergencies. For repair or maintenance, use a pillar jack or trolley jack under one of the jacking points. To raise both rear wheels it is permissible to jack up under the final drive casing, but use a block of wood as an insulator. To raise both front wheels, remove the engine undertray and place the jack below the centre of the front axle crossmember.

If the whole vehicle is being raised on a four-point workshop lift, place the lifting arms as shown in the illustration. If the front lifting arms are placed below the jacking points, the vehicle may become nose heavy.

Never venture under a vehicle supported solely by a jack – always supplement the jack with axle stands. These may be placed under the jacking points, the rear axle tube or the front crossmember. Do not use four axle stands under the jacking points, however – as with the four-point light, the vehicle could become nose heavy.

For many procedures, an alternative to jacking is to use ramps or to place the vehicle over a pit. Make sure that ramps and axle stands are adequately rated for the weight of the vehicle and in good condition.

Towing

Towing eyes are provided at both front and rear of the vehicle (photos). The rear towing eye should be used only for emergency towing of another vehicle: for trailer towing a properly fitted towing bracket is required.

Vehicles with automatic transmission must not be towed further than 20 miles (30 km) or faster than 12 mph (20 km/h). If these conditions cannot be met, or if transmission damage has already occurred, the propeller shaft must be removed or the vehicle towed with its rear wheels off the ground.

When being towed, insert the ignition key and turn it to position II. This will unlock the steering and allow lights and direction indicators to be used. If the engine is not running, greater effort will be required to operate steering and brakes.

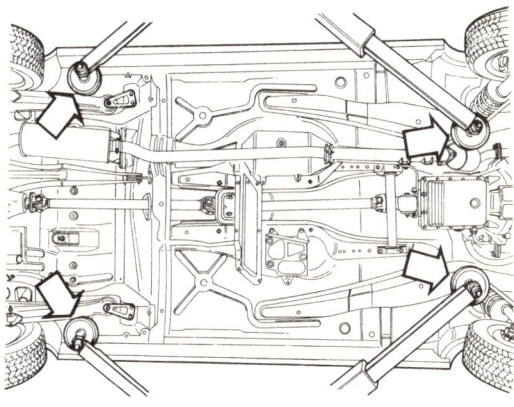

Support points (arrowed) for a four-point lift, or when using four axle stands

Prise out the cover plate ...

... for access to the front towing eye

Jacking, towing and wheel changing

Wheel changing

Park on a firm flat surface if possible. Apply the handbrake and engage reverse gear or P. Chock the wheel diagonally opposite the one being removed.

Remove the wheel trim, when applicable, for access to the wheel nuts. Prise the trim off if necessary using a screwdriver. Slacken the wheel nuts half a turn each using the wheelbrace (photos).

If the car is fairly new, the wheels and tyres will have been balanced on the vehicle during production. To maintain this relationship, mark the position of the wheel relative to the hub. (This is not necessary if the tyre is to be removed for repair or renewal, since the balance will inevitably be altered). Some wheels are positively located by a spigot on the hub, again making marking unnecessary.

Engage the jack lead in the jacking point nearest the wheel being removed (photo). Turn the jack handle clockwise to lower the foot of the jack to the ground. If the ground is soft or uneven, place a plank or block under the foot of the jack to spread the load.

Jack up the vehicle until the wheel is clear of the ground. Remove the wheel nuts and lift the wheel off the studs. Fit the new wheel onto the studs and secure it with the nuts. Tighten the nuts until they are snug, but do not tighten them fully yet.

Lower the vehicle and remove the jack. Carry out the final tightening of the wheel nuts in criss-cross sequence. The use of a torque wrench is strongly recommended, especially when light alloy wheels are fitted. See Chapter 10 Specifications for the recommended tightening torque.

Refit the wheel trim, when applicable, and stow the tools. If a new wheel has been brought into service, check the tyre pressure at the first opportunity.

Prising off the wheel trim

Slackening the wheel nuts

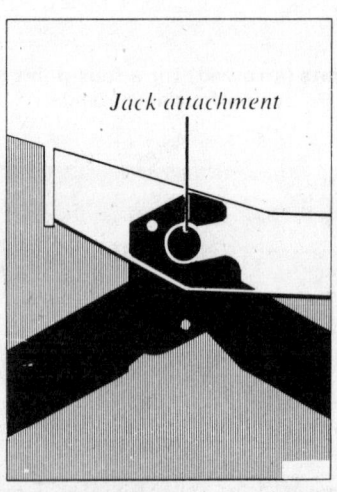
Jack head engagement with the jacking point

Jack head engaged in the jacking point

Buying spare parts and vehicle identification numbers

Buying spare parts

Spare parts are available from many sources, for example: Volvo garages, other garages and accessory shops, and motor factors. Our advice regarding spare parts sources is as follows:

Officially appointed garages – This is the best source for parts which are peculiar to your car and are not generally available (eg complete cylinder heads, internal gearbox components, badges, interior trim etc). It is also the only place at which you should buy parts if your vehicle is still under warranty – non Volvo – components may invalidate the warranty. To be sure of obtaining the correct parts it will always be necessary to give the storeman your car's vehicle identification number, and if possible, to take the 'old' part along for positive identification. Remember that some parts are available on a factory exchange scheme – any parts returned should always be clean! It obviously makes good sense to go straight to the specialists on your car for this type of part for they are best equipped to supply you.

Other garages and accessory shops – These are often very good places to buy materials and components needed for the maintenance of your car (eg oil filters, spark plugs, bulbs, drivebelts, oils and greases, touch-up paint, filler paste, etc). They also sell general accessories, usually have convenient opening hours, charge lower prices and can often be found not far from home.

Motor factors – Good factors will stock all of the more important components which wear out relatively quickly (eg clutch components, pistons, valves, exhaust systems, brake cylinders/pipes/hoses/seals and pads, etc). Motor factors will often provide new or reconditioned components on a part exchange basis – this can save a considerable amount of money.

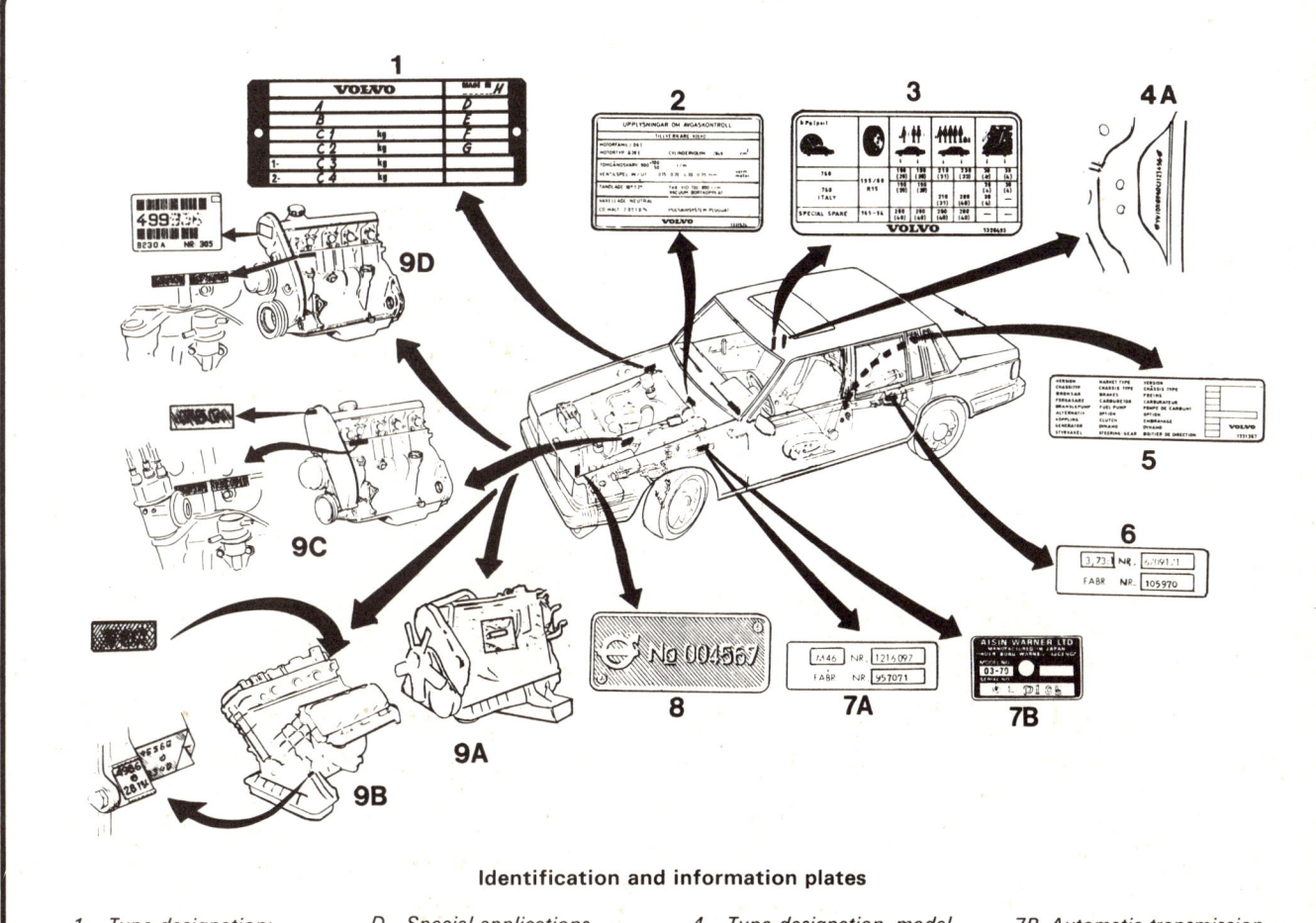

Identification and information plates

1	Type designation:	D	Special applications	4	Type designation, model year and chassis number	7B	Automatic transmission number
A	Type approval	E	Market code	5	Service plate	8	Body number
B	Vehicle identification number	F	Colour code	6	Final drive ratio, type and serial number	9A	Engine number (Diesel)
C1	Gross vehicle weight	G	Trim level			9B	Engine number (B 28)
C2	Gross train weight	H	Country of origin	7A	Manual gearbox number	9C	Engine number (B 23)
C3	Front axle weight	2	Emission control decal (not UK)			9D	Engine number (B 230)
C4	Rear axle weight	3	Tyre pressure information				

Buying spare parts and vehicle identification numbers

Vehicle identification numbers

When ordering spare parts, always give as much information as possible. Quote the car model, year of manufacture and if necessary the chassis number.

The locations of the various identification plates are shown in the accompanying diagram. The information to be found on the service plate is normally sufficient for routine maintenance and repair requirements (photo). It is interpreted as follows:

Part	Manufacturer	Code
Brakes	Girling front and rear	1
	Girling front and ATE rear	2
	DBA front, ATE rear	3
Carburettor	SU	2
	Pierburg	3
	Solex	5
Fuel pump	Bosch	3
	AC-Delco	4
	Sofabex	5
Clutch	Fichtel & Sachs	2
	Verto/Valeo	3
Alternator	Bosch	1
Steering box	Cam Gear	2
	Zahnrad Fabrik (ZF)	3

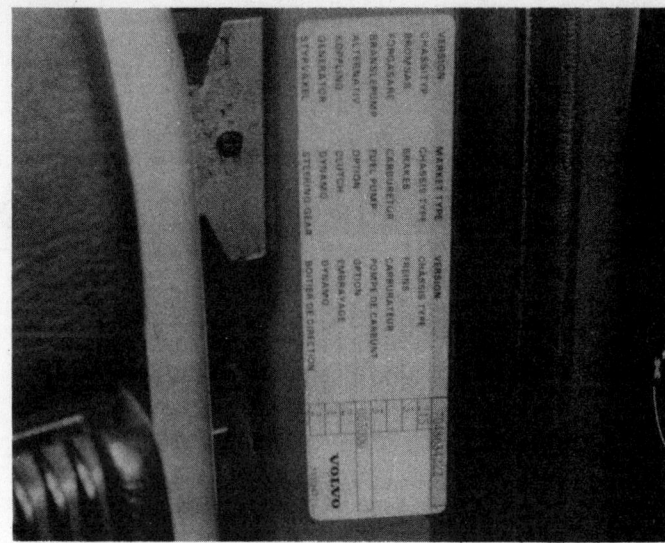

The service plate provides information needed when buying certain spares

Some models do not have a service plate, and the information required for maintenance and repair is on the type designation plate.

A to C4 As before
D1 Main type, number of doors, engine, equipment
D2 Body type, gearbox, left- or right-hand drive
D3 Market code
E Chassis type
F Emission control
G Steering gear:
   Cam gear — 2
   Zahnrad Fabrik (ZF) — 3
H Brakes:
   Girling front, ATE rear — 2
   DBA front, ATE rear — 3
   Girling front and rear — 4
   DBA front, Girling rear — 5
K Interior fittings code
L Colour code
M Special applications

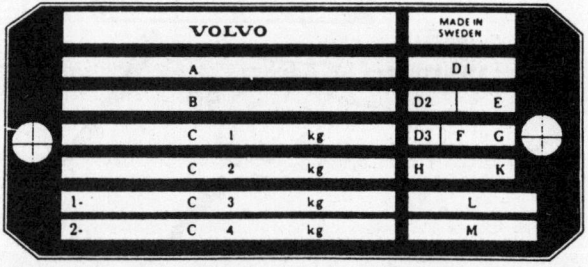

Type designation plate on later models

General repair procedures

Whenever servicing, repair or overhaul work is carried out on the car or its components, it is necessary to observe the following procedures and instructions. This will assist in carrying out the operation efficiently and to a professional standard of workmanship.

Preparation

Read through the relevant Section(s) of the manual before starting work, to make sure that the necessary tools and spare parts are to hand. If it is not possible to tell in advance what spare parts will be required, check the availability and cost of the appropriate parts so that time is not wasted dismantling an item for which no spares are available, or which would be cheaper to exhange than to repair.

Joint mating faces and gaskets

Where a gasket is used between the mating faces of two components, ensure that it is renewed on reassembly, and fit it dry unless otherwise stated in the repair procedure. Make sure that the mating faces are clean and dry with all traces of old gasket removed. When cleaning a joint face, use a tool which is not likely to score or damage the face, and remove any burrs or nicks with an oilstone or fine file.

Make sure that tapped holes are cleaned with a pipe cleaner, and keep them free of jointing compound if this is being used unless specifically instructed otherwise.

Ensure that all orifices, channels or pipes are clear and blow through them, preferably using compressed air.

Oil seals

Whenever an oil seal is removed from its working location, either individually or as part of an assembly, it should be renewed.

The very fine sealing lip of the seal is easily damaged and will not seal if the surface it contacts is not completely clean and free from scratches, nicks or grooves. If the original sealing surface of the component cannot be restored, the component should be renewed.

Protect the lips of the seal from any surface which may damage them in the course of fitting. Use tape or a conical sleeve where possible. Lubricate the seal lips with oil before fitting and, on dual lipped seals, fill the space between the lips with grease.

Unless otherwise stated, oil seals must be fitted with their sealing lips toward the lubricant to be sealed.

Use a tubular drift or block of wood of the appropriate size to install the seal and, if the seal housing is shouldered, drive the seal down to the shoulder. If the seal housing is unshouldered, the seal should be fitted with its face flush with the housing top face.

Screw threads and fastenings

Always ensure that a blind tapped hole is completely free from oil, grease, water or other fluid before installing the bolt or stud. Failure to do this could cause the housing to crack due to the hydraulic action of the bolt or stud as it is screwed in.

When tightening a castellated nut to accept a split pin, tighten the nut to the specified torque, where applicable, and then tighten further to the next split pin hole. Never slacken the nut to align a split pin hole unless stated in the repair procedure.

When checking or retightening a nut or bolt to a specified torque setting, slacken the nut or bolt by a quarter of a turn, and then retighten to the specified setting.

Locknuts, locktabs and washers

Any fastening which will rotate against a component or housing in the course of tightening should always have a washer between it and the relevant component or housing.

Spring or split washers should always be renewed when they are used to lock a critical component such as a big-end bearing retaining nut or bolt.

Locktabs which are folded over to retain a nut or bolt should always be renewed.

Self-locking nuts can be reused in non-critical areas, providing resistance can be felt when the locking portion passes over the bolt or stud thread.

Split pins must always be replaced with new ones of the correct size for the hole.

Special tools

Some repair procedures in this manual entail the use of special tools such as a press, two or three-legged pullers, spring compressors etc. Wherever possible, suitable readily available alternatives to the manufacturer's special tools are described, and are shown in use. In some instances, where no alternative is possible, it has been necessary to resort to the use of a manufacturer's tool and this has been done for reasons of safety as well as the efficient completion of the repair operation. Unless you are highly skilled and have a thorough understanding of the procedure described, never attempt to bypass the use of any special tool when the procedure described specifies its use. Not only is there a very great risk of personal injury, but expensive damage could be caused to the components involved.

Tools and working facilities

Introduction

A selection of good tools is a fundamental requirement for anyone contemplating the maintenance and repair of a motor vehicle. For the owner who does not possess any, their purchase will prove a considerable expense, offsetting some of the savings made by doing-it-yourself. However, provided that the tools purchased meet the relevant national safety standards and are of good quality, they will last for many years and prove an extremely worthwhile investment.

To help the average owner to decide which tools are needed to carry out the various tasks detailed in this manual, we have compiled three lists of tools under the following headings: *Maintenance and minor repair*, *Repair and overhaul*, and *Special*. The newcomer to practical mechanics should start off with the *Maintenance and minor repair* tool kit and confine himself to the simpler jobs around the vehicle. Then, as his confidence and experience grow, he can undertake more difficult tasks, buying extra tools as, and when, they are needed. In this way, a *Maintenance and minor repair* tool kit can be built-up into a *Repair and overhaul* tool kit over a considerable period of time without any major cash outlays. The experienced do-it-yourselfer will have a tool kit good enough for most repair and overhaul procedures and will add tools from the *Special* category when he feels the expense is justified by the amount of use to which these tools will be put.

It is obviously not possible to cover the subject of tools fully here. For those who wish to learn more about tools and their use there is a book entitled *How to Choose and Use Car Tools* available from the publishers of this manual.

Maintenance and minor repair tool kit

The tools given in this list should be considered as a minimum requirement if routine maintenance, servicing and minor repair operations are to be undertaken. We recommend the purchase of combination spanners (ring one end, open-ended the other); although more expensive than open-ended ones, they do give the advantages of both types of spanner.

> *Combination spanners - 10, 11, 12, 13, 14 & 17 mm*
> *Adjustable spanner - 9 inch*
> *Engine sump drain plug key (V6 only)*
> *Spark plug spanner (with rubber insert)*
> *Spark plug gap adjustment tool*
> *Set of feeler gauges*
> *Brake bleed nipple spanner*
> *Screwdriver - 4 in long x $^{1}/_{4}$ in dia (flat blade)*
> *Screwdriver - 4 in long x $^{1}/_{4}$ in dia (cross blade)*
> *Combination pliers - 6 inch*
> *Hacksaw (junior)*
> *Tyre pump*
> *Tyre pressure gauge*
> *Oil can*
> *Fine emery cloth (1 sheet)*
> *Wire brush (small)*
> *Funnel (medium size)*

Repair and overhaul tool kit

These tools are virtually essential for anyone undertaking any major repairs to a motor vehicle, and are additional to those given in the *Maintenance and minor repair* list. Included in this list is a comprehensive set of sockets. Although these are expensive they will be found invaluable as they are so versatile - particularly if various drives are included in the set. We recommend the ½ in square-drive type, as this can be used with most proprietary torque wrenches. If you cannot afford a socket set, even bought piecemeal, then inexpensive tubular box spanners are a useful alternative.

The tools in this list will occasionally need to be supplemented by tools from the *Special* list.

> *Sockets (or box spanners) to cover range in previous list*
> *Reversible ratchet drive (for use with sockets)*
> *Extension piece, 10 inch (for use with sockets)*
> *Universal joint (for use with sockets)*
> *Torque wrench (for use with sockets)*
> *'Mole' wrench - 8 inch*
> *Ball pein hammer*
> *Soft-faced hammer, plastic or rubber*
> *Screwdriver - 6 in long x $^{5}/_{16}$ in dia (flat blade)*
> *Screwdriver - 2 in long x $^{5}/_{16}$ in square (flat blade)*
> *Screwdriver - 1$^{1}/_{2}$ in long x $^{1}/_{4}$ in dia (cross blade)*
> *Screwdriver - 3 in long x $^{1}/_{8}$ in dia (electricians)*
> *Pliers - electricians side cutters*
> *Pliers - needle nosed*
> *Pliers - circlip (internal and external)*
> *Cold chisel - $^{1}/_{2}$ inch*
> *Scriber*
> *Scraper*
> *Centre punch*
> *Pin punch*
> *Hacksaw*
> *Valve grinding tool*
> *Steel rule/straight-edge*
> *Allen keys (inc. splined/Torx type if necessary)*
> *Selection of files*
> *Wire brush (large)*
> *Axle-stands*
> *Jack (strong trolley or hydraulic type)*

Special tools

The tools in this list are those which are not used regularly, are expensive to buy, or which need to be used in accordance with their manufacturers' instructions. Unless relatively difficult mechanical jobs are undertaken frequently, it will not be economic to buy many of these tools. Where this is the case, you could consider clubbing together with friends (or joining a motorists' club) to make a joint purchase, or borrowing the tools against a deposit from a local garage or tool hire specialist.

The following list contains only those tools and instruments freely available to the public, and not those special tools produced by the vehicle manufacturer specifically for its dealer network. You will find occasional references to these manufacturers' special tools in the text of this manual. Generally, an alternative method of doing the job without the vehicle manufacturers' special tool is given. However, sometimes, there is no alternative to using them. Where this is the case and the relevant tool cannot be bought or borrowed, you will have to entrust the work to a franchised garage.

- Valve spring compressor
- Piston ring compressor
- Balljoint separator
- Universal hub/bearing puller
- Impact screwdriver
- Micrometer and/or vernier gauge
- Dial gauge
- Stroboscopic timing light
- Tachometer
- Universal electrical multi-meter
- Cylinder compression gauge
- Lifting tackle
- Trolley jack
- Light with extension lead

Buying tools

For practically all tools, a tool factor is the best source since he will have a very comprehensive range compared with the average garage or accessory shop. Having said that, accessory shops often offer excellent quality tools at discount prices, so it pays to shop around.

There are plenty of good tools around at reasonable prices, but always aim to purchase items which meet the relevant national safety standards. If in doubt, ask the proprietor or manager of the shop for advice before making a purchase.

Care and maintenance of tools

Having purchased a reasonable tool kit, it is necessary to keep the tools in a clean serviceable condition. After use, always wipe off any dirt, grease and metal particles using a clean, dry cloth, before putting the tools away. Never leave tools lying around after they have been used. A simple tool rack on the garage or workshop wall, for items such as screwdrivers and pliers is a good idea. Store all normal wrenches and sockets in a metal box. Any measuring instruments, gauges, meters, etc, must be carefully stored where they cannot be damaged or become rusty.

Take a little care when tools are used. Hammer heads inevitably become marked and screwdrivers lose the keen edge on their blades from time to time. A little timely attention with emery cloth or a file will soon restore items like this to a good serviceable finish.

Working facilities

Not to be forgotten when discussing tools, is the workshop itself. If anything more than routine maintenance is to be carried out, some form of suitable working area becomes essential.

It is appreciated that many an owner mechanic is forced by circumstances to remove an engine or similar item, without the benefit of a garage or workshop. Having done this, any repairs should always be done under the cover of a roof.

Wherever possible, any dismantling should be done on a clean, flat workbench or table at a suitable working height.

Any workbench needs a vice: one with a jaw opening of 4 in (100 mm) is suitable for most jobs. As mentioned previously, some clean dry storage space is also required for tools, as well as for lubricants, cleaning fluids, touch-up paints and so on, which become necessary.

Another item which may be required, and which has a much more general usage, is an electric drill with a chuck capacity of at least 5/16 in (8 mm). This, together with a good range of twist drills, is virtually essential for fitting accessories such as mirrors and reversing lights.

Last, but not least, always keep a supply of old newspapers and clean, lint-free rags available, and try to keep any working area as clean as possible.

Spanner jaw gap comparison table

Jaw gap (in)	Spanner size
0.250	1/4 in AF
0.276	7 mm
0.313	5/16 in AF
0.315	8 mm
0.344	11/32 in AF; 1/8 in Whitworth
0.354	9 mm
0.375	3/8 in AF
0.394	10 mm
0.433	11 mm
0.438	7/16 in AF
0.445	3/16 in Whitworth; 1/4 in BSF
0.472	12 mm
0.500	1/2 in AF
0.512	13 mm
0.525	1/4 in Whitworth; 5/16 in BSF
0.551	14 mm
0.563	9/16 in AF
0.591	15 mm
0.600	5/16 in Whitworth; 3/8 in BSF
0.625	5/8 in AF
0.630	16 mm
0.669	17 mm
0.686	11/16 in AF
0.709	18 mm
0.710	3/8 in Whitworth; 7/16 in BSF
0.748	19 mm
0.750	3/4 in AF
0.813	13/16 in AF
0.820	7/16 in Whitworth; 1/2 in BSF
0.866	22 mm
0.875	7/8 in AF
0.920	1/2 in Whitworth; 9/16 in BSF
0.938	15/16 in AF
0.945	24 mm
1.000	1 in AF
1.010	9/16 in Whitworth; 5/8 in BSF
1.024	26 mm
1.063	1 1/16 in AF; 27 mm
1.100	5/8 in Whitworth; 11/16 in BSF
1.125	1 1/8 in AF
1.181	30 mm
1.200	11/16 in Whitworth; 3/4 in BSF
1.250	1 1/4 in AF
1.260	32 mm
1.300	3/4 in Whitworth; 7/8 in BSF
1.313	1 5/16 in AF
1.390	13/16 in Whitworth; 15/16 in BSF
1.417	36 mm
1.438	1 7/16 in AF
1.480	7/8 in Whitworth; 1 in BSF
1.500	1 1/2 in AF
1.575	40 mm; 15/16 in Whitworth
1.614	41 mm
1.625	1 5/8 in AF
1.670	1 in Whitworth; 1 1/8 in BSF
1.688	1 11/16 in AF
1.811	46 mm
1.813	1 13/16 in AF
1.860	1 1/8 in Whitworth; 1 1/4 in BSF
1.875	1 7/8 in AF
1.969	50 mm
2.000	2 in AF
2.050	1 1/4 in Whitworth; 1 3/8 in BSF
2.165	55 mm
2.362	60 mm

Safety first!

Professional motor mechanics are trained in safe working procedures. However enthusiastic you may be about getting on with the job in hand, do take the time to ensure that your safety is not put at risk. A moment's lack of attention can result in an accident, as can failure to observe certain elementary precautions.

There will always be new ways of having accidents, and the following points do not pretend to be a comprehensive list of all dangers; they are intended rather to make you aware of the risks and to encourage a safety-conscious approach to all work you carry out on your vehicle.

Essential DOs and DON'Ts

DON'T rely on a single jack when working underneath the vehicle. Always use reliable additional means of support, such as axle stands, securely placed under a part of the vehicle that you know will not give way.

DON'T attempt to loosen or tighten high-torque nuts (e.g. wheel hub nuts) while the vehicle is on a jack; it may be pulled off.

DON'T start the engine without first ascertaining that the transmission is in neutral (or 'Park' where applicable) and the parking brake applied.

DON'T suddenly remove the filler cap from a hot cooling system – cover it with a cloth and release the pressure gradually first, or you may get scalded by escaping coolant.

DON'T attempt to drain oil until you are sure it has cooled sufficiently to avoid scalding you.

DON'T grasp any part of the engine, exhaust or catalytic converter without first ascertaining that it is sufficiently cool to avoid burning you.

DON'T allow brake fluid or antifreeze to contact vehicle paintwork.

DON'T syphon toxic liquids such as fuel, brake fluid or antifreeze by mouth, or allow them to remain on your skin.

DON'T inhale dust – it may be injurious to health (see *Asbestos* below).

DON'T allow any spilt oil or grease to remain on the floor – wipe it up straight away, before someone slips on it.

DON'T use ill-fitting spanners or other tools which may slip and cause injury.

DON'T attempt to lift a heavy component which may be beyond your capability – get assistance.

DON'T rush to finish a job, or take unverified short cuts.

DON'T allow children or animals in or around an unattended vehicle.

DO wear eye protection when using power tools such as drill, sander, bench grinder etc, and when working under the vehicle.

DO use a barrier cream on your hands prior to undertaking dirty jobs – it will protect your skin from infection as well as making the dirt easier to remove afterwards; but make sure your hands aren't left slippery. Note that long-term contact with used engine oil can be a health hazard.

DO keep loose clothing (cuffs, tie etc) and long hair well out of the way of moving mechanical parts.

DO remove rings, wristwatch etc, before working on the vehicle – especially the electrical system.

DO ensure that any lifting tackle used has a safe working load rating adequate for the job.

DO keep your work area tidy – it is only too easy to fall over articles left lying around.

DO get someone to check periodically that all is well, when working alone on the vehicle.

DO carry out work in a logical sequence and check that everything is correctly assembled and tightened afterwards.

DO remember that your vehicle's safety affects that of yourself and others. If in doubt on any point, get specialist advice.

IF, in spite of following these precautions, you are unfortunate enough to injure yourself, seek medical attention as soon as possible.

Asbestos

Certain friction, insulating, sealing, and other products – such as brake linings, brake bands, clutch linings, torque converters, gaskets, etc – contain asbestos. *Extreme care must be taken to avoid inhalation of dust from such products since it is hazardous to health.* If in doubt, assume that they *do* contain asbestos.

Fire

Remember at all times that petrol (gasoline) is highly flammable. Never smoke, or have any kind of naked flame around, when working on the vehicle. But the risk does not end there – a spark caused by an electrical short-circuit, by two metal surfaces contacting each other, by careless use of tools, or even by static electricity built up in your body under certain conditions, can ignite petrol vapour, which in a confined space is highly explosive.

Always disconnect the battery earth (ground) terminal before working on any part of the fuel or electrical system, and never risk spilling fuel on to a hot engine or exhaust.

It is recommended that a fire extinguisher of a type suitable for fuel and electrical fires is kept handy in the garage or workplace at all times. Never try to extinguish a fuel or electrical fire with water.

Note: *Any reference to a 'torch' appearing in this manual should always be taken to mean a hand-held battery-operated electric lamp or flashlight. It does NOT mean a welding/gas torch or blowlamp.*

Fumes

Certain fumes are highly toxic and can quickly cause unconsciousness and even death if inhaled to any extent. Petrol (gasoline) vapour comes into this category, as do the vapours from certain solvents such as trichloroethylene. Any draining or pouring of such volatile fluids should be done in a well ventilated area.

When using cleaning fluids and solvents, read the instructions carefully. Never use materials from unmarked containers – they may give off poisonous vapours.

Never run the engine of a motor vehicle in an enclosed space such as a garage. Exhaust fumes contain carbon monoxide which is extremely poisonous; if you need to run the engine, always do so in the open air or at least have the rear of the vehicle outside the workplace.

If you are fortunate enough to have the use of an inspection pit, never drain or pour petrol, and never run the engine, while the vehicle is standing over it; the fumes, being heavier than air, will concentrate in the pit with possibly lethal results.

The battery

Never cause a spark, or allow a naked light, near the vehicle's battery. It will normally be giving off a certain amount of hydrogen gas, which is highly explosive.

Always disconnect the battery earth (ground) terminal before working on the fuel or electrical systems.

If possible, loosen the filler plugs or cover when charging the battery from an external source. Do not charge at an excessive rate or the battery may burst.

Take care when topping up and when carrying the battery. The acid electrolyte, even when diluted, is very corrosive and should not be allowed to contact the eyes or skin.

If you ever need to prepare electrolyte yourself, always add the acid slowly to the water, and never the other way round. Protect against splashes by wearing rubber gloves and goggles.

When jump starting a car using a booster battery, for negative earth (ground) vehicles, connect the jump leads in the following sequence: First connect one jump lead between the positive (+) terminals of the two batteries. Then connect the other jump lead first to the negative (–) terminal of the booster battery, and then to a good earthing (ground) point on the vehicle to be started, at least 18 in (45 cm) from the battery if possible. Ensure that hands and jump leads are clear of any moving parts, and that the two vehicles do not touch. Disconnect the leads in the reverse order.

Mains electricity and electrical equipment

When using an electric power tool, inspection light etc, always ensure that the appliance is correctly connected to its plug and that, where necessary, it is properly earthed (grounded). Do not use such appliances in damp conditions and, again, beware of creating a spark or applying excessive heat in the vicinity of fuel or fuel vapour. Also ensure that the appliances meet the relevant national safety standards.

Ignition HT voltage

A severe electric shock can result from touching certain parts of the ignition system, such as the HT leads, when the engine is running or being cranked, particularly if components are damp or the insulation is defective. Where an electronic ignition system is fitted, the HT voltage is much higher and could prove fatal.

Routine maintenance

For modifications, and information applicable to later models, see Supplement at end of manual

The maintenance schedules below are basically those recommended by the manufacturer. Servicing intervals are determined by mileage or time elapsed – this is because fluids and systems deteriorate with age as well as with use. Follow the time intervals if the appropriate mileage is not covered within the specified period.

Vehicles operating under adverse conditions need more frequent maintenance. 'Adverse conditions' include climatic extremes, full-time towing or taxi work, driving on unmade roads, and a high proportion of short journeys. In such cases the engine oil and filter should be changed every 3000 miles (5000 km) or three months, whichever comes first.

Weekly, every 250 miles (400 km), or before a long journey

    Check tyre pressures and inspect tyres (Chapter 10, Sec 32)
    Check engine oil level (Chapter 1, Sec 2 or 48)
    Check coolant level (Chapter 2, Sec 2)
    Check brake fluid level (Chapter 9, Sec 3)
    Top up washer reservoir(s), adding a screen wash such as Turtle Wax High Tech Screen Wash (Chapter 12, Sec 28)
    Inspect engine bay and under vehicle for leaks
    Check function of lights, horn, wipers, etc.

Every 6000 miles (10 000 km) or six months, whichever comes first

    Renew engine oil and filter (Chapter 1, Secs 2 and 3/48 and 49)
    Check power steering fluid level (Chapter 10, Sec 3)
    Check coolant antifreeze concentration (Chapter 2, Sec 2)
    Renew spark plugs (Chapter 4, Sec 3)
    Inspect battery (Chapter 12, Sec 3)
    Check idle speed and CO level (Chapter 3, Sec 8)
    Inspect tyres thoroughly (Chapter 10, Sec 32)
    Check brake pad wear (Chapter 9, Secs 4 and 5)
    ABS: check that spare wheel well drain is clear

Every 12 000 miles (20 000 km) or twelve months, whichever comes first

In addition to the work previously specified
    Check operation of brake servo (Chapter 9, Sec 6)
    Check handbrake adjustment (Chapter 9, Sec 7)
    Check automatic transmission selector adjustment (Chapter 6, Sec 39)
    Grease bonnet hinges
    Check front wheel bearing adjustment (Chapter 10, Sec 4)
    Check security and condition of steering and suspension components (Chapter 10, Sec 2)
    Inspect brake hydraulic pipes and hoses (Chapter 9, Sec 8)
    Check tightness of trailing arm nuts, control arm, strut and steering gear fastenings, and front axle crossmember bolts. (First 12 000 miles only). (Chapter 10, Sec 2)
    Inspect clutch hydraulic components (Chapter 5, Sec 2) or check adjustment of cable (Chapter 5, Sec 3)
    Check transmission oil level (Chapter 6, Sec 2 or 28)
    Inspect propeller shaft, centre bearing and universal joints (Chapter 7, Sec 2)
    Check condition and security of exhaust system (Chapter 3, Sec 19)
    Check rear axle oil level (Chapter 8, Sec 3)
    Check condition of fuel lines
    Thoroughly inspect engine for fluid leaks
    Check condition of underseal and paintwork (Chapter 11, Sec 2)
    Inspect in-line fuel filter (carburettor models) (Chapter 3, Sec 2)
    Check condition and tension of accessory drivebelts(s) (Chapter 4, Sec 2)
    Lubricate distributor felt pad (B28E engine only) (Chapter 4, Sec 2)
    Inspect distributor cap, rotor arm and HT leads (Chapter 4, Sec 2)
    Check turbo boost pressure switches (when applicable) (Chapter 3, Sec 45)
    Check turbo tamperproof seals (when applicable) on wastegate actuator rod
    Check operation of kickdown cable (automatic transmission) and adjust if necessary (Chapter 6, Sec 31)

Every 24 000 miles (40 000 km) or two years, whichever comes first

In addition to the work previously specified
    Renew automatic transmission fluid (Chapter 6, Sec 29)
    Renew fuel filter (Chapter 3, Sec 7)
    Renew air cleaner element (Chapter 3, Sec 5)
    Clean crankcase ventilation hoses, flame trap, etc (Chapter 1, Sec 5 or 51)
    Check valve clearances, (Chapter 1, Sec 4 or 50)
    Perform a compression test (Chapter 1, Sec 45 or 96)
    Renew coolant (Chapter 2, Secs 3 to 5)
    Renew brake fluid by bleeding (Chapter 9, Sec 9); at the same time consider renewing rubber seals and flexible hoses as a precautionary measure

Every 48 000 miles (80 000 km) or four years, whichever comes first

In addition to the work previously specified
    Renew camshaft drivebelt (B 23/B 230 engines) (Chapter 1, Sec 6)

Under-bonnet view of a Volvo 760 GLE

1 Battery
2 Ignition control unit
3 Ignition coil
4 Air conditioning compressor
5 Engine oil dipstick
6 Suspension turrets
7 Identification plate
8 Ignition vacuum advance valve
9 Brake fluid reservoir
10 Brake servo
11 Engine oil filler cap
12 Air control valve
13 Fuel distributor
14 Automatic transmission dipstick
15 Vacuum pump
16 Air conditioner receiver/drier
17 Fuel filter
18 Coolant expansion tank
19 Air cleaner
20 Screen washer filler cap
21 Bonnet catches
22 Radiator
23 Power steering reservoir
24 Left-hand rocker cover
25 Inlet manifold
26 Air intake
27 Compressor drivebelt
28 Radiator top hose
29 Throttle cable

Under-bonnet view of a Volvo 760 Turbo

1 Screen washer filler cap
2 Air cleaner
3 Airflow meter
4 Coolant expansion tank
5 Suspension turrets
6 Identification plate
7 Brake and clutch fluid reservoir
8 Brake servo
9 Turbocharger
10 Turbo air outlet
11 Bypass valve
12 Bypass valve hose
13 Engine oil filler cap
14 HT leads
15 Clutch master cylinder
16 Auxiliary air valve
17 Vacuum delay valve
18 Engine oil dipstick
19 Throttle linkage
20 Air conditioner receiver/drier
21 Ignition coil
22 Power steering pump and reservoir
23 Fuel pressure regulator
24 Ignition distributor
25 Radiator top hose
26 Battery
27 Bonnet catches
28 Intercooler
29 Radiator

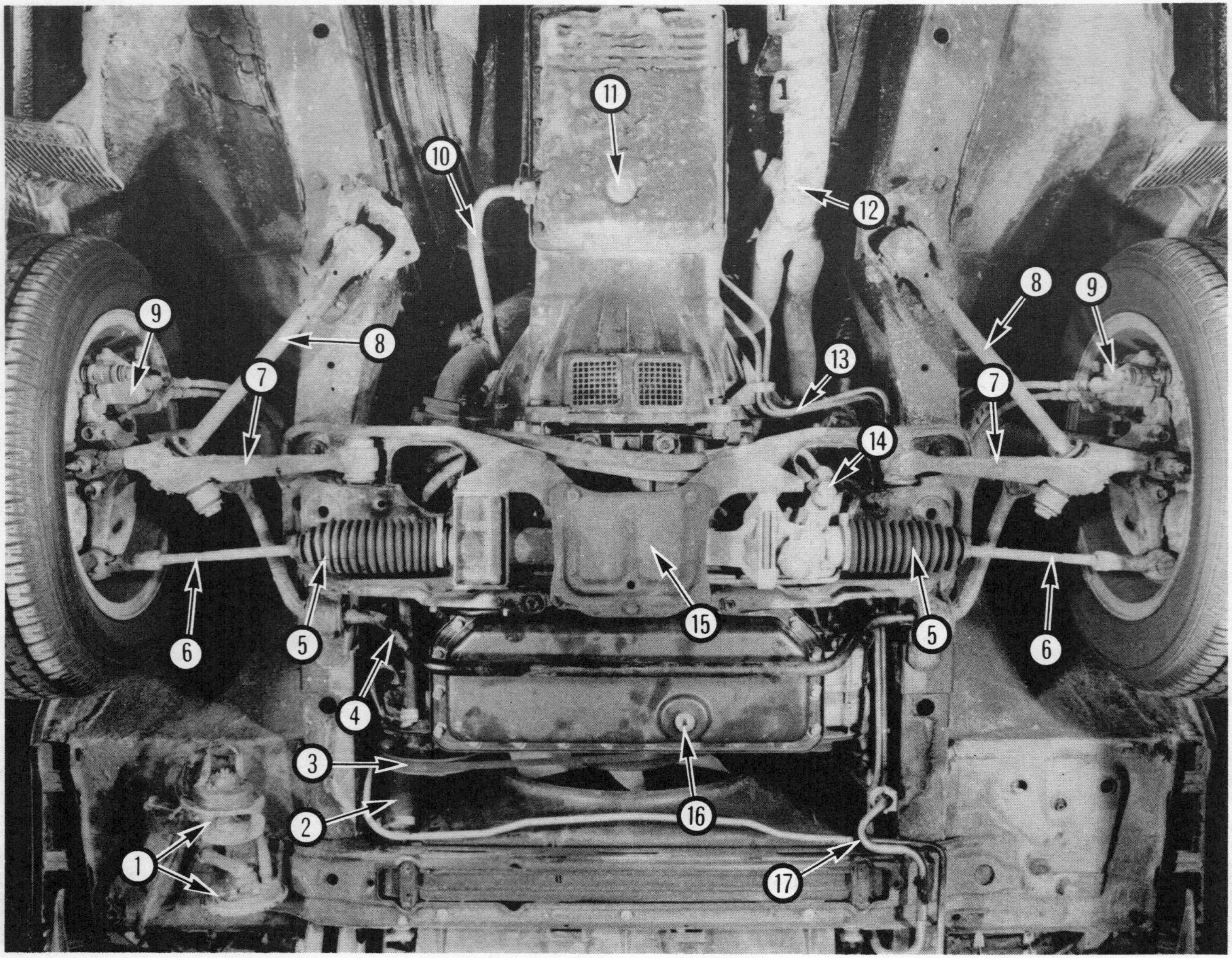

Front underside view of a Volvo 760 GLE

1 Horns
2 Radiator bottom hose
3 Steering pump
4 Anti-roll bar
5 Steering rack bellows
6 Track rods
7 Control arms
8 Radius rods
9 Brake calipers
10 Transmission dipstick/filler tube
11 Transmission drain plug
12 Exhaust downpipe
13 Transmission fluid cooler lines
14 Steering hydraulic unions
15 Jacking plate
16 Engine oil drain plug
17 Refrigerant lines (air conditioning)

Front underside view of a Volvo 760 Turbo

1 Horn
2 Refrigerant lines (air conditioning)
3 Oil cooler hoses
4 Vacuum tank
5 Anti-roll bar
6 Steering rack bellows
7 Track rods
8 Control arms
9 Radius rods
10 Brake calipers
11 Engine oil drain plug
12 Brace
13 Clutch slave cylinder
14 Gearbox filler/level plug
15 Gearbox drain plug
16 Gearbox mounting
17 Overdrive solenoid
18 Overdrive
19 Damper
20 Propeller shaft flange
21 Exhaust downpipe
22 Steering intermediate shaft
23 Steering hydraulic unions
24 Jacking plate
25 Radiator bottom hose

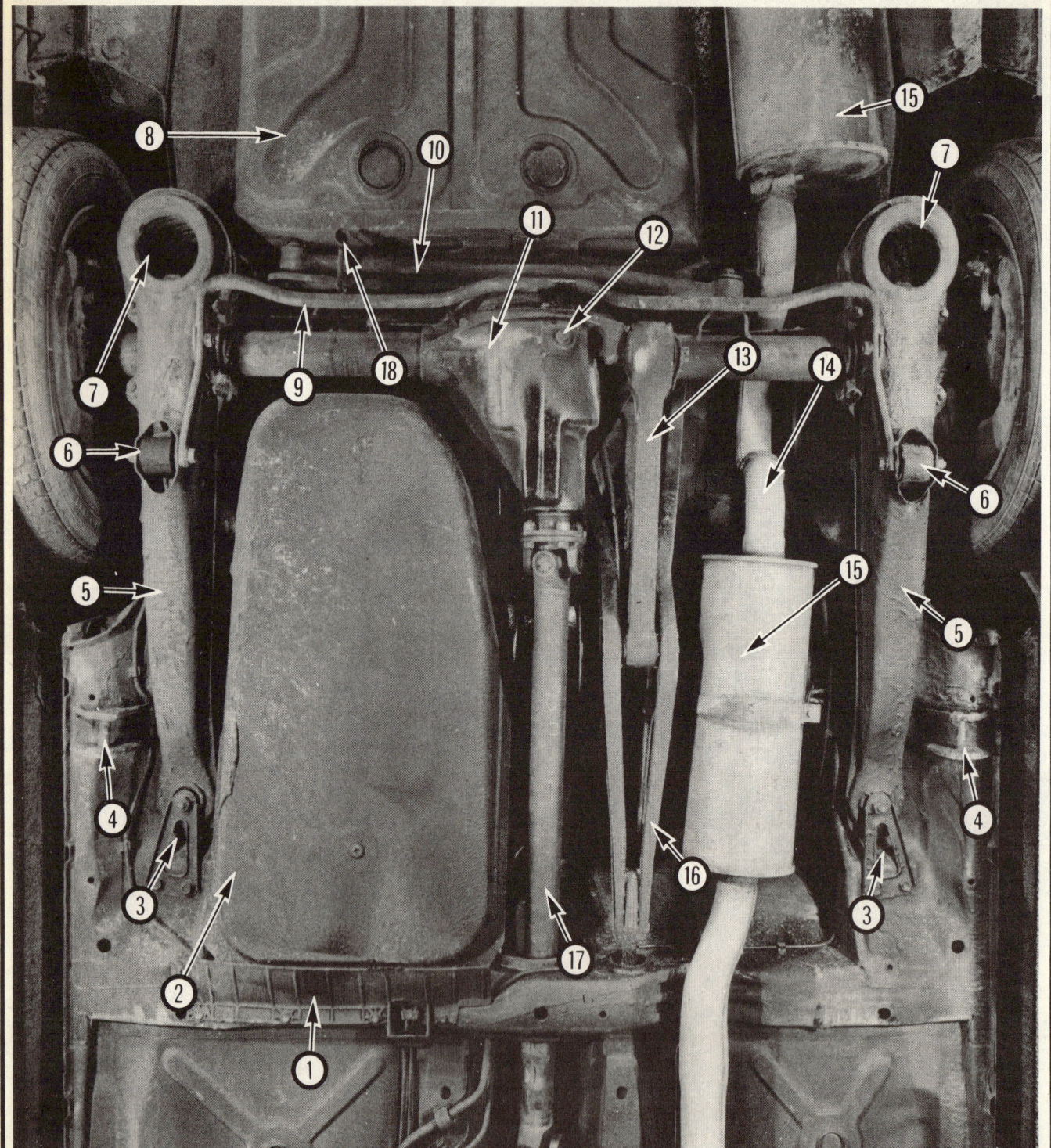

Rear underside view of a Volvo 760 Turbo

1	Mud deflector	6	Shock absorber lower mountings
2	Main fuel tank	7	Spring pans
3	Trailing arm brackets	8	Spare wheel well
4	Rear jacking points	9	Anti-roll bar
5	Trailing arms	10	Panhard rod
11	Rear axle	15	Silencers
12	Rear axle drain plug	16	Subframe
13	Torque rod	17	Propeller shaft
14	Exhaust pipe	18	Fuel tank breather

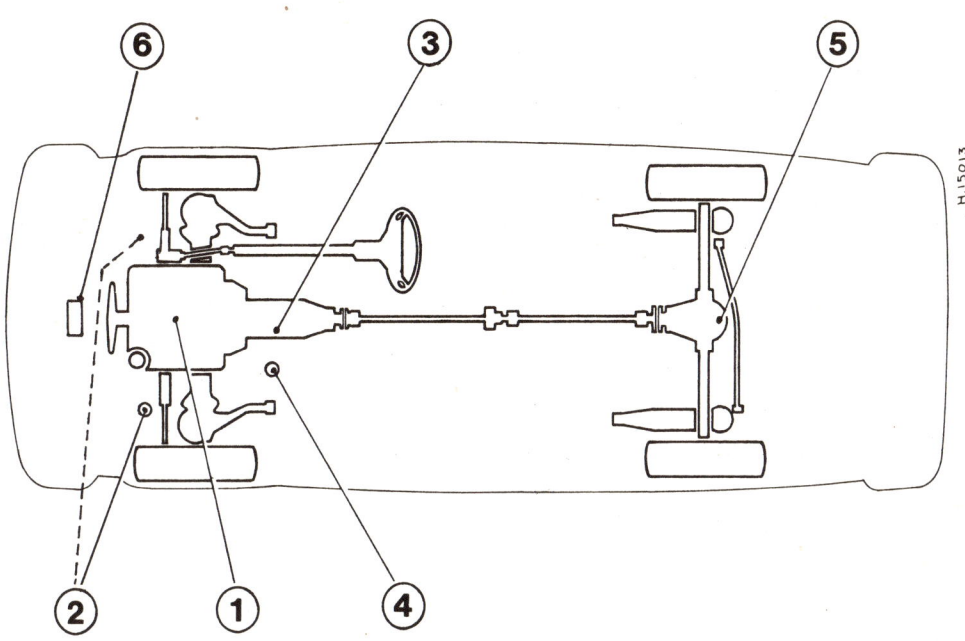

Recommended lubricants and fluids

Component or system	Lubricant type/specification	Duckhams recommendation
1 Engine	Multigrade engine oil, viscosity range SAE 10W/30 to 15W/50, to API SF/CC, SF/CD or better	Duckhams QXR, QS, Hypergrade Plus or Hypergrade
2 Cooling system	Volvo coolant type C and clean water	Duckhams Universal Antifreeze and Summer Coolant
3 Manual gearbox	Volvo thermo oil	Duckhams QXR
4 Automatic transmission		
Models up to 1983	ATF type F or G	Duckhams Uni-Matic
1984 and later models	Dexron IID type ATF	Duckhams Uni-Matic
5 Rear axle		
Except limited slip differential	Hypoid gear oil, viscosity SAE 90 EP, to to API GL 5 or 6	Duckhams Hypoid 80W/90S
Limited slip differential	Special Volvo oil (No 1 161 276-9), or gear oil as above with Volvo additive (No 1 161 129-0)	Duckhams Hypoid 90DL
6 Power steering	ATF type A, F or G	Duckhams Uni-Matic
Brake hydraulic system	Hydraulic fluid to DOT 4	Duckhams Universal Brake and Clutch Fluid

Conversion factors

Length (distance)
Inches (in)	X	25.4	= Millimetres (mm)	X	0.0394	= Inches (in)
Feet (ft)	X	0.305	= Metres (m)	X	3.281	= Feet (ft)
Miles	X	1.609	= Kilometres (km)	X	0.621	= Miles

Volume (capacity)
Cubic inches (cu in; in^3)	X	16.387	= Cubic centimetres (cc; cm^3)	X	0.061	= Cubic inches (cu in; in^3)
Imperial pints (Imp pt)	X	0.568	= Litres (l)	X	1.76	= Imperial pints (Imp pt)
Imperial quarts (Imp qt)	X	1.137	= Litres (l)	X	0.88	= Imperial quarts (Imp qt)
Imperial quarts (Imp qt)	X	1.201	= US quarts (US qt)	X	0.833	= Imperial quarts (Imp qt)
US quarts (US qt)	X	0.946	= Litres (l)	X	1.057	= US quarts (US qt)
Imperial gallons (Imp gal)	X	4.546	= Litres (l)	X	0.22	= Imperial gallons (Imp gal)
Imperial gallons (Imp gal)	X	1.201	= US gallons (US gal)	X	0.833	= Imperial gallons (Imp gal)
US gallons (US gal)	X	3.785	= Litres (l)	X	0.264	= US gallons (US gal)

Mass (weight)
Ounces (oz)	X	28.35	= Grams (g)	X	0.035	= Ounces (oz)
Pounds (lb)	X	0.454	= Kilograms (kg)	X	2.205	= Pounds (lb)

Force
Ounces-force (ozf; oz)	X	0.278	= Newtons (N)	X	3.6	= Ounces-force (ozf; oz)
Pounds-force (lbf; lb)	X	4.448	= Newtons (N)	X	0.225	= Pounds-force (lbf; lb)
Newtons (N)	X	0.1	= Kilograms-force (kgf; kg)	X	9.81	= Newtons (N)

Pressure
Pounds-force per square inch (psi; lbf/in^2; lb/in^2)	X	0.070	= Kilograms-force per square centimetre (kgf/cm^2; kg/cm^2)	X	14.223	= Pounds-force per square inch (psi; lbf/in^2; lb/in^2)
Pounds-force per square inch (psi; lbf/in^2; lb/in^2)	X	0.068	= Atmospheres (atm)	X	14.696	= Pounds-force per square inch (psi; lbf/in^2; lb/in^2)
Pounds-force per square inch (psi; lbf/in^2; lb/in^2)	X	0.069	= Bars	X	14.5	= Pounds-force per square inch (psi; lbf/in^2; lb/in^2)
Pounds-force per square inch (psi; lbf/in^2; lb/in^2)	X	6.895	= Kilopascals (kPa)	X	0.145	= Pounds-force per square inch (psi; lbf/in^2; lb/in^2)
Kilopascals (kPa)	X	0.01	= Kilograms-force per square centimetre (kgf/cm^2; kg/cm^2)	X	98.1	= Kilopascals (kPa)
Millibar (mbar)	X	100	= Pascals (Pa)	X	0.01	= Millibar (mbar)
Millibar (mbar)	X	0.0145	= Pounds-force per square inch (psi; lbf/in^2; lb/in^2)	X	68.947	= Millibar (mbar)
Millibar (mbar)	X	0.75	= Millimetres of mercury (mmHg)	X	1.333	= Millibar (mbar)
Millibar (mbar)	X	0.401	= Inches of water (inH$_2$O)	X	2.491	= Millibar (mbar)
Millimetres of mercury (mmHg)	X	0.535	= Inches of water (inH$_2$O)	X	1.868	= Millimetres of mercury (mmHg)
Inches of water (inH$_2$O)	X	0.036	= Pounds-force per square inch (psi; lbf/in^2; lb/in^2)	X	27.68	= Inches of water (inH$_2$O)

Torque (moment of force)
Pounds-force inches (lbf in; lb in)	X	1.152	= Kilograms-force centimetre (kgf cm; kg cm)	X	0.868	= Pounds-force inches (lbf in; lb in)
Pounds-force inches (lbf in; lb in)	X	0.113	= Newton metres (Nm)	X	8.85	= Pounds-force inches (lbf in; lb in)
Pounds-force inches (lbf in; lb in)	X	0.083	= Pounds-force feet (lbf ft; lb ft)	X	12	= Pounds-force inches (lbf in; lb in)
Pounds-force feet (lbf ft; lb ft)	X	0.138	= Kilograms-force metres (kgf m; kg m)	X	7.233	= Pounds-force feet (lbf ft; lb ft)
Pounds-force feet (lbf ft; lb ft)	X	1.356	= Newton metres (Nm)	X	0.738	= Pounds-force feet (lbf ft; lb ft)
Newton metres (Nm)	X	0.102	= Kilograms-force metres (kgf m; kg m)	X	9.804	= Newton metres (Nm)

Power
Horsepower (hp)	X	745.7	= Watts (W)	X	0.0013	= Horsepower (hp)

Velocity (speed)
Miles per hour (miles/hr; mph)	X	1.609	= Kilometres per hour (km/hr; kph)	X	0.621	= Miles per hour (miles/hr; mph)

*Fuel consumption**
Miles per gallon, Imperial (mpg)	X	0.354	= Kilometres per litre (km/l)	X	2.825	= Miles per gallon, Imperial (mpg)
Miles per gallon, US (mpg)	X	0.425	= Kilometres per litre (km/l)	X	2.352	= Miles per gallon, US (mpg)

Temperature

Degrees Fahrenheit = (°C x 1.8) + 32   Degrees Celsius (Degrees Centigrade; °C) = (°F - 32) x 0.56

It is common practice to convert from miles per gallon (mpg) to litres/100 kilometres (l/100km), where mpg (Imperial) x l/100 km = 282 and mpg (US) x l/100 km = 235

Fault diagnosis

Introduction

The vehicle owner who does his or her own maintenance according to the recommended schedules should not have to use this section of the manual very often. Modern component reliability is such that, provided those items subject to wear or deterioration are inspected or renewed at the specified intervals, sudden failure is comparatively rare. Faults do not usually just happen as a result of sudden failure, but develop over a period of time. Major mechanical failures in particular are usually preceded by characteristic symptoms over hundreds or even thousands of miles. Those components which do occasionally fail without warning are often small and easily carried in the vehicle.

With any fault finding, the first step is to decide where to begin investigations. Sometimes this is obvious, but on other occasions a little detective work will be necessary. The owner who makes half a dozen haphazard adjustments or replacements may be successful in curing a fault (or its symptoms), but he will be none the wiser if the fault recurs and he may well have spent more time and money than was necessary. A calm and logical approach will be found to be more satisfactory in the long run. Always take into account any warning signs or abnormalities that may have been noticed in the period preceding the fault – power loss, high or low gauge readings, unusual noises or smells, etc – and remember that failure of components such as fuses or spark plugs may only be pointers to some underlying fault.

The pages which follow here are intended to help in cases of failure to start or breakdown on the road. There is also a Fault Diagnosis Section at the end of each Chapter which should be consulted if the preliminary checks prove unfruitful. Whatever the fault, certain basic principles apply. These are as follows:

Verify the fault. This is simply a matter of being sure that you know what the symptoms are before starting work. This is particularly important if you are investigating a fault for someone else who may not have described it very accurately.

Don't overlook the obvious. For example, if the vehicle won't start, is there petrol in the tank? (Don't take anyone else's word on this particular point, and don't trust the fuel gauge either!) If an electrical fault is indicated, look for loose or broken wires before digging out the test gear.

Cure the disease, not the symptom. Substituting a flat battery with a fully charged one will get you off the hard shoulder, but if the underlying cause is not attended to, the new battery will go the same way. Similarly, changing oil-fouled spark plugs for a new set will get you moving again, but remember that the reason for the fouling (if it wasn't simply an incorrect grade of plug) will have to be established and corrected.

Don't take anything for granted. Particularly, don't forget that a 'new' component may itself be defective (especially if it's been rattling round in the boot for months), and don't leave components out of a fault diagnosis sequence just because they are new or recently fitted. When you do finally diagnose a difficult fault, you'll probably realise that all the evidence was there from the start.

Electrical faults

Electrical faults can be more puzzling than straightforward mechanical failures, but they are no less susceptible to logical analysis if the basic principles of operation are understood. Vehicle electrical wiring exists in extremely unfavourable conditions – heat, vibration and chemical attack – and the first things to look for are loose or corroded connections and broken or chafed wires, especially where the wires pass through holes in the bodywork or are subject to vibration.

All metal-bodied vehicles in current production have one pole of the battery 'earthed', ie connected to the vehicle bodywork, and in nearly all modern vehicles it is the negative (–) terminal. The various electrical components – motors, bulb holders etc – are also connected to earth, either by means of a lead or directly by their mountings. Electric current flows through the component and then back to the battery via the bodywork. If the component mounting is loose or corroded, or if a good path back to the battery is not available, the circuit will be incomplete and malfunction will result. The engine and/or gearbox are also earthed by means of flexible metal straps to the body or subframe; if these straps are loose or missing, starter motor, generator and ignition trouble may result.

Assuming the earth return to be satisfactory, electrical faults will be due either to component malfunction or to defects in the current supply. Individual components are dealt with in Chapter 12. If supply wires are broken or cracked internally this results in an open-circuit, and the easiest way to check for this is to bypass the suspect wire temporarily with a length of wire having a crocodile clip or suitable connector at each end. Alternatively, a 12V test lamp can be used to verify the presence of supply voltage at various points along the wire and the break can be thus isolated.

If a bare portion of a live wire touches the bodywork or other earthed metal part, the electricity will take the low-resistance path thus formed back to the battery: this is known as a short-circuit. Hopefully a

Fault diagnosis

short-circuit will blow a fuse, but otherwise it may cause burning of the insulation (and possibly further short-circuits) or even a fire. This is why it is inadvisable to bypass persistently blowing fuses with silver foil or wire.

Spares and tool kit

Most vehicles are supplied only with sufficient tools for wheel changing; the *Maintenance and minor repair* tool kit detailed in *Tools and working facilities,* with the addition of a hammer, is probably sufficient for those repairs that most motorists would consider attempting at the roadside. In addition a few items which can be fitted without too much trouble in the event of a breakdown should be carried. Experience and available space will modify the list below, but the following may save having to call on professional assistance:

Spark plugs, clean and correctly gapped
HT lead and plug cap – long enough to reach the plug furthest from the distributor
Distributor rotor
Drivebelt(s) – emergency type may suffice
Spare fuses
Set of principal light bulbs
Tin of radiator sealer and hose bandage
Exhaust bandage
Roll of insulating tape
Length of soft iron wire
Length of electrical flex
Torch or inspection lamp (can double as test lamp)
Battery jump leads
Tow-rope
Ignition water dispersant aerosol
Litre of engine oil
Sealed can of hydraulic fluid
Emergency windscreen
Worm drive clips

If spare fuel is carried, a can designed for the purpose should be used to minimise risks of leakage and collision damage. A first aid kit and a warning triangle, whilst not at present compulsory in the UK, are obviously sensible items to carry in addition to the above.

When touring abroad it may be advisable to carry additional spares which, even if you cannot fit them yourself, could save having to wait while parts are obtained. The items below may be worth considering:

Throttle cable
Cylinder head gasket
Alternator brushes
Tyre valve core

One of the motoring organisations will be able to advise on availability of fuel etc in foreign countries.

Engine will not start

Engine fails to turn when starter operated
Flat battery (recharge, use jump leads, or push start)
Battery terminals loose or corroded
Battery earth to body defective
Engine earth strap loose or broken
Starter motor (or solenoid) wiring loose or broken
Automatic transmission selector in wrong position, or inhibitor switch faulty
Ignition/starter switch faulty
Major mechanical failure (seizure)
Starter or solenoid internal fault (see Chapter 12)

Starter motor turns engine slowly
Partially discharged battery (recharge, use jump leads, or push start)
Battery terminals loose or corroded
Battery earth to body defective
Engine earth strap loose

Carrying a few spares can save you a long walk

Fault diagnosis

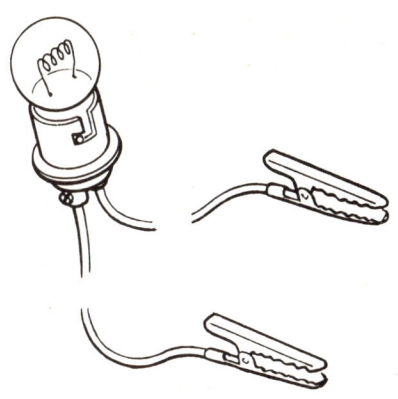

A simple test lamp is useful for tracing electrical faults

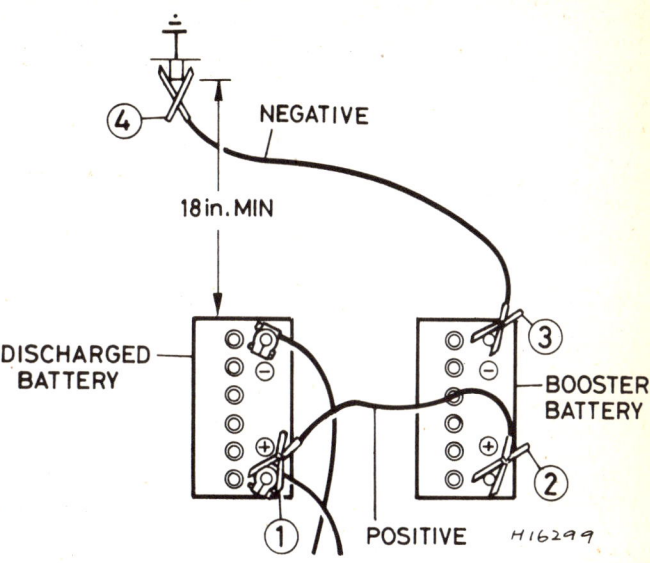

Jump start lead connections for negative earth – connect leads in order shown

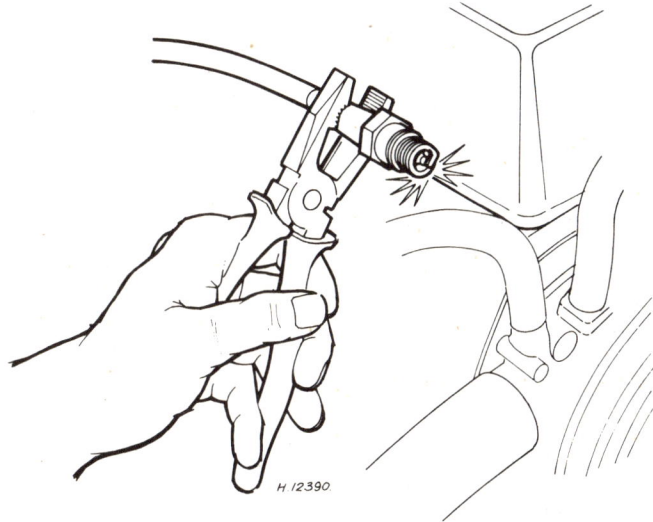

Hold plug firmly against engine metal, crank engine and check for spark. Note use of insulated tool. Use a spare plug, not one removed from the engine (fire hazard)

Starter motor (or solenoid) wiring loose
Starter motor internal fault (see Chapter 12)

Starter motor spins without turning engine
Flywheel gear teeth damaged or worn
Starter motor mounting bolts loose

Engine turns normally but fails to start
Damp or dirty HT leads and distributor cap (crank engine and check for spark – try moisture dispersant such as Holts Wet Start)
No fuel in tank (check for delivery at carburettor or start injector)
Fouled or incorrectly gapped spark plugs (remove, clean and regap)
Other ignition system fault (see Chapter 4)
Other fuel system fault (see Chapter 3)
Poor compression (see Chapter 1)
Major mechanical failure (eg camshaft drive)

Engine fires but will not run
Air leaks at carburettor or inlet manifold
Fuel starvation (see Chapter 3)
Ballast resistor defective, or other ignition fault (see Chapter 4)

Engine cuts out and will not restart

Engine cuts out suddenly – ignition fault
Loose or disconnected LT wires
Wet HT leads or distributor cap (after traversing water splash)
Coil failure (check for spark)
Other ignition fault (see Chapter 4)

Engine misfires before cutting out – fuel fault
Fuel tank empty
Fuel pump defective or filter blocked (check for delivery)
Fuel tank filler vent blocked (suction will be evident on releasing cap)
Carburettor needle valve sticking
Carburettor jets blocked (fuel contaminated)
Other fuel system fault (see Chapter 3)

Engine cuts out – other causes
Serious overheating
Major mechanical failure (eg camshaft drive)

Engine overheats

Slack or broken drivebelt – retension or renew (Chapter 2)
Coolant loss due to internal or external leakage (see Chapter 2)
Thermostat defective
Low oil level
Brakes binding
Radiator clogged externally or internally
Electric cooling fan not operating correctly
Engine waterways clogged
Ignition timing incorrect or automatic advance malfunctioning
Mixture too weak

Note: *Do not add cold water to an overheated engine or damage may result*

Low engine oil pressure

Warning light illuminated with engine running
Oil level low or incorrect grade
Defective sender unit

Fault diagnosis

Wire to sender unit earthed
Engine overheating
Oil filter clogged or bypass valve defective
Oil pressure relief valve defective
Oil pick-up strainer clogged
Oil pump worn or mountings loose
Worn main or big-end bearings

Note: *Low oil pressure in a high-mileage engine at tickover is not necessarily a cause for concern. Sudden pressure loss at speed is far more significant. In any event, check the warning light sender before condemning the engine.*

Engine noises

Pre-ignition (pinking) on acceleration
Incorrect grade of fuel
Ignition timing incorrect
Distributor faulty or worn
Worn or maladjusted carburettor
Excessive carbon build-up in engine
Crankcase ventilation system blocked (can also cause oil leaks)

Whistling or wheezing noises
Leaking vacuum hose
Leaking carburettor or manifold gasket
Blowing head gasket

Tapping or rattling
Incorrect valve clearances
Worn valve gear
Worn timing chain or belt
Broken piston ring (ticking noise)

Knocking or thumping
Unintentional mechanical contact (eg fan blades)
Worn drivebelt
Peripheral component fault (generator, water pump etc)
Worn big-end bearings (regular heavy knocking, perhaps less under load)
Worn main bearings (rumbling and knocking, perhaps worsening under load)
Piston slap (most noticeable when cold)

Chapter 1 Engine

For modifications, and information applicable to later models, see Supplement at end of manual

Contents

Part A: In-line engine

Auxiliary shaft – examination and renovation	33
Auxiliary shaft – refitting	39
Auxiliary shaft – removal	25
Camshaft – removal and refitting	9
Camshaft and tappets – examination and renovation	32
Camshaft drivebelt – removal, refitting and tensioning	6
Compression test – description and interpretation	45
Crankcase ventilation system – general	5
Crankshaft and bearings – examination and renovation	29
Crankshaft and main bearings – refitting	38
Crankshaft and main bearings – removal	26
Crankshaft spigot bearing – removal and refitting	15
Cylinder block and bores – examination and renovation	30
Cylinder head – decarbonising, valve grinding and renovation	36
Cylinder head – dismantling	35
Cylinder head – reassembly	37
Cylinder head – removal and refitting	10
Engine – dismantling for overhaul	24
Engine – reassembly after overhaul	40
Engine – reconnection to transmission	41
Engine – refitting (alone)	43
Engine – refitting (with transmission)	42
Engine – removal (alone)	21
Engine – removal (with transmission)	22
Engine – separation from transmission	23
Engine dismantling and reassembly – general	8
Engine mountings – removal and refitting	19
Examination and renovation – general	27
Fault diagnosis – engine	46
Flywheel/driveplate – examination and renovation	34
Flywheel/driveplate – removal and refitting	12
Flywheel ring gear – renewal	14
General description	1
Initial start-up after overhaul or major repair	44
Maintenance and inspection	2
Major operations possible with the engine installed	7
Methods of engine removal	20
Oil filter – renewal	3
Oil pump – examination and renovation	28
Oil pump – removal and refitting	17
Oil seal (crankshaft rear) – renewal	13
Oil seals (front) – renewal	11
Pistons and connecting rods – examination and renovation	31
Pistons and connecting rods – removal and refitting	18
Sump – removal and refitting	16
Valve clearances – checking and adjustment	4

Part B: V6 engine

Camshaft – removal and refitting	57
Camshafts and rocker gear – examination and renovation	81
Compression test – description and interpretation	96
Crankcase ventilation system – general	51
Crankshaft and bearings – examination and renovation	76
Crankshaft and main bearings – refitting	86
Crankshaft and main bearings – removal	72
Crankshaft oil seal (front) – renewal	58
Crankshaft oil seal (rear) – renewal	60
Crankshaft spigot bearing – removal and refitting	62
Cylinder block and liners – examination and renovation	77
Cylinder head – decarbonising, valve grinding and renovation	83
Cylinder head – dismantling	82
Cylinder head – reassembly	84
Cylinder heads – refitting (engine removed)	88
Cylinder heads – removal and refitting (engine installed)	56
Cylinder heads – removal (engine removed)	70
Cylinder liners – refitting and checking protrusions	85
Cylinder liners – removal	73
Engine – dismantling for overhaul	69
Engine – reassembly after overhaul	90
Engine – reconnection to transmission	91
Engine – refitting (alone)	93
Engine – refitting (with transmission)	92
Engine – removal (alone)	66
Engine – removal (with transmission)	67
Engine – separation from transmission	68
Engine dismantling and reassembly – general	53
Engine mountings – removal and refitting	64
Examination and renovation – general	74
Fault diagnosis – engine	97
Flywheel/driveplate – examination and renovation	79
Flywheel/driveplate – removal and refitting	59
Flywheel ring gear – renewal	61
General description	47
Initial start-up after overhaul or major repair	95
Maintenance and inspection	48
Major operations possible with the engine installed	52
Methods of engine removal	65
Oil filter – renewal	49
Oil level sensor – testing, removal and refitting	94
Oil pump – examination and renovation	75
Oil pump – removal and refitting	55
Pistons and connecting rods – examination and renovation	78
Pistons and connecting rods – refitting	87
Pistons and connecting rods – removal	71
Sump – removal and refitting	63
Timing chains and sprockets – examination and renovation	80
Timing chains and sprockets – removal and refitting	54
Timing scale – checking and adjusting	89
Valve clearances – checking and adjustment	50

Chapter 1 Engine

Specifications

Part A: In-line engine
General

Engine type	4-stroke, 4-cylinder in-line, ohc, spark ignition, water-cooled
Identification:	
B23ET	Fuel injection, turbocharged, up to 1984
B230E	Fuel injection, normally aspirated, from 1985 model year
B230ET	Fuel injection, turbocharged, from 1985 model year
B230 K	Carburettor, normally aspirated, from 1985 model year
Bore	96 mm (3.780 in) nominal
Stroke	80 mm (3.150 in)
Cubic capacity	2316 cc (141.4 cu in)
Compression ratio:	
B23ET and B230ET	9.0:1
B230E and B230 K	10.3:1
Compression pressure:	
Overall value	9 to 11 bar (131 to 160 lbf/in^2)
Variation between cylinders	2 bar (29 lbf/in^2) max
Maximum power:	
B23ET	127 kW (170 bhp) @ 5700 rpm
B230E	96 kW (129 bhp) @ 5500 rpm
B230ET	134 kW (180 bhp) @ 5800 rpm
B230K	84 kW (113 bhp) @ 5200 rpm
Maximum torque:	
B23ET	250 Nm (184 lbf ft) @ 3400 rpm
B230E	190 Nm (140 lbf ft) @ 3300 rpm
B230ET	260 Nm (192 lbf ft) @ 3400 rpm
B230K	192 Nm (142 lbf ft) @ 2500 rpm
Firing order	1-3-4-2 (No 1 at front)

Cylinder head

Warp limit – acceptable for use:	
Lengthwise	0.50 mm (0.020 in)
Across	0.25 mm (0.010 in)
Warp limit – acceptable for refinishing:	
Lengthwise	1.00 mm (0.039 in)
Across	0.50 mm (0.020 in)
Height:	
New	146.1 mm (5.752 in)
Minimum after refinishing	145.6 mm (5.732 in)

Cylinder bores

Standard sizes:	
C	96.00 to 96.01 mm (3.7795 to 3.7799 in)
D	96.01 to 96.02 mm (3.7799 to 3.7803 in)
E	96.02 to 96.03 mm (3.7803 to 3.7807 in)
G	96.04 to 96.05 mm (3.7811 to 3.7815 in)
First oversize	96.30 mm (3.7913 in)
Second oversize	96.60 mm (3.8031 in)
Wear limit	0.1 mm (0.004 in)

Pistons

Height:	
B23	75.4 mm (2.9685 in)
B230	64.7 mm (2.5472 in)
Weight:	
B23	562 ± 7 g (19.8 ± 0.25 oz)
B230	535 ± 7 g (18.9 ± 0.25 oz)
Weight variation in same engine	12 g (0.42 oz) max
Running clearance in bore:	
B23	0.05 to 0.07 mm (0.0020 to 0.0028 in)
B230	0.01 to 0.03 mm (0.0004 to 0.0012 in)

Piston rings

Height:	
Top compression (B23/B230) and second compression (B230)	1.728 to 1.740 mm (0.0680 to 0.0685 in)
Second compression (B23)	1.978 to 1.990 mm (0.0779 to 0.0783 in)
Oil control (B23)	3.975 to 3.990 mm (0.1565 to 0.1571 in)
Oil control (B230)	3.475 to 3.490 mm (0.1368 to 0.1374 in)
Clearance in groove:	
Top compression	0.060 to 0.092 mm (0.0024 to 0.0036 in)
Second compression	0.040 to 0.072 mm (0.0016 to 0.0028 in)
Oil control	0.030 to 0.065 mm (0.0012 to 0.0026 in)

Chapter 1 Engine

End gap (in 96.00 mm/3.7795 in bore):
   Compression rings, B23 .... 0.40 to 0.65 mm (0.016 to 0.026 in)
   Compression rings, B230 .... 0.30 to 0.55 mm (0.012 to 0.022 in)
   Oil control .... 0.30 to 0.60 mm (0.012 to 0.024 in)

Gudgeon pins
Diameter, standard:
   B23 .... 24.00 mm (0.9449 in)
   B230 .... 23.00 mm (0.9055 in)
Oversize available .... + 0.05 mm (0.0020 in)
Fit in connecting rod .... Light thumb pressure
Fit in piston .... Firm thumb pressure

Valve clearances (inlet and exhaust)
Checking value:
   Cold engine .... 0.30 to 0.40 mm (0.012 to 0.016 in)
   Warm engine .... 0.35 to 0.45 mm (0.014 to 0.018 in)
Setting value:
   Cold engine .... 0.35 to 0.40 mm (0.014 to 0.016 in)
   Warm engine .... 0.40 to 0.45 mm (0.016 to 0.018 in)
Adjusting shims available .... 3.30 to 4.50 mm (0.1299 to 0.1772 in) in steps of 0.05 mm (0.0020 in)

Inlet valves
Head diameter .... 44 mm (1.732 in)
Stem diameter:
   New .... 7.955 to 7.970 mm (0.3132 to 0.3138 in)
   Wear limit .... 7.935 mm (0.3124 in)
Valve head angle .... 44° 30'

Exhaust valves
Head diameter .... 35 mm (1.378 in)
Stem diameter (B230 E and K):
   New .... 7.945 to 7.960 mm (0.3128 to 0.3134 in)
   Wear limit .... 7.925 mm (0.3120 in)
Stem diameter (B23 and 230 ET):
   32 mm (1.26 in) from head .... Same as B230 E and K
   16 mm (0.63 in) from tip:
      New .... 7.965 to 7.980 mm (0.3136 to 0.3142 in)
      Wear limit .... 7.945 mm (0.3128 in)
Valve head angle .... 44° 30'

Valve seat inserts
Diameter (standard):
   Inlet .... 46.00 mm (1.8110 in)
   Exhaust .... 38.00 mm (1.4961 in)
Oversizes available .... + 0.25 and 0.50 mm (0.0098 and 0.0197 in)
Fit in cylinder head .... 0.17 mm (0.0067 in) interference
Valve seat angle .... 45° 00'

Valve guides
Length .... 52 mm (2.047 in)
Internal diameter .... 8.000 to 8.022 mm (0.3150 to 0.3158 in)
Height above cylinder head:
   Inlet .... 15.4 to 15.6 mm (0.606 to 0.614 in)
   Exhaust .... 17.9 to 18.1 mm (0.705 to 0.713 in)
Stem-to-guide clearance:
   New (inlet) .... 0.030 to 0.060 mm (0.0012 to 0.0024 in)
   New (exhaust) .... 0.060 to 0.090 mm (0.0024 to 0.0035 in)
   Wear limit (inlet and exhaust) .... 0.15 mm (0.0059 in)
Fit in head .... Press (9000 N/2025 lbf minimum)
External oversizes available .... 3 (marked by grooves)

Valve springs
Diameter .... 32.5 mm (1.28 in)
Free length .... 45.0 mm (1.77 in)
Length under load of:
   280 to 320 N (63 to 72 lbf) .... 38.0 mm (1.50 in)
   710 to 790 N (160 to 178 lbf) .... 27.0 mm (1.06 in)

Tappets (cam followers)
Diameter .... 36.975 to 36.995 mm (1.4557 to 1.4565 in)
Height .... 30.000 to 31.000 mm (1.1811 to 1.2205 in)
Shim clearance in tappet .... 0.009 to 0.064 mm (0.0004 to 0.0025 in)
Tappet clearance in cylinder head .... 0.030 to 0.075 mm (0.0012 to 0.0030 in)

Camshaft
Identification letter (stamped on end):
    B23ET .... B
    B230E V
    B230ET .... A
    B230K X
Maximum lift:
    A .... 10.50 mm (0.4134 in)
    B .... 10.60 mm (0.4172 in)
    V .... 11.37 mm (0.4476 in)
    X .... 10.65 mm (0.4193 in)
Bearing journal diameter .... 29.950 to 29.970 mm (1.1791 to 1.1799 in)
Bearing running clearance:
    New .... 0.030 to 0.071 mm (0.0012 to 0.0028 in)
    Wear limit .... 0.15 mm (0.0059 in)
Endfloat .... 0.1 to 0.4 mm (0.004 to 0.016 in)

Auxiliary shaft
Bearing journal diameter:
    Front .... 46.975 to 47.000 mm (1.8494 to 1.8504 in)
    Centre .... 43.025 to 43.050 mm (1.6939 to 1.6949 in)
    Rear .... 42.925 to 42.950 mm (1.6900 to 1.6909 in)
Bearing running clearance .... 0.020 to 0.075 mm (0.0008 to 0.0030 in)
Endfloat .... 0.20 to 0.46 mm (0.008 to 0.018 in)

Crankshaft – B23
Run-out .... 0.05 mm (0.0019 in) max
Endfloat .... 0.25 mm (0.0098 in) max
Main bearing journal diameter:
    Standard .... 63.451 to 63.464 mm (2.4981 to 2.4986 in)
    First undersize .... 63.197 to 63.210 mm (2.4881 to 2.4886 in)
    Second undersize .... 62.943 to 62.956 mm (2.4781 to 2.4786 in)
Main bearing running clearance .... 0.028 to 0.083 mm (0.0011 to 0.0033 in)
Main bearing out-of-round .... 0.07 mm (0.0028 in) max
Main bearing taper .... 0.05 mm (0.0020 in) max
Connecting rod bearing journal diameter:
    Standard .... 53.987 to 54.000 mm (2.1255 to 2.1260 in)
    First undersize .... 53.733 to 53.746 mm (2.1155 to 2.1160 in)
    Second undersize .... 53.479 to 53.492 mm (2.1055 to 2.1060 in)
Connecting rod bearing running clearance .... 0.024 to 0.070 mm (0.0009 to 0.0028 in)
Connecting rod bearing out-of-round .... 0.5 mm (0.0019 in) max
Connecting rod bearing tape .... 0.05 mm (0.0019 in) max

Crankshaft – B230
Run-out .... 0.025 mm (0.0010 in) max
Endfloat .... 0.080 to 0.270 mm (0.003 to 0.011 in)
Main bearing journal diameter:
    Standard .... 54.987 to 55.000 mm (2.1648 to 2.1654 in)
    First undersize .... 54.737 to 54.750 mm (2.1550 to 2.1555 in)
    Second undersize .... 54.487 to 54.500 mm (2.1452 to 2.1457 in)
Main bearing running clearance .... 0.024 to 0.072 mm (0.0010 to 0.0028 in)
Main bearing out-of-round .... 0.004 mm (0.0002 in) max
Main bearing taper .... 0.004 mm (0.0002 in) max
Connecting rod bearing journal diameter:
    Standard .... 48.984 to 49.005 mm (1.9285 to 1.9293 in)
    First undersize .... 48.734 to 48.755 mm (1.9187 to 1.9195 in)
    Second undersize .... 48.484 to 48.505 mm (1.9088 to 1.9096 in)
Connecting rod bearing running clearance .... 0.023 to 0.067 mm (0.0009 to 0.0026 in)
Connecting rod bearing out-of-round .... 0.004 mm (0.0002 in) max
Connecting rod bearing taper .... 0.004 mm (0.0002 in) max

Connecting rods
Length between centres:
    B23 .... 145 mm (5.709 in)
    B230 .... 152 mm (5.984 in)
Endfloat on crankshaft:
    B23 .... 0.15 to 0.35 mm (0.006 to 0.014 in)
    B230 .... 0.25 to 0.45 mm (0.010 to 0.018 in)
Weight variation in same engine:
    B23 .... 10 g (0.35 oz) max
    B230 .... 20 g (0.71 oz) max

Flywheel
Run-out .... 0.02 mm (0.0008 in) per 100 mm (3.9 in) diameter

Chapter 1 Engine

Lubrication system
Oil capacity (drain and refill):
    Engine only .... 3.35 litres (5.9 pints)
    Engine and oil filter .... 3.85 litres (6.8 pints)
    For Turbo oil cooler add .... 0.60 litre (1.1 pints)
Oil type/specification .... Multigrade engine oil, viscosity range SAE 10W/30 to 15W/50, to API SF/CC, SF/CD or better (Duckhams QXR, QS, Hypergrade Plus or Hypergrade)

Oil pressure (warm engine @ 2000 rpm) .... 2.5 to 6.0 bar (36 to 87 lbf/in²)
Oil filter... Champion C102

Oil pump
Type .... Gear, driven from intermediate shaft
Clearances:
    Endfloat .... 0.02 to 0.12 mm (0.0008 to 0.0047 in)
    Gear side clearance .... 0.02 to 0.09 mm (0.0008 to 0.0035 in)
    Backlash .... 0.15 to 0.35 mm (0.0059 to 0.0138 in)
    Driving gear bearing clearance .... 0.032 to 0.070 mm (0.0013 to 0.0028 in)
    Idler gear bearing clearance .... 0.014 to 0.043 mm (0.0006 to 0.0017 in)
Relief valve spring free length .... 39.20 mm (1.543 in)
Relief valve spring length under load of:
    46 to 54 N (10 to 12 lbf) .... 26.25 mm (1.033 in)
    62 to 78 N (14 to 18 lbf) .... 21.00 mm (0.827 in)

Torque wrench settings*

	Nm	lbf ft
Cylinder head bolts:		
Stage 1	20	15
Stage 2	60	44
Stage 3	Tighten 90° further	Tighten 90° further
Main bearing caps	110	81
Connecting rod bearing caps (B23):		
New bolts	70	52
Used bolts	63	47
Connecting rod bearing caps (B230)†:		
Stage 1	20	15
Stage 2	Tighten 90° further	Tighten 90° further
Flywheel/driveplate (use new bolts)	70	52
Spark plugs (dry threads)	25 ± 5	18 ± 4
Camshaft sprocket	50	37
Intermediate shaft sprocket	50	37
Camshaft bearing caps	20	15
Crankshaft pulley/sprocket bolt (B23)	165	122
Crankshaft pulley/sprocket bolt (B230):		
Stage 1	60	44
Stage 2	Tighten 60° further	Tighten 60° further
Sump bolts	11	8

*Oiled threads unless otherwise stated
†Renew bolts if length exceeds 55.5 mm (2.185 in)

Part B: V6 engine

General
Engine type .... 4-stroke, V6, twin ohc, spark ignition, water-cooled
Maker's designation B28E
Bore .... 91 mm (3.583 in) nominal
Stroke .... 73 mm (2.874 in)
Cubic capacity .... 2849 cc (173.8 cu in)
Compression ratio 9.5:1
Compression pressure:
    Overall value .... 8 to 11 bar (116 to 160 lbf/in²)
    Variation between cylinders .... 2 bar (29 lbf/in²) max
Maximum power .... 115 kW (154 bhp) @ 5700 rpm
Maximum torque .... 235 Nm (173 lbf ft) @ 3000 rpm
Firing order .... 1-6-3-5-2-4 (No 1 LH rear)

Cylinder head
Warp limit – acceptable for use .... 0.05 mm (0.002 in) per 100 mm (3.9 in) length
Warp limit – acceptable for refinishing .... No refinishing allowed
Height (new) .... 111.07 mm (4.373 in)

Cylinder liners
Bore:
    Grade 1 (takes grade A piston) .... 91.00 to 91.01 mm (3.5827 to 3.5831 in)
    Grade 2 (takes grade B piston) .... 91.01 to 91.02 mm (3.5831 to 3.5835 in)
    Grade 3 (takes grade C piston) .... 91.02 to 91.03 mm (3.5835 to 3.5839 in)
Liner protrusion above block (depends on gasket; consult your dealer):
    Checking value (used seals) .... 0.14 to 0.23 mm (0.0055 to 0.0091 in)
    Setting value (new seals) .... 0.16 to 0.23 mm (0.0063 to 0.0091 in)

Liner seal thickness:
  Blue mark .... 0.070 to 0.105 mm (0.0028 to 0.0041 in)
  White mark .... 0.085 to 0.120 mm (0.0034 to 0.0047 in)
  Red mark .... 0.105 to 0.140 mm (0.0041 to 0.0055 in)
  Yellow mark .... 0.130 to 0.165 mm (0.0051 to 0.0065 in)

Pistons
Diameter (matched to liners):
  Grade A .... 90.970 to 90.980 mm (3.5815 to 3.5819 in)
  Grade B .... 90.980 to 90.990 mm (3.5819 to 3.5823 in)
  Grade C .... 90.990 to 91.000 mm (3.5823 to 3.5827 in)
Clearance in bore .... 0.020 to 0.040 mm (0.0008 to 0.0016 in)
Height .... 65.3 mm (2.571 in)
Weight .... 455 ± 3 g (16.05 ± 0.11 oz)
Gudgeon pin bore:
  Blue mark .... 23.510 to 23.573 mm (0.9256 to 0.9257 in)
  White mark .... 23.507 to 23.510 mm (0.9255 to 0.9256 in)
  Red mark .... 23.504 to 23.507 mm (0.9254 to 0.9255 in)

Gudgeon pins
Diameter:
  Blue mark .... 23.497 to 23.500 mm (0.9251 to 0.9252 in)
  White mark .... 23.494 to 23.497 mm (0.9250 to 0.9251 in)
  Red mark .... 23.491 to 23.494 mm (0.9248 to 0.9250 in)
Clearance in connecting rod .... 0.020 to 0.041 mm (0.0008 to 0.0016 in)
Clearance in piston .... 0.010 to 0.016 mm (0.0004 to 0.0006 in)

Piston rings
Clearance in groove:
  Top compression .... 0.045 to 0.074 mm (0.0018 to 0.0029 in)
  Second compression .... 0.025 to 0.054 mm (0.0010 to 0.0021 in)
  Oil control .... 0.009 to 0.233 mm (0.0004 to 0.0092 in)
End gap (in 91.00 mm/3.5827 in bore):
  Top and second compression .... 0.40 to 0.60 mm (0.016 to 0.024 in)
  Oil control .... 0.40 to 1.45 mm (0.016 to 0.057 in)

Valve clearances
Inlet:
  Cold engine .... 0.10 to 0.15 mm (0.004 to 0.006 in)
  Warm engine .... 0.15 to 0.20 mm (0.006 to 0.008 in)
Exhaust:
  Cold engine .... 0.25 to 0.30 mm (0.010 to 0.012 in)
  Warm engine .... 0.30 to 0.35 mm (0.012 to 0.014 in)

Inlet valves
Head diameter .... 44 mm (1.73 in)
Stem diameter:
  26.5 mm (1.04 in) from head .... 7.965 to 7.980 mm (0.3136 to 0.3142 in)
  Just below collet groove .... 7.975 to 7.990 mm (0.3140 to 0.3146 in)
Valve head angle .... 29° 30'

Exhaust valves
Head diameter .... 37 mm (1.46 in)
Stem diameter:
  32 mm (1.26 in) from head .... 7.945 to 7.960 mm (0.3128 to 0.3134 in)
  Just below collet groove .... 7.965 to 7.980 mm (0.3136 to 0.3142 in)
Valve head angle .... 44° 30'

Valve seat inserts
Fit in cylinder head .... 0.070 to 0.134 mm (0.0028 to 0.0053 in) interference
Oversizes available 3
Valve seat angles:
  Inlet .... Compound (60° – 30° – 15°)
  Exhaust .... 45°

Valve guides
Internal diameter .... 8.000 to 8.022 mm (0.3150 to 0.3158 in)
Fit in cylinder head .... 0.052 to 0.095 mm (0.0020 to 0.0037 in) interference
External oversizes available .... 3 (marked by grooves)

Valve springs
Free length .... 47.1 mm (1.8543 in)
Length under load:
  230 to 266 N (52 to 60 lbf) .... 40.0 mm (1.5748 in)
  613 to 689 N (138 to 155 lbf) .... 30.0 mm (1.811 in)

Are your plugs trying to tell you something?

Normal.
Grey-brown deposits, lightly coated core nose. Plugs ideally suited to engine, and engine in good condition.

Heavy Deposits.
A build up of crusty deposits, light-grey sandy colour in appearance.
Fault: Often caused by worn valve guides, excessive use of upper cylinder lubricant, or idling for long periods.

Lead Glazing.
Plug insulator firing tip appears yellow or green/yellow and shiny in appearance.
Fault: Often caused by incorrect carburation, excessive idling followed by sharp acceleration. Also check ignition timing.

Carbon fouling.
Dry, black, sooty deposits.
Fault: over-rich fuel mixture.
Check: carburettor mixture settings, float level, choke operation, air filter.

Oil fouling.
Wet, oily deposits. Fault: worn bores/piston rings or valve guides; sometimes occurs (temporarily) during running-in period.

Overheating.
Electrodes have glazed appearance, core nose very white – few deposits. Fault: plug overheating. Check: plug value, ignition timing, fuel octane rating (too low) and fuel mixture (too weak).

Electrode damage.
Electrodes burned away; core nose has burned, glazed appearance. Fault: pre-ignition. Check: for correct heat range and as for 'overheating'.

Split core nose.
(May appear initially as a crack). Fault: detonation or wrong gap-setting technique.
Check: ignition timing, cooling system, fuel mixture (too weak).

WHY DOUBLE COPPER IS BETTER FOR YOUR ENGINE.

Unique Trapezoidal Copper Cored Earth Electrode — 50% Larger Spark Area — Copper Cored Centre Electrode

Champion Double Copper plugs are the first in the world to have copper core in both centre <u>and</u> earth electrode. This innovative design means that they run cooler by up to 100°C – giving greater efficiency and longer life. These double copper cores transfer heat away from the tip of the plug faster and more efficiently. Therefore, Double Copper runs at cooler temperatures than conventional plugs giving improved acceleration response and high speed performance with no fear of pre-ignition.

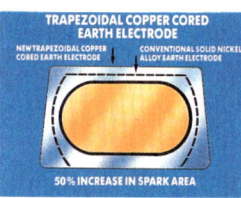 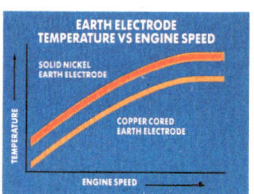

Champion Double Copper plugs also feature a unique trapezoidal earth electrode giving a 50% increase in spark area. This, together with the double copper cores, offers greatly reduced electrode wear, so the spark stays stronger for longer.

 FASTER COLD STARTING

 FOR UNLEADED OR LEADED FUEL

 ELECTRODES UP TO 100°C COOLER

 BETTER ACCELERATION RESPONSE

 LOWER EMISSIONS

 50% BIGGER SPARK AREA

 THE LONGER LIFE PLUG

Plug Tips/Hot and Cold.
Spark plugs must operate within well-defined temperature limits to avoid cold fouling at one extreme and overheating at the other.
Champion and the car manufacturers work out the best plugs for an engine to give optimum performance under all conditions, from freezing cold starts to sustained high speed motorway cruising.
Plugs are often referred to as hot or cold. With Champion, the higher the number on its body, the hotter the plug, and the lower the number the cooler the plug.

Plug Cleaning
Modern plug design and materials mean that Champion no longer recommends periodic plug cleaning. Certainly don't clean your plugs with a wire brush as this can cause metal conductive paths across the nose of the insulator so impairing its performance and resulting in loss of acceleration and reduced m.p.g.
However, if plugs are removed, always carefully clean the area where the plug seats in the cylinder head as grit and dirt can sometimes cause gas leakage.
Also wipe any traces of oil or grease from plug leads as this may lead to arcing.

This photographic sequence shows the steps taken to repair the dent and paintwork damage shown above. In general, the procedure for repairing a hole will be similar; where there are substantial differences, the procedure is clearly described and shown in a separate photograph.

First remove any trim around the dent, then hammer out the dent where access is possible. This will minimise filling. Here, after the large dent has been hammered out, the damaged area is being made slightly concave.

Next, remove all paint from the damaged area by rubbing with coarse abrasive paper or using a power drill fitted with a wire brush or abrasive pad. 'Feather' the edge of the boundary with good paintwork using a finer grade of abrasive paper.

Where there are holes or other damage, the sheet metal should be cut away before proceeding further. The damaged area and any signs of rust should be treated with Turtle Wax Hi-Tech Rust Eater, which will also inhibit further rust formation.

For a large dent or hole mix Holts Body Plus Resin and Hardener according to the manufacturer's instructions and apply around the edge of the repair. Press Glass Fibre Matting over the repair area and leave for 20-30 minutes to harden. Then ...

... brush more Holts Body Plus Resin and Hardener onto the matting and leave to harden. Repeat the sequence with two or three layers of matting, checking that the final layer is lower than the surrounding area. Apply Holts Body Plus Filler Paste as shown in Step 5B.

For a medium dent, mix Holts Body Plus Filler Paste and Hardener according to the manufacturer's instructions and apply it with a flexible applicator. Apply thin layers of filler at 20-minute intervals, until the filler surface is slightly proud of the surrounding bodywork.

For small dents and scratches use Holts No Mix Filler Paste straight from the tube. Apply it according to the instructions in thin layers, using the spatula provided. It will harden in minutes if applied outdoors and may then be used as its own knifing putty.

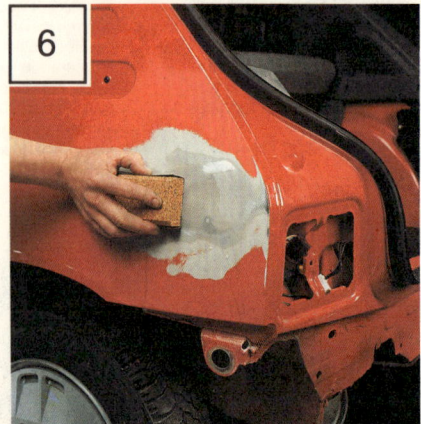

Use a plane or file for initial shaping. Then, using progressively finer grades of wet-and-dry paper, wrapped round a sanding block, and copious amounts of clean water, rub down the filler until glass smooth. 'Feather' the edges of adjoining paintwork.

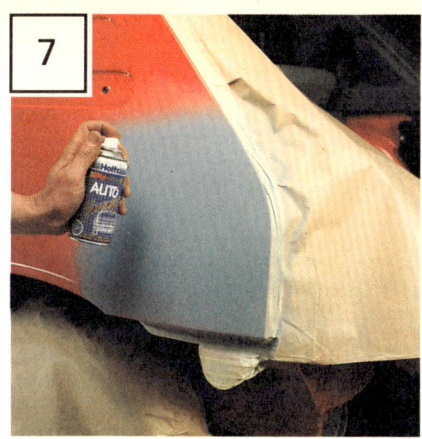

Protect adjoining areas before spraying the whole repair area and at least one inch of the surrounding sound paintwork with Holts Dupli-Color primer.

Fill any imperfections in the filler surface with a small amount of Holts Body Plus Knifing Putty. Using plenty of clean water, rub down the surface with a fine grade wet-and-dry paper – 400 grade is recommended – until it is really smooth.

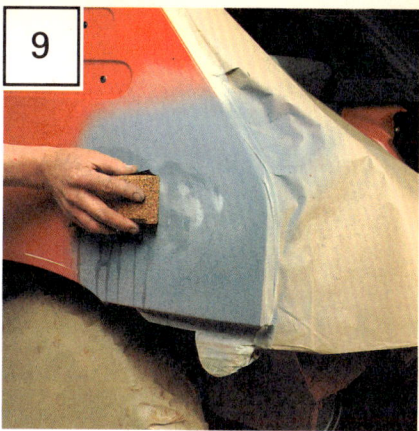

Carefully fill any remaining imperfections with knifing putty before applying the last coat of primer. Then rub down the surface with Holts Body Plus Rubbing Compound to ensure a really smooth surface.

Protect surrounding areas from overspray before applying the topcoat in several thin layers. Agitate Holts Dupli-Color aerosol thoroughly. Start at the repair centre, spraying outwards with a side-to-side motion.

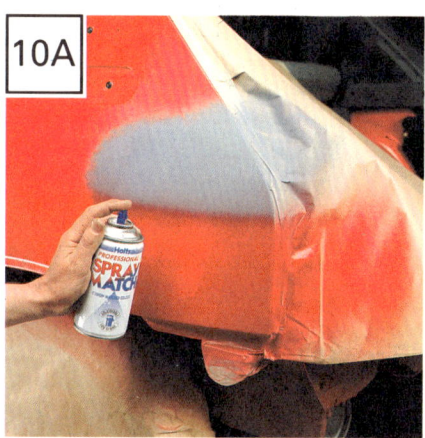

If the exact colour is not available off the shelf, local Holts Professional Spraymatch Centres will custom fill an aerosol to match perfectly.

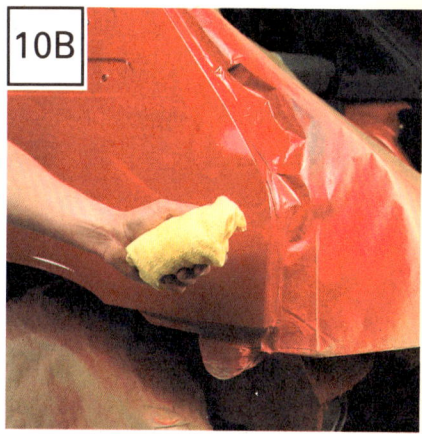

To identify whether a lacquer finish is required, rub a painted unrepaired part of the body with wax and a clean cloth.

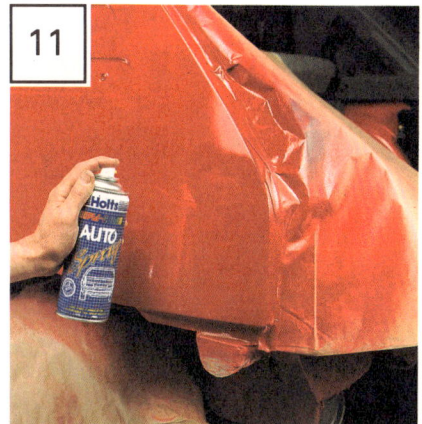

If *no* traces of paint appear on the cloth, spray Holts Dupli-Color clear lacquer over the repaired area to achieve the correct gloss level.

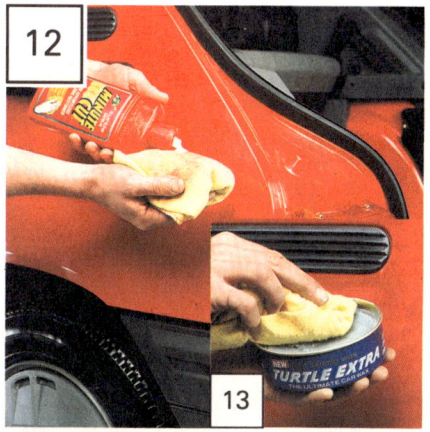

The paint will take about two weeks to harden fully. After this time it can be 'cut' with a mild cutting compound such as Turtle Wax Minute Cut prior to polishing with a final coating of Turtle Wax Extra.

When carrying out bodywork repairs, remember that the quality of the finished job is proportional to the time and effort expended.

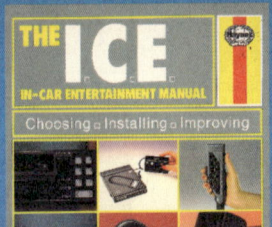

HAYNES No1 for DIY

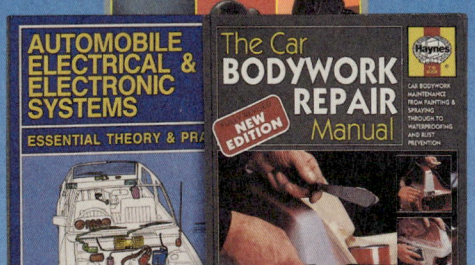

Haynes publish a wide variety of books besides the world famous range of **Haynes Owners Workshop Manuals**. They cover all sorts of DIY jobs. Specialist books such as the **Improve and Modify** series and the **Purchase and DIY Restoration Guides** give you all the information you require to carry out everything from minor modifications to complete restoration on a number of popular cars. In addition there are the publications dealing with specific tasks, such as the **Car Bodywork Repair Manual** and the **In-Car Entertainment Manual**. The **Household DIY** series gives clear step-by-step instructions on how to repair everyday household objects ranging from toasters to washing machines.

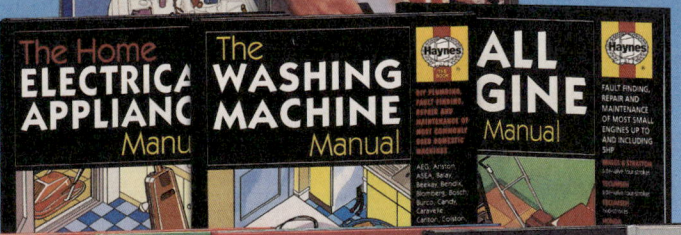

Whether it is under the bonnet or around the home there is a Haynes Manual that can help you save money. Available from motor accessory stores and bookshops or direct from the publisher.

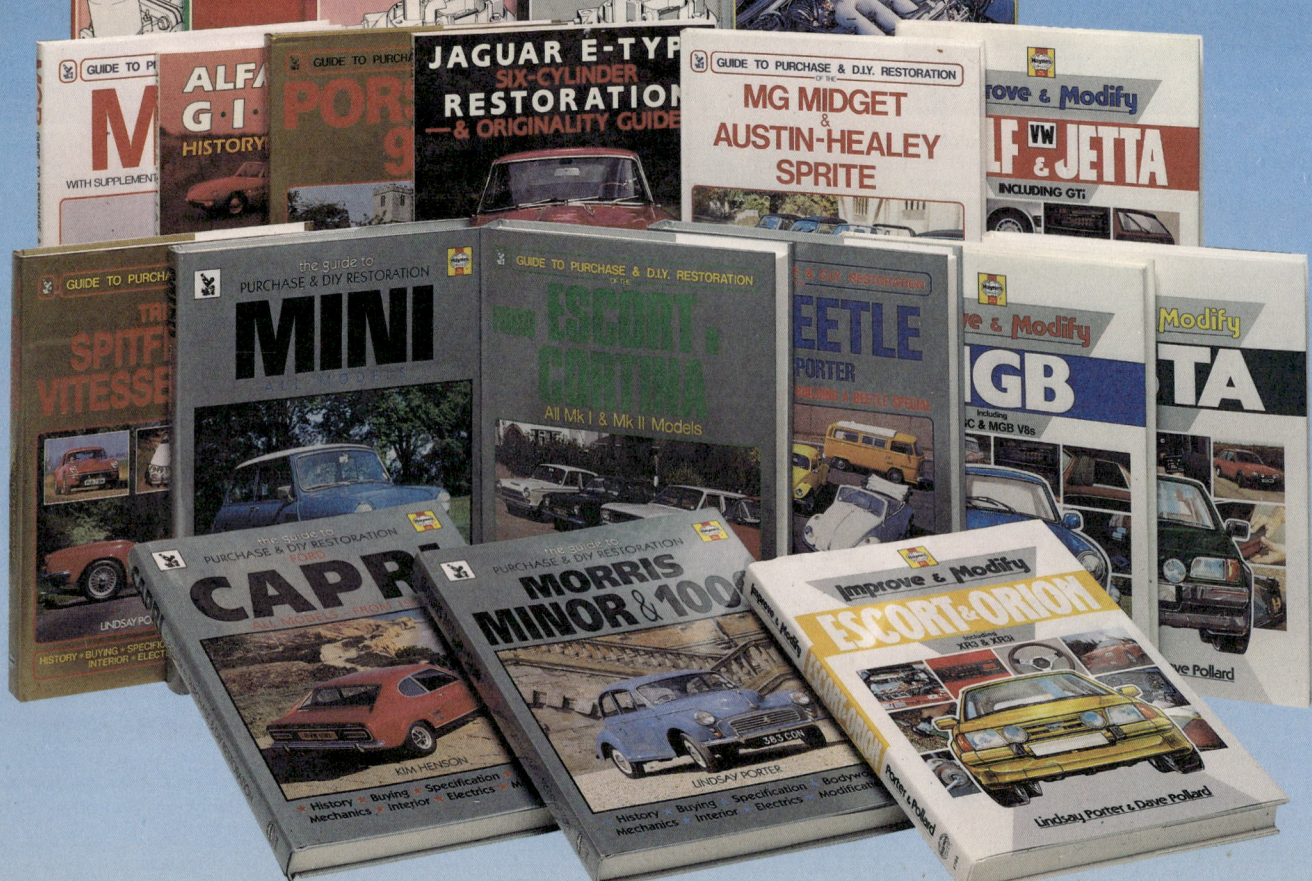

Rocker gear
Rocker arm clearance on shaft .... 0.012 to 0.054 mm (0.0005 to 0.0021 in)

Camshafts
Identification letter .... F
Identification number:
   Left .... 615 or ... 977
   Right .... 616 or ... 978
Maximum lift (at lobe) .... 5.96 mm (0.2347 in)
Bearing journal diameters:
   1 .... 40.440 to 40.465 mm (1.5921 to 1.5931 in)
   2 .... 41.040 to 41.065 mm (1.6158 to 1.6167 in)
   3 .... 41.640 to 41.665 mm (1.6394 to 1.6404 in)
   4 .... 42.240 to 42.265 mm (1.6630 to 1.6640 in)
Bearing running clearance .... 0.035 to 0.085 mm (0.0014 to 0.0034 in)
Endfloat:
   New .... 0.070 to 0.144 mm (0.0028 to 0.0057 in)
   Wear limit .... 0.5 mm (0.020 in)

Crankshaft
Run-out (measured on centre journals) .... 0.02 mm (0.0008 in)
Endfloat .... 0.070 to 0.270 mm (0.0028 to 0.0106 in)
Main bearing running clearance .... 0.038 to 0.088 mm (0.0015 to 0.0035 in)
Connecting rod bearing running clearance .... 0.030 to 0.080 mm (0.0012 to 0.0031 in)
Rear seal diameter:
   Standard .... 79.926 to 80.000 mm (3.1467 to 3.1496 in)
   Undersize .... 79.726 to 79.800 mm (3.1388 to 3.1417 in)
Main bearing journal diameter:
   Standard .... 70.043 to 70.062 mm (2.7576 to 2.7583 in)
   Undersize .... 69.743 to 69.762 mm (2.7458 to 2.7465 in)
Main bearing out-of-round .... 0.007 mm (0.0003 in) max
Main bearing taper .... 0.01 mm (0.0004 in) max
Main bearing shell thickness:
   Standard .... 1.961 to 1.967 mm (0.0772 to 0.0774 in)
   Oversize .... 2.111 to 2.117 mm (0.0831 to 0.0833 in)
Rear main bearing journal width:
   Standard .... 29.20 to 29.25 mm (1.1496 to 1.1516 in)
   First oversize .... 29.40 to 29.45 mm (1.1575 to 1.1594 in)
   Second oversize .... 29.50 to 29.55 mm (1.1614 to 1.1634 in)
   Third oversize .... 29.60 to 29.65 mm (1.1654 to 1.1673 in)
Thrust washer thickness:
   Standard .... 2.30 to 2.35 mm (0.0906 to 0.0925 in)
   First oversize .... 2.40 to 2.45 mm (0.0945 to 0.0965 in)
   Second oversize .... 2.45 to 2.50 mm (0.0965 to 0.0984 in)
   Third oversize .... 2.50 to 2.55 mm (0.0984 to 0.1004 in)
Connecting rod bearing journal diameter:
   Standard .... 52.267 to 52.286 mm (2.0578 to 2.0585 in)
   Undersize .... 51.967 to 51.986 mm (2.0459 to 2.0467 in)
Connecting rod bearing out-of-round .... 0.007 mm (0.0003 in) max
Connecting rod bearing taper .... 0.01 mm (0.0004 in) max
Connecting rod bearing shell thickness:
   Standard .... 1.842 to 1.848 mm (0.0725 to 0.0728 in)
   Oversize .... 1.992 to 1.998 mm (0.0784 to 0.0787 in)

Connecting rods
Length between centres .... 146.15 mm (5.7539 in)
Endfloat on crankshaft (between each pair of rods) ...... 0.20 to 0.38 mm (0.008 to 0.015 in)
Weight variation in same engine .... 2.5 g (0.09 oz) max

Flywheel
Run-out .... 0.05 mm (0.0020 in) max

Lubrication system
Oil capacity (drain and refill):
   Engine only .... 6.0 litres (10.6 pints)
   Engine and oil filter .... 6.5 litres (11.4 pints)
Oil type/specification .... Multigrade engine oil, viscosity range SAE 10W/30 to 15W/50, to API SF/CC, SF/CD or better (Duckhams QXR, QS, Hypergrade Plus or Hypergrade)
Oil pressure (warm engine):
   At 900 rpm .... 1 bar (14.5 lbf/in^2) minimum
   At 3000 rpm .... 4 bar (58 lbf/in^2)
Oil filter .... Champion C102

Oil pump
Type .... Gear, chain-driven from crankshaft sprocket
Clearances:
   Endfloat .... 0.025 to 0.084 mm (0.0010 to 0.0033 in)
   Gear side clearance .... 0.110 to 0.185 mm (0.0043 to 0.0073 in)

Backlash	0.17 to 0.27 mm (0.0067 to 0.0106 in)
Driving gear bearing clearance	0.015 to 0.053 mm (0.0006 to 0.0021 in)
Idler gear bearing clearance	0.015 to 0.051 mm (0.0006 to 0.0020 in)
Relief valve spring free length	89.5 mm (3.524 in)
Relief valve spring length under load of 88 N (20 lbf)	56.5 to 60.5 mm (2.224 to 2.382 in)

Torque wrench settings*

	Nm	lbf ft
Connecting rod bearing caps	45 to 50	33 to 37
Crankshaft pulley nut	240 to 280	177 to 207
Camshaft sprocket	70 to 90	52 to 66
Flywheel (use new bolts)	45 to 50	33 to 37
Spark plugs (dry threads)	12 ± 2	9 ± 1.5
Rocker cover	15	11
Cylinder head bolts (see text):		
Stage 1	60	44
Slacken, then Stage 2A	20	15
Stage 2B	Tighten 106° further	Tighten 106° further
Stage 3 (after warm-up and cooling)	Tighten 45° further	Tighten 45° further
Main bearing nuts (see text):		
Stage 1	30	22
Slacken, then Stage 2A	30 to 35	22 to 26
Stage 2B	Tighten 73 to 77° further	Tighten 73 to 77° further
Oil pump to block	10 to 15	7 to 11
Timing cover bolts	10 to 15	7 to 11

*Oiled threads unless otherwise stated

PART A: IN-LINE ENGINE

1 General description

The four-cylinder engine is of the overhead camshaft type. The cylinders are in line and the engine is mounted vertically and is in a 'north-south' attitude in the engine bay. Cooling is by water.

Drive to the camshaft is by toothed belt and sprockets. The camshaft drivebelt also drives an auxiliary shaft, which in turn drives the oil pump and (on B23 engines) the distributor. Other accessories are driven from the crankshaft pulley by V-belts.

The cylinder block is of cast iron and the cylinder head of aluminium alloy, with pressed-in valve guides and valve seats. The cylinder head is of the crossflow type, the inlet ports being on the left-hand side and the exhaust ports on the right.

The crankshaft runs in five shell type main bearings; the connecting

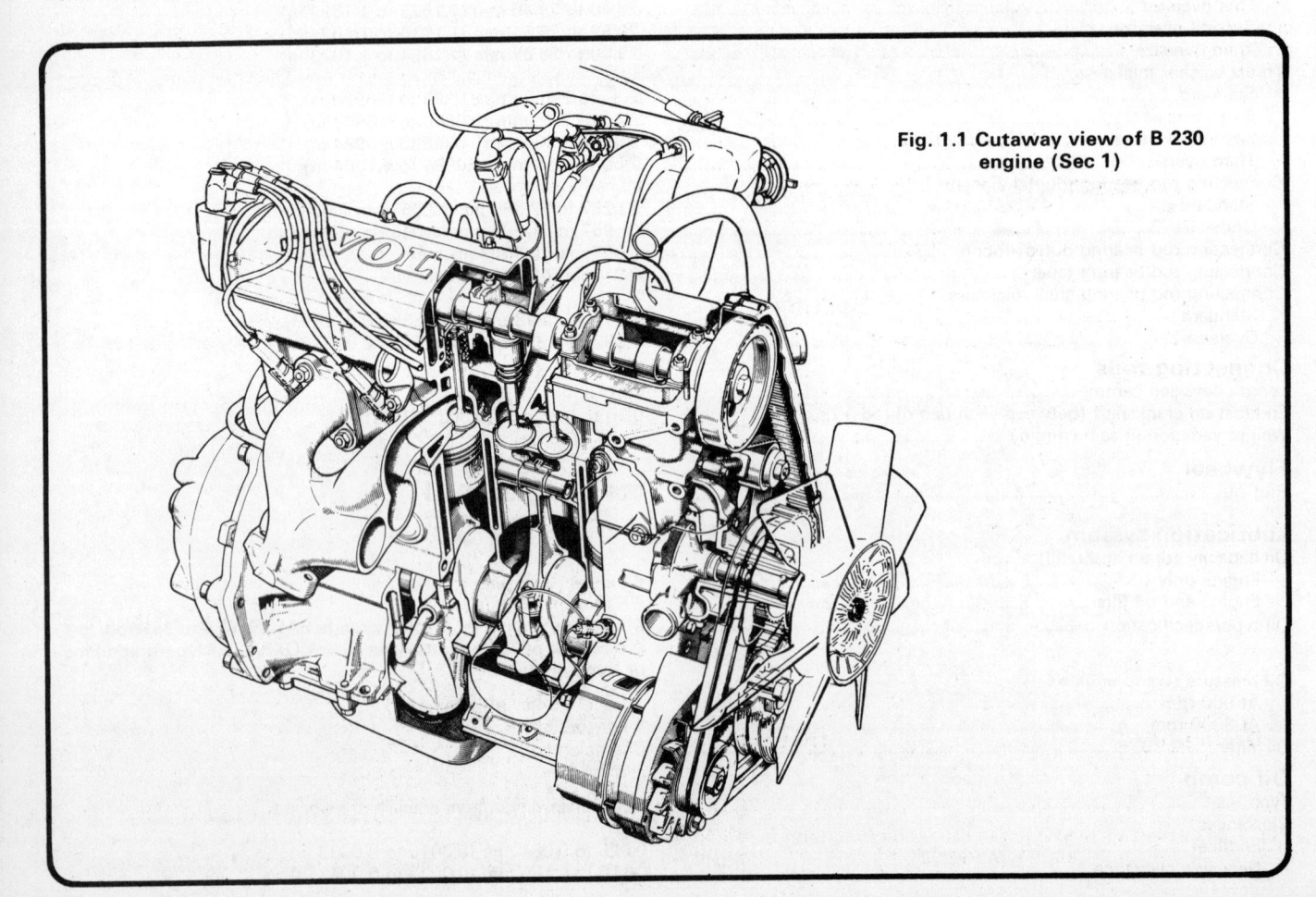

Fig. 1.1 Cutaway view of B 230 engine (Sec 1)

rod big-end bearings are also of the shell type. Crankshaft endfloat is taken by thrust flanges on No 5 main bearing (B23) or by separate thrust washers on No 3 main bearing (B230). The camshaft runs in plain bearings machined directly in the cylinder head.

Valve actuation is direct, the camshaft being located above the valves. The cam lobes depress bucket type tappets; valve clearance is determined by the thickness of the shim in the recess in the top of each tappet.

The lubrication system is of the full-flow, pressure-feed type. Oil is drawn from the sump by a gear type pump, driven from the auxiliary shaft. Oil under pressure passes through a full-flow filter before being fed to the various shaft bearings and to the valvegear. On some models an external oil cooler is fitted, mounted next to the radiator. Turbo models also have an oil feed and return for the turbocharger bearings.

Although the B230 engine series represents a considerable advance on the B23, with many engine components having been redesigned, from the mechanic's point of view the two engine types are almost identical. Significant differences will be found in the Specifications or in the appropriate section of text.

2 Maintenance and inspection

1 Every 250 miles (400 km), weekly, or before a long journey, check the oil level as follows.
2 With the vehicle parked on level ground, and with the engine having been stopped for a few minutes, open and prop the bonnet. Withdraw the dipstick, wipe it on a clean rag and re-insert it fully. Withdraw it again and read the oil level relative to the marks on the end of the stick.
3 The oil level should be in between the 'MAX' and 'MIN' marks on the dipstick. If it is at or below the 'MIN' mark, top up (via the oil filler cap) without delay. The quantity of oil required to raise the level from 'MIN' to 'MAX' on the dipstick is approximately 1 litre. Do not overfill.
4 The rate of oil consumption depends on leaks and on the quantity of oil burnt. External leakage should be obvious. Oil which is burnt may enter the combustion chambers through the valve guides or past the piston rings; excessive blow-by past the rings can also force oil out via the crankcase ventilation system. Driving conditions also affect oil consumption.
5 Every 6000 miles (10 000 km) or six months, whichever comes first, drain the engine oil immediately after a run. Park the vehicle on level ground, position a drain pan of adequate capacity under the sump and remove the drain plug. Allow the oil to drain for at least 15 minutes.
6 Clean the drain plug, the plug washer (if fitted) and the drain plug seat on the sump. Refit and tighten the drain plug, then fill the engine with the correct grade and quantity of oil.
7 Before running the engine, renew the oil filter as described in Section 3. When the engine is next started, there may be a delay in the extinguishing of the oil pressure warning light while the new filter fills with oil. Run the engine and check for leaks from the filter and drain plug, then stop the engine and check the oil level.
8 Every 24 000 miles (40 000 km nominal) or two years, check the valve clearances and adjust if necessary. See Section 4.
9 At the same interval, inspect the crankcase ventilation system hoses, breathers etc and clean or renew them as necessary. See Section 5.
10 Every 48 000 miles (80 000 km) or four years, renew the camshaft drivebelt (Section 6).
11 Regularly inspect the engine for leaks of oil, fuel or coolant, and rectify as necessary.

3 Oil filter – renewal

1 The oil filter is located low down on the right-hand side of the engine block. Access is particularly bad on Turbo models.
2 Raise and support the front of the vehicle, and remove the engine undertray if necessary. Place a drain pan below the filter.
3 Unscrew the filter from its seat using a chain or strap wrench (photo). If a suitable wrench is not available, drive a large screwdriver through the filter casing and use that as a lever to unscrew it. Be prepared for oil spillage.
4 Discard the old filter, remembering it is full of oil. Wipe clean the filter seat on the block.
5 Smear clean engine oil or grease onto the sealing ring of the new filter. Screw the new filter into position (photo).
6 Tighten the new filter by hand only, three-quarters of a turn beyond the point where the sealing ring contacts the seat (or as instructed by the manufacturer). If the filter is overtightened, it will be extremely difficult to remove.
7 Run the engine and check for oil leaks from around the base of the filter. Tighten a little further if necessary.
8 Stop the engine. Refit the undertray, then lower the vehicle and check the oil level.

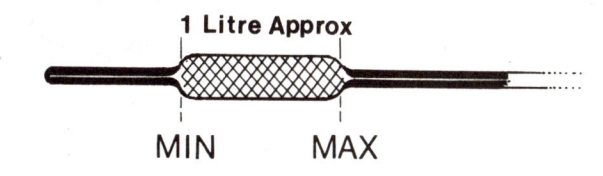

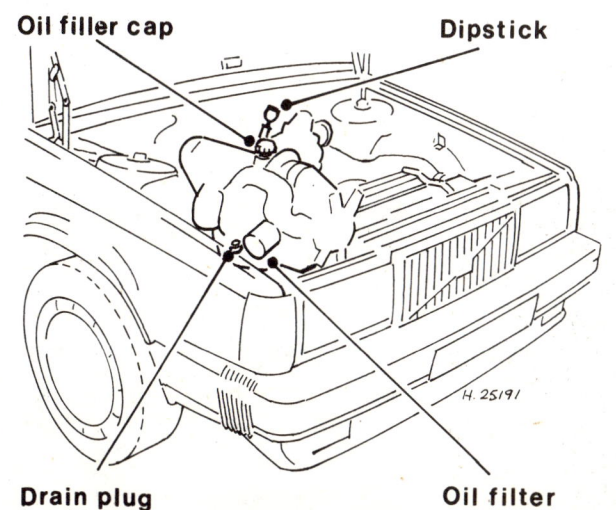

Fig. 1.2 Lubrication system maintenance points and dipstick markings (Sec 2)

Fig. 1.3 Engine oil drain plug (arrowed) (Sec 2)

Chapter 1 Engine

3.3 Removing the oil filter with a strap wrench (looking from below)

3.5 Fitting a new oil filter

4.2 Camshaft cover, showing all ten securing nuts

4 Valve clearances – checking and adjustment

1 Disconnect or remove items such as HT leads, vacuum/breather/boost pressure hoses, and if necessary the throttle cable, in order to gain access to the camshaft cover. Also unbolt the auxiliary air valve from the camshaft cover, where applicable.
2 Remove the securing nuts and lift off the camshaft cover. Note the location of the earth strap, the HT lead clip and similar items (photo). Recover the gasket.
3 Using a spanner on the crankshaft pulley centre bolt, bring the engine to TDC, No 1 cylinder firing. (This will be easier if the spark plugs are removed.) No 1 piston is at TDC when the notch on the crankshaft pulley is in line with the figure 'O' on the timing scale, and the cam lobes for No 1 cylinder (at the front) are both pointing obliquely upwards.
4 With the engine in this position, measure and record the clearance between the base of the front cam lobe and the tappet shim beneath it. Insert various thicknesses of feeler blade until a firm sliding fit is obtained (photo). This thickness is the clearance for No 1 exhaust valve. Write it down.
5 Repeat the measurement and recording on the second cam from the front. This gives No 1 inlet valve clearance.
6 Turn the crankshaft 180° (half a turn) clockwise so that the cam lobes for No 3 cylinder are pointing obliquely upwards. Measure and record the clearances for these two valves. The exhaust valve is always nearer the front.
7 Turn the crankshaft a further 180° and deal with No 4 cylinder, then 180° again for No 2.
8 Compare the clearances recorded with those given in the Specifications. If the recorded clearances are within limits, commence reassembly (paragraph 16). Otherwise, adjust the clearances as follows.
9 Gather together a small screwdriver or scriber, a pair of long-nosed pliers and a stout C-spanner or square section screwdriver. These will

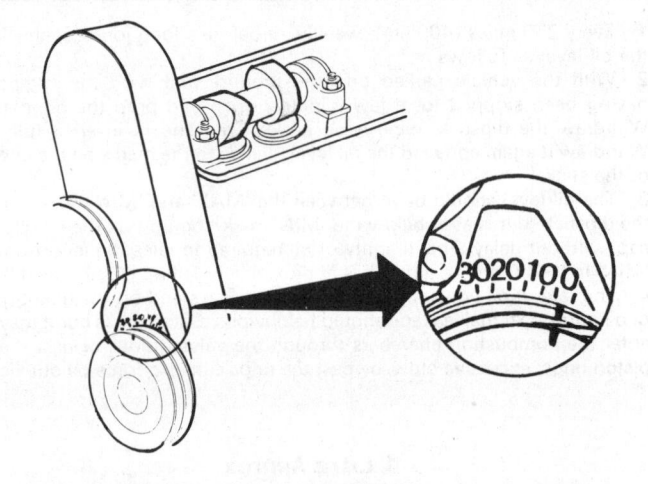

Fig. 1.4 Pulley and cam lobe positions – No 1 at TDC and firing (Sec 4)

substitute for the special tools normally required to change the shims with the camshaft in position. (Alternatively, the camshaft can be removed, but this involves much extra work.)
10 With the cam lobes in the same position as for checking, depress the tappet with the C-spanner or screwdriver. Only press on the edge of the tappet. Flick the shim out of the top of the tappet with the small screwdriver and remove it with the long-nosed pliers. Release the tappet (photos).

4.4 Measuring No 1 exhaust valve clearance

4.10A Freeing a tappet shim – the tappet is being held down with a square section screwdriver

4.10B Extracting the shim. Here a C-spanner is being used to depress the tappet

Chapter 1 Engine

4.11 Original shim thickness is engraved on the underside

4.16 Fitting the rubber plug at the rear of the camshaft

4.17 A new camshaft cover gasket in position

11 The correct thickness of shim must now be calculated. First the thickness of the old shim must be known. It may be engraved on the underside (photo), but ideally the actual thickness should be measured with a micrometer or vernier gauge. This will take account of any wear.

12 The required shim thickness can now be calculated as shown in this example:

Specified clearance (A) = 0.40 mm (0.016 in)
Measured clearance (B) = 0.28 mm (0.11 in)
Original shim thickness (C) = 3.95 mm (0.156 in)
Shim thickness required = C – A + B = 3.83 mm (0.151 in)

In this example the shim to be fitted would have to be 3.85 mm or 3.80 mm, giving clearances of 0.38 mm or 0.43 mm respectively.

13 Lubricate a new shim of the required thickness. Depress the tappet and insert the shim, marked side downwards. Release the tappet and check that the shim is properly located.

14 Repeat the operations on the adjacent tappet, if necessary, then proceed to the other valves, each time turning the crankshaft to position the cam lobes upwards. **Do not** turn the crankshaft whilst shims are missing from tappets, as the cam lobes may jam in them.

15 When all the required shims have been fitted, turn the crankshaft through several complete turns, then check all the clearances again.

16 Make sure that the rubber plug to the rear of the camshaft (B 23 only) is securely fitted and in good condition. Renew it if necessary (photo).

17 Refit the camshaft cover, using a new gasket (photo). Fit and tighten the nuts, remembering to fit the HT lead bracket and earth strap.

18 Reconnect the HT leads, vacuum hoses etc, then run the engine and check that there are no oil leaks from the camshaft cover.

5 Crankcase ventilation system – general

1 The crankcase ventilation system directs oil fumes and blow-by gases from the lower crankcase into the induction system, so that they may be drawn into the engine and burnt. Typical hose layouts are shown in Figs. 1.5 and 1.6

2 It is important to keep the hoses, flame traps, restrictors etc, clean and in good condition. Obstruction of the system can cause a build-up of pressure within the crankcase, with subsequent failure of oil seals. Leaks in the system can cause rough, or erratic idling, besides being a source of pollution.

3 Do not attempt to remove the long oil drain hose from below the oil trap, as the sump must be removed in order to refit it correctly.

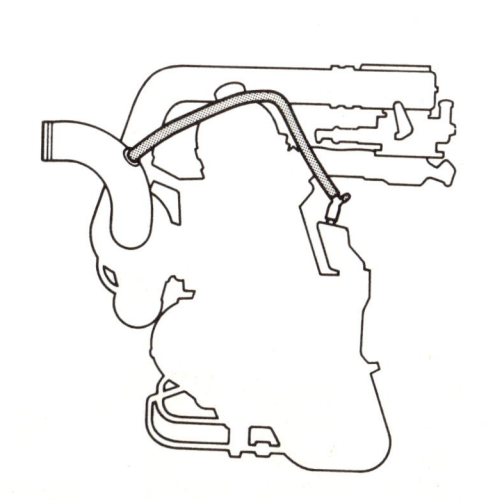

Fig. 1.5 Typical crankcase ventilation hose routing – Turbo (Sec 5)

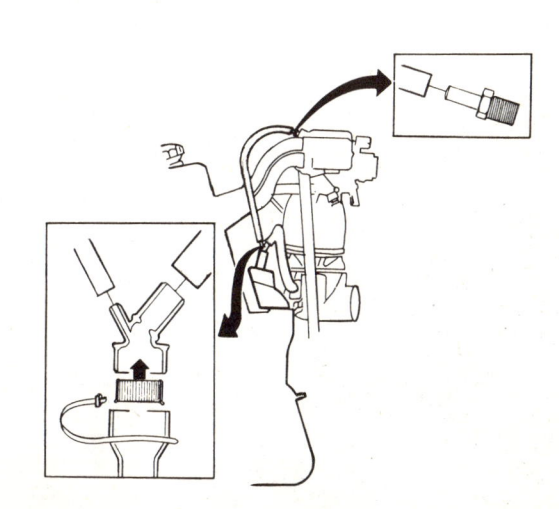

Fig. 1.6 Typical crankcase ventilation hose routing – normally aspirated (Sec 5)

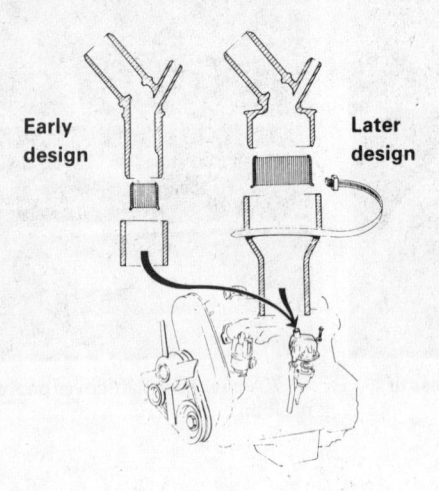

Fig. 1.7 Details of crankcase ventilation oil and flame traps (Sec 5)

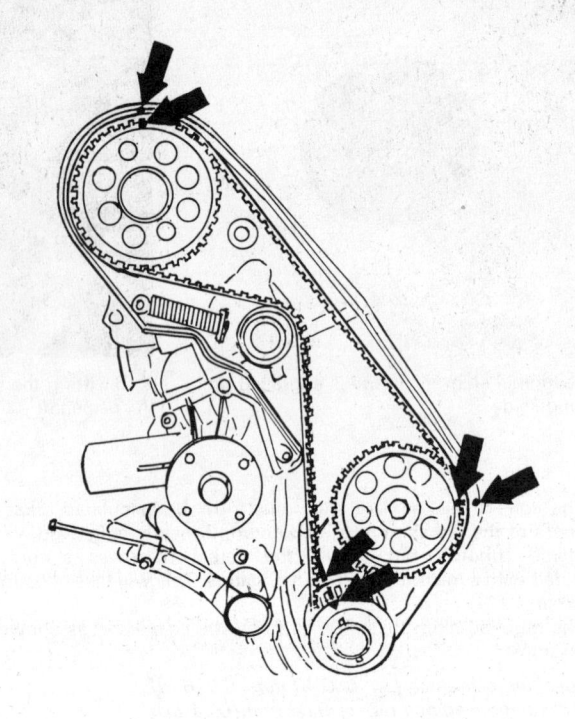

Fig. 1.8 Sprocket alignment marks (arrowed) – No 1 at TDC and firing. Auxiliary shaft marks are not critical (Sec 6)

6 Camshaft drivebelt – removal, refitting and tensioning

1 Remove the viscous coupled fan, the fan shroud, and the accessory drivebelts. Refer to Chapter 2 if necessary. On B230 engines, also remove the water pump pulley.
2 Unbolt and remove the camshaft drivebelt cover. (On B230 engines, just remove the top half of the cover.)
3 Using a spanner on the crankshaft pulley centre bolt, bring the engine to TDC, No 1 firing. This is indicated when the mark on the camshaft sprocket is in line with the mark on the camshaft cover or the drivebelt backplate. At the same time the marks on the crankshaft sprocket guide plate and the oil seal housing will be in line. (The pulley mark cannot be used even if the pulley is still in place, since the timing scale is on the drivebelt cover.) Although it is not critical, the position of the auxiliary shaft sprocket timing mark should also be noted (photos).
4 On B230 engines, remove the starter motor or the flywheel bottom cover plate. Have an assistant jam the ring gear teeth, then slacken the crankshaft pulley bolt without disturbing the set position of the crankshaft. Remove the bolt and the pulley, then remove the lower half of the camshaft drivebelt cover.
5 On B23 engines, if the crankshaft pulley was not removed with the accessory drivebelts, remove it now (photo).
6 Slacken the belt tensioner nut. Pull on the belt to compress the tensioner spring. Lock the tensioner in this position, either by tightening the nut again or by inserting a nail or similar into the hole in the tensioner shaft (photos).
7 Mark the running direction of the belt if it is to be re-used, then slip it off the sprockets and tensioner roller and remove it. **Do not** rotate the crankshaft, camshaft or auxiliary shaft with the bolt removed.
8 Spin the tensioner roller and check for roughness or shake; renew if necessary.
9 Do not contaminate the drivebelt with oil, nor kink it or fold it sharply.
10 Before refitting, make sure that all three sprockets are in the correct positions (paragraph 3). Slip the belt over the sprockets and round the roller, observing the correct running direction if the old belt is being re-used.
11 Recheck the alignment of the sprocket marks, then release the belt tensioner by slackening the nut or pulling out the nail (photo). Tighten the tensioner nut.
12 On B230 engines, refit the drivebelt lower cover and the crankshaft pulley. Make sure that the dowel (guide pin) on the sprocket engages with the hole in the pulley. Jam the ring gear teeth and tighten the pulley bolt to the specified torque. Refit the starter motor or flywheel cover.
13 On all engines, rotate the crankshaft two full turns clockwise. Stop

6.3A Timing marks – camshaft (A), auxiliary shaft (B) and crankshaft (C)

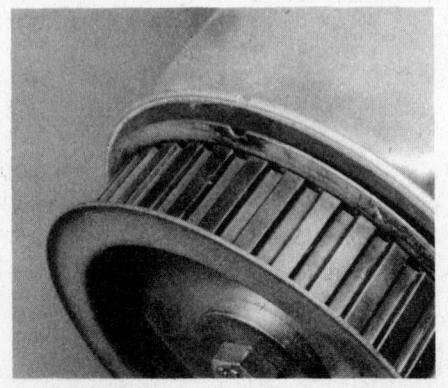

6.3B Close-up of the camshaft sprocket and cover marks

6.5 Removing the crankshaft pulley

Chapter 1 Engine   39

6.6A Slacken the tensioner nut ...

6.6B ... and insert a nail or rivet (arrowed) to restrain the spring

6.11 Pulling out the rivet to release the tension

at TDC, No 1 firing, and check that the various timing marks still align. Slacken and retighten the tensioner nut.
14 Refit the drivebelt cover (or the top section), then refit the accessory drivebelts, pulleys, fan etc.
15 Run the engine to operating temperature, then switch it off. Bring the engine to TDC, No 1 firing. Remove the access plug from the front of the drivebelt cover, slacken the tensioner nut once more and then retighten it. Refit the access plug (photo).
16 If a new belt has been fitted, repeat paragraph 15 after approximately 600 miles (1000 km).

7 Major operations possible with the engine installed

1 The following items may be relatively easily removed with the engine in place:

(a) Camshaft drivebelt
(b) Camshaft
(c) Cylinder head and valvegear
(d) Crankshaft/auxiliary shaft front oil seals
(e) Flywheel/clutch (after removal of gearbox)
(f) Crankshaft rear oil seal (after removal of flywheel)
(g) Engine mountings

2 It is possible to remove the sump with the engine installed, but the amount of preparatory work is formidable – see Section 16. With the sump removed, the oil pump, pistons and connecting rods may be dealt with.
3 The engine must be removed for removal of the crankshaft and main bearings.

8 Engine dismantling and reassembly – general

1 If the engine has been removed from the car for major overhaul, or if individual components have been removed for repair or renewal, observe the following general hints on dismantling and reassembly.
2 Thoroughly clean the exterior of the engine using a degreasing solvent or paraffin. Clean away as much of the external dirt and grease as possible before dismantling.
3 As parts are removed, clean them in a paraffin bath. However, do not immerse parts with internal oilways in paraffin as it is difficult to remove, usually requiring a high pressure hose. Clean oilways with nylon pipe cleaners.
4 Avoid working with the engine or any of the components directly on a concrete floor, as grit presents a real source of trouble.
5 Wherever possible, work should be carried out with the engine or individual components on a strong bench. If the work must be done on the floor, cover it with a board or sheets of newspaper.
6 Have plenty of clean, lint-free rags available and also some containers or trays to hold small items. This will help during reassembly and also prevent possible losses.

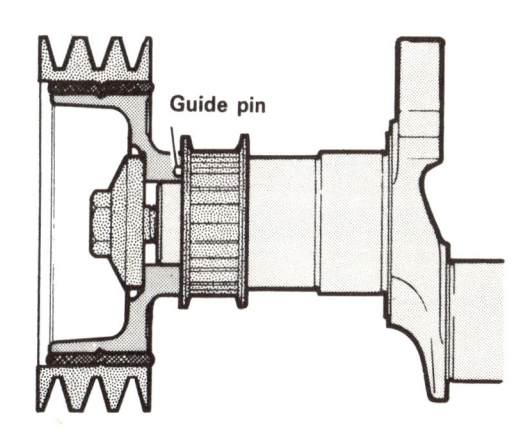

Fig. 1.9 Crankshaft pulley and sprocket details – B 230 engine (Sec 6)

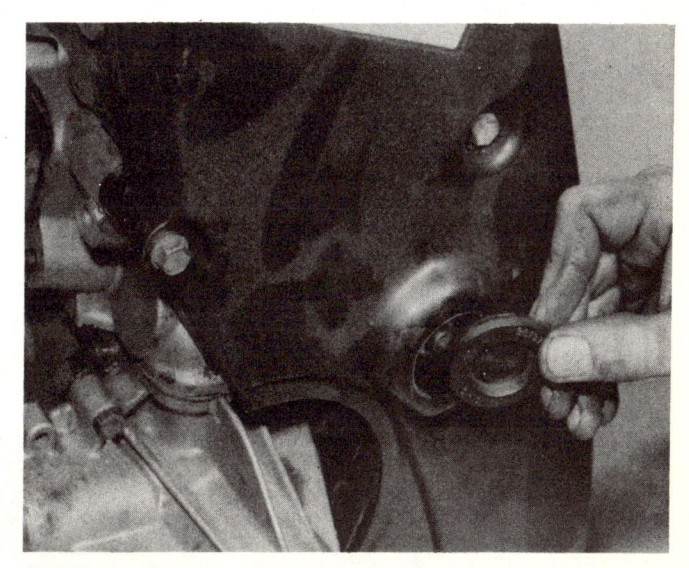

6.15 The tensioner nut access plug

7 Always obtain a complete set of gaskets if the engine is being completely dismantled, or all those necessary for the individual component or assembly being worked on. Keep the old gaskets with a view to using them as a pattern to make a replacement if a new one is not available.
8 When possible refit nuts, bolts and washers in their locations after removal as this helps to protect the threads and avoids confusion or loss.
9 During reassembly thoroughly lubricate all components, where appropriate, with engine oil, but avoid contaminating the gaskets and joint mating faces.
10 Besides gaskets, seals and so on, a good many cable ties will be needed to replace those which have to be cut when working on and around the engine.
11 Engine undertray removal is necessary for virtually all operations requiring access to the underside of the engine.

9 Camshaft – removal and refitting

Note: *If a new camshaft is to be fitted, the lubrication system must be flushed with two consecutive oil and filter changes **before** removing the old camshaft. Drain the oil and renew the filter, then run the engine for 10 minutes. Fresh oil and a new filter must be provided for the new camshaft. Failure to observe this may cause rapid wear of the new camshaft.*

1 Remove the camshaft drivebelt (Section 6). The belt can stay on the lower sprockets if wished.
2 Restrain the camshaft sprocket with a suitable tool through the holes in its face, or by clamping an old drivebelt around it. Slacken the camshaft sprocket bolt. **Do not** allow the camshaft to move, or piston/valve contact may occur (photo).
3 Remove the sprocket bolt and the sprocket itself. Note the position of any front plates, backplates and washers (photo).
4 On B230 engines, remove the distributor (Chapter 4, Section 4).
5 Remove the camshaft cover, noting the position of the HT lead bracket and the earth strap. Recover the gasket.
6 Make identification marks if necessary, then progressively slacken the camshaft bearing cap nuts (photo). The camshaft will rise up under the pressure of the valve springs – be careful that it does not stick and then suddenly jump up. Remove the bearing caps.
7 Lift out the camshaft complete with front oil seal. Be careful of the lobes, which may have sharp edges (photo).
8 If it is wished to measure the camshaft endfloat, temporarily remove all the tappets and shims, keeping them in order for refitting. Refit the camshaft and the rear bearing cap; measure the endfloat between the cap and the camshaft flange (photo). Excessive endfloat, if not due to wear of the camshaft itself, can be corrected by renewing the rear bearing cap. Refit the tappets and shims to their original locations on completion.
9 Commence refitting by liberally oiling the tappets, shims, the camshaft bearings and caps and the cam lobes. Use clean engine oil, or special camshaft lubricant if supplied with a new shaft.
10 Fit the camshaft in approximately the correct position for No 1 firing (No 1 lobes both pointing obliquely upwards). Apply sealant to the head mating surfaces of the front and rear bearing caps (photo). Fit all the bearing caps in their correct positions and pull them down by tightening the nuts a little at a time. When all the caps are seated, tighten the nuts to the specified torque.
11 Lubricate a new oil seal and fit it to the front of the camshaft, lips inwards. Tap it home with a piece of tube (photo).
12 Refit the camshaft sprocket and associated components. Restrain the sprocket and tighten the bolt to the specified torque.
13 Refit the distributor if it was removed.
14 Refit and tension the camshaft drivebelt (Section 6).
15 If any new parts have been fitted, check the valve clearances (Section 4).
16 Refit the camshaft cover, using a new gasket.
17 If a new camshaft has been fitted, run it in at moderate engine speeds for a few minutes (neither idling nor racing), or as directed by the manufacturer.

9.2 Restrain the sprocket and slacken the bolt

9.3A Removing the bolt, washer and front plate ...

9.3B ... the camshaft sprocket itself ...

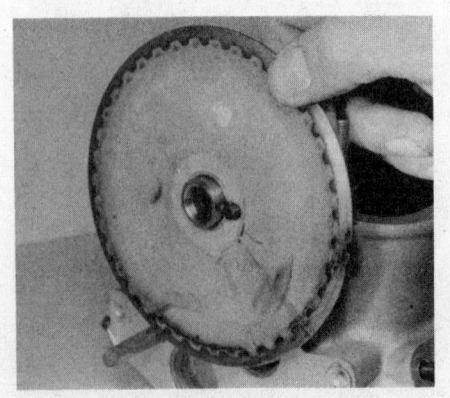

9.3C ... and the sprocket backplate. Other engines may differ slightly

9.6 Slackening a camshaft bearing cap nut

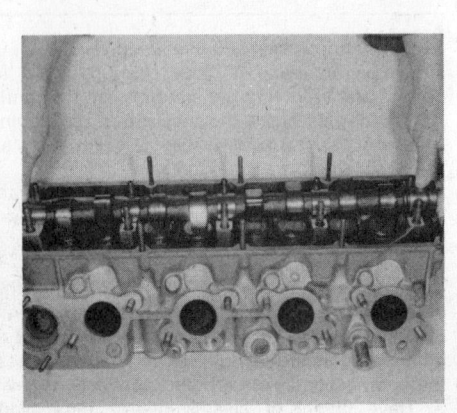

9.7 Removing the camshaft

Chapter 1 Engine

9.8 Measuring camshaft endfloat

9.10 Fitting the camshaft front bearing cap. Sealant is on shaded areas

9.11 Fitting a new oil seal

10 Cylinder head – removal and refitting

1 Disconnect the battery negative lead.
2 Drain the cooling system (Chapter 2, Section 3).
3 Disconnect or remove HT leads, vacuum/breather hoses etc, to gain access to the camshaft cover.
4 Disconnect the radiator top hose from the thermostat housing.
5 Remove the fan and fan shroud.
6 Remove all accessory drivebelts and the water pump pulley (Chapter 2, Section 7).
7 On B230 engines, remove the distributor cap.
8 Remove the camshaft drivebelt cover. (If the cover is in two sections, just remove the upper section.)
9 Bring the engine to TDC, No 1 firing, then remove the camshaft drivebelt and tensioner – see Section 6. (The drivebelt can stay on the lower sprockets.) **Do not** rotate the crankshaft or camshafts from now on.
10 On B230 engines, unbolt and remove the camshaft sprocket and the spacer washer. Also remove the camshaft drivebelt tensioner stud.
11 Remove the bolts which secure the drivebelt backplate to the cylinder head.
12 Remove the nuts which secure the inlet and exhaust manifolds to the cylinder head. Pull the manifolds off their studs to the sides of the engine bay, supporting them if necessary. Recover the gaskets.
13 Remove the camshaft cover. Recover the gasket.
14 Slacken the cylinder head bolts, half a turn at a time to begin with, in the order shown in Fig. 1.10. Remove the bolts. Obtain new bolts for reassembly if the old ones appear stretched.
15 Lift off the cylinder head. On B230 engines it will be necessary to bend the drivebelt backplate forwards a little.
16 Set the head down on a couple of wooden blocks to avoid damage to protruding valves. Recover the old head gasket.
17 Commence refitting by placing a new head gasket on the cylinder block. Make sure it is the right way up – all the bolt holes, oilways etc must line up (photo).
18 Make sure that the camshaft to set to the 'No 1 firing' position – both cam lobes for No 1 cylinder pointing obliquely upwards. Lower the head into position.
19 Oil the threads of the cylinder head bolts. Fit the bolts and tighten them, in the sequence shown in Fig. 1.11, to the specified Stage 1 torque.
20 In the same sequence tighten the bolts to the Stage 2 torque, then go round again and tighten the bolts through the angle specified for Stage 3. (If the engine is on the bench, it may be preferable to leave Stage 3 tightening until after engine refitting.) No further tightening is required.
21 The remainder of refitting is a reversal of the removal procedure. Use new gaskets etc where necessary.
22 Check the valve clearances (Section 4) before starting the engine. Also refer to Section 44, most of which applies.

10.17 A new head gasket in position

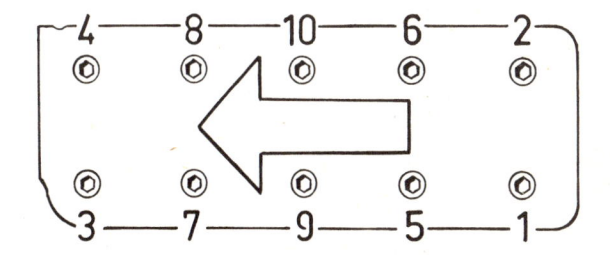

Fig. 1.10 Cylinder head bolt slackening sequence (Sec 10)

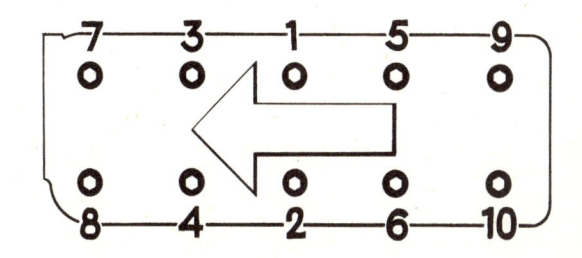

Fig. 1.11 Cylinder head bolt tightening sequence (Sec 10)

11.2 Unbolting the auxiliary shaft sprocket

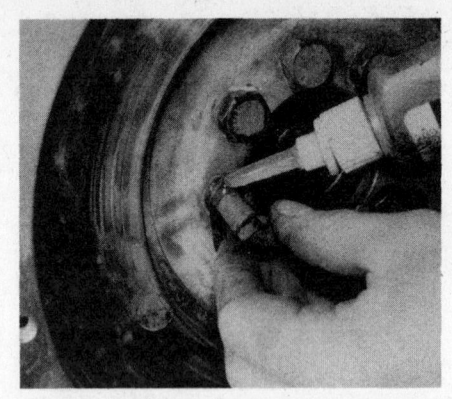

12.5A Applying thread locking compound to a flywheel bolt

12.5B Jam the ring gear and tighten the bolts

11 Oil seals (front) – renewal

1 Remove the camshaft drivebelt (Section 6)
2 Restrain and unbolt the appropriate sprockets for access to the failed seal (photo). If necessary, also remove the drivebelt tensioner and backplate.
3 Carefully extract the seal by prising it out with a small screwdriver or hooked tool. Do not damage the shaft sealing face.
4 Clean the seal seat. Examine the shaft sealing face for wear or damage which could cause premature failure of the new seal.
5 Lubricate the new oil seal. Fit the seal over the shaft, lips inwards, and tap it home with a piece of tube.
6 Refit the other disturbed components. Fit a new camshaft drivebelt if the old one was oil-soaked.

12 Flywheel/driveplate – removal and refitting

Flywheel
1 Remove the gearbox (Chapter 6, Section 4).
2 Remove the clutch pressure plate and driven plate (Chapter 5, Section 10).
3 Make alignment marks so that the flywheel can be refitted in the same position relative to the crankshaft. (If the flywheel is refitted incorrectly, the Motronic ignition system will not work.)
4 Unbolt the flywheel and remove it. Do not drop it, it is heavy. Obtain new bolts for reassembly.
5 Refit by reversing the removal operations. Use new bolts, tightened to the specified torque, with thread locking compound (photos).
6 Refit the clutch as described in Chapter 5.
7 Before refitting the gearbox, check that the Motronic ignition trigger 'slug' in the rim of the flywheel passes the (so-called) TDC sensor when No 1 piston is at 90° BTDC.

Driveplate
8 Remove the automatic transmission (Chapter 6, Section 37).
9 Proceed as above from paragraph 3, ignoring the references to the clutch. Note the location and orientation of the large washers on each side of the driveplate.

13 Oil seal (crankshaft rear) – renewal

1 Remove the flywheel or driveplate (Section 12).
2 Note whether the old seal is flush with the end of its carrier, or recessed into it.
3 Carefully prise out the old oil seal. Do not damage the carrier or the surface of the crankshaft.
4 Clean the oil seal carrier and the crankshaft. Inspect the crankshaft for a wear groove or ridge left by the old seal (photo).
5 Lubricate the carrier, the crankshaft and the new seal. Fit the seal, lips inwards, and use a piece of tube (or the old seal, inverted) to tap it home. If there is any wear on the crankshaft sealing surface, fit the new

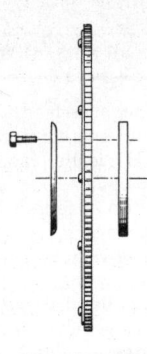

Fig. 1.12 Driveplate and washers (Sec 12)

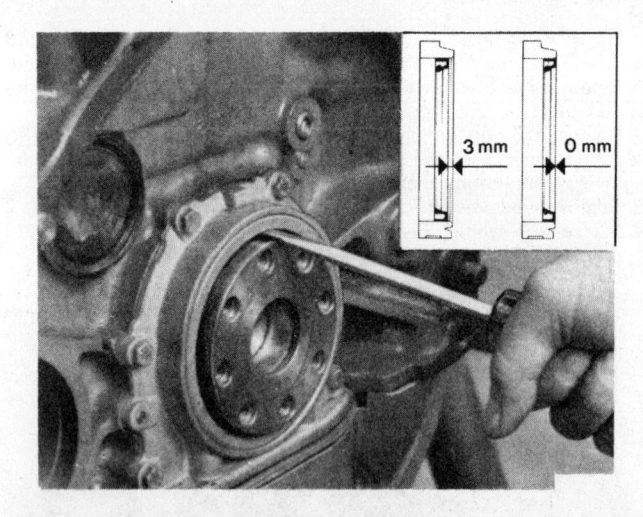

Fig. 1.13 Levering out the rear oil seal. Note recessed depth (inset) (Sec 13)

seal more deeply recessed than the old one. The seal may be recessed up to 6 mm (0.24 in) within the carrier.
6 Refit the flywheel or driveplate.

14 Flywheel ring gear – renewal

The driveplate ring gear cannot be renewed separately.
1 Remove the flywheel (Section 12).
2 Drill through the old ring gear at the root of two teeth. Do not drill into the flywheel.

Chapter 1 Engine

13.4 Wear groove left by the old oil seal

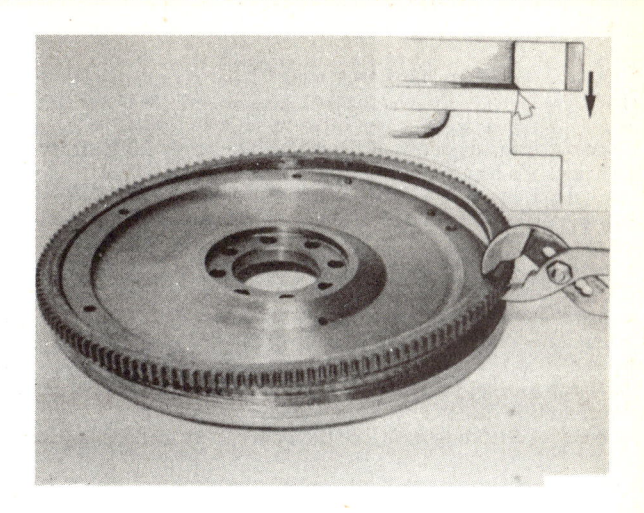

Fig. 1.14 Fitting the flywheel ring gear. Inset shows chamfer (Sec 14)

3 Split the ring gear, using a chisel above the drilled hole. Wear eye protection when doing this. Lever the ring gear off the flywheel and clean up its seat.
4 Heat the new ring gear to 230°C (446°F) in an oven or oil bath, or with a flame. If using a flame, place scraps of 40/60 solder (40% tin, 60% lead) on the ring gear to show when the correct temperature has been reached. The solder will melt at 220 to 230°C (428 to 446°F). Do not overheat the gear, or its temper will be lost.
5 Using tongs, fit the gear to the flywheel, with the inner chamfer towards the flywheel. Tap the gear home if necessary with a brass drift, then allow it to cool.
6 Refit the flywheel.

15 Crankshaft spigot bearing – removal and refitting

A spigot bearing is only fitted to manual gearbox models. With automatic transmission, a plain bush is used.
1 Remove the clutch pressure and driven plates (Chapter 5).
2 Prise out the bearing retaining circlip (photo).
3 Extract the bearing with a slide hammer or similar tool (photo).

4 When refitting, drive the bearing home with a piece of tubing which acts on its outer race. Refit the circlip.
5 Refit the clutch components.

16 Sump – removal and refitting

Note: *Read through this procedure first to see what is involved. Some readers will prefer to remove the engine*
1 Raise the front of the vehicle on ramps, or drive it over a pit.
2 Drain the engine oil. Refit the drain plug for safe keeping.
3 Disconnect the battery negative lead.
4 Remove the splash guard from below the engine.
5 Disconnect the exhaust downpipe from the silencer.
6 Remove the nuts which secure the engine mountings to the crossmember.
7 Release the clamp bolts and disconnect the shaft from the steering gear.
8 Support the engine from above, either with a hoist or with an adjustable support resting on the inner wings or suspension turrets. Satisfy yourself as to the security of the support arrangements.
9 Release the fan shroud and remove the engine oil dipstick, then raise the engine slightly to take the weight off the mountings. On the

15.2 Removing the spigot bearing circlip

15.3 Extracting the spigot bearing

B230 engine, be careful not to crush the distributor against the bulkhead.
10 Remove the engine mounting on the left-hand side. Cut the cable tie which secures the power steering hose nearby. Also remove the inlet manifold bracing strut (not on carburettor engines).
11 Remove the bolts which secure the front crossmember to the body.
12 Remove the flywheel/driveplate bottom cover plate.
13 Pull the front crossmember downwards to give adequate clearance below the sump. Disconnect or move aside power steering hoses as necessary.
14 Remove the sump securing bolts. Separate the sump from the block – if it is stuck, tap it with a soft-faced hammer. Do not lever between the mating faces.
15 Lower the sump, twist it to free it from the oil pump pick-up and remove it.
16 Clean the sump internally. Remove all traces of gasket from the sump and block faces.
17 Commence refitting by sticking a new gasket to the sump with grease.
18 Offer the sump to the block, being careful not to displace the gasket. Secure the sump with two bolts in opposite corners.
19 Fit all the sump bolts and tighten them progressively to the specified torque.
20 The remainder of refitting is a reversal of the removal procedure. Check that the drain plug is tight, then refill the engine with oil on completion.

Fig. 1.15 Removing the sump (Sec 16)

17 Oil pump – removal and refitting

1 Remove the sump (Section 16).
2 Remove the two bolts which secure the oil pump. Note that one of these bolts also secures the oil trap drain hose guide. Remove the pump and the guide.
3 Separate the oil pump delivery pipe from the pump. Recover the seals from each end of the pipe.
4 Commence refitting by fitting the delivery pipe, with new seals, to the pump (photo).
5 Fit the pump to the block, engaging the pump drivegear and the delivery pipe at the same time (photo).
6 Fit the two bolts and the drain hose guide. Tighten the bolts.
7 Make sure that the oil drain hose is correctly positioned, then refit the sump.

18 Pistons and connecting rods – removal and refitting

1 Remove the cylinder head (Section 10), the sump (Section 16) and the oil pump (Section 17).

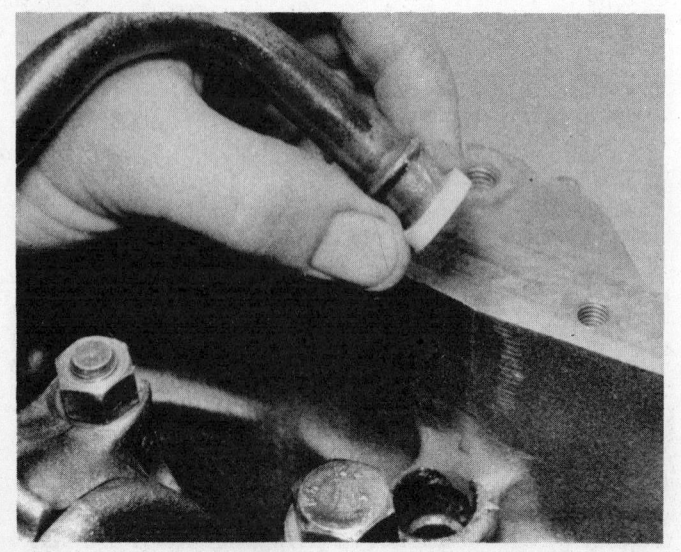

17.4 Fitting the delivery pipe to the oil pump. New seal (arrowed) is in the pump.

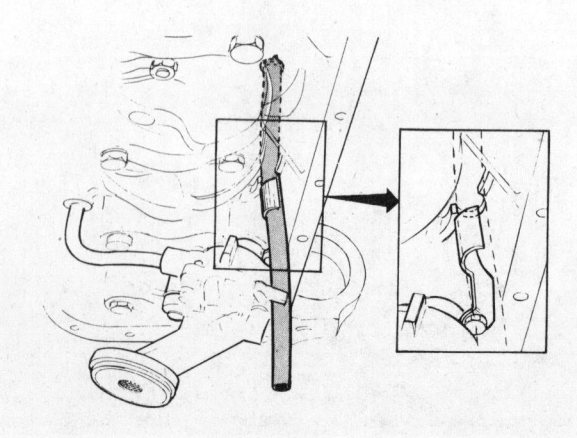

17.5 Fitting the delivery pipe and a new seal to the block

Fig. 1.16 Correct positioning of oil drain hose (Sec 17)

Chapter 1 Engine 45

2 Feel inside the tops of the bores for a pronounced wear ridge. Some authorities recommend that such a ridge be removed (with a scraper or ridge reamer) before attempting to remove the pistons. However, a ridge big enough to damage the pistons will almost certainly mean that a rebore and new pistons are needed anyway.
3 Turn the crankshaft to bring a pair of connecting rod caps into an accessible position. Check that there are identification numbers or marks on each connecting rod and cap; paint or punch suitable marks if necessary, so that each rod can be refitted in the same position and the same way round.
4 Remove two connecting rod nuts or bolts. Tap the cap with a soft-faced hammer to free it. Remove the cap and bearing shell.
5 Push the connecting rod and piston up and out of the bore. Recover the other half bearing shell if it is loose.
6 Refit the cap to the connecting rod so that they do not get mixed up. Keep the bearing shells in their original positions if there is any chance that they will be re-used.
7 Repeat the operations on the remaining connecting rods and pistons, turning the crankshaft as necessary to gain access to the connecting rod caps.
8 Examine the bores and the pistons (Sections 30 and 31). Also inspect the crankshaft journals (Section 29). If it is decided that the engine will have to be removed for further attention, the cylinder head must be refitted first, using at least six bolts to secure it.
9 Commence refitting by deglazing the bores with coarse emery or a 'glaze buster' hone. (Stuff rags into the bores first to protect the crankshaft, and be sure to remove all traces of abrasive.) The object is to give a cross-hatch finish to the bore which will help new rings to bed in quickly.
10 Oil a piston and its bore. Make sure that the top half of the bearing shell is fitted to the connecting rod and the bottom half to the cap (photo). Also check that the piston ring gaps are evenly spaced.
11 Fit a piston ring compressor to the piston. Insert the piston and connecting rod into their bore, making sure that the arrow on the piston crown points towards the front of the engine. Tap the piston crown with a wooden hammer handle so that the piston enters the bore. If it sticks, do not force it, but release the ring compressor and check the rings (photos).
12 Check that the bearing shell is still secure in the connecting rod. Oil the crankpin and draw the connecting rod onto it. Fit the cap with its shell, and tighten the nuts or bolts to the specified torque (photos).
13 Repeat the operations on the other three pistons, turning the crankshaft as necessary to receive the crankpins.
14 Check that the crankshaft is still free to rotate. Some stiffness is to be expected if new shells have been fitted, but binding or tight spots should be investigated.
15 Refit the oil pump, the sump and the cylinder head.

19 Engine mountings – removal and refitting

1 Disconnect the battery negative lead.
2 Remove the nuts from the mounting to be removed (photo).
3 Fit lifting tackle to the engine, or support it in some other way. Do not jack up directly onto the sump, as damage may result.
4 Take the weight off the mounting and remove it. It may be necessary to move aside power steering hoses, and (on fuel injection engines) to remove the inlet manifold bracing strut.
5 Refit by reversing the removal operations.

20 Methods of engine removal

The engine is removed by lifting it out of the engine bay, with or without the transmission.

21 Engine – removal (alone)

1 Disconnect the battery negative lead. Either remove the bonnet, or open it to its widest setting.

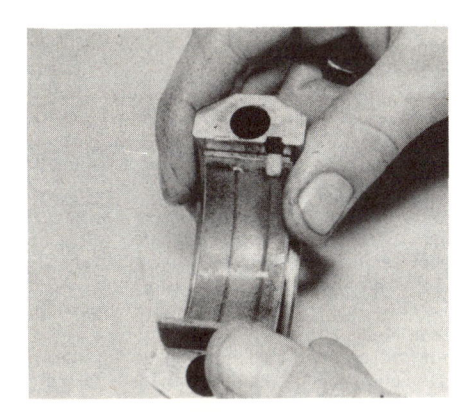

18.10 Fitting a bearing shell to a connecting rod cap

18.11A Fitting a piston to the bore

18.11B Arrow on piston crown must face front of engine. Letters 'D' are grade marks

18.12A Fitting the connecting rod cap ...

18.12B ... and tightening the nuts

19.2 The engine right-hand mounting – Turbo shown, others are more accessible

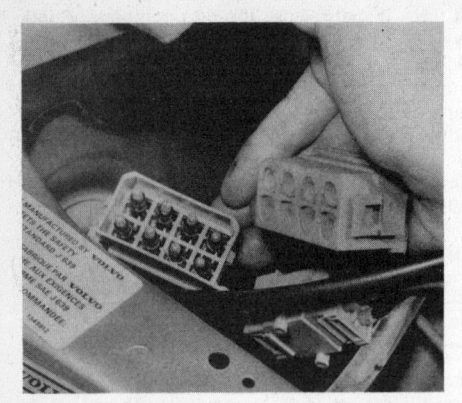

21.11 Disconnecting an engine wiring harness multi-plug

21.22 Disconnecting the oil cooler hoses

21.23 Disconnecting the vacuum tank

2. Drain the cooling system and remove the radiator. Refer to Chapter 2 if necessary.
3 On Turbo models, remove the intercooler and associated hoses (Chapter 3). Also remove the airflow meter-to-turbo hose.
4 On carburettor models, remove the air cleaner (Chapter 3).
5 On all models, remove the air cleaner hot air trunking.
6 On B230 engines, remove the distributor cap and HT leads. On B23 engines, remove the coil-to-distributor HT lead.
7 Disconnect the throttle cable.
8 Disconnect the brake servo vacuum hose.
9 Disconnect the fuel supply and return pipes. Be prepared for fuel spillage.
10 Disconnect the various crankcase ventilation, vacuum and pressure sensing hoses. Make notes or identifying marks if there is any possibility of confusion later.
11 Disconnect the engine wiring harness multi-plug(s), again making notes if necessary (photo).
12 Disconnect the starter motor feed lead, then remove the battery completely.
13 Disconnect the engine earth strap(s).
14 Disconnect the air conditioning compressor clutch lead.
15 Remove the power steering pump without disconnecting the hoses and wire it up out of the way. See Chapter 10 if necessary.
16 Disconnect the heater hoses at the rear of the engine.
17 Remove the starter motor (Chapter 12).
18 Disconnect the exhaust downpipe from the manifold or turbo exit.
19 If an oil cooler is fitted, unbolt its mounting bracket.
20 Raise and support the vehicle. Remove the engine undertray, if not already done.
21 Drain the engine oil and remove the oil filter.
22 If an oil cooler is fitted, disconnect the flexible hoses at their unions with the rigid pipes (photo). The contents of the oil cooler will drain out of the open unions. Remove the oil cooler.
23 Disconnect the hose from the vacuum tank (photo), then unbolt and remove it.
24 Remove those engine-to-transmission nuts and bolts which are accessible from below. Also remove the flywheel/driveplate bottom cover.
25 On automatic transmission models, unbolt the torque converter from the driveplate. Turn the crankshaft as necessary to gain access. Make alignment marks for reference when refitting.
26 Remove the air conditioning compressor drivebelt (Chapter 2, Section 7).
27 Remove the nuts which secure the air conditioning compressor bracket to the engine. Move the compressor aside without disconnecting the refrigerant hoses. It will rest in the space vacated by the battery.
28 Disconnect the two ignition sensor multi-plugs at the bulkhead. Identify the plugs to avoid confusion when refitting. (This only applies to the Motronic system.)
29 Support the gearbox from below, using a trolley jack for preference. Pad the jack head with rags or wood.
30 Attach the lifting tackle to the engine using the lifting eyes provided. Take the weight of the engine.
31 Remove the nuts which secure the engine bearers to the engine mountings.
32 Remove the remaining engine-to-transmission nuts and bolts.
33 Check that no wires, hoses etc have been overlooked. Raise the engine and draw it forwards off the transmission, at the same time raising the jack under the transmission. Do not allow the weight of the transmission to hang on the input shaft.
34 Once the engine is clear of the transmission, carefully lift it out of the engine bay and take it to the bench (photo).

22 Engine – removal (with transmission)

1 Proceed as in the previous Section, paragraphs 1 to 16, 18 to 23 and 26 to 28.
2 Disconnect the leads from the starter motor solenoid.
3 Remove the exhaust downpipe completely.

Manual gearbox
4 Remove the clutch slave cylinder (without disconnecting the hydraulic hose) or disconnect the clutch cable, as applicable. See Chapter 5.
5 Disconnect the gear lever – see Chapter 6, Sections 4 and 21.

Automatic transmission
6 Disconnect the kickdown and control linkages – see Chapter 6, Section 37.

All models
7 Disconnect electrical services from the transmission.
8 Unbolt the propeller shaft from the rear of the transmission.
9 Support the transmission. Unbolt the crossmember from the transmission and from the side rails and remove it.
10 Unbolt the bracing strut from below the bellhousing (when fitted).
11 Attach lifting tackle to the engine using the eyes provided. Take the weight of the engine.
12 Remove the nuts which secure the engine bearers to the engine mountings.
13 Check that no attachments have been overlooked. Raise the engine, at the same time lowering the support under the transmission, until the whole assembly can be lifted from the engine bay.

23 Engine – separation from transmission

1 Remove the starter motor.
2 Remove the bellhousing-to-engine nuts and bolts. Also remove the flywheel/driveplate bottom cover plate (if applicable).

Manual gearbox
3 With the aid of an assistant, draw the gearbox off the engine. Once it is clear of the dowels, do not allow it to hang on the input shaft.

Automatic transmission
4 Unbolt the torque converter from the driveplate, turning the crankshaft to gain access from below or through the starter motor hole. Make alignment marks for reference when refitting.

Chapter 1 Engine

21.34 Lifting out the engine

24.2A Removing the oil cooler adaptor nut ...

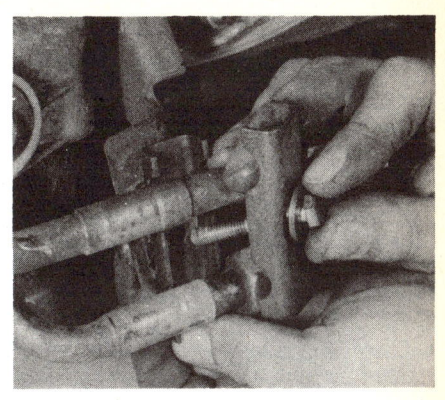

24.2B ... and the pipe bracket

5 With the aid of an assistant, draw the transmission off the engine. Make sure that the torque converter stays in the bellhousing.

24 Engine – dismantling for overhaul

1 With the engine removed for dismantling, remove ancillary components as follows, referring to other Chapters when necessary:

- (a) Alternator and brackets (Chapter 12)
- (b) Exhaust manifold, with turbocharger if fitted (Chapter 3)
- (c) Fan and water pump (Chapter 2)
- (d) Distributor (Chapter 4)
- (e) Inlet manifold with carburettor or injection components (Chapter 3)
- (f) Turbo bypass valve (Chapter 3, Section 46)
- (g) Spark plugs (Chapter 4)
- (h) Clutch pressure and driven plates (Chapter 5)
- (j) Ignition sensors and bracket (Chapter 4)
- (k) Auxiliary air valve or air control valve (not carburettor models) (Chapter 3, Section 34 or 46)

2 If an oil cooler is fitted, remove the pipework and the oil filter adaptor. The adaptor is secured by a large central nut and the pipework by a bracket (photos).
3 Unbolt the coolant distribution pipe.
4 Remove the dipstick. Unbolt the dipstick tube bracket and pull the tube out of the block (photo).
5 Unbolt the crankcase ventilation system oil trap and pull out the long drain hose. Working through that hole, lift up the oil pump drivegear/shaft and remove it (photo).
6 If not already done, remove the camshaft drivebelt cover. Jam the flywheel ring gear and slacken the crankshaft pulley/sprocket centre bolt.
7 Remove the camshaft drivebelt and tensioner (Section 6).
8 Unbolt and remove the camshaft and auxiliary shaft sprockets, then remove the camshaft drivebelt backplate.
9 Remove the cylinder head (Section 10).
10 Remove the crankshaft pulley (or pulley boss), the sprocket and guide plates.
11 Remove the flywheel or driveplate (Section 12).
12 Remove the sump and oil pump (Sections 16 and 17).
13 Remove the pistons and connecting rods (Section 18).
14 Remove any pressure switches, transducers etc, from the cylinder block.

25 Auxiliary shaft – removal

1 With the engine dismantled as described in the previous Section, unbolt and remove the front oil seal housing (photo). Note the cable clips attached to the bottom studs. Recover the gasket.
2 Withdraw the auxiliary shaft, being careful not to damage the bearings in the block (photo).

26 Crankshaft and main bearings – removal

1 With the pistons and connecting rods removed, position the engine with the crankshaft uppermost.
2 Remove the front and rear oil seal housings (if not already done) (photo).
3 If it is possible that the existing main bearing shells will be re-used, the crankshaft endfloat should be measured now to make sure that it is within limits. See Section 38, paragraph 9.
4 Inspect the main bearing caps for identifying numbers or marks. Paint or punch marks if necessary.

24.4 Removing the dipstick tube

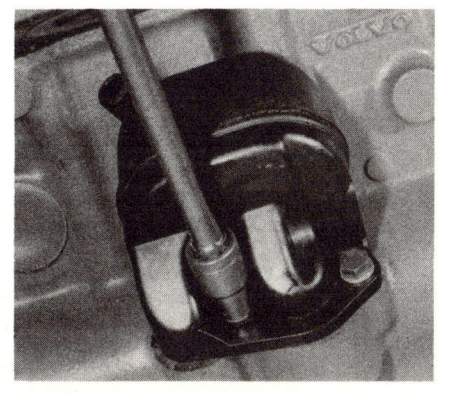

24.5 Unbolting the oil trap

25.1 Removing the front oil seal housing

48 Chapter 1 Engine

25.2 Removing the auxiliary shaft

26.2 Removing the rear oil seal housing

26.5 Removing the rear main bearing cap

5   Remove the main bearing cap bolts. Lift off the main bearing caps, tapping them with a soft-faced hammer if necessary to free them. Keep the bearing shells with their caps if they may be re-used (photo).
6   Lift out the crankshaft. Do not drop it, it is heavy (photo).
7   On B230 engines, recover the two half thrust washers from each side of the centre main bearing. (On B23 engines, No 5 bearing shells have integral thrust flanges.)
8   Remove the upper half main bearing shells from their seats in the crankcase by pressing the end of the shell furthest from the locating tab. Again, keep the shells in order if they are to be re-used.

27 Examination and renovation – general

With the engine completely stripped, clean all the components and examine them for wear. Each part should be checked, and where necessary renewed or renovated as described in the following Sections. Renew main and big-end shell bearings as a matter of course, unless you know that they have had little wear and are in perfect condition.

If in doubt as to whether to renew a component which is still just serviceable, consider the time and effort which will be incurred should it fail at an early date. Obviously the age and expected life of the vehicle must influence the standards applied.

Gaskets, oil seals and O-rings must all be renewed as a matter of routine. Flywheel bolts must be renewed because of the high stresses to which they are subjected. Some other bolts may be re-used if they do not show signs of stretching – see Specifications or the appropriate Section.

Take the opportunity to renew the engine core plugs while they are easily accessible. Knock out the old plugs with a hammer and chisel or punch. Clean the plug seats, smear the new plugs with sealant and tap them squarely into position.

28 Oil pump – examination and renovation

1   Remove the Allen screws which hold the two halves of the pump together (photo).
2   Remove the pick-up pipe and gear cover from the gear housing. Be prepared for the ejection of the relief valve spring (photo).
3   Remove the relief valve spring and plunger (or ball on early models) and the pump gears (photos).
4   Clean all components, paying particular attention to the pick-up screen, which is partly obscured by its housing. Inspect the gears, housing and gear cover for signs of wear or damage.
5   Measure the relief valve spring, and if possible compare its characteristics with those in the Specifications. Renew it if it is weak or distorted. Also inspect the plunger or ball for scoring or other damage.
6   Refit the gears to the casing. Using a straight-edge and feeler blades, check the gear side clearance and endfloat (photos). Also check the backlash between the teeth. If the clearances are outside the specified limits, renew the pump.
7   If the clearances are satisfactory, lubricate the gears. Lubricate and fit the relief valve plunger (or ball) and spring (photo).
8   Refit the pick-up pipe and gear cover. Fit and tighten the Allen screws.

29 Crankshaft and bearings – examination and renovation

1   Examine the crankpin and main journal surfaces for scoring or scratches. Measure the journals in several places, using a micrometer, to check for out-of-round and taper. Limits are given in the Specifications. If the journals are scored or out of limit, the crankshaft must be reground and undersize shells fitted.

26.6 Lifting out the crankshaft

28.1 Undoing the oil pump Allen screws

28.2 Separating the pick-up pipe and cover from the pump body

Chapter 1 Engine

28.3A Removing the oil pump idler gear ...

28.3B ... and the driving gear

28.6A Measuring oil pump gear side clearance ...

28.6B .. and gear endfloat

28.7 Fitting the relief valve plunger

2 Inspect the bearing shells for signs of general wear, pitting and scratching. Renew them as a matter of course unless they are in perfect condition. Refitting used shells is false economy.

3 On engines with separate thrust washers, these should be renewed if the crankshaft endfloat is excessive.

4 An accurate method of determining bearing wear is by the use of Plastigage. The crankshaft is located in the main bearings (and big-end bearings if necessary) and the Plastigage filament located across the journal or shell, which must be dry. The cap is then fitted and the bolts/nuts tightened to the specified torque. On removal of the cap the width of the filament is checked against a scale which shows the bearing running clearance. This clearance is then compared with that given in the Specifications (photos).

5 If the crankshaft is suspected of being cracked it should be taken to a suitably equipped workshop for crack testing. As a rough guide, an intact crankshaft will ring clearly if suspended from a cord and struck, whilst a cracked item will only give a dull note.

6 Renew the spigot bearing whilst it is accessible – see Section 15.

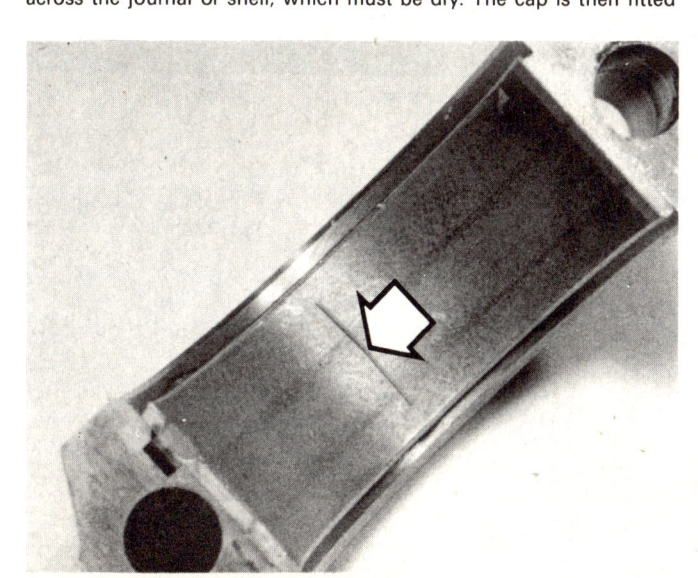

29.4A Plastigage filament (arrowed) on a bearing shell

29.4B Checking the width of the crushed filament

30 Cylinder block and bores – examination and renovation

1 The cylinder bores must be examined for taper, ovality, scoring and scratches. Start by examining the top of the bores; if these are worn, a slight ridge will be found which marks the top of the piston ring travel. If the wear is excessive, the engine will have had a high oil consumption rate accompanied by blue smoke from the exhaust.
2 If available, use an inside dial gauge to measure the bore diameter just below the ridge and compare it with the diameter at the bottom of the bore, which is not subject to wear. If the difference is more than 0.1 mm (0.004 in), the cylinders will normally require reboring with new oversize pistons fitted.
3 Proprietary oil control rings can be obtained for fitting to the existing pistons if it is felt that the degree of wear does not justify a rebore. However, any improvement brought about by such rings may be short-lived.
4 If new pistons or piston rings are to be fitted to old bores, deglaze the bores as described in Section 18.
5 If there is a ridge at the top of the bore and new piston rings are being fitted, either the top piston ring must be stepped ('ridge dodger' pattern) or the ridge must be removed with a ridge reamer. If the ridge is left, the piston ring may hit it and break.
6 Thoroughly examine the crankcase and cylinder block for cracks and damage and use a piece of wire to probe all oilways and waterways to ensure they are unobstructed.

31 Pistons and connecting rods – examination and renovation

1 Check that each piston and connecting rod carry identification and orientation marks, so that they may be reassembled correctly.
2 Remove one of the circlips which secure the gudgeon pin. Push the gudgeon pin out of the piston and connecting rod (photos).
3 Examine the pistons for scoring, burning and cracks. If new pistons of standard size are required, note that four grades are available – see Specifications. The grade letter is stamped on the piston crown and adjacent to each bore.
4 Measure the diameter of each piston, at 90° to the gudgeon pin and about 7 mm (0.28 in) above the bottom of the skirt. Subtract the diameter from the bore measurement (Section 30) to give the piston running clearance in the bore. Clearance limits are given in the Specifications.
5 If new rings are to be fitted to the existing pistons, expand the old rings over the top of the pistons. The use of two or three old feeler blades will be helpful in preventing the rings dropping into empty grooves. Note that the oil control ring is in three sections.
6 Before refitting the new rings to the pistons, insert them into the cylinder bore and use a feeler gauge to check that the end gaps are within the specified limits (photo).
7 Clean out the piston ring grooves using a piece of old piston ring as a scraper. Be careful not to scratch the aluminium surface of the pistons. Protect your fingers – piston ring edges are sharp (photo).
8 Check the vertical clearance of the new rings in their grooves. Renew the piston if the clearance is excessive.
9 Check the fit of the gudgeon pin in the connecting rod bush and in the piston. If there is perceptible play, a new bush or an oversize gudgeon pin must be fitted. Consult a Volvo dealer or other specialist.
10 Fit the rings to the piston, starting with the oil control ring sections. Observe the 'TOP' marking on the second compression ring. The other rings can be fitted either way up, unless the top ring is stepped, in which case the step must be uppermost (photos). Do not expand the compression rings too far or they will break – use feeler blades as described in paragraph 5.
11 Oil the gudgeon pin. Reassemble the connecting rod and piston, making sure the rod is the right way round, and secure the gudgeon pin with the circlip.

31.2A Removing a piston circlip

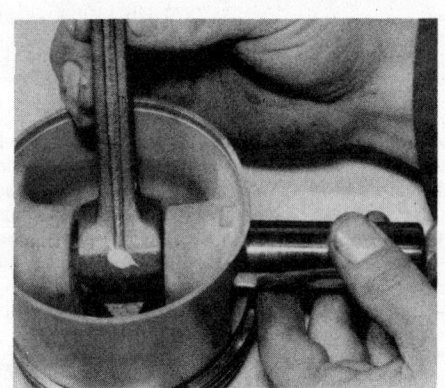

31.2B Pulling out the gudgeon pin

31.6 Measuring a piston ring end gap

31.7 Cleaning a piston ring groove

31.10A Fitting the oil control ring sections

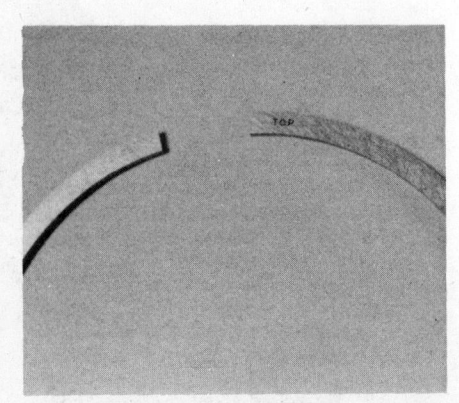

31.10B 'TOP' marking on second compression ring

Chapter 1 Engine

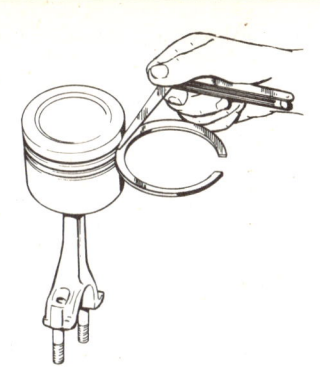

Fig. 1.17 Checking piston ring vertical clearance (Sec 31)

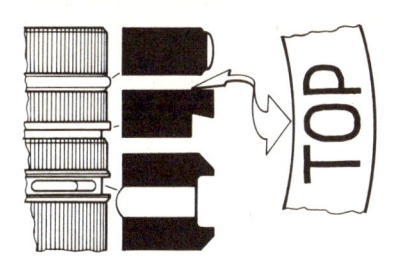

Fig. 1.18 Piston ring profiles (Sec 31)

32 Camshaft and tappets – examination and renovation

1 Inspect the cam lobes and the camshaft bearing journals for scoring or other visible evidence of wear. Once the surface hardening of the lobes has been penetrated, wear will progress rapidly.
2 Measure the bearing journals with a micrometer and check for ovality and taper. The bearing running clearances can be established by refitting the camshaft to the head and using Plastigage (Section 29). If the bearing caps and seats in the head are damaged, the head will have to be renewed.
3 Check the camshaft endfloat as described in Section 9.
4 Inspect the tappets for scuffing, cracking or other damage; measure their diameter in several places with a micrometer. Tappet clearance in the bore can be established by measuring bore diameter and subtracting tappet diameter from it. Renew the tappets if they are damaged or worn.
5 Inspect the tappet shims for visible damage; renew them if they are obviously worn. A selection of new shims should be available in any case for setting the valve clearances.

33 Auxiliary shaft – examination and renovation

1 Inspect the shaft bearing journals and gears for wear or damage. Measure the journals with a micrometer. Renew the shaft if it is worn or damaged.
2 If the auxiliary shaft bearings in the block are damaged, have them renewed by a Volvo dealer or other specialist.

34 Flywheel/driveplate – examination and renovation

1 Inspect the flywheel/driveplate for cracks or other damage. Pay particular attention to the clutch friction face of the flywheel – see Chapter 5, Section 12.
2 If the driveplate is damaged, it must be renewed. Small cracks or moderate scoring on the flywheel friction face can be removed by machining; this is specialist work. Large cracks or deep scores probably call for renewal.
3 Inspect the flywheel ring gear and renew it if necessary (Section 14).
4 Flywheel run-out may be checked during engine rebuilding, when the flywheel is mounted on the crankshaft.

35 Cylinder head – dismantling

1 If not done during removal, unbolt and remove the camshaft sprocket and backplate.
2 Unbolt and remove the thermostat housing and the engine lifting eye. Recover the thermostat and gasket.
3 Progressively slacken the camshaft bearing cap nuts. The camshaft will be lifted by valve spring pressure as this is done. Be careful that it does not spring up suddenly.
4 With the spring pressure relieved, remove the nuts and the camshaft bearing caps. Keep the caps in order: they are numbered 1 to 5, the numbers being on the thermostat side.
5 Lift out the camshaft with its oil seal(s).
6 Have ready a box divided into eight segments, or some other means of keeping matched components together.
7 Lift out the tappets and shims, keeping them identified for position (photo).
8 Recover the rubber rings from the valve stem tips.
9 Tap each valve stem smartly, using a light hammer and drift, to free the spring and associated items.
10 Fit a valve spring compressor to one valve. Compress the spring until the collets are exposed. Lift out the collets – a small screwdriver, a magnet or a pair of tweezers may be useful (photo). Carefully release the spring compressor and remove it.
11 Remove the valve spring upper seat and the valve spring. Pull the valve out of its guide (photo). If the valve stem tip has been 'hammered' or otherwise damaged so that it will not pass through the guide, do not force it, but dress the end with a fine file.
12 If it is an inlet valve which has been removed, pull off the valve

35.7 Removing a tappet

35.10 Extracting a collet with a magnet

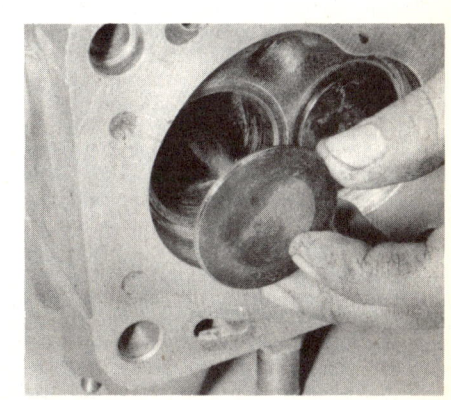

35.11 Removing a valve

35.12 Removing the inlet valve stem oil seal

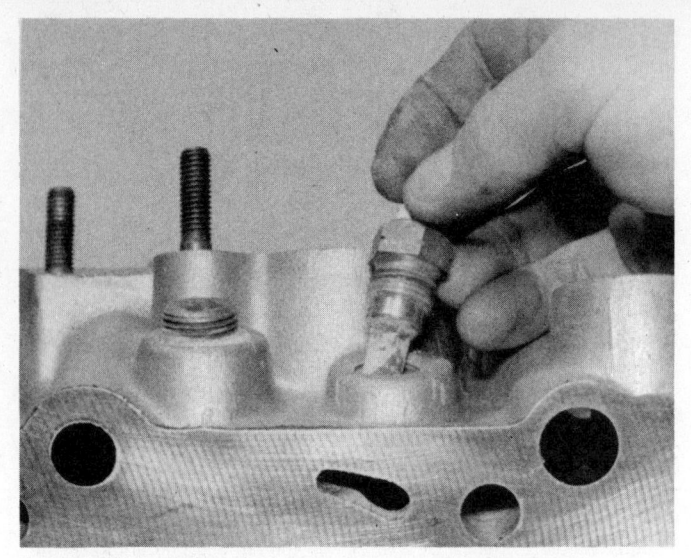

35.15 Removing a temperature sensor from the cylinder head

stem oil seal with a pair of long-nosed pliers. Recover the seal (photos).
13 Recover the valve spring lower seat. If there is much carbon build-up round the outside of the valve guide, this will have to be scraped off before the seat can be removed.
14 Repeat paragraphs 10 to 13 on the other valves. Remember to keep all matched components together if they are to be re-used.
15 Remove the temperature sensor(s) and thermotime switch (as applicable), identifying them if necessary (photo).

36 Cylinder head – decarbonising, valve grinding and renovation

1 With the cylinder head dismantled as described in the previous Section, use a scraper or wire brush to remove carbon from the combustion chambers and posts. Be careful not to scratch or gouge the head, especially on gasket mating faces.
2 Remove all traces of gasket from the mating faces and wash the head with paraffin.
3 Use a straight-edge and feeler blade to check that the cylinder head surface is not distorted. If it is, it must be resurfaced by a suitably equipped engineering works.
4 If the engine is still in the car, clean the piston crowns and cylinder bore upper edges, but make sure that no carbon drops between the pistons and bores. To do this, locate two of the pistons at the top of their bores and seal off the remaining bores with paper and masking tape. Press a little grease between the two pistons and their bores to collect any carbon dust; this can be wiped away when the piston is lowered. To prevent carbon build-up, polish the piston crown with metal polish, but remove all traces of the polish afterwards.
5 Examine the heads of the valves for pitting and burning, especially the exhaust valve heads. Renew any valve which is badly burnt.
Caution: *Exhaust valves on Turbo engines contain sodium and must not be mixed with other scrap metal. Consult a Volvo dealer for safe disposal of these valves.*
6 Inspect the valve seats for pitting, burning or cracks. Have the seats recut or renewed if necessary. Light pitting can be removed by valve grinding.
7 Check the fit of the valves in their guides, if possible using a dial gauge to measure play at the valve head when it is held 1 to 2 mm (0.04 to 0.08 in) off its seat. Limits are given in the Specifications. If renewing the valves does not improve the fit, the guides must be renewed. This is another specialist job.
8 If the old valves are fit for re-use, remove the carbon from them. This is most easily done by gripping the valve stem in an electric drill chuck, mounting the drill securely and spinning the valve against a wire brush or scraper. Wear eye protection.

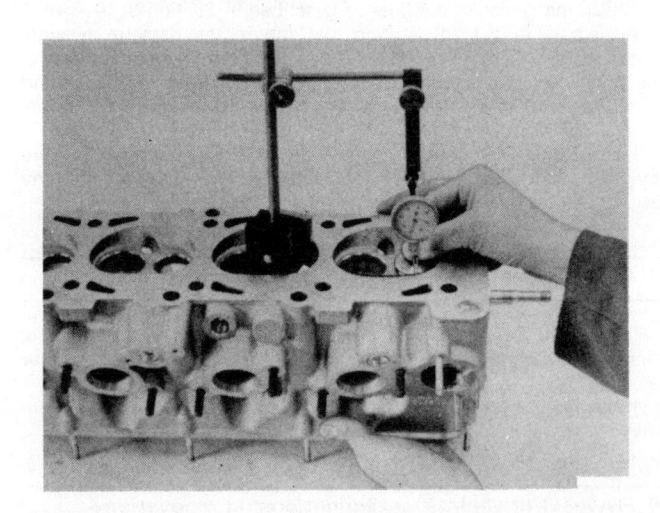

Fig. 1.19 Measuring valve play in the guide (Sec 36)

9 Valve grinding should now be carried out, either to mate new components together or to improve the fit of the old ones. Place the cylinder head upside down on a couple of wooden blocks.
10 Smear a trace of coarse carborundum paste on the seat face and press a suction grinding tool onto the valve head. With a semi-rotary action, grind the valve head to its seat, lifting the valve occasionally to redistribute the grinding paste. When a dull matt even surface is produced on both the valve seat and the valve, wipe off the paste and repeat the process with fine carborundum paste as before. A light spring placed under the valve head will greatly ease this operation. When a smooth unbroken ring of light grey matt finish is produced on both the valve and seat, the grinding operation is complete (photos).
11 Remove all traces of grinding paste from the valves, seats and guides, using first a paraffin-soaked rag, then a clean rag, finally compressed air (if available). It is important that no grinding paste is left in the engine.
12 Measure the free length of the valve springs, and if possible check their load/length characteristics – see Specifications. Renew the springs if they are obviously weak or distorted, or if they have seen much service.
13 Renew the inlet valve stem oil seals as a matter of course.

Chapter 1 Engine

36.10A Applying grinding paste to a valve

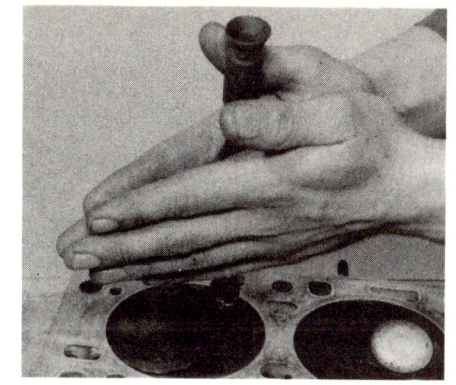

36.10B Grinding in a valve

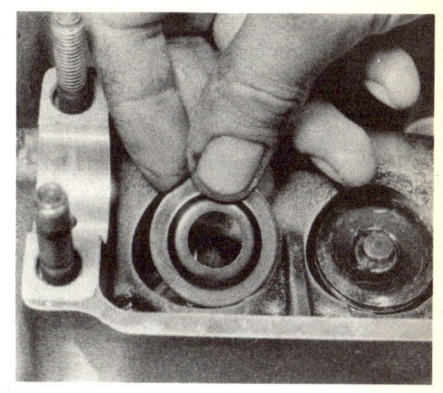

37.2 Fitting a valve spring lower seat

37 Cylinder head – reassembly

1 Refit the temperature sensors and similar items to their positions in the head. Use a smear of sealant on the threads.
2 Oil the stem of one valve and insert it into its guide. Fit the spring lower seat, dished side up (photo).
3 On an inlet valve, fit the valve stem oil seal, pushing it onto the valve guide with a piece of tube. Be careful not to damage the seal lips on the valve stem: if a protective sleeve is supplied with the seals, cover the collet grooves with it when fitting the seal.
4 Fit the valve spring and upper seat (photos). Compress the spring and fit the two collets, using a dab of grease on each to hold them in position. Carefully release the compressor.
5 Cover the valve stem with a cloth and tap it smartly with a light hammer to verify that the collets are properly seated.
6 Repeat paragraphs 2 to 5 on the other valves. (Stem oil seals are not fitted to exhaust valves.)
7 Fit new rubber rings to the valve stem tips (photo).
8 Oil the tappets and insert them into their bores. Fit a shim to each tappet (photo). Record the thickness of shims fitted for reference later.
9 Oil the camshaft lobes and bearings. Fit the camshaft and the bearing caps, applying sealant to the mating faces of the front and rear caps.
10 Fit the bearing cap nuts and pull down the caps by progressively tightening the nuts. When all the caps are seated, tighten the nuts to the specified torque.
11 Lubricate and fit a new front oil seal, lips inwards. Seat it with a piece of tube.
12 Fit the camshaft sprocket and plates. Secure the sprocket with the bolt and washer. Restrain the sprocket with a strap wrench or similar tool when tightening the bolt. (On the B230 engine the sprocket will have to be removed again when refitting the cylinder head, so do not bother to tighten the bolt fully.)
13 Check the valve clearances and adjust them if necessary as described in Section 4, turning the camshaft sprocket by hand or with a strap wrench.
14 Refit the thermostat, using a new gasket, and the engine lifting eye.

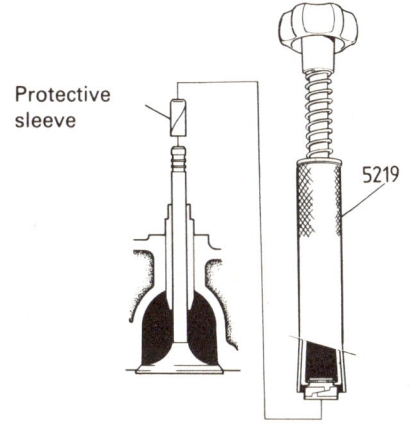

Fig. 1.20 Tool 5219 for fitting the valve stem oil seals – a piece of tube will do (Sec 37)

38 Crankshaft and main bearings – refitting

1 Wipe the bearing shell locations in the crankcase and main bearing caps with a clean non-fluffy rag.
2 Fit the main bearing shells to their locations (the same locations as previously occupied if they are being re-used). Press the shells home so that the tangs engage in the recesses provided. All the shells are the

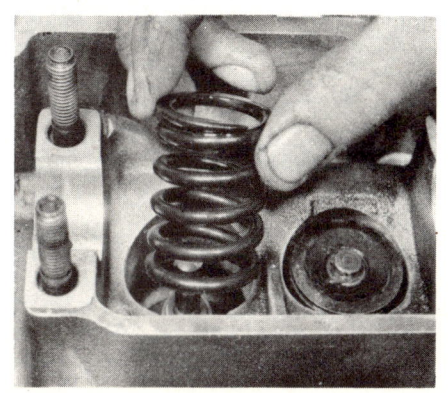

37.4A Fitting a valve spring ...

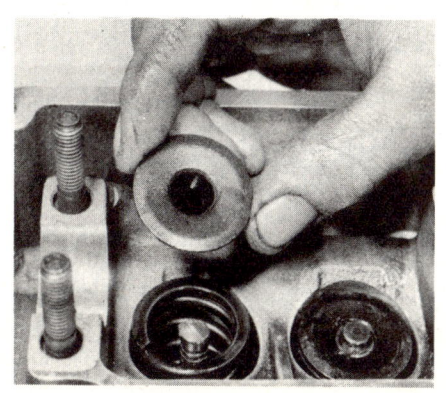

37.4B ... and the spring upper seat

37.7 Fitting a rubber ring to the valve stem tip

54 Chapter 1 Engine

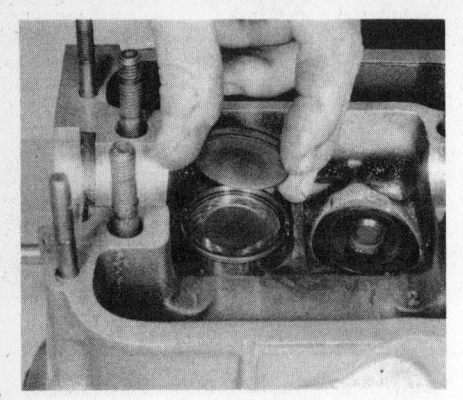

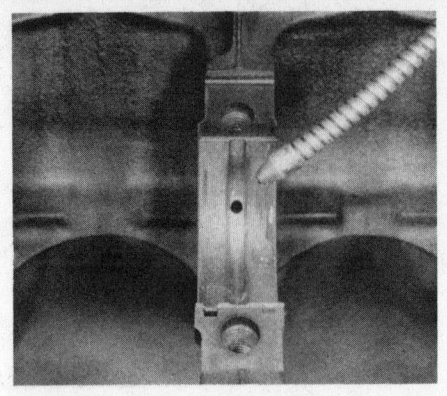

37.8 Fitting a shim to its tappet

38.4 Oiling a main bearing shell

38.7 Tightening a main bearing cap bolt

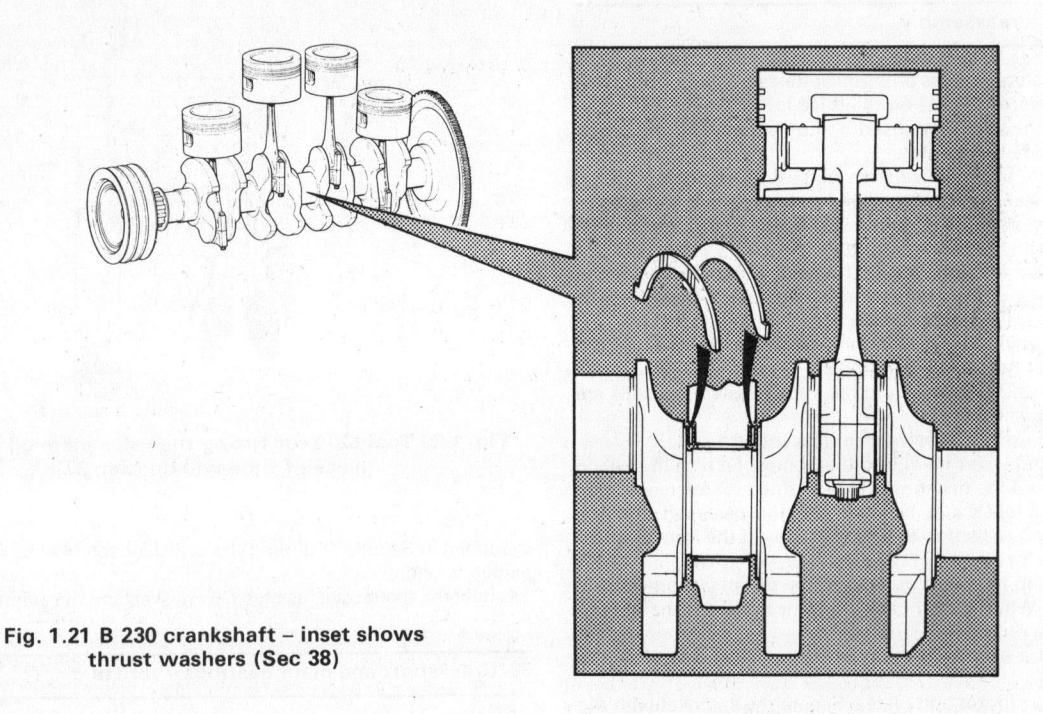

Fig. 1.21 B 230 crankshaft – inset shows thrust washers (Sec 38)

same, except for No 5 shells on the B23 engine which have thrust flanges.
3 On the B230 engine, smear some grease on the smooth sides of the half thrust washers. Place the washers in position on each side of the centre bearing in the crankcase. The slotted sides of the washers face outwards.
4 Oil the bearing shells in the crankcase (photo).
5 Wipe clean the crankshaft journals, then lower it into position with the help of an assistant. Make sure that the shells (and thrust washers, when applicable) are not displaced.
6 Inject oil into the crankshaft oilways. Oil the shells in the main bearing caps and fit the caps, each to its correct position and the right way round.
7 Fit the main bearing cap bolts and tighten them progressively to the specified torque (photo).
8 Rotate the crankshaft. Some stiffness is to be expected with new components, but there must be no tight spots or binding.
9 Check the crankshaft endfloat, levering it back and forth and placing a dial test indicator on the flywheel mounting face. On the B23 engine, endfloat can also be checked by inserting feeler blades between the crankshaft web and No 5 bearing flange.
10 Fit the rear oil seal carrier, using a new gasket. Trim the protruding ends of the gasket level with the sump mating face (photos).
11 Fit a new rear oil seal as described in Section 13.

39 Auxiliary shaft – refitting

1 Lubricate the auxiliary shaft bearing surfaces and feed the shaft into the block, being careful not to damage the bearings.
2 Fit the front oil seal housing, using a new gasket. Trim the ends of the gasket level with the sump mating face. Some of the housing bolts cannot be fitted yet because they also secure the camshaft drivebelt backplate.
3 Fit new oil seals in the front oil seal housing, lips inwards and lubricated. Use a piece of tube to seat the seals (photo).

40 Engine – reassembly after overhaul

1 Refit the oil pressure switch and any other senders or transducers to the block.
2 Refit the pistons and connecting rods (Section 18).
3 Refit the oil pump, delivery pipe and drain hose guide (Section 17).
4 Refit the oil trap drain hose, making sure that it is inserted fully into its hole and that it is secured towards its lower end by the guide (photo). It is impossible to fit this hose successfully after the sump has been fitted.
5 Refit the sump, using a new gasket (photos). Tighten the bolts progressively to the specified torque.

Chapter 1 Engine 55

38.10A Fitting a rear oil seal carrier gasket

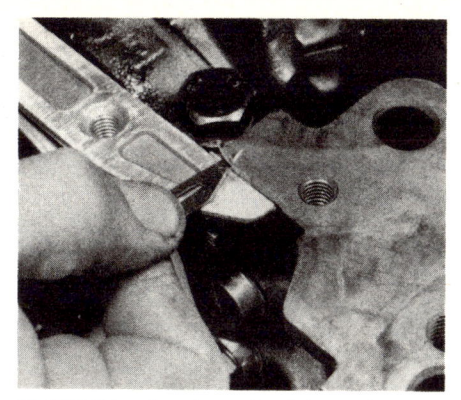

38.10B Trimming the ends of the gasket

39.3 Front oil seals and housing in position. Four bolt holes (arrowed) are shared with drivebelt backplate

40.4 Oil trap drain hose correctly fitted. The guide is secured by an oil pump bolt (arrowed)

40.5A Refitting the sump

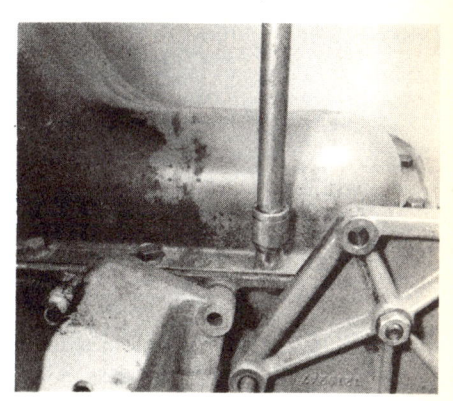

40.5B Tightening a sump bolt

6 Refit the flywheel or driveplate (Section 12), making sure that it is in the correct position.
7 Refit the crankshaft sprocket and guide plates, and the pulley or pulley boss. Fit the pulley bolt, jam the ring gear teeth and tighten the bolt to the specified torque (photos).
8 Refit the cylinder head (Section 10).
9 Refit the water pump, using a new gasket.
10 Refit the camshaft drivebelt backplate, the auxiliary shaft and camshaft sprockets, and the drivebelt and tensioner (Section 6). Refit the drivebelt cover.
11 Refit the oil pump drivegear/shaft, making sure it engages with the oil pump end with the auxiliary shaft gear (photo).
12 Fit a new O-ring to the crankcase ventilation system oil trap. Refit and secure the trap (photos).
13 Refit the dipstick tube and secure its bracket. Insert the dipstick.

14 Refit the coolant distribution pipe.
15 When applicable, refit the oil cooler adaptor, using a new O-ring (photo).
16 Refit the remaining ancillary components listed in Section 24, paragraph 1. If preferred, delicate items such as the distributor and alternator may be left until the engine has been refitted to the vehicle.
17 Make sure that the ignition sensor bracket is in place before mating the transmission to the engine, since it is impossible to fit it afterwards.

41 Engine – reconnection to transmission

Automatic transmission
1 Make sure that the torque converter is fully engaged in the

40.7A Fitting the crankshaft sprocket rear guide plate ...

40.7B ... the sprocket itself ...

40.7C ... the front guide plate and the pulley boss ...

56　　　　　　　　　　　　　　　　　　　Chapter 1 Engine

40.7D ... and the pulley bolt

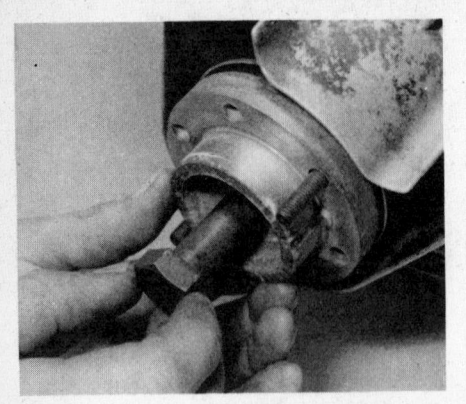

40.11 Fitting the oil pump drivegear

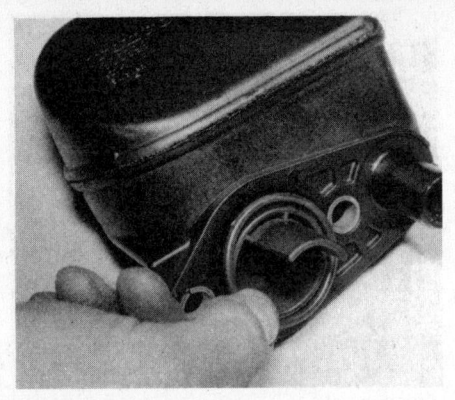

40.12A Fitting a new O-ring to the oil trap

40.12B Refitting the oil trap

40.15 Oil cooler adaptor – O-ring arrowed

transmission. Put a smear of grease or anti-seize compound on the torque converter locating spigot.
2　Offer the transmission to the engine, engaging the locating dowels. Fit a couple of bellhousing-to-engine nuts and bolts.
3　Insert the torque converter-to-driveplate bolts, turning the crankshaft to gain access. Just nip the bolts up at first, then tighten them in cross-cross sequence to the specified torque (Chapter 6 Specifications).

Manual gearbox
4　Make sure that the clutch is correctly centred and that the clutch release components are fitted to the bellhousing. Put a smear of grease or anti-seize compound on the input shaft splines.
5　Offer the gearbox to the engine. Rotate the crankshaft or the input shaft if necessary to align the input shaft and clutch driven plate splines. Do not allow the weight of the gearbox to hang on the input shaft.
6　Engage the gearbox on the engine dowels. Fit a couple of bellhousing-to-engine nuts and bolts.

All models
7　Fit the remaining bellhousing nuts and bolts, and (when applicable) the flywheel/driveplate bottom cover plate. Tighten the nuts and bolts progressively.
8　Refit the starter motor.

42 Engine – refitting (with transmission)

1　Refitting is essentially a reversal of the procedure in Section 22, with the following additional points.
2　On automatic transmission models, adjust the gear selector mechanism (Chapter 6, Section 30).
3　Refill the engine with oil and coolant.
4　Refill the manual gearbox or automatic transmission with lubricant if necessary.
5　Refer to Section 44 before starting the engine.

43 Engine – refitting (alone)

1　Make sure that the clutch is properly centred, or that the torque converter is fully engaged in the transmission. Put a smear of grease or anti-seize compound on the gearbox input shaft or the torque converter locating spigot.
2　Lower the engine into position; have an assistant watch to see that no pipes, wires etc, are trapped.
3　On manual gearbox models, rock the engine from side to side, or rotate the crankshaft slightly, to encourage the input shaft to enter the clutch driven plate. Do not allow the engine to hang on the input shaft.
4　When the bellhousing is engaged on the engine dowels, insert a couple of engine-to-bellhousing nuts and bolts and nip them up.
5　The remainder of refitting is now a reversal of the procedure in Section 21.
6　Refill the engine with oil and coolant.
7　Refer to Section 44 before starting the engine.

44 Initial start-up after overhaul or major repair

1　Make a final check to ensure that everything has been reconnected to the engine and that no rags or tools have been left in the engine bay.
2　Check that the accessory drivebelts are correctly tensioned.
3　Check that engine oil and coolant levels are correct.
4　Start the engine. This may take a little longer than usual as fuel is pumped up to the engine.
5　Check that the oil pressure light goes out when the engine starts.
6　Run the engine at a fast tickover and check for leaks of oil, fuel and coolant. Also check power steering and (when applicable) transmission fluid cooler unions for leaks. Some smoke and odd smells may be expected as assembly lubricant burns off exhaust components.
7　Bring the engine to operating temperature. Check that the throttle cable is correctly adjusted, then check the idle speed and mixture (Chapter 3, Section 8).
8　On automatic transmission models, check the adjustment of the kickdown cable (Chapter 6, Section 31).

Chapter 1 Engine

9 Stop the engine and allow it to cool. Recheck the oil and coolant levels and top up if necessary.
10 If new bearings, pistons etc, have been fitted, the engine should be run in at restricted speeds and loads for the first 600 miles (1000 km) or so. After this mileage the camshaft drivebelt should be re-tensioned (Section 6) and the valve clearances rechecked (Section 4). At the same time it will be beneficial to change the engine oil and filter. There is no need to retighten the cylinder head bolts.

45 Compression test – description and interpretation

1 When engine performance is down, or if misfiring occurs which cannot be attributed to the ignition or fuel system, a compression test can provide diagnostic clues. If the test is performed regularly it can give warning of trouble before any other symptoms become apparent.
2 The engine must be at operating temperature, the battery must be fully charged and the spark plugs must be removed. The services of an assistant will also be required.
3 Disable the ignition system by disconnecting the coil LT feed. Fit the compression tester to No 1 spark plug hole. (The type of tester which screws into the spark plug hole is to be preferred.)
4 Have the assistant hold the throttle wide open and crank the engine on the starter. Record the highest reading obtained on the compression tester.
5 Repeat the test on the remaining cylinders, recording the pressure developed in each.
6 Desired pressures are given in the Specifications. If the pressure in any cylinder is low, introduce a teaspoonful of clean engine oil into the spark plug hole and repeat the test.
7 If the addition of oil temporarily improves the compression pressure, this indicates that bore or piston wear was responsible for the pressure loss. No improvement suggests that leaking or burnt valves, or a blown head gasket, may be to blame.
8 A low reading from two adjacent cylinders is almost certainly due to the head gasket between them having blown.
9 On completion of the test, refit the spark plugs and reconnect the coil LT feed.

46 Fault diagnosis – engine

Symptom	Reason(s)
Engine fails to start	Discharged battery Loose battery connection Loose or broken ignition leads Moisture on spark plugs, distributor cap, or HT leads Incorrect spark plug gaps Cracked distributor cap or rotor Other ignition system fault Dirt or water in fuel Empty fuel tank Faulty fuel pump Other fuel system fault Faulty starter motor Low cylinder compressions
Engine idles erratically	Intake manifold air leak Leaking head gasket Incorrect valve clearances Worn camshaft lobes Faulty fuel pump Loose crankcase ventilation hoses Idle adjustment incorrect Uneven cylinder compressions
Engine misfires	Spark plugs worn or incorrectly gapped Dirt or water in fuel Idle adjustment incorrect Burnt out valve Leaking cylinder head gasket Distributor cap cracked Incorrect valve clearances Uneven cylinder compressions Worn carburettor Other fuel or ignition system fault
Engine stalls	Idle adjustment incorrect Intake manifold air leak Ignition timing incorrect
Excessive oil consumption	Worn pistons, cylinder bores or piston rings Valve guides and valve stem seals worn Oil leaks
Engine backfires	Idle adjustment incorrect Ignition timing incorrect Incorrect valve clearances Intake manifold air leak Sticking valve
Engine noises	See *'Fault diagnosis'* at beginning of manual

PART B: V6 ENGINE

47 General description

The V6 engine is a product of the PRV (Peugeot-Renault-Volvo) co-operative. It is well proven and is found in a wide selection of European cars destined for the upper end of the market.

The cylinder block, cylinder heads and crankcase are all made of aluminium alloy. The crankcase is split horizontally at the level of the crankshaft; a pressed steel sump is bolted onto the bottom of the lower crankcase.

A 90° 'V' configuration is used for the two banks of cylinders. Cast iron cylinder liners of the 'wet' type are used, as is common. French practise: coolant circulates freely between the liners and the block. The liners are sealed at the top by the head gasket and at the base by individually selected seals.

The crankshaft runs in four renewable shell bearings, with endfloat controlled by thrust washers at the flywheel end. Each crankpin is shared by two connecting rods. The connecting rod big-end bearings are also of the shell type.

The cylinder heads are of the crossflow type, the common inlet manifold sitting in the centre of the engine and the exhaust manifolds being on the outside. Each head carries valves, a camshaft and rocker gear. The camshafts are driven by separate chains with hydraulic tensioners.

The lubrication system is conventional. A gear type pump, driven from the crankshaft by a third chain, draws oil from the sump. Oil under pressure passes through an external canister filter before being supplied to the crankshaft, camshafts and rocker gear. The pistons and gudgeon pins are lubricated by splash. The timing chains are lubricated by spillage from the camshaft front bearings and the chain tensioner.

48 Maintenance and inspection

1 Check the engine oil level weekly, every 250 miles or before a long run. Top up if necessary (photos). See Section 2 for more details.
2 Change the oil every 6000 miles (10 000 km) or six months, also as described in Section 2. An 8 mm square drive key will be needed for the sump drain plug (photo).
3 At the same intervals renew the oil filter – see Sections 3, and 49.
4 Every 24 000 miles (40 000 km) or two years, or if valvegear noise becomes excessive, check the valve clearances (Section 50).
5 At the same intervals, inspect the crankcase ventilation system components (Sections 5 and 51).
6 There is no camshaft drivebelt to renew on the V6 engine. Although not specified by the makers, it would be prudent to inspect the timing chains at roughly 48 000 mile (80 000 km) or four-year intervals. See Section 80, paragraph 2.

49 Oil filter – renewal

Proceed as in Section 3, but note that the oil filter is located on the left-hand side of the block (photo).

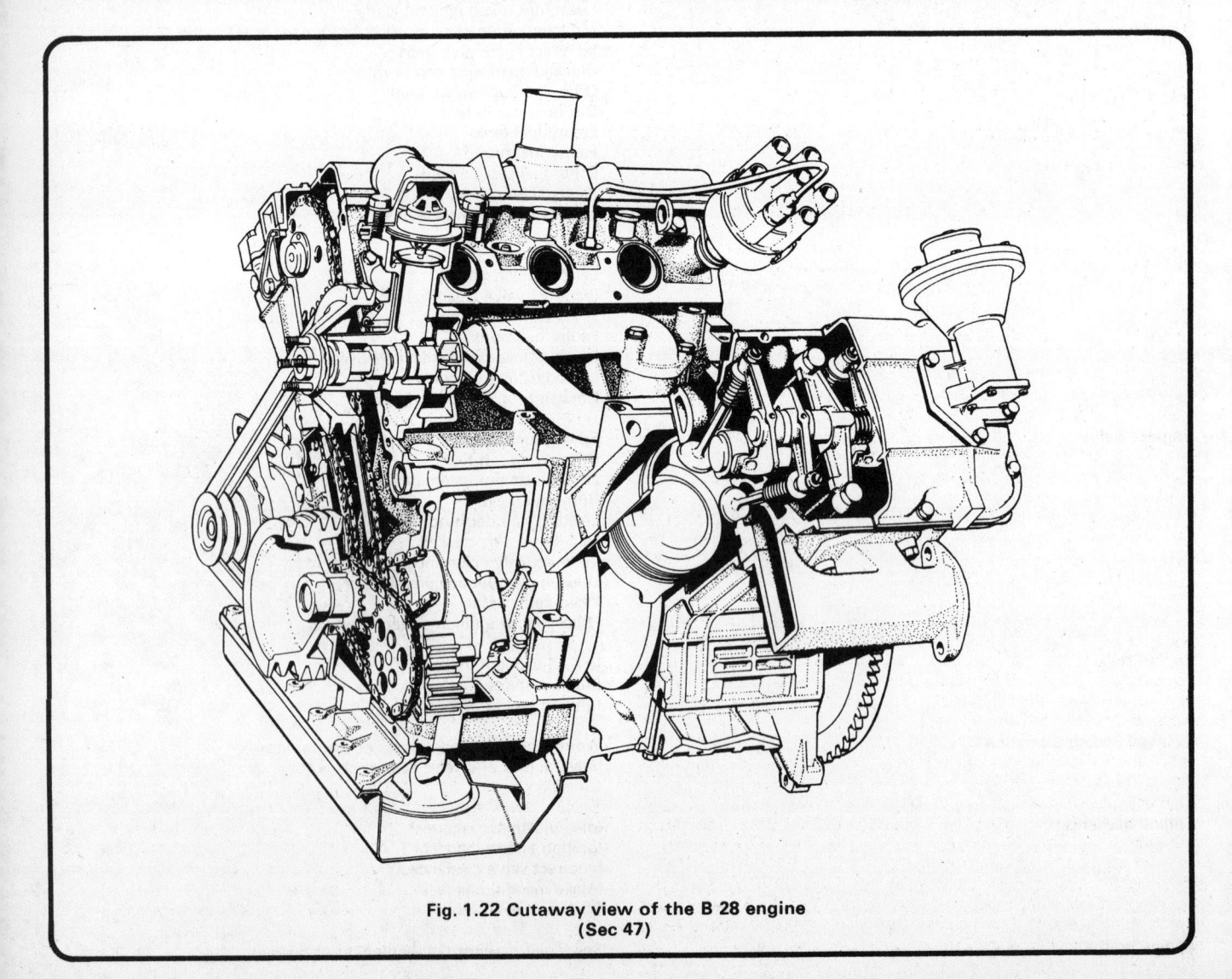

Fig. 1.22 Cutaway view of the B 28 engine
(Sec 47)

Chapter 1 Engine

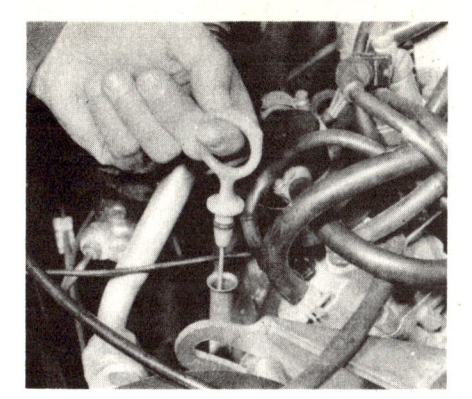

48.1A Removing the engine oil dipstick

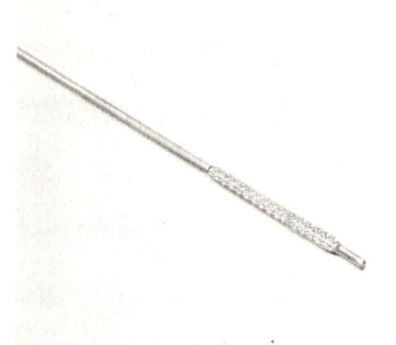

48.1B Dipstick tip – oil level should be in the hatched area

48.1C Topping-up the engine oil

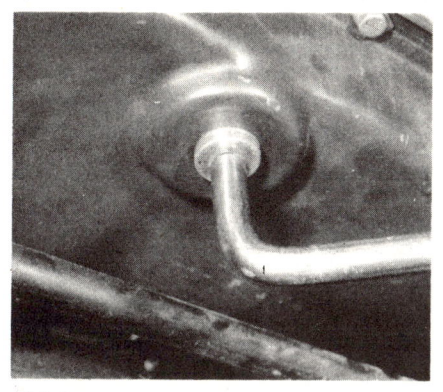

48.2 Removing the sump drain plug

49.0 Fitting a new oil filter

50.7 Removing a rocker cover bolt

50 Valve clearances – checking and adjustment

1 Disconnect the battery negative lead.
2 Disconnect the ignition harness connector on the right-hand inner wing.
3 Unbolt the control pressure regulator (without disconnecting it) and place it on the inlet manifold.
4 Remove the air inlet trunking, the oil filler cap and the crankcase ventilation hoses.
5 Unbolt and remove the vacuum pump.
6 Remove the air conditioning compressor drivebelt (Chapter 2, Section 7). Unbolt the compressor brackets from the engine and move the compressor and brackets to one side. **Do not** disconnect any refrigerant hoses, nor allow the weight of the compressor to hang on them.
7 Unbolt and remove both rocker covers. Recover the gaskets (photo).
8 Using a 36 mm socket on the crankshaft pulley nut, or (manual gearbox) by pushing the car along with a gear engaged, bring the engine to TDC, No 1 firing. This is achieved when the No 1 pulley notch is aligned with the 'O' mark on the timing scale, and both rocker arms for No 1 cylinder (LH rear) have a small amount of free play, showing that the valves are closed.
9 Insert a feeler blade of the specified thickness between the adjuster screw and valve stem on No 1 cylinder exhaust valve (the rearmost valve). The blade should be a firm sliding fit, neither tight nor slack.
10 If adjustment is required, slacken the locknut and turn the adjuster screw until the clearance is correct. Hold the adjuster screw stationary and tighten the locknut without disturbing the position of the screw. Recheck the fit of the feeler blade (photo).
11 Similarly check the clearances of the exhaust valves on cylinders No 3 and 6, and the inlet valves on cylinders No 1, 2 and 4. Remember that inlet and exhaust clearances are different. Inlet valves are nearest the centre of the engine, exhaust valves nearest the outside.

Fig. 1.23 No 1 cylinder notch in the TDC position (Sec 50)

1 No 1 cylinder notch   2 No 6 cylinder notch

12 Turn the crankshaft 360° (one full turn) clockwise, so that the No 1 pulley notch is again aligned with the 'O' mark, but this time No 1 cylinder rocker arms have no free play. In this position check the clearances of the exhaust valves on cylinders No 2, 4 and 5, and the inlet valves on cylinders No 3, 5 and 6.
13 Recheck all clearances, turning the crankshaft as necessary.
14 Refit the rocker covers, using new gaskets (photo).
15 Refit the remaining components in the reverse order to removal.

Chapter 1 Engine

50.10 Adjusting a valve clearance

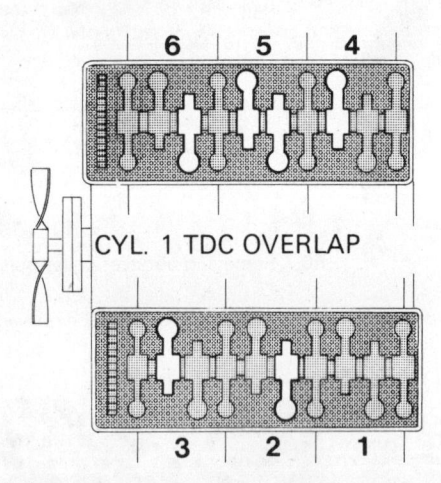

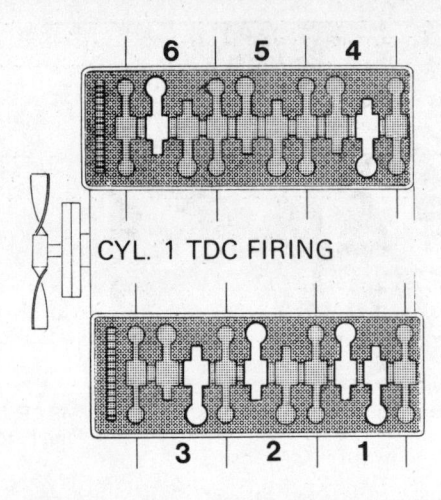

Fig. 1.24 With No 1 in firing position, adjust the valves in the light coloured positions (Sec 50)

Fig. 1.25 With No 1 ending exhaust stroke (valves overlapping) adjust the other valves shown (Sec 50)

50.14 Fitting a new rocker cover gasket

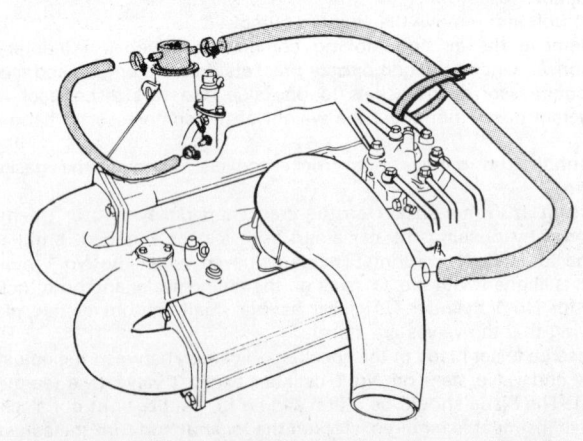

Tension the air conditioning compressor drivebelt (Chapter 2, Section 7).

16 Run the engine and check that there are no oil leaks from the rocker covers.

51 Crankcase ventilation system – general

Refer to Section 5 for general advice, and to Fig. 1.26 for component location.

52 Major operations possible with the engine installed

1 The following components can be removed with the engine in place:

(a) Timing chains, sprockets etc
(b) Cylinder heads and camshafts
(c) Crankshaft front oil seal
(d) Oil pump
(e) Flywheel/clutch (after removal of transmission)
(f) Crankshaft rear oil seal (after removal of flywheel)
(g) Engine mountings

Fig. 1.26 Crankcase ventilation components (Sec 51)

2 The sump can be removed with the engine installed, but this does not give access to any major components.
3 The engine must be removed for access to the pistons, cylinder liners and crankshaft.

Chapter 1 Engine

53 Engine dismantling and reassembly – general

Refer to Section 8.

54 Timing chains and sprockets – removal and refitting

1 If the engine is in the vehicle, carry out the following preliminary work:

 (a) Disconnect the battery
 (b) Remove the radiator, fan shroud and fan. When fitted, also remove the ATF cooler (Chapter 2)
 (c) Remove all accessory drivebelts (Chapter 2)
 (d) Unbolt and move aside the air conditioning compressor and the steering pump, both with their brackets
 (e) Remove the air inlet trunking, the oil filler cap and the crankcase ventilation hoses

2 Unbolt and remove the vacuum pump.
3 Unbolt the control pressure regulator from the right-hand valve cover and move it aside.
4 Remove the ten bolts which secure each rocker cover, noting the locations of the various lengths of bolt.
5 Remove the rocker covers and recover the gaskets.
6 Bring the engine to TDC, No 1 firing. See Section 50, paragraph 8.
7 Remove the blanking plate from the unused starter motor opening. Jam the starter ring gear through this opening, either by having an assistant brace a tyre lever or a large screwdriver in the gear teeth, or (preferably) by bolting a suitably shaped metal segment to engage with the teeth.
8 Using a 36 mm socket, undo the crankshaft pulley centre nut. This nut is very tight. Remove the ring gear jamming device.
9 Check that the keyway in the crankshaft pulley is upwards, then remove the pulley. (If the keyway were downwards, the key might fall into the sump).
10 Remove the 25 bolts which secure the timing cover (photo). Also remove the drivebelt idler pulleys, which share some of the timing cover bolts. Note the locations of the various lengths of bolt.
11 Move aside the wiring harness which passes in front of the timing cover.
12 Pull the timing cover off its dowels and remove it. Cover the holes leading to the sump with paper or rag, then recover the timing cover gasket.
13 Restrain the camshaft sprockets and slacken their centre bolts with a 10 mm Allen key (photo).
14 Retract each timing chain tensioner by turning the locking device anti-clockwise with a small screwdriver, at the same time pushing in the plunger (photo).
15 Unbolt the oil pump sprocket. Remove the sprocket and chain.
16 Unbolt and remove the timing chain tensioners. Identify them if they are to be re-used. Recover the oil strainer from behind each tensioner.
17 Unbolt and remove the timing chain guides and dampers.

Fig. 1.27 Metal segment (5112) jamming the ring gear teeth (Sec 54)

18 Check whether any markings are visible on the timing chain links. If there are none, and the existing timing chain is to be re-used, initially read paragraphs 25, 26, 27, 29 and 30 of this Section and make alignment marks on the chains when the crankshaft and camshaft sprockets are in the stated positions.
19 Remove the camshaft sprocket centre bolts, the camshaft sprockets and the timing chains. Identify left-hand and right-hand components.
20 Remove the oil pump drive sprocket, the outer Woodruff key, the spacer, the twin sprocket and the inner key. A pulley may be needed (photo).
21 Examine components as described in Section 80. Clean the gasket mating faces.
22 Commence reassembly by fitting new oil strainers to the chain tensioner recesses in the block. Refit and secure the chain tensioners, using thread locking compound on the bolts (photos).
23 Refit and secure the chain guides and dampers, again using thread locking compound (photos).
24 Oil the crankshaft nose and fit the inner Woodruff key to it (photo).
25 Fit the twin drive sprocket to the crankshaft. The mark on the sprocket must face outwards. Drive the sprocket home with a piece of tube if it is tight (photo). Be careful not to dislodge the Woodruff key.
26 Prepare to fit the left-hand timing chain. (Remember, left and right refer to the engine, not to the mechanic.) Temporarily refit the crankshaft pulley nut and turn the crankshaft until the keyway points to the left-hand camshaft. Turn the left-hand camshaft so that the sprocket locating groove points directly upwards (Fig. 1.28).
27 Fit the left-hand timing chain to the camshaft sprocket so that the two marked links on the chain are on each side of the mark on the sprocket. Offer the chain to the innermost section of the crankshaft sprocket so that the single marked link is in line with the mark on the sprocket. Tension the chain on the driving side (next to the straight

54.10 Removing a timing cover bolt

54.13 Slackening a camshaft sprocket bolt

54.14 Retracting a timing chain tensioner

54.20 Pulling off the twin sprocket

54.22A Fitting a new oil strainer ...

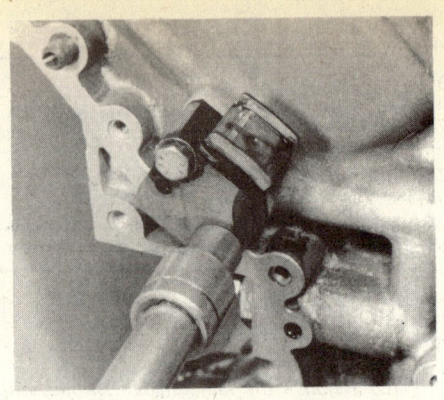

54.22B ... and bolting on the chain tensioner

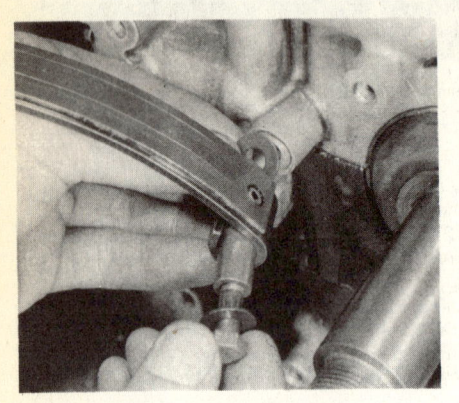

54.23A Refitting a chain damper

54.23B Tightening a chain guide bolt

54.24 Fit the inner Woodruff key (arrowed)

54.25 Driving the inner sprocket home

54.27A Single marked link aligned with crankshaft sprocket mark ...

54.27B ... and twin marked links straddling the camshaft sprocket mark

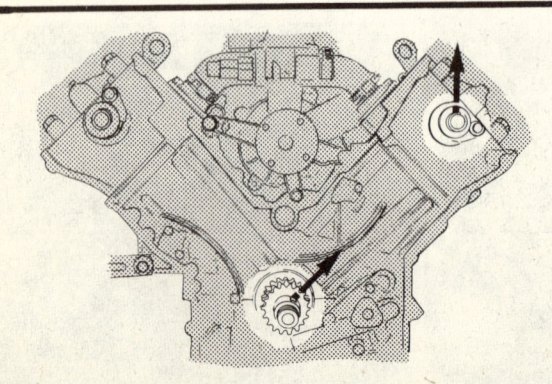

Fig. 1.28 Keyway and groove positions for fitting the left-hand timing chain (Sec 54)

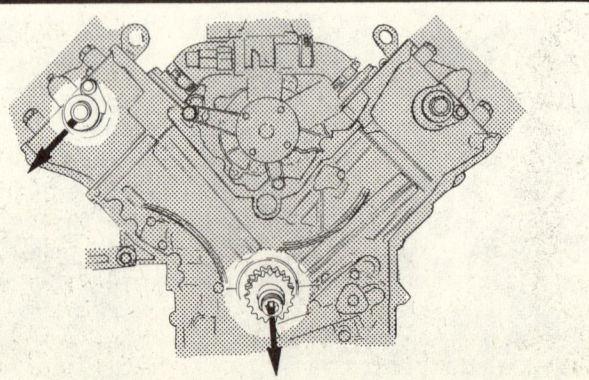

Fig. 1.29 Keyway and groove positions for fitting the right-hand timing chain (Sec 54)

Chapter 1 Engine

guide) and fit the camshaft sprocket to the camshaft. The sprocket must locate in the groove (photos).

28 Fit the bolt to the left-hand camshaft sprocket, but do not tighten it fully yet.

29 Prepare to fit the right-hand chain. Turn the crankshaft clockwise until the keyway points vertically downwards. Turn the right-hand camshaft until the sprocket locating groove is facing outwards and parallel with the head mating face (Fig. 1.29).

30 Fit the right-hand chain and sprockets in the same way as the left-hand one, with the twin marked links straddling the camshaft sprocket mark and the single marked link aligned with the crankshaft sprocket mark (photos). Turn the crankshaft a little if necessary to achieve alignment.

31 Fit the bolt to the right-hand camshaft sprocket. Tighten both camshaft sprocket bolts to the specified torque, restraining the sprockets with a screwdriver.

32 Release the timing chain tensioners by turning the locking devices a quarter turn clockwise. Do not force the plungers out.

33 Rotate the crankshaft through two full turns clockwise to set the chain tension. (The marks will no longer align – see paragraph 18.) Turn the crankshaft a further half turn clockwise so that the keyway points upwards again.

34 Remove the crankshaft pulley nut. Fit the spacer, the outer Woodruff key and the oil pump driving sprocket (photos).

35 Fit the oil pump driven sprocket and chain. Use thread locking compound on the sprocket bolts.

36 Oil the chains and remove the rag or paper from the sump holes. Check that nothing has been overlooked.

37 Refit the timing cover, using a new gasket (photo). Insert and tighten the 25 bolts, applying thread locking compound to the four bottom bolts. Remember to fit the wiring harness behind the idler pulleys.

38 Fit a new oil seal to the timing cover if necessary (Section 58), then refit the crankshaft pulley. Be careful not to dislodge the Woodruff key.

39 Turn the starter ring gear. Fit the crankshaft pulley nut and tighten it to the specified torque (photo). Remove the jamming device.

40 Refit the starter motor blanking plate.

41 Trim the protruding ends of the timing cover gasket flush with the cylinder heads.

42 Refit the rocker covers, using new gaskets.

43 Refit the remaining components by reversing the removal procedure, referring to other Chapters as necessary.

44 Check the ignition timing (Chapter 4, Section 6) and the idle speed and mixture (Chapter 3, Section 8).

55 Oil pump – removal and refitting

1 Proceed as for timing chain removal (Section 54, paragraphs 1 to 12).

2 Unbolt the oil pump sprocket. Remove the sprocket and chain (photo).

3 Remove the four bolts which secure the oil pump to the block. Withdraw the pump and recover the idler gear (photos).

4 Clean the pump and block mating faces, and clean out the pump recess in the block.

5 Commence refitting by oiling all components liberally. Insert the idler gear into the recess, fit the pump and secure it with the four bolts. Tighten the bolts to the specified torque.

6 Refit the sprocket and chain. Use thread locking compound on the sprocket bolts.

7 Refit the timing cover and associated components (Section 54).

56 Cylinder heads – removal and refitting (engine installed)

Note: *Read through this procedure before starting work to understand what is involved. In particular, note that if the cylinder liners are accidentally disturbed, the engine may have to be removed and completely dismantled to put things right.*

1 Disconnect the battery negative lead.

2 Drain the cooling system (Chapter 2, Section 3).

54.30A Right-hand chain marking at the crankshaft sprocket ...

54.30B ... and at the right-hand camshaft sprocket

54.34A Fit the spacer ...

54.34B ... and the outer Woodruff key

54.37 Fitting a new timing cover gasket

54.39 Tightening the crankshaft pulley nut

64   Chapter 1 Engine

55.2A Unbolt the oil pump sprocket ...

55.2B ... and remove it with the chain

55.3A Unbolt the oil pump ...

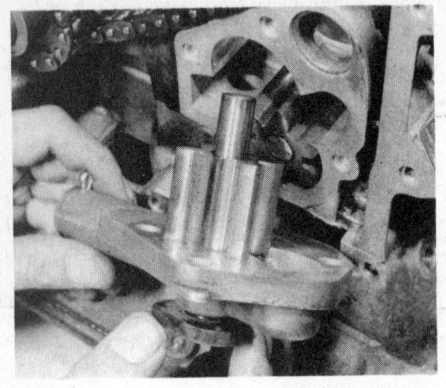

55.3B ... and remove it from the block. Idler gear will stay behind

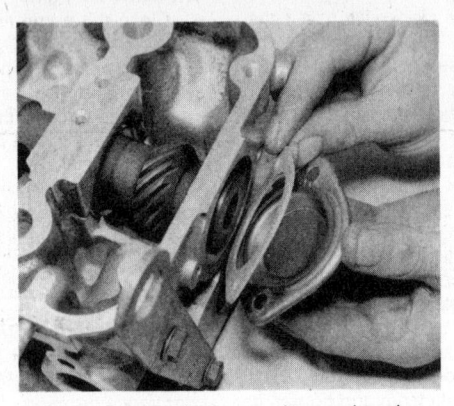

56.10 Removing the cover plate and gasket from the rear of the head

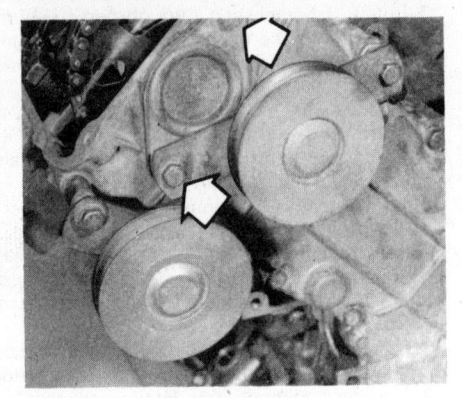

56.12A Cover plate bolts (arrowed) for access to right-hand sprocket bolt. Idler pulley may need to be removed

56.12B Threaded plug (arrowed) for access to left-hand sprocket bolt

56.14 Slackening a cylinder head bolt

3 Remove the inlet manifold and associated components (Chapter 3, Section 22).
4 Disconnect the coolant hose(s) from the head(s) to be removed. Also remove the radiator top and/or bottom hoses from the water pump and thermostat housing.
5 When removing the right-hand head, disconnect or remove the following items:

  (a) TDC sender and cable lead
  (b) Distributor (Chapter 4, Section 4)
  (c) Air conditioning compressor (without disconnecting the hoses)
  (d) Engine oil dipstick and tube

6 When removing the left-hand head, disconnect or remove the following items:

  (a) Vacuum pump
  (b) Hot air trunking

7 Disconnect the exhaust downpipes from both manifolds. Disconnect the exhaust mounting from the transmission and move the exhaust system rearwards so that the downpipes are clear of the manifold studs.
8 Perform the following operations on one head at a time.
9 Unbolt and remove the rocker cover.
10 Remove the cover plate from the rear of the head (photo).
11 Remove the four timing cover bolts which enter the cylinder head.
12 Remove the cover plate (RHS) or threaded plug (LHS) which gives access to the camshaft sprocket bolt (photos).
13 Restrain the camshaft sprocket. Slacken the sprocket centre bolt with a 10 mm Allen key.
14 Slacken the cylinder head/rocker gear bolts progressively in the sequence shown in Fig. 1.30. Remove the bolts and the rocker gear; identify them if both heads are to be removed (photo).
15 Slacken the camshaft thrust plate bolt. Move the thrust plate aside.
16 It is now necessary to fit a tool (Volvo tool No 5213, or equivalent) to retain the camshaft sprocket and to keep the chain taut. **If the**

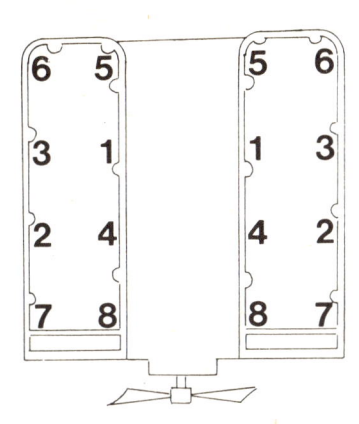

Chapter 1 Engine

Fig. 1.30 Cylinder head bolt slackening and tightening sequence (Sec 56)

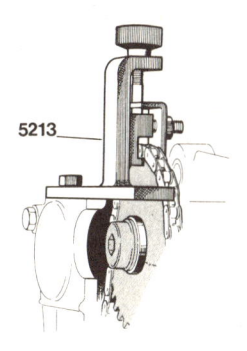

Fig. 1.31 Volvo tool 5213 for retention of the camshaft sprocket (Sec 56)

timing chain is allowed to slacken, the timing cover will have to be removed (Section 54) to reset the tensioner. If the tool is not available, proceed by removing the timing chains.

17 With the sprocket securely supported and the chain tension assured, unscrew the camshaft sprocket centre bolt. Remove the bolt from the right-hand camshaft, being careful not to drop it into the timing case. On the left-hand side there is not room to remove the bolt completely.

18 Move the camshaft rearwards so that it is clear of the sprocket.

19 Place a wooden or plastic lever between the head and the water branch pipe. Lever the head away from the block with a rocking motion and remove it. **Do not** try to lift the head straight off the block, or the cylinder liners will be disturbed.

20 Remove the gasket from the cylinder head or block. Recover the dowels if they are loose.

21 Fit cylinder liner retaining clamps, using some head bolts, spacers, and large washers or suitable pieces of scrap metal. The precise dimensions of the retainers are not important unless the pistons are to be removed, in which case they must not foul the bores (photo).

22 If it is wished to turn the crankshaft whilst the head is removed, or if the other head is to be removed, the camshaft sprocket retaining tool must be changed for one which will allow the sprocket to rotate. Volvo tool No 5105 is suitable.

23 Repeat the operations from paragraph 9 to remove the other cylinder head.

24 If the head has been removed to rectify a blown gasket, check the head for distortion (Section 36) and measure the liner protrusion (Section 85) to establish the reason for the gasket failure.

25 Commence refitting by removing the liner retaining clamps. Refit the camshaft sprocket retaining tool, if it was removed, being careful to keep the chain taut.

26 Fit the dowels in the cylinder block and keep them raised by

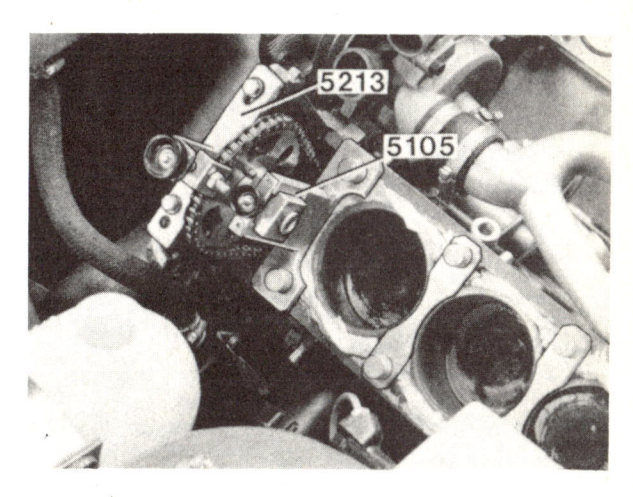

Fig. 1.32 Tool 5105 is needed in addition if the sprocket must be rotated (Sec 56)

inserting a nail or twist drill (3 mm/0.12 in diameter) in the holes beneath the dowels (photo).

27 Make sure that the exposed sections of the timing cover gasket are in good condition. If not, repair them with fragments cut from a new gasket. Apply jointing compound to the gasket sections.

28 With the gasket mating face clean and dry, fit a new head gasket to the block. Make sure that it is the right way round and the right way up (photo).

29 Lower the cylinder head and camshaft into position. Turn the camshaft until its hole lines up with the locating pin in the sprocket,

56.21 Home-made liner retaining clamps in position

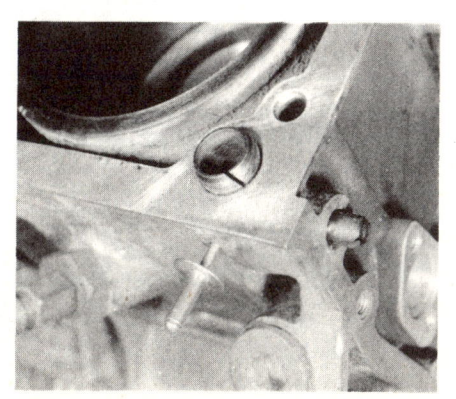

56.26 A rivet inserted below the dowel

56.28 Fitting a cylinder head gasket

then push the camshaft forwards to engage with the sprocket. Insert the sprocket centre bolt and tighten it lightly.
30 Remove the nails or drills from below the dowels. Fit the rocker gear and the cylinder head bolts. The bolts must be clean and have oiled threads.
31 Tighten the cylinder head bolts progressively, in the correct sequence (Fig. 1.30) and to the Stage 1 specified torque.
32 Slacken the first bolt, then retighten it to the Stage 2A specified torque. Tighten the bolt further through the angle specified for Stage 2B. Make up a cardboard template to indicate the angle required (photo).
33 Repeat the Stage 2 tightening process on each bolt in turn.
34 Remove the camshaft sprocket retaining tool. Move the camshaft thrust plate into position and tighten its retaining bolt.
35 Restrain the camshaft sprocket and tighten the sprocket centre bolt to the specified torque.
36 Refit the cover plate or access plug to the front of the timing cover. Use a new O-ring under the cover plate.
37 Fit and tighten the four timing cover bolts.
38 Refit the head rear cover plate, using a new gasket.
39 Refit the other cylinder head if it was removed.
40 Check and adjust the valve clearances (Section 50).
41 Provisionally fit the rocker covers, using new gaskets. Only secure them with a bolt at each corner for the time being, as they will have to come off again soon.
42 Refit the exhaust downpipes to the manifolds and secure the exhaust mounting.
43 Refit the items listed in paragraphs 5 and 6, with the exception of the air conditioning compressor.
44 Refit the coolant hoses. Refill the cooling system (Chapter 2, Section 5).
45 Refit the inlet manifold and fuel injection equipment (Chapter 3, Section 22).
46 Reconnect the battery. Run the engine and bring it up to normal operating temperature.
47 Stop the engine and allow it to cool for two hours.
48 Remove the rocker covers again. Tighten each cylinder head bolt, in the correct sequence, through the angle specified for Stage 3 tightening.
49 Refit the rocker covers, this time using all the bolts. Refit any other disturbed components.
50 Refit the air conditioning compressor.
51 Check the ignition timing (Chapter 4, Section 6) and the idle speed and mixture (Chapter 3, Section 8).

57 Camshaft – removal and refitting

1 Remove the cylinder head and rocker gear (Section 56 or 70).
2 If not done during removal, unbolt the camshaft thrust plate and remove the cover plate from the rear of the head (photo).
3 Withdraw the camshaft through the hole in the rear of the head, being careful not to damage the bearing surfaces (or your fingers) with the sharp edges of the cam lobes (photo).

4 Refit by reversing the removal operations, applying plenty of oil to the bearing journals and the cam lobes. If special lubricant is supplied with a new camshaft, use it.
5 Run in a new camshaft at moderate engine speeds (in the range 1500 to 2500 rpm) for a few minutes, or as specified by the manufacturer.

58 Crankshaft oil seal (front) – renewal

1 Disconnect the battery negative lead.
2 Remove the radiator, fan shroud and fan. When fitted, also remove the ATF auxiliary cooler (Chapter 2).
3 Remove all the accessory drivebelts (Chapter 2, Section 7).
4 Turn the crankshaft until No 1 pulley notch is aligned approximately with the 20° BTDC mark on the timing scale. This will position the keyway correctly.
5 Remove the crankshaft pulley (Section 54, paragraphs 7 to 9).
6 Carefully prise the oil seal from its location. Do not damage the seal seat.
7 Clean the seal seat in the timing cover and inspect the seal rubbing surface on the pulley. If the rubbing surface is damaged, it must be cleaned up or the pulley must be renewed, otherwise the new seal will fail prematurely.
8 Grease the lips of the new seal. Fit the seal, lips inwards, and tap it home using a large socket or a piece of tube.
9 Refit the crankshaft pulley, being careful not to dislodge the Woodruff key.
10 Jam the ring gear and tighten the crankshaft pulley nut to the specified torque. Remove the jamming device.
11 Refit the other components by reversing the removal sequence.

59 Flywheel/driveplate – removal and refitting

Proceed as in Section 12, but disregard the references to the ignition system. Note also that the arrangement of the driveplate washers differ (Fig. 1.33).

60 Crankshaft oil seal (rear) – renewal

Refer to Section 13, but disregard references to the fitted depth of the seal.

61 Flywheel ring gear – renewal

Refer to Section 14.

56.32 Angle-tightening a cylinder head bolt

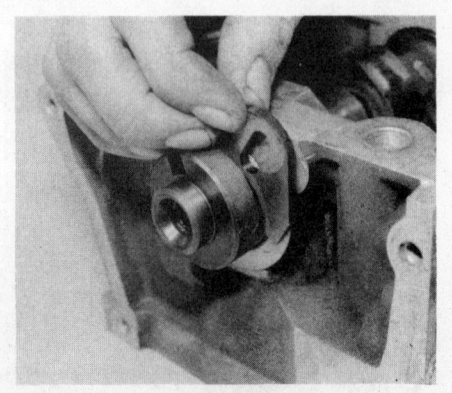

57.2 Removing a camshaft thrust plate

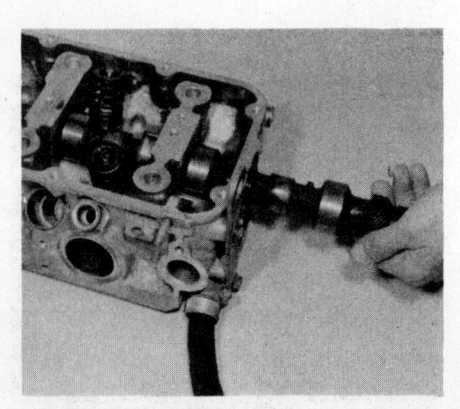

57.3 Removing a camshaft

Chapter 1 Engine

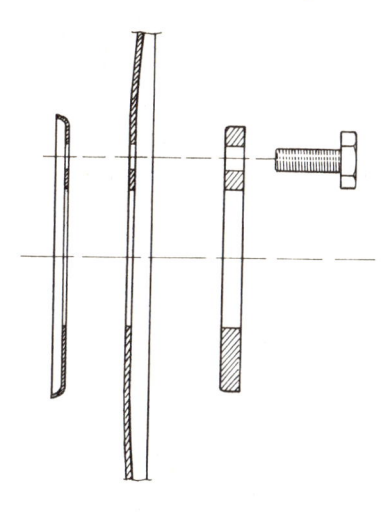

Fig. 1.33 Driveplate and washers (Sec 59)

63.2 Removing a sump bolt

63.5 Fitting a new sump gasket

62 Crankshaft spigot bearing – removal and refitting

Refer to Section 15, but note that there is no bearing retaining circlip on the V6 engine.

63 Sump – removal and refitting

The sump can be removed with the engine installed, but this does not give access to any major components

1 If the engine is in the vehicle, drain the oil. If it is on the bench, invert it.
2 Remove the 23 bolts and washers which secure the sump (photo).
3 Remove the sump, tapping or levering it gently if need be to free it. Recover the gasket.
4 The oil baffle and oil pump pick-up strainer are now accessible. If the strainer is removed, renew its O-ring.
5 Refit by reversing the removal operations, using a new gasket (photo). Tighten the sump bolts evenly.

64 Engine mountings – removal and refitting

1 Disconnect the battery negative lead.
2 Raise and support the front of the vehicle. From below, remove the through-bolt and nut and the spigot nuts from the lower part of the mounting to be removed.
3 Fit lifting tackle to the engine. Raise the engine slightly to unload the mountings and remove the lower part.
4 The upper part of the mounting, complete with rubber block, can now be removed.
5 Refit by reversing the removal operations.

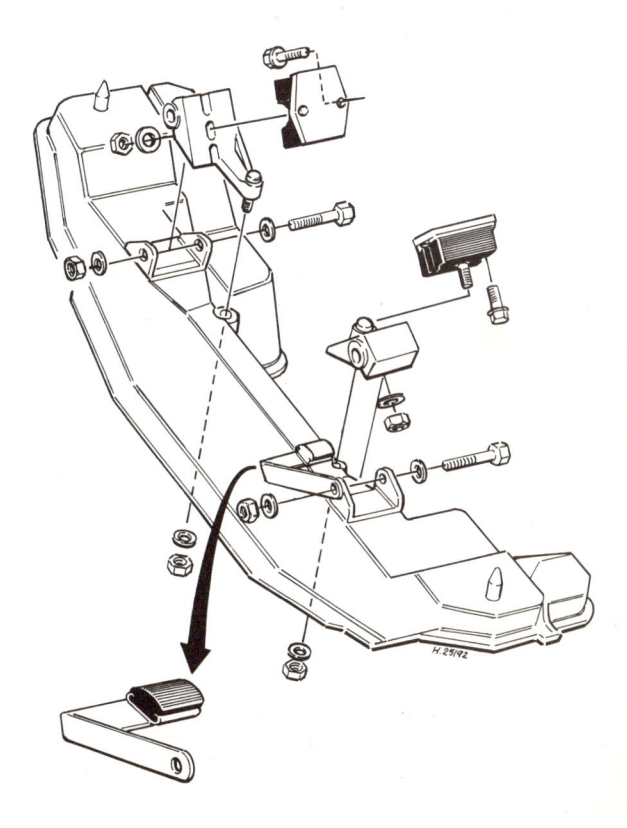

Fig. 1.34 Engine mounting and front crossmember (Sec 64)

65 Methods of engine removal

The makers recommend that the engine and transmission be removed together. This method is preferred if adequate lifting tackle is available.

Removal of the engine alone is easy enough on models with automatic transmission. This was done in our workshop. Refitting proved quite difficult because the angle at which the engine is suspended has to be just right in order to mate with the transmission. Thought should be given to this when attaching the lifting tackle: ideally the chains should be attached to a bar running lengthways and pivoted about its centre so that the angle can easily be adjusted.

Removal of the engine alone on manual gearbox models is not recommended by the makers and has not been tried. This is not to say that it is impossible, but the problems mentioned above will be magnified because of the need to align the clutch with the gearbox input shaft.

66 Engine – removal (alone)

1 Open the bonnet to the vertical position, or remove it completely.
2 Remove the battery (Chapter 12, Section 5).
3 Drain the cooling system and remove the radiator, the fan shroud and the viscous coupled fan (Chapter 2, Sections 3, 8 and 9). Remove the radiator top and bottom hoses from the engine.
4 Remove the air intake trunking from the air cleaner and the airflow sensor, together with the oil filler cap and the crankcase ventilation hoses.
5 Disconnect the exhaust downpipes from the manifolds. (If preferred, these can be tackled from below later on.) Recover the gaskets.
6 When fitted, remove the transmission fluid auxiliary cooler (Chapter 2, Section 17).
7 Unbolt the crossmember which carries the radiator top mounting brackets. Four of the bonnet lockbolts pass through this crossmember. Disengage the crossmember from the bonnet release cable and remove it.
8 Unbolt the air conditioning condenser bottom mountings. Move the condenser forwards, being careful not to strain the refrigerant pipes. **Do not** disconnect the pipes.
9 Remove the air conditioning compressor drivebelt (Chapter 2, Section 7). Disconnect the wires from the compressor clutch. Unbolt the compressor and its mounting brackets; place the compressor in the battery tray, being careful not to strain the refrigerant hoses. **Do not** disconnect the hoses.
10 Disconnect the engine and fuel injection wiring harness multi-plugs on top of the engine and next to the expansion tank. Unclip the impulse relay from the expansion tank so that it stays with the harness. Place the harness on the engine.
11 Unbolt the air conditioning harness earth tag from the inlet manifold. Move the harness out of the way.
12 Slacken the fuel tank filler cap to release any pressure. Disconnect the fuel feed pipe from the top of the fuel filter. Be prepared for fuel spillage.
13 Disconnect the fuel return pipe from the union on the left-hand inner wing, or from the fuel distributor.
14 Disconnect the throttle, kickdown and cruise control cables (as applicable) and move them aside.
15 Remove the hot air trunking from the air cleaner and the downpipe shroud.
16 Unbolt one end of the earth strap which joins the fuel distributor to the bulkhead.
17 Disconnect the distributor-to-coil HT lead and the distributor LT multi-plug.
18 Disconnect the distributor vacuum advance pipes from the control valve behind the right-hand suspension turret (when fitted). Make identifying marks for reference when refitting.
19 Disconnect the brake servo/heater control vacuum feed from the T-piece near the vacuum pump.
20 Disconnect the heater hoses at the rear of the engine
21 Remove the oil filter (Sec 49).
22 Raise and support the vehicle. Remove the ending undertray (if not already done).
23 Drain the engine oil. Refit and tighten the drain plug afterwards for safekeeping.
24 Unbolt the torque converter or clutch cover plate from the bottom of the bellhousing (photo).
25 Remove the starter motor (Chapter 12, Section 11).
26 Remove the steering pump drivebelt (Chapter 2, Section 7). Unbolt the pump and its brackets – some of the mounting bolts are more easily reached from above – and support the pump below the engine.
27 Unbolt and remove the blanking plate from the unused starter motor aperture.
28 Unbolt the earth strap from behind the engine right-hand mounting
29 On automatic transmission models, make alignment marks between the torque converter and driveplate. Working through the cover plate or starter motor aperture, remove the torque converter-to-driveplate bolts, turning the crankshaft to gain access. Lever the torque converter rearwards to check that it is free from the driveplate.
30 On all models, remove the bellhousing-to-engine bolts which are accessible from below (photo).
31 Slacken the exhaust mounting bracket at the rear of the transmission.
32 Remove the through-bolt from each engine mounting, and the single nut from each mounting which secures it to the front crossmember.
33 Lower the vehicle. Remove the alternator and its drivebelts (Chapter 12, Section 8). Recover the water pump pulley, which is now free.
34 Free the remaining bellhousing nuts and bolts. On automatic transmission models, one of these bolts also secures the transmission dipstick/filler tube.
35 Attach the lifting tackle to the engine, using the eyes provided.
36 Support the transmission with a trolley jack under the bellhousing.
37 Check that no attachments have been overlooked.
38 Lift the engine until the mounting studs are clear of the crossmember. Raise the trolley jack so that the transmission is still supported, then draw the engine forwards and off the transmission. With a manual gearbox, do not allow the weight to be taken by the input shaft.
39 Lift the engine out of the bay (photo). As assistant should guide it out and make sure that items such as the air conditioning compressor are not snagged. Be careful not to damage the air conditioning condenser, which will be punctured by the water pump pulley studs if the engine lurches forwards.
40 Set the engine down on the bench or on blocks of wood, making sure it is securely supported.
41 On automatic transmission models, secure a bar or a block of wood across the mouth of the bellhousing to keep the torque converter in place (photo).

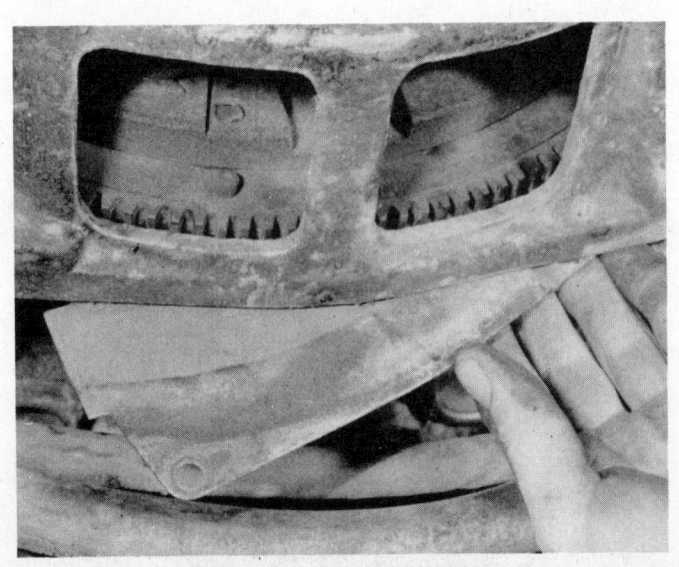

66.24 Removing the torque converter cover plate

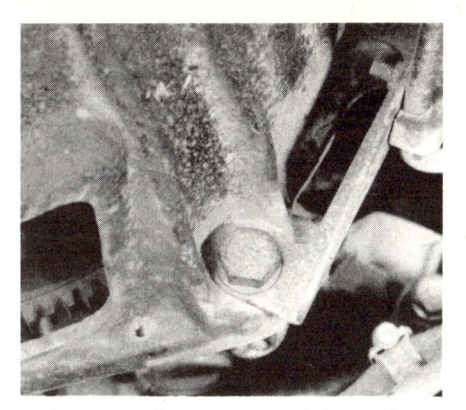

Chapter 1 Engine

66.30 One of the bellhousing bolts – this one retains a fluid cooler pipe bracket as well

66.39 Lifting out the engine

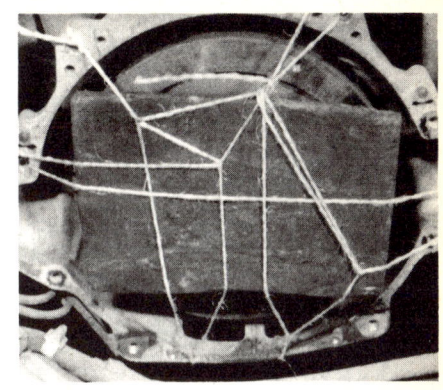

66.41 Retain the torque converter with a block of wood

67 Engine – removal (with transmission)

1 Proceed as described in Section 66, paragraphs 1 to 4 and 6 to 20.
2 Raise and support the vehicle. Remove the engine undertray.
3 Remove the steering pump (Section 66, paragraph 26).
4 Unbolt the earth strap from behind the engine right-hand mounting.
5 On manual gearbox models, disconnect the clutch cable or remove the clutch slave cylinder (without disconnecting the hydraulic hose) and remove the gear lever. See Chapter 5, Section 3 or 7, and Chapter 6, Section 21.
6 On automatic transmission models, unbolt the selector linkage from below at the shift lever end, and separate the wiring connectors.
7 On all models, disconnect the exhaust downpipe from the rest of the system.
8 Disconnect the propeller shaft from the transmission flange, making alignment marks for reference when refitting.
9 Support the rear of the engine from above, or use a trolley jack and a piece of wood from below, then remove the transmission rear mounting crossmember and associated components.
10 Remove the through-bolt for each engine mounting, and the single nut from each mounting which secures it to the front crossmember.
11 Lower the vehicle. Support the transmission with a trolley jack and a piece of wood (if not already done).
12 Attach the lifting tackle to the four lifting eyes.
13 Take the weight of the engine/transmission unit. Check that no attachments have been overlooked, then lift the unit out of the engine bay, lowering the jack under the transmission as the lifting commences. Have an assistant guide the unit and check that nothing is trapped.
14 Set the unit down on the bench or on blocks of wood, making sure it is securely supported.

68 Engine – separation from transmission

Refer to Section 23, and note the following points:

(a) It will be necessary to remove the oil filter before the starter motor can be removed
(b) Remove the blanking plate from the unused starter motor aperture
(c) Remove the exhaust downpipes

69 Engine – dismantling for overhaul

1 With the engine removed for dismantling, remove ancillary components as follows:

(a) Engine mountings (photo)
(b) Exhaust manifolds
(c) Dipstick tube and dipstick (photos)
(d) Clutch pressure and driven plates (Chapter 5)
(e) Alternator, accessory drivebelts and pulleys
(f) Inlet manifold (Chapter 3)
(g) Distributor (Chapter 4)
(h) Water pump with hoses and distribution pipe (Chapter 2)
(j) Ignition sensor and diagnostic socket
(k) Vacuum pump (Chapter 9)
(m) Oil level sensor (Section 94)
(n) Sensors, brackets, sender units, etc

2 Complete dismantling of the engine can now proceed as described in the following Sections.

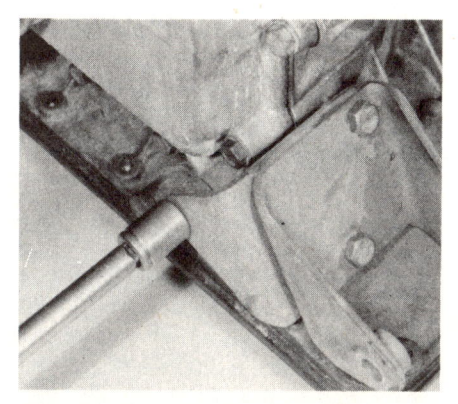

69.1A Unbolting an engine mounting bracket

69.1B Unbolting the dipstick tube bracket ...

69.1C ... and the tube base nut

Chapter 1 Engine

3 The oil pump should be removed at some stage after the removal of the timing chains and sprockets.

70 Cylinder heads – removal (engine removed)

1 Remove the timing chains and sprockets as described in Section 54, paragraph 4 onwards.
2 Slacken the cylinder head/rocker gear bolts progressively in the sequence shown in Fig. 1.30. Remove the bolts.
3 Remove the rocker gear. Keep left-hand and right-hand components separate.
4 Remove the cylinder heads, using a couple of metal bars inserted in two bolt holes to rock them away from the centre of the engine. (This is to avoid disturbing the liners. If the liners will be removed anyway, just lift the heads straight off.)
5 Recover the cylinder head gaskets.
6 If the cylinder liners are not to be disturbed, fit liner retaining clamps (Section 56, paragraph 27).

71 Pistons and connecting rods – removal

1 With the engine on the bench, remove the cylinder heads and the sump (Sections 63 and 70).
2 Remove the oil baffle and the oil pump pick-up strainer. Recover the O-ring (photos).
3 Remove the 14 small bolts and the eight main bearing nuts which secure the lower crankcase (photo). Lift off the lower crankcase; recover the oil pick-up tube O-ring.
4 Fit spacers to the main bearing cap studs and then refit the main bearing nuts so that the main bearings and crankshaft are secured for subsequent operations. This is particularly important if main bearings are not to be disturbed. In any event it is undesirable to have the crankshaft fall out unexpectedly.
5 If it is hoped to re-use the existing cylinder liners, make sure that they are securely clamped.
6 Place the engine on one side, or stand it on its rear face, so that both top and bottom are accessible and the crankshaft can be turned.
7 Inspect the connecting rods and caps to see that they carry identification marks or numbers. Note that the numbering system used here by the makers does **not** correspond to the cylinder numbering (Fig. 1.35).
8 Check the endfloat between each pair of connecting rods (photo). If it is outside the specified limits, all six connecting rods must be renewed.
9 Remove the nuts from one connecting rod cap, tapping it with a soft-faced hammer if it is stiff. Recover the bearing shell if it is loose and keep it with the cap.
10 Push the piston and rod up the bore and out of the top. Have an assistant catch the piston in a rag as it emerges. Tap the rod with a hammer handle if it is stiff. Recover the bearing shell if it is loose and keep it with the cap.
11 Fit the rod and cap back together and secure them with the nuts. Keep the bearing shells identified if they are to be re-used.

Cylinder		1	4	2	5	3	6
Marking of con rod and cap	early type	A	B	C	D	E	F
	late type	1	2	3	4	5	6
Crank webs (from rear)		1		2		3	

Fig. 1.35 Relationship of cylinder and connecting rod numbering. Letter ('L' shown here) is arbitrary (Sec 71)

12 Repeat the process to remove the other five pistons and rods, turning the crankshaft as necessary to gain access.

72 Crankshaft and main bearings – removal

1 Remove the pistons and connecting rods (Section 71). Also remove the flywheel or driveplate, if not already done.
2 Unbolt and remove the rear oil seal housing (photo).
3 Remove the main bearing nuts and spacers.
4 Remove the main bearing caps, making identifying marks if necessary so that they can be refitted in the same position. Tap them free with a soft-faced hammer. Keep the main bearing shells with their caps; recover the thrust washers from each side of the rear cap (photo).
5 Lift out the crankshaft. Do not drop it, it is heavy.
6 Remove the upper half main bearing shells from their seats in the

71.2A Unbolting the oil baffle and pick-up strainer

71.2B Removing the oil pick-up strainer

71.3 Removing a lower crankcase bolt

Chapter 1 Engine

71.8 Measuring the endfloat between a pair of connecting rods

72.2 Removing the rear oil seal housing

72.4 Removing the rear main bearing cap with thrust washers

75.1A Oil pump, showing relief valve split pin

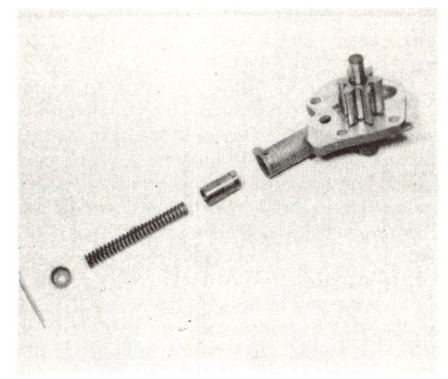

75.1B Oil pump and relief valve components

crankcase by pressing the end of the shell furthest from the locating tab. Keep the shells in order if they may be re-used.
7 Recover the upper halves of the thrust washers from each side of the rear bearing seat.

73 Cylinder liners – removal

1 With the cylinder heads and pistons removed, mark the position of each liner relative to the block. Also mark the cylinder number on each liner.
2 Remove the liner retaining clamps (if fitted), then lift the liners out of the block.
3 Clean the sealing lip on the outside of the liner and the sealing surface in the block.

74 Examination and renovation – general

Refer to Section 27.

75 Oil pump – examination and renovation

1 Remove the relief valve components by depressing the cap and extracting the split pin. Remove the cap, spring and plunger (photos).
2 Clean the pump body and the gears and inspect them for wear and damage. If evident, the pump must be renewed complete. Although clearances and wear limits are specified, they are not easy to measure because of the design of the pump. If in doubt, renew the pump.
3 Inspect the relief valve plunger for scoring. Measure the spring free length and if possible check its load/length characteristic – see Specifications. Relief valve spares are available.
4 Reassemble the pump and relief valve components. Use a new split pin.

76 Crankshaft and bearings – examination and renovation

Refer to Section 29.

77 Cylinder block and liners – examination and renovation

1 Remove the various blanking plugs from the cylinder block. Clean the block inside and out, not forgetting the oil and water channels. Blow through the channels with compressed air.
2 Inspect the block for cracks, distortion of mating faces or other

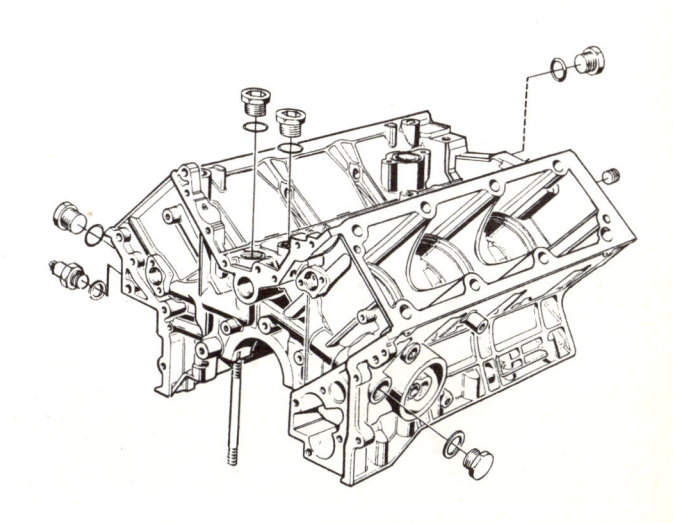

Fig. 1.36 Cylinder block blanking plugs (Sec 77)

damage. Seek professional advice if damage is found. Also check the condition of threaded holes.
3 Refit the blanking plugs to the block using new seals or copper washers.
4 Inspect the liners for cracks, internal scoring or other visible damage.
5 Check the liners for wear and ovality by measuring the internal diameter at several points, using an internal micrometer. Alternatively, measure a piston ring end gap at the bottom end of the liner (where it is unworn) and towards the top (where wear is greatest). Liners cannot be rebored and must be renewed, complete with pistons and in sets of six, if excessively worn. Refer to the next Section.

78 Pistons and connecting rods – examination and renovation

1 Examine the pistons for scoring, burning and cracks. Renew the complete set of pistons and liners if such damage is found.
2 Measure the piston diameter. If satisfactory, remove the piston rings to clean out the grooves and to check the gaps. See Section 31, paragraphs 4 to 7, and Fig. 1.37.
3 Inspect the connecting rod bolts for evidence of stretching or damaged threads. Renew the bolts if necessary by clamping the connecting rod in a soft-jawed vice and driving out the old bolts with a plastic mallet. Support the piston when doing this. Press in the new bolt using the vice, the connecting rod cap and a 12 mm socket.
4 The gudgeon pins are an interference fit in the connecting rod small-end bushes. Press tools are needed to remove and refit them; this must be done by specialists.
5 Accurate inspection of the connecting rods for twisting or other distortion requires them to be separated from the pistons.
6 New pistons and liners are supplied with gudgeon pins. The old connecting rods must be removed from the old pistons and fitted to the new ones by a Volvo dealer or other specialist.

79 Flywheel/driveplate – examination and renovation

Refer to Section 34.

80 Timing chains and sprockets – examination and renovation

1 Worn timing chains make a characteristic thrashing sound. Of itself chain wear is not serious, but if the chain stretches too far, a plunger may fall out of one of the tensioners. This will cause loss of oil pressure and possibly serious damage to the engine.
2 Chain wear may be checked without too much dismantling by removing the left-hand valve cover. The left-hand chain tensioner can now be inspected with the aid of a torch. If the tensioner plunger protrudes by four notches (8mm/0.32 in) or more, wear is excessive and the chains must be renewed.
3 Chains and sprockets wear together and should always be renewed together. To do otherwise is risking noise and rapid wear.
4 Inspect the chain guides and dampers. Renew them if they are badly grooved or otherwise damaged.
5 Inspect the chain tensioners, but do not dismantle them. If the plunger is removed from a tensioner, the tensioner must be renewed. Check that the tensioner oilways are not blocked.

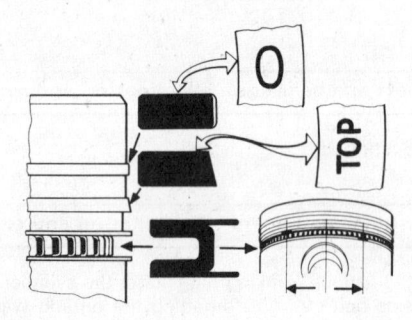

Fig. 1.37 Piston ring profiles. Stagger oil control rail gaps as shown (Sec 78)

Fig. 1.38 Removing a connecting rod bolt (Sec 78)

Fig. 1.39 Fitting a new connecting rod bolt (Sec 78)

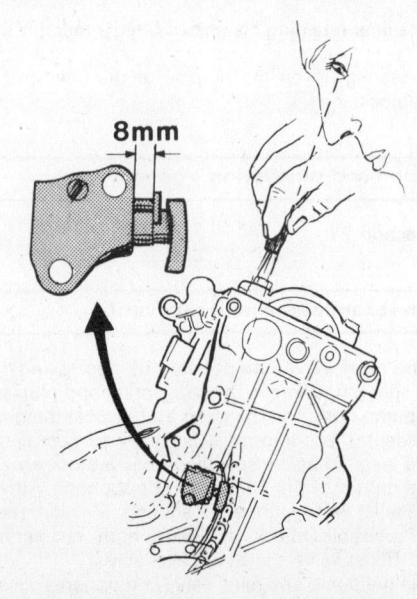

Fig. 1.40 Checking timing chain wear (Sec 80)

Chapter 1 Engine

6 Renew the tensioner oil strainers, the timing cover gasket and the camshaft cover gaskets as a matter of course. Also renew the crankshaft front oil seal unless it is known to be in perfect condition.

81 Camshafts and rocker gear – examination and renovation

1 Inspect the cam lobes and the camshaft bearing journals for scoring, scuffing or other damage. Once the surface hardening of the lobes has been penetrated, wear will progress rapidly.
2 Measure the bearing journals with a micrometer and check them for ovality and taper. To establish the bearing running clearance, an internal micrometer must be used to measure the bearings in the cylinder heads. Excessive wear or damage can only be corrected by renewing the camshaft and/or head.
3 Measure camshaft endfloat with the camshaft installed in the head (photo). If endfloat is excessive, renew the thrust plate.
4 Dismantle the rocker gear for examination as follows.
5 Remove the bolt from the pedestal furthest from the circlip. Hold the pedestal against spring pressure as the bolt is removed (photo).
6 Slide the pedestals, rocker arms, springs and spacers off the shaft, being careful to keep them all in order and the right way round (photo). Remove the circlip from the end of the shaft.
7 Clean all components, paying particular attention to the oilways. Renew worn or damaged items. The camshaft rubbing faces of the rocker arms are surface hardened and must not be machined.
8 Reassemble in the reverse order to dismantling, oiling all components liberally. Note that the flat faces of the pedestals face the circlip end of the shaft, and that the shaft oilways face downwards.

82 Cylinder head – dismantling

1 Remove the camshaft (Section 57).
2 Unbolt and remove the exhaust manifold. Recover the gasket sections.
3 Remove the spark plugs and any coolant unions, blanking plates, etc.
4 Have ready a box divided into six segments. Remove the valves as described in Section 35, paragraphs 9 to 13. Note that valve stem oil seals are fitted both to inlet and to exhaust valves on this engine (photos).
5 Keep the components of the left-hand and right-hand heads separate if both are being dismantled.

83 Cylinder head – decarbonising, valve grinding and renovation

1 Refer to Section 36, and note the following additional points.
2 If decarbonising the piston *in situ*, stuff paper or clean rags into the water jacket around the cylinder liners to keep dirt out. Remove the paper or rags on completion.
3 Renew both the inlet and the exhaust valve stem oil seals.
4 Renew the core plugs if they are obviously rusty or have seen much service.

84 Cylinder head – reassembly

1 Refit the valves as described in Section 37, paragraphs 2 to 6, noting that stem oil seals are fitted to all valves (photo).
2 Refit the exhaust manifold, using new gaskets. Separate the gasket sections by cutting, **not** by tearing.
3 Refit the camshaft (Section 57).
4 Refit the spark plugs, coolant unions, etc.

85 Cylinder liners – refitting and checking protrusion

Note: *Some cylinder gaskets require a different liner protrusion; consult your dealer.*

1 The protrusion of the cylinder liners above the top of the block must be accurately set to ensure good sealing of the head gaskets and liner

81.3 Measuring camshaft endfloat

81.5 Removing the rocker pedestal bolt

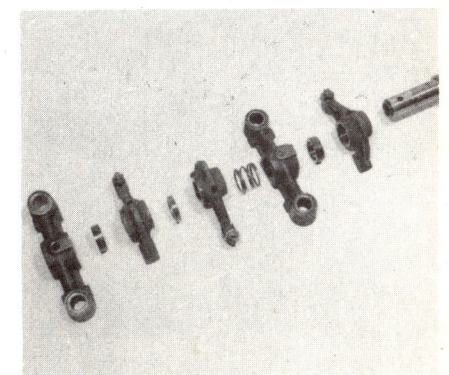

81.6 Rocker shaft components removed

82.4A Valve spring lower seat is not dished on the V6 engine

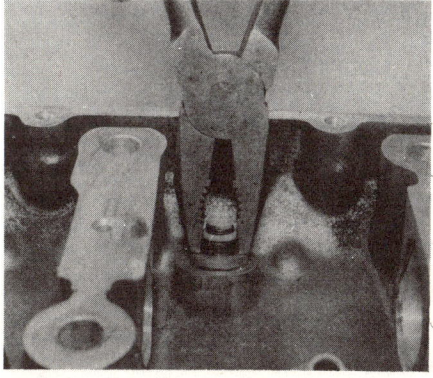

82.4B Removing a valve stem oil seal

84.1 Pressing home a new valve stem oil seal

74 Chapter 1 Engine

bases. Base seals are available in various thicknesses; a selection should be obtained for this procedure.
2 Make sure that the liners and their seats are perfectly clean.
3 Fit one liner to the block without a seal. Observe position and alignment marks if re-using the old liners. Clamp the liner lightly (Section 56, paragraph 21).
4 Using a straight-edge and feeler blades, or (preferably) a dial test indicator and a suitable bracket, measure the protrusion of the top of the liner relative to the top of the block (photo). Measure in three different places and record the results.
5 The difference between the three measurements must not exceed 0.05 mm (0.002 in). If it does, remove the liner and check for dirt on the liner or seat. If the difference is within limits, use the largest of the three measurements as the basis for calculation. For example:

Measurement 1	=	0.10 mm
Measurement 2	=	0.06 mm
Measurement 3	=	0.07 mm
Maximum difference	=	0.04 mm
Largest measurement	=	0.10 mm

6 Select a base seal of thickness such that the final protrusion of the liner will be within the specified range. Aim for the maximum allowable protrusion. For example:

Desired protrusion	=	0.23 mm (max)
Measurement 1 (above)	=	0.10 mm
Difference (seal thickness required)	=	0.13 mm

7 The seal closest in thickness to that required carries a red mark (see Specifications). It might prove marginally too thick, in which case a white marked seal would have to be used.
8 Remove the liner from the block.
9 Fit the thickness of seal just calculated to all the liners. The coloured identification tab should be positioned so that it will be visible when the liner is fitted. The tongues round the inside of the seal must fit into the liner base groove (photo).
10 Fit all the liners to the block, again observing position and alignment marks if applicable (photo). Clamp the liners.

11 Working on one cylinder bank at a time, measure the protrusion of each liner relative to the block and relative to the adjacent liner(s). The protrusion relative to the block must be as specified, and the difference between adjacent liners must not exceed 0.04 mm (0.0016 in).
12 Fit different thickness seals to individual liners if necessary to achieve the desired result. (No more than one seal may be used per liner). New liners may be rotated or swapped around if wished. Clamp the liners securely on completion.

86 Crankshaft and main bearings – refitting

1 Wipe clean the main bearing shell locations in the cylinder block and bearing caps.
2 Fit the bearing shells to their locations, observing the originally fitted positions if re-using the old shells. Press the shells home so that the tangs engage in the grooves. Make sure that the oil holes in the shells and block are aligned. The shells in the caps are plain (photo).
3 Fit the upper half thrust washers to each side of the rear bearing seat in the block. The grooves must face outwards. Use a smear of grease to hold them in position (photo).
4 Oil the bearing shells in the block. Lower the crankshaft into position, being careful not to dislodge the thrust washers.
5 Inject oil into the crankshaft oilways. Oil the shells in the main bearing caps and fit the thrust washers to the rear cap, grooves outwards. Fit the caps to their locations and the right way round. Tap them home if necessary.
6 Secure each cap with a spacer and a nut (photo). Fit two spacers to the rear cap and tighten its nuts to 40 Nm (30 lbf ft).
7 Position a dial test indicator probe against the crankshaft rear flange (photo). Lever the crankshaft back and forth and measure the endfloat. If it is outside the specified limits, different thickness thrust washers will have to be fitted.
8 Check that the crankshaft is free to rotate. Some stiffness is to be expected with new components, but there must be no light spots or binding.

85.4 Measuring cylinder liner protrusion

85.9 Liner with base seal fitted. Identification tab is arrowed

85.10 Fitting a liner to the block

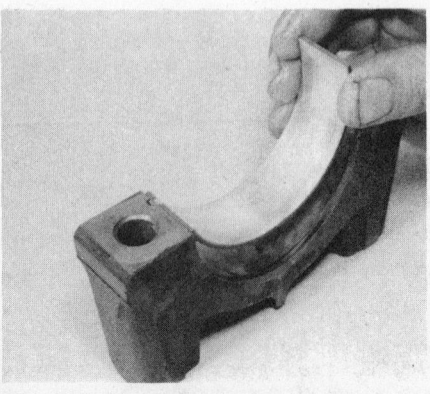

86.2 Fitting a main bearing shell to its cap

86.3 Rear main bearing seat and thrust washers

86.6 Main bearing caps retained with spacers and nuts

Chapter 1 Engine

86.7 Measuring crankshaft endfloat

86.9 Trimming the oil seal housing gasket

87.2A Fitting a piston to its liner

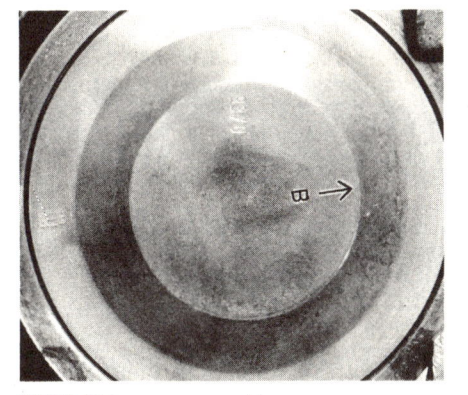

87.2B Piston crown markings – arrow points to front, 'B' is grade mark

87.2C Fitting a connecting rod cap

87.3 Tightening a connecting rod cap nut

9   Refit the rear oil seal housing, using a new gasket. Make sure that the flat edge of the housing is flush with the block, then tighten the five Allen screws to secure it. Trim the gasket flush (photo).
10 Lubricate a new rear oil seal, the crankshaft and the seal seat. Fit the seal, lips inwards, and seat it with a piece of tubing.
11 If the spigot bearing was removed (manual gearbox only), refit it now.

87 Pistons and connecting rods – refitting

1   Deglaze the liner bores if new rings are being fitted to old pistons and liners. See Section 18, paragraph 9.
2   Refit one piston and connecting rod as described in Section 18, paragraphs 10 to 12, but do not tighten the nuts yet. Remember that the connecting rod numbers do not correspond to the cylinder numbers (photos).

3   Refit the other five pistons. Tighten the bearing cap nuts to the specified torque when each pair of rods is fitted to a crankpin (photo). (If the nuts are tightened when only one rod is fitted, the twisting of the rod may produce a false result.)
4   Check that the crankshaft is still free to rotate.
5   Remove the main bearing cap nuts and spacers. Make sure that the old pick-up tube is in place and fit a new O-ring to it (photo).
6   Clean the upper and lower crankcase mating faces, then apply sealant to one face, including the areas round the main bearing studs (photo).
7   Fit the lower crankcase. Fit the main bearing nuts and the 14 small bolts, but only tighten them lightly at this stage.
8   Use a straight-edge to check that the rear edges of the crankcase halves are aligned (photo). Reposition the lower crankcase if necessary, first slackening the nuts and bolts.
9   Tighten the main bearing nuts in the sequence shown in Fig. 1.41 to the Stage 1 specified torque.

87.5 Oil pick-up tube O-ring (arrowed)

87.6 Applying sealant to the crankcase mating face

87.8 Checking the alignment of the crankcase halves

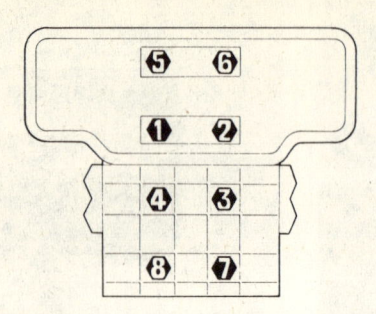

Fig. 1.41 Main bearing nut tightening sequence (Sec 87)

10 Recheck the alignment of the crankcase halves. Slacken the nuts if necessary and start again.
11 Slacken the first nut and retighten it to the Stage 2A specified torque. Tighten the nut further through the angle specified for Stage 2B. Use a protractor or a cardboard template to indicate the angle required (photo).
12 Repeat the slackening and retightening process on each nut in turn.
13 Check that the crankshaft is still free to rotate.
14 Tighten the 14 small bolts in the lower crankcase.
15 Refit the oil pump pick-up strainer, using a new O-ring, and the oil baffle.
16 Refit the sump, cylinder heads and other disturbed components.

88 Cylinder heads – refitting (engine removed)

1 Make sure that the head locating dowels are in position in the block. Insert a nail or similar item (approx 3 mm/0.12 in diameter) in the hole below each dowel to keep it raised.
2 Turn the crankshaft to bring No 1 cylinder to TDC.
3 Remove the liner retaining clamps from the left-hand cylinder bank. Make sure that the surface is clean, then fit a new gasket. The gasket must be for the left-hand side (they are not identical) and the right way up.
4 Position the left-hand camshaft so that its sprocket locating groove will point upwards when the head is fitted.
5 Fit the left-hand head and camshaft to the block.
6 Refit the left-hand rocker gear, making sure it is the correct way round.
7 Remove the nails from below the left-hand dowels. Insert the cylinder head bolts, clean and with oiled threads. Only hand tighten the bolts at present.
8 Similarly refit the right-hand head, but position the camshaft groove so that it will point outwards and parallel with the head mating face (photo).

9 Tighten the cylinder head bolts in the correct stages and sequence (Section 56, paragraphs 31 to 33).
10 Refit the timing chains, sprockets etc, as described in Section 54, paragraph 22 onwards. Remember to refit the oil pump if it was removed.
11 Check the valve clearances before refitting the valve covers (Section 50).
12 Only use four bolts to retain each valve cover for the time being, as they will have to be removed again after warm-up.

89 Timing scale – checking and adjusting

1 To check the accuracy of the timing seals, the timing cover and crankshaft pulley must be fitted, and the inlet manifold must be removed. It may also be necessary to remove the water pump.
2 Bring the crankshaft to approximately 20° BTDC on No 1 cylinder.
3 Remove the blanking plug from the checking hole, using an 8 mm square drive key (the same as for the sump drain plug). Recover the copper washer (photo).
4 Insert a drill shank or other rod, 8 mm (0.315 in) in diameter, into the hole (photo). Slowly turn the crankshaft clockwise until the rod drops into the slot in the crankshaft. This is TDC for No 1 cylinder.
5 In this position the '0' mark on the timing scale must be aligned exactly with the No 1 notch on the pulley. Slacken the scale securing bolts if necessary and adjust the scale position. Tighten the bolts and seal them with a dab of paint.
6 Remove the drill or rod. Refit the blanking plug, using a new copper washer, and tighten it.

90 Engine – reassembly after overhaul

1 Refit the flywheel or driveplate, using new bolts. Tighten the bolts to the specified torque (photos).
2 Refit the ancillary components listed in Section 69. If preferred, delicate items such as the alternator can be left until after refitting the engine.

91 Engine – reconnection to transmission

1 Refer to Section 41.
2 Refit the exhaust downpipes.
3 Refit the starter motor blanking plate.
4 Fit a new oil filter.

92 Engine – refitting (with transmission)

1 Refitting is essentially a reversal of the procedure in Section 67, with the following additional points.
2 On automatic transmission models, adjust the gear selector mechanism (Chapter 6, Section 30).

87.11 Angle-tightening a main bearing nut

88.8 Refitting the right-hand head (with exhaust manifold attached)

89.3 Removing the blanking plug from the TDC checking hole

Chapter 1 Engine

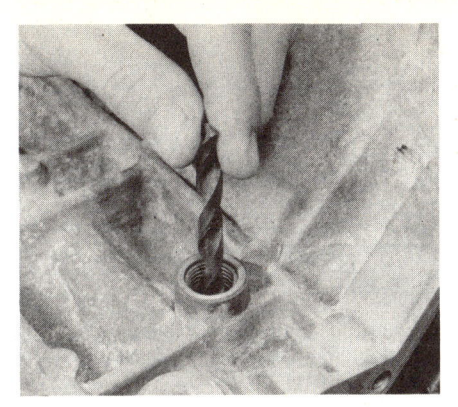

89.4 Inserting a drill shank into the hole

90.1A Fitting the driveplate front washer ...

90.1B ... the driveplate itself ...

90.1C ... the thick washer and bolts

94.4 Fitting the oil level sensor

94.5 Oil level warning control unit

3 Refill the engine with oil and coolant. Refill the transmission with lubricant if it was drained.
4 Refer to Section 95 before starting the engine.

93 Engine – refitting (alone)

1 Refer to Section 43, paragraphs 1 to 4. Also see Section 65.
2 Once the engine and transmission are mated, proceed by reversing the removal procedure (Section 66).
3 Refer to Section 95 before starting the engine.

94 Oil level sensor – testing, removal and refitting

1 When fitted, the oil level sensor is screwed into the right-hand side of the lower crankcase. The sensor protrudes into the sump.
2 To test the sensor, disconnect the multi-plug from it and measure the resistance across the sensor terminals. It should be 9Ω under all conditions.
3 To remove the sensor, unscrew it from its hole. Be careful with it, it is fragile.
4 When refitting the sensor, apply some sealant to its threads. Screw the sensor home, reconnect the multi-plug and check for correct operation (photo).

5 The oil level warning control unit is located under the rear console armrest/storage box (photo).

95 Initial start-up after overhaul or major repair

1 Refer to Section 44, and note the following points.
2 Check the ignition timing before adjusting the idle speed and mixture (Chapter 4, Section 6).
3 After the engine has warmed up, stop it and allow it to cool for at least two hours, then carry out the final tightening of the cylinder head bolts. Refit the valve covers using all the bolts.

96 Compression test – description and interpretation

Proceed as in Section 45, but first perform the following:

(a) Remove the air inlet trunking
(b) Unplug the auxiliary air valve or air control valve connector
(c) Disconnect the hose which joins the auxiliary air valve and the start injector

97 Fault diagnosis – engine

Refer to Section 46.

Chapter 2
Cooling, heating and air conditioning systems

For modifications, and information applicable to later models, see Supplement at end of manual

Contents

Accessory drivebelts – removal, refitting and tensioning	7
Air conditioning system – description and precautions	19
Air conditioning system – maintenance	20
Antifreeze mixture – general	6
Automatic climate control programmer – removal and refitting	30
Automatic climate control sensors – description	28
Automatic climate control sensors – testing, removal and refitting	29
Coolant level sensor – general	16
Cooling system – draining	3
Cooling system – filling	5
Cooling system – flushing	4
Electric fan – removal and refitting	10
Electric fan thermoswitch – removal, testing and refitting	11
Fault diagnosis – cooling system	31
Fault diagnosis – heating and air conditioning systems	32
General description	1
Heater/air conditioning control panel – removal and refitting	21
Heater blower motor – removal and refitting	25
Heater blower motor resistor – removal and refitting	26
Heater matrix – removal and refitting	23
Heater temperature control cable – removal, refitting and adjustment	22
Heater vacuum motors – removal and refitting	24
Heater water valve – removal and refitting	27
Heating and ventilation system – description	18
Maintenance and inspection	2
Oil coolers – removal and refitting	17
Radiator – removal and refitting	8
Temperature gauge – testing	15
Temperature gauge sender – removal and refitting	14
Thermostat – removal, testing and refitting	13
Viscous coupled fan – removal and refitting	9
Water pump – removal and refitting	12

Specifications

General
System type .... Water-based coolant, pump-assisted circulation, thermostatically controlled

Coolant capacity:
- B23/B230 .... 9.5 litres (16.7 pints)
- B28 .... 10.0 litres (17.6 pints)

Coolant type/specification .... Volvo coolant type C and clean water (Duckhams Universal Antifreeze and Summer Coolant)

Pressure cap rating
Overpressure:
- B23 .... 0.65 to 0.85 bar (9 to 12 lbf/in^2)
- B230 (early type) .... 0.75 bar (11 lbf/in^2)
- B230E and K (later type) .... 1.00 bar (15 lbf/in^2)
- B230ET (later type) .... 1.50 bar (22 lbf/in^2)
- B28 (early type) .... 0.75 bar (11 lbf/in^2)
- B28 (later type) .... 1.50 bar (22 lbf/in^2)

Underpressure .... 0.07 bar (1 lbf/in^2)

Thermostat
Opening commences:
- B23/B230 (type 1) .... 86 to 88°C (187 to 190°F)
- B23/B230 (type 2) and B 28 .... 91 to 93°C (196 to 199°F)

Fully open at:
- B23/B230 (type 1) .... 97°C (207°F)
- B23/B230 (type 2) and B 28 .... 102°C (216°F)

Torque wrench settings
	Nm	lbf ft
Water pump bolts:		
B28	15 to 20	11 to 15
B23/230	Not specified	

1 General description

The cooling system is conventional in operation. Water-based coolant is circulated around the cylinder block and head(s) by a belt-driven pump. A thermostat restricts circulation to the engine and heater matrix until operating temperature is achieved. When the thermostat opens, coolant circulates through the radiator at the front of the engine bay.

Cooling airflow through the radiator is provided by the forward motion of the vehicle, and by a viscous coupled fan on the water pump pulley. The design of the viscous coupling is such that fan speed remains low at low air temperatures, increasing as the temperature of the air coming through the radiator rises. In this way overcooling, unnecessary power loss and noise are minimised. On some models an electric cooling fan is placed in front of the air conditioning condenser, itself in front of the radiator, to supplement the airflow.

The cooling system is pressurised, which increases the efficiency of the system by raising the boiling point of the coolant. An expansion tank accommodates variations in coolant volume with temperature.

Chapter 2 Cooling, heating and air conditioning systems

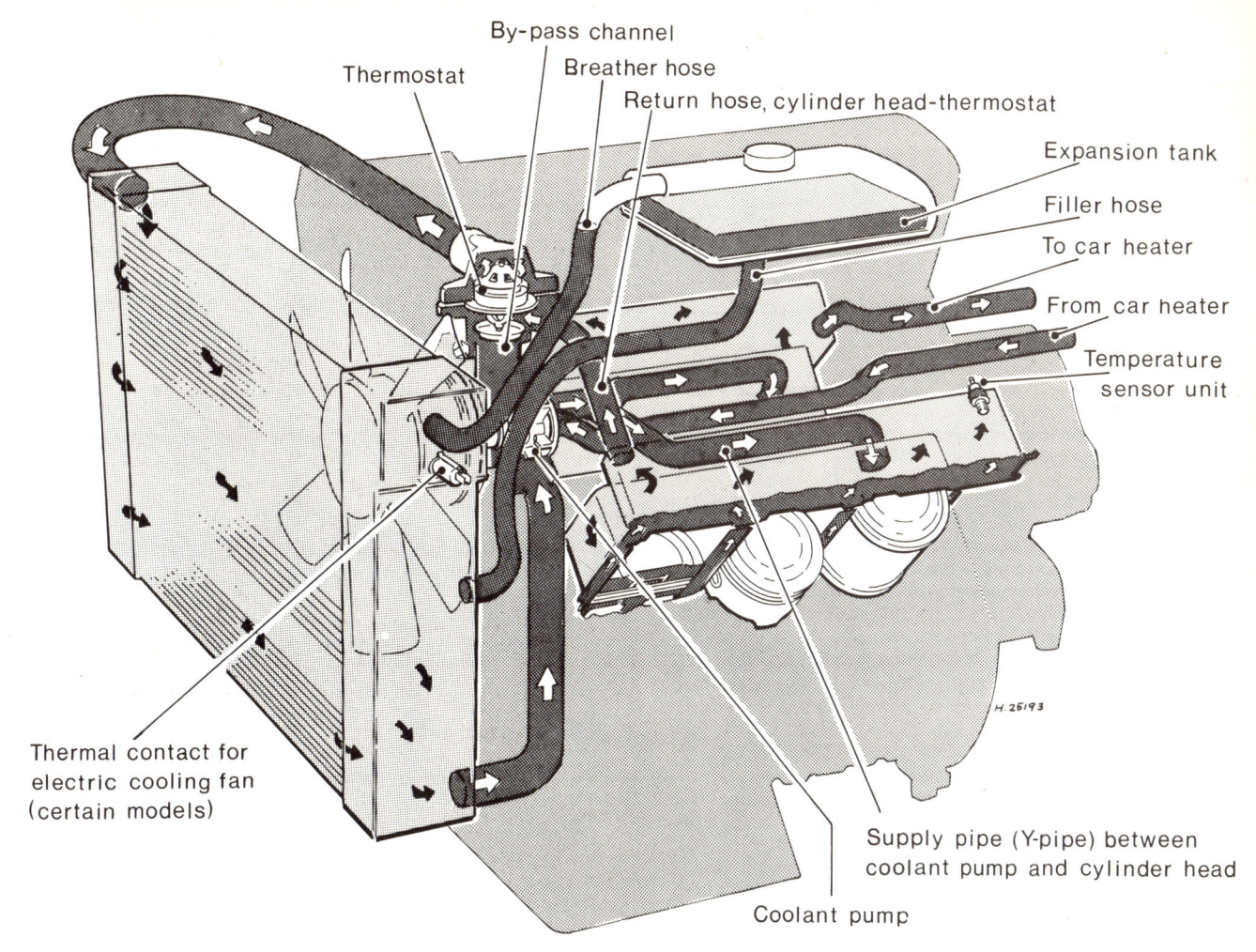

Fig. 2.1 Cooling system components and circulation – V6 shown (Sec 1)

Labels: By-pass channel; Thermostat; Breather hose; Return hose, cylinder head-thermostat; Expansion tank; Filler hose; To car heater; From car heater; Temperature sensor unit; Supply pipe (Y-pipe) between coolant pump and cylinder head; Coolant pump; Thermal contact for electric cooling fan (certain models)

Because the system is sealed, evaporative losses are minimal.

Heat from the coolant is used in the vehicle's heating system. The heating and air conditioning systems are described in Sections 18 and 19.

2 Maintenance and inspection

1 Every 250 miles, weekly, or before a long journey, check the coolant level as follows.
2 Open the bonnet. Observe the level of coolant through the translucent walls of the expansion tank (on one side of the engine bay). The level should be up to the 'MAX' mark when the engine is cold, and may be somewhat above the mark when hot.
3 If topping-up is necessary, wait for the system to cool down if it is hot. **There is a risk of scalding if the cap is removed whilst the system is hot.** Place a thick rag over the expansion tank cap and slacken it to release any pressure. When pressure has been released, carry on unscrewing the cap and remove it.
4 Top up to the 'MAX' mark with the specified coolant – see Section 6 (photo). In an emergency plain water is better than nothing, but remember that it is diluting the proper coolant. Do not add cold water to an overheated engine whilst it is still hot.
5 Refit the expansion tank cap securely when the level is correct. Check for leaks if there is a frequent need for topping up – losses from this type of system are normally minimal.
6 Every 6000 miles or six months, check the antifreeze concentration using a proprietary tester of the hydrometer or floating ball type. (If the owner is satisfied that no water has been added to the system, and that the concentration of antifreeze was originally correct, this check may safely be omitted.)
7 Every 12 000 miles or annually, inspect the accessory drivebelt(s) for fraying, glazing or other damage. Re-tension or renew as necessary – see Section 7.

2.4 Topping-up the coolant

8 Every two years, regardless of mileage, renew the coolant. The necessary information will be found in Sections 3 to 6. At the same time inspect all the coolant hoses and hose clips with a critical eye. It is worth renewing the hoses as a precautionary measure if suspect, rather than have one burst on the road. In an emergency, minor leaks from the radiator can be cured by using a radiator sealant such as Holts Radweld.
9 Occasionally clean insects and road debris from the radiator fins, using an air jet or a soft brush.
10 For maintenance of the air conditioning system see Section 20.

3 Cooling system – draining

The cooling system should not be drained when the coolant is hot, as there is a risk of scalding.
1 Remove the expansion tank filler cap. Also remove the engine undertray.
2 Move the heater temperature control to 'hot'. (Even though the heater is of the air mix type, the matrix water valve is closed in some conditions.)
3 Place a suitable drain pan under the engine and open the drain tap(s). V6 engines have two taps, one each side of the block; 4-cylinder engines have one tap on the right-hand side. Hoses can be fitted to the taps if wished to guide the coolant into the pan.
4 When coolant has finished flowing from the block, move the drain pan to below the radiator. Disconnect the bottom hose from the radiator and allow the rest of the coolant to drain.
5 Dispose of old coolant in a non-polluting fashion.

4 Cooling system – flushing

Flushing should not be necessary unless regular coolant changing has been neglected, or if plain water has been used as coolant.
1 Drain the system as previously described, and remove the thermostat (Section 13).
2 Using a garden hose, run clean water into the radiator top hose so that it flows through the radiator and out of the bottom hose stub. Carry on until clean water emerges.
3 Similarly feed water into the engine via the bottom hose so that it flows out of the thermostat housing. Protect electrical components from water spillage.
4 Flushing in the reverse direction to normal flow can be beneficial in some cases. The radiator should be removed for this, inverted and shaken gently as the water flows through it to dislodge any sediment.
5 If, after a reasonable period, the water still does not run clear, the radiator may be flushed using a chemical cleaner such as Holts Radflush or Holts Speedflush. It is important that the manufacturer's instructions are followed precisely.
6 When flushing is complete, refit the thermostat, hoses and any other disturbed components.

5 Cooling system – filling

1 Close the drain taps and make sure that all hoses and hose clips are securely fitted and in good condition.
2 Fill the system through the expansion tank until the level in the tank is up to the 'MAX' mark. Massage the large coolant hoses to disperse air pockets.
3 Fit the expansion tank cap. Run the engine up to operating temperature, checking for coolant leaks. Switch off the engine and allow it to cool.
4 Check the coolant level and top up again to the 'MAX' mark if necessary.

6 Antifreeze mixture – general

Warning: *Antifreeze mixture is poisonous. Keep it out of reach of children and pets. Wash splashes off skin and clothing with plenty of water. Wash splashes off vehicle paintwork, too, to avoid discoloration.*

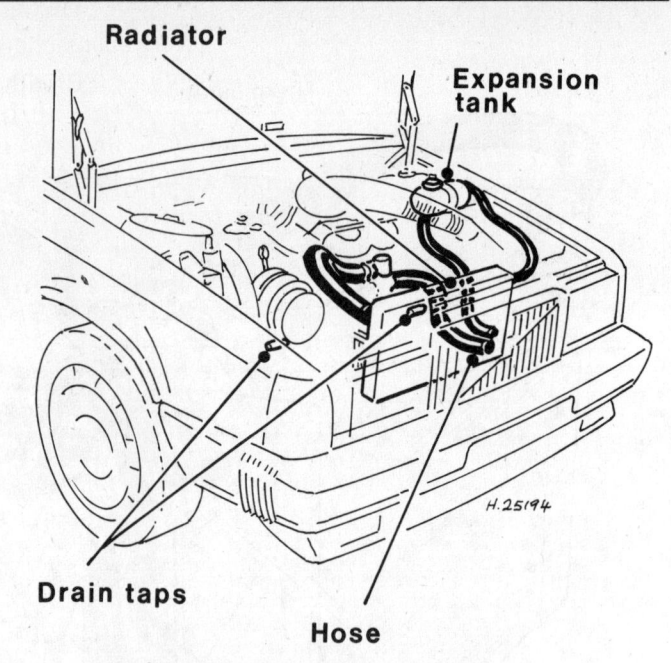

Fig. 2.2 Draining points – V6 engine (Sec 3)

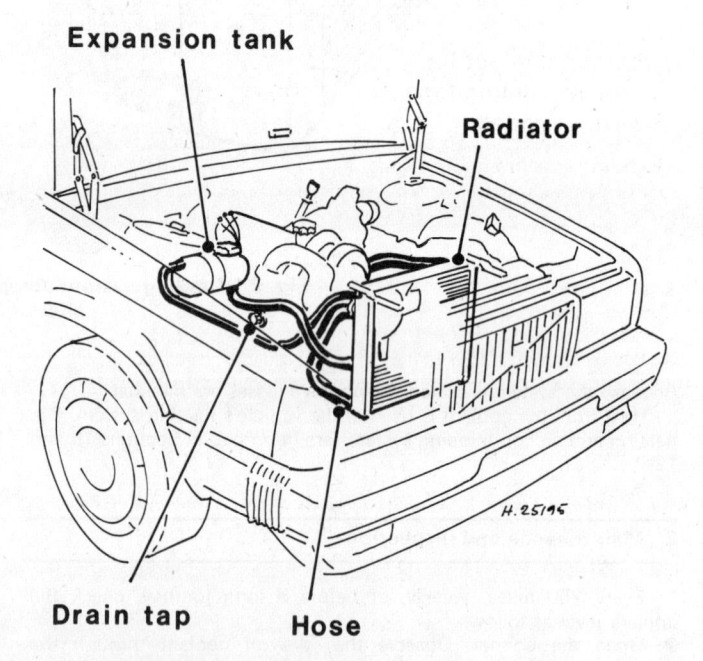

Fig. 2.3 Draining points – in-line engine (Sec 3)

1 The antifreeze/water mixture must be renewed every two years to preserve its anti-corrosive properties. (In some owner's literature a period of three years between changes is specified. The latest recommendation is for two years, however.) Never run the engine for long periods with plain water as coolant.
2 Only use the specified antifreeze (see *'Recommended lubricants and fluids'*). Inferior brands may not contain the necessary corrosion inhibitors, or may break down at high temperatures. Antifreeze containing methanol is particularly to be avoided, as the methanol evaporates.
3 The specified mixture is 50% antifreeze and 50% clean soft water (by volume). Mix the required quantity in a clean container and then fill the system as described in Section 5. Save any surplus mixture for topping-up.

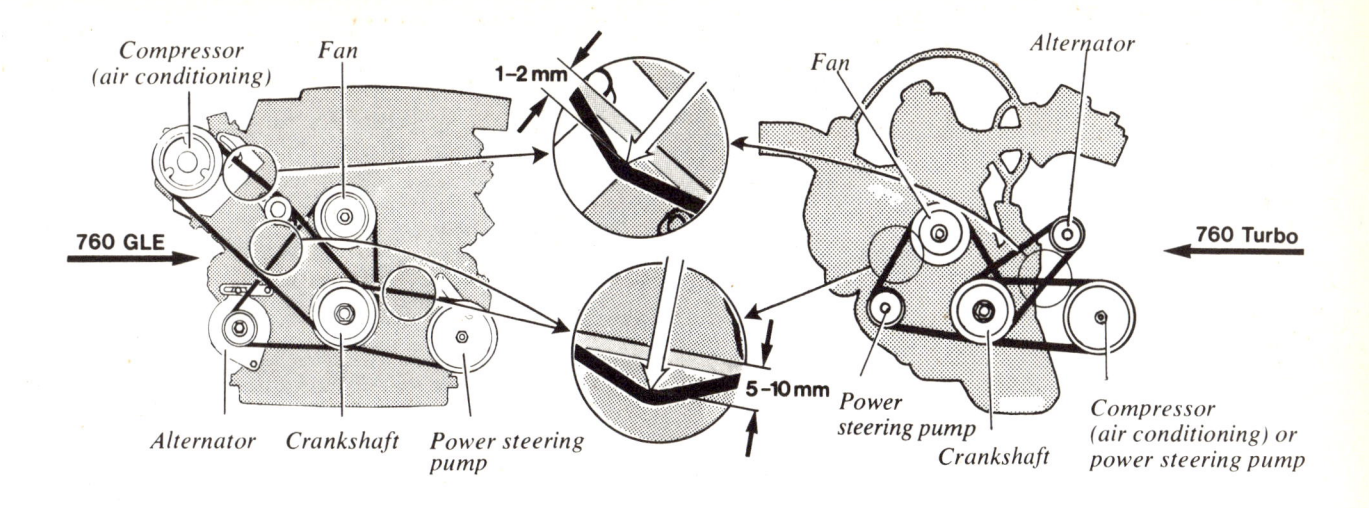

Chapter 2 Cooling, heating and air conditioning systems

Fig. 2.4 Some typical accessory drivebelt layouts (Sec 7)

7 Accessory drivebelts – removal, refitting and tensioning

1 The accessory drivebelts transmit power from the crankshaft pulley to the alternator, water pump/viscous fan, steering pump and air conditioning compressor (as applicable). A variety of belt arrangements and tensioning methods will be found, according to equipment and engine type. A representative selection is given here (Fig. 2.4).
2 When removing a particular drivebelt, it will obviously be necessary to remove those in front of it first.
3 Twin belts should always be renewed in pairs, even if only one is broken.

Water pump/alternator drivebelt(s)

4 Slacken the alternator pivot and adjusting strap nuts and bolts (photo).
5 Move the alternator towards the engine to release the belt tension. On some models a positive tensioning device is used: undo the tensioner screw to move the alternator inwards (photo).
6 Slip the belts of the pulleys and remove them (photo).
7 When refitting, move the alternator away from the engine until the belts can be deflected 5 to 10 mm (0.2 to 0.4 in) by firm thumb pressure in the middle of the longest run. Tighten the pivot and adjusting strap nuts and bolts in this position and recheck the tension.
8 On models with a positive tensioning device, be careful not to overtension the belt. On models without such a device, it may be helpful to lever the alternator away from the engine to achieve the desired tension. Only use a wooden or plastic lever, and only lever at the pulley end.

Steering pump drivebelt

9 Proceed as for the water pump/alternator drivebelt, noting the location of the pivot and adjuster strap nuts and bolts (photos). On V6 models, access to the steering pump is easier from below.

Air conditioning compressor drivebelt

10 On models where the compressor mountings allow it to be pivoted, proceed as for the water pump/alternator drivebelt (photo). Note, however, that the desired deflection of the belt is only 1 to 2 mm (0.04 to 0.08 in).
11 On models where the compressor mounting is rigid, belt tension is controlled by varying the number of shims between segments of the crankshaft pulley (photo).
12 To remove the drivebelt, unbolt the pulley from its hub. Remove the pulley segments and the shims. The drivebelt can now be removed (photo).
13 When refitting, experiment with the number of shims between the segments until belt tension is correct. Inserting shims decreases the tension, and *vice versa*. Fit unused shims in front of the pulley for future use.

All drivebelts

14 Recheck the tension of a new belt after a few hundred miles.

7.4 Alternator adjusting strap nut (arrowed) – without positive tensioner

7.5 Alternator adjusting strap with positive tensioner. Adjusting screw is arrowed

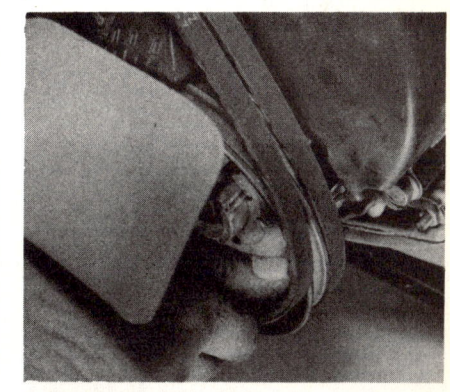

7.6 Removing a drivebelt

Chapter 2 Cooling, heating and air conditioning systems

7.9A Steering pump adjusting strap – in-line engine

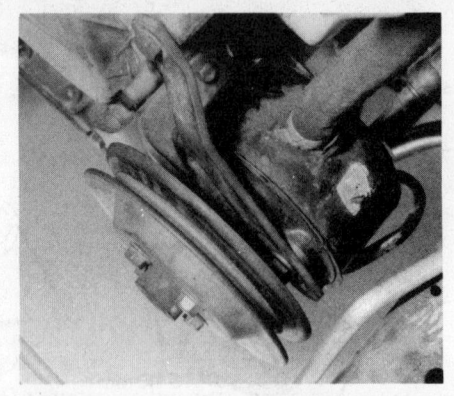

7.9B Steering pump adjusting strap – V6 engine

7.10 Air conditioning compressor adjusting straps – V6 engine

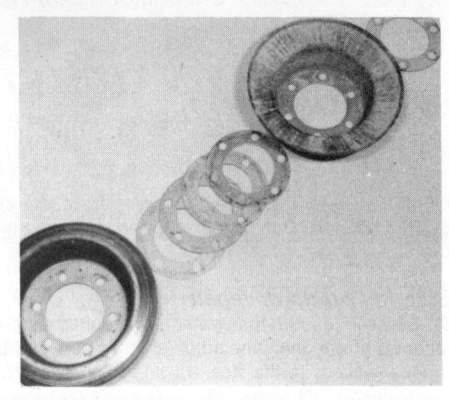

7.11 Crankshaft pulley segments and shims

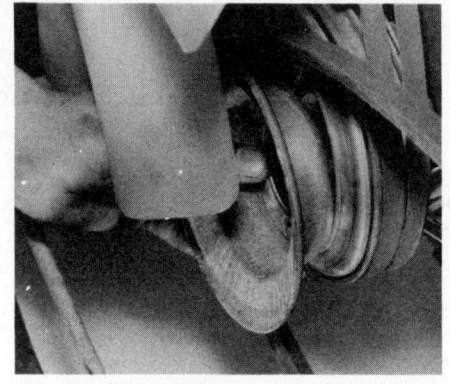

7.12 Removing the pulley front segment

8 Radiator – removal and refitting

1 Drain the cooling system by disconnecting the radiator bottom hose. Take precautions against scalding if the coolant is hot.
2 Disconnect the top hose, expansion tank hose and vent hose from the radiator (photos).
3 On automatic transmission models, disconnect the fluid cooler lines from the radiator. Be prepared for fluid spillage. Plug or cap the lines to keep dirt out.
4 Disconnect the leads from any thermal switches, sensors etc, in the radiator.
5 Unbolt the power steering fluid reservoir (when located on the radiator) and move it aside.
6 Unbolt the fan shroud and move it rearwards (photo).
7 Unbolt the radiator top mounting brackets (photo).

8 Lift out the radiator. Recover the bottom mountings if they are loose (photo).
9 Refit by reversing the removal operations. Refill the cooling system on completion. On automatic transmission models, check the fluid level (Chapter 6, Section 28).

9 Viscous coupled fan – removal and refitting

1 Remove the nuts which secure the viscous coupling to the water pump pulley studs.
2 Pull the fan and coupling off the studs (photo). Manipulate the assembly past the fan shroud and remove it. It may be necessary to release the shroud.
3 The fan and viscous coupling may now be separated if required.
4 Refit by reversing the removal operations.

8.2A Disconnect the radiator top hose ...

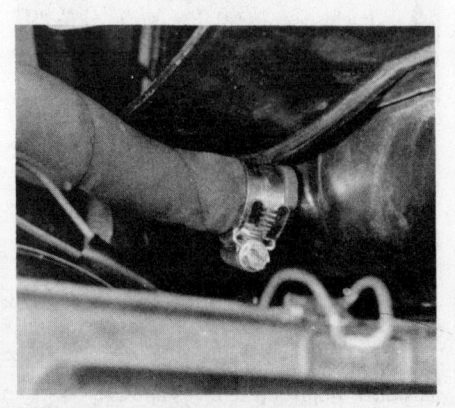

8.2B ... the expansion tank hose ...

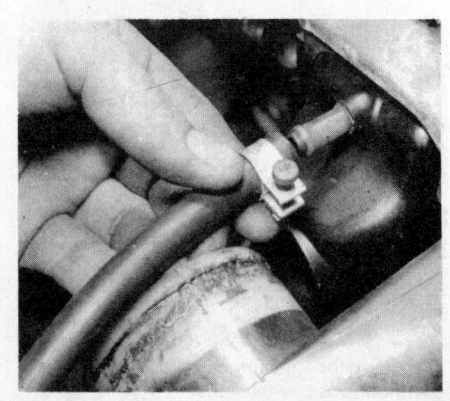

8.2C ... and the vent hose

Chapter 2 Cooling, heating and air conditioning systems

8.6 Unbolting the fan shroud

8.7 Removing a radiator mounting bracket

8.8 Radiator bottom mounting

10 Electric fan – removal and refitting

1 Remove the front grille panel (Chapter 11, Section 42).
2 Remove the four screws which secure the fan mounting bars. Disconnect the wiring multi-plug.
3 Remove the fan complete with mounting bars. The motor can be unbolted from the bars if wished.
4 Refit by reversing the removal operations.

11 Electric fan thermoswitch – removal, testing and refitting

1 Partially drain the cooling system to below the level of the thermoswitch (Section 3).
2 Disconnect the thermoswitch leads, unscrew it and remove it (photo).
3 To test the switch, connect a battery and test light to its terminals. Heat the switch in hot water. The switch should close (test light comes on) at approximately the temperature stamped on it, and open again (test light goes off) as it cools down. If not, renew it.
4 Refit the thermoswitch, using sealant on the threads, and reconnect its leads.
5 Refill the cooling system (Section 5).

12 Water pump – removal and refitting

1 Disconnect the battery negative lead.
2 Drain the cooling system (Section 3).
3 Remove the radiator and the fan shroud (Sec 8).
4 Remove the fan from the water pump (Section 9). Remove the pump drivebelt(s) and the pump pulley (photo).

In-line engine

5 Disconnect the radiator bottom hose and the heater pipe from the pump (photo).
6 Unbolt the water pump, slide it downwards and remove it (photo).
7 Before refitting the pump, renew its top sealing ring and body gasket (photos). When refitting, keep the pump pressed up against the cylinder head whilst tightening the nuts and bolts. Use a new seal on the heater pipe.

V6 engine

8 Remove the inlet manifold (Chapter 3, Section 22).
9 Remove the two hoses which connect the pump to the cylinder heads. Disconnect the remaining hoses from the pump and thermostat housing. Also disconnect the sensor and switch at the sides of the pump (photos).
10 Remove the three bolts which secure the pump to the block. Lift off the pump (photo).
11 If a new pump is to be fitted, transfer the rear housing, thermostat and housing, sender unit, blanking plugs etc, from the old pump to the new. Use new gaskets and seals. Also renew the pump hoses unless they are in perfect condition.
12 Fit the pump to the block and secure it with the three bolts, tightened to the specified torque.

All models

13 The remainder of refitting is a reversal of the removal procedure. Tension the alternator drivebelt(s) (Section 7) and refill the cooling system (Section 5) to complete.

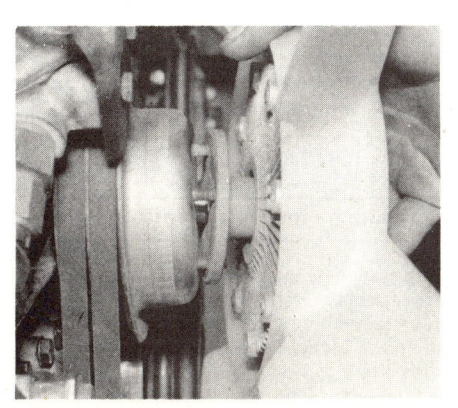

9.2 Removing the fan and viscous coupling

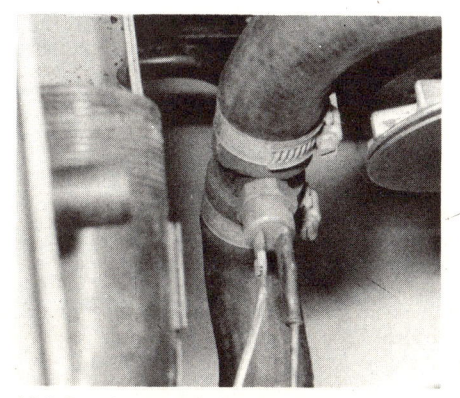

11.2 Fan thermoswitch in a hose adaptor. It may also be located in the radiator side tank

12.4 Removing the water pump pulley

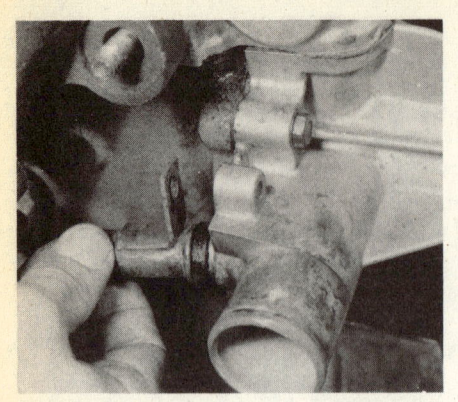

84   Chapter 2 Cooling, heating and air conditioning systems

12.5 Disconnecting the heater pipe from the pump

12.6 Unbolting the water pump

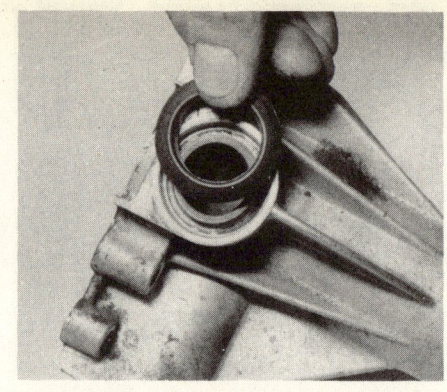

12.7A Fitting a new top sealing ring

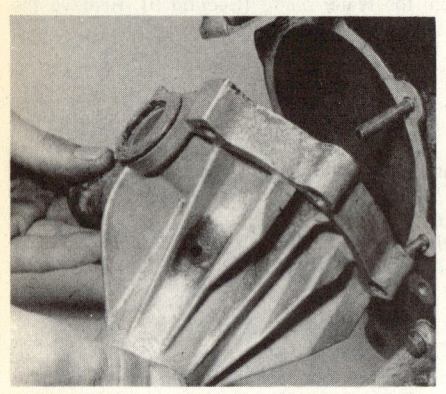

12.7B Refitting the pump, using a new gasket

12.9A One of the pump-to-head hoses

12.9B Two hoses at the rear of the pump

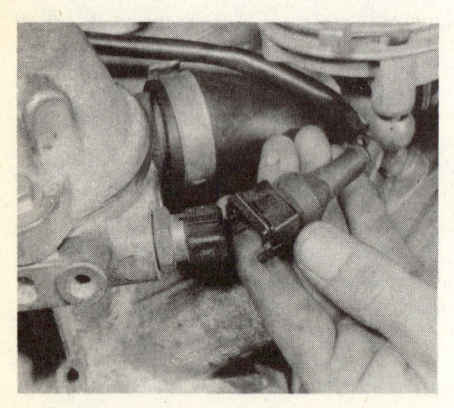

12.9C Disconnecting a temperature sensor

12.10 Unbolting the water pump

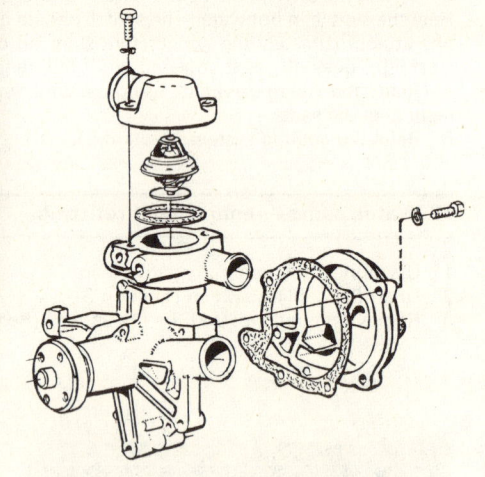

Fig. 2.5 V6 engine water pump and peripheral components (Sec 12)

13 Thermostat – removal, testing and refitting

1   Drain the cooling system (Section 3) until the coolant level is below the thermostat.
2   Unbolt the thermostat housing from the cylinder head or water pump. On some models an engine lifting eye may also be attached here; on V6 engines it will be necessary to unbolt a throttle cable bracket. Lift out the thermostat (photos).
3   Test the thermostat by heating it in a pan of water. If the thermostat fails to open in boiling water, or fails to close as the water cools, renew it. (Refer to the Specifications for precise temperature.)
4   Fit a sealing ring to the thermostat (photo).
5   Refit the thermostat and housing (and the engine lifting eye, when applicable). Fit and tighten the housing nuts.
6   Refill the cooling system (Section 5).

14 Temperature gauge sender – removal and refitting

1   Depressurize the cooling system by unscrewing the expansion tank cap. Take precautions against scalding if the system is hot. There is no need to drain the cooling system if a new sender unit is to hand.
2   Disconnect the lead from the sender unit and unscrew it from the cylinder head or water pump (photo).
3   Screw in the new sender unit, using a smear of sealant on the threads. Reconnect the leads.
4   Top up the coolant level if much was lost, then tighten the expansion tank cap.

Chapter 2 Cooling, heating and air conditioning systems

13.2A Unbolting the thermostat housing ...

13.2B ... and removing the thermostat (in-line engine)

13.2C Unbolting the thermostat housing ...

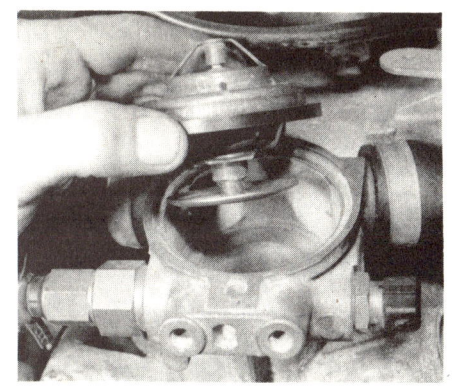

13.2D ... and removing the thermostat (V6 engine)

13.4 Fitting a thermostat sealing ring

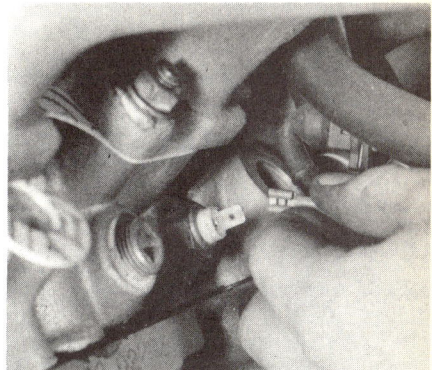

14.2 Disconnecting the temperature gauge sender unit

15 Temperature gauge – testing

Note: *If both the fuel gauge and the temperature gauge are inaccurate, the fault is probably in the instrument voltage stabiliser on the instrument panel printed circuit.*

1 Disconnect the lead from the temperature gauge sender. Connect a resistor, value approximately 68 Ω, between the lead and earth (vehicle metal). Switch on the ignition: the gauge should rise to roughly 75% of full scale deflection. If not, either the lead is broken or the gauge is defective. Switch off the ignition.
2 Measure the resistance of the temperature gauge sender with it immersed in a water bath of known temperature. The resistance should vary as follows:

Temperature	Resistance
60°C (140°F)	217 ± 35 Ω
90°C (194°F)	87 ± 15 Ω
100°C (212°F)	67 ± 11 Ω

3 If the gauge or the sender does not behave as described, renew it.
4 Reconnect the sender lead on completion.

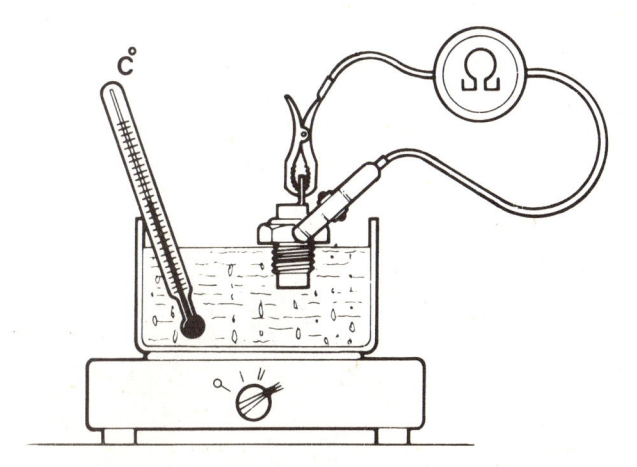

16 Coolant level sensor – general

1 A coolant level sensor is fitted in the expansion tank on later models. It consists of a float-operated switch connected to an instrument panel warning light.
2 If the sensor malfunctions, unscrew it from the expansion tank (with the system cold) and renew it.

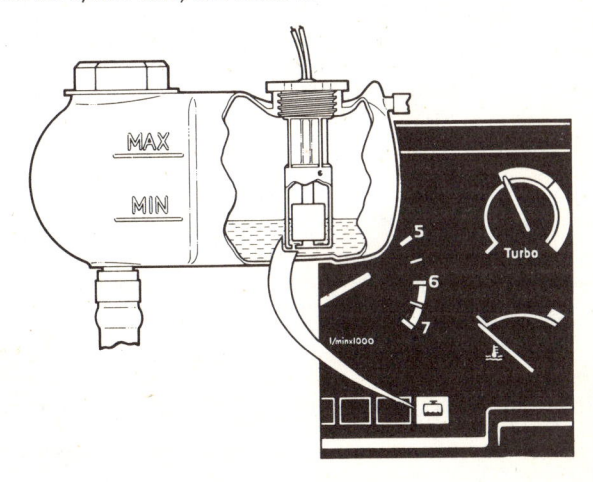

Fig. 2.6 Testing the temperature gauge sender (Sec 15)

Fig. 2.7 Coolant level sensor (Sec 16)

17 Oil coolers – removal and refitting

Engine oil cooler

1 When fitted, the engine oil cooler is mounted behind and to one side of the radiator.
2 Disconnect the oil cooler unions, either at the cooler itself or at the flexible hoses (photo). Be prepared for oil spillage.
3 Unbolt the oil cooler brackets and remove it (photo). The oil cooler can then be separated from the brackets if required.
4 If the oil cooler is to be re-used, flush it internally with solvent and then blow compressed air through it. Also clean it externally.
5 Refit by reversing the removal operations. Run the engine and check for oil leaks, then switch off and check the oil level.

ATF auxiliary cooler

6 When fitted, the ATF auxiliary cooler is mounted between the radiator and the air conditioning condenser.
7 Disconnect the flexible hoses from the union on the radiator (photo). Be prepared for fluid spillage. Cap open unions to keep dirt out.
8 Remove the radiator (Section 8).
9 Unbolt the cooler (photo). Feed the hoses through the side panel grommets and remove the cooler and hoses together.
10 Clean the cooler fins and flush it internally with clean ATF. Renew the hoses if necessary.
11 Refit by reversing the removal operations. Refill the cooling system, run the engine and check the transmission fluid level (Chapter 6, Section 28).

18 Heating and ventilation system – description

Depending on model and options selected, the heater may be fitted alone or in conjunction with an air conditioning unit. The same housings and heater components are used in all cases. The air conditioning system is described in Section 19.
The heater is of the fresh air type. Air enters through a grille in front of the windscreen. On its way to the various vents a variable proportion of the air passes through the heater matrix, where it is warmed by engine coolant flowing through the matrix.
Distribution of air to the vents, and through or around the matrix, is controlled by flaps or shutters. These are operated by vacuum motors

17.2 Oil cooler flexible hose unions

17.3 Unbolting an oil cooler bracket

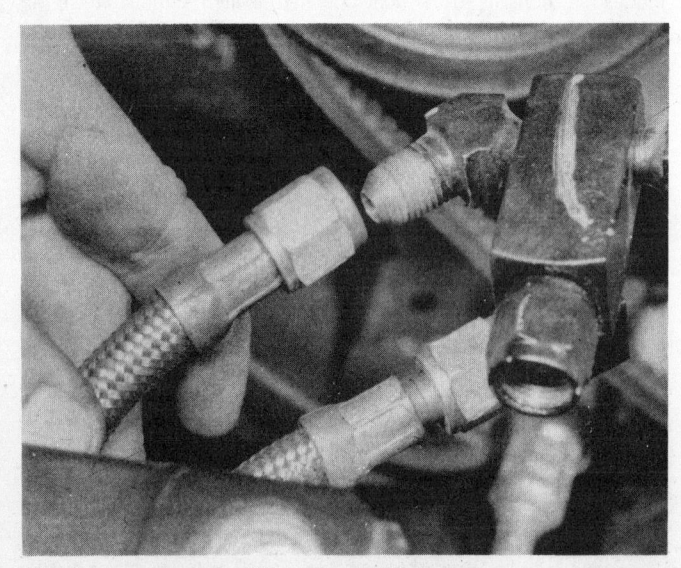

17.7 Disconnecting an ATF cooler flexible hose

17.9 An ATF cooler mounting bolt

87

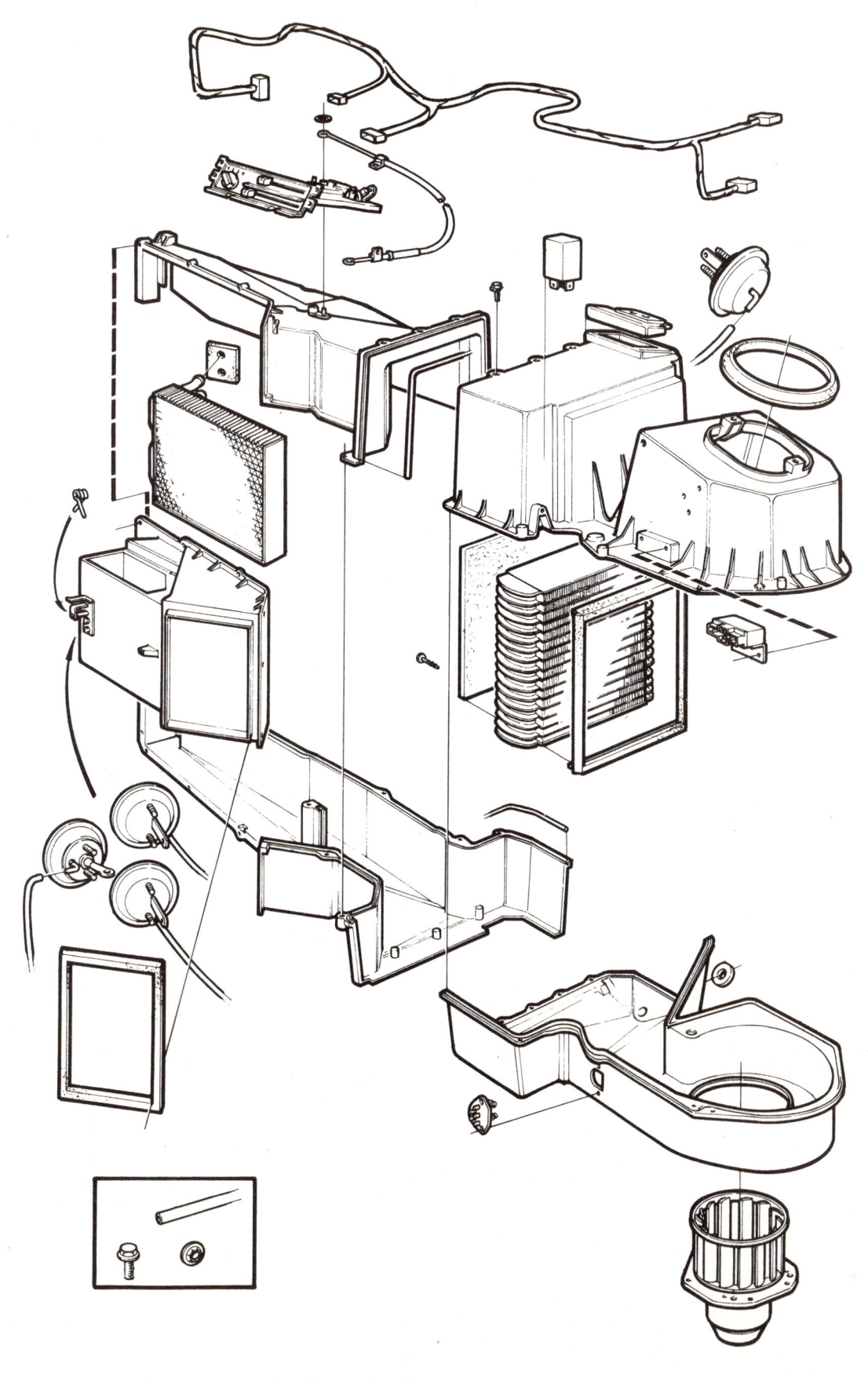

Fig. 2.8 Exploded view of heater/air conditioning distribution unit. LHD shown, RHD is mirror image (Sec 18)

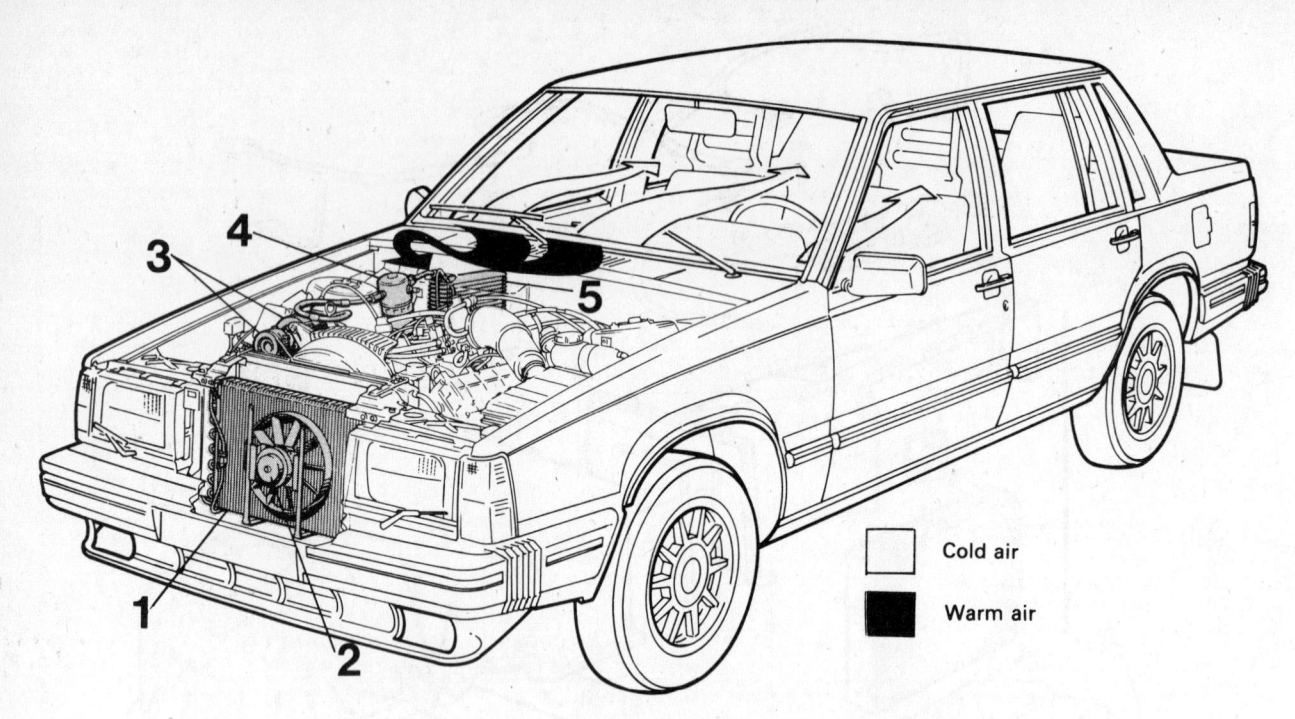

Fig. 2.9 Main components of the air conditioning system
(Sec 19)

1 Condenser
2 Electric fan
3 Compressor
4 Receiver/drier
5 Evaporator

(except for the air mix shutter on heater-only models, which is operated by cable). A vacuum tank is fitted under the vehicle on some models.

A four-speed electric blower is fitted to boost the airflow through the heater. On early models the blower is always running at low speed when the ignition is on.

19 Air conditioning system – description and precautions

1 Air conditioning is fitted as standard on most models and is optionally available on others. In conjunction with the heater the system enables any reasonable air temperature to be achieved inside the car. It also reduces the humidity of the incoming air, aiding demisting even when cooling is not required.
2 The refrigeration side of the air conditioning system functions in a similar way to a domestic refrigerator. A compressor, belt-driven from the crankshaft pulley, draws refrigerant in its gaseous phase from an evaporator. The compound refrigerant passes through a condenser where it loses heat and enters its liquid phase. After dehydration the refrigerant returns to the evaporator where it absorbs heat from air passing over the evaporator fins. The refrigerant becomes a gas again and the cycle is repeated.
3 Various subsidiary controls and sensors protect the system against excessive temperature and pressures. Additionally, engine idle speed is increased when the system is in use to compensate for the additional load imposed by the compressor.
4 On models with automatic climate control (ACC), the temperature of the incoming air is automatically regulated to maintain the cabin temperature at the level selected by the operator. An electro-mechanical programmer controls heater, air conditioner and blower functions to achieve this. See Section 28 for more details.

Precautions

5 Although the refrigerant is not itself toxic, in the presence of a naked flame (or a lighted cigarette) it forms a highly toxic gas. Liquid refrigerant spilled on the skin will cause frostbite. If refrigerant enters the eyes, rinse them with a dilute solution of boric acid and seek medical advice immediately.

6 In view of the above points, and of the need for specialised equipment for evacuating and recharging the system, any work which requires the disconnection of a refrigerant line must be left to a specialist.
7 Do not allow refrigerant lines to be exposed to temperatures above 110°C (230°F) – eg during welding or paint drying operations.
8 Do not operate the air conditioning system if it is known to be short of refrigerant, or further damage may result.

20 Air conditioning system – maintenance

1 Every 12 000 miles or annually, inspect the compressor drivebelt for correct tension and good condition. Adjust or renew if necessary as described in Section 7.
2 At the same intervals it is a good idea to clean the condenser fins. Remove the radiator grille and clean leaves, insects etc, from the fins, using an air jet or a soft brush. Be careful not to damage the condenser by over-vigorous cleaning.
3 Operate the air conditioning system for at least 10 minutes each month, even during cold weather, to keep the seals etc, in good condition.
4 Regularly inspect the refrigerant pipes, hoses and unions for security and good condition. Also inspect the vacuum hoses.
5 The air conditioning system will lose a proportion of its charge through normal seepage – typically up to 100 g (4 oz) per year – so it is as well to regard periodic recharging as a maintenance operation. Recharging must be done by a Volvo dealer or an air conditioning specialist.
6 Renewal of air conditioning system components, except for the sensors and other peripheral items covered in subsequent Sections, must also be left to a specialist.

21 Heater/air conditioning control panel – removal and refitting

1 For best access, remove the centre console side panels (Chapter 12, Section 35).

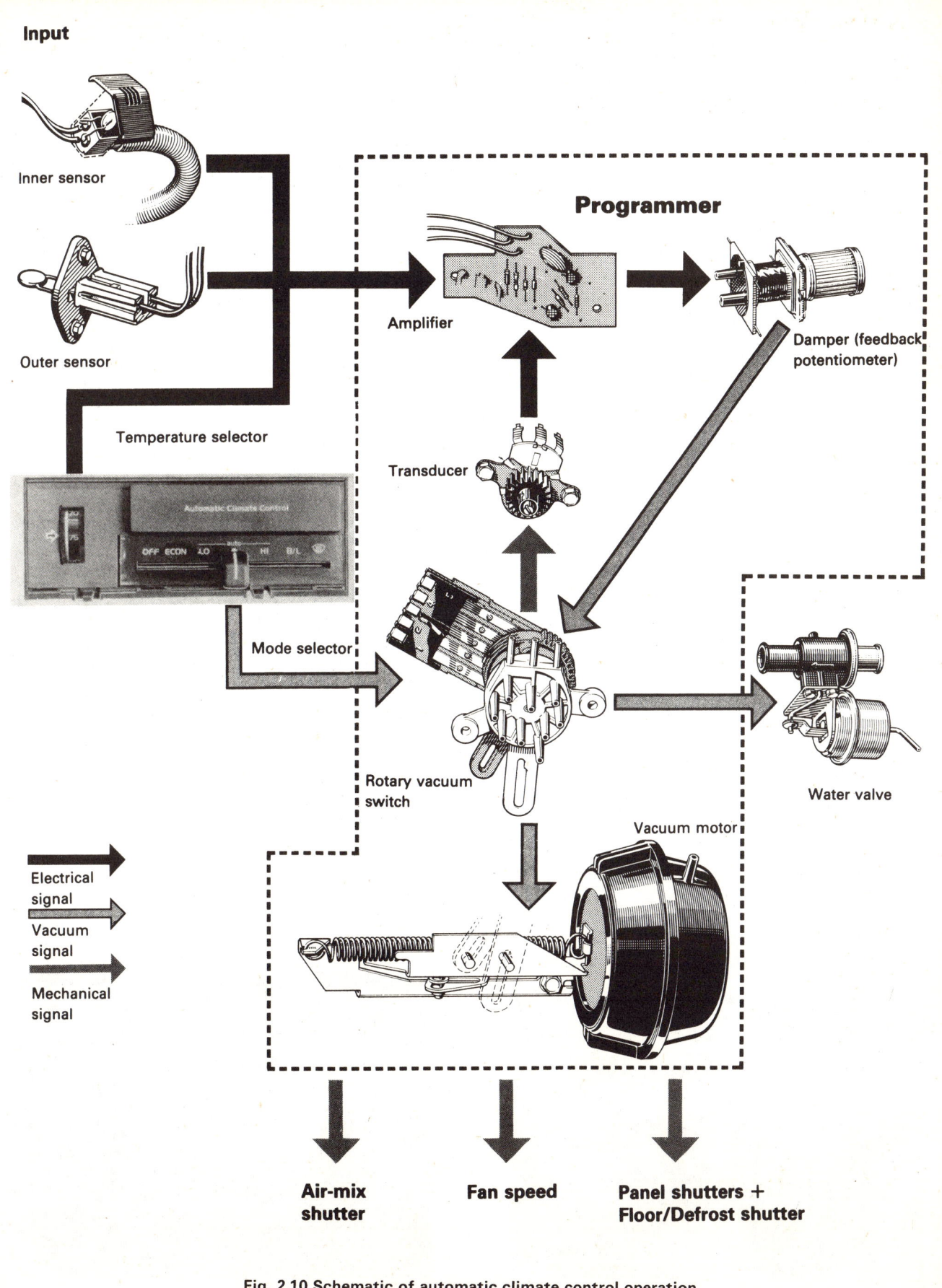

Fig. 2.10 Schematic of automatic climate control operation
(Sec 19)

Chapter 2 Cooling, heating and air conditioning systems

2 Remove the trim from around the control panel, if not already done. Remove the panel securing screws (photo).
3 Withdraw the panel and disconnect the control cables, multi-plugs and vacuum unions from it. Make notes or identifying marks if necessary for reference when refitting.
4 Refit by reversing the removal operations. Where a mechanical temperature control cable is fitted, see Section 22 for adjustment.

22 Heater temperature control cable – removal, refitting and adjustment

A mechanical temperature control cable is only fitted to vehicles without air conditioning.
1 Remove the glovebox (Chapter 11, Section 34).
2 With the temperature control in the 'WARM' position, disconnect the far end of the cable from the air mix shutter lever.
3 Remove the trim from around the heater control panel. Remove the screws which hold the panel to the centre console.
4 Ease the heater control panel away from the centre console until the cable is accessible. Disconnect the cable from the control panel, using a screwdriver to prise free the cable sleeve.
5 The cable can now be removed.
6 Refit by reversing the removal operations, noting the following points:
  (a) If the cable sleeve was damaged during removal, use a self-tapping screw to secure it (Fig. 2.13)
  (b) Adjust the position of the cable sleeve so that the air mix shutter travels over its full range of movement when the temperature control is operated

23 Heater matrix – removal and refitting

1 Disconnect the battery negative lead.
2 Depressurise the cooling system by removing the expansion tank cap. Take precautions against scalding if the coolant is hot.
3 Clamp the coolant hoses which lead to the heater matrix stubs on the bulkhead. Release the hose clips and disconnect the hoses from the stubs. Be prepared for coolant spillage.
4 Remove the centre console, rear console and glovebox. See Chapter 11, Sections 34 to 36.
5 Unclip the central electrical unit and move it aside.
6 Remove the centre panel vent. Remove the screw from the distribution unit and disconnect all the air ducts from the unit. Also remove the rear vent distribution ducts.
7 Disconnect the vacuum hoses from the vacuum motors (photo). On models with automatic climate control, also remove the hose which leads to the inner sensor (the aspirator hose).

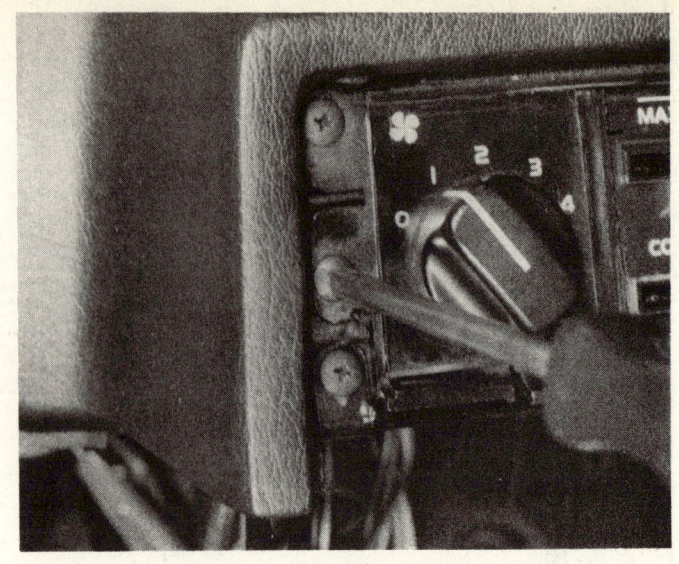

21.2 Removing a control panel securing screw

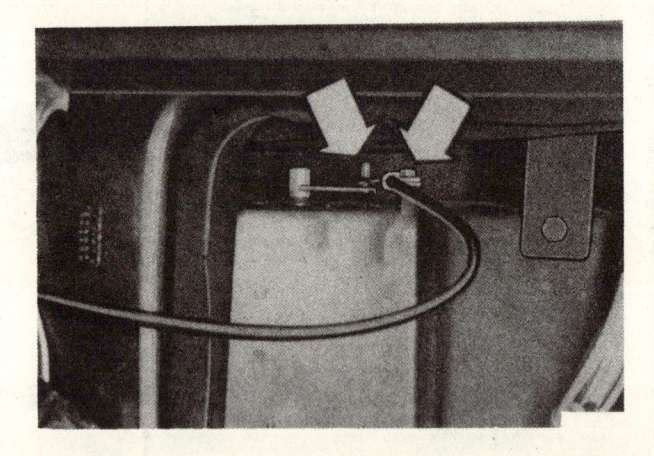

Fig. 2.11 Temperature control cable connections (arrowed) at shutter lever (Sec 22)

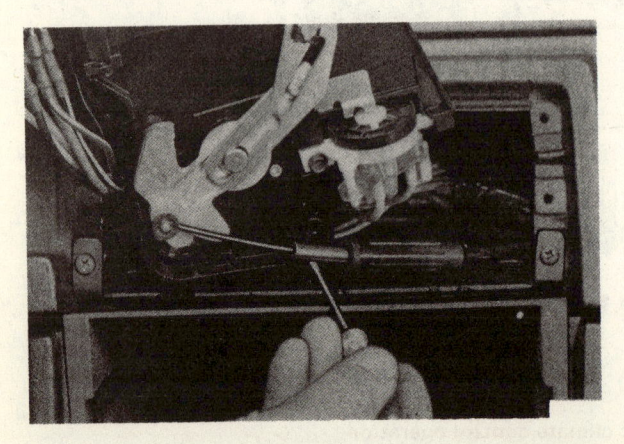

Fig. 2.12 Prising out the temperature control cable sleeve (Sec 22)

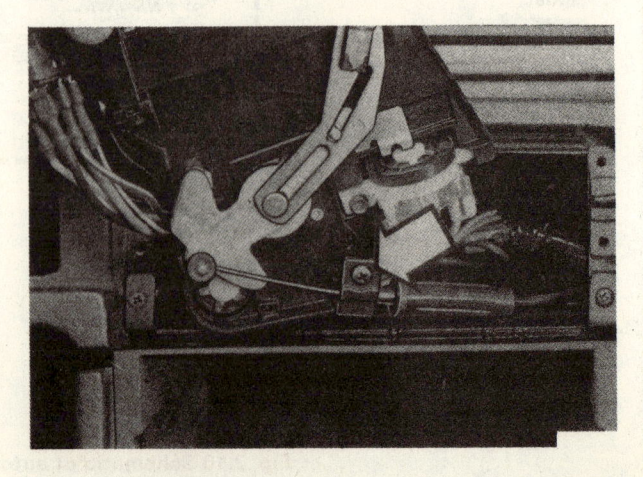

Fig. 2.13 Use a self-tapping screw (arrowed) to secure the sleeve (Sec 22)

Chapter 2 Cooling, heating and air conditioning systems

8 Remove the distribution unit.
9 Remove the heater matrix clips. Pull the matrix out and remove it; be prepared for coolant spillage.
10 Refit by reversing the removal operations. Refer to Fig. 2.14 or 2.15 for guidance when connecting the vacuum motor hoses.
11 Top up the cooling system on completion. Run the engine and check that there are no coolant leaks, then allow it to cool and recheck the coolant level.

24 Heater vacuum motors – removal and refitting

Motors in distribution unit
1 Remove the distribution unit. Refer to Section 23, but do not disconnect the coolant pipes from the matrix.
2 Remove the appropriate panel from the distribution unit for access to the motors. Remove the motors.
3 Refit by reversing the removal operations.

Recirculation shutter motor
4 Remove the glovebox (Chapter 11, Section 34). Also remove the outer panel vent and air duct.

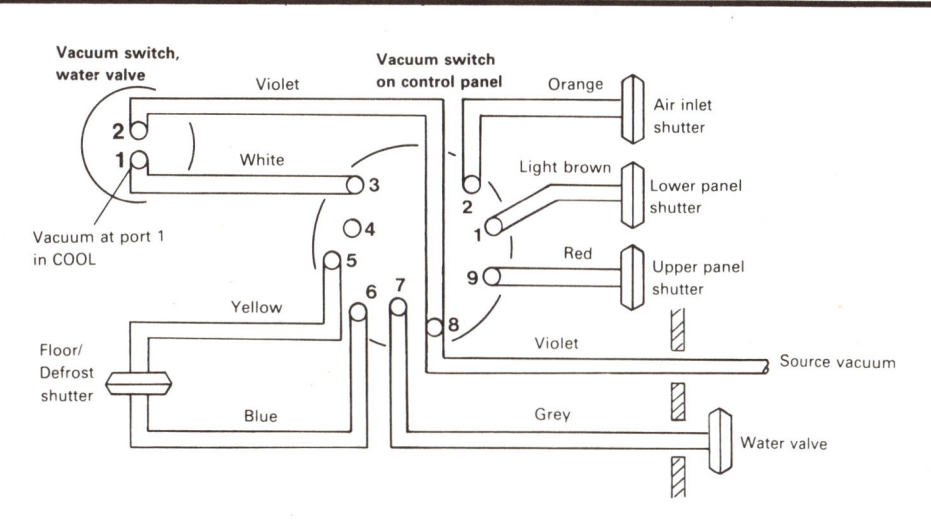

23.7 Distribution unit vacuum motors

Fig. 2.14 Vacuum hose connections – models without climate control (Sec 23)

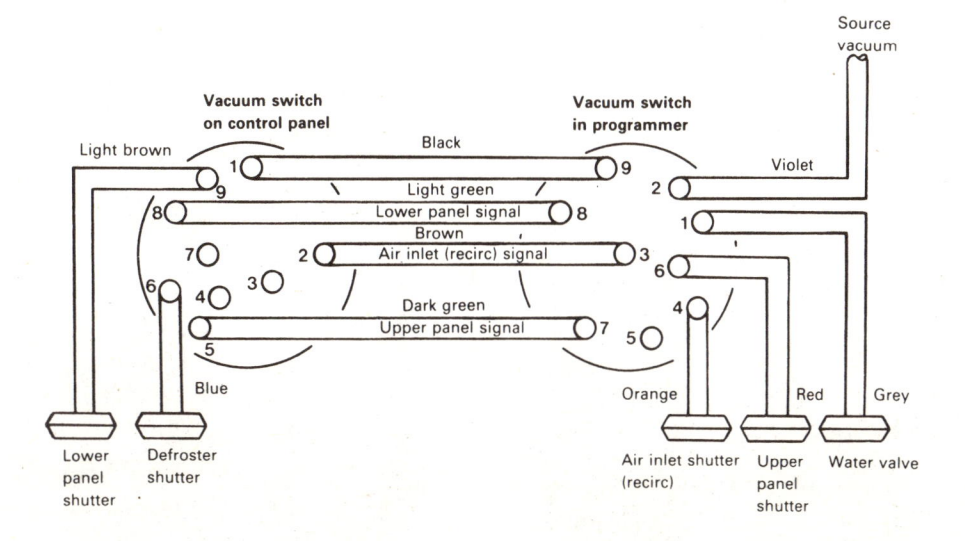

Fig. 2.15 Vacuum hose connections – models with climate control (Sec 23)

Chapter 2 Cooling, heating and air conditioning systems

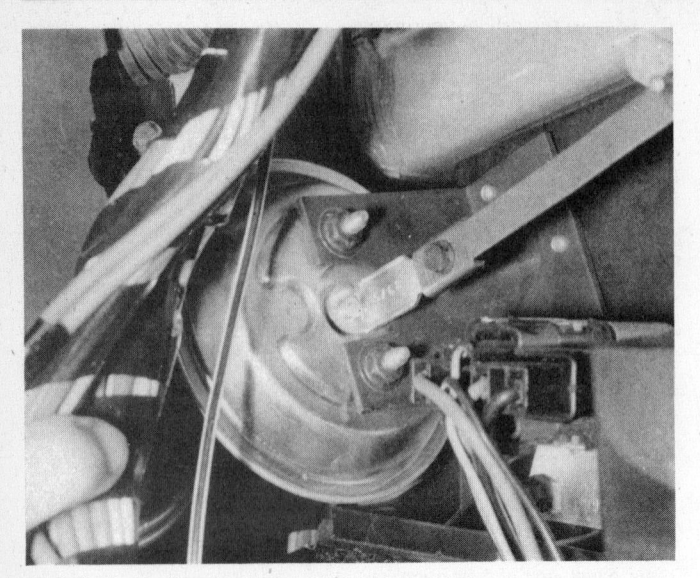

24.5 Recirculation shutter vacuum motor

5 Unbolt the control rod from the motor. Undo the two securing nuts, withdraw the motor and disconnect the vacuum hose (photo).
6 When refitting, make sure that both the shutter and the vacuum motor are in the resting position before tightening the control rod bolt.
7 The remainder of refitting is a reversal of the removal procedure.

25 Heater blower motor – removal and refitting

1 Remove the trim panel from below the glovebox.
2 Remove the screws which secure the motor to the housing.
3 Lower the motor and disconnect the cooling hose. Disconnect the wiring and remove the motor complete with centrifugal fan (photo).
4 Do not disturb any steel clips on the fan blades. They have been fitted for balancing purposes.
5 When refitting, apply sealant between the motor flange and housing. Connect the wiring and secure the motor.
6 Refit the motor cooling hose – this is important if premature failure is to be avoided.
7 Check for correct operation of the motor, then refit the trim panel.

26 Heater blower motor resistor – removal and refitting

1 Remove the glovebox (Chapter 11, Section 34).
2 Disconnect the multi-plug from the resistor.
3 Remove the two screws from the resistor and withdraw it (photo). Be careful not to damage the coils of resistance wire.

4 Refit by reversing the removal operations, but check the operation of the blower on all four speeds before refitting the glovebox.

27 Heater water valve – removal and refitting

1 Depressurise the cooling system by removing the expansion tank cap. Take precautions against scalding if the coolant is hot.
2 Clamp the coolant hoses on each side of the valve.
3 Disconnect the vacuum and coolant hoses from the valve and remove it (photo).
4 Refit by reversing the removal operations. Top up the coolant if much was lost.

28 Automatic climate control sensors – description

The four sensors peculiar to the ACC are the control panel sensor, the coolant thermal switch, the inner sensor and the outer sensor.
The control panel sensor operates in conjunction with the coolant thermal switch. If cabin temperature is below 18°C (64°F) and coolant temperature is below 35°C (95°F), the blower is prevented from operating (unless 'defrost' is selected). This prevents the ACC from blowing cold air into the cabin whilst the coolant warms up.
The inner sensor is located above the glovebox. It reads cabin air temperature.
The outer sensor is located in the blower housing and reads the temperature of the incoming air.
Acting on information received from the inner and outer sensors, the programmer applies the appropriate heating/cooling and blower speed settings to achieve the selected temperature.

29 Automatic climate control sensors – testing, removal and refitting

Control panel sensor
1 Remove the trim from around the control panel. Remove the control panel securing screws and pull the panel out.
2 Disconnect the sensor multi-plug. Use an ohmmeter to check the sensor for continuity. Continuity should be displayed above 18°C (64°F), and no continuity (open-circuit) at lower temperatures. Cool the sensor with some ice cubes, or warm it in the hands, to check that it behaves as described.
3 To remove the sensor, insert a thin screwdriver or a stiff wire into the multi-plug and prise out the sensor terminals.
4 Fit the new sensor by pressing its terminals into the multi-plug. Refit the control panel and trim.

Coolant thermal switch
5 The coolant thermal switch is located under the bonnet. It is screwed into a T-piece inserted in the heater supply hose.
6 Unplug the electrical connector and unscrew the thermal switch from the T-piece.
7 Test the thermal switch using an ohmmeter, or a battery and test

25.3 Removing the heater blower motor

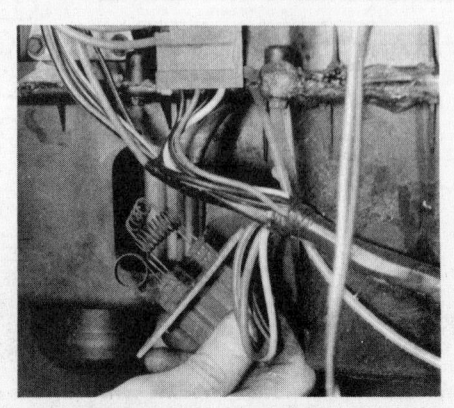

26.3 Removing the motor resistor

27.3 Disconnecting the heater water valve

Chapter 2 Cooling, heating and air conditioning systems

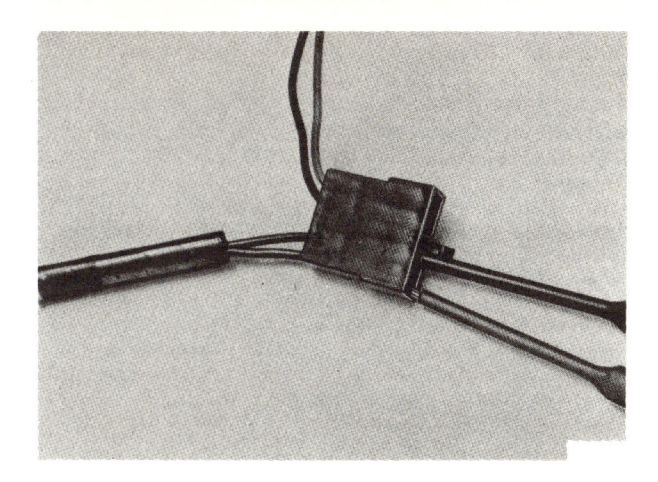

Fig. 2.16 Checking the control panel sensor for continuity (Sec 29)

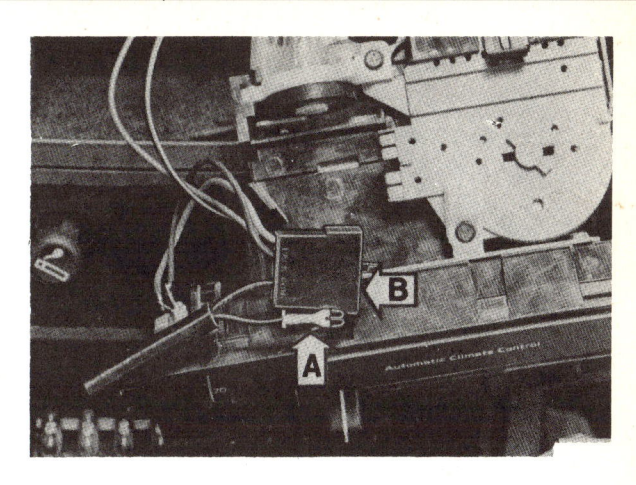

Fig. 2.17 Prising out the control panel sensor terminals (A) by inserting a stiff wire (B) (Sec 29)

Fig. 2.18 The coolant thermal switch (Sec 29)

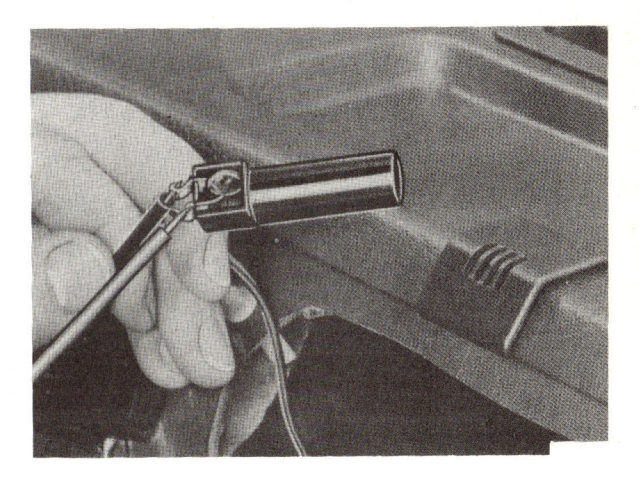

Fig. 2.19 Checking the inner sensor for continuity (Sec 29)

lamp, immersing the switch in a heated water bath. The switch should show continuity at temperatures of 30 to 40°C (86 to 104°F) upwards. As the water cools, continuity should be broken before the temperature reaches 10°C (50°F).
8 Note that if the lead to the thermal switch is accidentally disconnected, the heater blower will not work at cabin temperatures below 18°C (64°F), regardless of coolant temperature.
9 Refit by reversing the removal operations.

Inner sensor
10 Remove the glovebox (Chapter 11, Section 34).
11 Pull the air hose off the sensor, unclip the sensor and remove it.
12 The only test specified for this sensor is that it should display continuity. No resistance values are given.
13 Refit by reversing the removal operations.

Outer sensor
14 Access to the sensor for testing can be gained by removing the windscreen wiper arms, the scuttle panel and the air intake cover.
15 Measure the resistance of the sensor. At 20 to 23°C (68 to 73°F) the resistance should be 30 to 40 Ω. The higher the temperature, the lower the resistance.
16 For removal and refitting of the sensor, proceed by removing the air recirculation shutter motor (Section 24). The sensor can then be removed and a new one fitted.
17 On 1985 and later models, the sensor is mounted lower down in the fan housing. Access for testing and removal should therefore be possible without much dismantling.

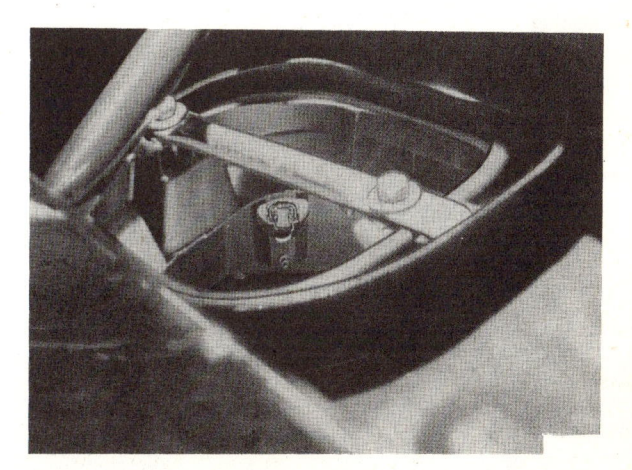

Fig. 2.20 The outer sensor location on pre-1985 models (Sec 29)

Chapter 2 Cooling, heating and air conditioning systems

30 Automatic climate control programmer – removal and refitting

1 Remove the glovebox (Chapter 11, Section 34). Also remove the outer panel vent and duct.
2 Disconnect the air mix shutter control rod, the electrical multi-plug and the vacuum pipe cluster from the programmer.
3 Remove the three screws which secure the programmer and remove it.
4 When refitting, secure the programmer with the three screws. Connect the vacuum cluster and the multi-plug, then adjust the shutter control rod as follows.
5 Run the engine to provide vacuum. Select maximum heat on the temperature control dial. Pull the control rod until it reaches its stop and secure it to the programmer arm.
6 Refit the duct, vent and glovebox.

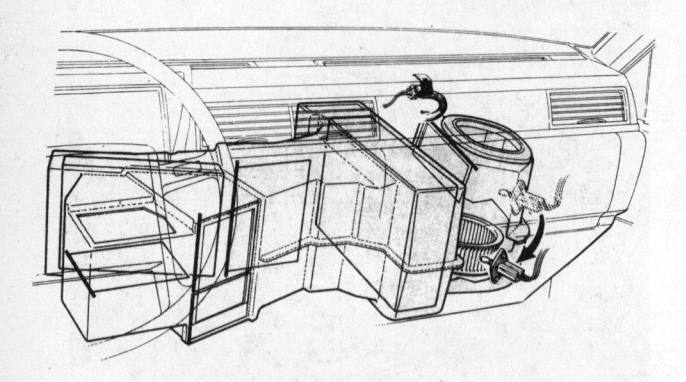

Fig. 2.21 Outer sensor location (arrowed) on later models (Sec 29)

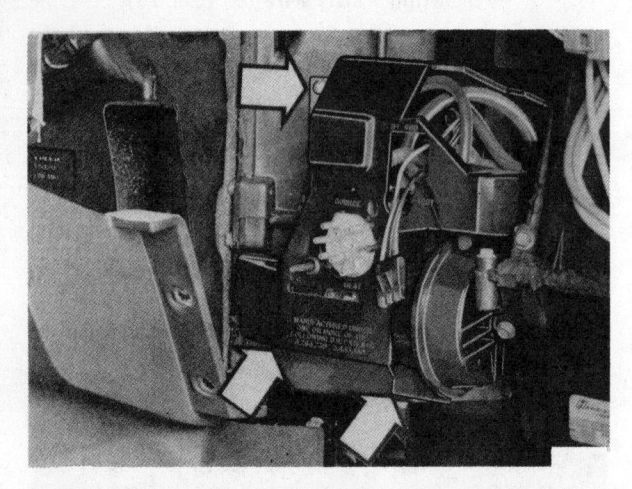

Fig. 2.22 ACC programmer disconnection points (arrowed) (Sec 30)

Fig. 2.23 ACC programmer securing screws (arrowed) (Sec 30)

31 Fault diagnosis – cooling system

Symptom	Reason(s)
Overheating	Coolant level low
	Drivebelt slipping or broken
	Radiator blocked
	Coolant hose collapsed
	Thermostat stuck shut
	Viscous-coupled fan not working
	Electric fan not working
	Ignition timing incorrect
	Fuel system fault (weak mixture)
	Exhaust system restricted
	Engine oil level low
	Cylinder head gasket blown
	Brakes binding
	New engine not yet run-in
Overcooling	Thermostat missing, jammed open or incorrect rating
Loss of coolant*	External leakage
	Overheating
	Pressure cap defective
	Cylinder head gasket blown
	Cylinder head or block cracked

Chapter 2 Cooling, heating and air conditioning systems

Symptom	Reason(s)
Oil and/or combustion gases in coolant	Cylinder head gasket blown Cylinder head or block cracked Cylinder liner base seal(s), failed (V6 only)

If the reason for loss of coolant is not obvious, have the cooling system pressure tested

32 Fault diagnosis – heating and air conditioning systems

Symptom	Reason(s)
Heater output insufficient	Overcooling (see Section 31) Water valve shut Control cable broken or maladjusted Vacuum motor broken or disconnected Heater matrix blocked internally
Air conditioner does not work	Compressor drivebelt slack or broken Fuse blown Loss of refrigerant
ACC gives full hot or cold with maximum fan speed	Sensor(s) disconnected or defective Programmer disconnected
ACC fan delay malfunction	Control panel sensor or coolant thermal switch defective Fan relay defective
ACC fan speed does not alter	Vacuum supply failure Programmer power supply failure

Chapter 3 Fuel and exhaust systems

For modifications, and information applicable to later models, see Supplement at end of manual

Contents

Air cleaner element – renewal	5
Air cleaner unit – removal and refitting	6
Auxiliary fuel tank – removal and refitting	12
Carburettor – adjustments on the vehicle	26
Carburettor – automatic choke adjustment	29
Carburettor – description	25
Carburettor – overhaul	28
Carburettor – removal and refitting	27
Constant idle speed system – adjustments	10
Constant idle speed system – description	9
Continuous injection system – description	32
Continuous injection system – testing procedures	33
Continuous injection system components – removal and refitting	34
Continuous injection system control unit – overhaul	35
Cruise control – general	21
Exhaust manifold(s) – removal and refitting	23
Exhaust system – inspection and repair	19
Exhaust system – removal and refitting	20
Fault diagnosis – carburettor	31
Fault diagnosis – continuous injection system	36
Fault diagnosis – fuel system (general)	24
Fault diagnosis – Motronic system	47
Fuel filter – renewal	7
Fuel gauge – testing	16
Fuel gauge senders – removal, testing and refitting	15
Fuel tank pump – removal and refitting	13
General description	1
Hot start valve – testing	30
Idle speed and mixture – adjustment	8
Inlet manifold – removal and refitting	22
Intercooler – removal and refitting	43
Main fuel pump – removal and refitting	14
Main fuel tank – removal and refitting	11
Maintenance and inspection	2
Motronic system – description	44
Motronic system – testing procedures (fuel side)	45
Motronic system components – removal and refitting	46
Tamperproof adjustment screws – caution	3
Throttle cable – removal, refitting and adjustment	17
Throttle pedal – removal and refitting	18
Turbo boost pressure – checking and adjustment	41
Turbocharger – description	37
Turbocharger – inspection and repair	40
Turbocharger – precautions	38
Turbocharger – removal and refitting	39
Turbo wastegate actuator – inspection and renewal	42
Unleaded fuel – general	4

Specifications

General

System type:
- B23ET and B230ET .... Motronic fuel injection, turbocharger
- B230E and B28E .... Continuous injection, normally aspirated
- B230K .... Pierburg 2B5 carburettor

Fuel tank capacity .... 60 or 82 litres (13.2 or 18.0 gallons) depending on model and year

Fuel octane rating .... 98 RON (UK 4-star)

Motronic fuel injection (B23ET and B230ET)

Air filter element .... Champion U532

Fuel filter .... Champion L204

Idle adjustments

Idle speed:
- B23ET .... 850 rpm
- B230ET .... 900 rpm (controlled by constant idle speed system – adjust to 800 rpm)

CO level at idle:
- B23ET .... $1.5 ^{+1.0}_{-0.5}$ %
- B230ET .... $1.0 ^{+1.0}_{-0.5}$ %

Pressure regulator

Line pressure .... 3.0 bar (43.5 lbf/in^2) above inlet manifold pressure

Fuel injectors

Delivery (at line pressure):
- Regular injectors .... 300 ml (10.6 fl oz) per minute
- Start injector .... 165 ml (5.8 fl oz) per minute

Injector ballast resistors (B230ET only) .... 5.5 to 6.5 Ω each

Thermal time switch

Injection time (± 2 sec):
- At −20°C (−4°F) .... 7.5 sec
- At 20°C (68°F) .... 2.0 sec
- At 35°C (95°F) and above .... Zero

Chapter 3 Fuel and exhaust systems

Auxiliary air valve (B23ET only)
Resistance	40 to 60 Ω
Fully open at	−30°C (−22°F)
Fully closed at	70°C (158°F)
Time to close from 20°C (68°F)	5 minutes

Air control valve (B230ET only)
Resistance (between terminals 3 and 4, and 4 and 5)	20 Ω approx

Fuel pumps
Main pump capacity at line pressure, +20°C (68°F) and input voltage:
12V	130 litres (28.6 gallons) per hour
11V	108 litres (23.8 gallons) per hour
10V	86 litres (18.9 gallons) per hour

Current consumption:
Main pump	6.5A max
Tank pump	1 to 2A

Turbocharger
Boost pressure (at full load and 3500 engine rpm)	0.45 to 0.53 bar (6.5 to 7.7 lbf/in²) over atmospheric
Overpressure switch opens at	0.65 to 0.75 bar (9.4 to 10.9 lbf/in²)
Boost pressure switch closes at	0.15 to 0.25 bar (2.2 to 3.6 lbf/in²)

Torque wrench settings
	Nm	lbf ft
Exhaust downpipe to wastegate housing	25	18
Wastegate housing to turbocharger	20	15
Turbocharger to exhaust manifold:		
Stage 1	1	0.7
Stage 2	45	33
Stage 3	Tighten 45° further	Tighten 45° further

Continuous injection system (B230E and B28E)

Air filter element
B230E	Champion U547
B28E	Champion U554

Fuel filter
Champion L204

Idle adjustments
Idle speed	900 rpm (adjust to 850 rpm on B28E with constant idle system)
CO level at idle:	
B230E*	$1 ^{+1.0}_{-0.5}$ %
B28E	2 ± 1%

* Automatic transmission only – measure CO level at 800 rpm

Fuel pressure
Line pressure:
B230E	4.5 to 5.3 bar (65 to 77 lbf/in²)
B28E	4.7 to 5.5 bar (68 to 80 lbf/in²)

Rest pressure:
B230E	1.5 to 2.4 bar (22 to 35 lbf/in²)
B28E	2.4 to 3.2 bar (35 to 46 lbf/in²)
Control pressure (warm regulator)	3.0 to 3.4 bar (44 to 49 lbf/in²)

Control pressure regulator
Resistance:
B230E	20 to 30 Ω
B28E	20 to 24 Ω

Fuel injectors
Opening pressure	3.5 to 4.1 bar (51 to 60 lbf/in²)
No leakage below	2.9 bar (42 lbf/in²)

Start injector delivery:
B230E	85 ml (3 fl oz) per minute
B28E	115 ml (4 fl oz) per minute

Thermal time switch
Injection time (± 2 sec):
- At −20°C (−4°F) .... 7.5 sec
- At 20°C (68°F) .... 2.0 sec
- At 35°C (95°F) and above .... Zero

Auxiliary air valve (not constant idle speed system)
- Resistance .... 40 to 60 Ω
- Fully open at .... −30°C (−22°F)
- Fully closed at .... 70°C (158°F)
- Time to close from +20°C (68°F) .... 5 minutes

Air control valve (constant idle speed system)
- Resistance (between terminals 3 and 4, and 4 and 5) .... 20 Ω approx

Coolant temperature sensor (constant idle speed system)
Resistance at:
- −10°C (14°F) .... 32,000 to 53,000 Ω
- 20°C (68°F) .... 8,500 to 11,500 Ω
- 80°C (176°F) .... 770 to 1,320 Ω

Fuel pumps
Main pump capacity at 5.0 bar (73 lbf/in²), 20°C (68°F) and input voltage:
- 12V .... 120 litres (26.4 gallon) per hour
- 11V .... 96 litres (21.1 gallons) per hour
- 10V .... 75 litres (16.5 gallons) per hour

Current consumption:
- Main pump .... 9.5A max
- Tank pump .... 1 to 2A

Torque wrench settings
	Nm	lbf ft
Inlet manifold bolts (B28E)	10 to 15	7 to 11

Pierburg 2B5 carburettor (B230K)

Air cleaner element
Champion U547

Idle adjustments
Idle speed:
- Manual .... 800 rpm
- Automatic .... 900 rpm
- CO level at idle .... 1.0 $^{+1.0}_{-0.5}$ %

Calibration
- Float height .... See text
- Accelerator pump delivery .... 10 to 14 ml (0.35 to 0.49 fl oz) per 10 strokes

Jet sizes

	Primary	Secondary
Air correction	140	65
Idle (air/fuel)	47.5/120	−
Auxiliary (air/fuel)	45/145	−
Enrichment	85	−
Main	X 112.5	X 137.5
Air bypass	−	140
Fuel bypass	−	100

Overhaul data
- Fast idle gap .... 4.0 mm (0.158 in)

Throttle linkage clearance (see text):
- Gap 'A' .... 0.10 to 0.50 mm (0.004 to 0.020 in)
- Gap 'B' .... 0.15 to 0.85 mm (0.006 to 0.034 in)

Choke flap opening under vacuum:
- Upper connection blocked (see text) .... 3.35 to 3.65 mm (0.132 to 0.144 in)
- Upper connection open .... 1.35 to 1.65 mm (0.053 to 0.065 in)
- Choke flap opening at full throttle .... 5.2 to 6.2 mm (0.205 to 0.244 in)
- Choke flap opening at 20°C (68°F) .... 0.55 to 2.05 mm (0.022 to 0.081 in)
- Clearance between choke flap and full load enrichment tube .... 0.5 mm (0.020 in)

Chapter 3 Fuel and exhaust systems

1 General description

All models have one or two rear-mounted fuel tanks and one or two electric fuel pumps. These (and the exhaust system) are the only common features.

The B230K engine has a Pierburg 2B5 carburettor. This is covered in Sections 25 to 31. The B230E and B28E engine have a fuel injection system of the continuous type; this is covered in Sections 32 to 36. B23ET and B230ET have a turbocharger and electronically controlled (Motronic) fuel injection; these are covered in Sections 37 to 47. The preliminary Sections cover all models.

Warning: Fuel hazards

Petrol offers multiple hazards – fire, explosion and toxicity. Do not smoke when working on the fuel system, or allow naked flames nearby. Ensure adequate ventilation if fuel vapour is likely to be produced. Refer to 'Safety first!' at the beginning of the manual for more details.

2 Maintenance and inspection

1 Keep an adequate supply of fuel in the tank at all times. A full tank is less likely to suffer from rust and condensation, both of which can cause problems by contaminating the fuel.
2 Every 6000 miles (10 000 km) or six months, whichever comes first, check the idle speed and mixture (CO level). See Section 8.
3 Every 12 000 miles (20 000 km) or annually, check the condition and security of the exhaust system (Section 19). Also inspect the fuel lines and hoses.
4 At the same interval on carburettor models, inspect the in-line fuel filter and renew it if it appears internally dirty (Section 7).
5 At the same interval on Turbo models, check the integrity of the tamperproof seals, and check the boost pressure switches (Section 45).
6 Every 24 000 miles (40 000 km) or two years, renew the fuel filter (Section 7) and the air cleaner element (Section 5). More frequent renewal of the air cleaner element may be necessary in dusty conditions.
7 Lubricate the throttle linkage occasionally and check that the cable is not frayed or kinked.

3 Tamperproof adjustment screws – caution

1 Certain adjustment points in the fuel system (and elsewhere) are protected by 'tamperproof' caps, plugs or seals. The purpose of such tamperproofing is to discourage, and to detect, adjustment by unqualified operators.
2 In some EEC countries (though not yet in the UK) it is an offence to drive a vehicle with missing or broken tamperproof seals.
3 Before disturbing a tamperproof seal, satisfy yourself that you will not be breaking local or national anti-pollution regulations by doing so. Fit a new seal when adjustment is complete when this is required by law.
4 Do not break tamperproof seals on a vehicle which is still under warranty.

4 Unleaded fuel – general

1 Unleaded fuel will theoretically be available in all EEC countries as time progresses, and may eventually replace leaded fuel completely.
2 It is generally believed that continuous use of unleaded fuel can cause rapid wear of conventional valve seats.
3 The vehicles covered by this manual should not be run on unleaded fuel without some modification being made. Consult a Volvo dealer for further details.

5 Air cleaner element – renewal

1 Release the clips which secure the air cleaner lid (photo).
2 On Turbo models, disconnect the airflow meter multi-plug and the meter-to-turbo trunking (photos). The multi-plug is released by levering out the wire clip.
3 Lift off the lid, with airflow meter when applicable, and remove the air cleaner element (photo).

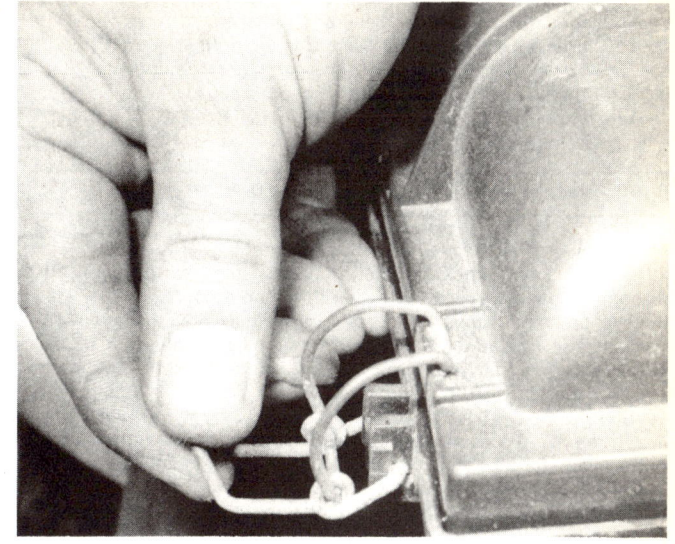

5.1 Releasing an air cleaner lid clip

5.2A Disconnecting the airflow meter multi-plug

5.2B Releasing the meter-to-tube trunking

5.3 Removing the air cleaner element

Chapter 3 Fuel and exhaust systems

4 Wipe clean inside the housing and lid with a cloth. Be careful not to sweep debris into the airflow meter or air intake.
5 Fit the new element, making sure it is the right way up. Press the seal on the rim of the element into the groove on the housing.
6 Refit the lid and secure it with the clips.
7 When applicable, reconnect the airflow meter.

6 Air cleaner unit – removal and refitting

1 Remove the air cleaner element (Section 5).
2 Disconnect the hot air inlet trunking from the housing. Also disconnect the crankcase ventilation hose (photos).
3 The preheating thermostat and shutter may now be removed if wished. The operation of the thermostat may be checked using a refrigerator or a hair dryer. The marking on the thermostat gives the approximate temperature at which it will be in the 'half-way' position (photos).
4 To remove the complete air cleaner housing, release the securing bolt or clip (photo). Lift out the housing, disengaging the cold air intake from the grommet in the inner wing.
5 Refit by reversing the removal operations.

7 Fuel filter – renewal

1 Disconnect the battery negative lead.

Carburettor models
2 The fuel filter is located under the bonnet, in the carburettor fuel feed line (photo). Slacken the hose clips and disconnect the hoses from the filter; be prepared for fuel spillage.
3 Fit the new filter, observing any direction of flow markings. Renew hoses and clips if necessary.

Fuel injection models
4 The filter is located under the bonnet, or with the main fuel pump, under the car.

Under-bonnet filter
5 Slacken the fuel outlet union on top of the filter. Be prepared for fuel spillage. Disconnect the unions (photo).
6 Slacken the filter clamp. Lift up the filter and disconnect the inlet union. Be prepared for further spillage.
7 Remove the old filter.
8 Fit the new filter, making sure it is the same way round as the old one. Use new copper washers on the unions (when applicable) and secure the filter in the clamp.

Under-car filter
9 Proceed as for fuel pump removal (Section 14), but remove the filter instead of the pump. Observe the arrow on the new filter showing the direction of fuel flow (photo).

All models
10 Reconnect the battery. Run the engine and check that there are no leaks.
11 Dispose of the old filter safely.

8 Idle speed and mixture – adjustment

An accurate tachometer (rev counter) and an exhaust gas analyser (CO meter) will be needed for this procedure. Otherwise, have the job done by a Volvo dealer or other specialist.
1 Refer to Section 3.
2 The engine valve clearances must be correct, the crankcase ventilation system hoses connected, the air cleaner element and the ignition system in good condition. Air conditioning and major electrical loads must be switched off. The throttle cable must be correctly adjusted (Section 17).
3 Bring the engine to normal operating temperature. Connect the

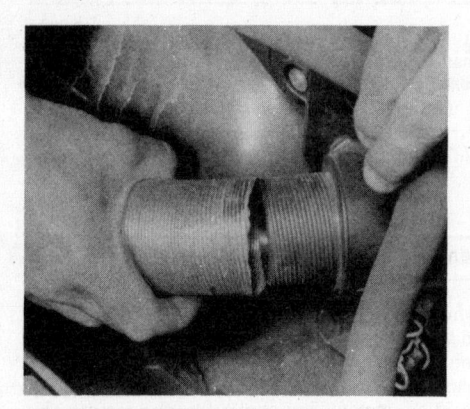

6.2A Disconnecting the hot air trunking

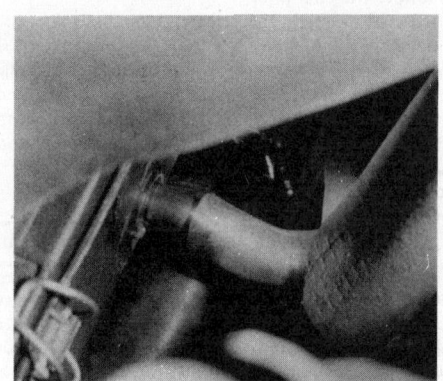

6.2B An air cleaner crankcase ventilation hose

6.3A Removing the air cleaner preheating thermostat assembly

6.3B Preheating thermostat assembly set to take in hot air. Thermostat itself is arrowed

6.4 An air cleaner securing bolt

7.2 Carburettor type fuel filter. Arrow shows direction of flow

Chapter 3 Fuel and exhaust systems

7.5 Slackening the fuel filter outlet union – continuous injection type

7.9 Fuel filter (on right) and pump – Turbo models. Arrows show direction of flow

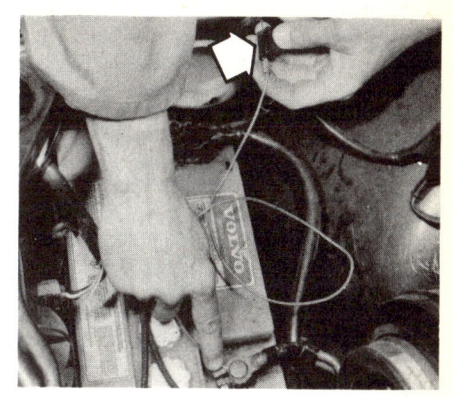

8.4 Earthing the constant idle speed system test wire. One end of the earth wire must fit the test connector (arrowed)

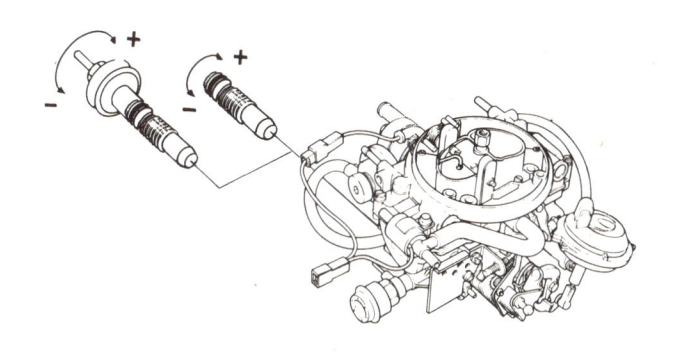

Fig. 3.1 Carburettor idle speed adjustment screw – compensation device fitted with air conditioning is on left (Sec 8)

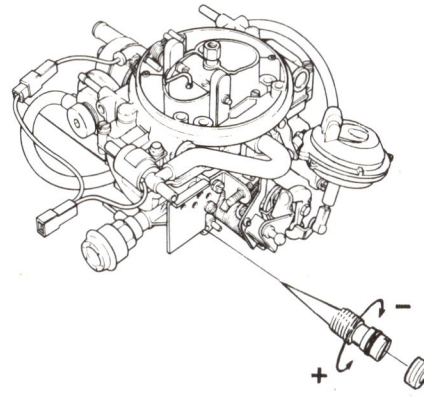

Fig. 3.2 Carburettor idle mixture adjustment screw and tamperproof plug (Sec 8)

tachometer and exhaust gas analyser as instructed by the manufacturers. Allow the engine to idle.

4 On models with a constant idle speed system (most B28E and all B230ET engines), disable the system by earthing the black and white test wire near the right-hand suspension turret (photo). The idle speed is then adjusted to a lower value than that normally maintained (see Specifications).

Carburettor models

5 Adjust the idle speed if necessary by turning the idle speed adjustment screw (models without air conditioning) or the idle compensation device (models with air conditioning) (photo).

6 Read the CO level and adjust if necessary by turning the idle mixture adjustment screw. This may be covered by a tamperproof plug (photo).

7 Readjust the idle speed if necessary.

Fuel injection models (not Turbo)

8 Adjust the idle speed using the adjustment screw on the inlet manifold (photo). (On B28E models, do not touch the other two screws).

9 Read the CO level and adjust if necessary by turning the mixture adjustment screw next to the fuel distributor (photo). A long Allen key will be needed. After each adjustment, remove the Allen key, rev the

8.5 Carburettor idle speed adjustment screw (arrowed)

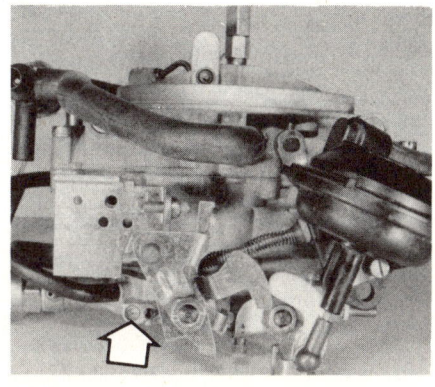

8.6 Carburettor idle mixture adjustment screw is underneath the plug (arrowed)

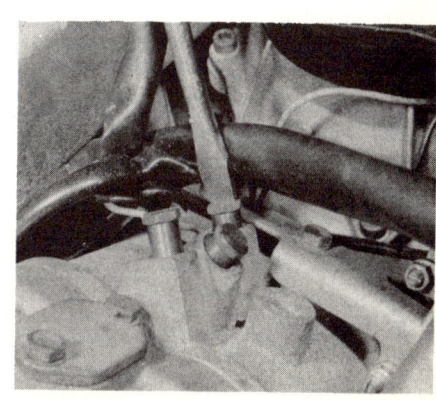

8.8 Idle speed adjustment – B 28 E. Do not touch the other two screws

102   Chapter 3 Fuel and exhaust systems

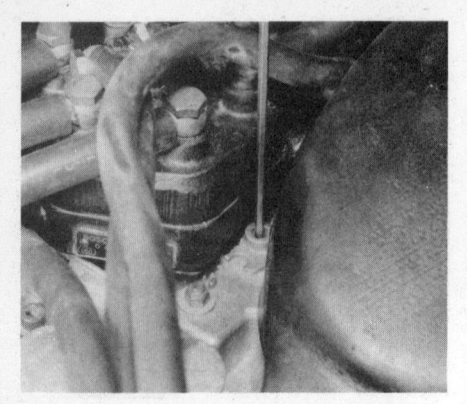

8.9 Idle mixture adjustment using a long Allen key. The hole may be tamperproofed

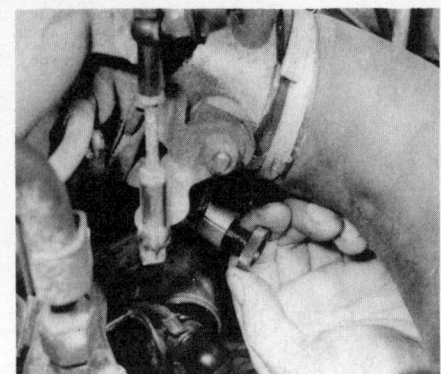

8.11 Idle speed adjustment – Turbo

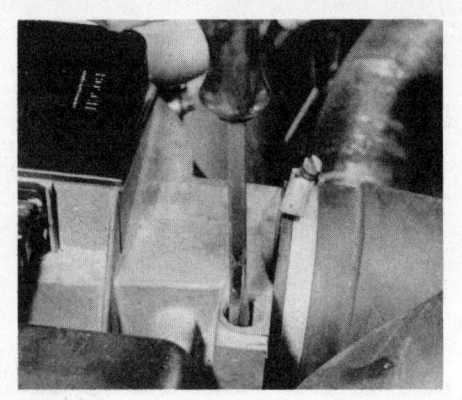

8.12 Idle mixture adjustment – Turbo. The hole may be tamperproofed

engine briefly and allow the CO reading to stabilise before re-checking.
10 Readjust the idle speed if necessary.

Fuel injection models (Turbo)
11 Adjust the idle speed if necessary by turning the knurled adjustment screw next to the throttle housing (photo).
12 Read the CO level and adjust if necessary by turning the adjustment screw on the airflow meter (photo).
13 Readjust the idle speed if necessary.

All models
14 If the CO reading is initially too high, but falls substantially when the crankcase ventilation hose is disconnected from the oil trap (in-line engines) or when the oil filler cap is removed (V6 engines), this suggests that fuel contamination of the oil is affecting the readings. Change the engine oil before proceeding.
15 When the idle speed and mixture are within the specified limits, stop the engine and disconnect the test gear.
16 On models with a constant idle speed system, disconnect the test wire earth lead.
17 Fit new tamperproof seals when required.

9 Constant idle speed system – description

All B230ET and most B28E engines are equipped with a constant idle speed system. The system operates to keep idle speed more or less constant regardless of varying accessory loads (air conditioning, alternator, etc).
The components of the system are a control unit, an air control valve, a throttle switch and a temperature sensor (photo). Additionally the system receives an input from the ignition coil (B28E) or the Motronic control unit (B230ET).
The air control valve allows a certain amount of air to bypass the throttle butterfly. The amount is regulated by the control unit, acting on data received indicating engine speed, throttle position and coolant temperature.
If the constant idle speed system malfunctions, and the adjustments described in the next Section are ineffective, consult a Volvo dealer or other specialist.

10 Constant idle speed system – adjustments

Before attempting to adjust the constant idle speed system, make sure that the ignition system and the rest of the fuel system are in good condition and correctly adjusted.

B28E
1 Connect an accurate tachometer and an exhaust gas analyser (CO meter) to the engine as instructed by their makers.

2 With the engine warmed up, turn all three idle adjustment screws fully clockwise. From this position unscrew the balance screw indicated in Fig. 3.4 by five complete turns. Do not touch the other screws.
3 Disconnect the link rod from the throttle cable drum. Check that the drum moves freely.
4 Connect the constant idle speed test wire to earth (Section 8, paragraph 4).
5 Disconnect the orange wire from terminal 2 of the throttle switch. Do not allow the end of the wire to touch metal.
6 Run the engine. Adjust the idle speed to 700 rpm by means of the lower adjustment screw on the throttle lever.
7 Adjust the idle speed to 850 rpm by means of the idle speed adjustment screw on the manifold.
8 Disconnect the test wire from earth (paragraph 4). The idle speed should rise to 900 rpm.
9 Remove the oil filler cap to prevent crankcase fumes affecting the CO level. Adjust the CO level to 1.8% (Section 8, paragraph 9).
10 Repeat the procedure from paragraph 4 until the idle speeds and CO level are as specified, without further adjustment. Switch off the engine.
11 Disconnect the test gear, remake the original electrical connections and refit the oil filler cap.
12 Connect a 12 volt test light between terminal 4 of the throttle switch and the battery positive terminal.

9.0 Air control valve (left) on constant idle speed system

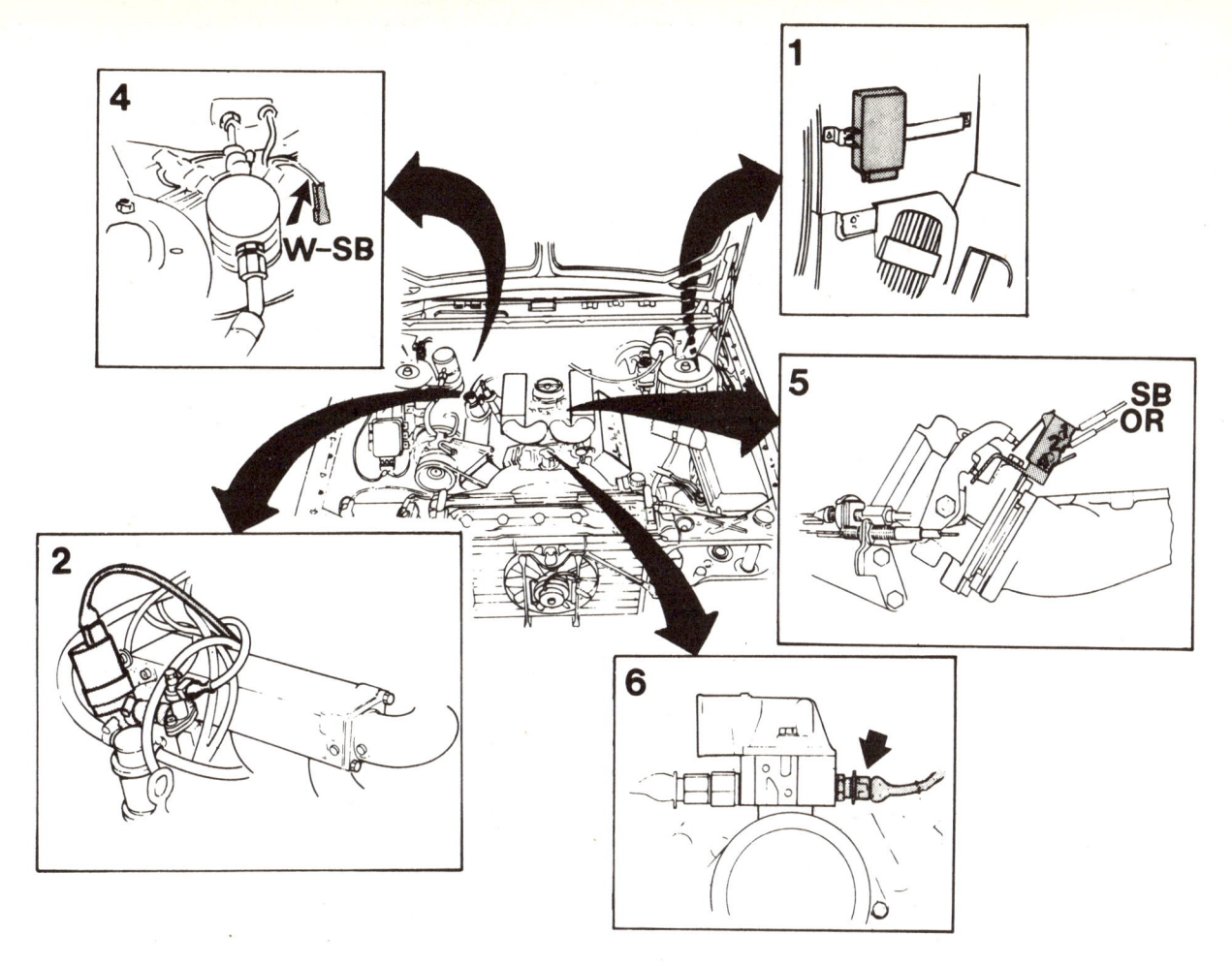

Fig. 3.3 Constant idle speed system components (Sec 9)

1 Control unit
2 Air control valve
4 Test wire
5 Throttle switch
6 Temperature sensor

OR Orange
SB Black
W White

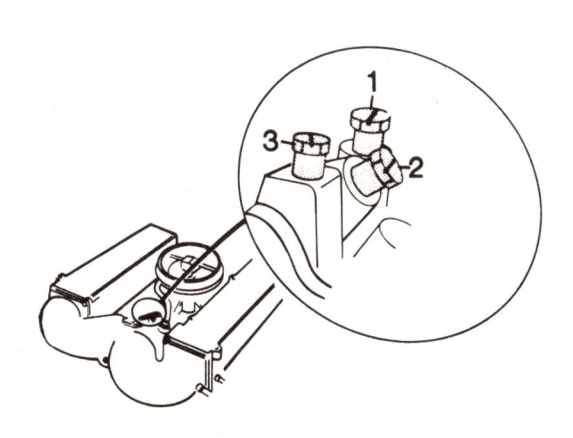

Fig. 3.4 Idle adjustment screws on manifold – B28 E (Sec 10)

1 Idle speed adjustment screw
2 Balance screw – do not unscrew
3 Balance screw – unscrew 5 turns

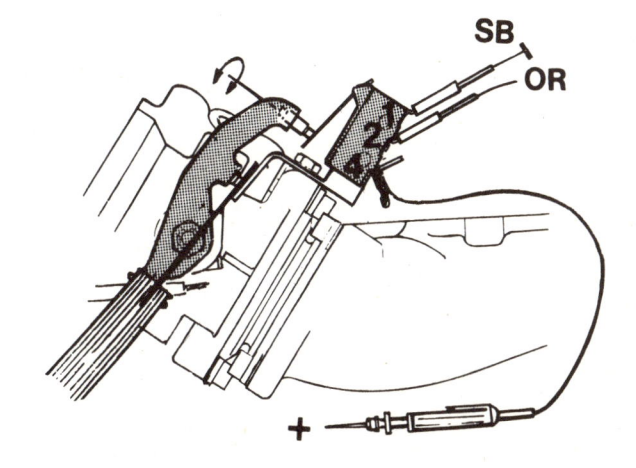

Fig. 3.5 Adjusting the throttle switch with a feeler blade and a test light (Sec 10)

OR Orange

SB Black

Chapter 3 Fuel and exhaust systems

10.13 Adjusting the constant idle speed throttle switch

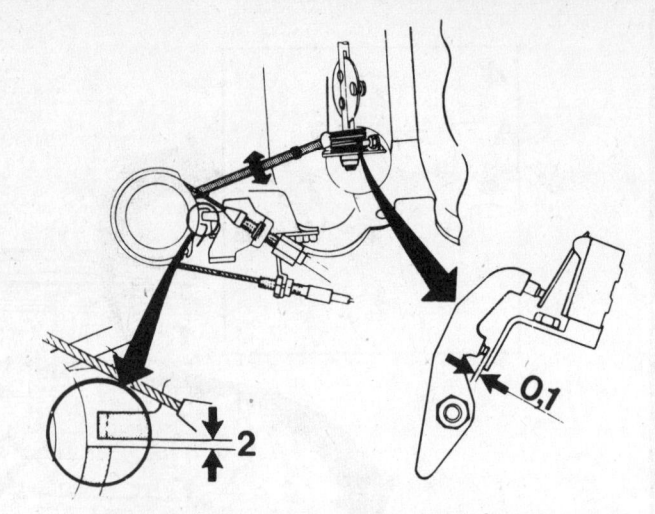

Fig. 3.6 Insert a 2 mm feeler blade (inset, left) and adjust the rod for a lever gap of 0.1 mm (right) (Sec 10)

13 Insert a feeler blade 0.30 mm (0.012 in) thick below the switch lower adjustment screw (photo). Unscrew the upper adjustment screw until the light just goes out, then screw it in again until the light just comes on.
14 Substitute feeler blades of 0.20 mm (0.008 in) and 0.60 mm (0.024 in) thickness. The test light should stay on with the thinner blade, and go out with the thicker one. Disconnect the test light.
15 Check the adjustment of the throttle cable (and cruise control cable, if applicable). The cable should be just taut at idle, without holding the throttle drum off its stop.
16 Have the assistant hold the throttle pedal down. Check that the throttle drum contacts the full throttle stop. Also check the adjustment of the kickdown cable (when applicable) in this position – see Chapter 6, Section 31. Release the throttle.
17 Insert a feeler blade 2 mm (0.080 in) thick between the throttle drum and the idle stop. Reconnect the throttle link rod, adjusting its length if necessary to achieve a clearance of 0.10 mm (0.004 in) between the lower adjustment screw and the stop (Fig. 3.6). Remove the feeler blade.
18 This completes adjustment of the constant idle speed system. Continued malfunction may be due to poor electrical connections; broken, kinked or blocked air or vacuum hoses; wrongly fitted air control valve (Fig. 3.7); other component fault.

B230ET
19 No specific information is available.

11 Main fuel tank – removal and refitting

1 Disconnect the battery negative lead. Observe strict fire precautions throughout the operation.
2 Drain the tank, storing the fuel in suitable sealed containers.
3 On models with an auxiliary tank, remove it (Section 12). On models without an auxiliary tank, remove the access hatch from the luggage area floor.
4 Disconnect the fuel hoses, filler and breather hoses (when applicable) and the tank pump connector. See Section 13, paragraph 5.
5 Raise and support the rear of the vehicle. Support the fuel tank, remove the securing nuts, bolts and reinforcing plates and lower the tank.
6 Repair of a leaking tank must only be undertaken by professionals. Even when the tank is empty, it may still contain explosive vapours. **Do not** attempt to weld or solder the tank. 'Cold' repair compounds are available and these are suitable for DIY use.

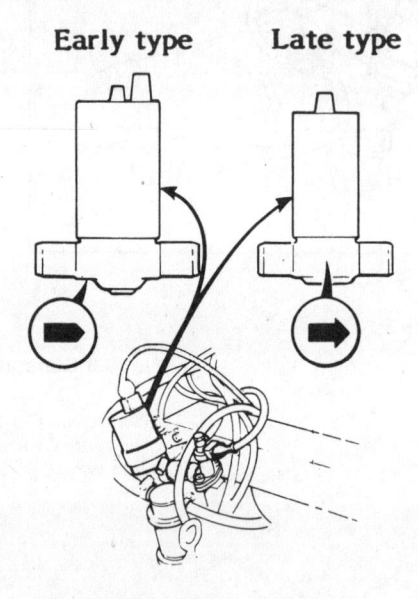

Fig. 3.7 Correct fitting of air control valve. Arrow shows direction of air flow (Sec 10)

7 If a new tank is to be fitted, apply rustproofing compound to it beforehand.
8 Refit by reversing the removal operations, using new hoses and clips as necessary.

12 Auxiliary fuel tank – removal and refitting

1 The auxiliary tank is empty when there is less than 60 litres (13.2 gallons) of fuel remaining. Drain some fuel from the main tank if necessary to empty the auxiliary tank.
2 Disconnect the battery negative lead. Observe strict fire precautions throughout the operation.
3 Unclip and remove the boot lining.
4 Unbolt and remove the tank cover plate and the filler pipe cover plate (photo).
5 Disconnect the wires from the fuel gauge auxiliary sender unit.
6 Disconnect the breather hose from the auxiliary tank (photo).
7 Remove the four bolts which secure the auxiliary tank.
8 Carefully lift the tank as far as possible, without straining the

Chapter 3 Fuel and exhaust systems

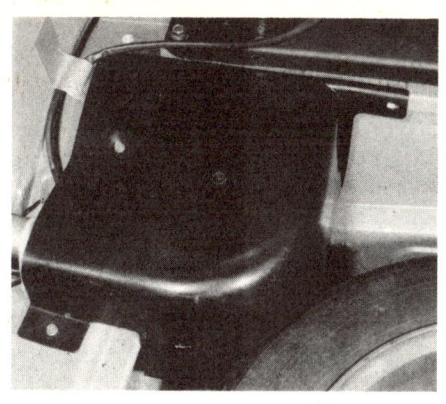

11.5 Fuel tank strap nuts – Estate

12.4 Fuel filler pipe cover plate – Saloon

12.6 Auxiliary tank breather hose

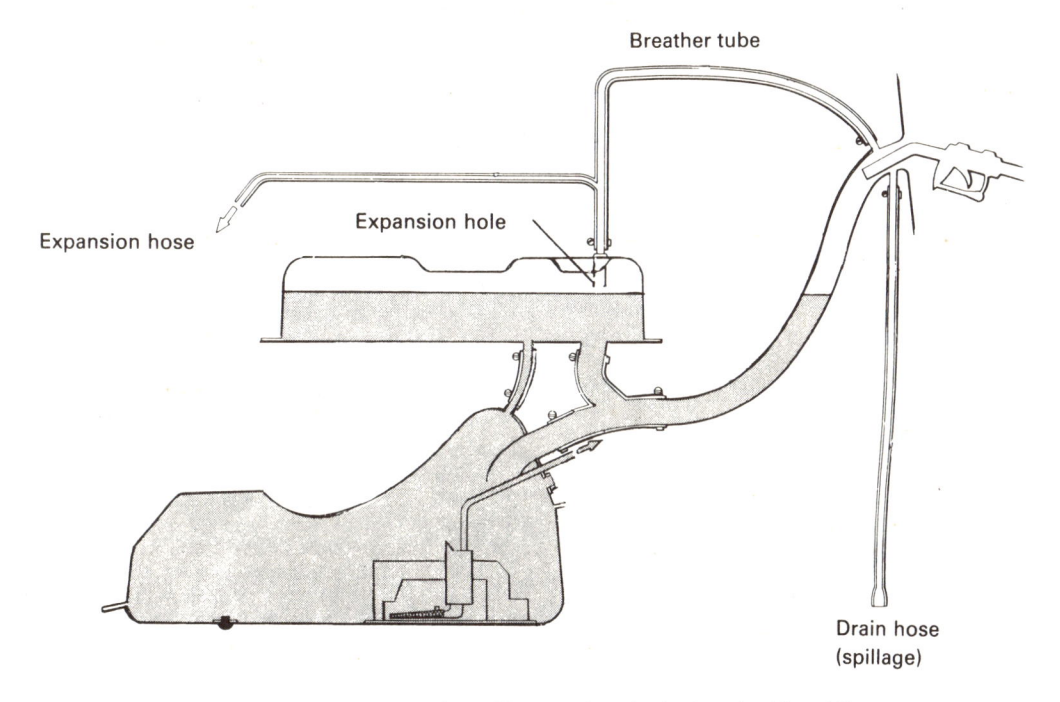

Fig. 3.8 Relationship of auxiliary and main fuel tanks (Sec 12)

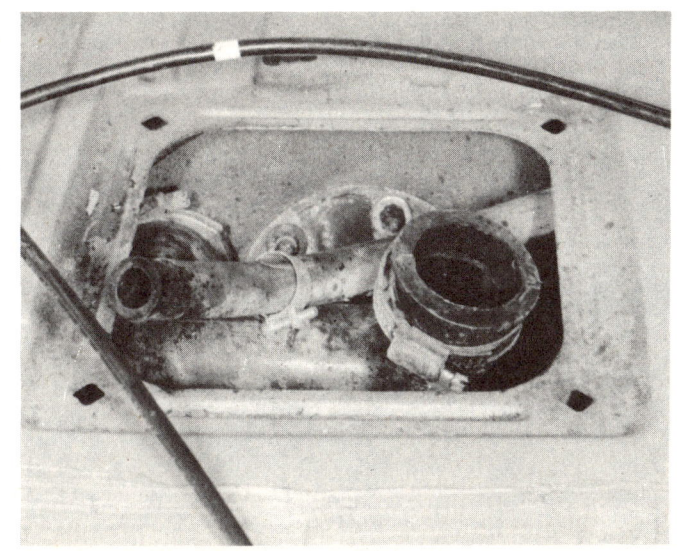

12.9 The two hoses which join the main and auxiliary tanks

connecting hoses beneath it. Prop the tank in the raised position with a couple of wooden blocks.
9 Working under the left-hand end of the auxiliary tank, disconnect the two hoses which join it to the main tank (photo).
10 Lift out the auxiliary tank, unclipping the thin breather pipe from its forward edge.
11 Refit by reversing the removal operations. Renew hoses and clips as necessary.

13 Fuel tank pump – removal and refitting

1 Disconnect the battery earth lead. Observe strict fire precautions throughout the operation.
2 Gain access to the top of the main fuel tank, either by removing the auxiliary tank (Section 12), or by removing the access hatch when no auxiliary tank is fitted. For Estate models see Figs. 3.9 and 3.10.
3 Clean around the tank pump/sender unit cover plate.
4 Disconnect and plug the fuel supply and return hoses. Also disconnect the breather hose (when applicable).
5 Follow the electrical lead back to the nearest multi-plug and disconnect it. If the plug will not pass through the holes in the bodywork *en route* to the tank, prise the connectors out of the plug. Unbolt the earth tag and pull the lead into the same compartment as the tank.

Chapter 3 Fuel and exhaust systems

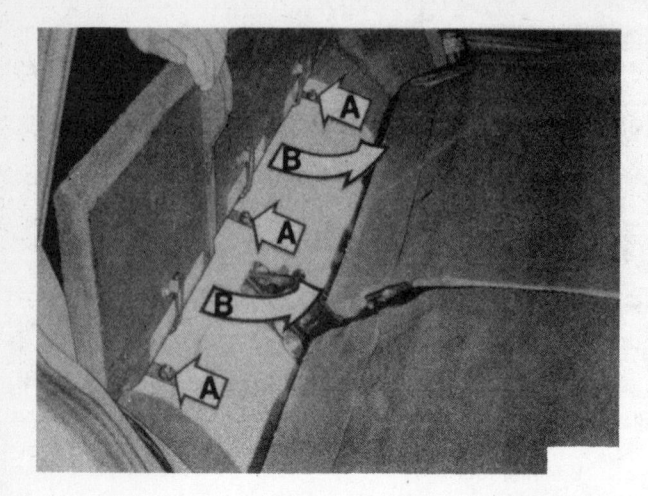

Fig. 3.9 Access to fuel tank hatch on Estates: remove screws (A) and pull floor forwards and upwards (B) (Sec 13)

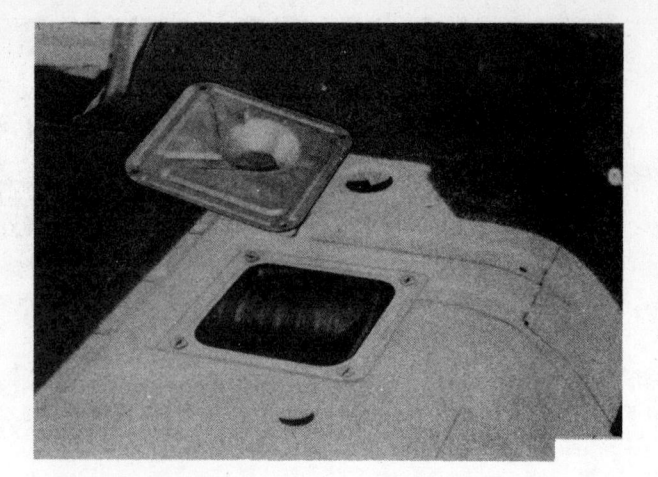

Fig. 3.10 Fuel tank hatch removed – Estate (Sec 13)

6 Remove the nuts which secure the cover plate to the tank.
7 The tank pump/sender unit can now be removed from the tank. Some manipulation will be needed. Do not force the unit out, it is delicate (photo).
8 Unclip the pick-up screen and tube from the pump. Remove the clamp screw, disconnect the electrical leads and remove the pump (photos).
9 If the pump is defective it must be renewed.
10 Fit the pump to the sender unit and secure the pick-up components. Make sure that the O-ring is in good condition and that the pick-up screen is clean (photo). Connect the electrical leads and tighten the clamp screw.
11 The pump is spring-loaded against the cover plate to ensure that the pick-up screen sits at the very bottom of its well. Temporarily defeat the spring by pressing the pump towards the plate and wedging it in this position with a length of matchstick on a piece of string. Feed the string out through the breather hole (photo).
12 Check the condition of the sealing ring on the tank and renew it if necessary.
13 Offer the pump/sender unit to the tank and work it into position. Fit and tighten the securing nuts.
14 Pull the piece of string to release the matchstick. The spring will force the pick-up screen to the bottom of the tank. Withdraw the string and matchstick through the breather hole. (No harm will result if the matchstick is lost.)
15 On models with an auxiliary tank, refit and secure the blanking strut to the breather pipe.

16 Reconnect the electrical lead, not forgetting the earth tag.
17 Reconnect the fuel hoses and (when applicable) the breather hose (photo).
18 Refit the auxiliary tank or the access hatch.
19 Reconnect the battery.

14 Main fuel pump – removal and refitting

Fuel injection models

1 Raise the vehicle on ramps or drive it over a pit.
2 Disconnect the battery negative lead.
3 Unbolt the fuel pump cradle from the underside of the vehicle (photo). Pull the cradle off the grommets.
4 Disconnect the electrical leads from the pump, noting the colours of the wires and the corresponding terminals (photo).
5 Disconnect the fuel supply and outlet pipes from the pump. Be prepared for fuel spillage. Plug or cap the open pipe unions.
6 Unbolt the pump brackets and remove the pump.
7 Refit by reversing the removal operations, using new sealing washers as necessary. Run the engine and check for leakage before lowering the vehicle.

Carburettor models

8 Disconnect the battery negative lead.
9 Clean around the hose unions on the pump, then slacken the hose clamps and disconnect the hoses (photo). Be prepared for fuel spillage. Plug the tank hose with a bolt or metal rod.

13.7 Removing the tank pump

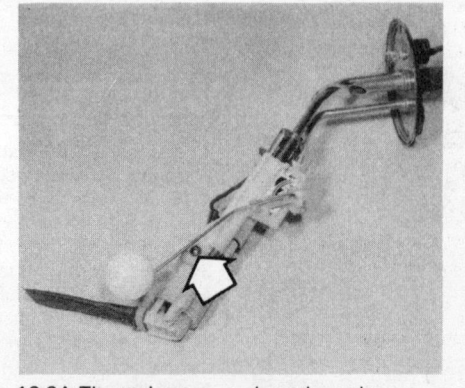

13.8A The tank pump and sender unit. Pump clamp screw arrowed.

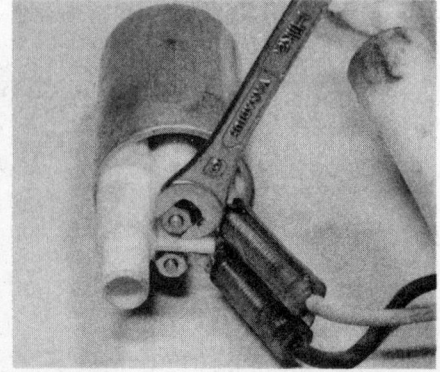

13.8B Disconnecting a lead from the pump

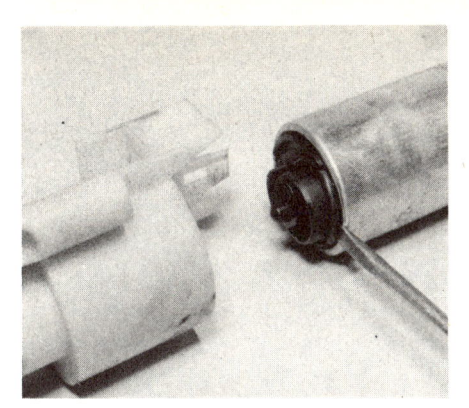

Chapter 3 Fuel and exhaust systems

13.10 The screwdriver shows the pump O-ring

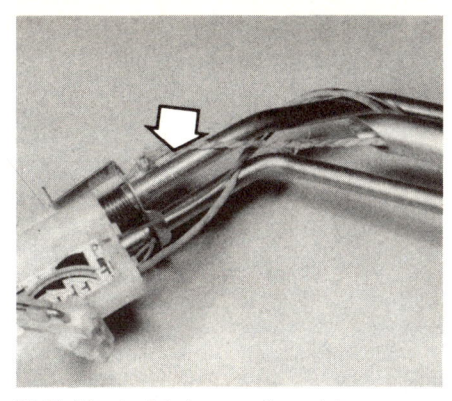

13.11 Matchstick (arrowed) wedging pump in raised position. String goes through breather tube

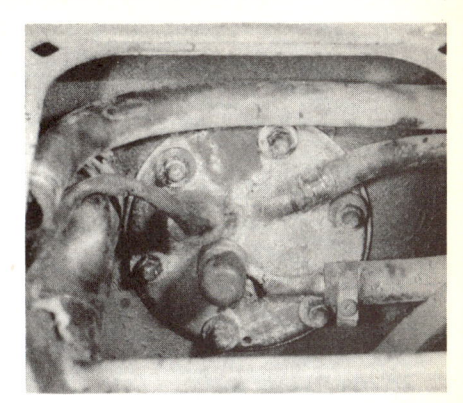

13.17 Fuel hoses and breather blanking stub reconnected

14.3 Unbolting the fuel pump cradle

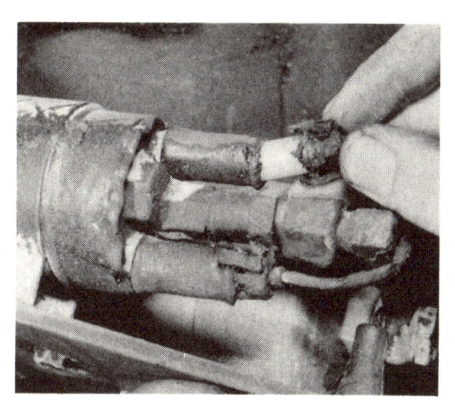

14.4 Disconnecting a fuel pump lead

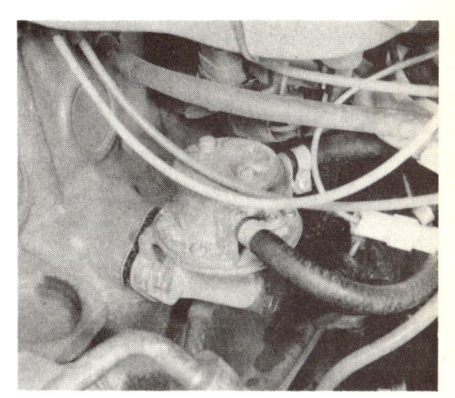

14.9 Mechanical fuel pump with hoses connected

10 Unbolt the fuel pump from the block and remove it. Recover the gasket and any spacers.
11 The pump cover may be removed for cleaning of the filter screen if wished. Further dismantling should not be attempted unless a repair kit can be obtained. Various makes of pump may be fitted.
12 Refit by reversing the removal operations, using a new gasket.
13 Run the engine and check for leakage.

15 Fuel gauge senders – removal, testing and refitting

Main tank sender unit
1 The removal and refitting procedure is as given in Section 13 for the fuel tank pump.
2 To test the sender unit, connect an ohmmeter between the black and the grey/white wires. Move the float up and down and verify that the resistance changes with float position as shown in Fig. 3.11.
3 A defective sender unit must be renewed.

Auxiliary tank sender unit
4 Disconnect the battery negative lead. Observe strict fire precautions throughout the operation.
5 Unclip and remove the boot lining.
6 Disconnect the leads from the top of the sender unit (photo).
7 Using a large screwdriver or tyre lever, release the sender unit from the auxiliary tank by twisting it anti-clockwise. Remove the sender unit (photo).
8 Connect an ohmmeter across the sender unit terminals and check that the resistance varies smoothly with float movement between 1 and 18 Ω.
9 Refit by reversing the removal operations, using a new sealing ring if necessary.

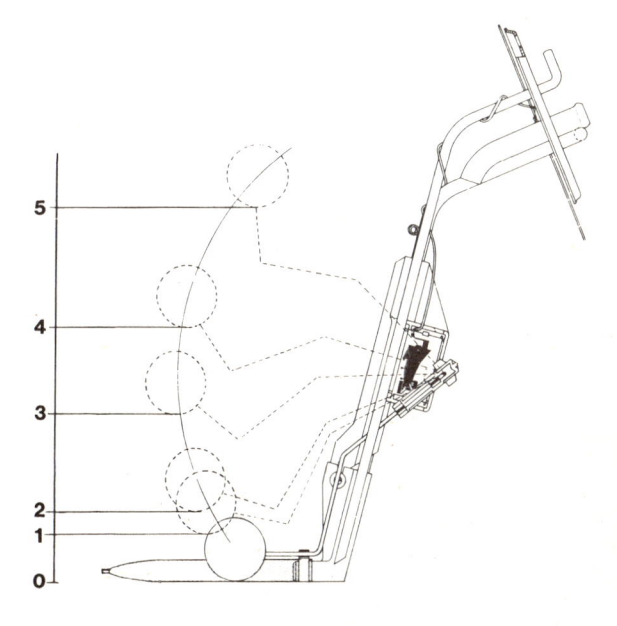

Fig. 3.11 Fuel gauge sender position/resistance characteristic (Sec 15)

Position	Resistance (approx)
0	296 Ω
1	196 Ω
2	145 Ω
3	98 Ω
4	68 Ω
5	36 Ω

15.6 Auxiliary tank sender unit with wires connected

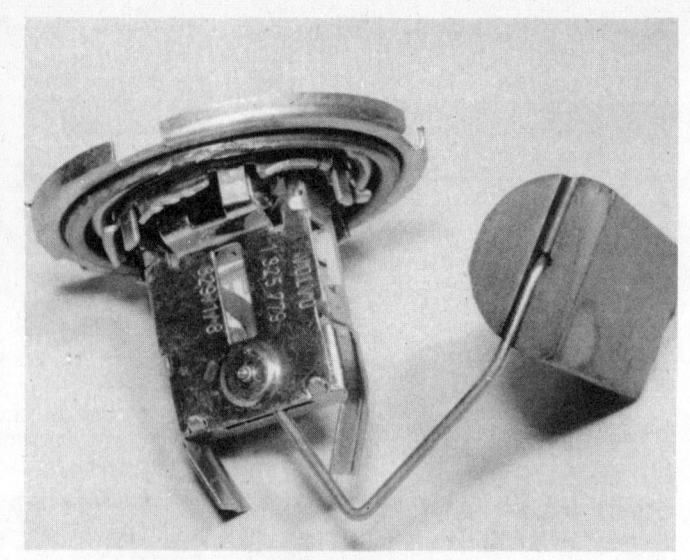

15.7 Auxiliary tank sender unit removed

16 Fuel gauge – testing

1 If both the fuel gauge and the temperature gauge are malfunctioning, the fault is probably in the instrument voltage stabiliser on the instrument panel printed circuit.
2 Connect a resistor, of value 68 Ω, between the fuel gauge connector near the tank and earth. (The wire leading to this connector is coloured white and green.) Switch on the ignition: the gauge should show approximately three-quarters full. If not, the fault lies in the gauge, its power supply or the wiring.
3 Testing of the sender units is covered in Section 15. Note that when two sender units are fitted, they are connected in series, so if either one fails the gauge will read zero.

17 Throttle cable – removal, refitting and adjustment

1 Release the cable outer by extracting the spring clip and unhook the inner from the drum (photos).
2 Inside the vehicle, remove the trim from below the steering column. Pull the cable inner through the end of the pedal and slide the split bush off the end of the cable (photo).
3 Release the cable grommet from the bulkhead and pull the cable into the engine bay. Note the routing of the cable, release it from any clips or ties and remove it.

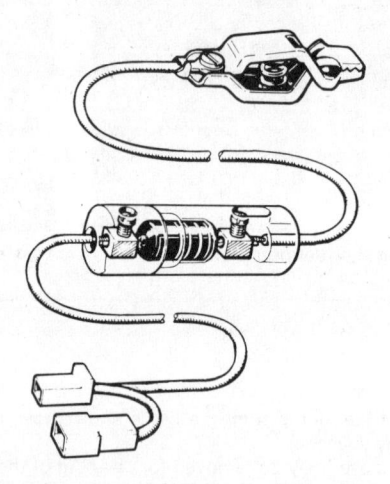

Fig. 3.12 Fuel gauge test lead and resistor (Sec 16)

4 Refit by reversing the removal operations, then adjust the cable as follows.
5 Disconnect the link rod which joins the cable down to the throttle valve(s) by levering off a balljoint (photo).
6 With the throttle pedal released, the cable inner should be just taut,

17.1A Remove the spring clip ...

17.1B ... and unhook the cable inner from the drum

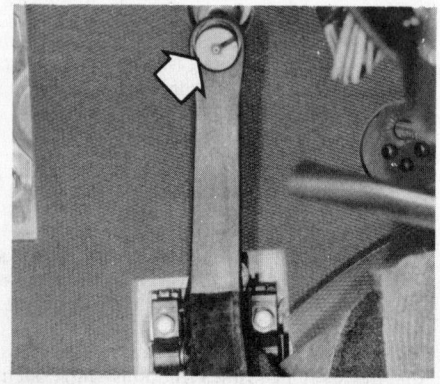

17.2 Split bush (arrowed) secures cable at pedal end

Chapter 3 Fuel and exhaust systems

and the cable drum must be resting against the idle stop. With the pedal fully depressed the drum must contact the full throttle stop. Adjust if necessary by means of the threaded sleeve.

7 On automatic transmission models, check the adjustment of the kickdown cable (Chapter 6, Section 31).

8 Reconnect the link rod, adjusting its length if necessary as follows.

B28E

9 The length of the link rod should be such that neither the cable drum nor the throttle valves are disturbed from their idle positions.

B23/230E/ET

10 Reconnect the link rod and place a 1 mm (0.040 in) feeler blade between the cable drum and the idle stop. In this position the clearance between the throttle lever and the adjustment screw must be 0.1 mm (0.004 in). Adjust the link rod (**not** the adjustment screw) if necessary to achieve this.

B230K

11 Make sure that the fast idle screw is on the lowest part of the cam (in the normal idling position). In this position the throttle valve lever should contact the stop screw.

12 Reconnect the link rod: the cable drum should be held off the idle stop by 0.5 to 1.0 mm (0.020 to 0.040 in). Adjust the link rod if necessary to achieve this.

18 Throttle pedal – removal and refitting

1 Remove the trim from below the steering column.
2 Depress the pedal fully. Grip the cable inner with pliers and release the pedal. Separate the cable inner from the split bush.
3 Remove the pedal bracket bolts and remove the pedal and bracket.
4 Refit by reversing the removal operations. Check the cable adjustment on completion (Section 17).

19 Exhaust system – inspection and repair

1 Periodically inspect the exhaust system for freedom from corrosion and security of mountings. Large holes will be obvious; small holes may be found more easily by letting the engine idle and partly obstructing the tailpipe with a wad of cloth.

2 Check the condition of the rubber mountings by applying

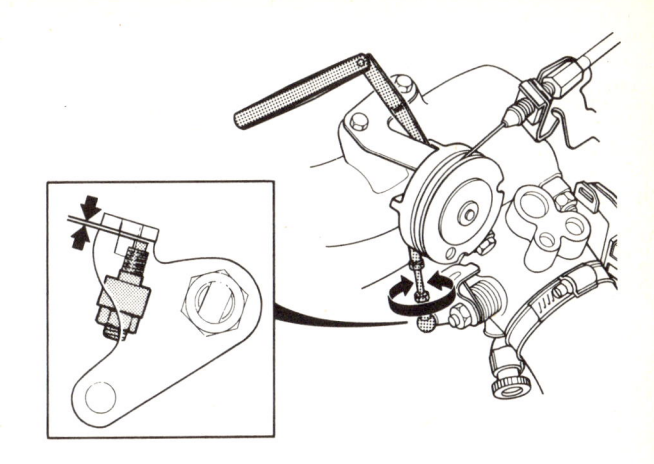

Fig. 3.13 Throttle link rod adjustment – B23/230 E/ET. For lever-to-screw clearance (inset) see text (Sec 17)

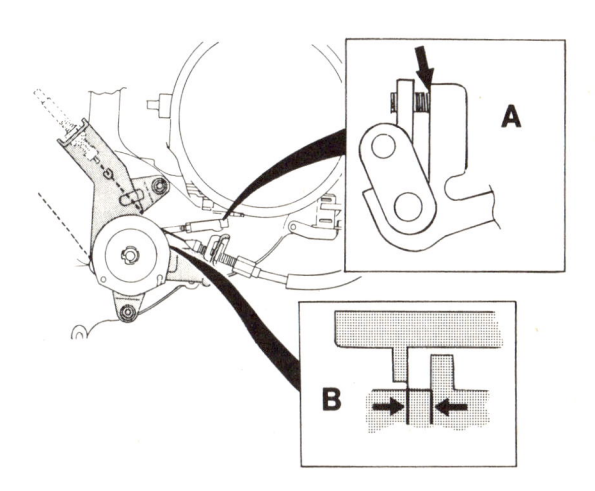

Fig. 3.14 Throttle link rod adjustment – B 230 K (Sec 17)

A Throttle valve lever contacts screw
B Gap between cable drum and idle stop

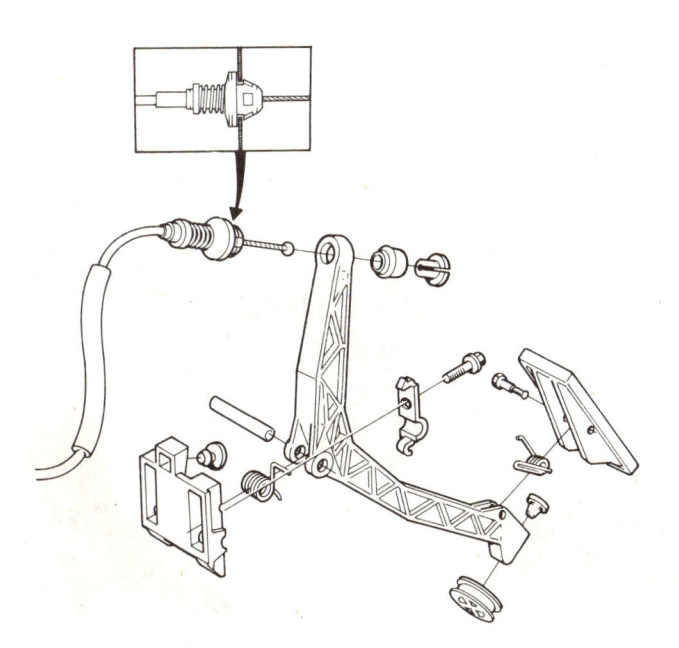

Fig. 3.15 Throttle pedal and associated components (Sec 18)

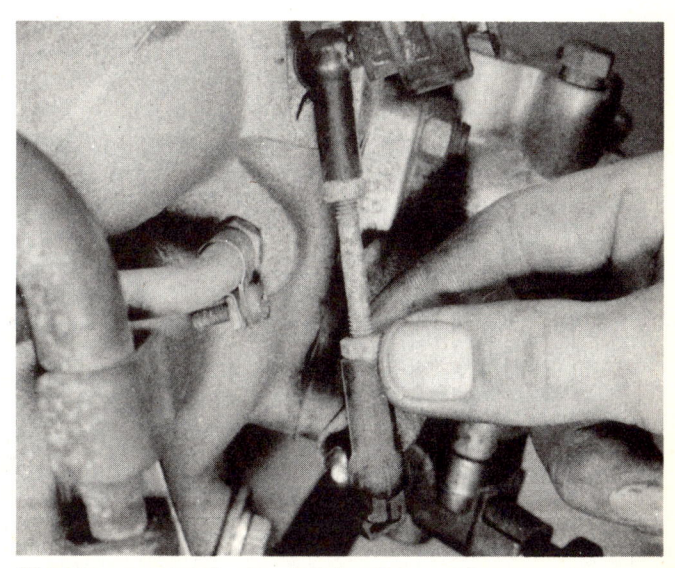

17.5 Disconnecting the throttle link rod

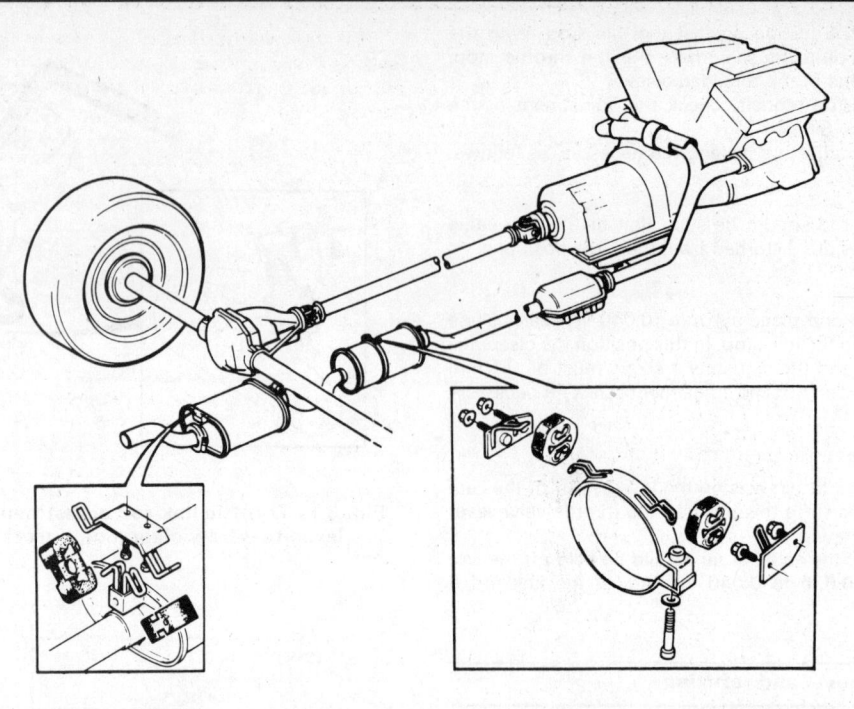

Fig. 3.16 Exhaust system mountings – B 28 E. On most models the system passes above the rear axle (Sec 20)

downward pressure on the exhaust system and observing the mountings for splits or cracks. Renew deteriorated mountings (photo).
3 Small holes or splits in the exhaust system may be repaired using a proprietary compound such as Holts Flexiwrap and Holts Gun Gum. Holts Flexiwrap is an MOT-approved permanent exhaust repair. For best results the section in question should be renewed.

20 Exhaust system – removal and refitting

1 Details of exhaust system routing and mounting will vary with model and year, but the principles of removal and refitting remain the same.
2 In many cases it will be found easier to remove the complete system from the downpipe(s) rearwards and then to renew individual sections on the bench. One exception is on models where the system passes over the rear axle; here it is better to separate the joints, or to cut the pipe if it is rusty anyway.
3 To remove the complete system, raise and support the vehicle at a convenient working height. Apply penetrating oil to the nuts, bolts and clamps which will have to be undone.

4 Unbolt the flanged joint at the union of the exhaust system with the downpipe(s) (photo).
5 If the system passes over the rear axle, remove one of the U-pipe clamps and separate the system there (photo).
6 With the aid of an assistant, unhook the system from its mountings and remove it.
7 To remove the downpipe(s), release the mounting clamp from the bellhousing and separate the joints from the manifold(s) or turbocharger. Also disconnect the hot air trunking, and if necessary unbolt the hot air shroud (photo).
8 Commence refitting with the downpipe(s), using new gaskets. Apply anti-seize compound to the threads. Fit the bellhousing mounting clamp bolt but do not tighten it yet.
9 Sling the rest of the system on its mountings and couple it up, using a new sealing ring at the flanged joint. Apply exhaust jointing compound to the sliding joints and anti-seize compound to all threads.
10 Tighten all the joints from the front rearwards, but leave the bellhousing clamp loose until everything else has been tightened. (On models with the B23ET engine, follow the tightening sequence shown in Fig. 3.17). Twist the sliding joints slightly if necessary so that the system hangs easily and without touching the body.

19.2 An exhaust system rubber mounting

20.4 Exhaust flanged joint

20.5 Exhaust U-pipe clamp

Chapter 3 Fuel and exhaust systems

11 Run the engine for a few minutes and check the system for leaks. Allow it to cool and retighten the joints.
12 Lower the vehicle.

21 Cruise control – general

When fitted, the cruise control allows the vehicle to maintain a steady speed selected by the driver, regardless of gradients or prevailing winds.

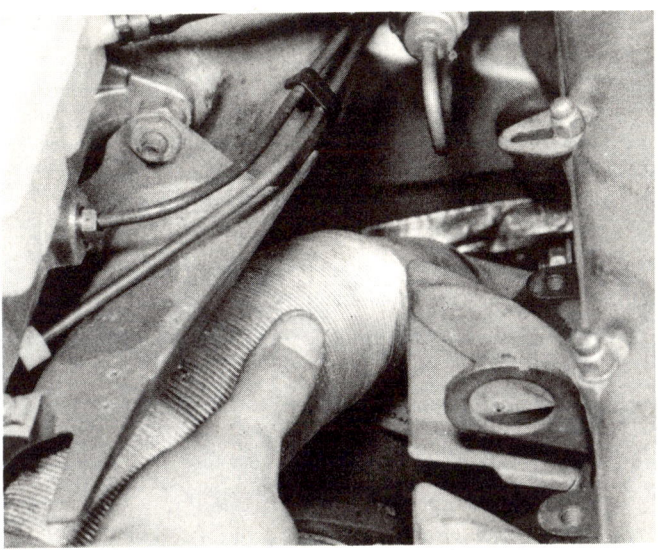

20.7 Disconnecting the hot air trunking – Turbo shown

The main components of the system are a control unit, a control switch, a vacuum servo and a vacuum pump. Brake and (when applicable) clutch pedal switches protect the engine against excessive speeds or loads should a pedal be depressed whilst the system is in use.

In operation, the driver accelerates to the desired speed and then brings the system into use by means of the switch. The control unit then monitors vehicle speed (from the speedometer pulses) and opens or closes the throttle by means of the servo to maintain the set speed. If the switch is moved to 'OFF', or the brake or clutch pedal is depressed, the servo immediately closes the throttle. The set speed is stored in the control unit memory and the system can be reactivated by moving the switch to 'RESUME', provided that vehicle speed has not dropped below 25 mph (40 km/h).

The driver can override the cruise control for overtaking simply by depressing the throttle pedal. When the pedal is released, the set speed will be resumed.

The cruise control cannot be engaged at speeds below 25 mph (40 km/h), and should not be used in slippery or congested conditions.

No specific removal, refitting or adjustment procedures were available at the time of writing. Problems should be referred to a Volvo dealer or other specialist.

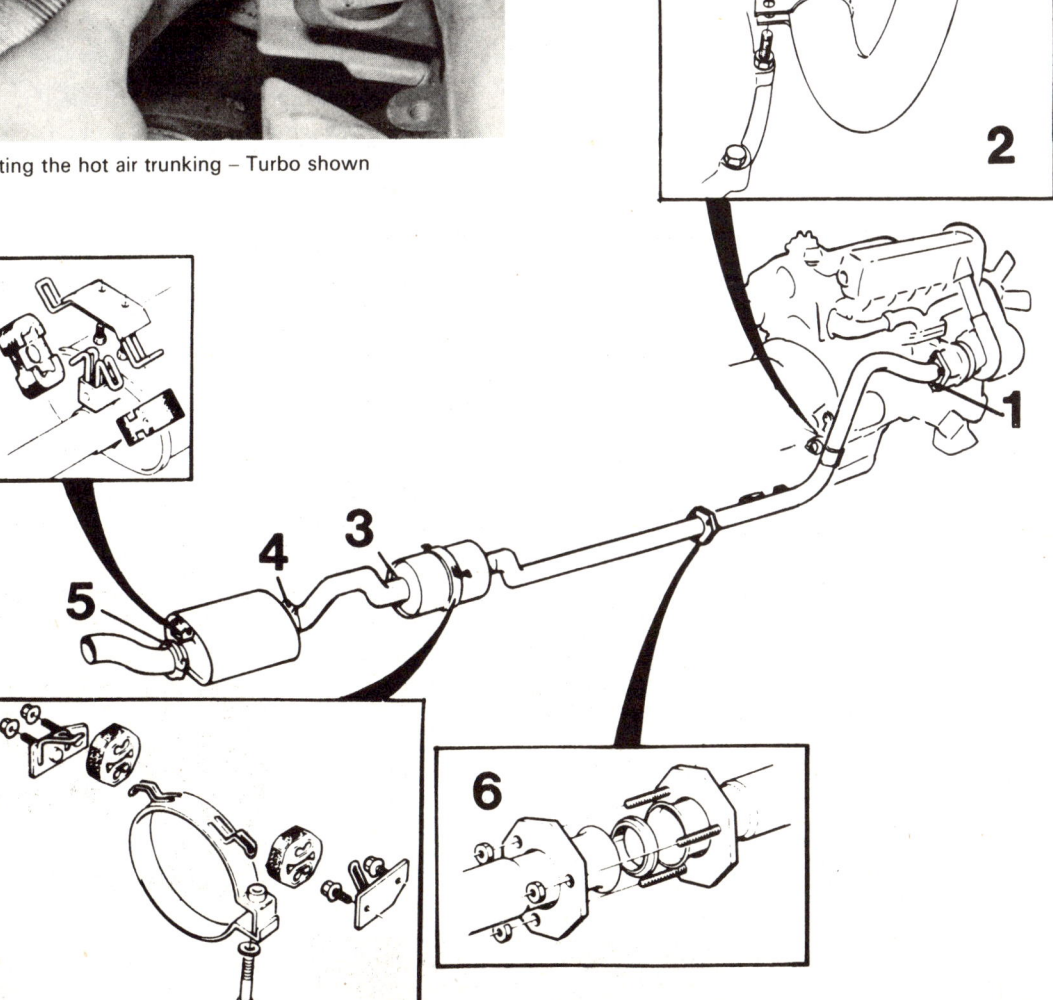

Fig. 3.17 Exhaust system details – B 23 ET. Tighten in numerical sequence (Sec 20)

112   Chapter 3 Fuel and exhaust systems

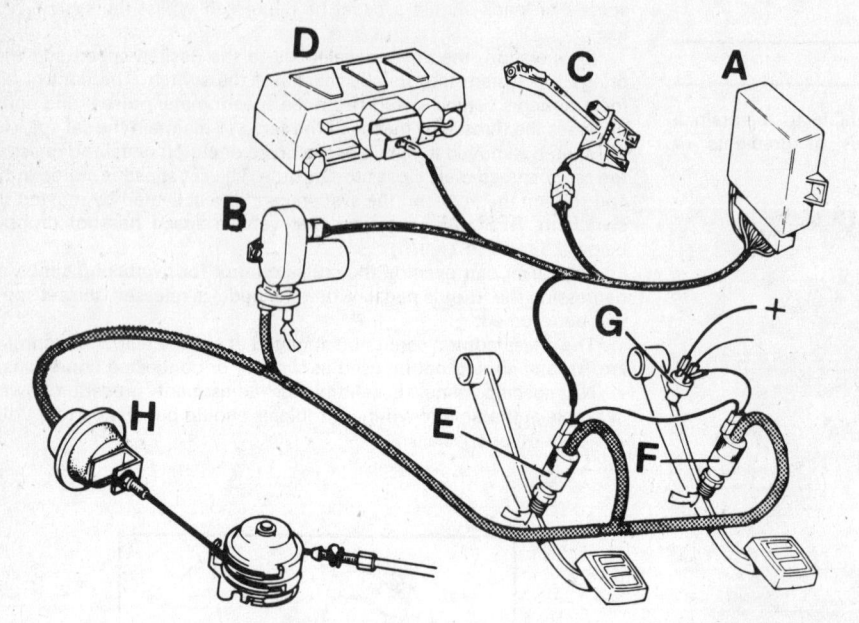

Fig. 3.18 Cruise control components (Sec 21)

A Control unit
B Vacuum pump
C Control switch
D Speedometer
E Clutch pedal switch/valve
F Brake pedal switch/valve
G Stop-light switch
H Vacuum servo

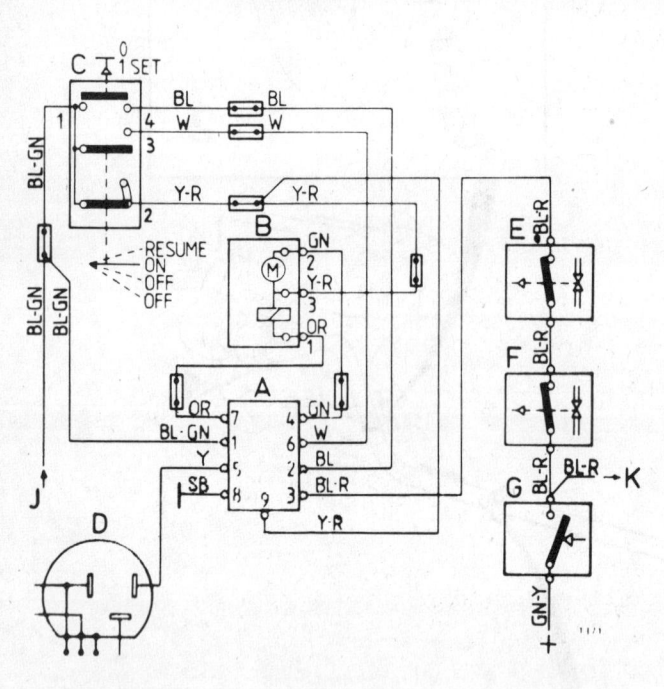

Fig. 3.19 Cruise control wiring diagram. For A to H see Fig. 3.18 (Sec 21)

J   From fuse No 11
K   To stop-lights
BL Blue
BN Brown
GN Green
OR Orange
R   Red
SB Black
W   White
Y   Yellow

3 Disconnect the vacuum hoses from the manifold, making identifying marks if necessary. Clamp and disconnect coolant hoses also.
4 Unbolt and remove the manifold. Recover the gasket.

In-line engine (fuel injection)

5 If the manifold is coolant-heated, drain the cooling system.
6 Disconnect the air inlet trunking from the throttle housing.
7 Disconnect the control cable(s) from the throttle drum.
8 Disconnect the injector multi-plugs (Turbo models) and any other electrical services obstructing removal.
9 Disconnect vacuum, pressure, breather and coolant hoses from the manifold, making identifying marks if necessary (photo).
10 Disconnect the fuel feed and return pipes from the fuel rail or pressure regulator. Be prepared for fuel spillage.
11 Disconnect or move aside the cold start injector and the auxiliary air valve (as applicable).
12 Check that nothing has been overlooked, then unbolt and remove the manifold complete with throttle housing and (on Turbo models) injection equipment (photos). Recover the gasket.

V6 engine

13 Slacken the fuel tank cap to release any residual pressure.
14 Remove the oil filler cap, the crankcase ventilation hoses and the air intake trunking.
15 Disconnect the control cables attached to the throttle drum and move them out of the way.
16 Disconnect the injection wiring harness at the multi-plugs next to the expansion tank, and from the injection system components on and around the manifold. Move the harness out of the way.
17 Disconnect the vacuum hoses from the manifold, making identifying marks if necessary (photos).
18 Disconnect the HT leads from the spark plugs, unclip the leads and move them aside.
19 Release the fuel injector retaining clips and remove the injectors from their bores.
20 Disconnect the fuel feed line from the top of the fuel filter, and the fuel return line from the union on the left-hand inner wing. Be prepared for fuel spillage; plug or cap open lines.
21 Unbolt the control pressure regulator and place it on the manifold.
22 Remove the four bolts which secure the manifold. Lift off the manifold complete with the throttle housing and fuel injection components. Recover the O-rings (photos).

All models

23 Refit by reversing the removal operations, using new gaskets or O-rings. It may be necessary to cut the gasket to clear adjacent components (photo).

22 Inlet manifold – removal and refitting

1 Disconnect the battery negative lead.

In-line engine (carburettor)

2 Remove the carburettor (Section 27), or disconnect all services from it but leave it attached to the manifold.

Chapter 3 Fuel and exhaust systems

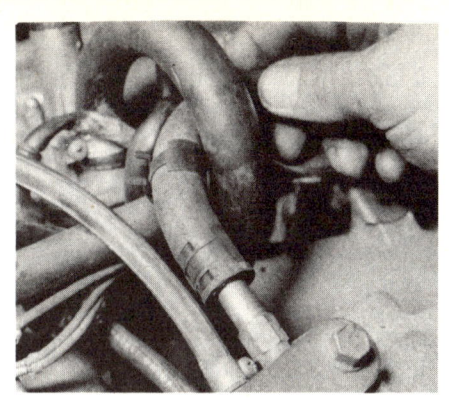

22.9 Disconnecting the manifold vacuum hoses – Turbo shown

22.12A Remove the bracing strut top bolt (arrowed) ...

22.12B ... and the manifold-to-head nuts

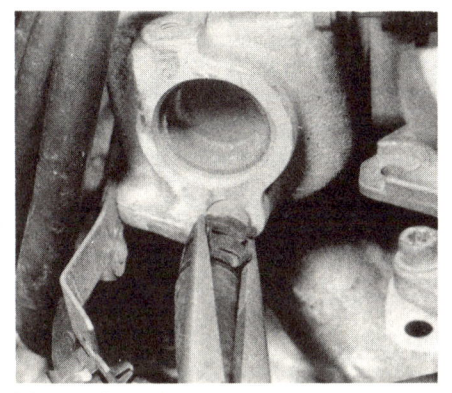

22.17A V6 manifold removal: disconnecting a large vacuum hose ...

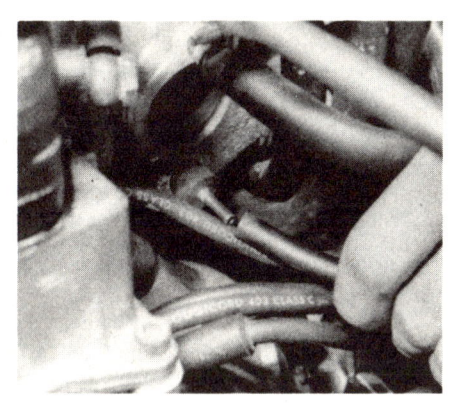

22.17B ... and a small one

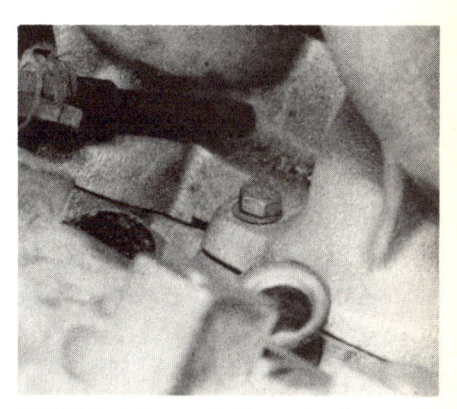

22.22A An inlet manifold securing bolt ...

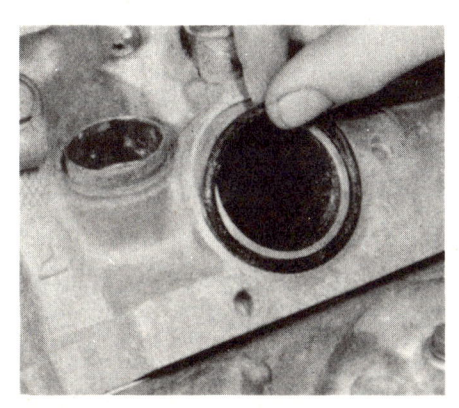

22.22B ... and an O-ring

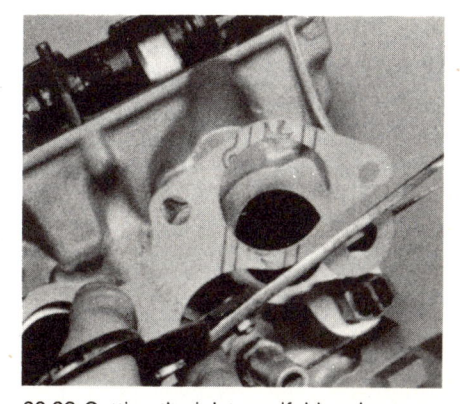

22.23 Cutting the inlet manifold gasket to clear the thermal timer

24 Adjust the throttle drum cable(s), run the engine and check the idle speed and mixture on completion.
25 Top up the cooling system if necessary.

23 Exhaust manifold(s) – removal and refitting

In-line engines (except Turbo)
1 Remove the hot air trunking (when so equipped).
2 Disconnect the exhaust downpipe from the manifold.
3 Unbolt the manifold from the cylinder head and remove it. Recover the gaskets.
4 When refitting, use new gaskets. The marking 'UT' must face away from the cylinder head.

5 Apply anti-seize compound to the threads. Fit the manifold to the head and tighten the nuts evenly.
6 Reconnect the exhaust downpipe and the hot air trunking.
7 Run the engine and check for leaks.

In-line engines (Turbo)
8 See Section 39.

V6 engines
9 Raise and support the front of the vehicle. Disconnect the battery negative lead.
10 Disconnect the exhaust downpipes from the manifold flanges (photo). (Both must be disconnected even if only one manifold is to be removed.)
11 Disconnect the hot air trunking if it is in the way.

114 Chapter 3 Fuel and exhaust systems

23.10 An exhaust manifold-to-downpipe flanged joint

23.15 Exhaust manifold gaskets (head removed)

23.16 Fitting an exhaust manifold (head removed)

12 Disconnect the exhaust system front mounting from the transmission. Move the exhaust system rearwards until the pipes are clear of the manifolds. Support the exhaust system if necessary so that it is not strained.
13 Remove the securing nuts and take off the manifolds. Recover the gaskets; obtain new ones for use when refitting.
14 Manifold gaskets are supplied in packets of three. Separate the gasket sections by cutting, **not** by folding or tearing.
15 Fit the gasket sections over the cylinder head studs, with the reinforced metallic edge facing the head (photo).
16 Fit the manifolds over the studs (photo). Secure them with the nuts; use a little anti-seize compound on the threads. Tighten the nuts evenly.
17 Fit the exhaust pipes to the manifold, using new gaskets, metallic edges facing the manifolds. Fit and tighten the nuts, again using anti-seize compound.
18 Refit the exhaust system front mounting.
19 Refit the hot air trunking.
20 Lower the vehicle and reconnect the battery.
21 Run the engine and check for leaks.

24 Fault diagnosis – fuel system (general)

Excessive fuel consumption, poor performance or difficult starting are not necessarily caused by fuel system faults. Ignition system faults and engine maladjustment or wear may also cause such symptoms. Simple causes of high consumption are under-inflated tyres, binding brakes, an unsympathetic driving style or unfavourable conditions of use. A clogged air cleaner element or fuel filter can also adversely affect performance and consumption.

The specific fault diagnosis Sections later in this Chapter assume that the above points have been checked, and that the fuel used is of good quality and of the specified octane rating.

25 Carburettor – description

The Pierburg 2B5 carburettor fitted to the B230K engine is a twin barrel, fixed jet, downdraught instrument. The choke (cold start device) is semi-automatic, being brought into operation by depressing and releasing the throttle pedal before starting.

Operation of the two barrels is sequential. At idle and light running only the primary barrel is used. Opening of the throttle valve in the secondary barrel is controlled by vacuum developed in the primary barrel, with a mechanical interlock preventing opening of the secondary throttle valve until the primary valve is at least half open. This system gives better performance from low speeds than a simple mechanical linkage.

Choke operation after start-up is controlled by a bi-metallic spring, which gradually opens the choke flap as it is heated by an electric element and by circulated coolant. Vacuum and mechanical devices

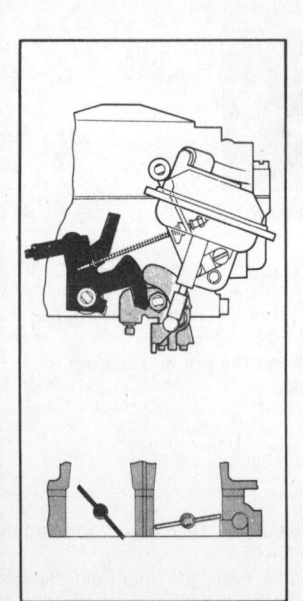

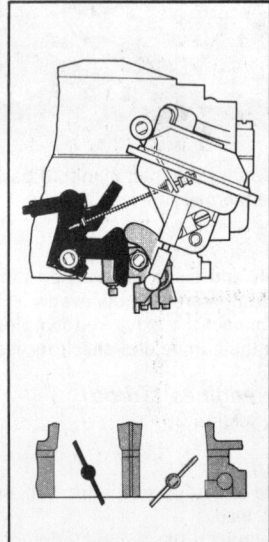

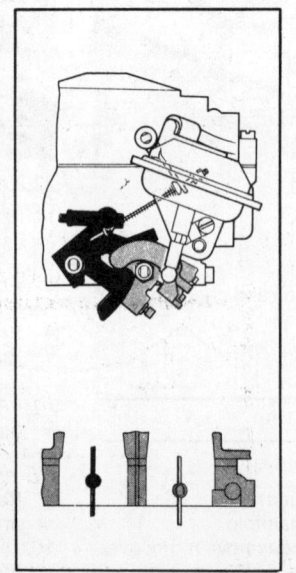

Fig. 3.20 Sequential operation of carburettor throttle valves (Sec 25)

Chapter 3 Fuel and exhaust systems

influence the degree of opening of the flap during the warm-up phase to avoid over-enrichment of the mixture.

When the engine is running, the float chamber is vented into the carburettor air inlet. This ensures that restriction of the air intake (due for instance to a blocked air cleaner element) affects both the air and the fuel systems equally and does not cause over-enrichment. When the ignition is switched off, a hot start valve vents the float chamber to atmosphere, so avoiding a build-up of fumes in the air inlet which could make hot starting difficult.

A fuel shut-off system improves fuel economy by cutting off the fuel supply when the throttle is released, the engine is warm and engine speed exceeds 1610 rpm. The system uses a vacuum valve, a solenoid valve and a throttle switch, with overall control from the ignition control unit. The solenoid and vacuum valves are also used to prevent run-on (dieseling) after switching off the ignition.

Twin floats and float chambers maintain a more or less constant fuel level, even on gradients or when cornering hard. A fuel return system is used to keep fuel temperature steady and avoid vapour locks.

The carburettor seen in the photographs in this Chapter is in fact a development of the 2B5 known as a 2B7. Apart from a three-stage choke vacuum unit and the absence of a fuel return connection, it is the same as the 2B5.

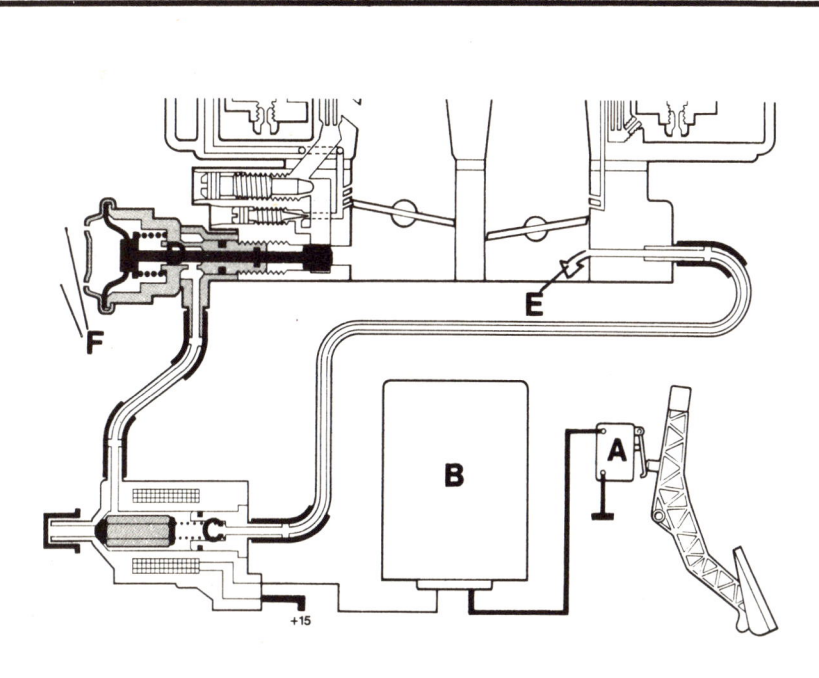

Fig. 3.21 Operation of fuel shut-off system (Sec 25)

A Throttle switch C Solenoid valve E Manifold vacuum
B Control unit   D Vacuum valve   F Atmospheric pressure

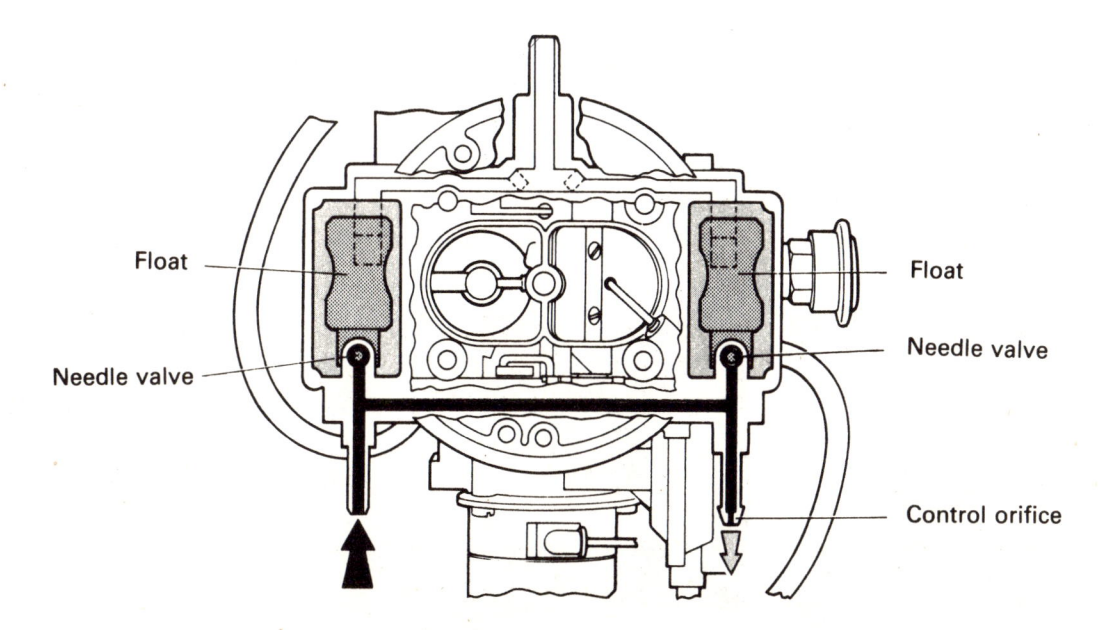

Fig. 3.22 Carburettor fuel supply and return system (Sec 25)

26 Carburettor – adjustments on the vehicle

Apart from normal idle adjustment (Section 8), the only other adjustment performed on the vehicle is to the fast idle. Proceed as follows.

1 Disconnect the air inlet trunking from the top of the carburettor.
2 Separate the cable drum-to-carburettor link rod by prising off its balljoint.
3 Open the primary throttle valve fully by hand. Turn the choke flap so that the fast idle adjustment screw is resting on the sixth (highest but one) step of the fast idle cam. Release the throttle valve.
4 In this position measure the fast idle gap between the throttle valve lever and the stop screw. If the gap is not as specified, turn the fast idle adjustment screw to correct it.
5 When the gap is correct, refit the link rod and the air trunking.
6 Check the idle speed and mixture (Section 8).

27 Carburettor – removal and refitting

1 Disconnect the battery negative lead.
2 Remove the air inlet trunking from the top of the carburettor by undoing the knurled nut and disconnecting the breather and vacuum pipes. Recover the O-ring (photos).
3 Disconnect the throttle link rod.
4 Disconnect the carburettor electrical feed (photo).
5 Disconnect the vacuum hoses from the carburettor, making identifying marks if necessary. (Coloured washers, corresponding to the vacuum hose colours, may be found on the connecting stubs.)
6 Slacken the expansion tank cap to release residual pressure, taking precautions against scalding if the coolant is hot. Clamp the coolant hoses which feed the choke and disconnect them (photo). Be prepared for coolant spillage.
7 Disconnect the fuel supply and return hoses. Be prepared for fuel spillage. Plug the hoses to keep dirt out.

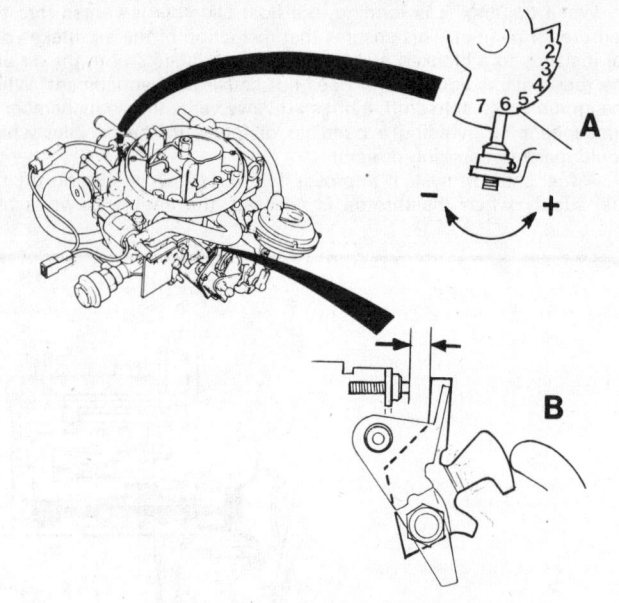

Fig. 3.23 Fast idle adjustment (Sec 26)

A Adjustment screw on 6th step   B Fast idle gap

8 Remove the four Allen screws which secure the carburettor to the manifold (photo). Lift off the carburettor and recover the spacer and gaskets.
9 Refit by reversing the removal operations, using new gaskets and (if necessary) a new inlet trunking O-ring.

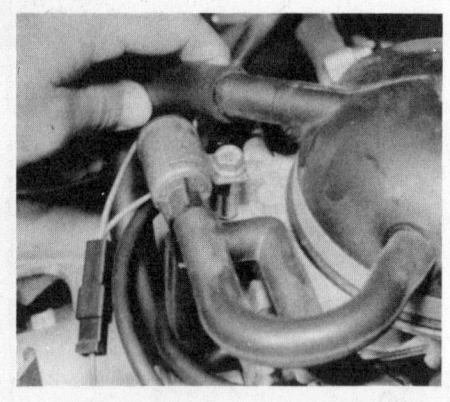

27.2A Disconnecting a carburettor breather pipe

27.2B Undoing the knurled nut

27.2C Carburettor inlet O-ring

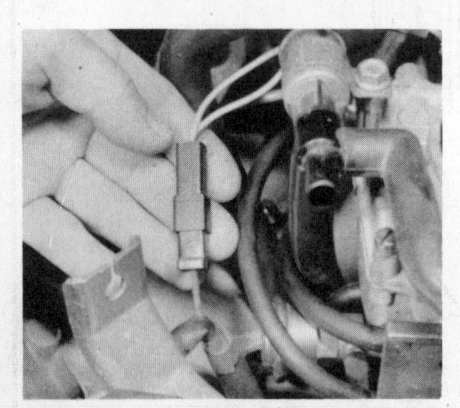

27.4 Carburettor electrical connector

27.6 The automatic choke coolant hoses. Fuel supply hose is in foreground

27.8 Removing a carburettor securing screw

Chapter 3 Fuel and exhaust systems

10 Check the fast idle setting (Section 26) and adjust the idle speed and mixture (Section 8).
11 Top up the cooling system if necessary.

28 Carburettor – overhaul

The operations described here should be regarded as the limit of possible overhaul. It may be more satisfactory to renew a well worn carburettor

1 With the carburettor removed from the vehicle, empty the petrol from it. Clean the carburettor externally using paraffin and an old toothbrush, then wipe it dry.
2 Operate the throttle linkage and inspect the throttle spindles and valves. Renew the carburettor if they are worn or damaged. (The throttle housing cannot be renewed separately because special equipment is needed to match it to the rest of the carburettor.)

Top cover

3 Unbolt the hot start valve. Disconnect its hoses and electrical connector and remove it (photo).
4 Disconnect the secondary throttle vacuum unit hose, noting where it is fitted (photo).
5 Disconnect the accelerator pump link at the bottom end by prising it out (photo).
6 Disconnect the choke link at the top end by unclipping it from the plastic lever (photo).
7 Remove the central stud and the four screws which secure the top cover (photo).
8 Lift off the top cover and recover the gasket.
9 Check the part load enrichment valve spring and piston for freedom of movement in the cover (photo). If the piston sticks, renew the carburettor.
10 Inspect the choke and accelerator pump linkages and renew worn components.

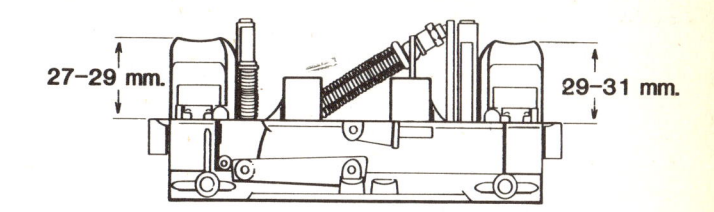

Fig. 3.24 Float height measurement (Sec 28)

27–29 mm.   29–31 mm.

11 Invert the top cover so that the floats are uppermost. Cover the fuel return connection and blow into the fuel supply connection: the valves must not leak. Lift the floats and the valves must open.
12 Still with the cover inverted, measure the float heights (Fig. 3.24). The gasket must be removed and the spring-loaded pins in the needle valves must not be depressed.
13 If a needle valve leaks or a float height is incorrect, renew the valve and float together. They are removed by tapping out the float pivot pin (photos).

Secondary throttle vacuum unit

14 Connect a vacuum pump to the secondary throttle vacuum unit.
15 Open the primary throttle wide and apply vacuum. The secondary throttle must open. Hold the vacuum to see that the unit does not leak, then release it.
16 Renew the vacuum unit if necessary. Separate its link rod from the balljoint in any case (photo).

Throttle housing and linkage

17 Disconnect the remaining hose from the choke vacuum unit, noting where it is connected.
18 Remove the screw from the underside of the throttle housing.

28.3 Hot start valve securing bolt (arrowed)

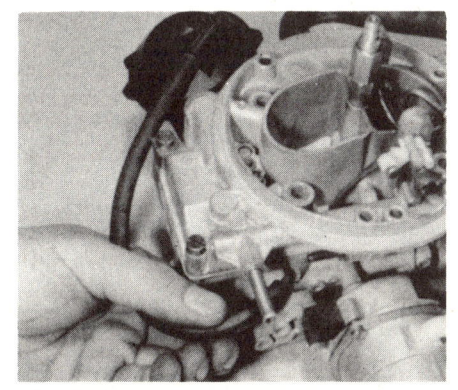

28.4 Disconnecting the secondary throttle unit vacuum hose

28.5 Prise out the accelerator pump link rod (arrowed)

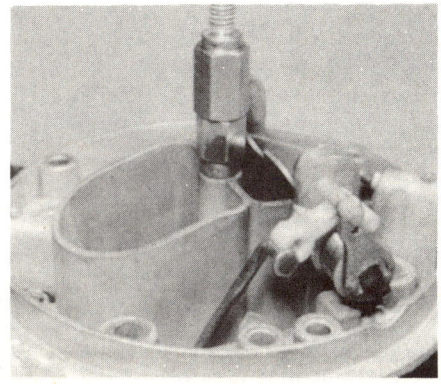

28.6 Choke link rod separated from the plastic lever

28.7 Top cover securing screws and stud (arrowed)

28.9 Part load enrichment valve (arrowed)

Chapter 3 Fuel and exhaust systems

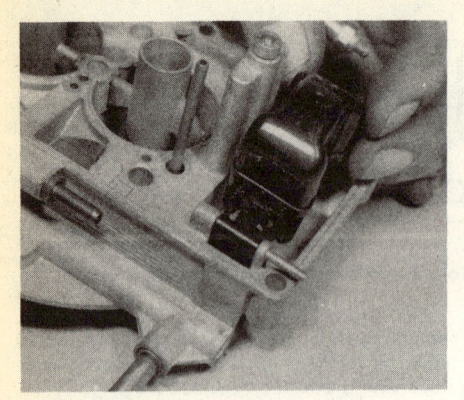

28.13A Removing a float ...

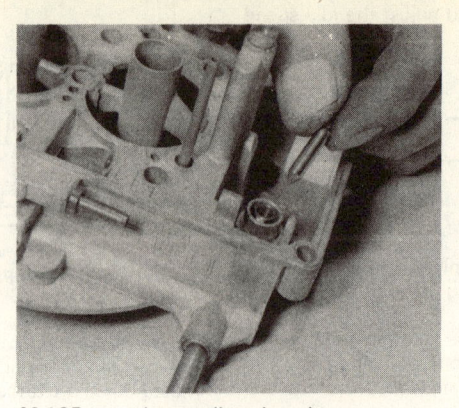

28.13B ... and a needle valve pin

28.16 Separating the secondary throttle unit balljoint

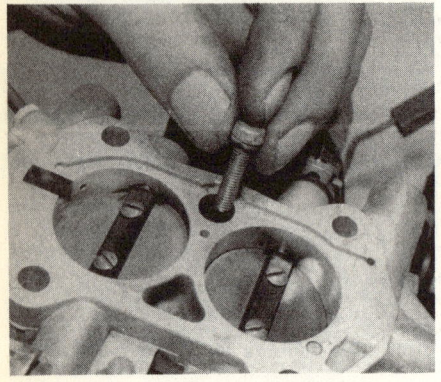

28.18A Removing the throttle housing screw

28.18B Remove the thick gasket

28.19 Throttle linkage – secondary throttle stop screw is arrowed. (Primary stop screw is on carburettor body)

Separate the carburettor body from the housing and recover the thick gasket (photos).
19 Inspect the throttle linkage on the side of the housing. Renew parts as necessary. **Do not** disturb the throttle stop screws (photo).
20 Check the throttle linkage clearance between the base of the plastic lever and the metal fork, pushing the lever one way and then the other to measure gaps 'A' and 'B' (Fig. 3.25). Bend the fork if the gaps are not as specified.

21 Apply vacuum to the fuel shut-off vacuum valve and check that the valve plunger moves in to block the idle mixture passage (photo). Dismantle the valve for inspection if necessary.

Choke components
22 Check for alignment marks between the bi-metallic spring housing and the choke carrier (photo). Make marks if necessary, then remove

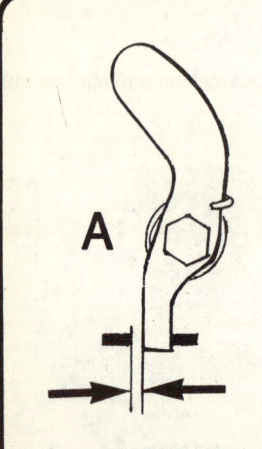

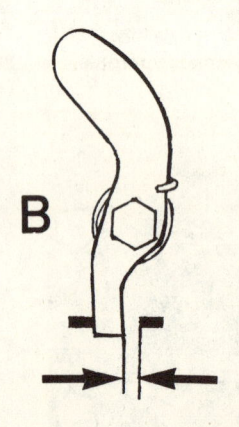

Fig. 3.25 Throttle linkage clearances. For 'A' and 'B' see text (Sec 28)

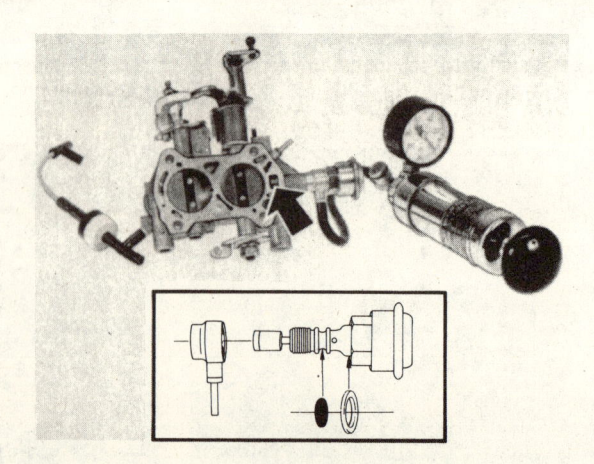

Fig. 3.26 Testing the fuel shut-off vacuum valve. Inset shows valve dismantled (Sec 28)

Chapter 3 Fuel and exhaust systems

28.21 Idle mixture passage (arrowed) and fuel shut-off valve (on right)

28.22 Alignment marks (arrowed) between bi-metallic spring housing and choke carrier

28.25A Choke carrier showing the three securing screws

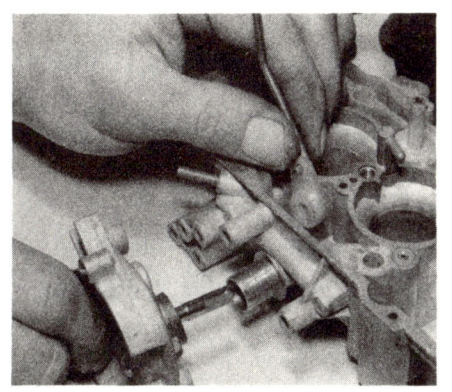

28.25B Removing the choke carrier

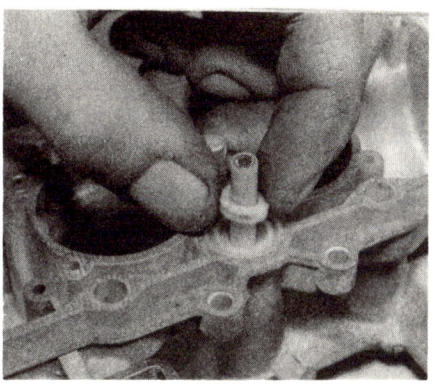

28.27 Accelerator pump felt washer

28.28 Prising up the piston retainer

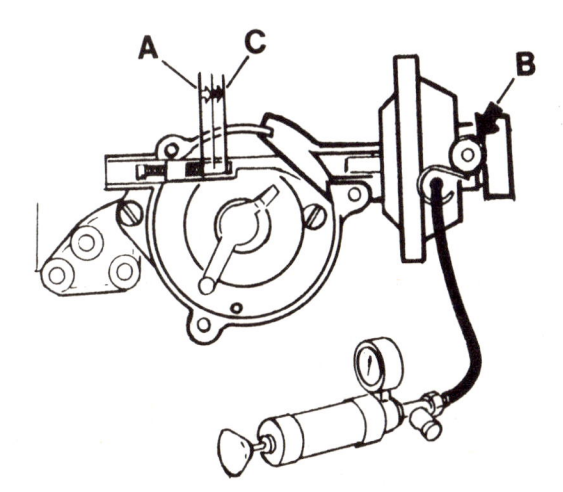

Fig. 3.27 Applying vacuum to the choke vacuum unit (Sec 28)

A First stage movement (upper connection exposed)
B Upper connection
C Second stage movement (upper connection covered)

the three screws which secure the clamp ring. Remove the clamp ring, water jacket and spring housing together. Recover the gasket.
23 The water jacket can be separated from the spring housing if wished after removing the central bolt. Recover the O-ring.
24 Apply vacuum to the lower connection on the choke vacuum unit. Cover the upper connection and apply vacuum until the pullrod contacts the far stop. Hold the vacuum to check for leakage, then release it.

25 Slide the vacuum unit and cover off the choke carrier. Remove the three screws which secure the carrier, noting that the short screw goes behind the vacuum unit pullrod. Prise off the inner plastic link and withdraw the choke carrier with shaft, spring, etc (photos).
26 Examine the choke components and renew worn or damaged items.

Accelerator pump
27 Remove the felt washer from the pump piston rod (photo).
28 Prise up the piston retainer (photo).
29 Lift out the piston and spring (photos).
30 Examine the components and renew as necessary.

Jets
31 Remove the jets and blow through them with compressed air (photo). **Do not** probe them with wire or their calibration will be upset. The air correct jets are pressed in place and cannot be removed.
32 Remove the idle speed and mixture adjustment screws.
33 Blow compressed air through the various channels and orifices, then refit the jets and adjustment screws. Screw the adjustment screws fully home, then back out the speed adjustment screw $2^{1}/_{2}$ turns and the mixture adjustment screw $3^{1}/_{2}$ turns. Final adjustment will be made after refitting.

Reassembly
34 Obtain a repair kit containing new gaskets, seals and other renewable items.
35 Refit the accelerator pump spring, plunger, retainer and felt washer.
36 Refit the throttle housing to the carburettor body, using a new thick gasket. Secure them with the screw.
37 Refit the secondary throttle unit (if removed) and reconnect the vacuum hose and link rod (photo). Set the link rod as short as possible without actually opening the throttle.
38 Refit the choke carrier, shaft and associated components. Slide the choke vacuum unit and cover into place (photo).
39 Refit the top cover, using a new gasket, and reconnect the choke and accelerator pump links (photo).

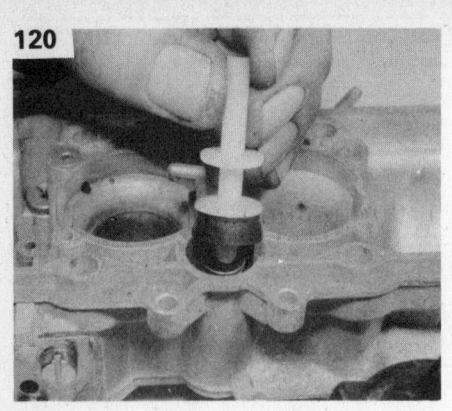

28.29A Removing the accelerator pump piston ...

28.29B ... and spring

28.31 Unscrewing a jet

28.37 Reconnecting the throttle link rod balljoint

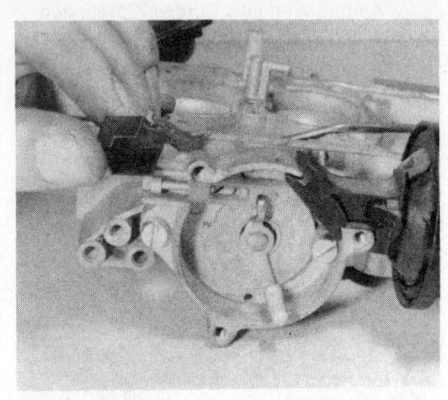

28.38 Fitting the plastic cover to the choke carrier

28.39 Fitting a new top cover gasket

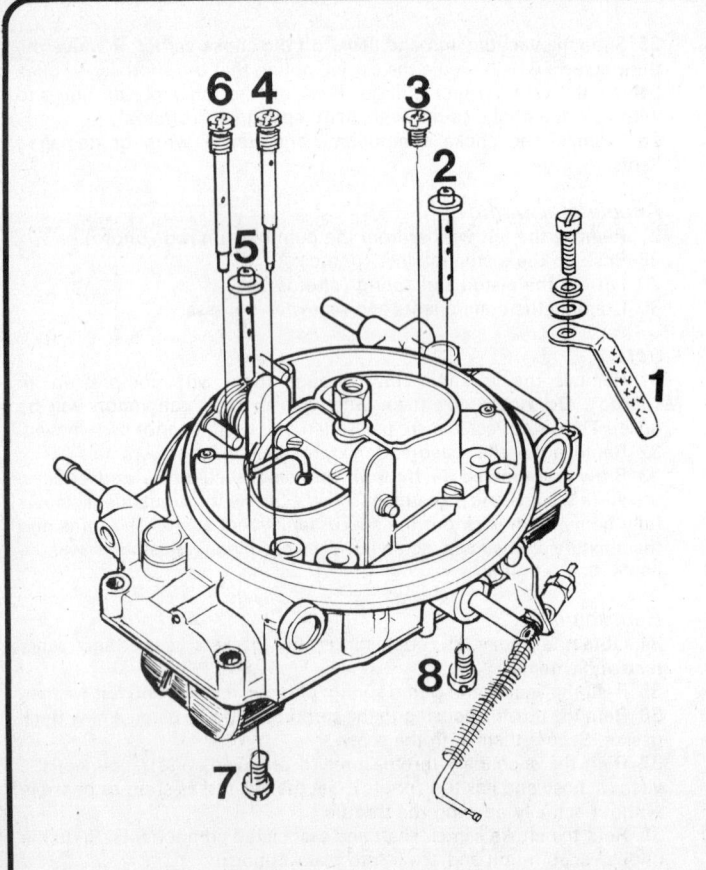

Fig. 3.28 Carburettor jet identification (Sec 28)

1 Identification plate
2 Air correction jet/emulsion tube
3 Air bypass jet
4 Idle air/fuel jet
5 Air correction jet/emulsion tube
6 Auxiliary air/fuel jet
7 Primary main jet
8 Secondary main jet
9 Enrichment jet
10 Bypass jet

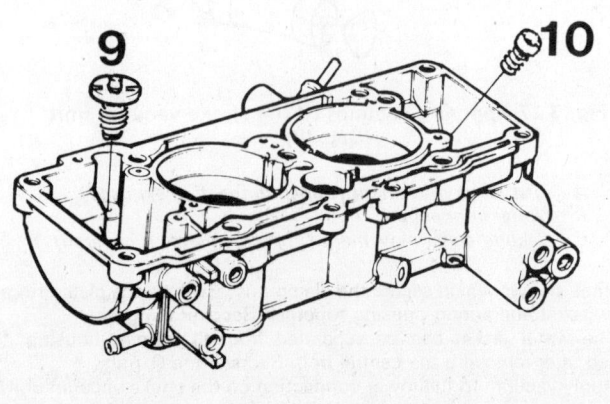

Chapter 3 Fuel and exhaust systems

40 Perform the automatic choke adjustments (Section 29, paragraph 3 onwards), then adjust the accelerator pump as follows.

Accelerator pump delivery

41 Taking appropriate fire precautions, fill the float chambers with petrol via the feed pipe until it starts to emerge from the return pipe.
42 Position the base of the carburettor over a funnel leading to a measuring cylinder. Holding the fast idle cam out of the way, open and close the throttle fully ten times, taking about one second per opening and allowing three seconds between openings. Measure the quantity of fuel delivered.
43 If the quantity of fuel is not as specified, turn the link rod adjusting nut in the appropriate direction and repeat the test.

Final assembly

44 Refit the hot start valve and hoses, and the fuel shut-off valve if it was removed.
45 Lubricate the linkages whilst they are accessible.
46 Reconnect any remaining vacuum hoses and wires.

29 Carburettor – automatic choke adjustment

1 Remove the carburettor (Section 27).
2 Make alignment marks if necessary, then remove the three screws which secure the bi-metallic spring housing clamp ring. Remove the clamp ring, spring housing with water jacket and gasket.
3 If not already done, set the fast idle gap (Section 26).
4 Subsequent adjustments must be done in the order given.
5 Close the choke flap by hooking a rubber band round the choke link and the choke vacuum unit.
6 Open and close the throttle. The fast idle adjustment screw should come to rest on the seventh (highest) step of the fast idle cam. There should be a small clearance (0.5 to 1.0 mm/0.02 to 0.04 in) between the choke link and the vacuum unit rod. Bend the tag on the link rod if necessary to achieve this.
7 Connect a vacuum pump to the lower connection on the choke vacuum unit. Cover the upper connection. Apply vacuum and measure the choke flap opening: it should correspond to gap 'A' (see Specifications). Adjust if necessary at screw 'A' (Fig. 3.30).
8 Expose the upper connection, maintaining vacuum on the lower. Gap 'B' must now be measured, and if necessary adjusted at screw 'B'.
9 Release the vacuum but leave the pump connected. Set the throttle so that the fast idle screw is again on the seventh cam step. Cover the choke vacuum upper connection and apply vacuum to the lower. Open and close the throttle: the fast idle screw should return to the sixth stop and be at least 1.0 mm (0.04 in) clear of the seventh step. Adjust if necessary by bending the link (Fig. 3.31).
10 Uncover the choke vacuum upper connection and remove the vacuum pump.
11 Move the choke link clockwise so that the choke flap is fully open and hold it in this position. Open and close the throttle: the fast idle adjustment screw should return to a position clear of all the cam steps, clearing the first (lowest) step by at least 0.5 mm (0.02 in). Bend the tag on the choke flap spindle if necessary to adjust.
12 Release the choke link so that the rubber band closes the flap again. Open the throttle wide and measure the choke flap opening. If it is not as specified, slacken the throttle spindle nut and reposition the cam, then tighten the nut. Remove the rubber band.
13 Refit the bi-metallic spring housing and associated components, engaging the eye of the spring with the end of the link and observing the alignment marks (photo). Insert the three screws into the clamp ring and just nip them up.
14 Check the choke flap opening at 20°C (68°F) – assuming this to be the ambient temperature – and adjust to within specification if necessary by turning the bi-metallic spring housing. Tighten the three screws when adjustment is correct.
15 With the choke flap still in the 20°C (68°F) position, check the clearance between the end of the full load enrichment tube and the

Fig. 3.29 With fast idle screw on 7th cam step (A), clearance (B) between link and rod must be as stated in text (Sec 29)

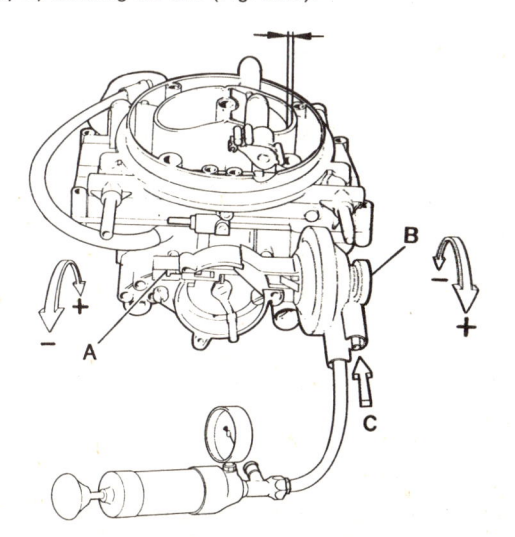

Fig. 3.30 Choke flap opening under vacuum. For A and B see text (Sec 29)

C Vacuum unit upper connection

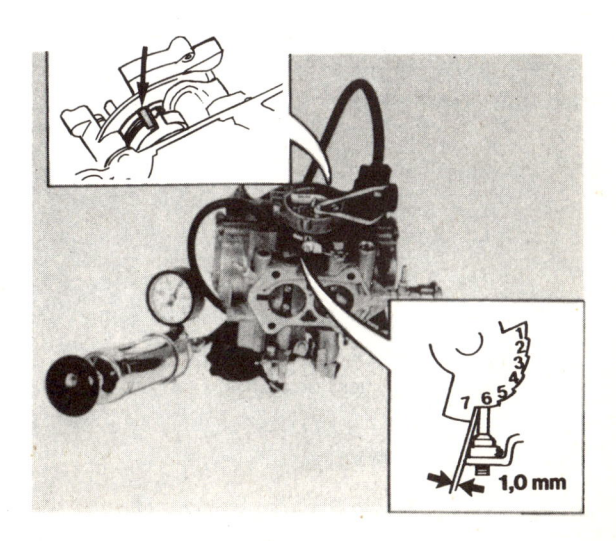

Fig. 3.31 Fast idle screw clearance from cam stop (inset, bottom) – adjust if necessary by bending link (inset, top) (Sec 29)

122   Chapter 3 Fuel and exhaust systems

29.13 Refitting the bi-metallic spring housing – spring eye and link (arrowed) must engage

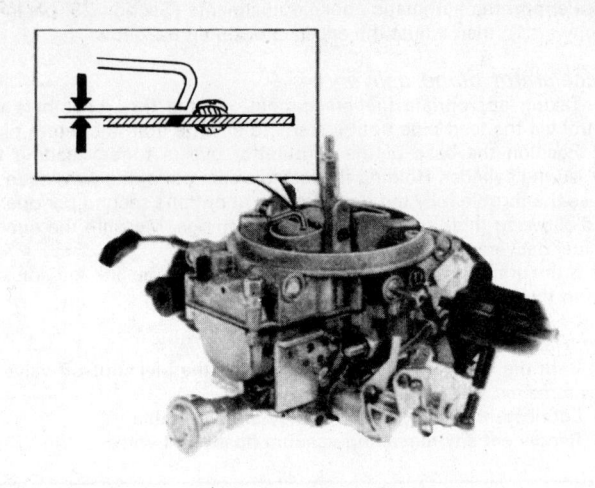

Fig. 3.32 Clearance between full load enrichment tube and choke flap (Sec 29)

flap. Bend the tube if the clearance is not as specified. Too low a setting will cause high fuel consumption, too high will cause cold starting problems.

30 Hot start valve – testing

1 Inspect the valve hoses to see that they are not kinked or perished. If the hoses between the valve and the carburettor are defective, rough running and poor fuel consumption may result.
2 Check that the breather hose is unobstructed. Its lower end is near the transmission crossmember. If this hose is blocked, difficult hot starting may result.
3 Check the valve by blowing through the hoses with the ignition on and off. With the ignition on, the valve must be closed (breather hose and carburettor hoses not connected). With the ignition off, the valve must open the breather hose to the carburettor. If not, either the valve or the power supply is at fault.

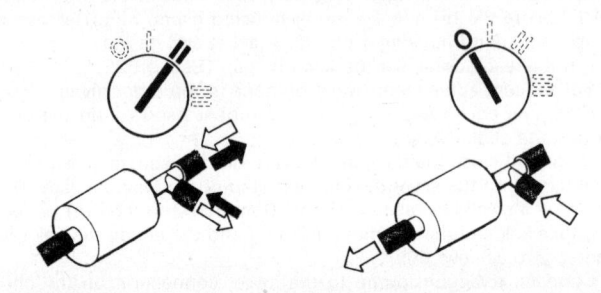

Fig. 3.33 Air flow through hot start valve with ignition on and off (Sec 30)

31 Fault diagnosis – carburettor

Symptom	Reason(s)
Difficult starting when cold	Incorrect procedure (depress and release throttle to set choke) Fast idle adjustment incorrect Induction air leak Automatic choke seized or wrongly adjusted Enrichment tube set too high
Difficult starting when hot	Hot start valve defective or hoses blocked Fuel return line blocked Induction air leak Float needle valve leaking Float level incorrect
Stalling or rough running after cold start	Fuel cut-off system fault Fast idle adjustment incorrect Idle speed adjustment incorrect Induction air leak Automatic choke seized or wrongly adjusted Choke vacuum unit leaking or wrongly adjusted
Idle rough or unstable	Incorrect idle adjustment Choke flap not opening Idle jet blocked Induction air leak

Chapter 3 Fuel and exhaust systems 123

Symptom	Reason(s)
Poor acceleration	Idle adjustment incorrect
	Accelerator pump defective or wrongly adjusted
	Secondary throttle vacuum unit defective
	Induction air leak
	Auxiliary jet blocked
	Part load enrichment piston stuck
Poor top speed	Throttle valve(s) not opening fully
	Secondary throttle vacuum unit defective
	Enrichment tube too high
	Part load enrichment piston stuck
	Float level incorrect
	Wrong jets fitted
Backfiring in exhaust	Fuel cut-off system fault
	Idle adjustment incorrect
	Idle or auxiliary jet blocked
	Wrong jets fitted
	Air leak into exhaust
Fuel consumption excessive	External leakage
	Fuel cut-off system fault
	Hot start valve or hoses defective
	Choke flap not opening fully
	Enrichment tube too low
	Part load enrichment piston stuck
	Float level incorrect
	Air jets blocked
	Wrong jets fitted

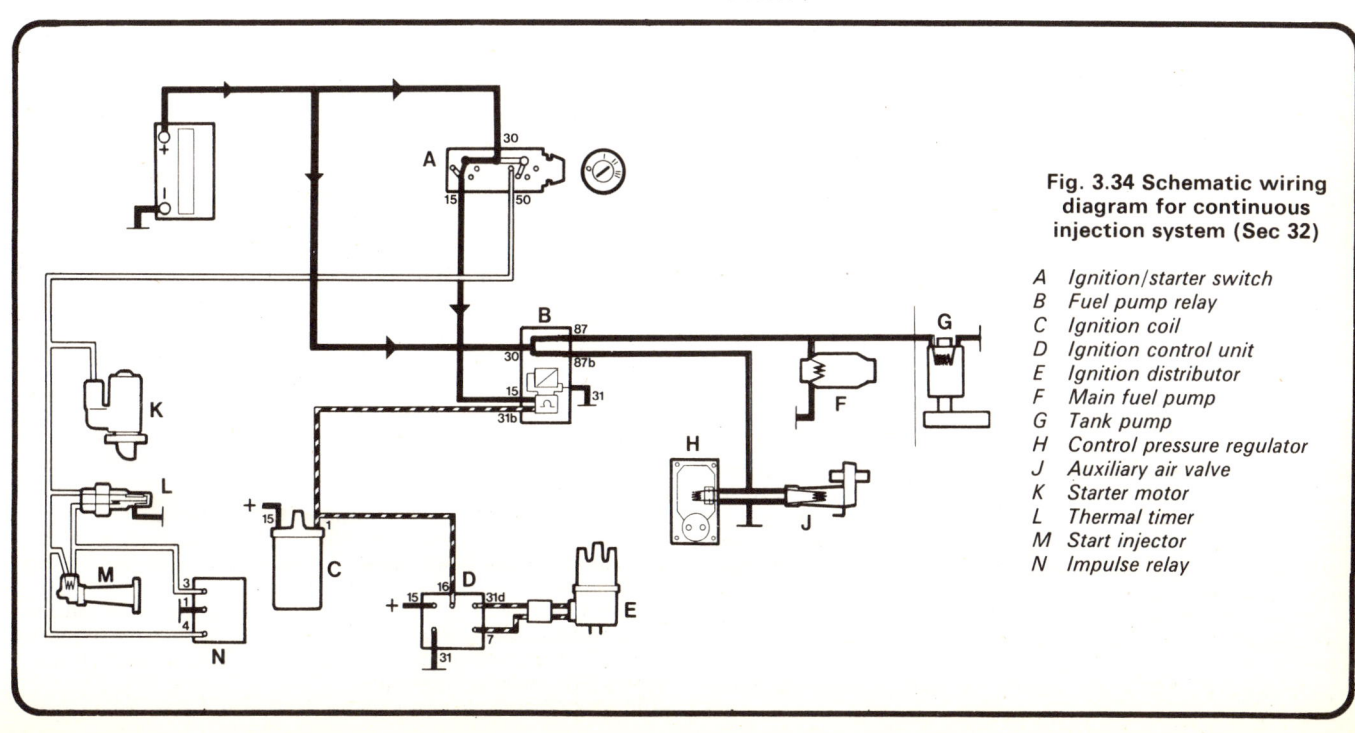

32 Continuous injection system – description

The continuous fuel injection system is found on the B28E and B230E engines. It is a well-proven system, with little to go wrong and no 'black boxes' to worry about. As the name implies, fuel injection takes place continuously whilst the engine is running. The rate of injection is varied to suit the prevailing speed and load.

Fuel is drawn from the tank by the tank pump. It passes to the main fuel pump where the line pressure of around 5 bar (73 lbf/in²) is established. An accumulator next to the pump provides a reservoir of pressure to improve hot starting. From the accumulator the fuel passes through a filter and then to the fuel distributor on top of the inlet manifold.

The fuel distributor looks a little like an ignition distributor, but it has fuel lines instead of HT leads. There is one fuel line per injector, with additional lines for the start injector and the control pressure regulator. The fuel distributor's main function is to regulate the fuel supply to the injectors in proportion to the incoming airflow. Incoming air deflects the airflow sensor plate, which moves the control plunger in the fuel distributor and so varies the supply to the injectors. The airflow sensor and the fuel distributor together are sometimes called the fuel control unit.

The control pressure regulator reduces the control pressure during warm-up and under condition of low manifold vacuum, and so enriches the mixture. (A lower control pressure means that the airflow sensor plate is deflected further, and the quantity of fuel injected is increased.)

Fig. 3.34 Schematic wiring diagram for continuous injection system (Sec 32)

A Ignition/starter switch
B Fuel pump relay
C Ignition coil
D Ignition control unit
E Ignition distributor
F Main fuel pump
G Tank pump
H Control pressure regulator
J Auxiliary air valve
K Starter motor
L Thermal timer
M Start injector
N Impulse relay

Chapter 3 Fuel and exhaust systems

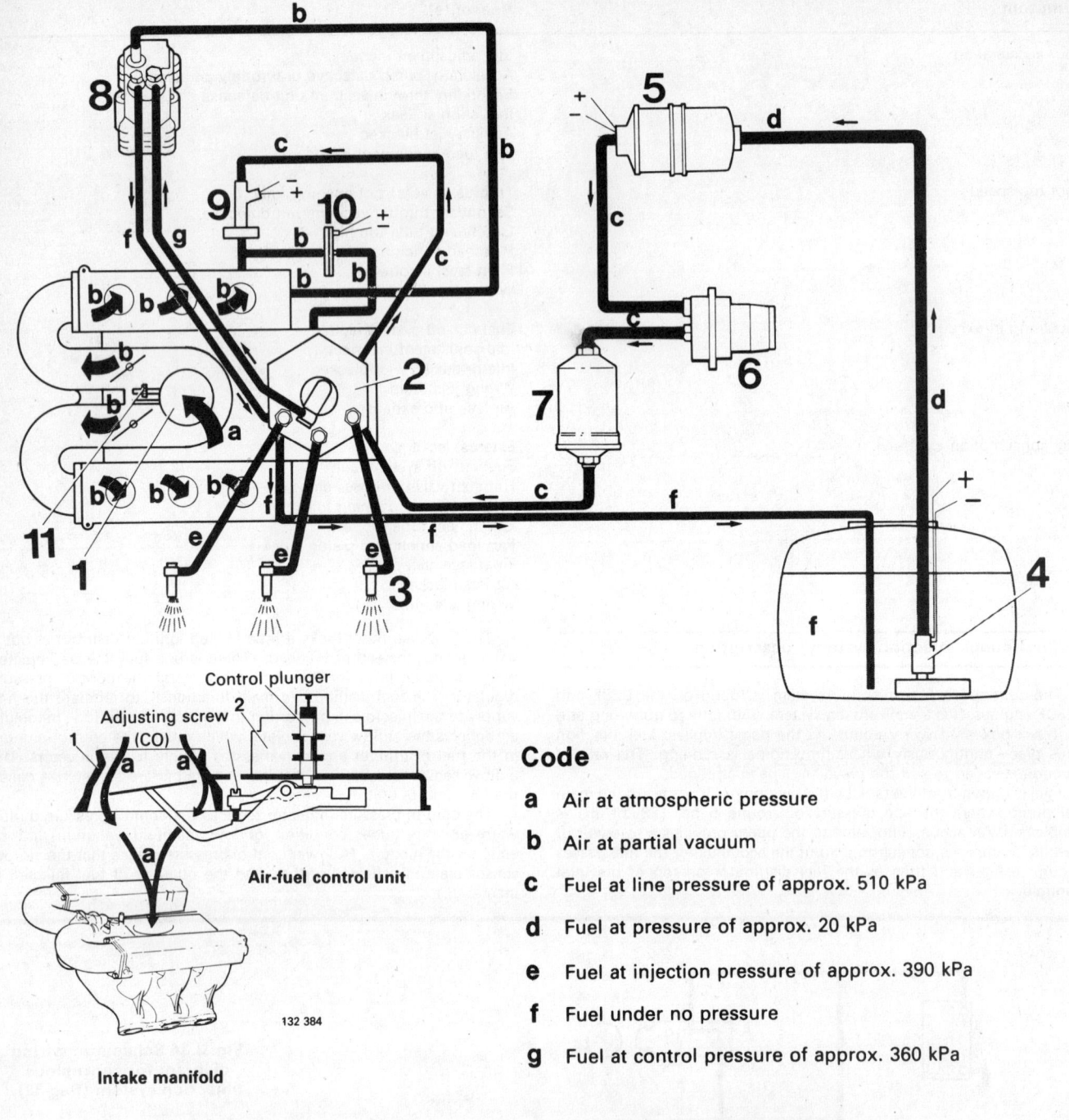

Code

- **a** Air at atmospheric pressure
- **b** Air at partial vacuum
- **c** Fuel at line pressure of approx. 510 kPa
- **d** Fuel at pressure of approx. 20 kPa
- **e** Fuel at injection pressure of approx. 390 kPa
- **f** Fuel under no pressure
- **g** Fuel at control pressure of approx. 360 kPa

Fig. 3.35 Continuous injection system components – V6 shown (Sec 32)

1 Airflow sensor	4 Tank pump	7 Filter	10 Auxiliary air valve
2 Fuel distributor	5 Main fuel pump	8 Control pressure regulator	11 Mixture adjustment screw
3 Injector	6 Accumulator	9 Start injector	

An electrically-controlled start injector provides extra fuel during engine starting. A thermal time switch controls the duration of start injector operation when the engine is cold; on a hot engine an impulse relay provides a smaller quantity of extra fuel to be injected. An auxiliary air valve provides the extra air needed to maintain idle speed when the engine is cold. On models with the constant idle speed system, an air control valve takes the place of the auxiliary air valve.

Most of the information in the following Sections relates to the continuous injection system as fitted to the B28E engine, since there is a lack of information for the B230E. Differences will be found to relate mainly to component access, besides the obvious difference of the number of cylinders.

33 Continuous injection system – testing procedures

1 All the tests described in this Section are within the capability of the competent home mechanic. Some extra equipment, such as a pressure gauge and a multi-meter, may have to be purchased for some of the tests.

2 Some tests result in fuel being sprayed from unions or injectors. Take stringent safety precautions, not only against fire but also against fume intoxication. Dispose of ejected fuel safely – only return it to the tank if its cleanliness is certain.

3 It is emphasised that it is no use attempting to diagnose faults in the

Chapter 3 Fuel and exhaust systems

injection system until all other engine-related systems are in good order. In particular, the battery must be fully charged when making electrical tests.
4 Disconnect the plug from the ignition control unit before starting the tests. Also disconnect the control pressure regulator and auxiliary air valve wiring plugs until they are to be tested, otherwise they will heat up during preliminary tests and give false results.

Start injector and associated components
5 The engine must be cold for this test.
6 Remove the two Allen screws which secure the start injector. Withdraw the injector and position it with its nozzle in a clean glass jar.
7 Unclip the impulse relay (next to the expansion tank) and disconnect it (photo).
8 Have ready a stop-watch. Time the duration of fuel injection from the start injector whilst an assistant cranks the engine on the starter motor. Duration will vary from 7.5 seconds at coolant temperatures of −20°C (−4°F) to zero at 35°C (95°F) and above. Duration at 20°C (68°F) should be about two seconds.

9 If no injection occurs at all, disconnect the start injector wiring plug. Connect a 12 volt test lamp or voltmeter across the plug terminals and repeat the cranking test. If voltage is present during cranking, but the injector did not operate, the injector is defective. No voltage suggests a fault in the thermal timer or wiring.
10 If injection occurs continuously, disconnect the injector wiring plug and repeat the test. Fuel issuing without the plug connected means that the injector is defective. If no fuel issues where it did before, the fault is probably in the thermal timer.
11 When satisfied with the results, reconnect the injector plug, then test the impulse relay as follows.
12 Reconnect the impulse relay plug. Again have an assistant crank the engine. After the initial period of continuous injection controlled by the thermal timer, the impulse relay should cause the injector to 'chatter' and deliver 0.1 sec bursts of fuel, with intervals of 0.3 sec between them. If this is not the case, test the relay by substitution.
13 Run the fuel pump by switching on the ignition and connecting Volvo test relay 5170. Feel the fuel filter to check that the pump is running – it will vibrate slightly. Leakage from the start injector must not exceed one drop per minute – if it does, renew it.
14 Switch off the ignition and refit the start injector and other disturbed components.

Air-fuel control unit
15 Remove the air intake trunking from the top of the airflow sensor.
16 Inspect the airflow sensor plate. Pass a slip of paper around its edge to check that it does not touch anywhere. Centre the plate if necessary by slackening its central screw, repositioning the plate and tightening the screw.
17 Run the fuel pump (paragraph 13). Depress the airflow sensor plate to simulate airflow into the manifold, at the same time listening to the injectors. The injectors should buzz when the plate is moved, and stop buzzing when it is released. If the injectors buzz with the plate at rest, the fuel distributor control plunger has jammed. If the injectors do not buzz at all, the line pressure is incorrect (perhaps because the fuel pump is not working). If the sensor plate jams, the airflow sensor needs to be reconditioned or renewed. Some resistance to motion is normal – do not mistake this for jamming.
18 Check the line and control pressures as described later in the test procedures, then measure the resting height of the airflow sensor plate. The fuel pump must be running and control pressure must be at its maximum value, so the control pressure regulator must have been energised for at least five minutes.
19 Measure the height of the sensor plate relative to the venturi waist. The plate should be flush with, or no more than 0.3 mm (0.012 in) above, the waist. Adjustment is made by tapping the pin to the left of the mixture screw hole up or down. To tap the pin upwards, the airflow sensor will have to be removed.
20 It is emphasised that the resting height of the sensor plate depends on the control pressure. If the control pressure is incorrect, it is no use trying to adjust the sensor plate.
21 Refit the disturbed components and remake the original electrical connections.

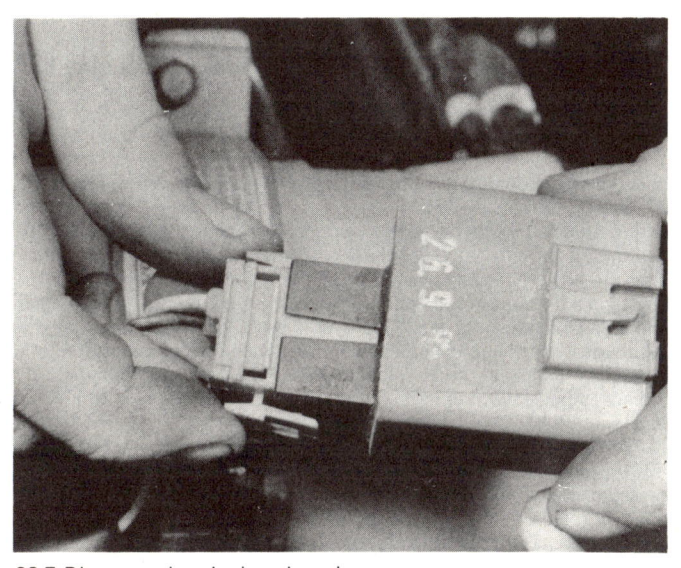

33.7 Disconnecting the impulse relay

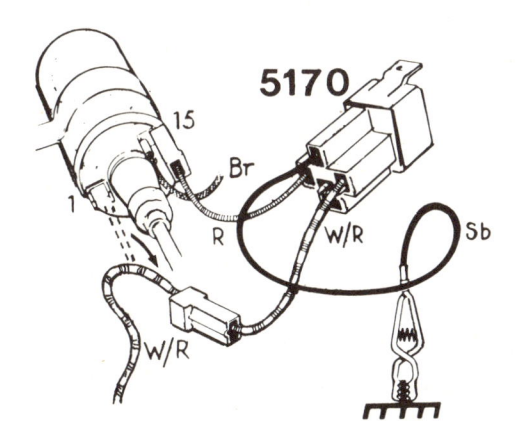

Fig. 3.36 Starting the fuel pumps with test relay 5170 (Sec 33)

1 Ignition coil (distributor side)
15 Ignition coil (battery side)
Br Brown
R Red
Sb Black
W White

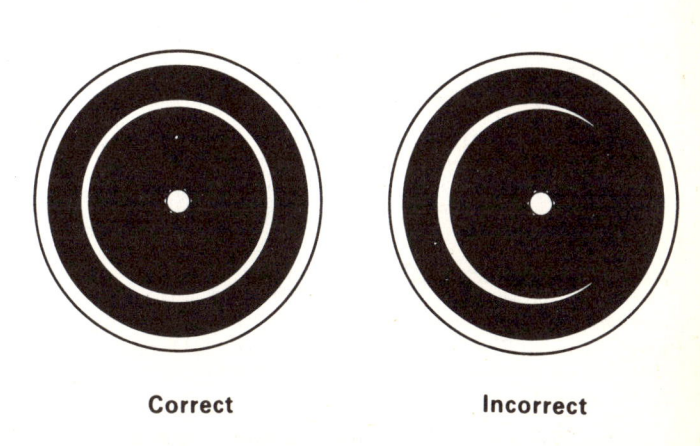

Correct   Incorrect

Fig. 3.37 Check the centralisation of the airflow sensor plate (Sec 33)

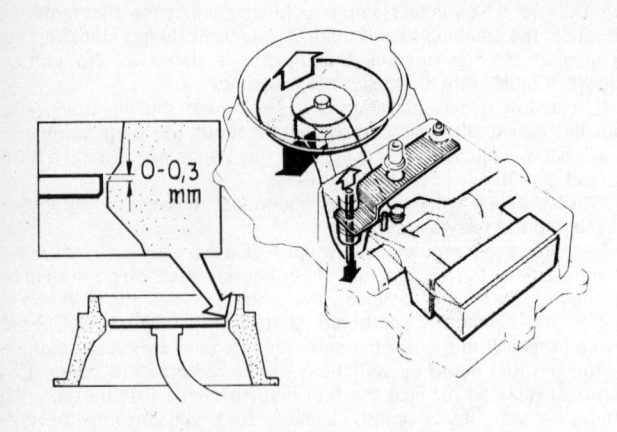

Fig. 3.38 Sensor plate height is adjusted by moving pin (small arrows) (Sec 33)

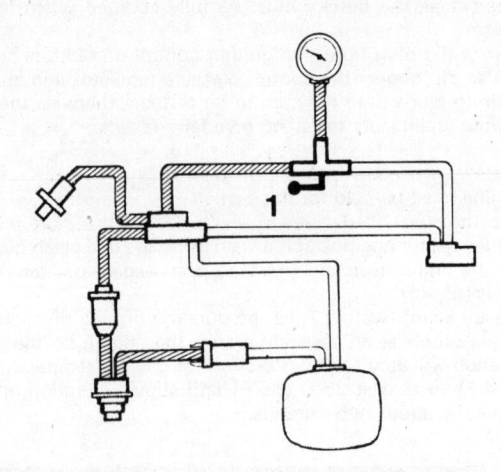

Fig. 3.39 Measuring line pressure – T-piece in position 1 (Sec 33)

Line, control and resting pressures
22 A pressure gauge of adequate range will be needed for these tests, together with a switchable T-piece and the appropriate pipes and unions. If these are not available, it will probably be quicker and cheaper to have a Volvo dealer or fuel injection specialist perform the tests.
23 With the ignition off, connect the gauge and T-piece between the control pressure regulator and the fuel distributor. Switch the T-piece so that the gauge is connected to the fuel distributor, and the control pressure regulator is isolated.
24 Run the fuel pump (paragraph 13).
25 Wait for the pressure gauge reading to stabilise, then record it and compare it with the specified line pressure. If the pressure is too high, either the fuel return line is blocked or the line pressure regulator is defective. Too low a pressure can be caused by leakage on the pressure side of the main pump; a defective accumulator; a defective tank pump or a clogged pick-up filter or main fuel filter.
26 With the line pressure correct, switch the T-piece to connect the control pressure regulator and the fuel distributor. When the pump is running, the gauge will show control pressure. Desired pressure from a cold regulator (ie the wiring plug is not connected) varies with ambient temperature as shown in the graph (Fig. 3.41). Too high a pressure can only be due to a blocked return line. If the pressure is too low, the test should be repeated with a new control pressure regulator.
27 Having achieved the correct cold control pressure, connect the wiring plug to the control pressure regulator. Leave the ignition on for five minutes to allow the regulator to warm up, then read the control pressure again. Desired warm pressure is given in the Specifications. Too high a pressure is still due to a blocked return line, and should have shown up in the cold test. Too low a pressure may mean that the regulator is not heating up – check for battery voltage at the wiring plug, and measure the resistance across the regulator terminals. If voltage is present at the regulator but it is not heating up, it should be renewed.
28 With line pressure and control pressure correct, finally check the resting pressure. With the gauge still connected to read control pressure, switch off the ignition. Pressure should stabilise at 2.4 to 3.2 bar (35 to 46 lbf/in²) and after 20 minutes should still be above 2.4 bar (35 lbf/in²). Pressure drop denotes external leakage, a leaking line pressure regulator, or a defective fuel pump non-return valve. Loss of resting pressure is a cause of hot start difficulties.
29 When all pressures are correct, disconnect the gauge and restore the original connections.

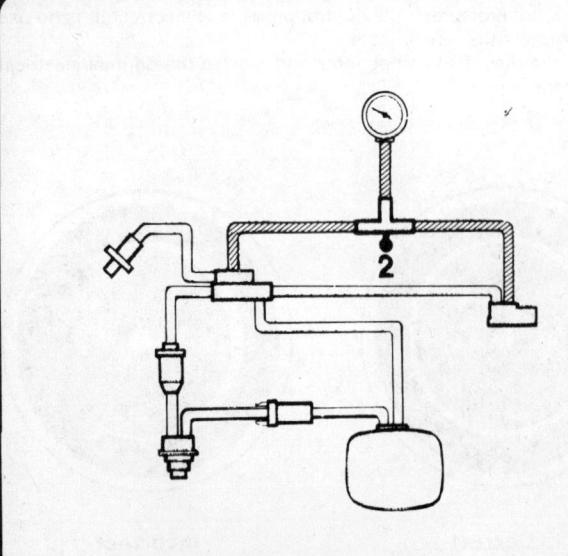

Fig. 3.40 Measuring control pressure – T-piece in position 2 (Sec 33)

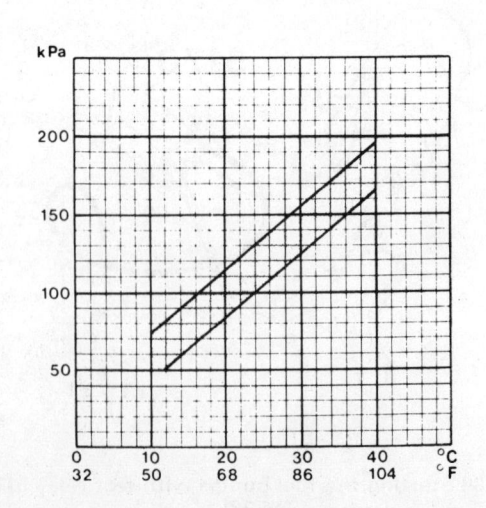

Fig. 3.41 Control pressure versus temperature (Sec 33)

Chapter 3 Fuel and exhaust systems

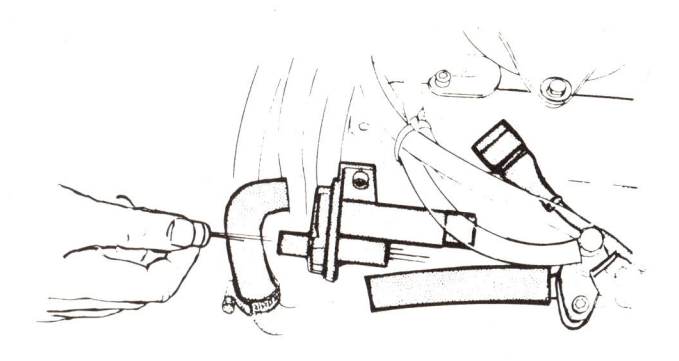

Fig. 3.42 Shining a torch through the auxiliary air valve (Sec 33)

Auxiliary air valve (without constant idle system)
30 With the auxiliary air valve cold (at room temperature), disconnect the air hoses from it. Shine a torch through the valve to establish that it is at least partly open. If it is not, and refuses to open when tapped or shaken, renew it.
31 Connect the wiring plug to the valve and run the fuel pump. The valve should close completely within five minutes. If it does not, tap it lightly (engine vibrations will normally cause the valve to close). If it still does not close, and there is battery voltage at the wiring plug, renew the valve.

Air control valve (with constant idle system)
32 With the ignition off, disconnect the multi-plugs from the constant idle system control unit (behind the side trim in the front passenger footwell).
33 Start the engine and allow it to idle. Using two pieces of wire, join together multi-plug terminals Nos 5 and 2, and 4 and 1. Engine speed should rise to between 1600 and 2400 rpm – if not, the valve (or its power supply) is at fault.

Injectors and fuel distributor
34 Release the injector securing clips and pull the injectors from their recesses. Place them on clean rags.
35 Run the fuel pump (paragraph 13). Observe the injectors: if they drip, the fuel distributor is leaking internally and should be renewed. (The injector tips may become moist with fuel – this is acceptable.)
36 Depress the airflow sensor plate slowly, observing the injectors. They should all start to discharge fuel at the same time. If one or more start to discharge before the others, have the suspect injector(s) cleaned and tested or renewed.
37 Injector delivery can be checked by leading each injector into an identical measuring tube or glass. (Since the object is to detect difference in delivery, it does not matter if the receptacles are not calibrated, so long as they are all the same). Depress the airflow sensor

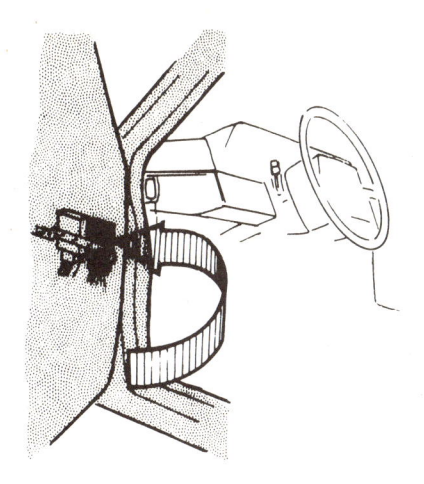

Fig. 3.43 Constant idle system control unit (arrowed) (Sec 33)

plate through half its travel, hold it in this position until the tubes are nearly full, then release it. All injectors should have delivered the same quantity of fuel, to within 20%.
38 If a difference in delivery greater than 20% is found, exchange 'good' and 'bad' injectors and repeat the test. If the difference follows the injector, the injector is defective. If the difference stays with the same line regardless of injector, the fault is in the fuel distributor.
39 Refit the injectors and other disturbed components when the test is complete.
40 Further testing and cleaning of fuel injectors should be left to a Volvo dealer or fuel injection specialist, since a special test rig is required.
41 A rough assessment of injector spray pattern can be made when performing the delivery test. Examples of good and bad spray patterns are given in Fig. 3.46.

Line pressure regulator – adjustment
42 Line pressure is adjusted by adding or removing shims at the pressure regulator spring. The pressure regulator is located in the fuel distributor. It is not recommended that any attempt be made at adjustment unless the necessary seals, shims and pressure gauges are available. Scrupulous cleanliness must be observed.
43 For every 0.1 mm (0.004 in) increase in shim thickness, line pressure will be increased by 0.15 bar (2.2 lbf/in²). Various shim thicknesses are available.

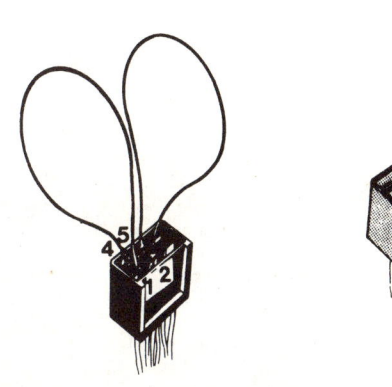

Fig. 3.44 Bridge terminals 5 and 2, and 4 and 1 (Sec 33)

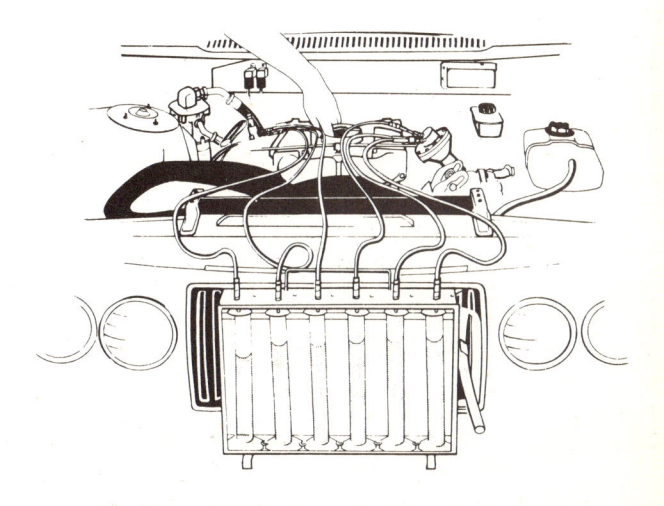

Fig. 3.45 Checking injector delivery with measuring cylinders (Sec 33)

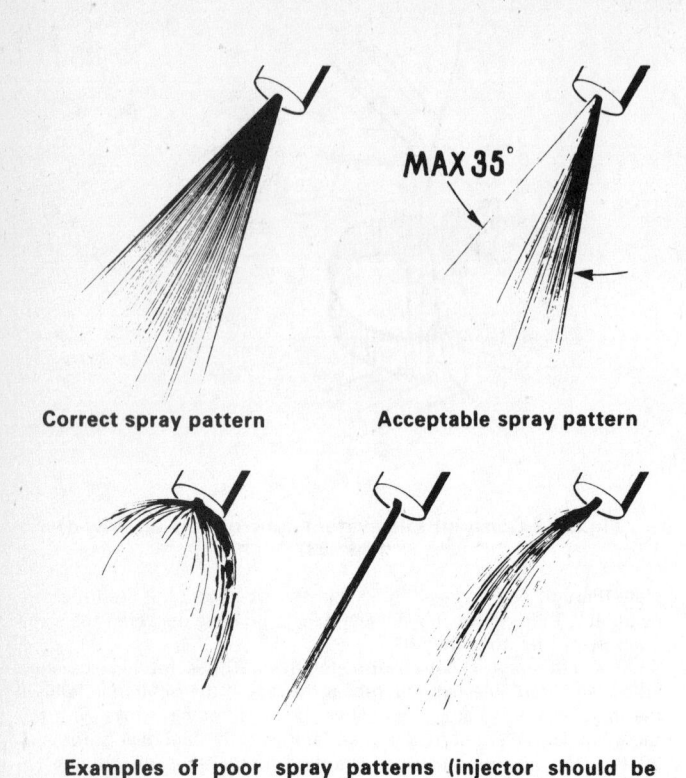

128 Chapter 3 Fuel and exhaust systems

Correct spray pattern

Acceptable spray pattern

Examples of poor spray patterns (injector should be replaced)

Fig. 3.46 Injector spray patterns (Sec 33)

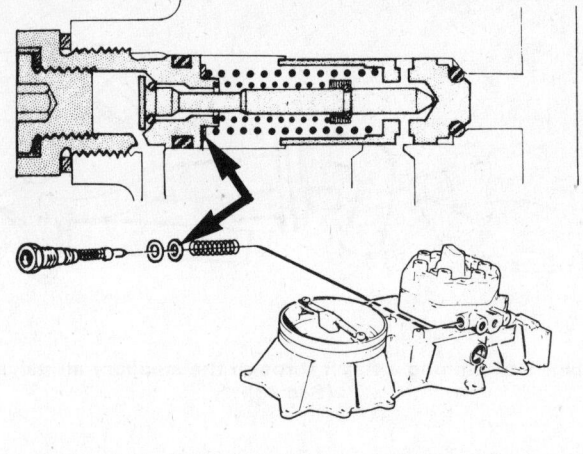

Fig. 3.47 Line pressure regulator – adjusting shims arrowed (Sec 33)

6 Refit by reversing the removal operations, using a new seal and union washers if necessary.
7 Note that new injectors are filled with a preservative wax before storage. This wax must be flushed out of the injector before it is fitted. Take advice from the seller of the injectors.

Fuel control unit
8 Slacken the fuel filler cap to release residual pressure.
9 Disconnect the various fuel unions from the fuel distributor, making notes or identifying marks if necessary (photos). Be prepared for fuel spillage.
10 Remove the air intake trunking from the top of the airflow sensor.
11 Disconnect the electrical leads from the constant idle system throttle switch (when so equipped).
12 Remove the twelve Allen screws which secure the upper half of the control unit. Note the location of the earth tags (photo).
13 Lift off the upper half of the control unit and recover the gasket.
14 Further dismantling is described in Section 35. Note that the fuel distributor can be removed independently of the airflow sensor if required.
15 When refitting, use a new gasket (photo). If necessary use new sealing washers on the fuel unions. Leave one of the injector unions disconnected.
16 Carry out the basic setting of the control unit as follows. Run the fuel pump (Section 33, paragraph 13). Turn the mixture adjustment screw clockwise until fuel just starts to drip from the open union, then back the screw off half a turn. Stop the fuel pump and reconnect the remaining union.
17 Check the line, control and testing pressures, and the resting height of the airflow sensor plate (Section 33). Adjust the idle speed and mixture on completion (Section 8).

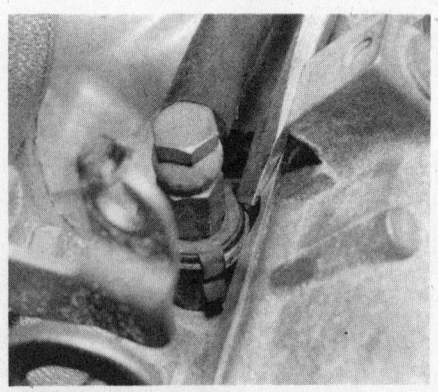

34 Continuous injection system components – removal and refitting

1 Disconnect the battery negative lead.

Fuel accumulator
2 This is removed in the same way as the main fuel pump (Section 14), except that there are no electrical connections to attend to.
3 In practise it is probably easier to remove the pump, accumulator and cradle all together. The accumulator can then be removed under relatively clean conditions on the bench.

Injectors
4 Release the spring clip and pull the injector from its recess (photos).
5 Disconnect the fuel line union from the injector.

34.4A Prise up the injector spring clip ...

34.4B ... and pull out the injector

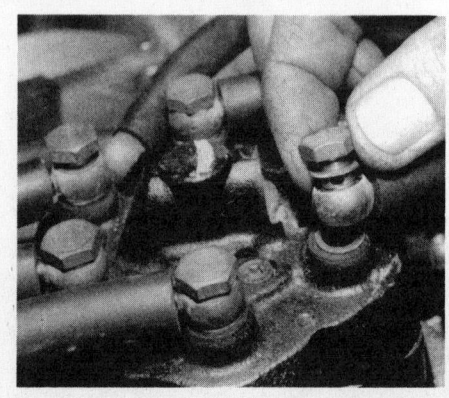

34.9A Disconnecting an injector union – note sealing washers

Chapter 3 Fuel and exhaust systems

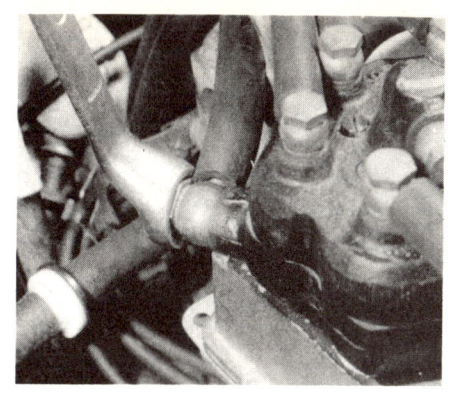

34.9B Disconnecting the fuel supply union from the distributor

34.12 Control unit showing some of the securing screws. Note earth tag (arrowed)

34.15 A new gasket fitted to the bottom half of the control unit

Start injector
18 Disconnect the fuel and electrical feeds from the injector. Be prepared for fuel spillage.
19 Remove the two Allen screws and withdraw the injector.
20 Refit by reversing the removal operations.

Air control/auxiliary air valve
21 Disconnect the electrical plug from the valve.
22 Disconnect the air hoses from the valve.
23 Remove the valve securing screws and remove the valve.
24 Refit by reversing the removal operations.

Control pressure regulator
25 Disconnect the electrical and vacuum feeds from the regulator (photo).
26 Disconnect the fuel unions from the regulator. They are different sizes so they cannot be confused. Be prepared for fuel spillage.
27 Unbolt the regulator and remove it.
28 Refit by reversing the removal operations. Use new union sealing washers if necessary.

Thermal timer
29 Unscrew the expansion tank filler cap to release any pressure in the cooling system. Take precautions against scalding if the coolant is hot.
30 Disconnect the electrical plug from the thermal timer (photo).
31 Unscrew the thermal timer and remove it. Plug the hole with a tapered wooden plug to minimise coolant loss.
32 When refitting, apply sealant to the timer threads and screw it into position. Reconnect the electrical plug.
33 Top up the cooling system if necessary.

Fuel pump relay
34 This relay is located in position 'E' of the central electrical unit. See Chapter 12, Section 24.

35 Continuous injection system control unit – overhaul

1 Remove the upper half of the control unit (Section 34).
2 Remove the three plain screws which secure the fuel distributor (photo). Lift off the fuel distributor, being careful not to drop the control plunger. Recover the O-ring.
3 Remove the control plunger (photo). Clean it with clean fuel and a soft brush or the fingers only – do not use any tools on it. Also clean the metering slots in the fuel distributor.
4 Unscrew the pressure regulator plug and withdraw the pressure regulator components, keeping them in order. Clean the plungers and renew the O-rings and sealing washers. Refit the components and tighten the plug (photos).
5 If any plungers are scored or otherwise damaged, or if the control plunger sticks in the distributor, the complete fuel distributor must be renewed. Do not attempt to dismantle it further.
6 To dismantle the airflow sensor, remove the two circlips which secure the pivot shaft end plugs. Remove the plugs, O-rings, spring and balls. Note which side the spring is fitted (photos).
7 Slacken the clamp screws on the sensor plate lever and tap out the pivot shaft. Lift out the lever, plate and arm (photo).
8 Examine all parts and renew as necessary. Make sure that the spring and pin which determine the resting height of the plate are not loose (photo).
9 Reassemble the airflow sensor, lightly greasing the shaft and balls. Align the lever so that the mixture adjustment screw is in line with its hole before tightening the clamp screws.
10 Centre the sensor plate in the venturi and tighten its clamp screw. Adjust the height of the plate by tapping the pin up or down so that the plate rests just above the venturi waist. After refitting, the plate can only be adjusted downwards (Section 33).
11 Refit the fuel distributor and control plunger, using a new O-ring, and tighten the securing screws (photo).

34.25 Control pressure regulator electrical and vacuum connections

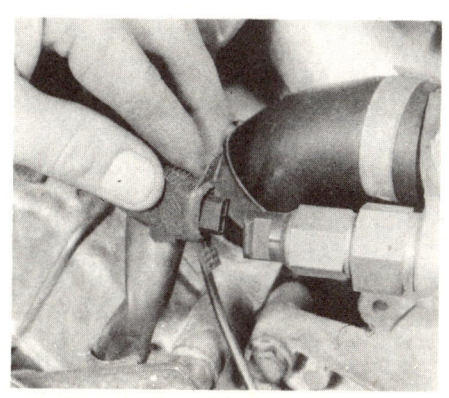

34.30 Disconnecting the thermal timer multi-plug

35.2 Removing a fuel distributor securing screw. The other two are arrowed

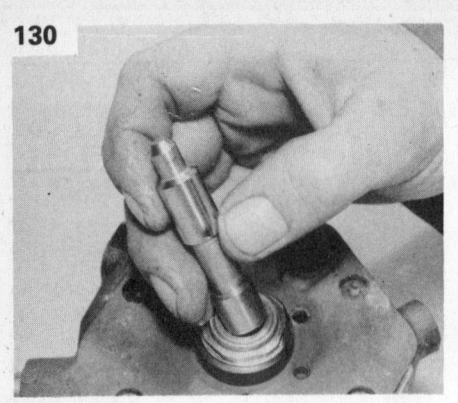

35.3 Removing the fuel distributor control plunger

35.4A Refitting the pressure regulator outer plunger ...

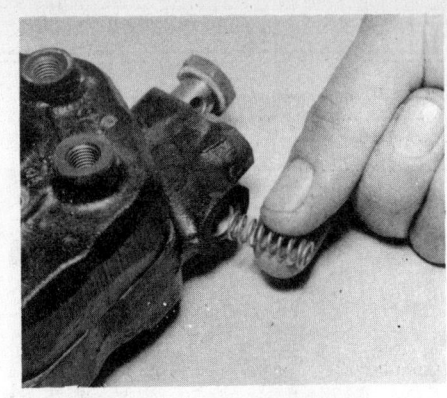

35.4B ... pressure regulator spring ...

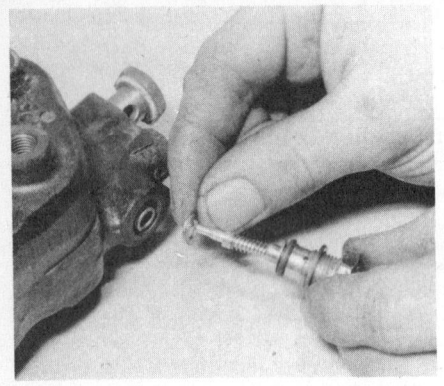

35.4C ... the inner plunger and plug. Shims control pressure

35.6A Remove the pivot shaft circlip ...

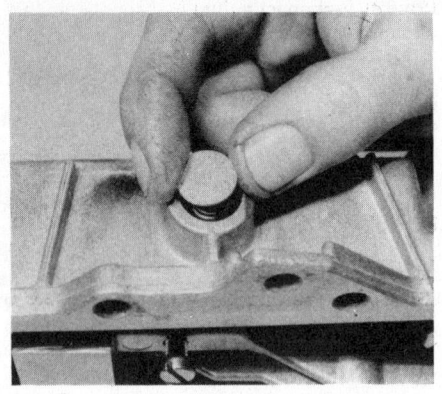

35.6B ... the end cover ...

35.6C ... the O-ring ...

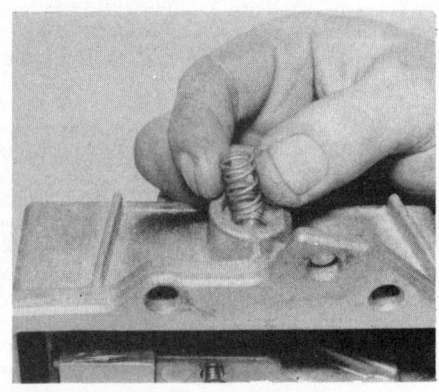

35.6D ... the spring (one side only) ...

35.6E ... and the ball

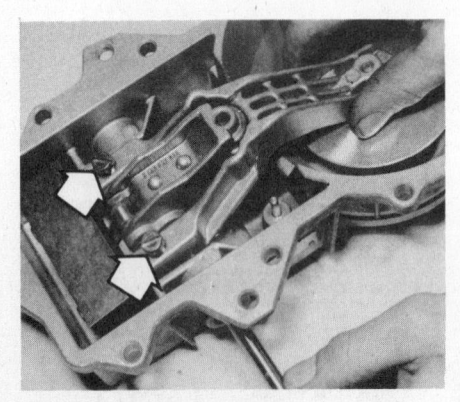

35.7 Slacken the clamp screws (arrowed) and withdraw the pivot shaft

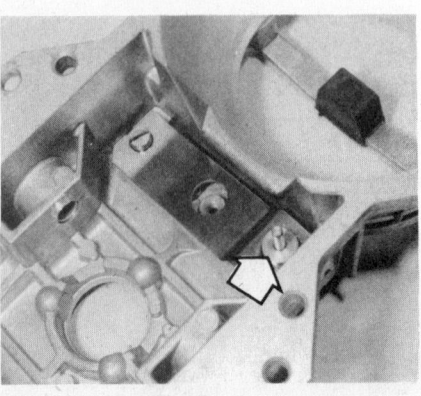

35.8 Spring and pin (arrowed) must not be loose

35.11 Distributor with control plunger and O-ring fitted

Chapter 3 Fuel and exhaust systems

36 Fault diagnosis – continuous injection system

Symptom	Reason(s)
Engine will not start	Tank empty Fuel pump or relay defective Induction air leak Fuel distributor control plunger seized Fuel pressure incorrect Airflow sensor plate height incorrect
Difficult starting when cold	Start injector defective Thermal timer defective Auxiliary air valve defective Constant idle system fault
Difficult starting when hot	Start injector leaking Impulse relay defective Fuel pressure lost at rest
Erratic running when cold and during warm-up	Cold control pressure incorrect
Erratic running when hot	Warm control pressure incorrect Line pressure incorrect
Erratic running, hot and cold	Induction air leak Control pressure incorrect Mixture adjustment incorrect Throttle valve(s) loose or worn
Excessive fuel consumption	External leakage Start injector leaking Control pressure incorrect Mixture adjustment incorrect
Poor performance	Throttle valve(s) not opening fully Control pressure incorrect Tank pump defective Main pump defective Fuel filter blocked Mixture adjustment incorrect
Erratic idle	Misfiring (ignition fault) Induction air leak Fuel distributor control plunger sticking Throttle valve(s) loose or worn Injector(s) leaking or poor spray pattern

37 Turbocharger – description

The turbocharger increases the efficiency of the engine by raising the pressure in the inlet manifold above atmospheric pressure. Instead of the air/fuel mixture being simply sucked into the cylinders it is actively forced in.

Energy for the operation of the turbocharger comes from the exhaust gas. The gas flows through a specially-shaped housing (the turbine housing) and in so doing spins the turbine wheel. The turbine wheel is attached to a shaft, at the other end of which is another vaned wheel known as the compressor wheel. The compressor wheel spins in its own housing and compresses the inducted air on the way to the inlet manifold.

After leaving the turbocharger, the compressed air passes through an intercooler, which is an air-to-air heat exchanger mounted in front of the radiator. Here the air gives up heat which it acquired when being compressed. This temperature reduction improves engine efficiency and reduces the risk of detonation.

Boost pressure (the pressure in the inlet manifold) is limited by a wastegate, which diverts the exhaust gas away from the turbine wheel in response to a pressure-sensitive actuator. As a further precaution, a pressure-sensitive switch cuts out the fuel pump if boost pressure becomes excessive. Boost pressure is displayed to the driver by a gauge on the instrument panel.

The turbo shaft is pressure-lubricated by means of a feed pipe from

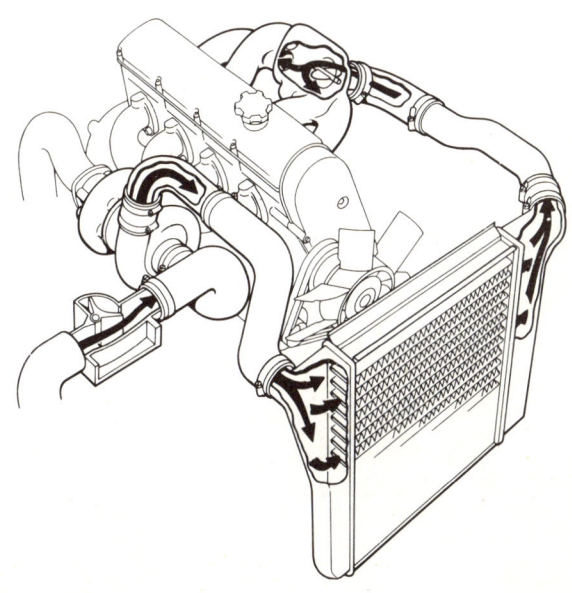

Fig. 3.48 Air flow through turbocharger and intercooler to inlet manifold (Sec 37)

132 Chapter 3 Fuel and exhaust systems

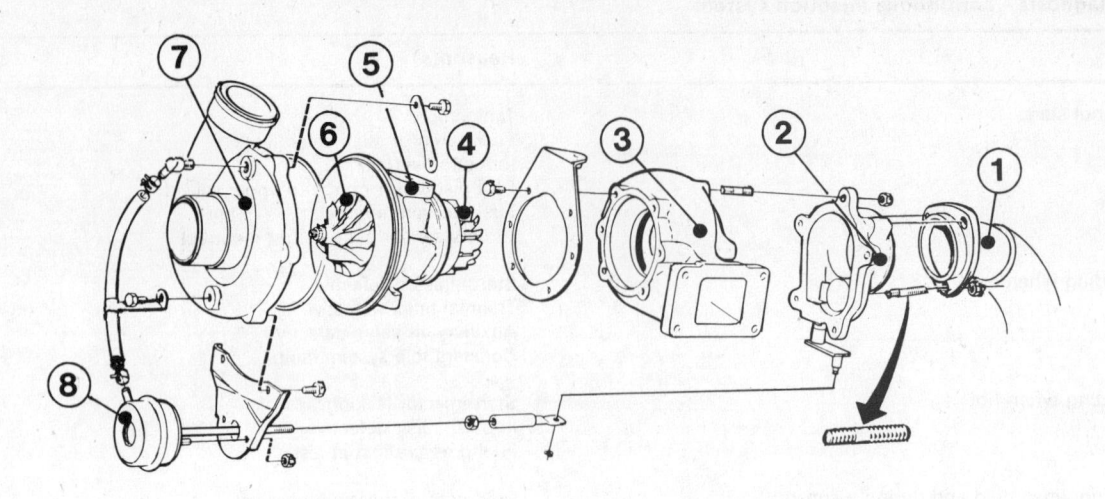

Fig. 3.49 Turbocharger and associated components (Sec 37)

1 Exhaust downpipe
2 Wastegate housing
3 Turbine housing
4 Turbine wheel
5 Shaft housing
6 Compressor wheel
7 Compressor housing
8 Wastegate actuator

the engine's main oil gallery. The shaft 'floats' on a cushion of oil. A drain pipe returns the oil to the sump.

38 Turbocharger – precautions

1 The turbocharger operates at extremely high speeds and temperatures. Certain precautions must be observed to avoid premature failure of the turbo or injury to the operator.
2 Do not operate the turbo with any parts exposed. Foreign objects falling onto the rotating vanes could cause extensive damage and (if ejected) personal injury.
3 Do not race the engine immediately after start-up, especially if it is cold. Give the oil a few seconds to circulate.
4 Always allow the engine to return to idle speed before switching it off – do not blip the throttle and switch off, as this will leave the turbo spinning without lubrication.
5 Allow the engine to idle for several minutes before switching off after a high-speed run.
6 Observe the recommended intervals for oil and filter changing, and use a reputable oil of the specified quality. Neglect of oil changing, or use of inferior oil, can cause carbon formation on the turbo shaft and subsequent failure.

39 Turbocharger – removal and refitting

1 The turbocharger and exhaust manifold are removed together. Begin by removing the turbo-to-intercooler and the airflow meter-to-turbo hoses. The airflow meter hose is also connected to the bypass valve (photo).
2 Remove the air cleaner hot air trunking.
3 Disconnect the exhaust downpipe from the turbo outlet. Remove the heat shield (photo).
4 Disconnect the oil drain pipe from the turbo. Be prepared for some oil spillage (photo).
5 Unbolt and remove the stiffener plate from below the manifold (photo).
6 Unbolt the oil feed pipe from the block (photo).
7 Remove the eight nuts which secure the exhaust manifold to the cylinder head. Note that one of the nuts secures a lifting eye (photo).
8 Lift off the manifold and turbocharger. Recover the exhaust post gaskets (photo).
9 Remove the oil feed pipe and recover the gasket.
10 Remove the locking plates from the four bolts which secure the turbocharger to the manifold. Bend up the plate tabs with a chisel and prise or drive them off the bolts. New plates will be needed for reassembly.
11 Clamp the manifold in a vice and remove the four bolts. Lift the turbo off the manifold. Recover the other halves of the locking plates.
12 Measure the length of the bolts and renew them if they have stretched to longer than 89 mm (3.50 in).
13 Fit the turbo to the manifold and secure it with the four bolts, applying anti-seize compound to their threads. Remember to fit the new locking plate sections.
14 Tighten the bolts to the specified torque, following the sequence in Fig. 3.50. Make up a template for the final angular tightening (photo).

39.1 Turbo connecting hoses

39.2 Turbo-to-downpipe bolts – two arrowed, third is hidden

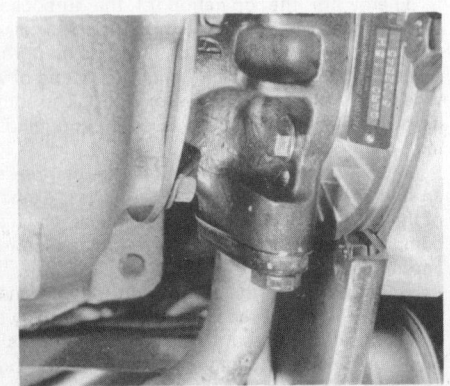

39.4 Turbo oil drain pipe connection

Chapter 3 Fuel and exhaust systems

39.5 Exhaust manifold stiffener plate

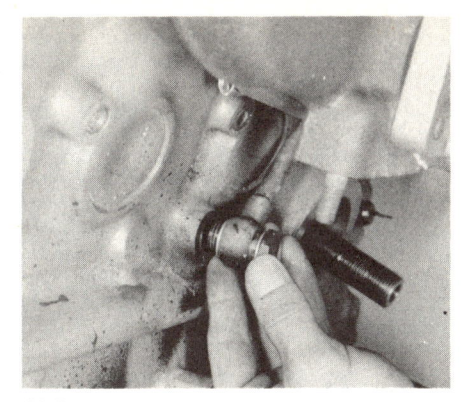

39.6 Turbo oil feed connection – note sealing washers

39.7 Removing an exhaust manifold nut

39.8 An exhaust port gasket

39.14 Angle-tightening the bolts

39.15A Fit the lockplate outer half ...

39.15B ... drive it over the bolt heads ...

39.15C ... crimp over the tabs ...

39.15D ... and knock the tabs down

15 Fit the outer halves of the locking plates. Drive them over the bolt heads with a hammer and a tube, crimp the tabs with pliers and finally knock the tabs down with a mallet (photos).

16 The remainder of refitting is a reversal of the removal procedure. Use new gaskets, oil pipe sealing washers, etc.

17 Before starting the engine, disconnect the ignition coil LT feed and crank the engine on the starter for six 10-second bursts. This will prime the turbocharger with oil.

18 Reconnect the ignition coil, run the engine and check that there are no oil leaks.

40 Turbocharger – inspection and repair

1 With the turbo removed from the manifold, inspect the housings for cracks or other visible damage.

2 Spin the turbine or compressor wheel to verify that the shaft is

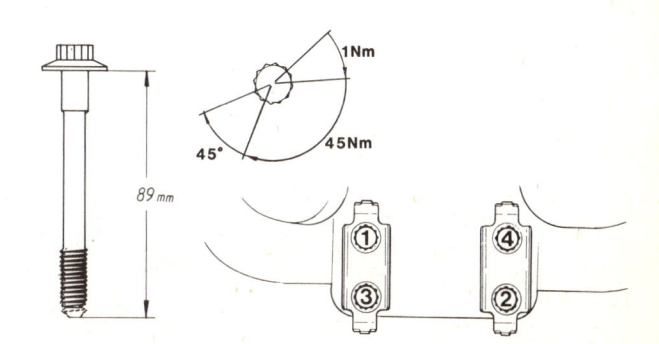

Fig. 3.50 Turbo bolt tightening sequence and stages. Renew bolts if they are longer than shown (Sec 39)

intact and to feel for excessive shake or roughness. Some play is normal since in use the shaft is 'floating' on a film of oil. Check that the wheel vanes are undamaged.
3 Inspect the wastegate actuator (see Section 42) and check that the seal and locking wire are intact.
4 If the exhaust or induction passages are obviously oil-contaminated, the shaft oil seals have probably failed. Oil on the induction side will collect in the intercooler, which should at least be drained and preferably flushed out before fitting a new turbo.
5 No DIY repair of the turbo is possible. The old unit may be accepted in part exchange for a new one.

41 Turbo boost pressure – checking and adjustment

Caution: *Unskilled or improper adjustment can cause serious engine damage.*
1 Using a T-piece, connect a pressure gauge (range approximately 0 to 1 bar/15 lbf/in²) into the boost pressure hose at the manifold. Position the gauge so that it can be read from inside the vehicle.
2 With the engine warmed up, have an assistant read the gauge. Accelerate at full throttle from 1500 rpm in 3rd gear, applying the brakes at the same time for a few seconds at 3500 rpm. (Do not apply the brakes for more than five seconds, or damage may result.) The assistant must note the highest pressure recorded.
3 If the boost pressure is not as specified, allow the engine to cool, then remove the seal from the wastegate actuator rod (photo). Unclip the rod from the wastegate lever and screw the threaded sleeve in or out after releasing the locknut. Lengthening the rod decreases the pressure, and *vice versa*; one complete turn of the sleeve will raise or lower pressure by approximately 0.03 bar (0.4 lbf/in²).
4 Secure the actuator rod with a new retaining clip and repeat the test.
5 When adjustment is correct, fit a new locking wire and seal if required. Tighten the actuator rod locknut.

42 Turbo wastegate actuator – inspection and renewal

1 If the boost pressure is incorrect and cannot be adjusted, the wastegate actuator may be defective. Check as follows.
2 Break the seal and unclip the actuator rod from the wastegate lever. The rod should retract into the actuator a little way. Mark the position of the rod where it enters the actuator, then reconnect the rod to the wastegate and measure the distance moved by the mark. It must be 2 to 6 mm (0.08 to 0.24 in); if not, renew the actuator.

3 To remove the actuator, unclip the link rod, remove the connecting hose and undo the two retaining nuts.
4 Commence refitting by securing the actuator to its bracket using two new nuts, then perform a preliminary adjustment as follows.
5 Connect a pump and pressure gauge to the actuator. Apply 0.48 bar (7 lbf/in²) pressure.
6 With the pressure applied, adjust the link rod so that it fits onto the wastegate lever (in the closed position) without being tight or slack. Tighten the adjuster locknut.
7 Secure the link rod to the wastegate lever using a new clip.
8 Remove the pump and gauge. Connect the actuator hose to the compressor housing.
9 Check the boost pressure (Section 41).

43 Intercooler – removal and refitting

1 Remove the radiator top mountings and carefully move the radiator rearwards.
2 Disconnect the hoses from the intercooler and lift it out (photo).
3 If turbocharger failure has occurred, the intercooler may contain a substantial quantity of oil. A drain plug is provided.
4 Refit by reversing the removal operations.

Fig. 3.51 Boost pressure gauge (5230) connected (left) and positioned for the road (right) (Sec 41)

41.3 Wastegate actuator rod and seal

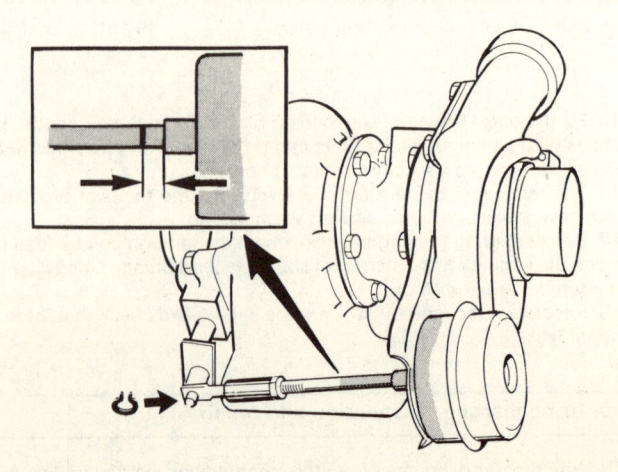

Fig. 3.52 Measuring wastegate actuator retraction (inset) (Sec 42)

Chapter 3 Fuel and exhaust systems

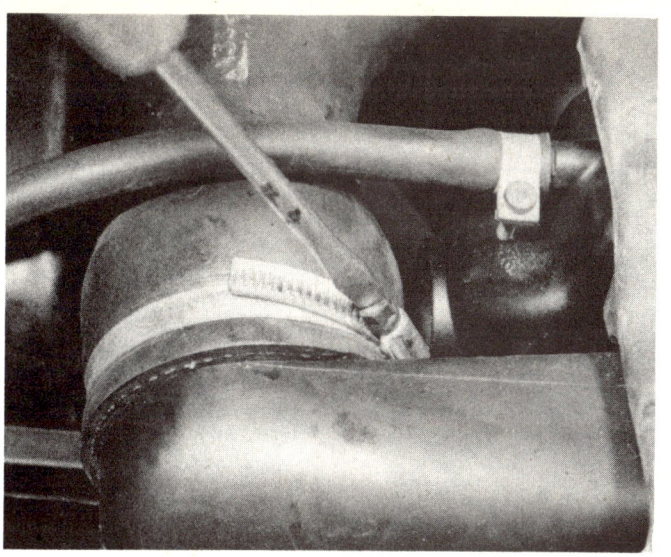

43.2 Disconnecting an intercooler hose

44 Motronic system – description

The Motronic system is basically Bosch L-Jetronic fuel injection with integrated ignition system control. Fuel injection and ignition functions are both controlled by the same unit in order to achieve optimum efficiency, driveability and power.

The control unit receives information of engine speed and crankshaft position, air flow into the induction system, coolant temperature and throttle position. The temperature of the inducted air after it has passed through the turbocharger and the intercooler is also measured. A signal relay informs the control unit if the air conditioning is in operation, necessitating a rise in idle speed. The unit also compensates for the effects of fluctuations in battery voltage, and provides extra enrichment via the start injector when the starter motor is operating. A thermal timer controls the duration of start injection.

Control unit outputs are to the fuel pump relay, the fuel injectors and the ignition system. The fuel pump is only allowed to run when the engine is running or when the starter motor is operating.

The fuel injectors are electrically controlled. Injection quantity is determined by the length of time for which the injectors are open. The quantity is increased in proportion to air flow, and also in response to sudden opening of the throttle and during warm-up.

Idle speed is raised during the warm-up phase by an auxiliary air valve, similar to that used in the continuous injection system, which allows a small quantity of air to bypass the throttle valve.

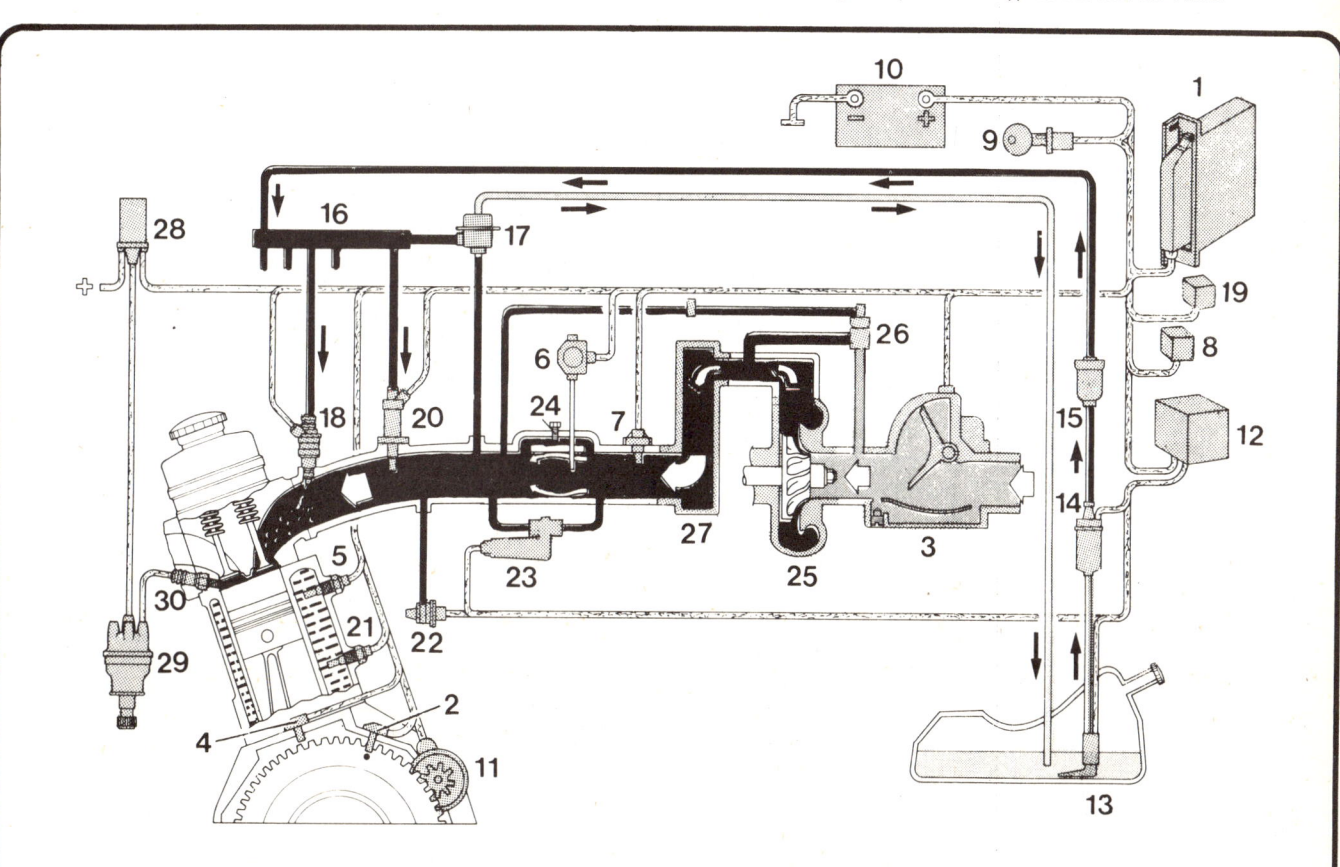

Fig. 3.53 A more detailed view of the Motronic system (Sec 44)

1 Control unit
2 Flywheel sensor (TDC)
3 Airflow meter
4 Flywheel sensor (engine speed)
5 Temperature sensor
6 Throttle position switch
7 Charge air temperature sensor
8 Air conditioning signal relay
9 Ignition/starter switch
10 Battery
11 Starter motor
12 System relay
13 Tank pump
14 Main fuel pump
15 Fuel filter
16 Fuel rail
17 Fuel pressure regulator
18 Fuel injector(s)
19 Torque limiter (overdrive) relay
20 Start injector
21 Thermal timer
22 Boost overpressure switch
23 Auxiliary air valve
24 Idle speed adjustment screw
25 Turbocharger
26 Bypass valve
27 Intercooler
28 Ignition coil
29 Ignition distributor
30 Spark plug

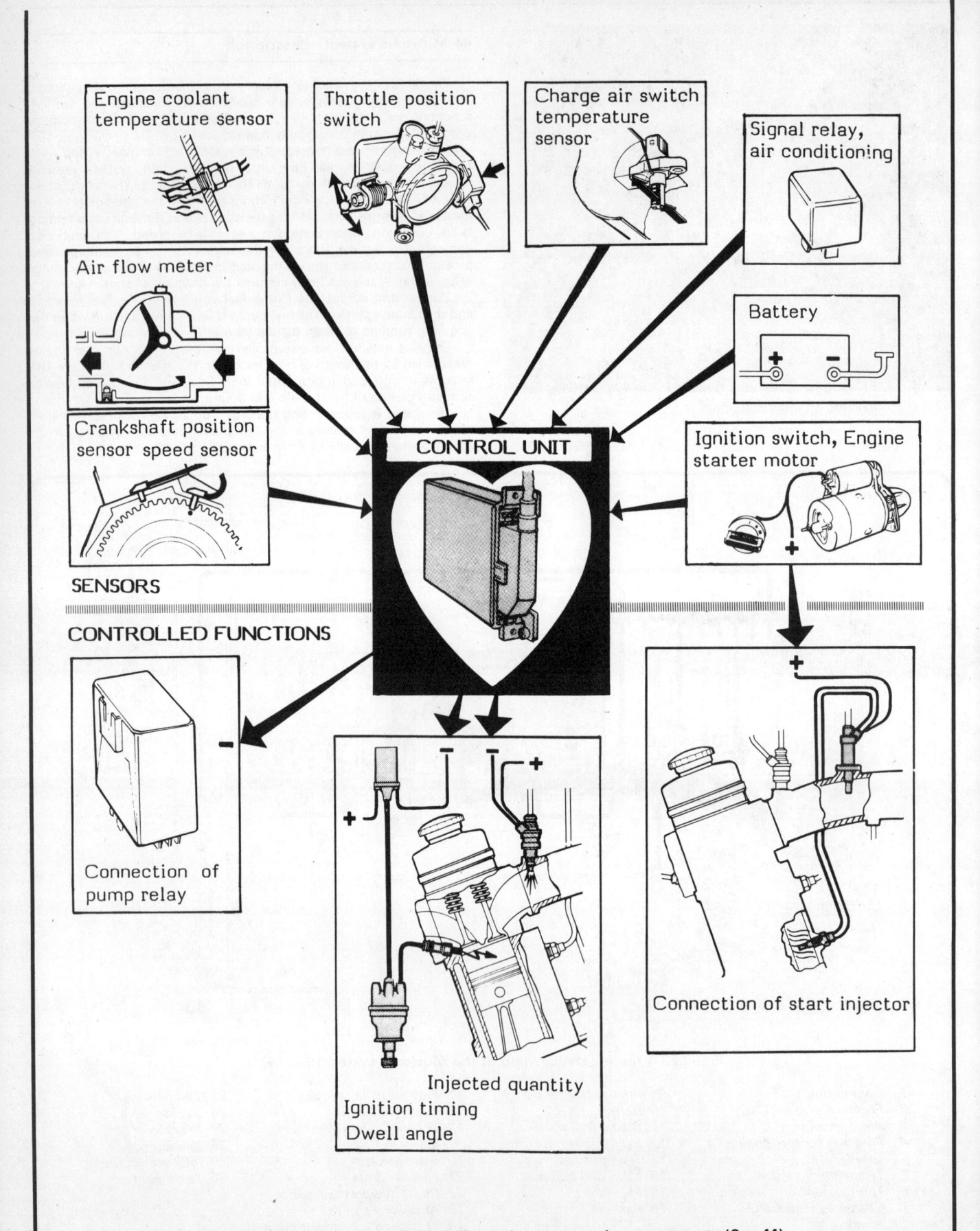

Fig. 3.54 Schematic relationship of Motronic system major components (Sec 44)

Chapter 3 Fuel and exhaust systems

Fuel pressure regulation is in proportion to manifold vacuum or boost pressure, and is achieved by varying the amount of fuel allowed out of the injector rail into the return line. The constant circulation of fuel makes for a steady fuel temperature and avoids vapour locks.

Besides regulating normal driving functions, the Motronic control unit also protects against overrevving by cutting out alternate injector pulses at speeds in excess of 6200 rpm.

In the event of a sensor failure or other malfunction, the control unit imposes preset values on injection and ignition systems to provide a 'limp home' facility.

Ignition-related aspects of the Motronic system, including a test procedure, are covered in Chapter 4.

45 Motronic system – testing procedures (fuel side)

To carry out this procedure a fuel pressure gauge, a manifold pressure/vacuum gauge with pump and a multi-meter will be needed. Testing of the electrical side is covered in Chapter 4.

This procedure applies specifically to the B23ET engine. The B230ET is believed to be similar.

Preliminary checks
1 Check the system fuses (Nos 1 and 15). Check the security of all under-bonnet electrical connectors, not forgetting the earth straps.
2 Inspect the air intake trunking for leaks. Leakage on the engine side of the airflow meter will seriously weaken the mixture.
3 Check the adjustment of the throttle switch (Chapter 4, Section 8).
4 Check the auxiliary air valve (Section 33, paragraphs 30 and 31).

Fuel pumps and fuel pressure
5 Remove the trim panel below the steering column for access to the pedals. Separate the wiring connector which feeds the boost overpressure switch (photo). With the ignition on, earthing one of the connectors should make the fuel pumps run. (Both terminals are fed by the same colour wires, so it is not possible to say which connector must be earthed.)
6 If the fuel pumps cannot be heard running, either the Motronic system relay or the wiring is defective. If one pump runs but not the other, the relay is OK and the fault must be in the pump or the wiring. Remove the fuel filler cap to hear if the tank pump is operating.
7 Switch off the ignition. Connect a fuel pressure gauge into the supply side at the fuel rail. Take appropriate fire precautions.
8 Run the fuel pumps (paragraph 5) and read the line pressure. Desired pressure is given in the Specifications.

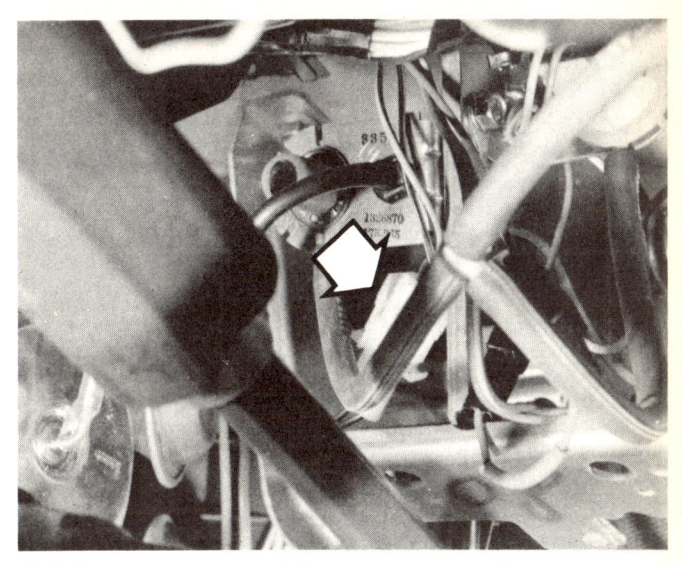

45.5 Boost overpressure switch connector (arrowed)

9 If line pressure is too high, switch off the ignition. Disconnect the fuel return line from the pressure regulator. Blow down the line to see if it is blocked. If the line is not blocked, the high pressure is being caused by the pressure regulator. Clear the line or renew the regulator, then reconnect the return line.
10 If line pressure is too low, pinch the return line (pump still running) and observe the rise in pressure. (Do not allow the pressure to exceed 6 bar/87 lbf/in^2). If pressure rises quickly, the low pressure is probably caused by the regulator. A slow rise suggests a blocked supply line or filter. No rise at all means that the main fuel pump is at fault.
11 With the line pressure correct, apply vacuum to the fuel pressure regulator (pump still running). The line pressure should drop by the same amount as the vacuum applied.
12 Apply pressure (not in excess of 0.7 bar/10 lbf/in^2) to the regulator instead of vacuum. The line pressure should rise by the same amount as the pressure applied.
13 Switch off the ignition, disconnect the pressure gauge and remake the original connections.

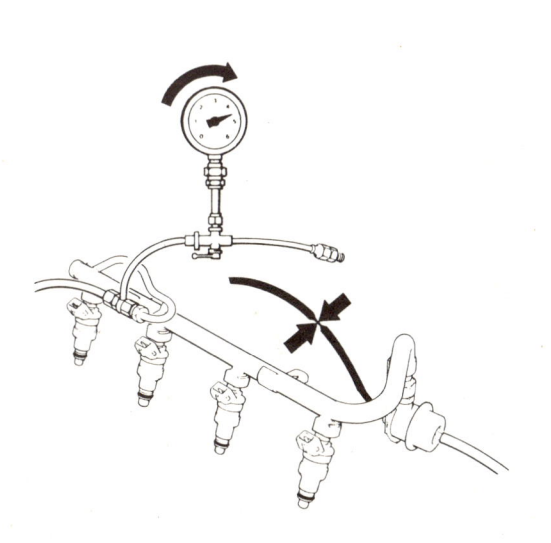

Fig. 3.55 Pinch the fuel return line and pressure should rise (Sec 45)

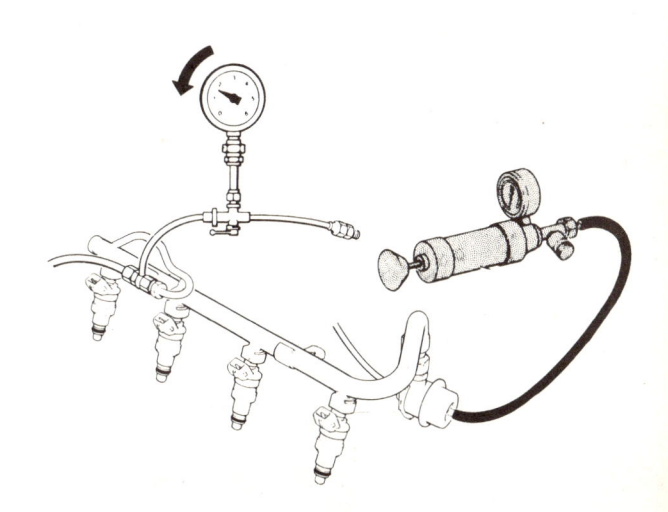

Fig. 3.56 Apply vacuum to the regulator and pressure should fall (Sec 45)

Chapter 3 Fuel and exhaust systems

14 Disconnect the earthing wire from the boost overpressure switch connector.

Boost pressure switches

15 Connect a self-powered continuity tester (buzzer, light or multi-meter) across the terminals of the boost pressure switch. Connect the pressure gauge and pump to the boost pressure hose at the inlet manifold. Apply pressure and note when the switch closes: desired values are given in the Specifications.

16 Transfer the continuity tester to the overpressure switch. This switch is normally closed. Increase the pressure and note when the switch opens.

17 If the pressure switches do not behave as specified, renew them.

18 Disconnect the test gear and remake the original connections.

Start injector and thermal timer

19 Remove the hose which connects the auxiliary air valve to the inlet manifold. The operation of the start injector can be viewed through the hose connector on the manifold.

20 Disconnect the multi-plug from the thermal timer (photo). Connect terminal 'W' of the plug to earth.

21 Disable the ignition system by disconnecting the coil LT feed. Have an assistant crank the engine on the starter: the start injector should deliver fuel all the time the starter is operating.

22 If no fuel emerges, disconnect the start injector multi-plug and check for battery voltage when the starter motor is in action. No voltage shows a wiring fault; voltage but no injection shows a faulty start injector.

23 Check that the start injector does not leak fuel under resting line pressure (ignition off). Leakage greater than one drop per minute is not acceptable.

24 Reconnect the air valve hose and the thermal timer multi-plug. Connect a voltmeter or test lamp to the start injector multi-plug. Have an assistant crank the engine on the starter. Voltage should appear at the multi-plug for a number of seconds, depending on coolant temperature – see Specifications.

25 Renew the thermal timer if it does not behave as described.

26 Disconnect the test gear and remake the original connections.

Bypass valve

27 Disconnect the large upper hose from the bypass valve. Look through the hose stub and check that the valve is closed.

28 Disconnect the valve vacuum hose from the inlet manifold. Apply vacuum to the hose and check that the valve opens. If not, apply vacuum directly to the bypass valve: if it opens now where it did not previously, the delay valve in the vacuum line is defective. The delay valve must be fitted with the white side towards the inlet manifold.

29 Disconnect the vacuum pump and remake the original hose connections.

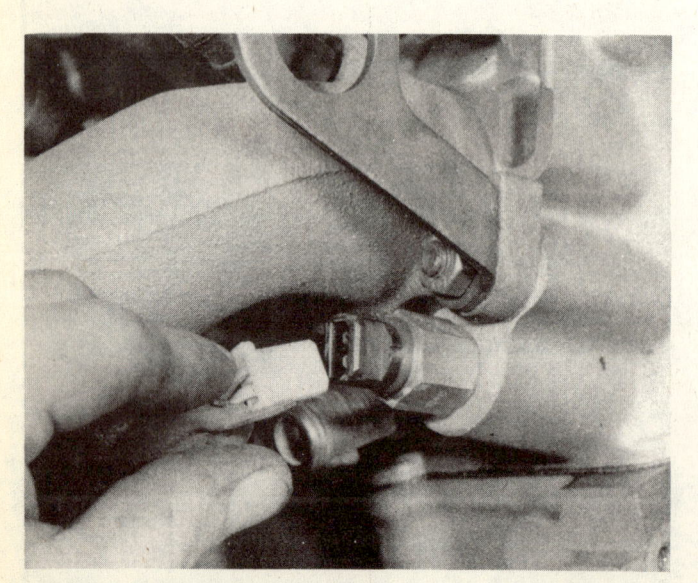

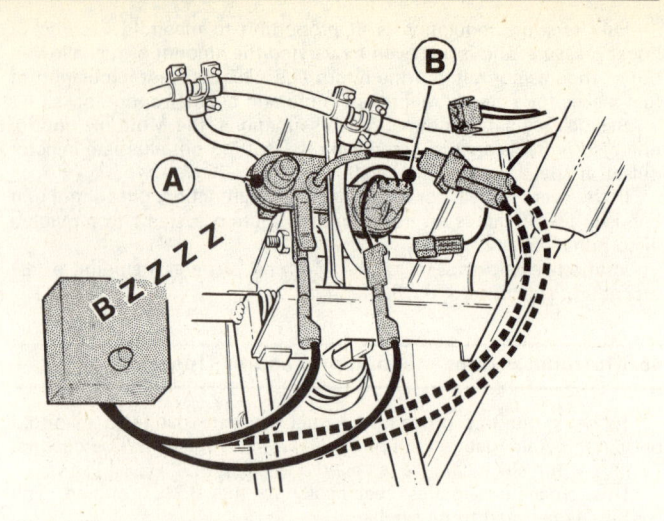

Fig. 3.57 Using a buzzer to test the boost pressure switches (Sec 45)

A Boost pressure   B Boost overpressure

Fig. 3.58 Operation of the start injector can be viewed through the hose connector (Sec 45)

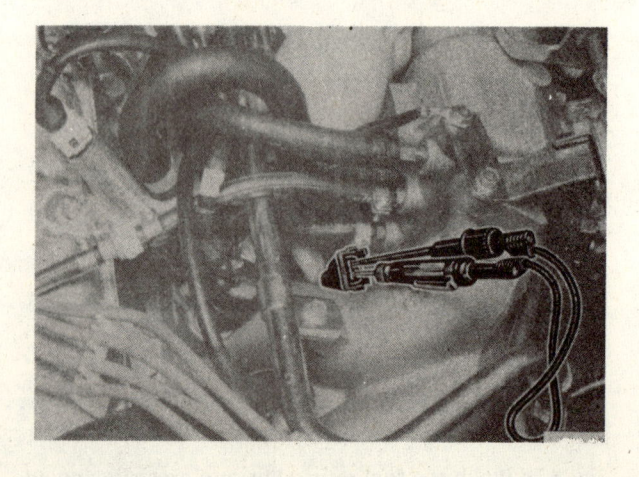

45.20 Disconnecting the thermal timer

Fig. 3.59 Using a test light to detect battery voltage at the start injector plug (Sec 45)

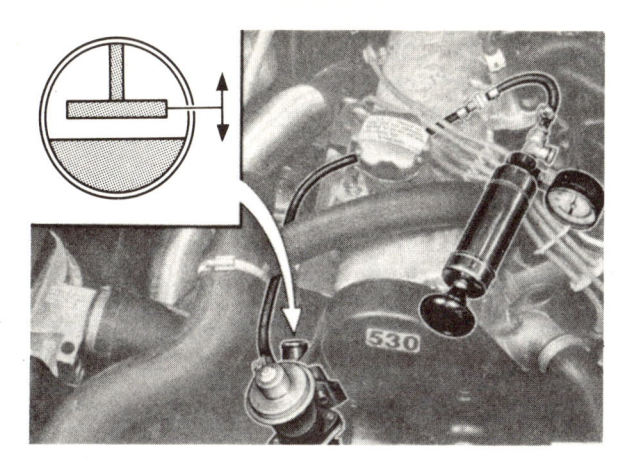

Chapter 3 Fuel and exhaust systems

Fig. 3.60 Checking the operation of the bypass valve (Sec 45)

46 Motronic system components – removal and refitting

1 Disconnect the battery negative lead.

Fuel injectors

2 Disconnect the fuel rail unions from the supply pipe, the start injector and the pressure regulator (photo). Be prepared for fuel spillage.
3 Disconnect the multi-plugs from the injectors. Prise out the wire clips to release the multi-plugs (photo).
4 Remove the injector securing clips by prising them out (photo).
5 Unbolt the fuel rail from the manifold, noting the location of any earth tags. Pull the rail off the injectors (photo).
6 Pull the injectors out of their locations in the manifold (photo).
7 Refit by reversing the removal operations. Use new sealing rings on the injectors and smear them with silicone grease. Renew the securing clips also.

Start injector

8 Disconnect the start injector union from the fuel rail (photo). Be prepared for fuel spillage.
9 Disconnect the start injector multi-plug.
10 Unbolt the start injector from underneath the inlet manifold and remove it.
11 Refit by reversing the removal operations.

Fuel pressure regulator

12 Disconnect the fuel return pipe and the fuel rail union from the regulator (photos). Be prepared for fuel spillage.
13 Disconnect the vacuum pipe from the regulator.
14 Undo the regulator securing nut and remove the regulator from its bracket.
15 Refit by reversing the removal operations.

Auxiliary air valve

16 Disconnect the multi-plug from the valve.
17 Disconnect the air hoses from the valve.
18 Unbolt the valve from the camshaft cover and remove it.
19 Refit by reversing the removal operations.

Airflow meter

20 Remove the air cleaner lid (Section 5).
21 Release the connecting hose, unbolt the airflow meter from the lid and remove it (photo). Do not drop it, it is fragile.
22 Refit by reversing the removal operations. If a new meter has been fitted, adjust the idle speed and mixture on completion (Section 8).

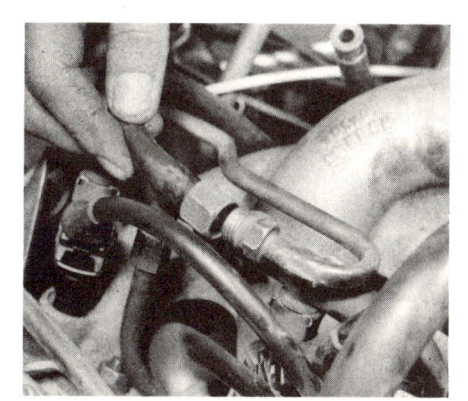

46.2 Fuel rail supply union

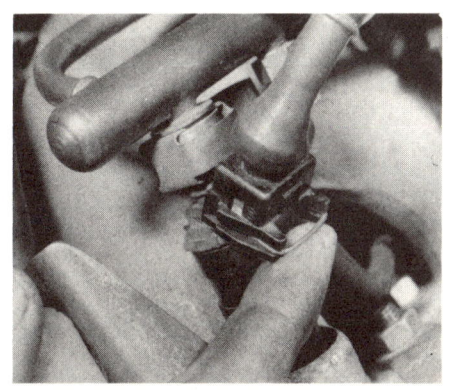

46.3 Prise out the wire clip to release the multi-plug

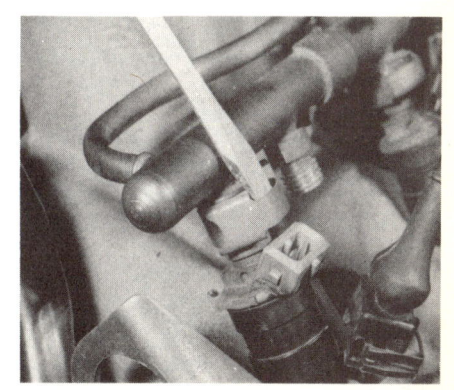

46.4 Prise out the injector securing clip

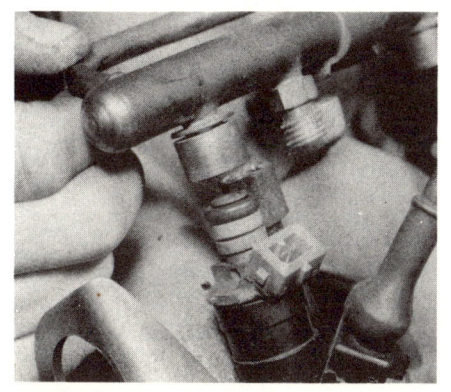

46.5 Removing the fuel rail

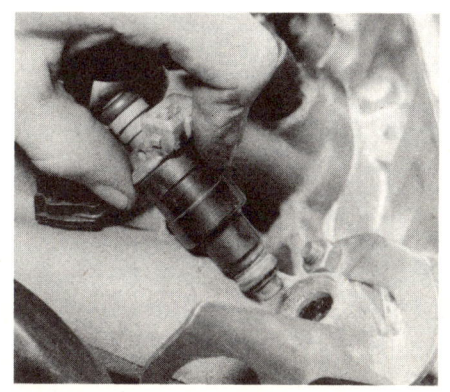

46.6 Removing a fuel injector

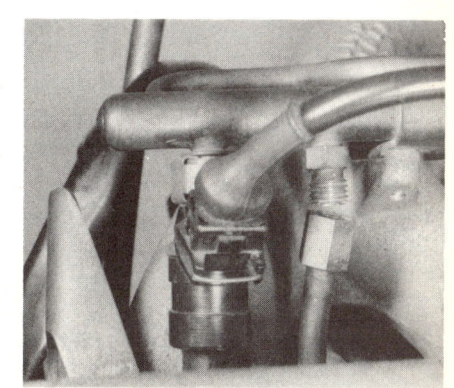

46.8 Start injector union

46.12A Disconnecting the regulator return pipe ...

46.12B ... and the fuel rail union

46.21 Releasing the airflow meter connecting hose

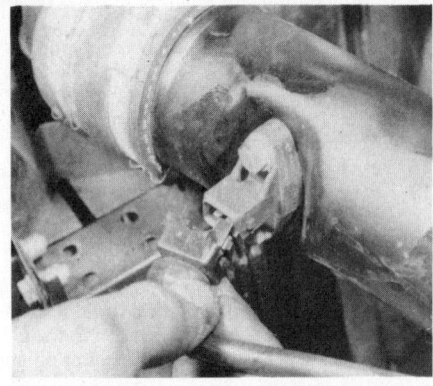

46.24 Disconnecting the charge air temperature sensor

46.28 Disconnecting a hose from the bypass valve

Thermal timer
23 See Section 34, paragraphs 29 to 33. The thermal timer on this engine is screwed into the cylinder head below the manifold, at the rear of the engine.

Charge air temperature sensor
24 Disconnect the multi-plug from the temperature sensor (photo).
25 Unbolt the sensor from the induction pipe and remove it.
26 Refit by reversing the removal operations.

Control unit
27 See Chapter 4, Section 12.

Bypass valve
28 Disconnect the hoses from the valve, making identifying marks if necessary (photo).
29 Unbolt the valve from its bracket and remove it.
30 Refit by reversing the removal operations. Do not disturb the adjustment screw on top of the valve, which is factory set.

47 Fault diagnosis – Motronic system

Symptom	Reason(s)
Difficult starting (hot or cold)	No 1 fuse blown Loose or dirty connections Induction air leak Fuel pump defective Fuel pressure incorrect Overpressure switch defective
Difficult starting when cold	Auxiliary air valve not opening Start injector or thermal timer defective Coolant temperature sensor defective
Irregular idle	Induction air leak Throttle linkage incorrectly adjusted Throttle switch incorrectly adjusted Coolant temperature sensor defective
Irregular running	Induction air leak Injector(s) defective Fuel pressure incorrect Control unit defective
Surge on overrun	Bypass valve not opening
Fuel consumption excessive	External leakage Fuel pressure incorrect Idle mixture adjustment incorrect
Poor performance	Throttle linkage incorrectly adjusted Air cleaner element clogged Induction air leak Tank pump not operating Fuel pressure incorrect Boost pressure incorrect Charge air temperature sensor defective Idle mixture adjustment incorrect

Chapter 4 Ignition system

For modifications, and information applicable to later models, see Supplement at end of manual

Contents

Ballast resistor – testing, removal and refitting	16	Ignition/engine management relays – location	15
Coil – testing, removal and refitting	17	Ignition timing – checking and adjustment	6
Distributor – overhaul	5	Knock sensor (EZ-K system) – removal and refitting	10
Distributor – removal and refitting	4	Maintenance and inspection	2
Electronic traction control – description	14	Motronic control unit – removal and refitting	12
Fault diagnosis – ignition system	19	Motronic system – testing procedure (electrical side)	18
Flywheel sensors (Motronic system) – removal and refitting	9	Spark plugs – removal, inspection and refitting	3
General description	1	Throttle position switch – adjustment	8
Ignition advance systems – description and testing	7	Torque limiter system – description and testing	13
Ignition control unit – removal and refitting	11		

Specifications

General
System type .... Electronic (breakerless); computerised control on all except B28E
Maker's designation:
  B23ET and B230ET .... Motronic
  B230E and B230K .... EZ-K
  B28E TSZ-4
Firing order:
  4-cylinder .... 1-3-4-2 (No 1 at front)
  6-cylinder .... 1-6-3-5-2-4 (No 1 LH rear)

Spark plugs
Make and type:
  B23ET .... Champion N7YCC or N7YC
  B230ET .... Champion RN7YCC or RN7YC
  B230K (to 1986) .... Champion N9YCC or N9YC
  B230E (to 1987) .... Champion N7YCC or N7YC
  B28E .... Champion S7YCC or S7YC

Electrode gap:
  N7YCC, RN7YCC, N9YCC and S7YCC .... 0.8 mm (0.032 in)
  N7YC, RN7YC, N9YC and S7YC .... 0.7 mm (0.028 in)

HT leads
740 and 760 Turbo (4-cylinder) .... Champion CLS12, boxed set

Ballast resistor
Resistance (total) .... 1.0 ± 0.1 Ω

Ignition timing*
B23ET .... 10° BTDC @ 750 rpm
B230E .... 12° BTDC @ 900 rpm
B230ET .... 10° BTDC @ 900 rpm
B230K .... 15° BTDC @ 800 rpm
B28E .... 10° BTDC @ 900 rpm; 25° to 29° BTDC @ 2500 rpm
Vacuum unit disconnected (when applicable)

Ignition coil
Primary resistance (typical) .... 0.5 to 0.9 Ω
Secondary resistance (typical) .... 6000 to 9000 Ω

Distributor (B28E)
Direction of rotation .... Clockwise
Impulse sender coil resistance .... 540 to 660 Ω
Rotor air gap .... 0.3 mm (0.012 in)
Rotor arm track resistance .... 4 to 6 kΩ

Motronic system test data
Airflow meter temperature sensor resistance:
  −10°C (+14°F) .... 8260 to 10 560 Ω
  +20°C (68°F) .... 2280 to 2720 Ω
  +50°C (122°F) .... 760 to 910 Ω

Chapter 4 Ignition system

Airflow meter potentiometer resistance:
   Terminals 6 and 9 .... 500 to 1100 Ω
   Terminals 6 and 7 .... 8 to 200 Ω
Flywheel sensor resistance .... 1000 Ω approx
Charge air temperature sensor resistance:
   +20°C (68°F) .... 985 to 1015 Ω
   +40°C (104°F) .... 1080 to 1110 Ω
   +130°C (266°F) .... 1550 to 1620 Ω

Coolant temperature sensor resistance:
   −10°C (+14°F) .... 8100 to 10 770 Ω
   +20°C (68°F) .... 2280 to 2720 Ω
   +80°C (176°F) .... 290 to 364 Ω

Torque wrench settings

	Nm	lbf ft
Spark plugs (dry threads):		
4-cylinder	25 ± 5	18 ± 4
6-cylinder	12 ± 2	9 ± 1.5
Knock sensor bolt	11	8

1 General description

The ignition system is responsible for igniting the compressed fuel/air charge in each cylinder in turn. This must be done at precisely the right moment for the prevailing engine speed and load. The various engines have slightly different systems, but the principles of operation are the same.

Low tension (LT) pulses are no longer produced by a contact breaker, but by the rotation of a toothed wheel in a magnetic field, or by sensors mounted near the rim of the flywheel. These pulses are received by a control unit where they are amplified to the level necessary to drive the ignition coil. The ignition coil converts the LT pulses into the high tension (HT) pulses needed to fire the spark plugs. HT pulses are conveyed to the appropriate spark plugs via the HT leads, distributor cap and rotor arm.

Ignition timing (the moment when the spark occurs) is varied according to engine speed and load. The faster the engine is turning, the earlier must the spark occur (more advance) in order to allow enough time for combustion. On V6 engines, ignition advance is controlled mechanically by centrifugal weights and springs in the distributor. A vacuum unit provides additional advance under conditions of high inlet manifold vacuum (denoting a light lead). On the in-line engines, the ignition advance is determined by the control unit, which receives information on engine temperature, throttle position and (on Turbo models) charge air temperature, as well as engine speed. The B230E/K control unit also receives a manifold vacuum signal.

The EZ-K system fitted to B230E/K engines incorporates an 'anti-knock' sensor mounted under the inlet manifold. This sensor

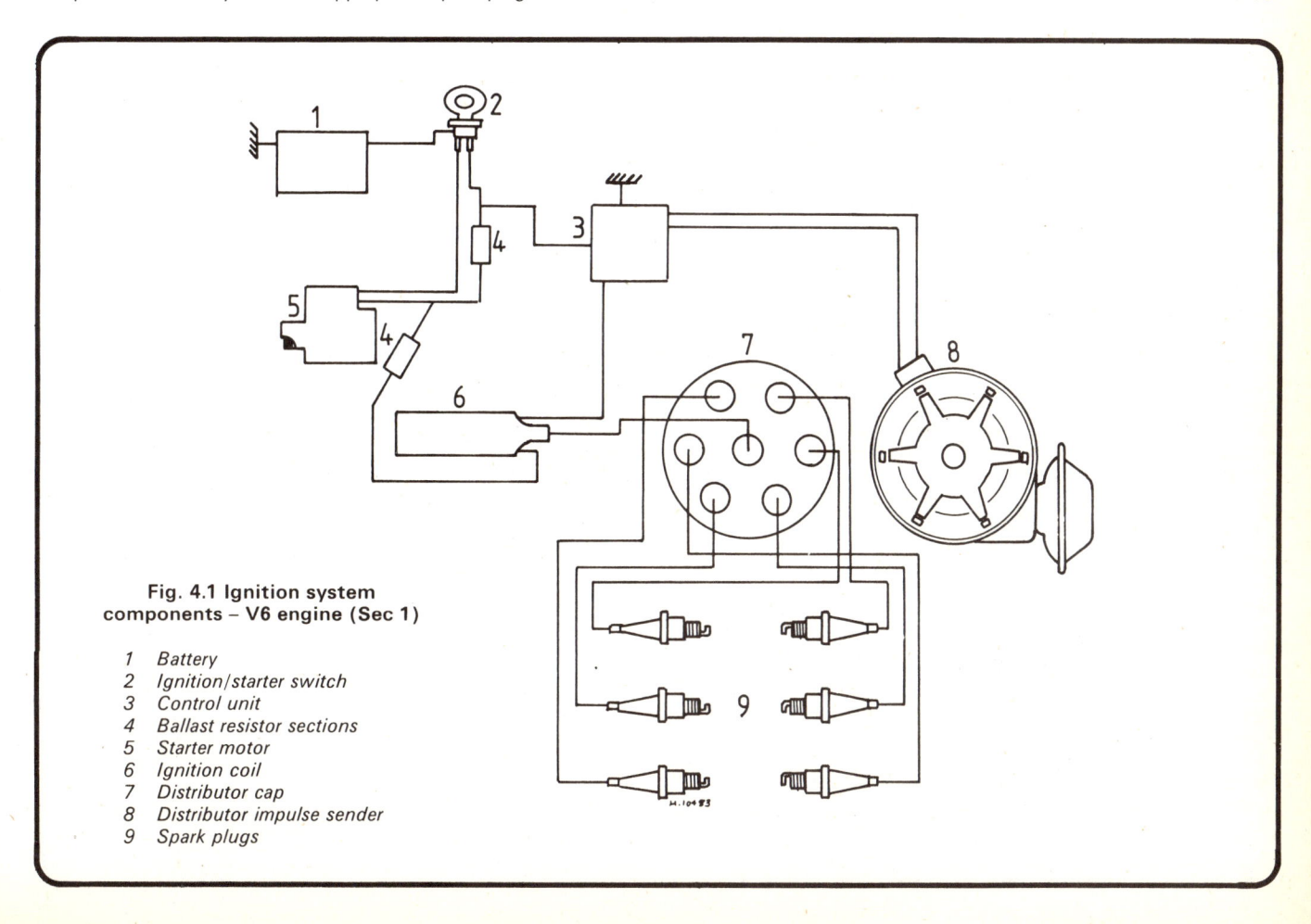

Fig. 4.1 Ignition system components – V6 engine (Sec 1)

1. Battery
2. Ignition/starter switch
3. Control unit
4. Ballast resistor sections
5. Starter motor
6. Ignition coil
7. Distributor cap
8. Distributor impulse sender
9. Spark plugs

Chapter 4 Ignition system

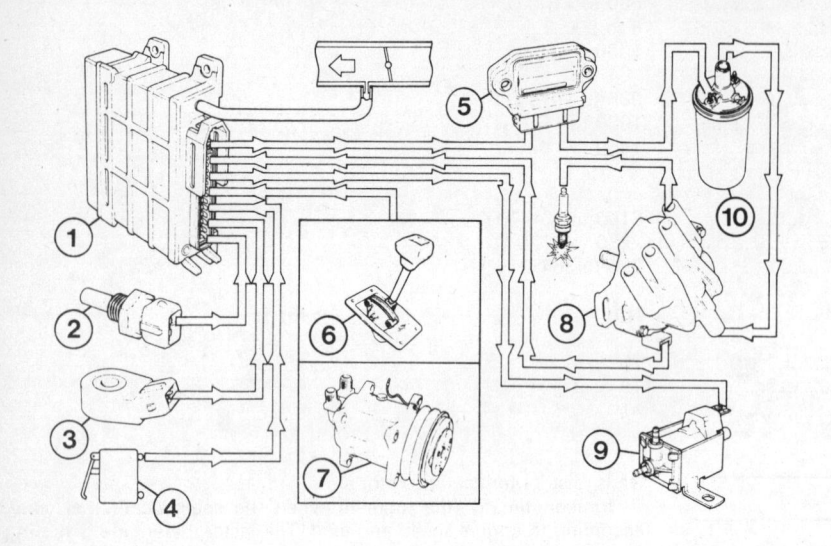

Fig. 4.2 EZ-K ignition system components (Sec 1)

1. Control unit
2. Temperature sensor
3. Knock sensor
4. Throttle switch
5. Power amplifier
6. Automatic transmission selector
7. Air conditioning compressor
8. Distributor
9. Fuel cut-off solenoid valve
10. Ignition coil

causes the control unit to retard ignition timing if pre-ignition occurs, so protecting the engine from damage caused by over-advanced timing or poor quality fuel.

The Motronic system fitted to Turbo models controls the fuel injection and ignition systems as an integrated whole; this is sometimes referred to as engine management. Special applications of this system are described in Sections 13 and 14. A more detailed description of the system is given in Chapter 3.

Ignition system HT voltage – warning

Take care to avoid receiving electric shocks from the HT side of the ignition system. Do not handle HT leads, or touch the distributor or coil, when the engine is running. When tracing faults in the HT system, use well insulated tools to manipulate live leads. **Electronic ignition HT voltage could prove fatal.**

2 Maintenance and inspection

1 Every 6000 miles (10 000 km) or six months, inspect the spark plugs as described in Section 3. Renew them or clean, regap and refit them. (The makers specifiy renewal at every service interval; the reader may follow this course if wished and so dispense with cleaning).
2 On B28E engines, lubricate the felt pad under the distributor rotor arm with a couple of drops of engine oil every 12 000 miles (20 000 km) or annually. Do not over-lubricate. (To remove the distributor cap, see Section 4, paragraphs 19 to 21).
3 On all models, inspect the HT leads, distributor cap and rotor arm every 12 000 miles (20 000 Km). Renew cracked, damaged or burnt items.
4 At all times keep the coil tower, HT leads and distributor cap clean and dry.

3 Spark plugs – removal, inspection and refitting

Note: *Spark plugs should not be removed when the engine is hot.*

1 The correct functioning of the spark plugs is vital for the correct running and efficiency of the engine. It is essential that the plugs fitted are appropriate for the engine, and the suitable type is specified at the beginning of this chapter. If this type is used and the engine is in good condition, the spark plugs should not need attention between scheduled replacement intervals. Spark plug cleaning is rarely necessary and should not be attempted unless specialised equipment is available as damage can easily be caused to the firing ends.
2 Open the bonnet. Make sure that the ignition is switched off. Remove any air cleaner trunking or similar items obstructing access to the spark plugs.
3 Inspect the HT leads to see if they carry their cylinder numbers – if not, number each lead, using sticky tape or paint.

Fig. 4.3 Take care to avoid electric shocks from the HT leads (Sec 1)

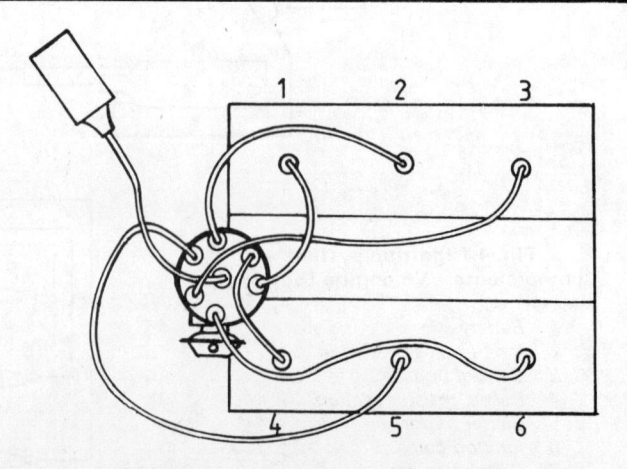

Fig. 4.4 HT leads and cylinder numbering – V6. Coil location may vary (Sec 3)

Chapter 4 Ignition system

4 Pull the HT lead connectors off the plugs. Pull on the connectors, not on the leads (photo).
5 Blow away any dirt from around the spark plug recesses in the cylinder head(s), using a bicycle pump or an air line.
6 Unscrew and remove the plugs, using a proprietary plug spanner or a spark plug socket, extension and ratchet. (For the V6 engine a deep socket spanner, 5/8in/16 mm AF, 3/8 in drive, will be required, together with a universal joint or other 'flexible' drive (photo). A suitable socket is available from the Champion Sparking Plug Company, part No CT-910).
7 Inspect the plugs and compare them with those illustrated by the coloured photographs in this Chapter. The condition of the plugs will tell much about the overall condition of the engine.
8 When fitting new spark plugs, check the electrode gap with a feeler gauge. Adjust if necessary by carefully bending the side electrode. Do not attempt to bend the centre electrode, or lever against it when bending the side electrode.
9 Make sure that the plug insulators are clean and that the screwed HT lead adaptors are tight. Pay particular attention to the plug seating surfaces on V6 engines, since these plugs have no sealing washers ('taper seat' type) and any dirt will cause a bad seal.
10 Screw each plug into its hole by hand. If a plug is reluctant to go in, do not force it with a spanner, but unscrew it and try again. If the plug is cross-threaded, it is the cylinder head which will be damaged.
11 Final tightening of the spark plugs should ideally be carried out using a torque wrench. The tightening torques are given in the Specifications. If a torque wrench is not available, tighten the plugs beyond the point where they contact the head as follows:

Taper seat plugs – One-sixteenth of a turn maximum
Plugs with washers – One-quarter of a turn maximum

12 If the taper seat type of plug is overtightened, the sealing faces will bite together and removal will be very difficult.

13 Refit the HT leads to the plugs, paying attention to the cylinder numbers. Push each connector firmly onto its plug.
14 Run the engine to verify that the HT leads have been refitted correctly.

4 Distributor – removal and refitting

B23

1 Unclip the distributor cap, or remove its securing screws, and move it aside.
2 Bring the engine to TDC, No 1 firing. Turn the engine with a spanner on the crankshaft pulley bolt until the notch on the pulley is aligned with the 'D' (TDC) mark on the timing seal, and the distributor rotor arm tip is pointing to the quarter corresponding to No 1 spark plug lead (photo).
3 Make alignment marks between the rotor arm tip and the rim of the distributor body, and between the base of the distributor body and the engine block.
4 Unbolt the clamp plate and lift out the distributor (photo). Notice how the rotor arm turns as the distributor is withdrawn.
5 When refitting, set the rotor arm in roughly the same position as it assumed after withdrawal. Insert the distributor so that the distributor-to-block marks are aligned. As the distributor is pushed home, the rotor arm will turn and should end up in the previously marked position for No 1 firing (photo). If not, withdraw the distributor and try again.
6 Refit and secure the distributor clamp plate.

B230

7 Identify the spark plug leads and disconnect them from the cap (photo).
8 Undo the three screws which secure the distributor cap. Access is

3.4 Pulling an HT lead connector off a spark plug

3.6 Unscrewing a spark plug with a socket and a universal joint

4.2 Timing marks on the in-line engine – pulley notch is at '0' (TDC)

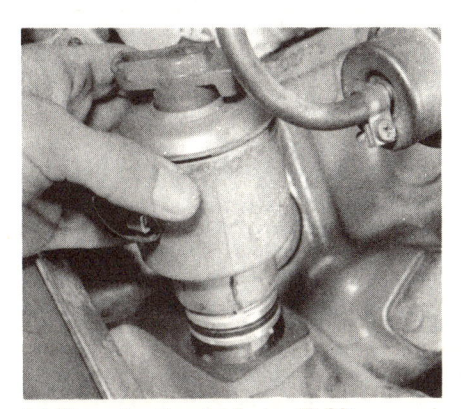

4.4 Removing the distributor (B 23)

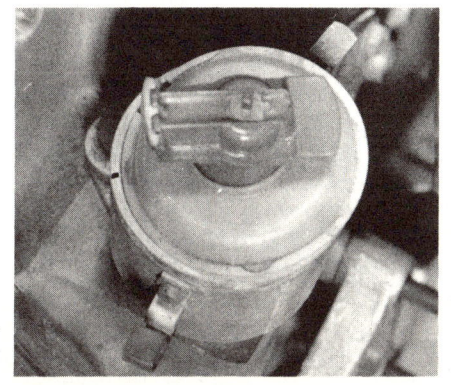

4.5 Rotor arm tip aligned with notch in distributor rim (B 23) – dust shield removed

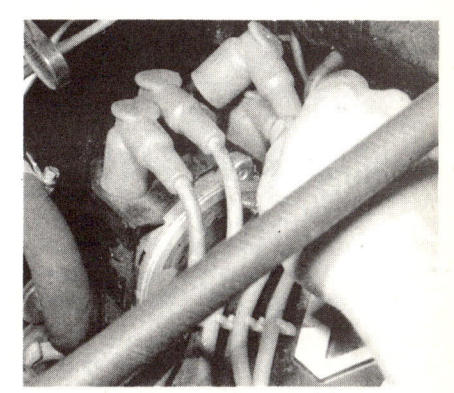

4.7 Disconnecting a spark plug lead (B 230)

4.8 Distributor cap securing screws (arrowed) – distributor removed

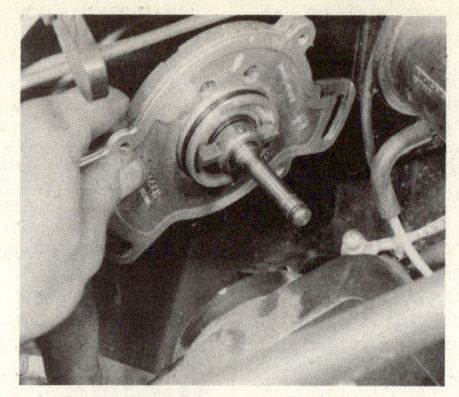

4.14 Removing the distributor (B 230)

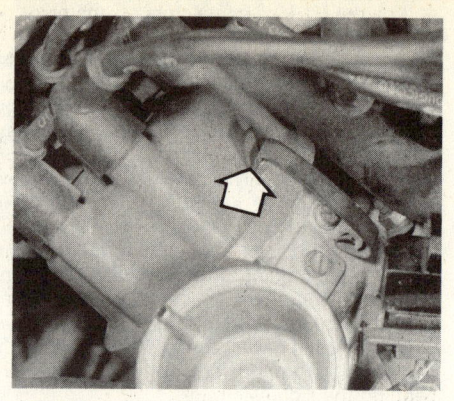

4.21 One of the distributor cap clips (arrowed) – B 28

4.22 Timing marks on the V6 engine. Pulley notch is at '0' (TDC), but there are two notches

4.23A Prise up the spring clip ...

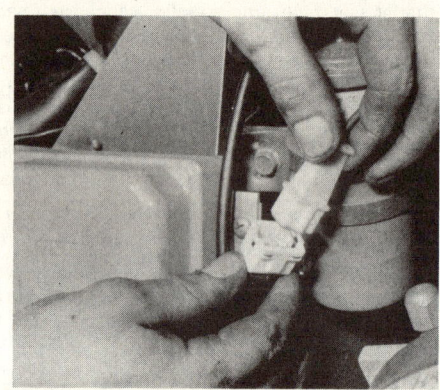

4.23B ... and unplug the LT connector

restricted. The screws are captive, so do not attempt to remove them from the cap (photo).
9 Lift off the distributor cap with the coil HT lead still attached.
10 Pull off the rotor arm and remove the dust shield.
11 Make alignment marks between the distributor flange and the cylinder head.
12 Remove the two bolts which secure the distributor.
13 Disconnect the LT connector from the distributor (when applicable).
14 Remove the distributor from the cylinder head (photo).
15 Renew the distributor O-rings if necessary.
16 When refitting, offer the distributor to the head, observing the alignment marks, and turn the shaft to align the drive dogs with the slots in the camshaft. The drive is offset, so there is no possibility of incorrect fitting.
17 The remainder of refitting is a reversal of the removal procedure.
18 On B230E and B230K engines, check the ignition timing (Section 6).

B28
19 Remove the air intake trunking from the airflow sensor.
20 Remove the twelve Allen screws which secure the fuel control unit top section. (If it is only wished to remove the distributor cap and rotor arm, this can just about be done without disturbing the fuel control unit. Some of the HT leads will have to be disconnected from the cap *in situ*).
21 Identify the HT leads. Unclip the distributor cap and remove it, disconnecting HT leads as necessary (photo). Lift the fuel control unit up, without straining the fuel lines, to provide clearance for the cap.
22 Bring the engine to TDC, No 1 firing (photo). (See paragraph 2 and Fig. 4.11).
23 Disconnect the LT connector between the distributor and the ignition control unit. Also disconnect the vacuum advance pipe (photos).

24 Make alignment marks between the distributor flange and the cylinder head. Remove the clamp nut and lift out the distributor, again lifting the fuel control unit to provide clearance.
25 Commence refitting by positioning the rotor arm approximately 30° clockwise of the No 1 reference notch in the rim of the distributor (Fig. 4.5).
26 Insert the distributor, observing the previously made alignment marks. The rotor arm should turn to align with the reference notch as

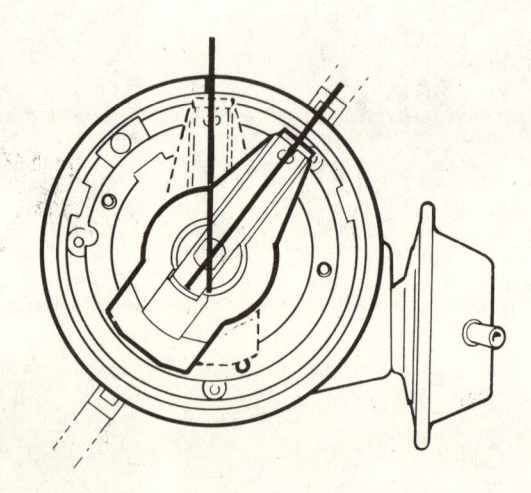

Fig. 4.5 Rotor arm position before refitting (shown solid) and after refitting (dotted) (Sec 4)

Chapter 4 Ignition system

4.26 Rotor arm tip aligned with No 1 cylinder notch in distributor rim (B 28) – dust shield removed

the distributor is pushed home (photo). Fit the clamp nut and tighten it lightly.
27 Reconnect the LT connector.
28 Refit the distributor cap and HT leads.
29 Secure the fuel control unit and refit the air trunking.
30 Check the ignition timing and adjust if necessary (Section 6), then fully tighten the clamp nut.

5 Distributor – overhaul

1 Check the availability and cost of spares before deciding to overhaul the distributor.

TSZ-4 system (B28E)

2 Pull off the rotor arm and remove the dust shield (photo).
3 Remove the circlip from the shaft. Carefully pull or prise off the rotor wheel. Recover the locating pin (photos).
4 Remove the LT connector securing screw. Pull out the connector (photo).
5 Remove the two screws which secure the vacuum unit. Unhook the vacuum unit from the baseplate and remove it (photos).
6 Slacken the screws which secure the cap clips. There is no need to remove the clips unless they are to be renewed.
7 Remove the three screws which secure the baseplate to the distributor body. Remove the remaining circlip from the shaft. Lift out the baseplate complete with coil and magnet (photo).
8 The centrifugal advance weights and springs are now accessible (photo).
9 To remove the distributor shaft itself, drive out the roll pin which secures the gear to the shaft. Remove the gear and lift out the shaft.
10 Examine all parts and renew as necessary. Measure the resistance of the impulse sender coil (see Specifications for the correct value). Apply vacuum to the vacuum unit to verify that the arm moves and that the diaphragm is intact. Renew the rotor arm and distributor cap unless they are known to be in perfect condition.
11 Reassemble by reversing the dismantling procedure. Lubricate the shaft bearing and the centrifugal mechanism with a little light oil. Tap the rotor wheel pin home with a punch or similar tool (photos).
12 After refitting the rotor wheel, check that the gaps between the rotor and station teeth are as specified. Bend the stator teeth carefully if necessary.

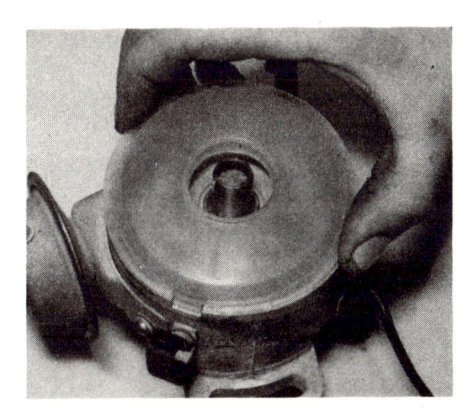

5.2 Removing the dust shield

5.3A Removing the upper circlip. Note locating pin (arrowed) in groove

5.3B Pulling off the rotor wheel

5.4 Undoing the LT connector screw

5.5A Removing the vacuum unit screws ...

5.5B ... and unhook the vacuum unit

5.7 Removing the baseplate with coil and magnet

5.8 Distributor body showing centrifugal advance weights and springs

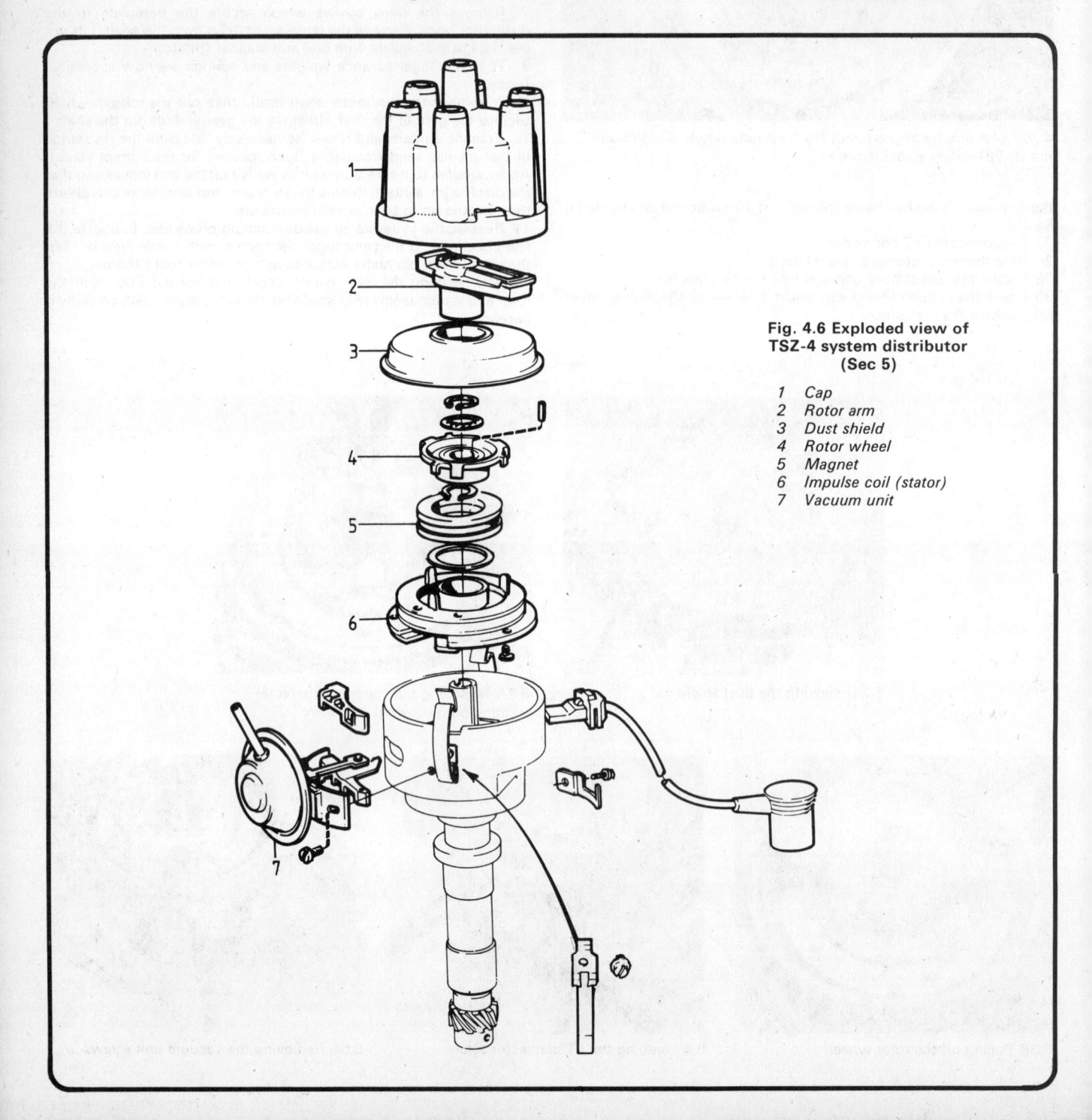

Fig. 4.6 Exploded view of TSZ-4 system distributor (Sec 5)

1. Cap
2. Rotor arm
3. Dust shield
4. Rotor wheel
5. Magnet
6. Impulse coil (stator)
7. Vacuum unit

Chapter 4 Ignition system

5.11A Align the grooves and insert the rotor wheel pins ...

5.11B ... and tap it home

5.14 Drive out the dog securing pin (arrowed) to release the shaft components

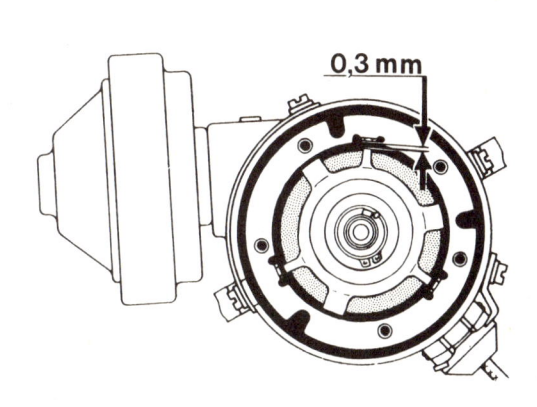

Fig. 4.7 Gap between rotor and stator teeth must be as shown (Sec 5)

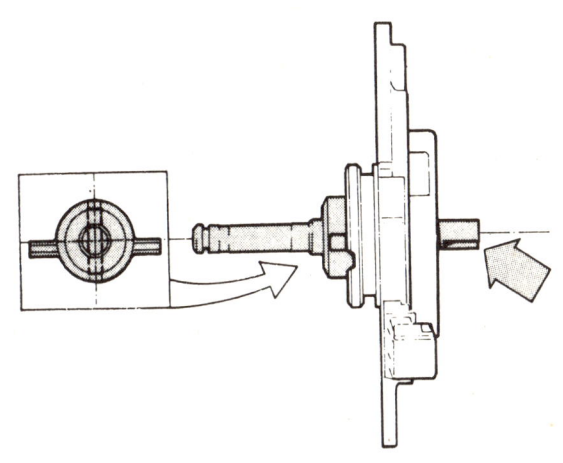

Fig. 4.8 Relationship between dog offset (left) and shaft cutaway (right) on EZ-K distributor (Sec 5)

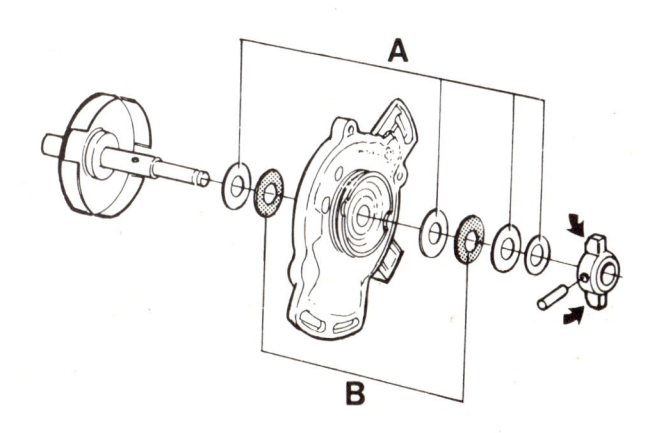

Fig. 4.9 EZ-K distributor shaft components (Sec 5)

A Steel washers
B Fibre washers

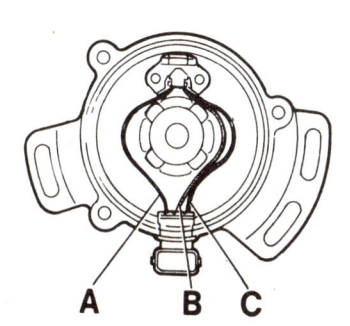

Fig. 4.10 EZ-K distributor – Hall sensor wiring (Sec 5)

A Black
B Green
C Red

Motronic system (B23/230ET)

13 Overhaul of the distributor is normally limited to renewal of the cap and rotor arm. There is nothing else to go wrong in the Motronic system distributor, apart from the shaft itself.

EZ-K system (B230E/K)

14 If spares are available, the Hall sensor on the distributor baseplate can be renewed. The shaft and associated components must be removed first; they are released by driving out the dog securing pin (photo). Note the relationship of the dog offset to the shaft.
15 Drill out the rivets which secure the old sensor. Remove the sensor and connector; transfer the connector and wiring to the new sensor.
16 Secure the sensor with new rivets. Reassemble the shaft components and secure them with a new pin.

6 Ignition timing – checking and adjustment

This is not a routine operation, since there is normally no reason why the timing should vary. On Turbo engines (with the Motronic

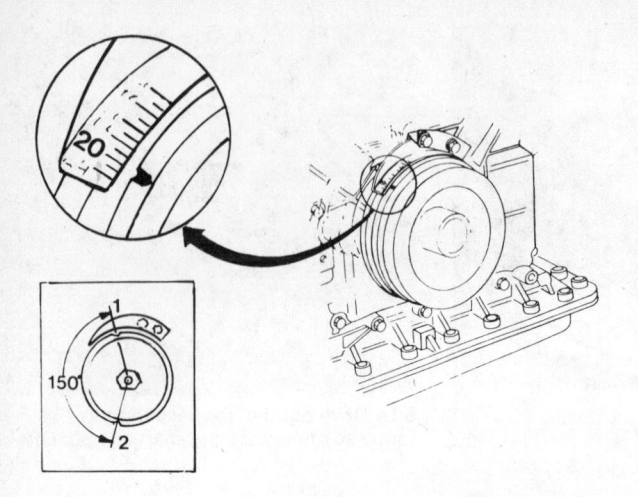

Chapter 4 Ignition system

Fig. 4.11 Timing marks for B 28 E – 20° BTDC highlighted. Note the two pulley notches (Sec 6)

1 No 1 cylinder notch   2 No 6 cylinder notch

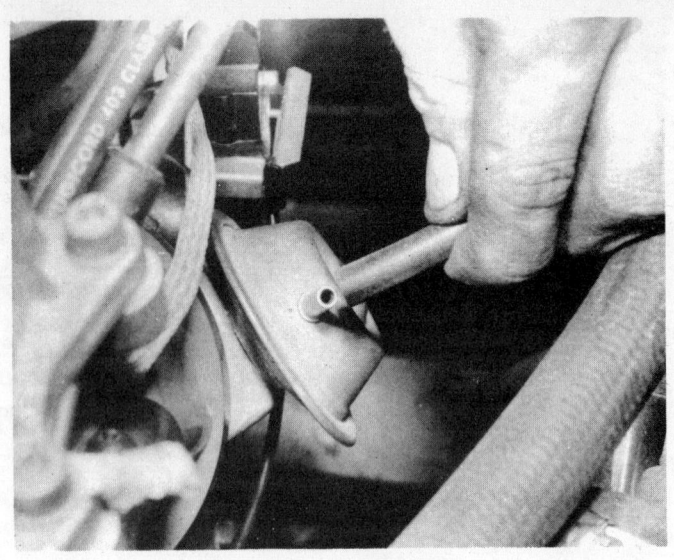

6.3 Disconnecting the distributor vacuum pipe

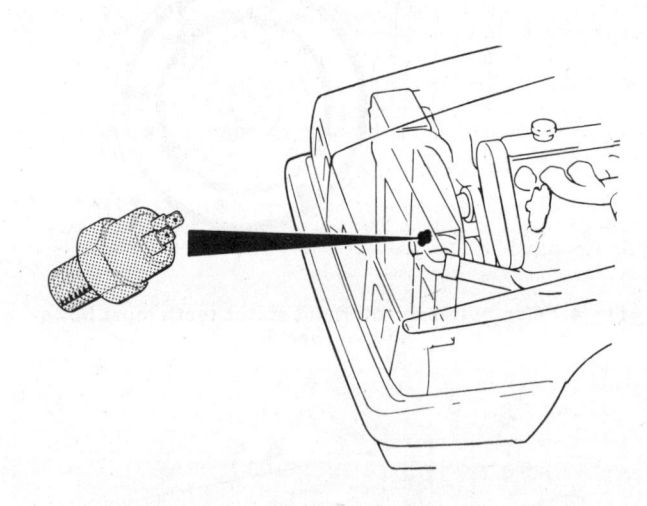

system) the timing cannot be adjusted; although it can be checked if wished.

1 Bring the engine to operating temperature with the air conditioning switched off. With the engine stopped, connect a timing light (stroboscope) and a tachometer as instructed by the manfuacturers.
2 Highlight the notch on the crankshaft pulley and the desired marks on the timing scale with white paint or typist's correction fluid. (See Specifications for the desired values). Be sure to use the correct pulley notch on the B28E engine (Fig. 4.11).
3 On B28E and B230E/K engines, disconnect and plug the distributor or control unit vacuum advance pipe (photo).
4 Run the engine at the specified idle speed and shine the timing light on the timing scale. **Caution:** *Do not get electrical leads, clothing, long hair etc, caught in the drivebelts or the fan.* The pulley notch will appear stationary and (if the timing is correct) in alignment with the appropriate mark on the timing scale.
5 If adjustment is necessary, stop the engine. Slacken the distributor mountings and turn the distributor a small amount. Turning the distributor against the direction of shaft rotation advances the timing, and *vice versa*. Tighten the distributor mountings after each adjustment, restart the engine and recheck the timing. (Do not attempt to turn the Motronic system distributor. It will not affect the timing).
6 When the timing at idle is correct, check the advance on non-Turbo models by observing the timing at the higher specified speed. If this is incorrect on B28E engines, it is caused by a defective mechanical advance mechanism; on other models it must be due to a fault in the control unit or its inputs. (Normal operation of the knock sensor, when applicable, will also retard the timing).
7 If a vacuum pump is available, apply vacuum to the advance unit (when so equipped) and verify that the timing is advanced.
8 Stop the engine, disconnect the test gear and remake the original electrical and vacuum connections.

7 Ignition advance systems – description and testing

1 Beside the speed, load and knock-related controls imposed on the ignition timing, some models are subject to ignition advance applied for the purpose of raising or maintaining engine speed. The advance may be imposed electrically (within the control unit) or mechanically (by switching a vacuum feed).

Temperature-related advance (overheating)

2 This system is fitted to certain vehicles with air conditioning and the EZ-K ignition system. By means of a radiator-mounted thermoswitch, ignition timing is advanced by 8 to 13° if the coolant temperature exceeds 103° (217°F). The throttle switch must be closed (denoting throttle pedal released).

Fig. 4.12 Radiator thermoswitch fitted for some temperature-related advance systems (Sec 7)

3 The system may be checked by bridging the thermoswitch terminals with the engine idling: the timing advance will be evidenced by a raised idle speed and a change in note, or may be verified accurately using a timing light.
4 The thermoswitch may be tested as described in Chapter 2, Section 11, but using oil instead of water to heat the switch.

Temperature-related advance (warm-up)

5 This system is fitted to the B230K engine to maintain idle speed and promote rapid warm-up. Acting on information received from the coolant temperature sensor, the control unit advances the timing by 15° below 5°C (41°F), and by 5 to 6° between 5 and 12°C (41 and 54°F).
6 The system may be checked by watching the ignition timing during warm-up after a cold start in winter. Alternatively, the sensor may be fooled by immersing it in iced water. (A spare sensor, or a blanking plug, would have to be fitted to the cylinder head).

Air conditioning advance (idle compensation)

7 This system is fitted to the B23ET engine to maintain idle speed when the air conditioning compressor is engaged at idle. It operates via a relay located behind the Motronic control unit in the driver's footwell. The amount of advance is approximately 17°.

Chapter 4 Ignition system

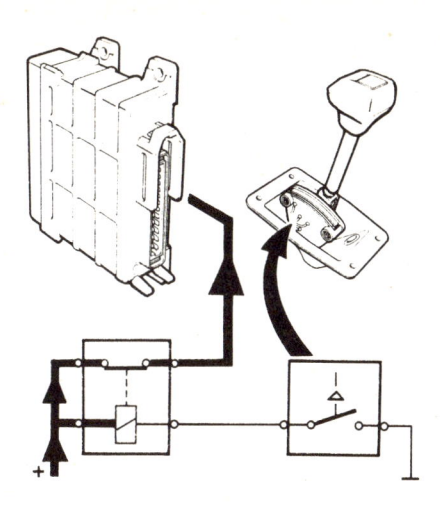

Fig. 4.13 Transmission-controlled advance system (Sec 7)

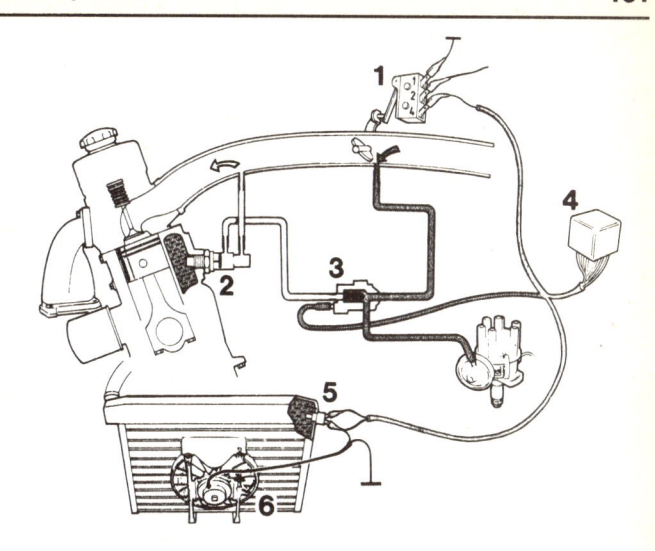

Fig. 4.14 Components of switched vacuum advance system (Sec 7)

1 Throttle switch
2 Thermal vacuum valve
3 Solenoid valve
4 Relay
5 Thermoswitch
6 Electric fan

8 Checking should be carried out by observing the ignition timing whilst an assistant switches the air conditioning on and off. The relay itself is best tested by substitution.

Transmission-controlled advance

9 This system is found on the B230K engine with automatic transmission. Its function is to prevent the engine stalling when a drive range is selected, by advancing the timing by 8°. A gear selector switch and a relay convey the information to the control unit.
10 Checking should be carried out by observing the ignition timing whilst an assistant moves the selector lever through its positions.
Caution: *Apply the handbrake firmly and chock the drive wheels so that the vehicle cannot move.*
11 Note that with this system the extra advance is applied when the relay is not energised, and is cancelled when the relay is energised.

Switched vacuum advance (temperature-related)

12 This system is fitted to some B28E engines. It is linked with the throttle microswitch (part of the constant idle system). Not all the features of the system described here are fitted to all models. The main components of the system are shown in Fig. 4.14.
13 At coolant temperature below 55°C (131°F), the thermal vacuum valve is closed and no vacuum is applied.
14 Between 55 and 100°C (137 and 212°F) the thermal vacuum valve is open. The throttle microswitch controls the solenoid valve so that manifold vacuum is applied above idle (photo). At idle the solenoid valve changes to a vacuum tapping above the throttle valve. The ignition is thus retarded, which makes for smooth idling.
15 Above 100°C (212°F) the radiator thermoswitch closes. This operates the electric cooling fan and also de-energises the relay. The solenoid valve is thus de-energised, regardless of throttle position, and manifold vacuum is applied all the time. This raises the idle speed and improves engine cooling.
16 Testing of the warm-up and normal running phases may be carried out by observing the shift in timing whilst an assistant operates the throttle microswitch. The overheating mode may be simulated by disconnecting the thermoswitch and bridging the connector plug terminals. The thermoswitch itself can be tested by immersing it in boiling water.

8 Throttle position switch – adjustment

On models with the constant idle speed system, see Chapter 3, Section 10.

Rotary switch

1 Listen to the switch whilst an assistant opens and closes the throttle (engine stopped). The switch should be heard to click as soon as the throttle starts opening.

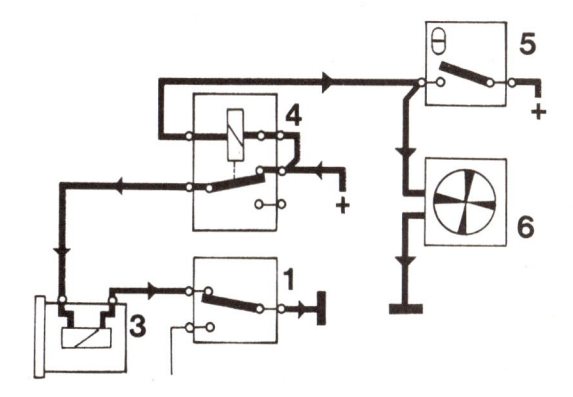

Fig. 4.15 Circuit diagram of switched vacuum advance system. For key see Fig. 4.14 (Sec 7)

7.14 Solenoid vacuum valve on B 28 engine

Chapter 4 Ignition system

2 If adjustment is necessary, slacken the switch mounting screws. Turn the switch clockwise slightly, then anti-clockwise until it clicks. Tighten the mounting screws in this position and recheck (photo).

Microswitch
3 This may be mounted near the throttle drum or below the throttle pedal. The principle of adjustment is as just described for the rotary switch, but the mountings are different (photo).

9 Flywheel sensors (Motronic system) – removal and refitting

1 Disconnect the sensor multi-plug near the bulkhead. Identify the plugs if both sensors are to be removed (photo).

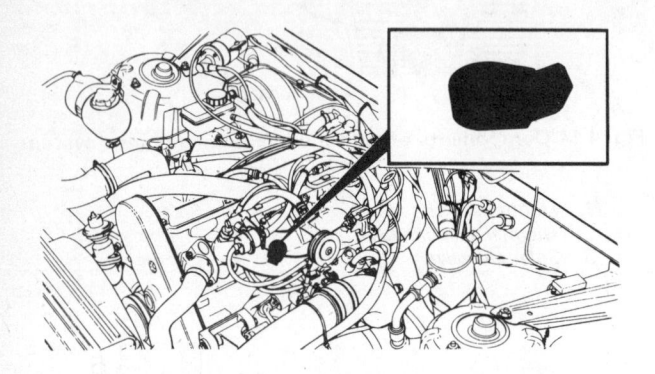

Fig. 4.16 Knock sensor location (Sec 10)

2 Remove the Allen screw which secures the sensor to the bracket. Access is restricted (photo).
3 Withdraw the sensor from the bracket and remove it.
4 Refit by reversing the removal operations.

10 Knock sensor (EZ-K system) – removal and refitting

1 Disconnect the multi-plug from the knock sensor.
2 Remove the sensor securing bolt and the sensor itself (photo). It is located under the inlet manifold.
3 Refit by reversing the removal operations. Apply thread locking compound to the bolt and tighten it to the specified torque.

11 Ignition control unit – removal and refitting

1 Disconnect the battery negative lead.

TSZ-4 (B28E)
2 The control unit is located on the right-hand inner wing, next to the coil. Disconnect the multi-plug from the base of the unit (photo).
3 Undo the securing screws and remove the control unit.
4 Refit by reversing the removal operations.

EZ-K (B230E/K)
5 Remove the trim from below the steering column.
6 Remove the four screws which secure the control unit to the right-hand end of the pedal bracket. Depress the throttle pedal for access to two of the screws (photo).
7 Disconnect the multi-plug and the vacuum pipe. Remove the control unit (photo).
8 Refit by reversing the removal operations.

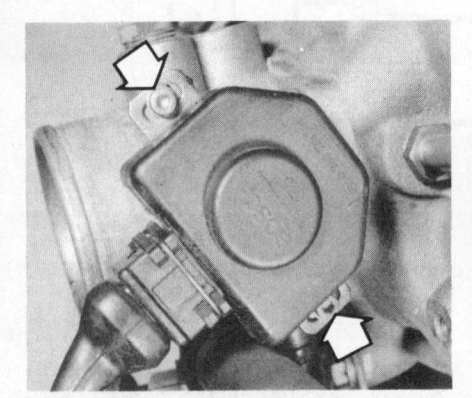

8.2 Rotary type throttle switch – mounting screws arrowed

8.3 Throttle microswitch at throttle drum – mounting nuts arrowed

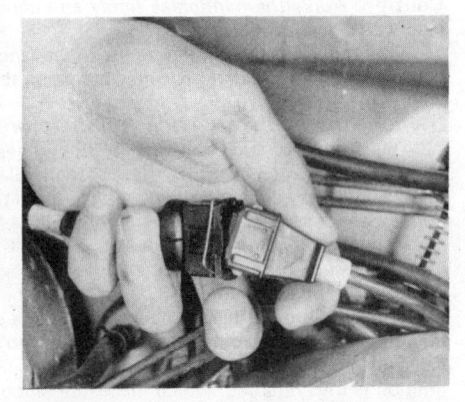

9.1 Disconnecting a flywheel sensor multi-plug

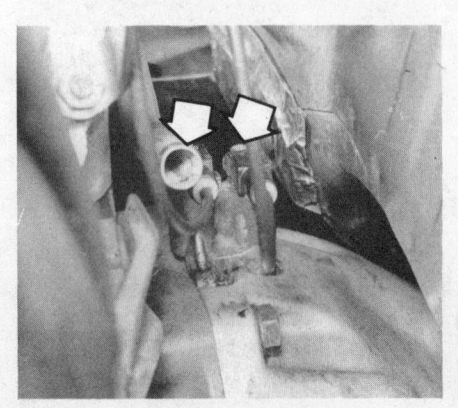

9.2 The two flywheel sensors (arrowed)

10.2 Knock sensor securing bolt (arrowed)

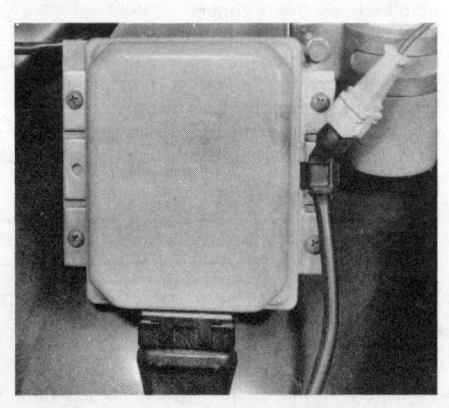

11.2 Ignition control unit – TSZ-4

Chapter 4 Ignition system

11.6 Removing an ignition control unit screw – EZ-K

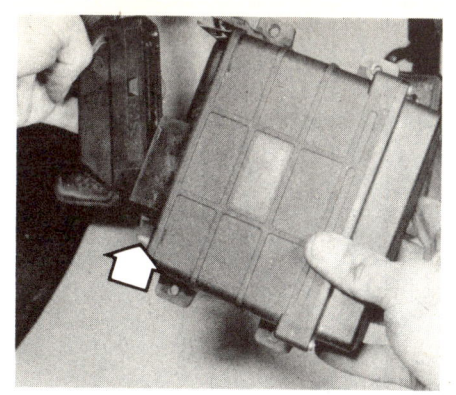

11.7 Disconnecting the control unit multi-plug. Vacuum pipe stub is arrowed

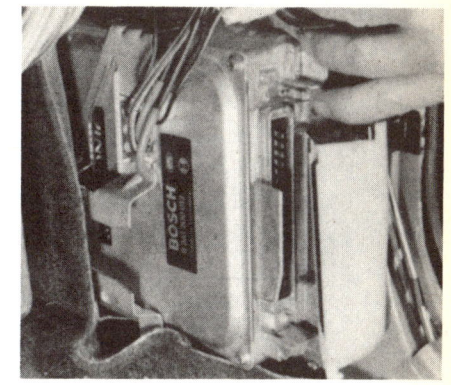

12.4 Disconnecting the Motronic control unit multi-plug

12 Motronic control unit – removal and refitting

1 Remove the steering column/pedal lower trim panel.
2 Release the forward end of the sill/seat belt reel trim by removing the screws securing it. These screws are concealed by covers.
3 Unclip the right-hand footwell trim panel and remove it.
4 Make sure the ignition is switched off, then disconnect the multi-plug from the control unit (photo).
5 Remove the two securing screws and slide the unit out of its bracket.
6 Refit by reversing the removal operations.
7 On some later models the power stage of the control unit is mounted separately under the bonnet, forward of the left-hand suspension turret.

13 Torque limiter system – description and testing

1 This system is fitted to Turbo models with manual transmission prior to 1986. Its function is to limit engine power during the engagement of overdrive. (For 1986 model year the construction of the overdrive was modified to withstand maximum torque without slipping).
2 The components of the system are a torque limiter relay, a turbo boost pressure switch and an overdrive oil pressure switch. The relay is located in position 'B' of the central electrical unit. The boost pressure switch is on the brake pedal bracket and the oil pressure switch is on the overdrive housing. The relay is also connected to the Motronic control unit and to the overdrive relay.
3 When the driver engages overdrive, if turbo boost pressure exceeds 0.3 bar (4.4 lbf/in²), the system momentarily interrupts the power supply to No 2 injector, so reducing engine power output slightly.
4 Malfunction of the system is indicated by slipping of the overdrive when engaged at high load, or by rough running at high load with overdrive engaged. Rough running suggests that the torque limiter relay is not closing: the testing procedure is included in Section 18 (paragraph 18). Slipping suggests that the relay is not opening, which may be checked as follows.
5 A 12 volt test light and a manifold pressure gauge with pump will be needed.
6 Chock the front wheels, raise the rear of the vehicle so that the rear wheels are free and support it securely. Damage or injury could result if the vehicle falls off the supports.
7 Connect the test light between earth and terminal 13 in the 20-pole connector. This connector is located behind the trim next to the right-hand side A-pillar, in a black plastic retainer. Terminal 13 has a green/black wire running to it.
8 Switch on the ignition: the test light should come on. If not, see Section 18, paragraph 18.
9 Disconnect the multi-plug from No 2 injector. (This will make the operation of the relay easier to detect).

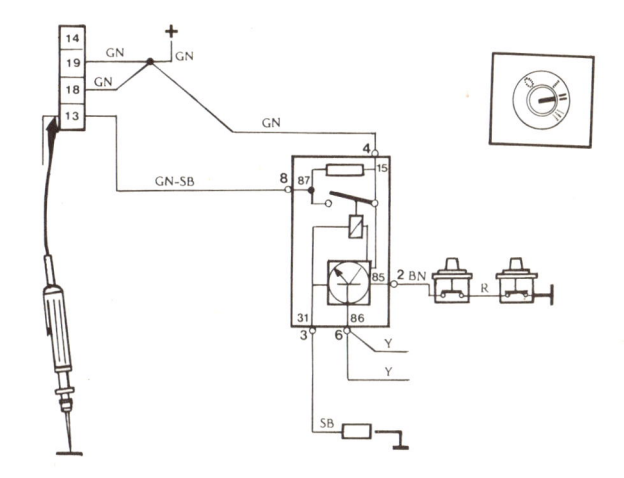

Fig. 4.17 Circuit diagram of torque limiter components – test light shown connected at left (Sec 13)

BN	Brown	SB	Black
GN	Green	Y	Yellow
R	Red		

10 Connect the manifold pressure gauge and pump to the boost pressure hose at the inlet manifold. Apply 0.3 bar (4.4 lbf/in²) to close the boost pressure switch.
11 Run the engine, taking appropriate precautions against carbon monoxide poisoning. Engage 4th gear and increase engine speed to indicate roughly 45 mph (70 km/h).
12 Observe the test light, which should still be lit. Engage the overdrive: the test light should dim for a fraction of a second, confirming the operation of the relay. If not, the relay, the wiring or the switches may be faulty. Note that the relay was modified in 1984; old and new relays are not interchangeable unless modifications are made to the wiring. Consult a Volvo dealer for details.
13 Stop the engine and lower the vehicle. Disconnect the test gear, remake the original connections and refit disturbed components.

14 Electronic traction control – description

Available on some later models, electronic traction control (ETC) monitors the rotation of all four wheels and limits engine torque if wheel slip is occurring. The driver can therefore accelerate hard on loose or slippery surfaces without danger of losing control. In many ways the system mirrors the anti-lock braking system (ABS); when both systems are fitted, some components are shared.
The main components of the system are the wheel sensors and the control unit. The front wheel sensors are mounted on the stub axles

154

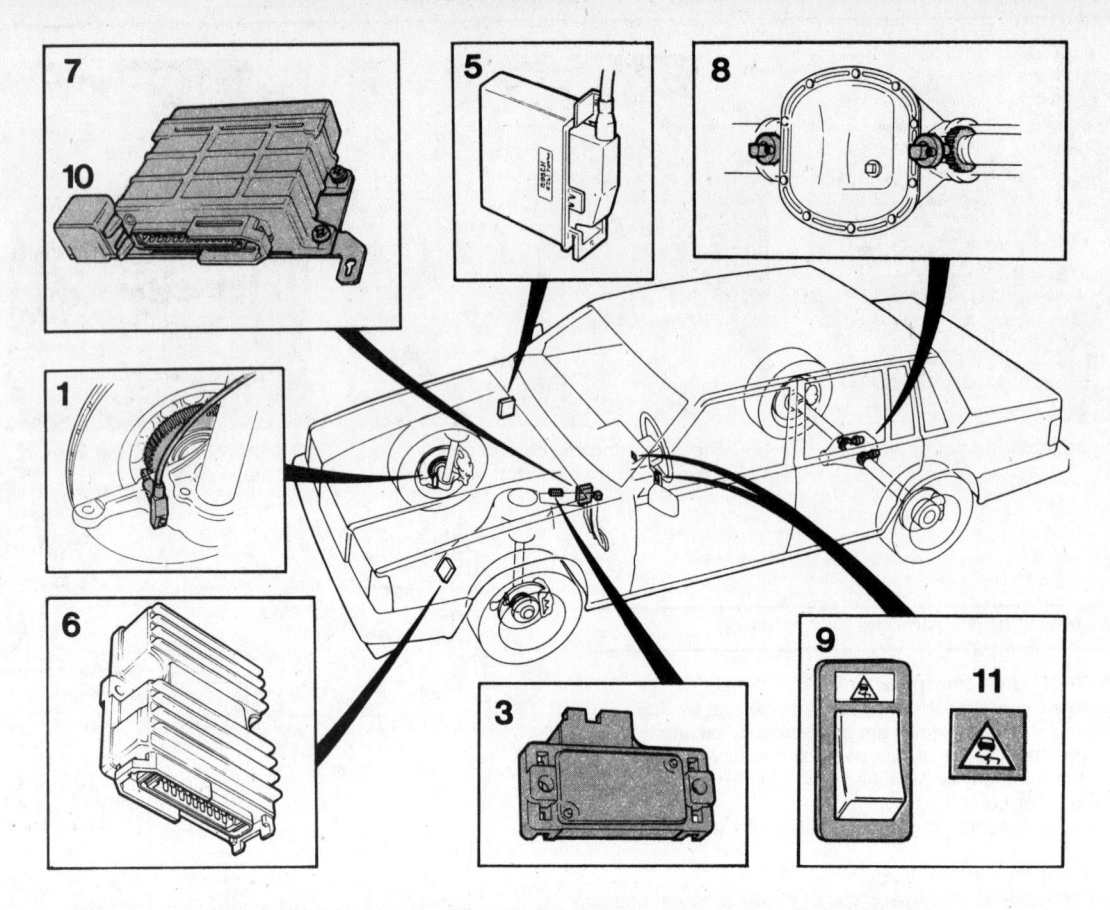

Fig. 4.18 Electronic traction components. For key see Fig. 4.19 (Sec 14)

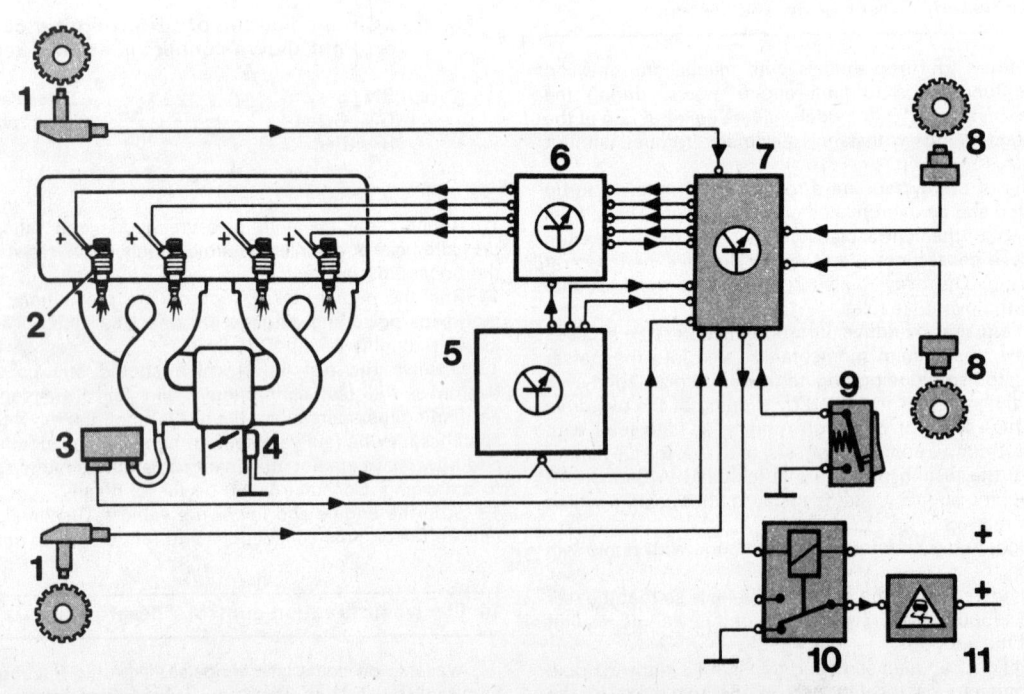

Fig. 4.19 Electronic traction control circuit diagram (schematic) (Sec 14)

1 Front wheel sensors	4 Throttle switch	7 ETC control unit	10 Relay
2 Fuel injectors	5 Motronic control switch	8 Rear wheel sensors	11 Warning light
3 Manifold pressure sensor	6 Motronic power stage	9 Control switch	

Chapter 4 Ignition system

and are shared with the ABS, when fitted. The rear wheel sensors are mounted on the rear axle tubes on each side of the differential housing. The control unit and relay are mounted on the brake pedal bracket.

Minor components include a manifold pressure sensor, a control switch and a warning light. The ETC components are connected to the Motronic control unit and power stage, by which the fuel injectors are controlled.

The driver can turn the system on or off as wished. When it is turned off, the warning light is constantly lit.

In use, with the system switched on, acceleration which would provoke wheelspin is countered by the ETC progressively cutting the power supply to the fuel injectors. In the extreme case, three injectors can be out completely and the remaining one by half. When the system is doing this, the warning light flashes and the relay ticks to show the driver that excessive throttle is being applied. As traction improves, the power supply to the injectors is gradually restored.

In the event of a fault in the system, three injectors will be cut and the warning light will come on. If the switch is then held depressed for ten seconds, a fault code will be flashed by the light (unfortunately details of this code are not yet available). If the switch is pressed momentarily, the ETC will be switched off and normal engine operation will be restored, but the fault code will be lost.

Faults in the ETC system should be referred to a Volvo dealer.

15 Ignition/engine management relays – location

1 The following relays are located on the central electrical unit:

Ignition advance (except B230E) – position 'O' or 'R'
Motronic – position 'E'
Torque limiter – position 'B'

2 The ignition advance relay on B230E models is under the bonnet, on the left-hand suspension turret.
3 The air conditioning ignition advance relay on B23ET engines is located behind the Motronic control unit.
4 The ETC relay is located on the brake pedal bracket.

16 Ballast resistor – testing, removal and refitting

1 A ballast resistor is only fitted to B28E engines. Its function is to assist starting. During normal running the resistor is in series with the coil LT windings, and approximately 6 volts are dropped across both resistor and coil. During start-up a section of the resistor is bypassed to compensate for the drop in battery voltage caused by the starter motor.
2 To test the ballast resistor, first disconnect the wiring from it. Use a suitable multi-meter to measure the resistances across the two uppermost terminals. The desired value is given in the Specifications.
3 To remove the ballast resistor, disconnect the wiring from it, remove the securing screw and remove the resistor complete with bracket (photo).
4 Refit by reversing the removal procedure.

17 Coil – testing, removal and refitting

1 Inspect the coil visually for cracks, leakage of insulating oil or other obvious damage. Renew it if such damage is evident.
2 Disconnect the LT and HT leads from the coil (photo). Use a multi-meter to measure the primary resistance (between the LT terminals). The desired value is given in the Specifications.
3 Measure the secondary resistance (between one of the LT terminals and the HT terminal) and compare it with that specified.
4 Renew the coil if the resistance of either winding differs widely from the correct value.
5 To remove the coil, disconnect the wires from it and release the clamp bracket. Slide the coil out of the bracket.
6 Refit by reversing the removal operations.

18 Motronic system – testing procedure (electrical side)

A multi-meter will be needed for this procedure. It is assumed that the fuel side is working correctly (Chapter 3, Section 45).

This procedure applies specifically to the B23ET engine. The B230ET is believed to be similar.

1 Gain access to the Motronic control unit and disconnect the multi-plug from it (Section 12).
2 Remove the securing screw and the cable clip and slide the cover off the multi-plug.
3 Tests will now be made across various terminals in the multi-plug. Insert the meter probes through the holes in the side of the plug, **not** into the mating face. The terminal numbers are marked on the side of the plug.
4 Check for continuity between earth and terminals 5, 16, 17 and 19. If continuity is not shown, check the earth tags on the inlet manifold.
5 Check for battery voltage between earth and terminals 18 and 35. Voltage should be present when the ignition is on and when the starter motor is operating.
6 Check the voltage between terminal 1 and earth with the ignition on. This is the earthy side of the coil LT winding, so less than full voltage may be obtained.

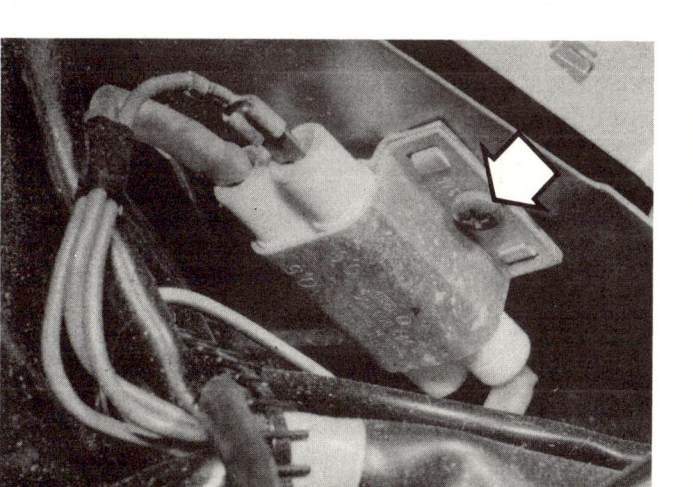

16.3 Ignition ballast resistor – securing screw arrowed

17.2 Disconnecting the coil HT lead

Chapter 4 Ignition system

Fig. 4.20 Motronic circuit diagram (Sec 18)

1. Airflow meter
2. TDC sensor
3. Engine speed sensor
4. Charge air temperature sensor
5. Throttle position switch
6. Coolant temperature sensor
7. Air conditioning signal relay
8. Starter motor
9. Start injector
10. Thermal timer
11. Control unit
12. Battery
13. Ignition/starter switch
14. Auxiliary air valve
15. System relay
16. Boost overpressure switch
17. Tank pump
18. Main fuel pump
19. Torque limiter relay
20. Overdrive pressure switch
21. Boost pressure switch
22. Fuel injectors
23. Ignition coil
24. Fuses (positions as numbered)

Connectors
A 4-pole, RH inner wing
B 3-pole, on bulkhead
C 3-pole, on bulkhead
D 8-pole, RH inner wing
E 20-pole, RH A-pillar
F 20-pole, LH A-pillar
G 11-pole, RH A-pillar
H Positive feed, RH A-pillar
J 20-pole, LH A-pillar
K Between gear lever and central electrical unit
L 8-pole, LH inner wing

Earthing points
I   Inlet manifold
II   Inlet manifold
III Control unit bracket
IV   Boot
V   Central electrical unit

Colour code
BL Blue
BN Brown
GN Green
GR Grey
OR Orange
P   Pink
R   Red
SB Black
VO Violet
W   White
Y   Yellow

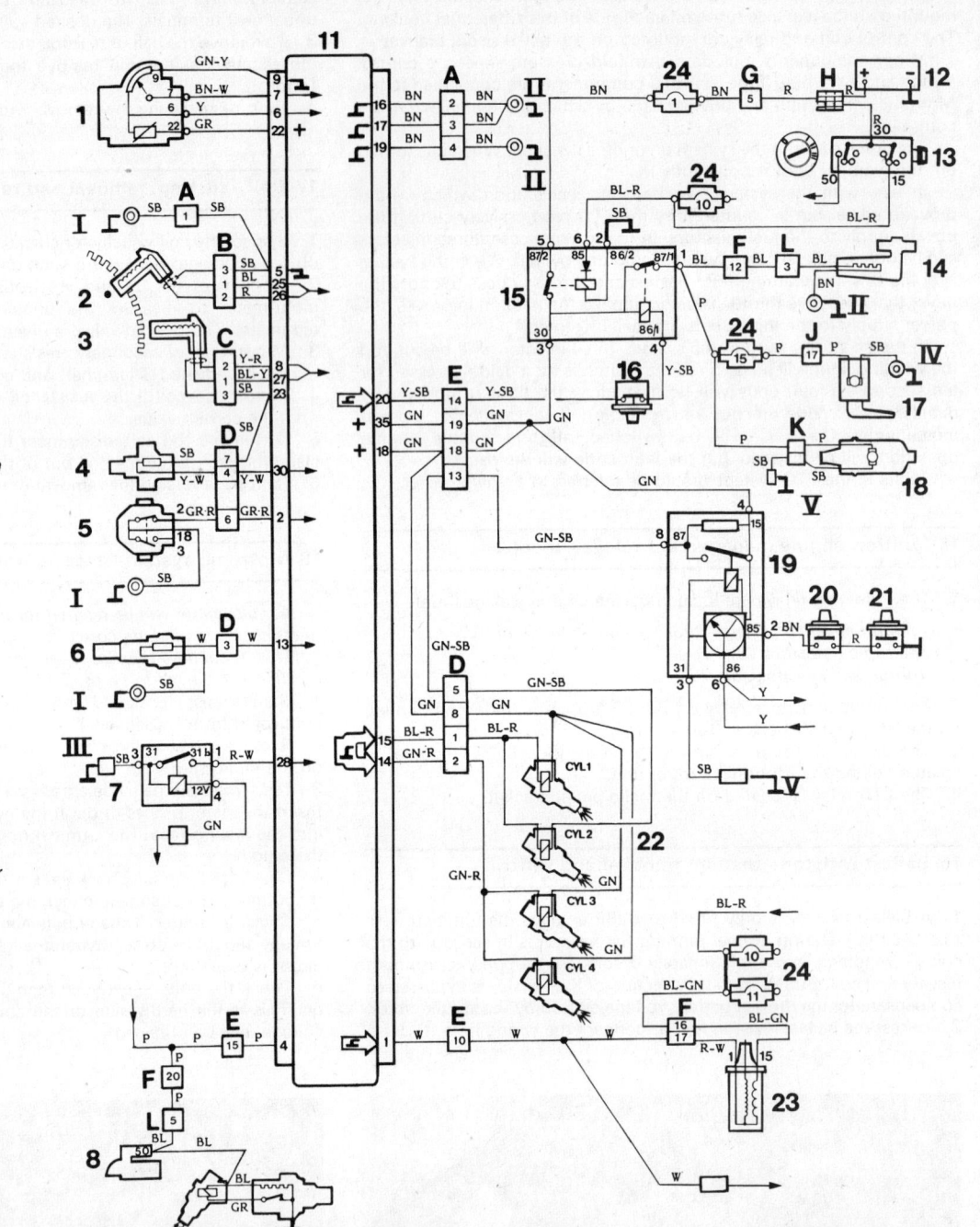

7 Connect the voltmeter between terminal 4 and earth. Voltage must only appear here when the starter motor is operating.
8 Earth terminal 20 and switch on the ignition. Both fuel pumps must run. (See also Chapter 3, Section 45).
9 Measure the resistance of the airflow meter temperature sensor across terminals 6 and 22 (see Specifications). If an incorrect result is obtained, repeat the measurement at the airflow meter multi-plug to see if the fault is in the sensor or the wiring.
10 Measure the airflow potentiometer resistance across terminals 6 and 9, and 6 and 7. Again, repeat the measurements at the meter if the initial results are not as specified.
11 Measure the resistance of the flywheel sensors across terminals 25 and 26, and 8 and 27. Repeat the measurements at the multi-plugs on the bulkhead if the initial readings are not as specified. Do not interchange the multi-plugs (they are the same physically) or the engine will not run.
12 Measure the resistance of the charge air temperature sensor across terminals 23 and 30. Resistance varies with temperature – see Specification. Repeat the measurement at the sensor if necessary.
13 Check for continuity between terminal 2 and earth. There should be continuity (zero resistance) with the throttle pedal released, and no continuity as soon as the throttle is depressed. If not, check the adjustment of the throttle switch (Section 8), and the inlet manifold earthing.
14 Measure the coolant temperature sensor resistance between earth and terminal 13. Repeat the measurement at the sensor if the resistance is not as specified.
15 Check the function of the air conditioning ignition advance relay as

follows. Measure the resistance between terminal 28 and earth: it should be infinite. Separate the relay feed wire connector (light green wire) near the bulkhead and apply 12 volts to the relay side: the relay should operate and the resistance fall to zero.

16 Measure the resistance of the fuel injectors. Injectors 3 and 4 are connected in parallel to terminals 14 and 18, so the desired resistance here is approximately 1 Ω. A reading of 2 Ω shows that one injector is defective or disconnected.

17 Measure the resistance of injector No 1 across terminals 15 and 18: 2 Ω is correct. The same applies to injector No 2, but this is measured between terminal 15 in the control unit plug and terminal 13 in the 20-pole connector. This connector is located behind the trim next to the right-hand side A-pillar, in a black plastic retainer.

18 The torque limiter system may now be checked. Measure the voltage between terminal 13 in the 20-pole connector and earth with the ignition on. Battery voltage should be indicated: if not, the torque limiter relay or its wiring is faulty. The effect of this would be to stop No 2 injector working – see Section 13.

19 Testing of the system is now complete. Remove the test gear, remake the original connections (with the ignition off) and refit disturbed trim.

19 Fault diagnosis – ignition system

Electronic ignition systems are normally very reliable. Faults are most likely to be due to loose or dirty connections, or 'tracking' of HT voltage due to dirt, dampness or damaged insulation.

The old practise of checking for a spark by holding the live end of an HT lead a short distance away from the block is not recommended by the makers, since there is a risk of damaging the coil insulation. For the same reason, diagnosing misfires by pulling off one plug cap at a time is also forbidden. In either case there is the risk of a powerful electric shock.

Engine will not start

1 If the ignition system seems completely dead, check the appropriate fuse (see Chapter 12 Specifications – on 1982/83 models the ignition system is not fused).
2 Inspect the coil, the distributor and the control unit for loose connections.
3 Make sure that the coil tower, the distributor cap and the HT leads are clean and dry. If they are wet, a moisture dispersant, such as Holts Wet Start, can be very effective. To prevent the problem recurring, Holts Damp Start can be used to provide a sealing coat, so excluding any further moisture from the ignition system. In extreme difficulty, Holts Cold Start will help to start a car when only a very poor spark occurs.
4 Inspect the inside of the distributor cap and the rotor arm for visible damage, 'tracking' (thin black lines) and burning. Check that the centre brush in the cap is intact and free to move. Test the rotor arm by substitution if possible.
5 On the B28E engine, measure the resistance of the distributor impulse sender coil. This can be done at the terminals of the multi-plug which connects the distributor with the control unit.
6 Further testing should be left to a Volvo dealer or other specialist.

Engine misfires

7 An irregular misfire suggest a loose connection or an intermittent fault on the LT side of the system, or an HT fault on the coil side of the rotor arm.
8 Regular misfiring is almost certainly due to a fault in the distributor cap, HT leads or spark plugs. Since disconnecting plug leads with the engine running is forbidden, the author suggests the use of a timing light (strobe) on each plug lead in turn to verify the presence of HT voltage.
9 If HT is not appearing on one particular lead, the fault is in the lead or the distributor cap. If HT is present on all leads, the fault is in a spark plug. Remove the plugs and renew them, or clean and re-gap them (Section 3).
10 Again, further testing should be done by a specialist.

Chapter 5 Clutch

Contents

Clutch – removal and refitting	10
Clutch cable – adjustment, removal and refitting	3
Clutch components – inspection	12
Clutch hydraulic system – bleeding	9
Clutch master cylinder – overhaul	6
Clutch master cylinder – removal and refitting	5
Clutch pedal – removal and refitting	4
Clutch release bearing – removal and refitting	11
Clutch slave cylinder – overhaul	8
Clutch slave cylinder – removal and refitting	7
Fault diagnosis – clutch	13
General description	1
Maintenance and inspection	2

Specifications

General
Clutch type .... Single dry plate, diaphragm spring
Actuation .... Hydraulic or cable, according to model and market
Plate diameter (nominal):
  B230E .... 216 mm (8.5 in)
  All other models .... 229 mm (9.0 in)

Hydraulic fluid
Type/specification .... Hydraulic fluid to DOT 4 (Duckhams Universal Brake and Clutch Fluid)

Adjustment
Hydraulic actuation .... Automatic in use
Cable actuation – free play at release fork .... 1 to 3 mm (0.04 to 0.12 in)

Driven plate
Lining wear limit .... Not specified

Pressure plate
Warp limit .... 0.2 mm (0.008 in)

1 General description

A single dry plate diaphragm spring clutch is fitted. Operation may be hydraulic or mechanical, according to model and market.

The main components of the clutch are the pressure plate, the driven plate (sometimes called the friction plate or disc) and the release bearing. The pressure plate is bolted to the flywheel, with the driven plate sandwiched between them. The centre of the driven plate carries female splines which mate with the splines on the gearbox input shaft. The release bearing is attached to the release fork and acts on the diaphragm spring fingers of the pressure plate.

When the engine is running and the clutch pedal is released, the diaphragm spring clamps the pressure plate, driven plate and flywheel firmly together. Drive is transmitted through the friction surfaces of the flywheel and pressure plate to the linings of the driven plate and thus to the gearbox input shaft.

When the clutch pedal is depressed, the pedal movement is transmitted (hydraulically or by cable) to the release fork. The fork moves the bearing to press on the diaphragm spring fingers. Spring pressure on the pressure plate is relieved, and the flywheel and pressure plate spin without moving the driven plate. As the pedal is released, spring pressure is restored and the drive is gradually taken up.

The clutch hydraulic system consists of a master cylinder, a slave cylinder and the associated pipes and hoses. The fluid reservoir is shared with the brake master cylinder.

Wear in the drive plate linings is compensated for automatically by the hydraulic system components. The cable needs periodic adjustment to compensate for wear and stretch.

2 Maintenance and inspection

1 Every 12 000 miles (20 000 km nominal) or annually, check the clutch adjustment on models with cable actuation – see Section 3.
2 Periodically check the clutch hydraulic components for leakage, and check the condition of the flexible hose.
3 When renewing the brake fluid, also renew the fluid in the clutch system by bleeding (Section 9).
4 Prompt investigation of clutch slip, screaching etc, may prevent further damage occurring.

3 Clutch cable – adjustment, removal and refitting

1 Cable adjustment is correct when the free play at the release fork is as given in the Specifications. Adjust if necessary by means of the locknuts and threaded adjuster at the end of the cable outer.
2 To remove the cable, slacken the adjuster as far as possible. Disconnect the return spring (if fitted) from the release fork and

Chapter 5 Clutch

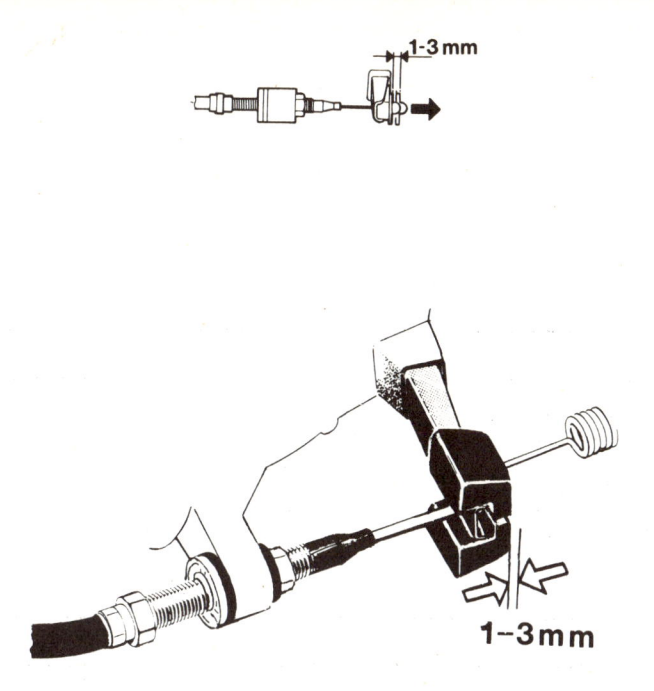

Fig. 5.1 Clutch cable adjustment at release fork – with return spring (bottom) or without (top) (Sec 3)

unhook the cable inner. If a rubber buffer is fitted at the end of the inner, note which way round it goes.
3 Remove the trim panel below the steering column for access to the pedals. Remove the retainer which secures the cable inner to the pedal.
4 Pull the cable into the engine bay and remove it.
5 Refit by reversing the removal operations, making sure the cable is correctly routed. Adjust the cable on completion.

4 Clutch pedal – removal and refitting

On RHD vehicles, the transmission tunnel blocks removal of the clutch pedal pivot bolt. The brake and clutch pedal box must therefore be removed as follows.
1 Disconnect the battery negative lead.
2 Remove the steering column (Chapter 10, Section 13).
3 Disconnect the brake pedal from the servo pushrod by removing the clevis pin.
4 Similarly disconnect the clutch pedal from the master cylinder pushrod, or (when applicable) disconnect the clutch cable.
5 Remove the three bolts which secure the top of the pedal box to the scuttle (photo).
6 Remove the six nuts which secure the pedal box to the bulkhead (photo). (These nuts also secure the clutch master cylinder and the brake servo.)
7 Disconnect the wiring from the stop-light switch. Also disconnect electrical and vacuum/pressure feeds from turbo boost and cruise control switches, or the ignition control unit (as applicable).
8 Remove the pedal box and pedals from the vehicle. Note how the brake pedal return spring bears against the scuttle.
9 Disengage the clutch pedal return spring. Remove the pivot nut and bolt, remove the clutch pedal and recover the bushes.
10 Refit by reversing the removal operations. Apply grease to the pedal bushes and to the pivot bolt.

5 Clutch master cylinder – removal and refitting

1 Disconnect the fluid supply hose from the master cylinder. Have ready a container to catch the fluid which will spill. Wash spilt fluid off paintwork immediately.

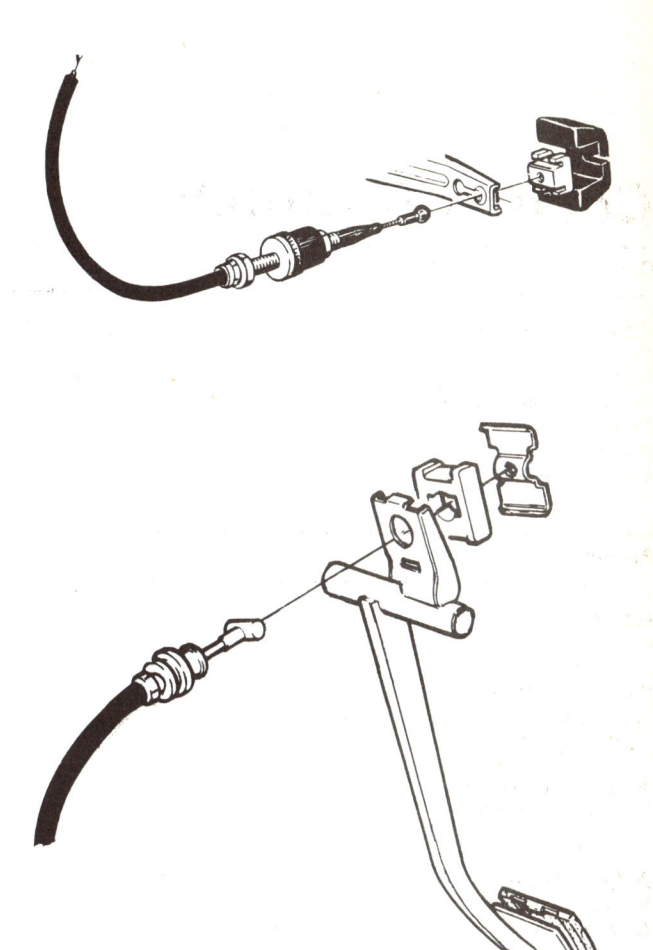

Fig. 5.2 Clutch cable attachment details (Sec 3)

2 Disconnect the pressure pipe union from the end of the cylinder. Be prepared for further fluid spillage. Cover the open pipe union with a piece of polythene and a rubber band to keep dirt out.
3 Remove the trim panel below the steering column.
4 Remove the clevis pin which secures the clutch pedal to the master cylinder pushrod (photo).
5 Remove the two nuts which secure the master cylinder to the bulkhead.
6 Remove the master cylinder, being careful not to drip fluid onto the paintwork.
7 Refit by reversing the removal operations, noting the following points:

   (a) *With the pedal released there should be 1 mm (0.04 in) clearance between the pushrod and the piston. Adjust if necessary by screwing the clevis up or down the pushrod*
   (b) *Bleed the hydraulic system on completion (Section 9)*

6 Clutch master cylinder – overhaul

Refer to Section 8. Overhaul of the master cylinder is basically the same, except that there is a washer under the piston retaining circlip, and the piston has two seals.

Chapter 5 Clutch

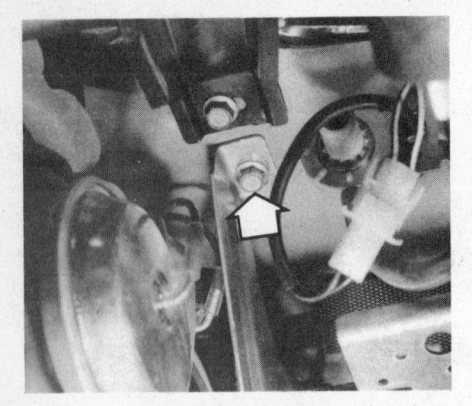

4.5 One of the bolts (arrowed) securing the pedal box to the scuttle

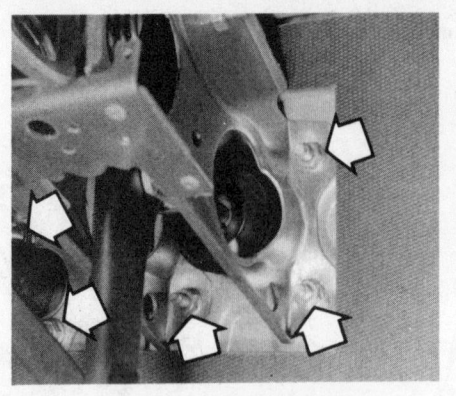

4.6 Five of the six nuts (arrowed) which secure the pedal box to the bulkhead

5.4 Clutch pedal clevis pin (arrowed)

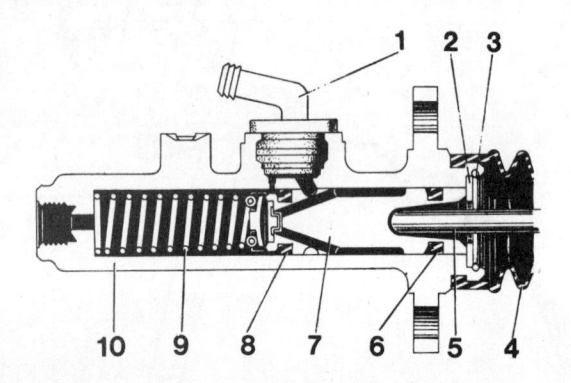

Fig. 5.3 Sectional view of clutch master cylinder (Sec 6)

1. Fluid inlet
2. Washer
3. Circlip
4. Dust boot
5. Pushrod
6. Outer seal
7. Piston
8. Inner seal
9. Spring
10. Cylinder body

7 Clutch slave cylinder – removal and refitting

1 Raise the vehicle on ramps, or drive it over a pit.
2 Slacken the flexible hose union on the slave cylinder (photo).
3 Unbolt the slave cylinder or remove its securing circlip, according to type (photo).
4 Withdraw the slave cylinder with pushrod. Unscrew the cylinder from the flexible hose. Plug or cap the open end of the hose to minimise fluid loss. Recover the sealing washer.
5 Refit by reversing the removal operations. Check the 'set' of the flexible hose after tightening; correct it if necessary by repositioning the hose-to-pipe union in the bracket.
6 Bleed the clutch hydraulic system (Section 9).

8 Clutch slave cylinder – overhaul

1 Empty the fluid out of the cylinder and clean it externally.
2 Remove the dust boot and pushrod.
3 Remove the circlip (if fitted) from the mouth of the cylinder.
4 Shake or tap out the piston and spring. If the piston is stuck, carefully blow it out with **low** air pressure (eg from a foot pump).
5 Remove the seal from the piston.

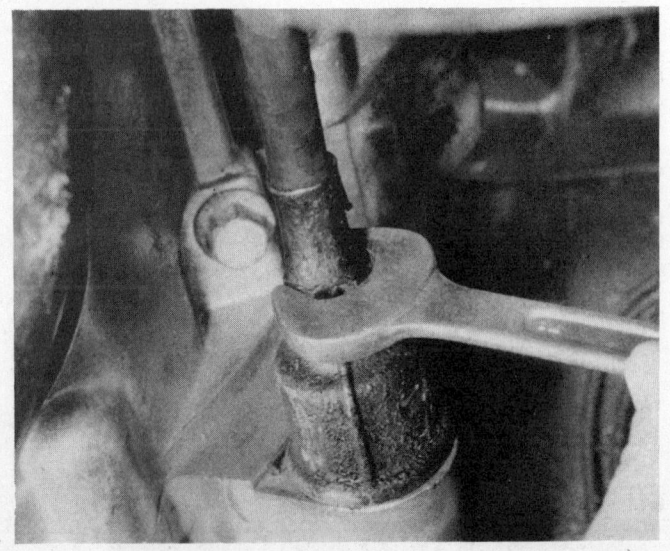

7.2 Undoing the clutch slave cylinder hydraulic union

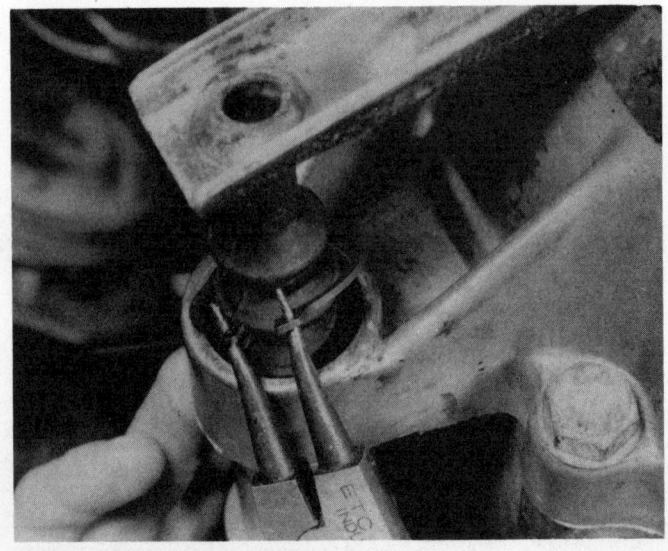

7.3 Removing the slave cylinder circlip

Chapter 5 Clutch   161

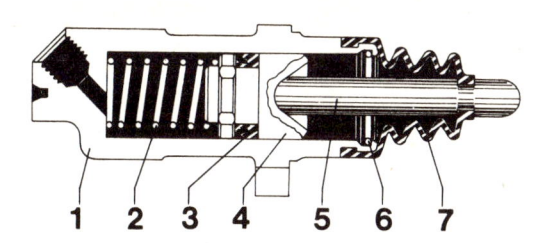

Fig. 5.4 Sectional view of clutch slave cylinder (Sec 8)

1 Cylinder body
2 Spring
3 Seal
4 Piston
5 Pushrod
6 Circlip
7 Dust boot

6 Clean the piston and bore with wire wool and methylated spirit. If either is badly rusted or scored, renew the complete cylinder. Otherwise, obtain a repair kit containing a new seal and dust boot.
7 Dip the new seal in clean hydraulic fluid and fit it to the piston, using the fingers only. Make sure it is the right way round.
8 Lubricate the piston and bore with clean hydraulic fluid. Insert the spring and the piston into the bore (photos).
9 When applicable, refit the circlip to the open end.
10 Fit the new dust boot over the pushrod. Place the pushrod in the cup of the piston and seat the dust boot on the cylinder (photo).

9 Clutch hydraulic system – bleeding

1 Top up the hydraulic fluid reservoir with fresh clean fluid of the specified type.
2 Slacken the bleed screw on the slave cylinder (photo). Fit a length of clear hose over the screw. Place the other end of the hose in a jar containing an inch or so of hydraulic fluid.
3 Have an assistant depress the clutch pedal. Tighten the bleed screw when the pedal is depressed. Have the assistant release the pedal, then slacken the bleed screw again.
4 Repeat the process until clean fluid, free of air bubbles, emerges from the bleed screw. Tighten the screw at the end of a pedal downstroke and remove the hose and jar.
5 Top up the hydraulic fluid reservoir.
6 Pressure bleeding equipment may be used if preferred – see Chapter 9, Section 9.

10 Clutch – removal and refitting

1 Remove the engine or gearbox, as wished.
2 Make alignment marks between the pressure plate and the flywheel.
3 Slacken the pressure plate bolts half a turn at a time until the spring pressure is released. Remove the bolts, the pressure plate and the driven plate (photo). Note which way round the driven plate is fitted.
4 Commence refitting by cleaning the friction surfaces of the flywheel and pressure plate with a non-greasy solvent, followed by a wipe with a clean cloth. Clean oil or grease off the hands before handling the clutch.
5 If a purpose-made centering tool is not to hand, make one from a length of wooden dowel which must be a snug fit in the crankshaft spigot bearing. Build up the dowel with masking tape so that when the tip of the tool is in the crankshaft bearing, the built-up section just fits through the splines in the driven plate.
6 Offer the driven plate to the flywheel, making sure it is the right way round (photo). It is probably marked 'SCHWUNGRAD' or 'FLYW-HEEL SIDE'.
7 Hold the driven plate in position with the centering tool or a

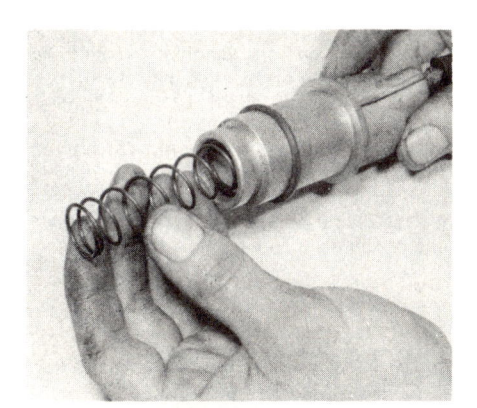

8.8A Insert the spring ...

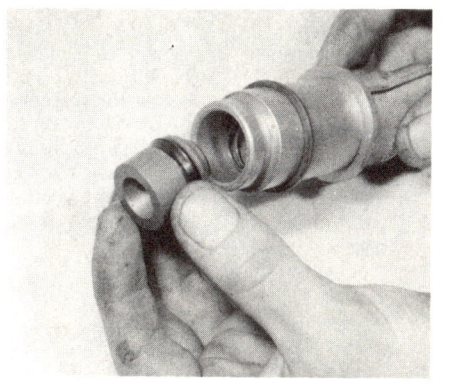

8.8B ... followed by the piston. Make sure it is the right way round

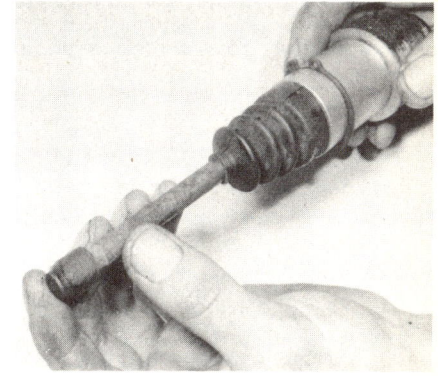

8.10 Fitting the dust boot and pushrod

9.2 Clutch slave cylinder bleed screw (arrowed)

10.3 Removing the clutch pressure plate

10.6 Fitting the clutch driven plate

10.8 Clutch centering tool in position

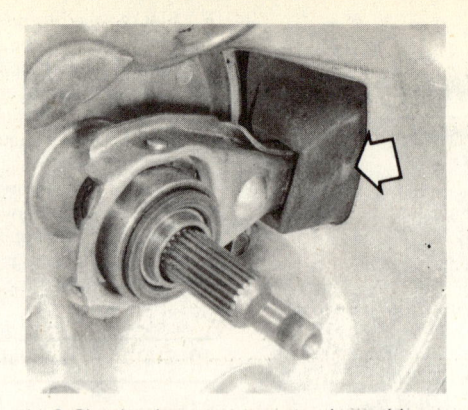

11.2 Clutch release components in position. Dust boot arrowed

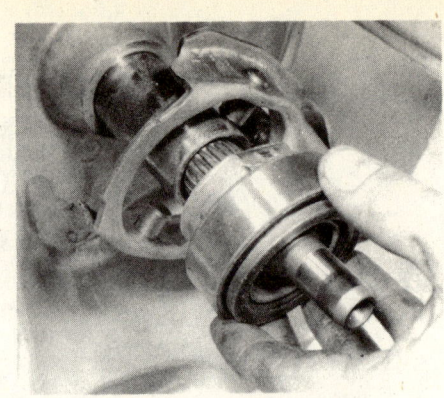

11.4 Removing the release bearing and fork

screwdriver and fit the pressure plate over it. Observe the alignment marks if the original plate is being refitted.
8 Fit the pressure plate bolts and tighten them evenly until the driven plate is being gripped but can still be moved. Insert the centering tool, if it is not already in position, and tighten the pressure plate bolts progressively (photo).
9 Remove the centering tool and check visually that the driven plate is central relative to the crankshaft pilot bearing. If the plate is not central, it will be impossible for the gearbox input shaft to enter it.
10 Refit the engine or gearbox.

11 Clutch release bearing – removal and refitting

1 Remove the engine or gearbox, as wished.
2 Free the release fork dust boot from the bellhousing (photo).
3 Disconnect the release fork from the pivot ball-stud. There may be a spring clip securing the fork to the stud, or there may be nothing.
4 Slide the bearing and fork off the guide sleeve and separate them (photo).

5 Clean the guide sleeve and smear a little grease on it. Lightly grease the fork pivot and tips too.
6 Refit by reversing the removal operations. When the release fork is secured by a spring clip, note that the clip should pass below the groove in the ball-stud.

12 Clutch components – inspection

1 Clean the flywheel, driven plate and pressure plate with a damp cloth, or an old damp paintbrush. *Take care not to disperse or inhale the dust, which may contain asbestos.*
2 Examine the friction surfaces of the flywheel and the pressure plate for scoring or cracks (photo). Light scoring may be ignored. Excessive scoring or cracks can sometimes be machined off the flywheel face – consult a specialist. The pressure plate must be renewed if it is badly scored or warped.
3 Inspect the pressure plate cover and the diaphragm spring for damage, or blue discoloration suggesting overheating. Pay attention to the tips of the spring fingers where the release bearing operates. Renew the pressure plate if in doubt.
4 Renew the driven plate if the friction linings are worn down to, or approaching, the rivets. If the linings are oil-soaked or have a hard black glaze, the source of oil contamination – the crankshaft rear oil seal or gearbox input shaft oil seal – must be dealt with before the plate is renewed. Also inspect the driven plate springs, hub and splines.

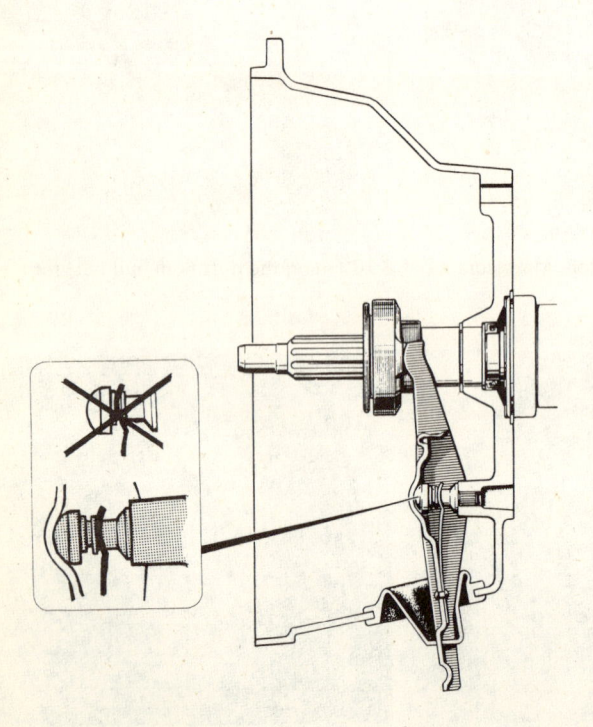

Fig. 5.5 Correct fitting of release fork clip below ball-stud groove (Sec 11)

12.2 Flywheel friction surface, showing light scoring and a small crack

Chapter 5 Clutch

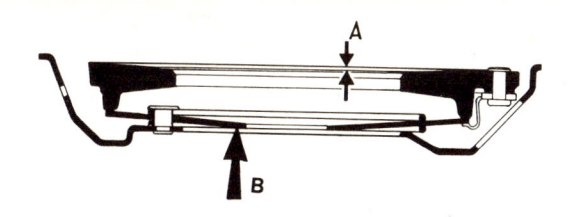

Fig. 5.6 Check the pressure plate for warping (A) and inspect the spring finger tips (B) (Sec 12)

5 Note that if the driven plate only is renewed, problems may be experienced related to the bedding-in of the driven plate and old pressure plate. It is certainly better practise to renew the driven plate and pressure plate together, if finances permit.

6 Try the fit of the driven plate (whether new or used) on the gearbox input shaft splines. It must neither bind nor be slack.

7 Spin the release bearing in the clutch bellhousing and feel for roughness or shake. The bearing should be renewed without question unless it is known to be in perfect condition. For renewal see Section 11.

8 Renewal of the crankshaft spigot bearing should also be considered at this stage. See Chapter 1, Section 15.

13 Fault diagnosis – clutch

Symptom	Reason(s)
Judder when taking up drive	Clutch friction surfaces worn Oil contamination of clutch Splines on driven plate or input shaft worn Pressure plate defective Engine/gearbox mountings worn
Clutch drag (failure to release)	Air in hydraulic system Driven plate sticking on splines Driven plate rusted to flywheel (after long periods of disuse) Cable free play excessive Input shaft seized in crankshaft spigot bearing
Clutch slip (engine speed increases without increasing vehicle speed)	Friction surfaces worn or oil contaminated Pressure plate defective Insufficient free play in cable
Noise when depressing clutch pedal (engine stopped)	Pedal shaft dry Clutch cable or release fork pivot dry
Noise when clutch pedal held depressed (engine running)	Release bearing dry or worn Pressure plate spring fingers damaged

Chapter 6 Manual gearbox, overdrive and automatic transmission

For modifications, and information applicable to later models, see Supplement at end of manual

Contents

Part A: Manual gearbox and overdrive

Fault diagnosis – manual gearbox and overdrive	25
Gearbox components – examination and renovation	10
Gearbox (M46) – dismantling into major assemblies	5
Gearbox (M46) – reassembly	15
Gearbox (M47) – dismantling into major assemblies	6
Gearbox (M47) – reassembly	14
Gearbox oil seals – renewal	3
Gear lever – removal and refitting	21
Gear lever pullrod – renewal	22
General description	1
Input shaft – dismantling	9
Input shaft – reassembly	11
Layshaft – dismantling	8
Layshaft – reassembly	12
Mainshaft – dismantling	7
Mainshaft – reassembly	13
Maintenance and inspection	2
Manual gearbox – removal and refitting	4
Overdrive – description	16
Overdrive – dismantling	18
Overdrive – overhaul	19
Overdrive – reassembly	20
Overdrive – removal and refitting	17
Overdrive switches – removal and refitting	23
Reversing light switch – removal and refitting	24

Part B: Automatic transmission

Automatic transmission – removal and refitting	37
Fault diagnosis – automatic transmission	38
Fluid level checking	28
Fluid renewal	29
Gear selector – checking and adjustment	30
General description	26
Kickdown cable – adjustment	31
Kickdown cable – renewal	32
Kickdown marker – adjustment	33
Maintenance and inspection	27
Overdrive switch (AW71) – removal and refitting	35
Starter inhibitor/reversing light switch – removal and refitting	34
Transmission oil seals – renewal	36

Specifications

Manual gearbox and overdrive

General
Gearbox type:
- M46 .... 4 forward gears, overdrive and one reverse. Synchro on all forward gears
- M47 .... 5 forward gears and one reverse. Synchro on all forward gears
- Overdrive type .... Laycock J or P

Lubrication
- Lubricant type/specification .... Volvo thermo oil (Duckhams QXR)
- Lubricant capacity:
  - M46 .... 2.3 litres (4.1 pints)
  - M47 .... 1.3 litres (2.3 pints)

Ratios
- 1st .... 4.03 : 1
- 2nd .... 2.16 : 1
- 3rd .... 1.37 : 1
- 4th .... 1 : 1
- Overdrive (M46) .... 0.79 : 1
- 5th (M47):
  - Up to 1985 .... 0.83 : 1
  - 1986 models .... 0.82 : 1
- Reverse .... 3.68 : 1

Overhaul data
- Reverse gear-to-selector clearance .... 0.1 to 1.0 mm (0.004 to 0.039 in)
- Shaft endfloat:
  - Input shaft .... 0.01 to 0.20 mm (0.0004 to 0.0079 in)
  - Layshaft – M46 .... 0.03 mm (0.0012 in) endfloat to 0.05 mm (0.0020 in) preload
  - Layshaft – M47 .... 0.01 to 0.10 mm (0.0004 to 0.0039 in)
  - Mainshaft .... 0.01 to 0.20 mm (0.0004 to 0.0079 in)
  - 5th gear synchro hub-to-circlip clearance (M47) ..... 0.01 to 0.20 mm (0.0004 to 0.0079 in)

Torque wrench settings

	Nm	lbf ft
Bellhousing nuts and bolts	35 to 50	26 to 37
Gear lever mounting bracket bolts	35 to 50	26 to 37
Gearcase cover bolts	15 to 25	11 to 18
Layshaft bolt (M47)	35 to 45	26 to 33

Chapter 6 Manual gearbox, overdrive and automatic transmission

Torque wrench settings (continued)	Nm	lbf ft
Drive flange nut:		
M46	175	129
M47, size M16	70 to 90	52 to 66
M47, size M20	90 to 110	66 to 87
Rear cover bolts:		
M46	12 to 18	9 to 13
M47	35 to 50	26 to 37
Bearing retaining plate bolts (M47)	15 to 25	11 to 18
Overdrive-to-intermediate case nuts	12	9
Overdrive main-to-rear case nuts	12	9
Overdrive bridge piece nuts	10	7
Overdrive solenoid	50	37

Automatic transmission

General

Type	4 forward speeds and one reverse. Torque converter, with high-speed lock-up on some models
Designation	AW71 or ZF4HP22

Lubrication

Lubricant type/specification:	
Models up to 1983	ATF type F or G (Duckhams Uni-Matic)
1984 and later models	Dexron IID type ATF (Duckhams Uni-Matic)
Lubricant capacity – drain and refill*:	
AW71	3.9 litres (6.9 pints)
ZF4HP22	2.0 litres (3.5 pints)
Lubricant capacity – from dry:	
AW71	7.5 litres (13.2 pints)
ZF4HP22	7.7 litres (13.6 pints)

* A greater quantity will be needed for fluid renewal. See Section 29

Ratios

	AW71	ZF4HP22
1st	2.45 : 1	2.48 : 1
2nd	1.45 : 1	1.48 : 1
3rd	1 : 1	1 : 1
4th (overdrive)	0.69 : 1	0.73 : 1
Reverse	2.21 : 1	2.09 : 1
Torque converter	1 to 2 : 1	1 to 2.3 : 1

Stall speed

AW71:	
B230K	2500 rpm
B28E	2100 rpm
ZF4HP22:	
B230E	2150 rpm
B230ET	2000 to 2450 rpm

Throttle cable setting

Stop-to-sleeve distance:	
Idling	0.25 to 1.00 mm (0.01 to 0.04 in)
Kickdown	51.0 ± 1.6 mm (2.01 ± 0.07 in)

Torque wrench settings

	Nm	lbf ft
AW71		
Converter housing to engine:		
M10	35 to 50	26 to 37
M12	55 to 90	41 to 66
Driveplate to torque converter	41 to 50	30 to 37
Centre support to gearcase (in steps of 5 Nm/4 lbf ft)	24 to 28	18 to 21
Oil pan	4 to 5	3 to 4
Oil cooler union	20 to 30	15 to 22
Dipstick tube nut	65 to 70	48 to 52
Drain plug	18 to 23	13 to 17
Drive flange nut	45	33
ZF4HP22		
Converter housing to engine:		
M10	35 to 50	26 to 37
M12	55 to 90	41 to 66
Driveplate to torque converter:		
M8	17 to 27	13 to 20
M10	41 to 50	30 to 37
Oil pan	7 to 9	5 to 7
Filler tube nut	85 to 115	63 to 85
Drive flange nut	100	74

Chapter 6 Manual gearbox, overdrive and automatic transmission

PART A: MANUAL GEARBOX AND OVERDRIVE

1 General description

Depending on model and year, the manual gearbox will be four-speed with overdrive (type M46) or five-speed (type M47). The two gearboxes are very similar. They are conventional in design and very sturdy.

Drive from the engine is transmitted to the input shaft by the clutch. The gear on the input shaft is permanently meshed with the front gear on the layshaft; the remaining layshaft gears (except reverse) are permanently meshed with their counterparts on the mainshaft. Only one mainshaft gear at a time is actually locked to the shaft, the others are freewheeling. The selection of gears is by sliding synchro units: movement of the gear lever is transmitted to selector forks, which slide the appropriate synchro unit towards the gear to be engaged and lock it to the mainshaft. In 4th gear the input shaft is locked to the mainshaft. In neutral, none of the mainshaft gears are locked.

Reverse gear is obtained by sliding an idler gear into mesh with the layshaft and mainshaft reverse gears. The introduction of the idler gear reverses the direction of rotation of the mainshaft.

A gear ratio higher than 4th is provided by the overdrive (described in Section 16) or by 5th gear. The 5th gear components are mounted at the rear of the gearbox, in a housing separate from the main gearcase.

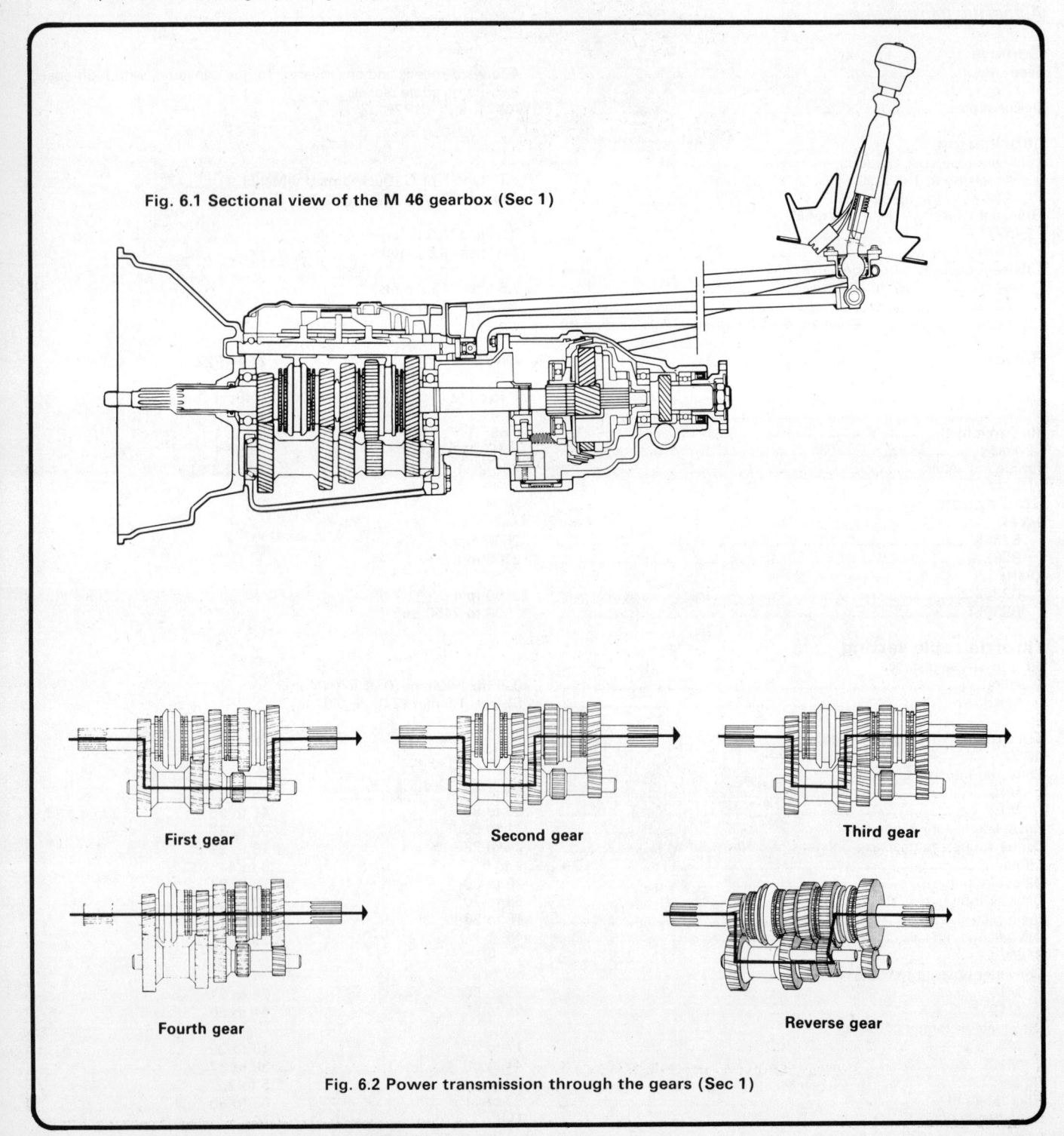

Fig. 6.1 Sectional view of the M 46 gearbox (Sec 1)

First gear

Second gear

Third gear

Fourth gear

Reverse gear

Fig. 6.2 Power transmission through the gears (Sec 1)

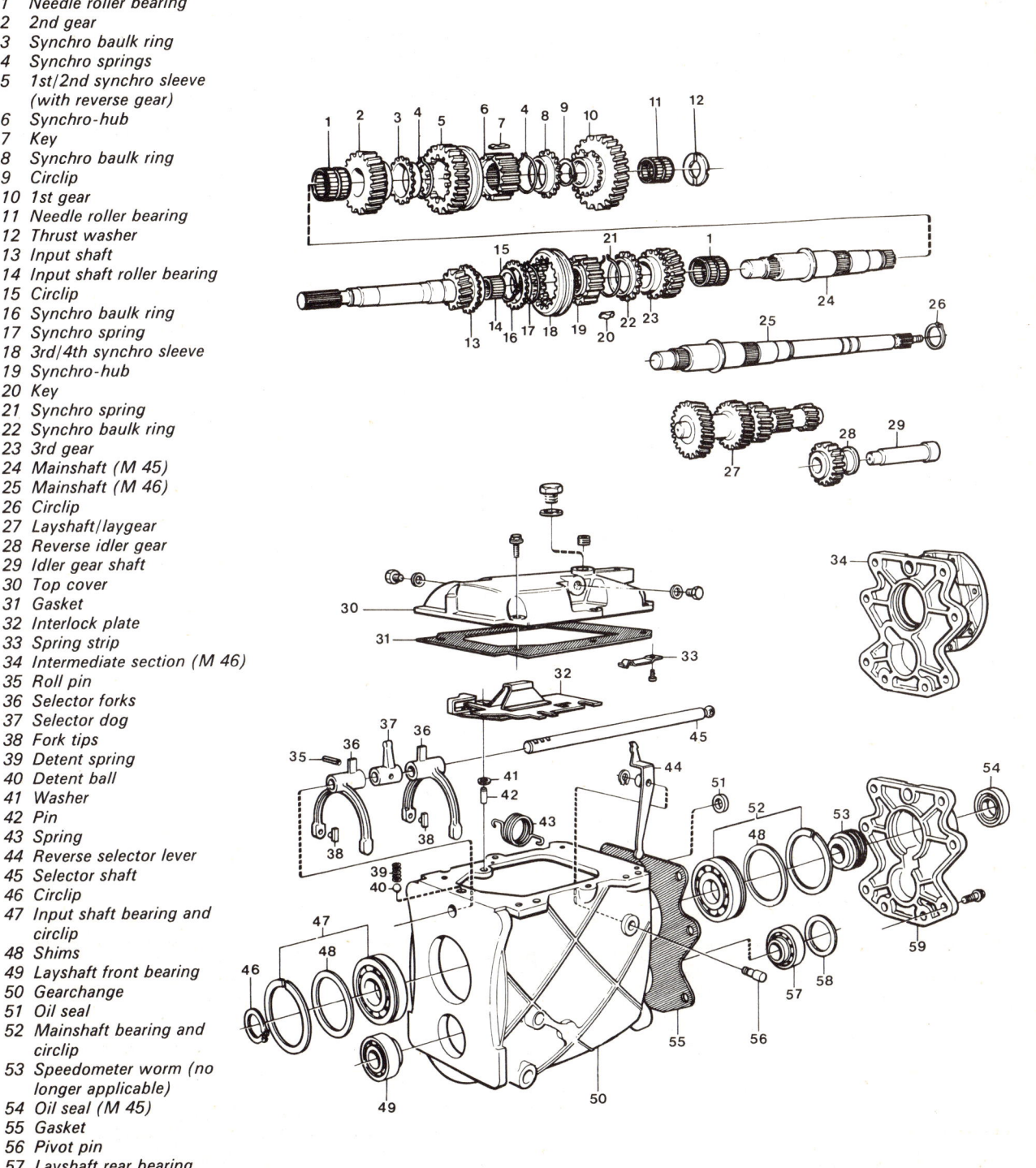

1 Needle roller bearing
2 2nd gear
3 Synchro baulk ring
4 Synchro springs
5 1st/2nd synchro sleeve (with reverse gear)
6 Synchro-hub
7 Key
8 Synchro baulk ring
9 Circlip
10 1st gear
11 Needle roller bearing
12 Thrust washer
13 Input shaft
14 Input shaft roller bearing
15 Circlip
16 Synchro baulk ring
17 Synchro spring
18 3rd/4th synchro sleeve
19 Synchro-hub
20 Key
21 Synchro spring
22 Synchro baulk ring
23 3rd gear
24 Mainshaft (M 45)
25 Mainshaft (M 46)
26 Circlip
27 Layshaft/laygear
28 Reverse idler gear
29 Idler gear shaft
30 Top cover
31 Gasket
32 Interlock plate
33 Spring strip
34 Intermediate section (M 46)
35 Roll pin
36 Selector forks
37 Selector dog
38 Fork tips
39 Detent spring
40 Detent ball
41 Washer
42 Pin
43 Spring
44 Reverse selector lever
45 Selector shaft
46 Circlip
47 Input shaft bearing and circlip
48 Shims
49 Layshaft front bearing
50 Gearchange
51 Oil seal
52 Mainshaft bearing and circlip
53 Speedometer worm (no longer applicable)
54 Oil seal (M 45)
55 Gasket
56 Pivot pin
57 Layshaft rear bearing
58 Shim
59 End casing (M 45)

Fig. 6.3 Exploded view of M 45/M 46 gearbox. (Type M 45, without overdrive, is not fitted to this range) (Sec 1)

Fig. 6.4 Exploded view of M 47 gearbox (Sec 1)

1. Rear end casing
2. Oil seal
3. Top cover
4. Gasket
5. Spring strip
6. Interlock plate
7. 5th gear housing
8. Gasket
9. Bearing race
10. Bearing track
11. Speedometer worm (no longer applicable)
12. Bearing track
13. Bearing race
14. Shim
15. Bearing retainer
16. Selector dog
17. Roll pins
18. Selector shaft
19. 5th gear selector fork
20. Reverse selector lever
21. Pivot pin
22. Selector forks
23. Selector dog
24. Fork tips
25. Detent spring
26. Detent ball
27. Selector shaft
28. Washer
29. Pin
30. Circlip
31. Shims
32. Input shaft bearing and circlip
33. Layshaft front bearing
34. Gearcase
35. Gasket
36. Oil seal
37. Mainshaft bearing and circlip
38. Spacer
39. Bearing retaining plate
40. Layshaft 5th gear
41. Layshaft rear bearing
42. Layshaft/laygear
43. Reverse idler gear
44. Idler gear shaft
45. 5th synchro-hub
46. Collar
47. Spring
48. Key
49. 5th synchro sleeve
50. Spring
51. Synchro baulk ring
52. Circlip
53. Spacer
54. Needle roller bearing
55. Mainshaft 5th gear
56. Spacer
57. Spigot bearing
58. Input shaft
59. Input shaft roller bearing
60. Circlip
61. Synchro baulk ring
62. Springs
63. 3rd/4th synchro sleeve
64. Synchro-hub
65. Synchro baulk ring
66. 3rd gear
67. Key
68. Mainshaft
69. 2nd gear
70. Synchro baulk ring
71. 1st/2nd synchro sleeve (with reverse gear)
72. Springs
73. Synchro-hub
74. Key
75. Synchro baulk ring
76. Circlip
77. Washer
78. 1st gear
79. Vibration damper ring
80. Damper spring
81. Damper washer
82. Thrust washer

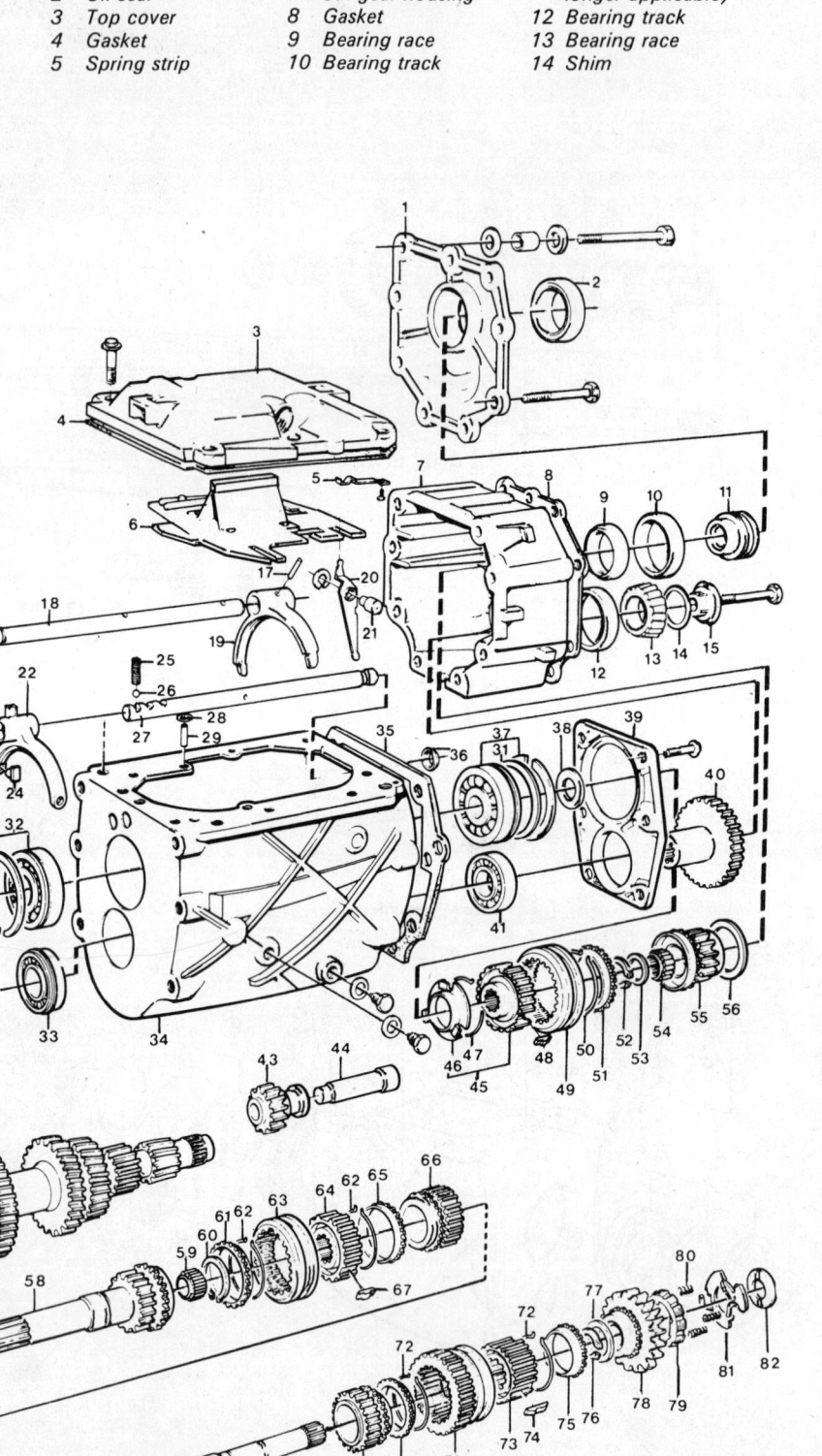

Chapter 6 Manual gearbox, overdrive and automatic transmission

2 Maintenance and inspection

1 Every 12 000 miles or annually, or if prompted by unusual noises or evidence of leakage, check the gearbox oil level as follows.
2 Drive the vehicle over a pit, or raise and support it so that it is still level.
3 Clean the area round the filler/level plug on the side of the gearbox. Unscrew and remove the plug (photo).
4 Oil should be up to the level of the bottom of the plug hole. Insert a clean piece of stiff wire if necessary to check.
5 Top up if necessary with the specified oil (photo). Do not overfill: allow any surplus to drip out of the plug hole.
6 Refit and tighten the filler/level plug, using a new sealing washer if necessary.
7 Frequent need for topping-up can only be due to leakage, which should be found and rectified.
8 Oil changing is no longer specified as a routine operation, although a drain plug is provided for the owner who wishes to do so.

3 Gearbox oil seals – renewal

Drive flange
1 Raise and support the vehicle.
2 Unbolt the propeller shaft from the drive flange and move it aside.
3 Counterhold the flange and undo its central nut (photo).
4 Draw off the flange, using a puller if necessary. Do not try to hammer it off. Be prepared for oil spillage.
5 Prise out the old oil seal and clean up its seat. Inspect the seal rubbing surface on the flange: clean it, or renew the flange if necessary, to avoid premature failure of the new seal.
6 Lubricate the new seal and fit it, lips inwards, using a piece of tube to tap it home. On the M47 gearbox the seal should be recessed by 2.5 mm (0.1 in).
7 On the M46 gearbox, apply locking compound to the output shaft splines. Be careful not to contaminate the seal.

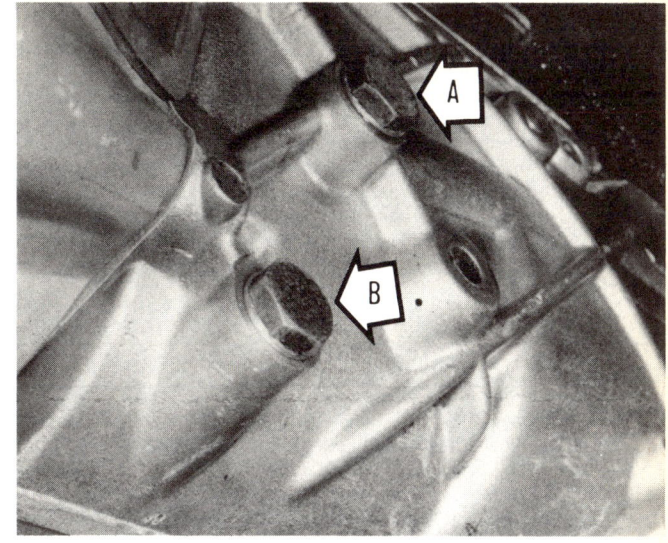

2.3 Gearbox filler/level plug (A) and drain plug (B)

8 Refit the flange and secure it with the nut, tightened to the specified torque.
9 Refit the propeller shaft.
10 Top up the gearbox oil.
11 Lower the vehicle. Check for leaks after the next run.

Input shaft
12 Renew the gearbox (Section 4).
13 Remove the clutch release components from the bellhousing.
14 Unbolt the bellhousing and remove it. Recover the input shaft bearing shim and clean off the old gasket.

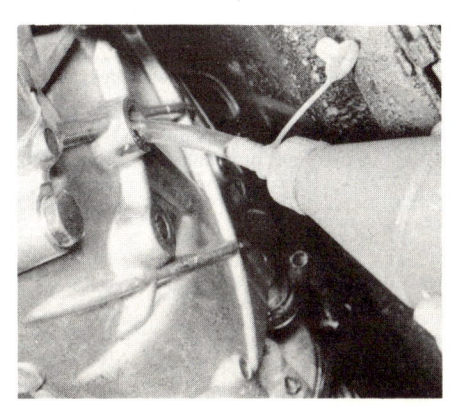

2.5 Topping-up the gearbox oil

3.3 Gearbox drive flange

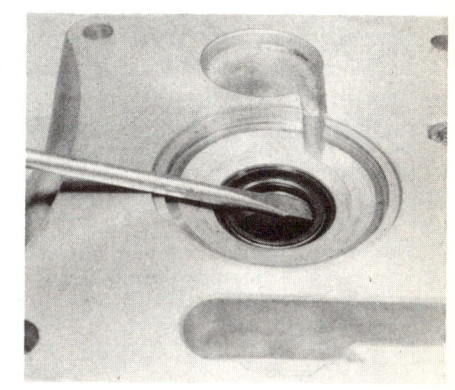

3.15 Removing the input shaft oil seal

3.17 Seating the input shaft oil seal

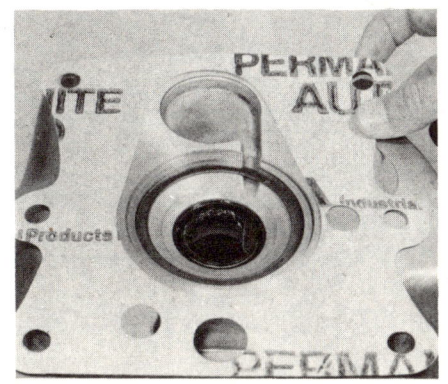

3.18 Fitting the bellhousing gasket

3.19 Tightening a bellhousing bolt

15 Lever the old oil seal out of the bellhousing and clean out its seat (photo).
16 Inspect the seal rubbing face of the input shaft. If it is damaged, a new shaft may be required.
17 Lubricate the new seal and fit it to the bellhousing, lips pointing to the gearcase side. Use a piece of tube to seat it (photo).
18 Refit the bellhousing to the gearcase, using a new gasket (photo). Remember to refit the input shaft bearing shim; use a smear of grease to hold it in position if necessary.
19 Fit the bellhousing bolts and tighten them to the specified torque (photo).
20 Refit the clutch release components.
21 Refit the gearbox.

4 Manual gearbox – removal and refitting

1 On overdrive gearboxes, if the overdrive is to be dismantled it should first be relieved as described in Section 17, paragraph 1.
2 Disconnect the battery negative lead.
3 On B230 engines, arrangements must be made to support the engine from above to prevent damage to the distributor due to engine movement. The best way to support the engine is with a bar resting in the bonnet channels with an adjustable hook appropriately placed.
4 Raise and support the vehicle.
5 Unbolt the propeller shaft from the gearbox output flange.
6 Remove the Allen screw which secures the gear lever to the selector rod. Push out the pin and separate the rod from the lever.
7 Slacken the exhaust downpipe-to-silencer joint so that some movement of the pipe is possible.
8 Unbolt the gearbox mounting crossmember from the gearbox and from the side rails. Remove the crossmember.
9 On B230 engines, adjust the engine support so that the distributor cap is 10 mm (0.4 in) from the bulkhead.
10 Disconnect the gearbox wiring harness multi-plugs.
11 Remove the gear lever (Section 21). Alternatively, the gear lever carrier can be unbolted now and left on the vehicle.
12 Remove the starter motor (Chapter 12). On B28 engines, also remove the cover plate from the unused starter motor mounting.
13 Remove the clutch slave cylinder without disconnecting the hydraulic pipe, or disconnect the clutch cable, as applicable. See Chapter 5.
14 Remove all but two of the engine-to-gearbox nuts and bolts. Note the position of cable clips, exhaust brackets, etc.
15 Support the gearbox, preferably with a cradle and trolley jack, or else with the aid of an assistant. It is too heavy for one person to remove alone.
16 Remove the remaining engine-to-gearbox nuts and bolts. Draw the gearbox off the engine. Do not allow the weight of the gearbox to hang on the input shaft.
17 Remove the gearbox from under the vehicle.
18 Refit by reversing the removal operations, noting the following points:

  (a) Apply a smear of molybdenum-based grease to the input shaft splines
  (b) Make sure that the clutch driven plate is properly centred, and that the clutch release components have been fitted in the bellhousing
  (c) Adjust the clutch cable (when applicable)
  (d) Refill or top up the gearbox oil

19 Check for correct operation on completion.

5 Gearbox (M46) – dismantling into major assemblies

1 Clean the gearbox and drain the oil. Unbolt the gear lever carrier if this was removed with the gearbox, disconnecting the associated wiring (photo).

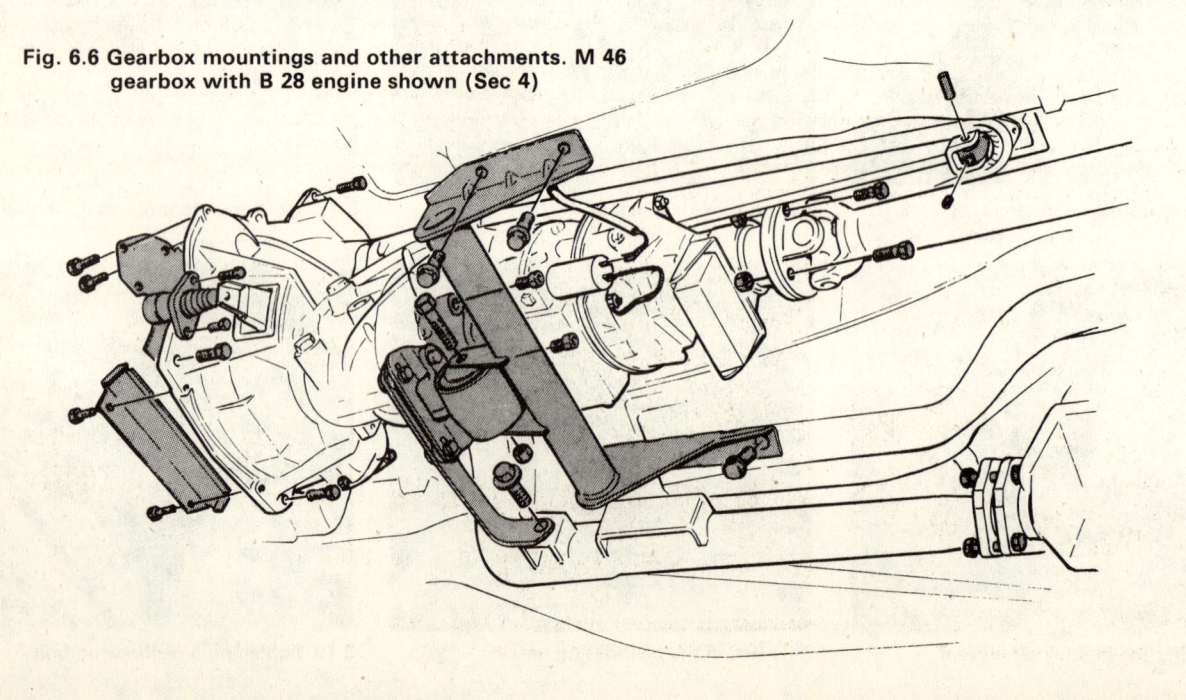

Fig. 6.5 Engine support bar (5006) resting in the bonnet channels (Sec 4)

Fig. 6.6 Gearbox mountings and other attachments. M 46 gearbox with B 28 engine shown (Sec 4)

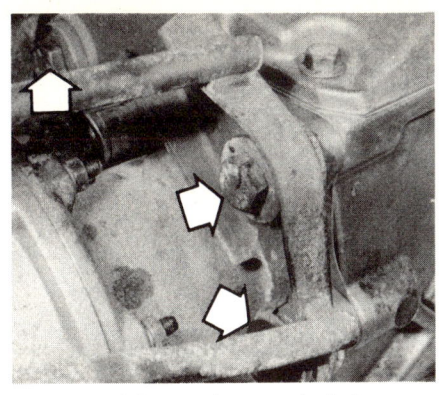

Chapter 6 Manual gearbox, overdrive and automatic transmission

5.1 Three of the gear lever carrier bolts (arrowed)

5.2A Remove the coupling pin ...

5.2B ... and remove the selector rod

5.3 Unbolting the gearbox top cover

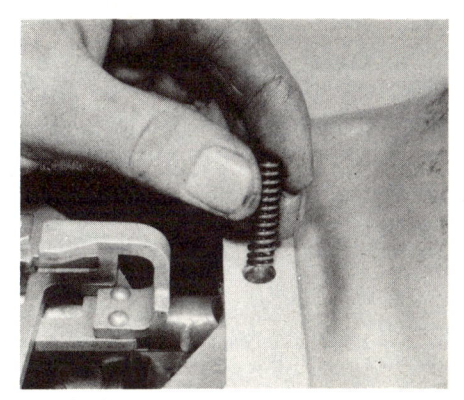

5.4A Remove the detent spring ...

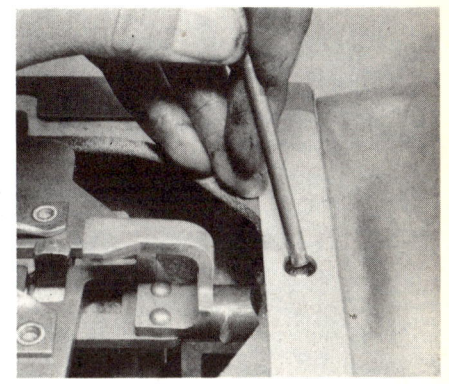

5.4B ... and extract the ball with a magnet

2 Separate the selector rod coupling by sliding back the coupling sleeve and pressing out the rear pin. Remove the selector rod (photos).
3 Unbolt and remove the top cover. Recover the gasket (photo).
4 Remove the detent spring and ball from the hole above the selector shaft (photos).
5 Remove the interlock plate and the large spring. Recover the washers from the interlock plate pegs (photos).
6 Remove the eight nuts which secure the overdrive unit to the intermediate section. Pull off the overdrive; if it is stuck, use a puller, not hammer blows.
7 Turn the selector shaft through 90°. Remove the other pin from the coupling and pull off the coupling (photo).
8 Remove the bolts which secure the intermediate section to the gearcase. Remove the intermediate section and recover the mainshaft and layshaft bearing shims which will be released (photos).
9 Remove the clutch release fork and bearing, referring if necessary to Chapter 5.

10 Remove the four bolts which secure the bellhousing. Lift off the bellhousing and recover the input shaft bearing shim.
11 Knock out the roll pin which secures the selector dog to the selector shaft.
12 Note the orientation of the selector dog and forks, making alignment marks if necessary. Withdraw the selector shaft, then lift out the dog and forks (photos).
13 Extract the layshaft bearing outer tracks from each end of the gearcase. An internal puller will be needed for this. Drive the layshaft back and forth if necessary to produce enough slack for the puller to grip.
14 Remove the circlip which secures the oil pump cam to the mainshaft. Pull off the cam. Recover the Woodruff key if it is loose, otherwise leave it in place (photos).
15 Remove the inner and outer circlips which secure the mainshaft ball-bearing. Lever the bearing out of the case using a couple of

5.5A Removing the interlock plate and spring

5.5B Removing an interlock plate washer

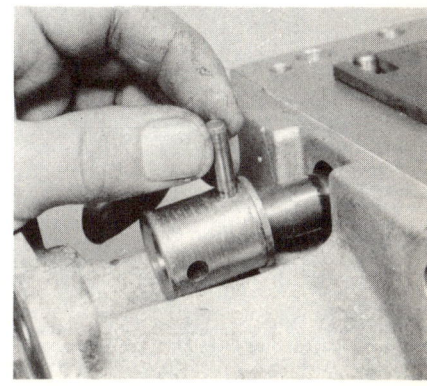

5.7 Removing the other pin from the coupling

5.8A Unbolting the intermediate section

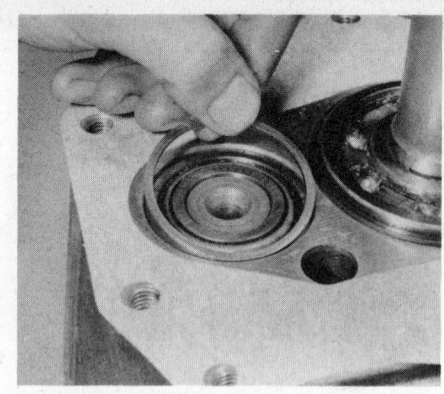

5.8B Removing the layshaft shim

5.12A Removing the selector dog ...

5.12B ... and a selector fork

5.14A Remove the circlip ...

5.14B ... and slide off the cam

5.15A Removing the mainshaft bearing inner circlip ...

5.15B ... and outer circlip

5.15C Pulling off the mainshaft bearing

5.16A Levering out the input shaft bearing

5.16B Removing the input shaft

5.16C Recover 4th synchro baulk ring

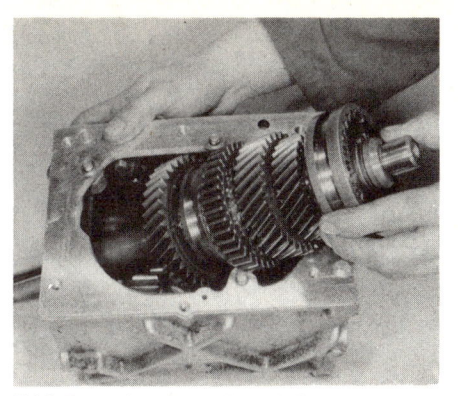

Chapter 6 Manual gearbox, overdrive and automatic transmission

5.17 Removing the mainshaft from the gearcase

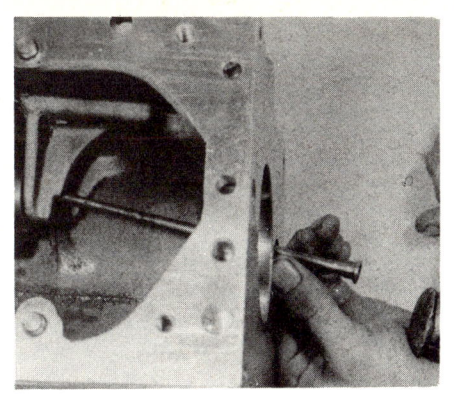

5.19 Driving out the reverse idler shaft

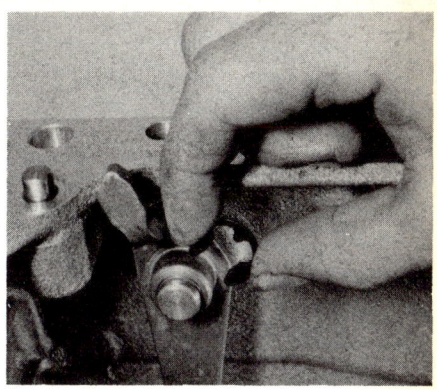

5.20 Reverse selector lever E-clip

screwdrivers in the outer circlip groove. The shaft will come out with the bearing until a gear meets the gearcase; at this point a long-legged puller will have to be used to draw the bearing off the shaft (photos).
16 Remove the input shaft and bearing by levering under the bearing outer circlip. Pull the shaft out of the gearcase. Recover the roller bearing and 4th synchro baulk ring (photos).
17 Work the mainshaft out of the top of the gearcase. It is a tight fit (photo).
18 Lift out the layshaft/laygear assembly.
19 Note the orientation of reverse idler gear. Drive the idler shaft rearwards and remove the shaft and gear (photo).
20 Remove the reverse selector lever, which is secured by an E-clip (photo).
21 The gearbox is now dismantled into its major assemblies for inspection or renewal.

6 Gearbox (M47) – dismantling into major assemblies

1 Remove the gear lever carrier, the selector rod and coupling, the top cover and the interlock plate. See Section 5, paragraphs 1 to 5 and 7.
2 Remove the clutch release components. Unbolt and remove the bellhousing; recover the input shaft shim.
3 Counterhold the drive flange and remove the flange nut. Pull the flange off the splines.
4 Unbolt and remove the rear end casing. Recover the gasket.
5 Remove the bolt from the rear end of the layshaft. Put aside the bearing retainer and shim which were secured by the bolt. Refit the bolt, screwing it in five or six turns only.
6 Pull off 5th gear housing, using a two-legged puller acting on the layshaft bolt as shown (Fig. 6.8). Recover the gasket.
7 Remove the washers and the roller bearing from the end of the mainshaft.

Pre-1986 models
8 Pull 5th gear off the layshaft, using a two-legged puller acting on the layshaft bolt.
9 From the mainshaft remove 5th gear, the two needle roller bearings, the spacer and 5th synchro baulk ring.
10 Remove the layshaft bolt.
11 Note the orientation of the selector dogs and forks. Drive out the roll pins and remove the selector shafts, dogs and forks.
12 Remove the circlip which secures 5th synchro-hub to the mainshaft. Unhook the synchro spring, pull off the sleeve and recover the sliding keys.
13 Pull 5th synchro-hub off the mainshaft. It may be necessary to remove some bolts from the rear bearing retainer plate to provide clearance for the puller. Recover the spacer.

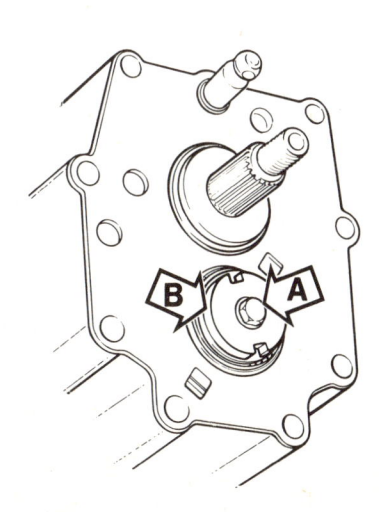

Fig. 6.7 Layshaft bolt (A) and bearing retainer (B) (Sec 6)

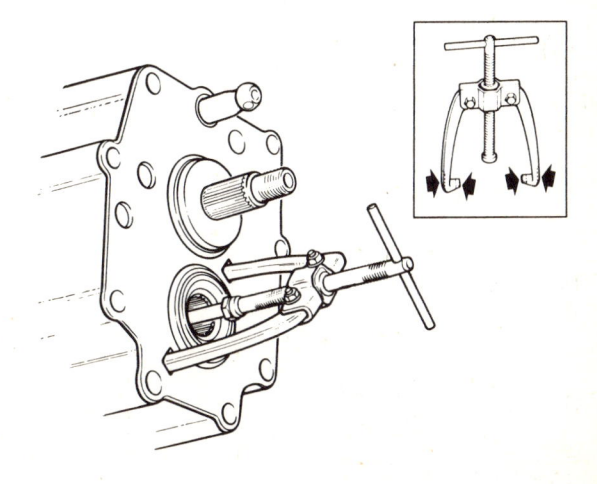

Fig. 6.8 Pulling off 5th gear housing. Puller feet may need to be ground down (inset) (Sec 6)

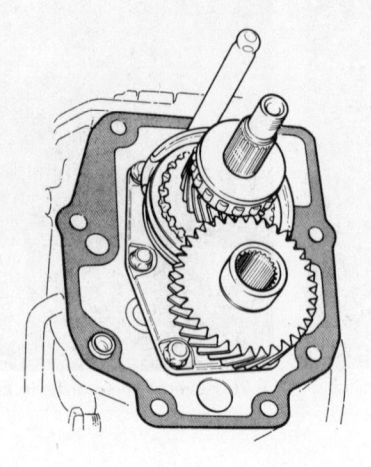

Fig. 6.9 5th gear components – pre-1986 models (Sec 6)

14 Remove the remaining bolts from the bearing retaining plate. Remove the plate and recover the mainshaft shim.

1986 models
15 Proceed as above, but making allowance for the fact that 5th gear synchro components are now on the layshaft instead of the mainshaft.

All models
16 Further dismantling is now as described in Section 5, paragraph 13 onwards.

7 Mainshaft – dismantling

1 Clamp the mainshaft in a soft-jawed vice, front end uppermost.
2 Remove the circlip which secures 3rd/4th synchro unit (photo).
3 Pull off 3rd/4th synchro and 3rd gear together.
4 Remove 3rd gear needle roller bearing (photo).
5 Invert the shaft. If a vibration damper is fitted to 1st gear, lever off the washer and recover the springs (photos).
6 Remove 1st gear, with the vibration damper ring if applicable. Use a puller if it is tight.
7 Remove 1st gear needle roller bearing and baulk ring (photos).
8 Remove the circlip which secures 1st/2nd synchro unit.

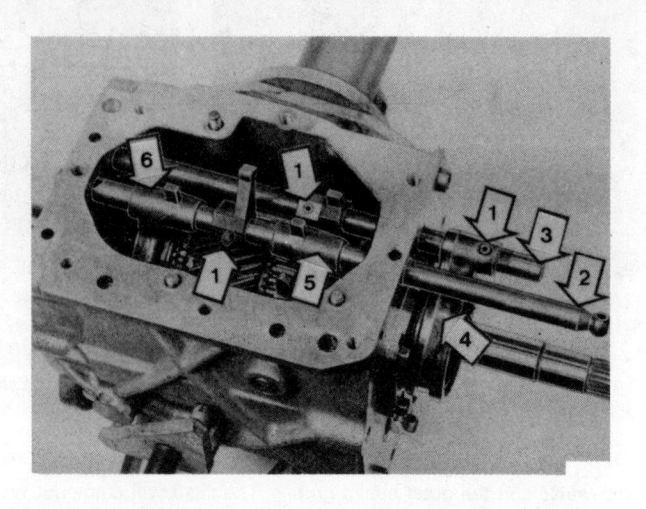

Fig. 6.10 Selector components – M 47 gearbox (Sec 6)

1 Roll pins
2 Main selector shaft
3 5th selector shaft
4 5th selector fork
5 1st/2nd selector fork
6 3rd/4th selector fork

9 Pull off 1st/2nd synchro unit (which includes reverse gear) and 2nd gear together.
10 Remove 2nd gear needle roller bearing (photo).
11 The synchro units and gears which were removed together may be dismantled for examination, but take care not to get the parts mixed up.

8 Layshaft – dismantling

1 If the layshaft bearings are to be renewed, pull the races off the ends of the shaft (photo).

9 Input shaft – dismantling

1 If the input shaft bearing is to be renewed, remove its securing circlip (photo).
2 Press or drive the shaft out of the bearing.

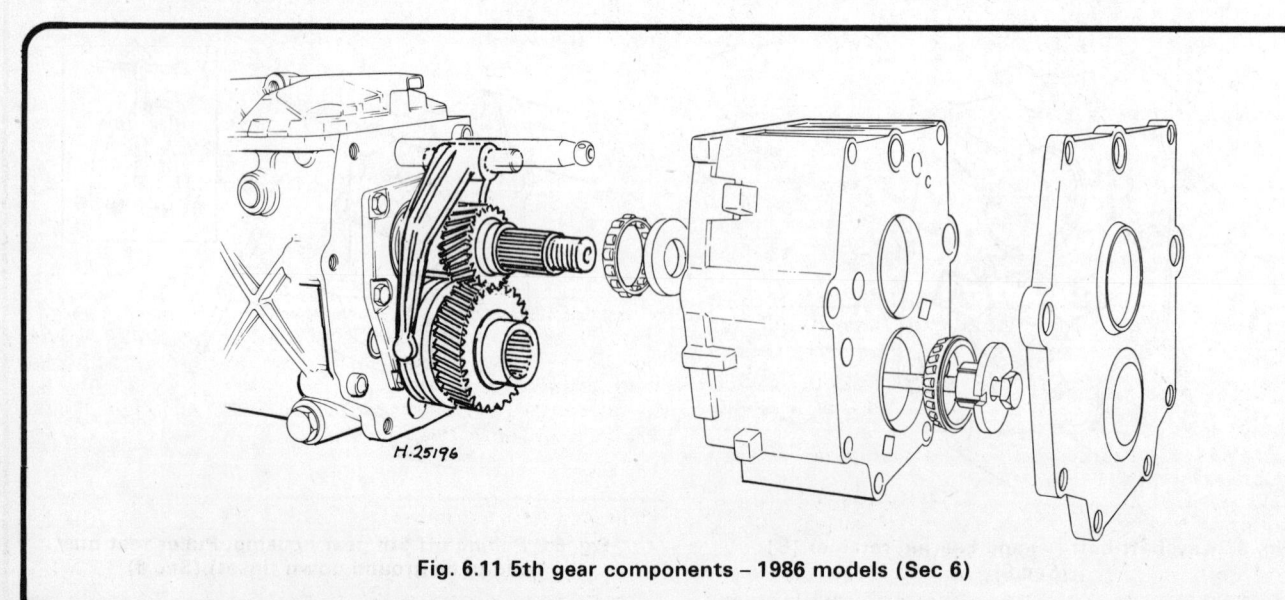

Fig. 6.11 5th gear components – 1986 models (Sec 6)

Chapter 6 Manual gearbox, overdrive and automatic transmission

7.2 Removing 3rd/4th synchro circlip

7.4 Removing 3rd gear needle roller bearing

7.5A Prise off the vibration damper washer ...

7.5B ... and remove the springs

7.7A Removing 1st gear needle roller bearing ...

7.7B ... and 1st gear baulk ring

7.10 Removing 2nd gear needle roller bearing

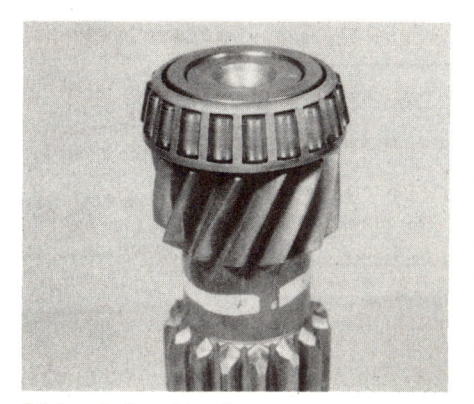

8.1 Layshaft and bearing

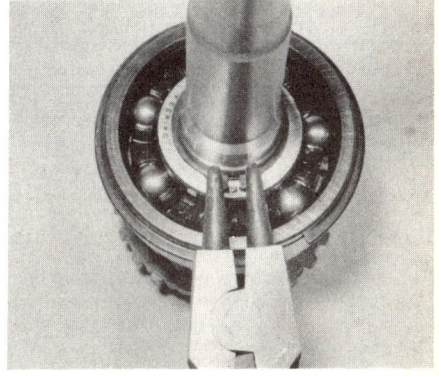

9.1 Removing the input shaft bearing circlip

10 Gearbox components – examination and renovation

1 Before commencing a comprehensive overhaul, check the availability and price of spare parts. Also check the price of a new or reconditioned unit.
2 Oil seals, gaskets, roll pins and similar items must be renewed as a matter of course (photo).
3 Bearings should be renewed unless they are known to be in perfect condition. A selection of bearing preload shims will then also be required: read through the reassembly procedure.
4 Synchro baulk rings should be renewed if they have seen much service. Examine the other synchro components, particularly the springs and sliding keys, and renew as necessary.
5 Inspect the gear teeth for chips or other damage. If such damage is found, check the mating gear also. Mainshaft gears can be renewed individually, but damage to a layshaft gear means that the complete layshaft must be renewed.
6 Examine the selector forks for wear, especially at the tips. The tips may be renewable separately (photo).

11 Input shaft – reassembly

1 If the input shaft bearing has been renewed, press the bearing onto the shaft, applying pressure only to the inner track. Fit the bearing circlips.

Chapter 6 Manual gearbox, overdrive and automatic transmission

10.2 Prising out the selector shaft oil seal (M 46 gearbox)

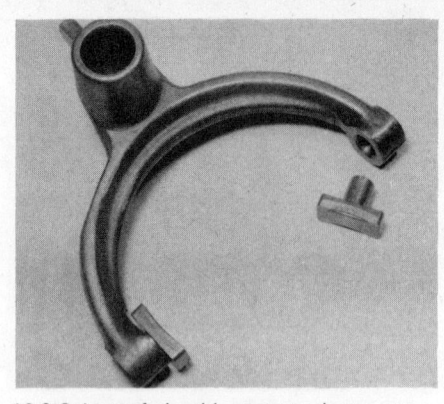

10.6 Selector fork with separate tips

11.2A Fitting the input shaft roller bearing ...

11.2B ... and 4th synchro baulk ring

13.2A Fitting mainshaft 2nd gear ...

13.2B ... and 2nd gear baulk ring

2 Fit the roller bearing and 4th synchro baulk ring to the shaft (photos), then put it aside until needed for reassembly.

12 Layshaft – reassembly

1 Fit the bearing races to the layshaft and tap them home, using a tube resting on the inner part of the bearing only.

13 Mainshaft – reassembly

1 Lubricate all components as they are refitted to the mainshaft.
2 With the mainshaft in the vice rear end upwards, start by fitting 2nd gear needle roller bearing, 2nd gear and its synchro baulk ring (photos).

3 If the synchro unit has been dismantled, it may either be reassembled on the bench, or on the shaft as shown here. Start by inserting the spring into the 2nd gear side of the synchro-hub, with the spring end in one of the cut-outs (photo). The tail of the spring should curve away from the hub centre rather than towards it.
4 Fit the synchro-hub to the mainshaft and drive it home with a piece of tube. Secure it with the circlip (photos).
5 Fit the circlip keys in the cut-outs (photo). Have an assistant hold the keys in place and fit 1st/2nd synchro sleeve, with the reverse gear teeth towards 2nd gear. If there are three bevelled teeth inside the sleeve, they must line up with the three cut-outs in the hub.
6 Fit the other synchro spring, engaging its free end in the same key as the lower one (photo). Again pay attention to the direction in which the tail curves (Fig. 6.13).
7 Fit 1st gear needle roller bearing, 1st gear baulk ring and 1st gear itself (photo).
8 When a vibration damper is fitted, insert the damper ring into 1st

13.3 End of synchro spring (arrowed) in cut-out

13.4A Fitting 1st/2nd synchro-hub

13.4B ... and its circlip

Chapter 6 Manual gearbox, overdrive and automatic transmission 177

13.5 Inserting a sliding key

13.6 Fitting the other synchro spring

13.7 Fitting mainshaft 1st gear

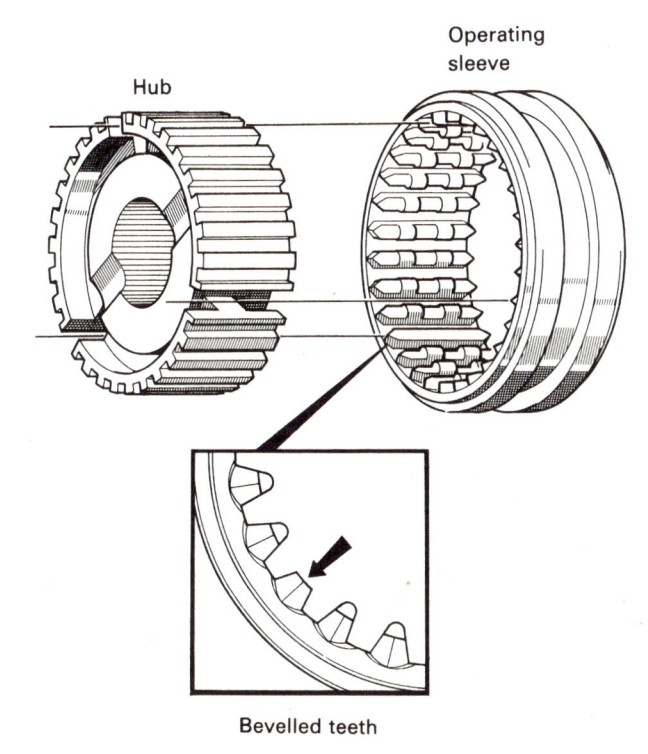

Fig. 6.12 Relationship of synchro-hub to sleeve (Sec 13)

Fig. 6.13 Synchro spring hooked into key (1). Tail (2) must curve away from hub (Sec 13)

gear (photo). Fit the springs and tap home the washer. The tangs in the washer go into the recesses in the ring.

9 Invert the shaft. Fit 3rd gear needle roller bearing, 3rd gear and 3rd gear baulk ring (photos).

10 Fit 3rd/4th synchro unit complete, or build it up on the shaft as was done for the other unit. Use a piece of tube to drive the hub home (photos).

11 Fit the 3rd/4th synchro circlip.

13.8 Fitting the vibration damper ring

13.9A Fitting mainshaft 3rd gear ...

13.9B ... and 3rd gear baulk ring

13.10A Fitting the assembled 3rd/4th synchro unit ...

13.10B ... and driving it home

14 Gearbox (M47) – reassembly

1   The first stages of reassembly, and calculation of the input shaft bearing shim, are carried out as described for the M46 gearbox (Section 15, paragraphs 8 to 24).
2   Mainshaft bearing shim calculation is also similar (Section 15, paragraphs 25 to 27), but the bearing seat is located in the bearing retaining plate.
3   Fit the bearing retaining plate with the selected mainshaft bearing shim. Tighten the bolts to the specified torque. Tap the plate whilst tightening the bolts to settle the bearings.

Pre-1986 models
4   Fit the spacer and 5th synchro-hub to the mainshaft. Tap the hub home and fit the circlip.
5   Measure the clearance between the hub and the circlip. If it is not as specified, remove the hub again and change the spacer for one of the required thickness.
6   Refit 5th synchro sliding keys, sleeve and spring. The three bevelled teeth in the sleeve must align with the keys.
7   Refit the selector shafts, dogs and forks, making sure that they are the right way round. Secure them with new roll pins. The 5th selector fork roll pin should be driven in flush; support the shaft when doing this.
8   Lubricate the roller bearings for mainshaft 5th gear. Place them in the gear with the spacer.
9   Fit 5th synchro baulk ring to the synchro-hub.
10   Fit mainshaft and layshaft 5th gears together. Draw layshaft 5th gear into position using the layshaft bolt and bearing retainer. Remove the bolt and retainer.
11   Fit the roller bearing and washers to the mainshaft; bearing taper facing rearwards.
12   Fit the bearing outer races and a new selector shaft oil seal to the 5th gear housing.
13   Fit the 5th gear housing, using a new gasket smeared with grease.
14   If the layshaft, its bearings or 5th gear housing have been replaced, calculate the shim thickness as follows.

Layshaft bearing shim calculation
15   Secure 5th gear housing with four bolts and spacers. Tighten the bolts to the specified torque.
16   Temporarily fit the layshaft bearing retainer and bolt, without the shim. Tighten the bolt to the specified torque. Measure the layshaft endfloat using a dial test indicator. Push the layshaft up and down, rotating it at the same time to settle the bearings.

Fig. 6.14 Drawing layshaft 5th gear into position (Sec 14)

Fig. 6.15 Measuring layshaft endfloat (Sec 14)

Chapter 6 Manual gearbox, overdrive and automatic transmission

17 Subtract the specified endfloat from the measured endfloat to give the shim thickness required. For example:

Measured endfloat = 0.72 mm
Specified endfloat = 0.01 to 0.10 mm
Difference = 0.62 to 0.71 mm

Shims are available in thicknesses from 0.10 to 0.75 mm.
18 Remove the dial test indicator, the layshaft bolt and the bearing retainer, the housing bolts and spacers.

Final reassembly

19 Fit the selected layshaft bearing shim, the bearing retainer and the bolt. Use thread locking compound on the bolt threads and tighten it to the specified torque.
20 Check that the vent hole in the end casing is not blocked. Fit a new oil seal and a new gasket to the end casing, smearing them with grease.
21 Secure the end casing with two lower bolts.
22 Refit the gear selector coupling and selector rod.
23 Refit the gear lever carrier and the remaining end casing/carrier bolts with washers and spacers. Tighten the bolts to the specified torque.
24 Refit the interlock plate and top cover (Section 15, paragraphs 34 to 38).
25 Refit the clutch release components.
26 Refit the drive flange. Counterhold the flange and tighten the nut to the specified torque.

1986 models

27 The operations are similar to those just described, making allowance for the different position of 5th gear synchro components.

15 Gearbox (M46) – reassembly

Layshaft bearing shim calculation

1 After renewal of the layshaft, layshaft bearings or the gearbox casing, the required shim thickness must be determined as follows.
2 Place the layshaft in the casing. Fit the front bearing race and drive it in from the outside of the casing, using a piece of tube. Leave the race protruding by a small amount (1 mm/0.04 in approx) – it will adopt its correct position in the next step.
3 Fit the bellhousing and its gasket. Secure it with the four bolts, tightened to the specified torque.
4 Invert the gearbox so that it stands on the bellhousing. Fit the layshaft rear bearing outer race and drive it into position. Rotate the layshaft while driving in the race; it is correctly fitted when a slight resistance to turning is felt.
5 Use a depth gauge, or a straight-edge and feeler blades, to measure accurately the distance from the rear end face of the casing (with gasket fitted) to the rear bearing outer race (photo). Select shims from the thicknesses available to make up a total thickness equal to the distance measured, or close enough to it to fall within the specified tolerance. For example:

Distance measured = 1.52 mm
Allowed tolerance = 0.03 mm endfloat to 0.05 mm preload
Acceptable shim thickness = (1.52 − 0.03) to (1.52 + 0.05)
= 1.49 to 1.57 mm

6 Shims are available in the following thicknesses:

0.05 mm
0.10 mm
0.15 mm
0.35 mm
0.50 mm
0.70 mm
1.00 mm

Therefore in the above example, shim packs of 1.50 or 1.55 mm could be made up, both of which fall within tolerance.
7 Remove the bellhousing and gasket, the layshaft and its races.

Fitting shafts and bearings (part 1)

8 Fit the reverse selector lever and secure it with the E-clip (photo).
9 Fit reverse idler gear and shaft, engaging the groove on the gear with the end of the selector lever (photo).
10 Tap the idler gear shaft home until it is flush with the end casing, or no more than 0.05 mm (0.002 in) proud of it (photo).
11 Check that clearance exists between the reverse idler gear groove and the tip of the lever (Fig. 6.16). Adjust if necessary by tapping the lever pivot pin.
12 Refit the layshaft (photo).
13 Refit the mainshaft into the gearcase. Fit the thrust washer to the rear of the mainshaft (models without a 1st gear vibration damper).
14 Fit the mainshaft rear bearing, with the outer circlip in the bearing groove. Do not try to drive the bearing home yet.
15 Refit the input shaft and bearing, remembering to insert 4th synchro baulk ring and the roller bearing between the input shaft and the mainshaft. Tap the outer race of the input shaft bearing into the case until the outer circlip contacts the case.
16 Refit the layshaft bearing races (photo). Tap them in gently.
17 If a new input shaft bearing or bellhousing has been fitted, determine the shim thickness as follows.

Input shaft bearing shim calculation

18 Measure the protrusion of the input shaft bearing from the front of the gearcase (photo).
19 Fit a new gasket to the bellhousing. Measure the depth of the bearing seat, including the thickness of the gasket (photo).
20 Subtract the bearing protrusion of the seat depth, then subtract the specified endfloat from the result, to determine shim thickness. For example:

Depth = 5.32 mm
Protrusion = 4.65 mm
Difference − 0.67 mm
Specified endfloat − 0.01 to 0.20 mm
Difference minus endfloat − 0.47 to 0.66 mm

The range of shims available is from 0.25 to 1.00 mm. Select a suitable thickness from the range.

15.5 Measuring the depth of the layshaft bearing race

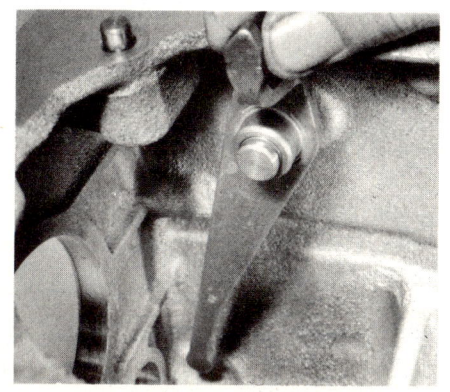

15.8 Fitting the reverse selector lever

15.9 Fitting reverse idler gear and shaft

180 Chapter 6 Manual gearbox, overdrive and automatic transmission

15.10 Checking the protrusion of the reverse idler shaft

15.12 Layshaft refitted to the gearcase

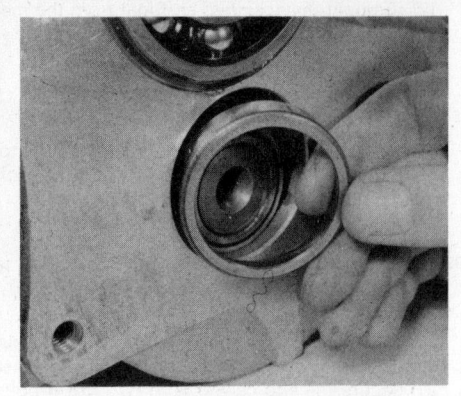

15.16 Fitting a layshaft bearing race

15.18 Measuring the protrusion of the input shaft bearing

15.19 Measuring the depth of the input shaft bearing seat

15.21 Fitting the input shaft bearing shim

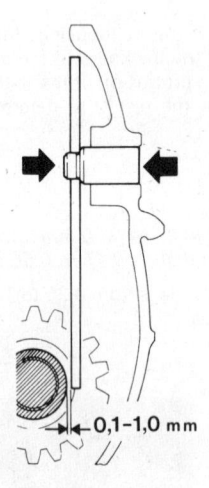

Fig. 6.16 Reverse gear-to-selector lever clearance. Adjust by tapping shaft in or out (arrows) (Sec 15)

Fitting shafts and bearings (part 2)
21 Fit the selected input shaft shim to the bellhousing, using a smear of grease to hold it in position (photo).
22 Fit the bellhousing, using a new gasket. Insert the securing bolts and tighten them to the specified torque.
23 Stand the gearbox on the bellhousing. Drive the mainshaft bearing home and fit the inner circlip (photo). (It may be necessary temporarily to remove the outer circlip to enable the bearing to sit slightly deeper whilst the inner circlip is fitted. The bearing can then be levered up and the outer circlip refitted.)
24 If the mainshaft bearing or the intermediate case section has been renewed, the bearing shim thickness must be calculated as follows.

Mainshaft bearing shim calculation
25 Fit a new gasket to the rear of the gearcase. Measure the protrusion of the mainshaft bearing above the gearcase.
26 Measure the depth of the bearing seat in the intermediate use section.
27 Calculate the shim thickness required as described in paragraph 20. The range of shims available is from 0.60 to 1.00 mm.

Final reassembly
28 Refit the overdrive oil pump cam and circlip. Make sure that the Woodruff key is secure.
29 Refit the selector forks, dog and shaft, making sure they are the right way round. Secure the dog with a new roll pin (photo).
30 Fit the selected layshaft bearing shim to the layshaft rear bearing, and the mainshaft bearing shim to the intermediate case (photo). Use a smear of grease to hold them in position.
31 Fit the intermediate case to the rear of the gearcase, using a new gasket. Secure it with the lower bolts. (The upper bolts also secure the gear lever carrier.)
32 Refit the selector shaft-to-rod coupling and the selector rod.
33 Refit the overdrive unit. Tighten the nuts to the specified torque.
34 Refit the washers to the interlock plate pegs. Fit the interlock plate and the large spring.
35 Fit a new top cover gasket.
36 Refit the detent ball and spring (photo).
37 Refit the top cover (photo). Fit and tighten its securing bolts.
38 Check the engagement of all gears, at the same time rotating the input shaft.
39 Refit the clutch release components.
40 Refit the gear lever carrier if it was removed with the gearbox.

Chapter 6 Manual gearbox, overdrive and automatic transmission

15.23 Mainshaft bearing with circlips fitted

15.29 Driving in the selector dog roll pin

15.30 Fitting the mainshaft bearing shim

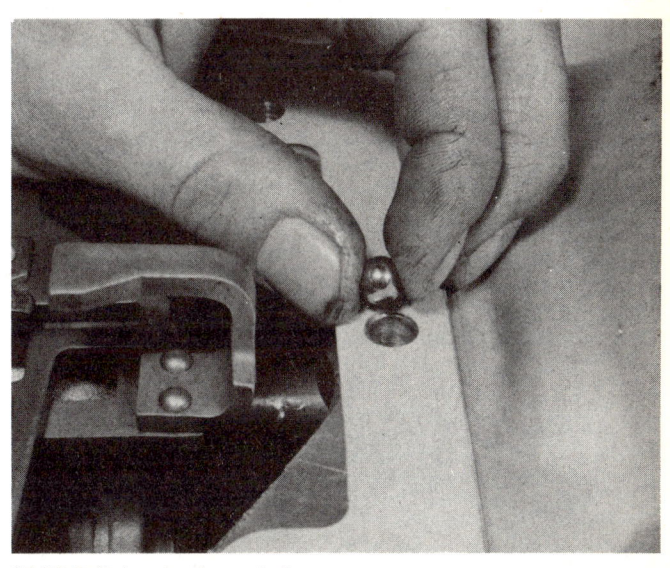

15.36 Refitting the detent ball

15.37 Refitting the top cover

16 Overdrive – description

The overdrive is essentially an extra gearbox, driven by the output shaft of the main gearbox and producing on its own output shaft a step-up ratio of 0.797 : 1. The 'gear change' is controlled hydraulically, the hydraulic control valve being operated by a solenoid. The electrical connections to the solenoid are taken through a switch on the cover of the main gearbox which ensures that overdrive can only be brought into operation when the car is in top gear. The activating switch for the whole system is mounted in the top of the gear lever knob.

A cutaway illustration of the overdrive and an exploded view are shown in Figs. 6.17 and 6.18.

The heart of the overdrive is the epicyclic gear system, whose components are shown in Fig. 6.18. These parts are assembled on the elongated mainshaft which extends from the main gearbox. Two of these parts, the planet carrier and the unidirectional clutch, are splined to the mainshaft and always revolve, therefore, at mainshaft speed. The unidirectional clutch sits inside the output shaft and ensures that if nothing else is driving the output shaft, it will be driven by the gearbox mainshaft. In this manner the 1 : 1 ratio is obtained. When this occurs, the planet carrier and the annulus on the output shaft are revolving at the same speed, so the planet gears within the planet carrier are not

Chapter 6 Manual gearbox, overdrive and automatic transmission

being driven forwards or backwards and remain stationary on their splines. This means that the sun wheel must also be revolving at the same speed as the planet carrier and the annulus. The sun wheel is splined to the sliding clutch member and this too is revolving at the mainshaft speed. In practise the sliding clutch member is held against the tapered extension of the mainshaft when the 1 : 1 ratio is required and the whole gear system is locked together (Fig. 6.19).

To obtain the step-up ratio, the sliding clutch member is drawn away from the output shaft annulus and comes up against the outer casing of the gearbox which holds it stationary. It is still splined to the sun wheel, so this, too, is prevented from turning.

The planet carrier continues to revolve at mainshaft speed, but because the sun wheel is stationary the planet wheels turn around their spindles in the planet carrier. This means that the outer teeth of the planet wheel (which mesh with the annulus) are moving relative to the planet carrier, and this makes the annulus move faster than the planet carrier.

The sliding member is bolted to bridge pieces. Behind these bridge pieces are hydraulically operated pistons which are able to push the bridge pieces away from the case against the action of the clutch return springs when the hydraulic pressure is great enough. This means that changing gear is simply a matter of raising the oil pressure applied to

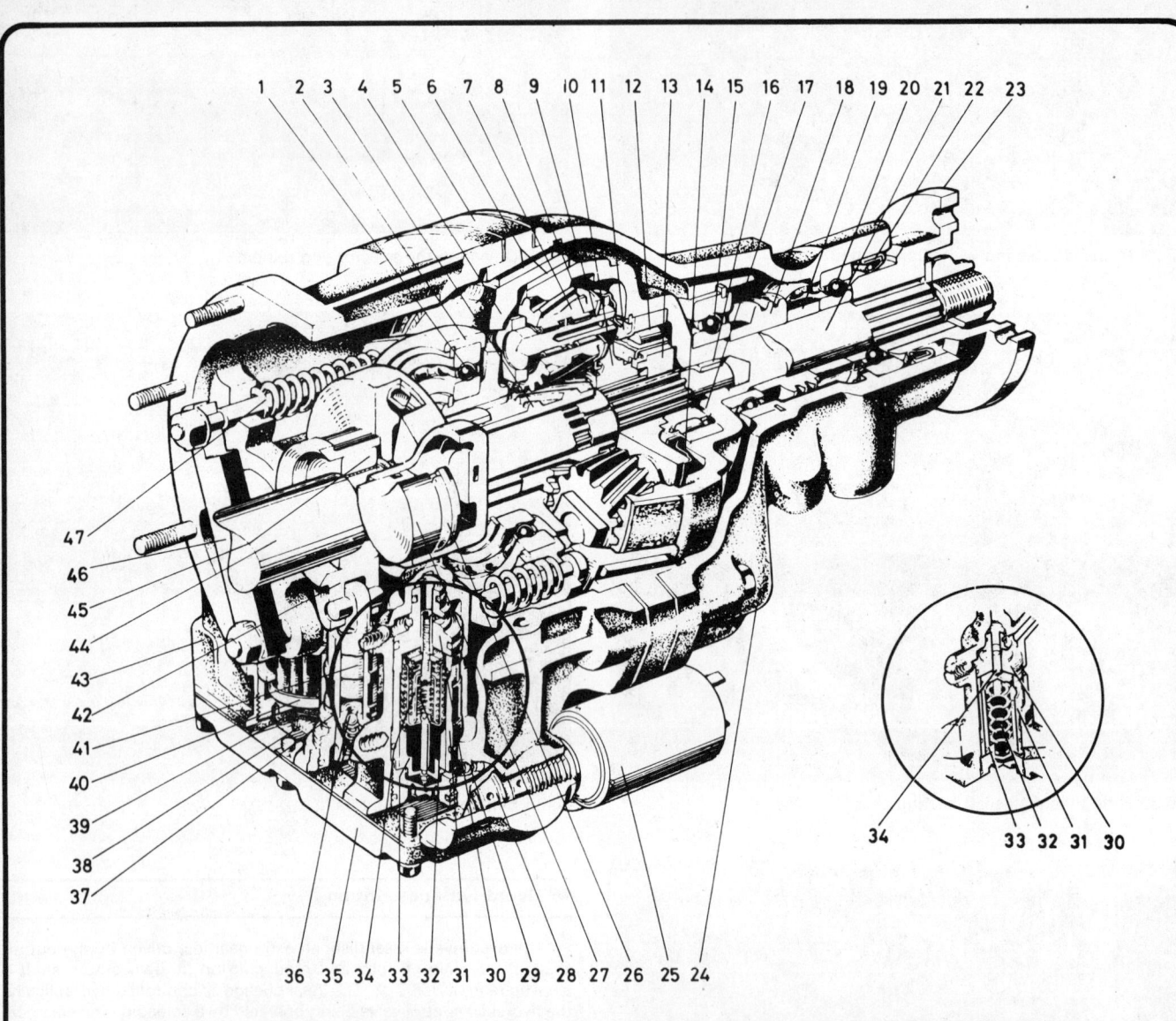

Fig. 6.17 Cutaway view of Type 'J' overdrive. Inset shows later type pressure relief valve (Sec 16)

#	Part	#	Part	#	Part	#	Part
1	Thrust bearing	13	Uni-directional clutch	25	Solenoid	37	Pump cylinder
2	Thrust bearing retainer	14	Oil trap	26	Piston seal	38	Magnet
3	Sun wheel	15	Ball-bearing	27	Piston	39	Pick-up strainer
4	Clutch sliding member	16	Bush	28	Operating valve	40	Pressure filter
5	Brake ring	17	Thrust washer	29	Orifice nozzle	41	Pump plunger
6	Clutch member linings	18	Speedometer worm (no longer applicable)	30	Cylinder top	42	Connecting rod
7	Planet gear	19	Spacer	31	Cylinder	43	Front casing
8	Needle bearing	20	Ball-bearing	32	Spring	44	Input shaft (gearbox mainshaft)
9	Shaft	21	Output shaft	33	Large piston	45	Cam
10	Planet carrier	22	Oil seal	34	Small piston	46	Bridge piece
11	Oil thrower	23	Coupling flange	35	Baseplate	47	Spring
12	Uni-directional clutch rollers	24	Rear casing	36	Check valve for oil pump		

Fig. 6.18 Exploded view of Type 'J' overdrive (Sec 16)

#	Part	#	Part	#	Part	#	Part
1	Nut	18	Circlip	36	Stud	52	Plug
2	Lockwasher	19	Circlip	37	Orifice nozzle	53	Nut
3	Bridge piece	20	Circlip	38	Seal	54	Piston
5	Breather	21	Stud	39	Plug	57	O-ring
6	Front casing	22	Piston seal	40	O-ring	58	Cylinder
7	Gasket	23	Piston	41	End piece	59	O-ring
8	Brake ring	24	Connecting rod	42	Piston	60	Plug
9	O-ring	25	Non-return ball	43	Washer	61	Spring
10	O-ring	26	Non-return valve spring	44	Spring	62	Ball
11	Seal	27	Plug	45	Retainer	63	Non-return body
12	Gasket	28	Key	46	Spring	64	O-ring
13	Solenoid	29	Resilient ring	47	Screw	65	Pump body
14	Bolt	30	Circlip	48	Screw	66	Pump plunger
15	Thrust bearing retainer	31	Cam	49	Holder	67	Washer
16	Spring	32	Piston pin	50	Spring	68	Pressure filter
17	Thrust bearing	33	Gasket	51	O-ring	69	Seal

70 Plug
71 Data plate
72 Screw
73 Planet gear and carrier
74 Sun wheel
75 Clutch sliding member
76 Pick-up strainer
77 Gasket
78 Magnet
79 Baseplate
80 Bolt
81 Resilient washer
84 Bush
85 Thrust washer
86 Oil thrower
87 Circlip
88 Uni-directional clutch
89 Stud
90 Resilient washer
91 Nut
95 Speedometer pinion*
96 O-ring*
97 Bush*
99 Bolt*
100 Retainer*
101 Oil seal*
102 Stud
106 Speedometer worm*
110 Output shaft
111 Ball bearing
112 Spacer
113 Rear casing
114 Ball bearing
115 Oil seal
116 Flange
117 Washer
118 Nut

* No longer applicable

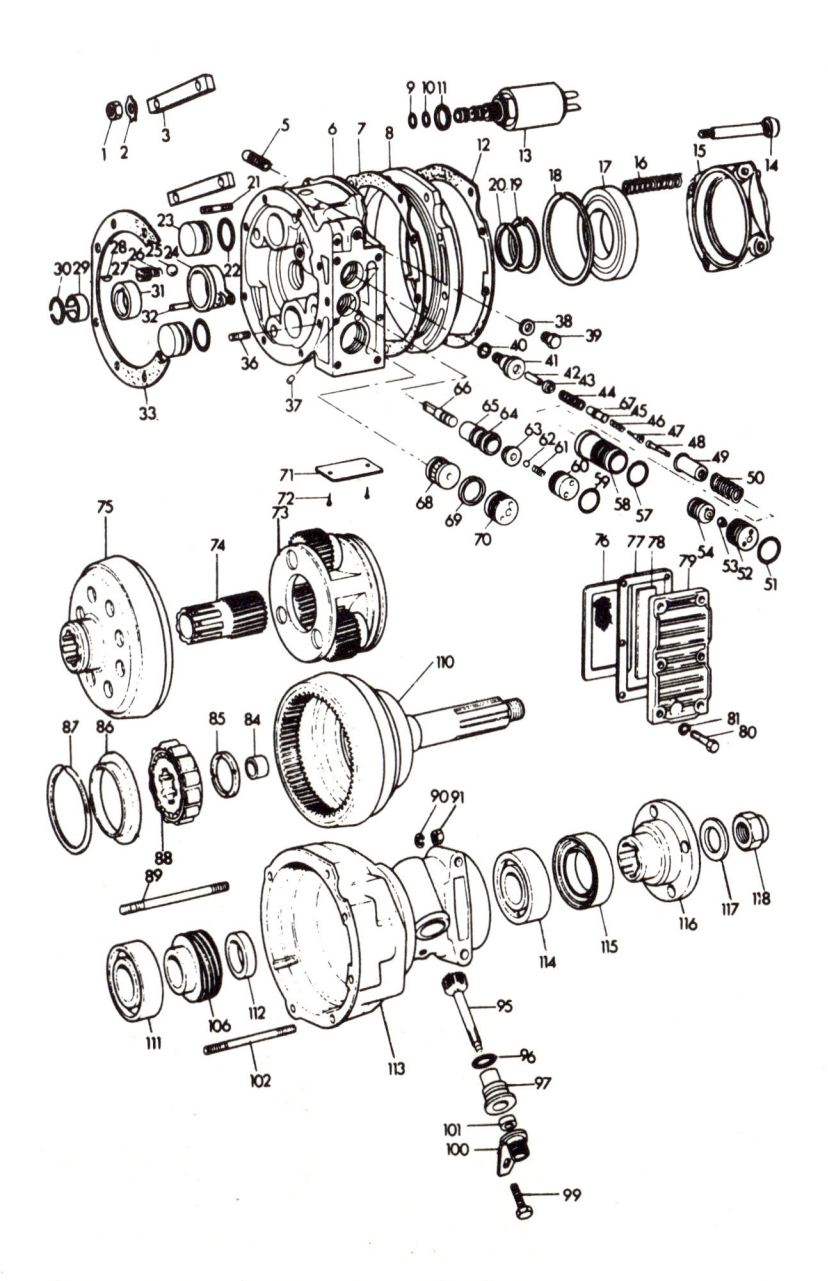

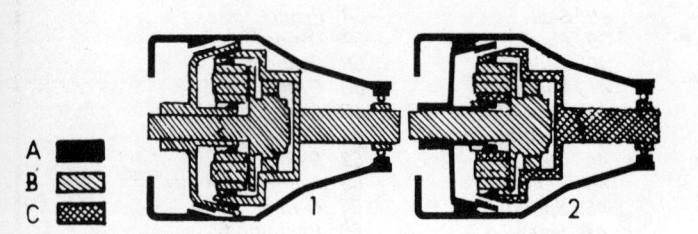

Fig. 6.19 Operating principle of overdrive (Sec 16)

1 Direct drive
2 Overdrive
A Non-rotating parts
B Parts rotating at speed of input shaft
C Parts rotating faster than input shaft

the pistons. The oil pressure is generated in the first instance by a piston pump which is driven by a cam on the mainshaft extension. A solenoid valve controls the application of oil pressure to the pistons.

The overdrive hydraulic system incorporates a pressure relief valve, a pick-up strainer and a pressure filter.

From 1986 a type 'P' overdrive is fitted to some models instead of the type 'J' previously fitted. The main difference between the two types is that the 'P' can handle a greater torque without slipping, due to the increased dimensions of the friction linings and some other components. This renders the torque limiting system (Chapter 4, Section 13) redundant.

17 Overdrive – removal and refitting

It is only necessary to remove the overdrive for overhaul or renewal. Work on the hydraulic system components can be carried out on the vehicle provided that scrupulous cleanliness is observed.

1 Relieve the pressure in the overdrive by driving the vehicle with overdrive engaged, then disengaging the overdrive with the clutch pedal depressed.
2 Raise and support the vehicle.
3 Unbolt the propeller shaft from the drive flange.
4 Support the gearbox and remove the crossmember. Lower the rear of the gearbox, being careful not to damage the distributor on the B230 engine.
5 Disconnect the wiring from the overdrive solenoid and (when applicable) the pressure switch (photo).
6 Remove the eight nuts which secure the overdrive to the gearbox intermediate section. Lift off the overdrive (photos); be prepared for oil spillage. If the overdrive will not come off, use a slide hammer on the drive flange; do not lever between the gearbox and overdrive casings.

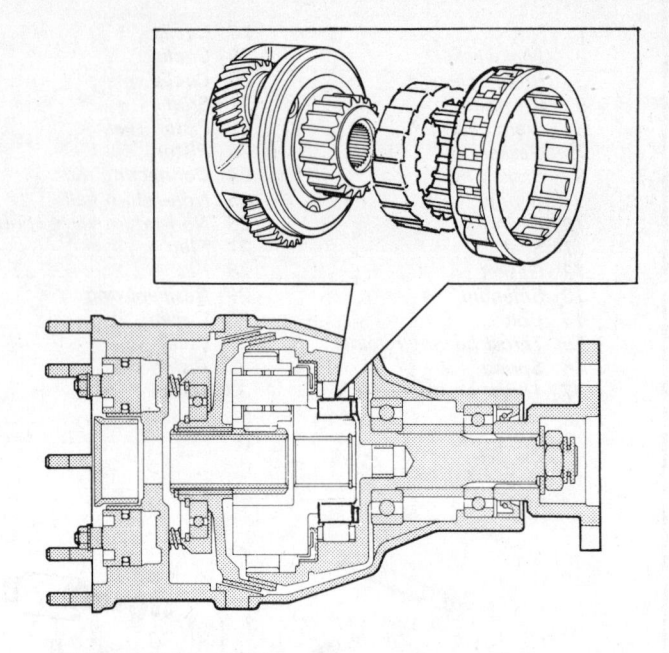

Fig. 6.20 Sectional view of Type 'P' overdrive. Inset shows later type unidirectional clutch and gear carrier (Sec 16)

7 Refit by reversing the removal operations, noting the following points:

(a) Use a new gasket between the gearbox and the overdrive
(b) Tighten the nuts progressively to the specified torque
(c) Top up the gearbox oil, road test the vehicle, then check the oil level again

18 Overdrive – dismantling

1 With the overdrive removed from the gearbox, clean it externally and clamp it in a soft-jawed vice.
2 Disconnect the earth lead fom the solenoid. Unscrew and remove the solenoid, using a 1 in AF open-ended spanner with thin jaws (photo).
3 Remove the overdrive pressure switch (Turbo models only) (photo).
4 Slacken the six nuts which secure the main and rear casings (photo). Remove four of the nuts, leaving two (opposite and on long studs) which should be slackened alternately a little at a time until the spring pressure separating the casings is relieved. Note the location of the solenoid earth lead under one of the nuts. Recover the special washers.

17.5 Disconnecting the overdrive solenoid

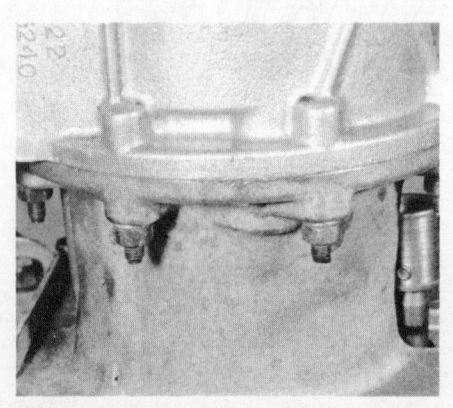

17.6A Four of the overdrive securing nuts

17.6B Lifting off the overdrive (gearbox on bench)

Chapter 6 Manual gearbox, overdrive and automatic transmission

18.2 Removing the overdrive solenoid

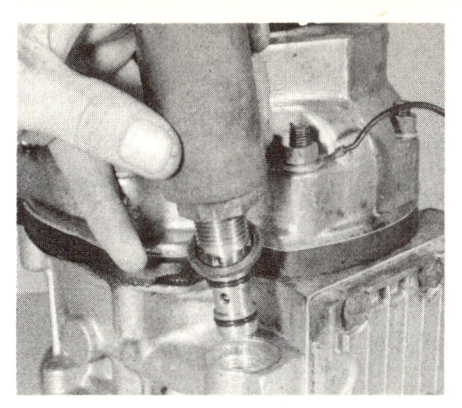

18.3 Removing the pressure switch

18.4 Undoing an overdrive casing nut

5 Remove the bridge piece nuts and the bridge pieces themselves (photo). The nuts will be under spring pressure at first.
6 Separate the main and rear casings.
7 Remove the planet gears and carrier, the sun wheel with cone clutch and springs, and the brake ring (photos).

8 Clamp the main casing in the vice. Unbolt and remove the valve cover and the oil pick-up strainer (photo).
9 Remove the two pistons, using long-nosed pliers if necessary (photo).
10 Using a peg spanner or (at a pinch) a pair of circlip pliers, unscrew

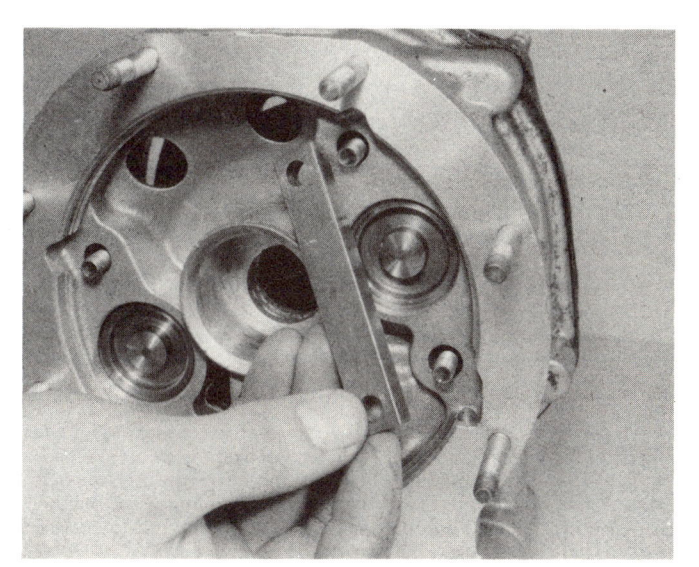

18.5 Removing a bridge piece

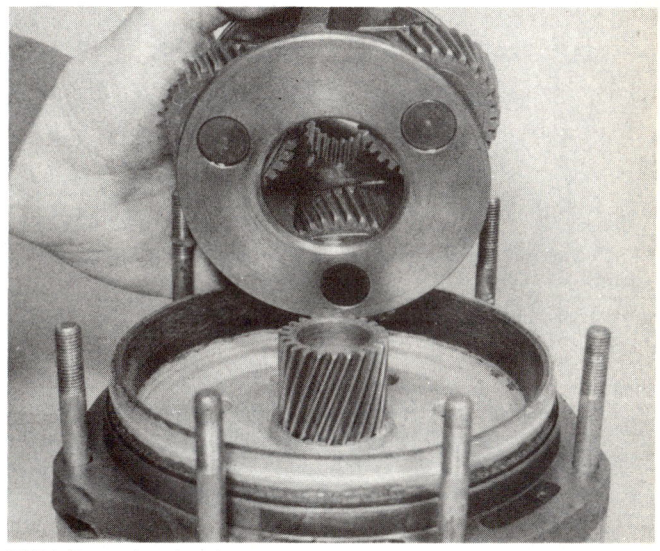

18.7A Removing the planet gear set ...

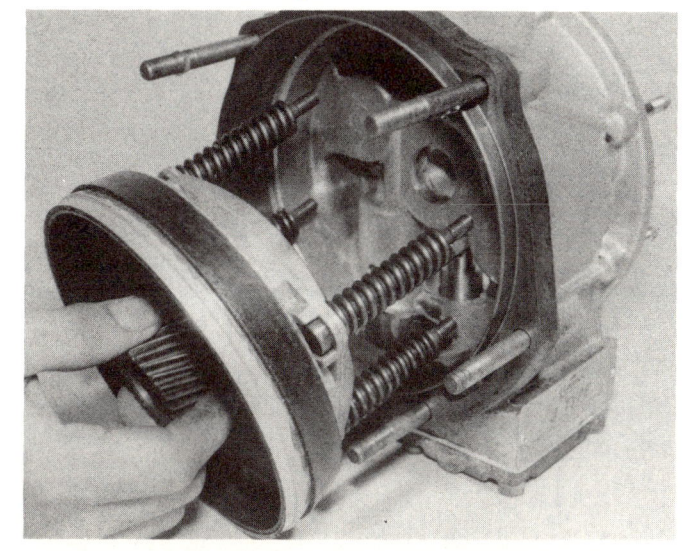

18.7B ... and the sun wheels, cone clutch and springs

18.8 Removing the overdrive valve cover

Chapter 6 Manual gearbox, overdrive and automatic transmission

18.9 Removing a piston

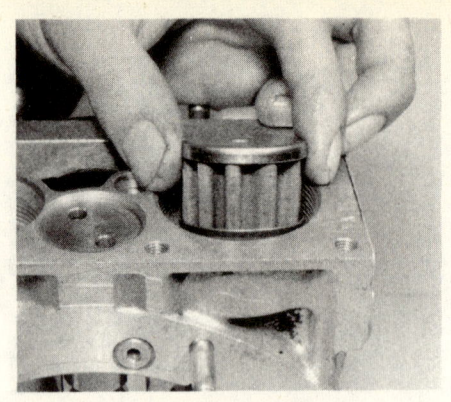

18.10 Removing the pressure filter

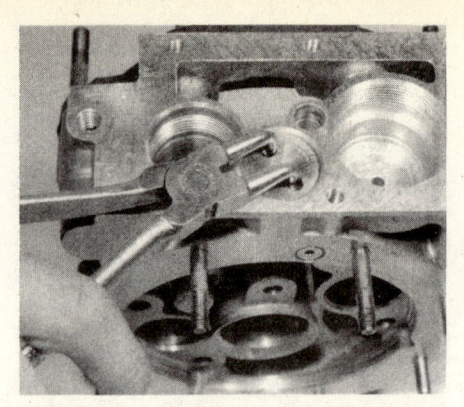

18.11A Unscrewing the non-return valve plug

18.11B Remove the non-return valve spring ...

18.11C ... the ball ...

18.11D ... and the valve seat

the pressure filter (large) plug. Remove the washer and filter (photo).
11 Similarly remove the non-return valve (small) plug. Remove the spring, ball and valve seat (photos). The sleeve and pump plunger can then be removed if wished.
12 Finally unscrew the relief valve plug. Remove the piston, springs and plunger, being careful not to lose any shims. The valve sleeve may be extracted if necessary (photos).

13 Further dismantling depends on the work to be undertaken. The thrust bearing can be removed from its cage after removing the two circlips and separating the cage from the sun wheel and cone clutch (photo).
14 The unidirectional clutch can be removed from the rear casing after removing the circlip and the oil thrower. Recover the thrust washer.
15 To remove the output shaft bearings, the flange must be removed and the shaft pressed out of the rear casing. The front bearing can then be pulled off the shaft and the rear bearing driven out of the casing. There is a spacer between the two bearings.

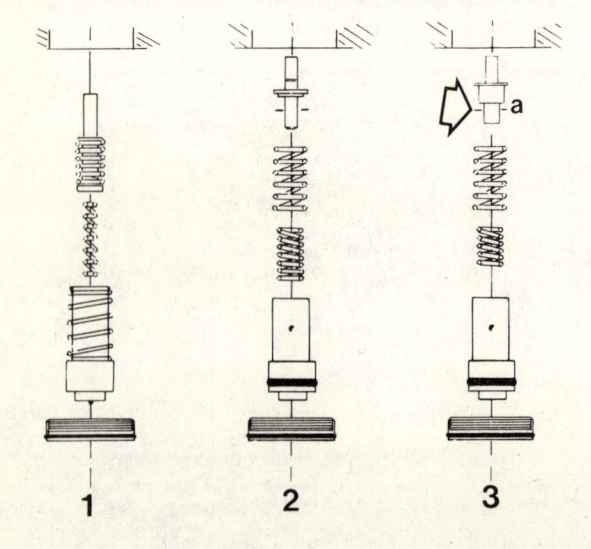

Fig. 6.21 Relief valve components. Types 1 and 2 are obsolete, type 3 is current (Sec 18)

a Shims

18.12A Removing the relief valve piston ...

Chapter 6 Manual gearbox, overdrive and automatic transmission

18.12B ... the small spring ...

18.12C ... the large spring ...

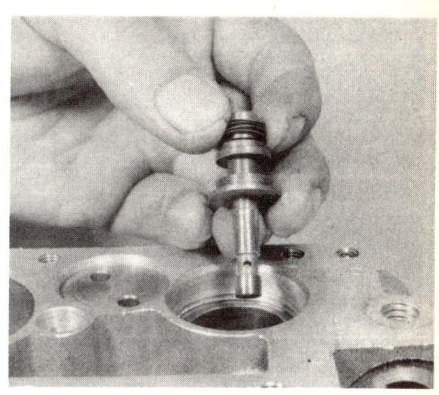

18.12D ... and the plunger with shims

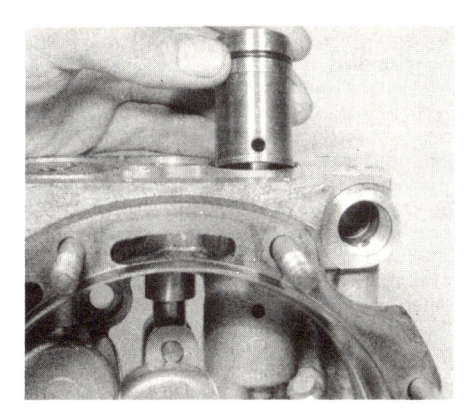

18.12E Removing the relief valve sleeve

18.13 The thrust bearing in its cage

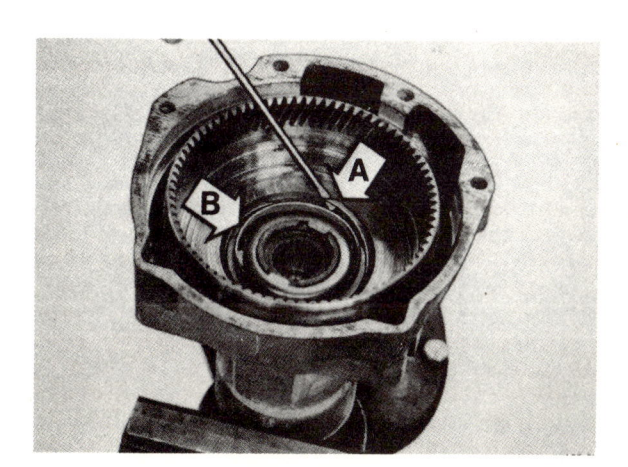

Fig. 6.22 Removing the circlip (A) which secures the unidirectional clutch (Sec 18)

B Oil thrower

Fig. 6.23 Pulling off the output shaft front bearing (Sec 18)

19 Overdrive – overhaul

1 See Section 10, paragraphs 1 and 2.
2 Flush oilways and control orifices with solvent and blow them dry.
3 Examine hydraulic system components such as pistons, valves and valve seats for wear and damage. Renew as necessary.
4 Wash the bearings and the one-way clutch in solvent. Check for roughness and visible damage.
5 Clean out the groove in front of the annulus (ring gear). Dirt will collect here as a result of centrifugal force.
6 Examine the gears for wear and damage. Do not try to dismantle the planet gear set.
7 Check the brake ring for cracks, and the clutch linings for burning or abrasion.
8 Check the condition of the clutch springs. They should be 55.5 ± 1.5 mm (2.185 ± 0.059 in) long.
9 Check the solenoid using a 12 volt battery and an ammeter. Current consumption should be approximately 2A and the solenoid plunger should move in and out without sticking.
10 Renew the oil pick-up strainer and the pressure filter unless they are in perfect condition.

Chapter 6 Manual gearbox, overdrive and automatic transmission

11 All parts must be perfectly clean, and reassembly must take place under clean conditions, to avoid subsequent trouble with the hydraulic system.

20 Overdrive – reassembly

1 Press the output shaft bearings into the rear casing and onto the shaft. Do not forget the spacer. Press the output shaft into the casing and fit a new flange oil seal.
2 Refit the thrust washer, the one-way clutch inner race and the roller cage. Fit the oil thrower and circlip, then check the operation of the one-way clutch.
3 Refit the drive flange, using locking compound on the splines. Do not contaminate the oil seal. Counterhold the flange and tighten the nut to the specified torque.
4 Refit the planet gear set to the output shaft, lining up the splines.
5 Fit the thrust bearing to its cage. Assemble the cage, sun wheel and cone clutch. Make sure the clutch linings are clean and dry, then lubricate them with ATF.
6 Refit the relief valve components and secure them with the plug, using a new O-ring.
7 Refit the oil pump sleeve and plunger, lubricating them with ATF (photo). The groove and bevel on the sleeve must align with the pressure filter recess, and the chamfer on the plunger ring must face forwards.
8 Refit the non-return valve seat and ball. Fit the spring into the plug and refit the plug, using a new O-ring.
9 Fit the pressure filter, washer and plug.
10 Tighten all the plugs with a peg spanner.
11 Fit the oil pick-up strainer, a new cover gasket and the valve cover (photo). Tighten the cover bolts.
12 Lubricate the two pistons and insert them into their bores. Make sure they are the right way round.
13 Refit the brake ring to the rear casing, using a new gasket. Fit another new gasket on the main casing side of the brake ring (photo).
14 Refit the sun wheel/cone clutch assembly. Fit the clutch springs over the bearing cage studs.
15 Fit the main and rear casings together, being careful not to displace the gasket. Fit the piston bridge pieces and the nuts, tightening the nuts against spring pressure.
16 Fit the main-to-rear casing nuts, using new nylon washers (photo). Remember the solenoid earth tag. Tighten the nuts progressively to the specified torque, then recheck the tightness of the bridge piece nuts.
17 Refit and tighten the solenoid. Also refit the pressure switch, when applicable.

21 Gear lever – removal and refitting

1 Raise and support the vehicle for access to the underside of the gear lever.
2 Remove the Allen screw which secures the pin at the end of the gear lever (photo). Press out the pin from the gear lever and the selector rod.
3 On 1986 and later models, remove the large circlip from the base of the gear lever (photo).
4 Inside the vehicle, remove the console section from around the gear lever.
5 Remove the outer gaiter. Undo the four screws which secure the inner gaiter clamp plate. Remove the clamp plate, noting which way round it is fitted, and peel the inner gaiter up the gear lever.
6 On pre-1986 models, remove the large circlip from the base of the gear lever (photo).
7 Pull the gear lever upwards and withdraw it. Disconnect the overdrive switch wires (when applicable). Do not disturb the reverse detent plate bolts.
8 Refit by reversing the removal operations. Check the clearance between the reverse detent plate and finger with 1st gear engaged: it should be between 0.5 and 1.5 mm (0.02 to 0.06 in). Adjust if necessary by slackening the detent plate bolts. When adjustment is

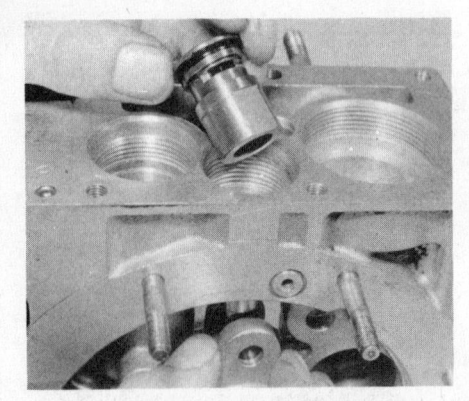

20.7 Refitting the oil pump sleeve and plunger

20.11 Fitting the oil pick-up strainer

20.13 Fitting a casing gasket

20.16 Fitting a casing stud nylon washer

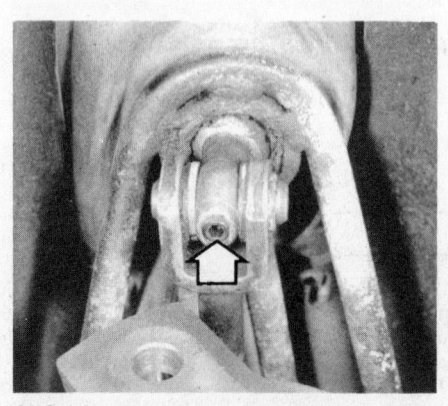

21.2 Allen screw (arrowed) securing gear lever pin

21.3 Circlip (arrowed) at base of gear lever on later models

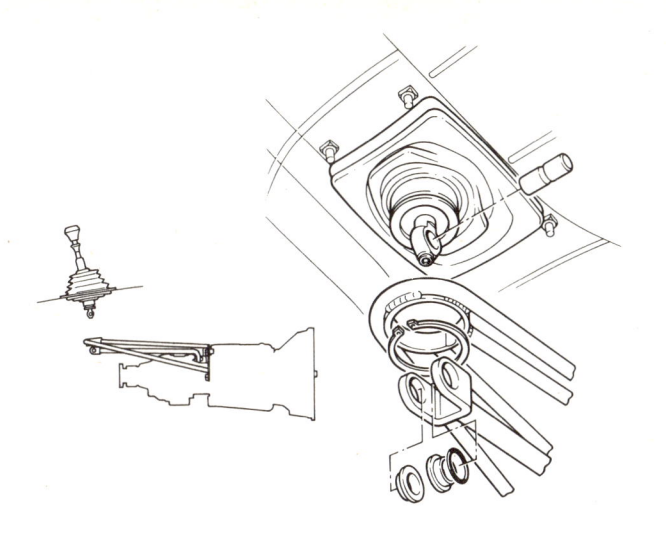

Chapter 6 Manual gearbox, overdrive and automatic transmission

Fig. 6.24 Base fixings of later type gear lever (Sec 21)

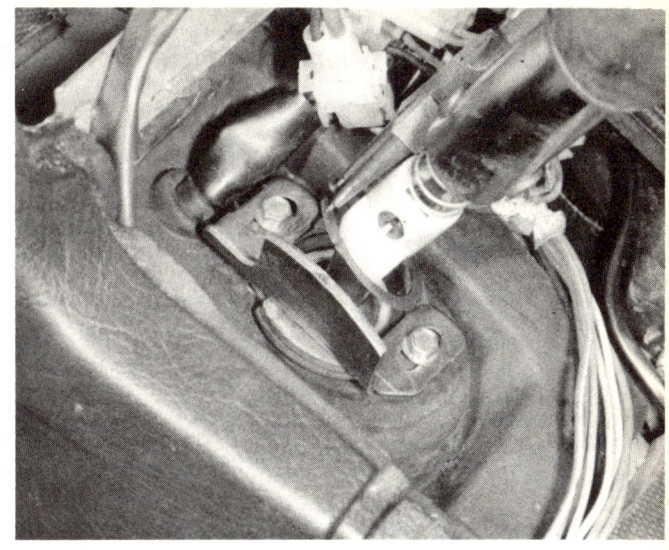

21.6 Removing the gear lever circlip – earlier models

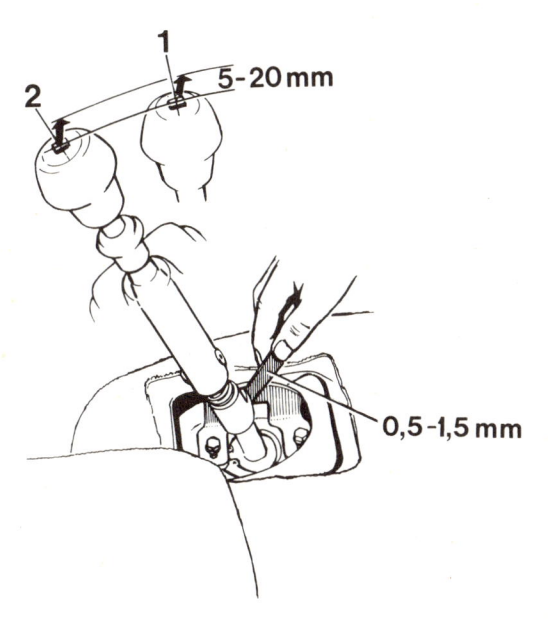

correct, side-to-side play of the gear lever knob in 1st or 2nd gear should be 5 to 20 mm (0.2 to 0.8 in).

22 Gear lever pullrod – renewal

1 The gear lever pullrod transmits the motion from the collar under the gear knob to the interlock sleeve at the base of the gear lever. If it breaks, it will not be possible to engage reverse gear. Proceed as follows.
2 On models with overdrive, remove the trim panel from the right-hand side of the centre console. Separate the overdrive wiring connector there and tie a piece of string to the wire leading to the gear lever.
3 On all models, remove the gear lever boot. Drive out the roll pin which secures the lever to the stub. Lift off the gear lever, at the same time pulling the overdrive wire and string through, when applicable. Untie the string.
4 Remove the overdrive switch, when applicable. Remove the gear lever knob by gripping the lever in a soft-jawed vice and tapping the knob with a soft hammer and an open-ended spanner (photo). The knob is glued onto splines and may not come off undamaged. Clean off the old glue.
5 Remove the old pullrod. It may be plastic or metal. The metal rod is removed by undoing the grub screw at the top of the lever, then withdrawing the pullrod, spring and interlock sleeve downwards. To remove the plastic rod, release the catch at the base and (on overdrive models) lift up the collar slightly to free the top of the rod (photos).

Fig. 6.25 Gear lever adjustment in 1st or 2nd gear positions (1 and 2) (Sec 21)

22.4 Removing the gear lever knob

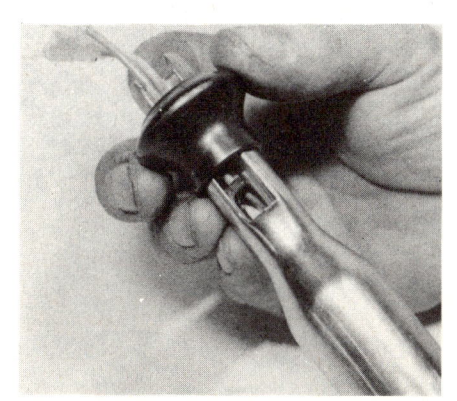

22.5A Lift up the collar ...

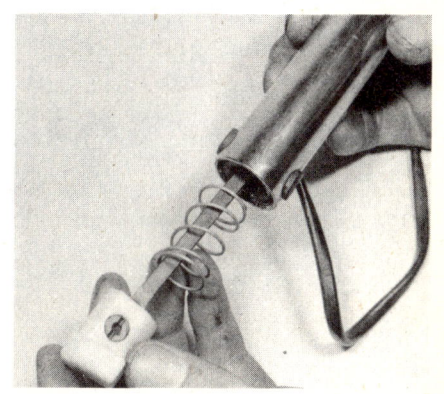

22.5B ... and remove the pullrod with the spring and interlock sleeve

190 Chapter 6 Manual gearbox, overdrive and automatic transmission

6 Soak the new pullrod in water for one hour before fitting it. Fit the new rod from below on overdrive models, being careful not to displace the rubber bushes and wiring. Make sure it engages with the collar. On models without overdrive, fit the pullrod to the collar and then insert them both from above. On all models, position the pullrod bottom lug to receive the interlock sleeve.
7 Fit the spring and interlock sleeve, engaging the pullrod lug in the catch on the sleeve.
8 Refit the gear lever over the splines, using a little glue if wished. Do not use powerful glue, or there will be a problem with any subsequent removal.
9 Refit the gear lever to the stub and secure it with the roll pin. On overdrive models, reattach the string to the wiring and draw the wire back to the centre console. Reconnect the wires and refit the trim panel.
10 Check the selection of all gears, including reverse. If adjustment is necessary, see Section 21, paragraph 8.
11 When adjustment is satisfactory, refit the gear lever boot, the overdrive switch and any other disturbed items.

23 Overdrive switches – removal and refitting

Control switch
1 Prise off the trim plate from the top of the gear lever knob (photo).
2 Prise out the switch and disconnect it (photos).
3 Refit by reversing the removal operations.

Inhibitor (gearbox) switch
4 Raise and support the vehicle.
5 Except on V6 models, support the gearbox, remove the crossmember and slacken the exhaust flanged joint. Lower the rear of the gearbox slightly for access to the top cover.
6 Clean around the switch, disconnect the wires and unscrew it (photos).
7 Refit by reversing the removal operations.

Pressure switch (Turbo only)
8 Raise and support the vehicle.
9 The switch is located in front of the overdrive solenoid. Clean around the switch, disconnect the wiring and unscrew it. Be prepared for oil spillage.
10 Refit by reversing the removal operations. Top up the gearbox oil if necessary.

24 Reversing light switch – removal and refitting

1 Gain access to the gearbox top cover (Section 23, paragraphs 4 and 5).
2 Clean around the switch, disconnect the wires and unscrew it (photo).
3 Refit by reversing the removal operations.

23.1 Removing the gear lever knob trim plate

23.2A Prise out the overdrive switch ...

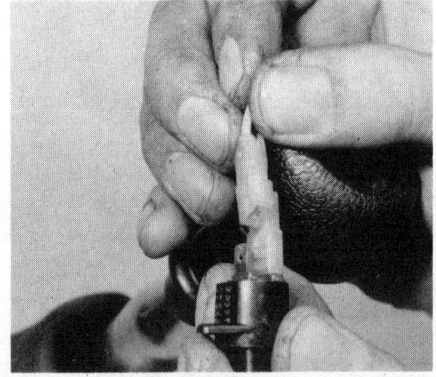

23.2B ... and disconnect it

23.6A Disconnect the inhibitor switch ...

23.6B ... and unscrew it from the cover

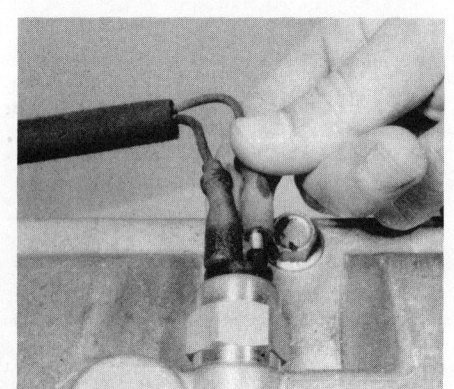

24.2 Disconnecting the reversing light switch

Chapter 6 Manual gearbox, overdrive and automatic transmission

25 Fault diagnosis – manual gearbox and overdrive

Note: *It is not necessarily a good idea to dismantle the gearbox in an attempt to rectify a minor fault. A gearbox which is noisy, or on which the synchro can be beaten by a quick change, may carry on for a long time in this state. Piecemeal renewal of components can actually increase noise by presenting different wear surfaces to each other*

Symptom	Reason(s)
Excessive noise	Oil level low or incorrect grade Gear lever gaiter torn or displaced Worn bearings or shafts Worn gear teeth
Difficult engagement of gears	Clutch fault Gear lever adjustment incorrect Synchro units worn or damaged Selector components worn or damaged
Jumps out of gear	Synchro units worn or damaged Selector components worn
Overdrive will not engage	Oil level low Solenoid or wiring faulty Gear lever switch faulty Inhibitor switch faulty Relay faulty Oil pump non-return valve stuck open Oil filter blocked
Overdrive will not release. (**Caution:** Do not reverse the vehicle or damage will result)	Control valve sticking Solenoid sticking Electrical fault (solenoid permanently energised) Cone clutch stuck (tap brake ring to release) Relief valve stuck
Overdrive slips when engaging	Oil level low Oil filter blocked Oil pump non-return valve leaking Relief valve piston sticking Clutch linings worn or glazed Torque limiter system malfunction (see Chapter 4)
Overdrive noisy in direct drive, quiet in overdrive	Thrust (cone clutch) bearing worn
Overdrive noisy except during engagement	Output shaft bearings worn

PART B: AUTOMATIC TRANSMISSION

26 General description

The automatic transmission has four forward speeds and one reverse. Gear changing between forward speeds is normally fully automatic, responding to speed and load, although the drive can prevent the selection of higher ratios. On the AW71 transmission the highest (4th) gear is provided by an overdrive unit fitted between the torque converter and the rest of the transmission. The ZF4HP transmission is an integrated four-speed unit.

Drive is taken from the engine to the transmission by a torque converter. This is a type of fluid coupling which under certain conditions has a torque multiplying effect. On later models the torque converter is mechanically locked at high speeds in 3rd and 4th gear, so eliminating losses due to slip and improving fuel economy.

The gear selector has six or seven positions: P, R, N, D, 3 (on some models), 2 and 1. The engine can only be started in positions P and N. In position P the transmission is mechanically locked: this position must only be engaged when the vehicle is stationary. In position R reverse is engaged, in N neutral. In position D gear changing is automatic throughout the range; positions 3, 2 and 1 prevent the selection of higher ratios when this is desired. These lower positions must not be selected at speeds so high as to cause engine overrevving.

When position 3 is missing from the selector, a button on the side of the selector knob serves to inhibit the engagement of 4th (overdrive) gear. A dashboard warning light reminds the driver when this has been done.

A 'kickdown' facility causes the transmission to shift down a gear (subject to engine speed) when the throttle is fully depressed. This is useful when extra acceleration is required. Kickdown is controlled by a cable linkage from the throttle cable drum.

The transmission fluid is cooled by a heat exchanger built into one of the radiator side tanks, and (on some models) by an auxiliary cooler mounted in front of the radiator. See also Chapter 2, Section 17.

The automatic transmission is a complex unit, but if it is not abused it is reliable and long-lasting. Repair or overhaul operations are beyond the scope of many dealers, let alone the home mechanic; specialist advice should be sought if problems arise which cannot be solved by the procedures given in this Chapter.

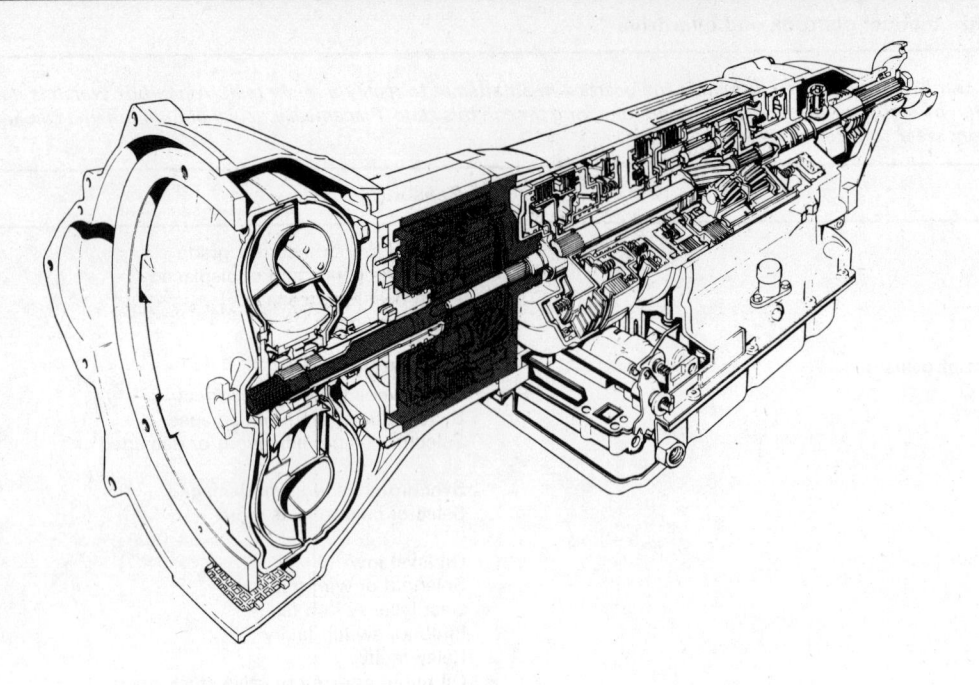

Fig. 6.26 Cutaway view of the AW 71 transmission (Sec 26)

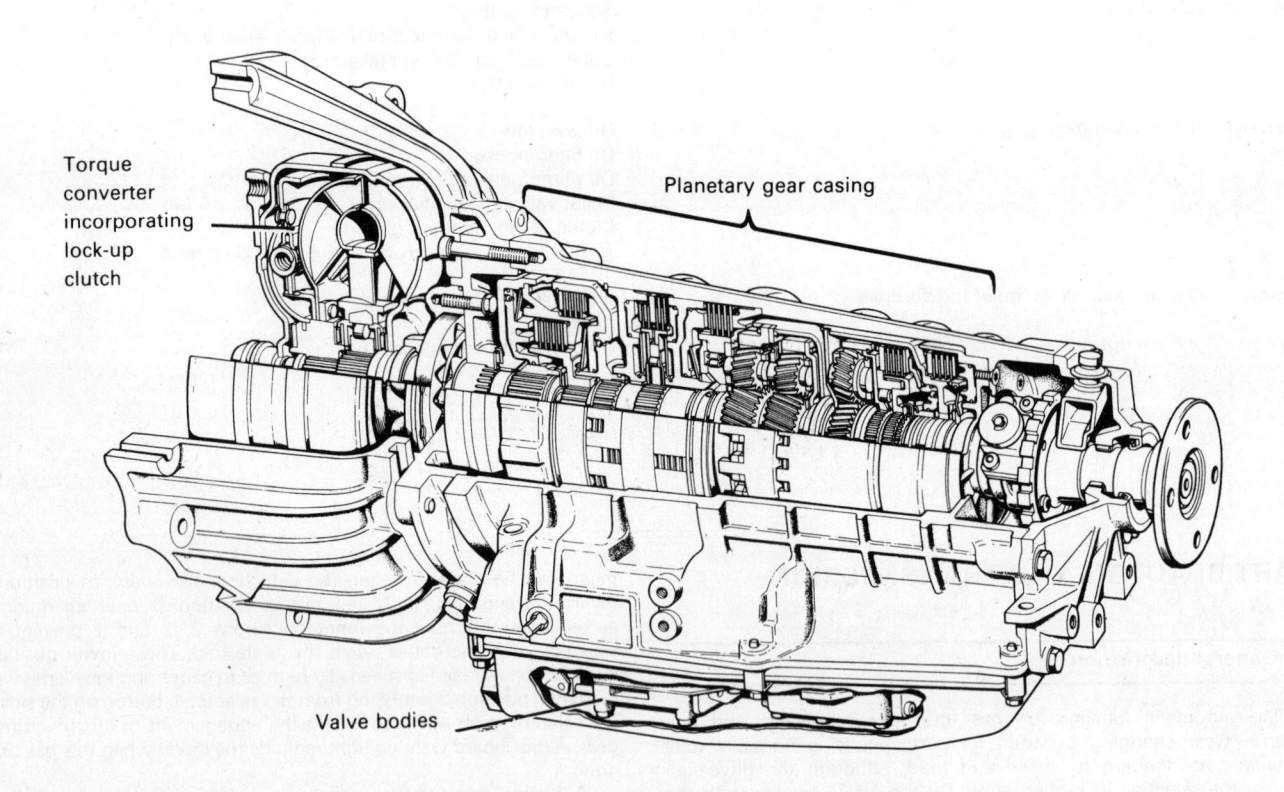

Fig. 6.27 Cutaway view of the ZF 4HP transmission (Sec 26)

Labels: Torque converter incorporating lock-up clutch; Planetary gear casing; Valve bodies

27 Maintenance and inspection

1 Every 12 000 miles or annually, or at the first sign of any malfunction, check the transmission fluid level (Section 28). Investigate the cause of persistent fluid loss.
2 At the same intervals check the adjustment of the selector mechanism (Section 30) and the kickdown cable (Section 31).
3 Every 24 000 miles or two years renew the transmission fluid as described in Section 29.
4 If contamination of the fluid occurs, or the transmission overheats and the fluid smells burnt, flush the cooling system by following the procedure in Section 29; also flush the auxiliary cooler (when fitted) using a hand pump. The auxiliary cooler is thermostatically controlled and will not be flushed during the fluid renewal procedure.

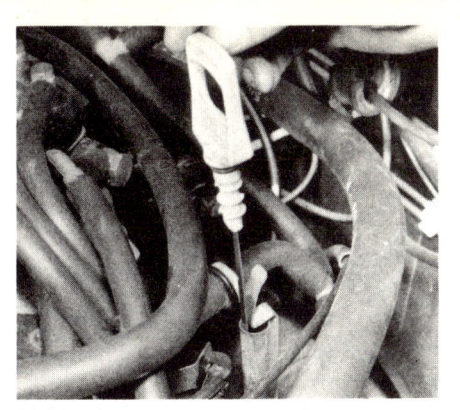

Chapter 6 Manual gearbox, overdrive and automatic transmission

28.3A Withdrawing the transmission dipstick

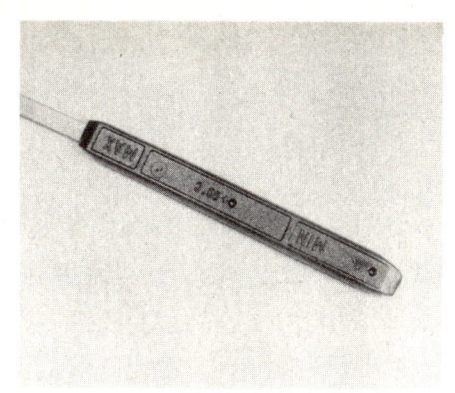

28.3B Transmission dipstick markings

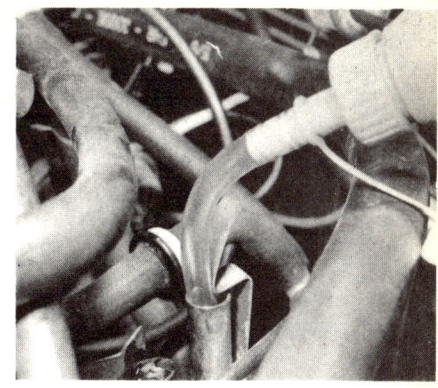

28.4 Topping-up the transmission fluid

28 Fluid level checking

1 Warm up the transmission fluid by driving the vehicle for at least 10 miles. Park on level ground and allow the engine to idle.
2 Move the gear selector through all positions (apply the footbrake) and finish in position P. Wait two minutes.
3 Still with the engine iding, withdraw the transmission dipstick. Wipe it with a clean lint-free cloth, re-insert it fully, withdraw it again and read the level. Read the side marked 'HOT' or '+90°C' (photos).
4 If topping-up is necessary, this is done via the dipstick tube (photo). Only use fresh transmission fluid of the specified type, and take great care not to introduce dirt into the transmission. Do not overfill: the distance from 'MIN' to 'MAX' on the dipstick represents half a litre (not quite a pint) of fluid.
5 Refit the dipstick, making sure it is fully home, and stop the engine.

29 Fluid renewal

Caution: *If the vehicle has just been run, the transmission fluid may be very hot.*
1 Raise and support the vehicle.

AW71
2 If no drain plug is fitted, proceed from paragraph 3. If a drain plug is fitted, remove it and allow the contents of the sump to drain into a drain pan. Refit and tighten the drain plug (photo). Add 2.0 litres (3.5 pints) of fresh ATF via the dipstick tube.
3 Clean the oil cooler return union (the rearmost one) on the side of the transmission (photo). Disconnect the union and attach a clear plastic hose to the line from the cooler. Lead the hose into the drain pan.
4 Raise the engine at idle. Fluid will flow into the drain pan. When bubbles appear in the fluid, stop the engine.
5 Add 2.0 litres (3.5 pints) of fresh ATF via the dipstick tube.
6 Repeat paragraph 4, then remove the plastic hose and reconnect the oil cooler union.
7 Add a further 2.0 litres (3.5 pints) of fresh ATF.
8 Lower the vehicle. Check the oil level as described in Section 28, paragraph 2 onwards, but use the 'COLD' or '+40°C' side of the dipstick. Top up as necessary.
9 Dispose of the old fluid safely.

ZFHP422
10 Proceed as above, but note the following points:

  (a) Top up in increments of 2.5 litres (4.4 pints)
  (b) The oil cooler return union is the lower of the two

30 Gear selector – checking and adjustment

1 Check that the gear lever is vertical in position P (not touching the centre console). Adjust from below if necessary by slackening the actuator lever nut.

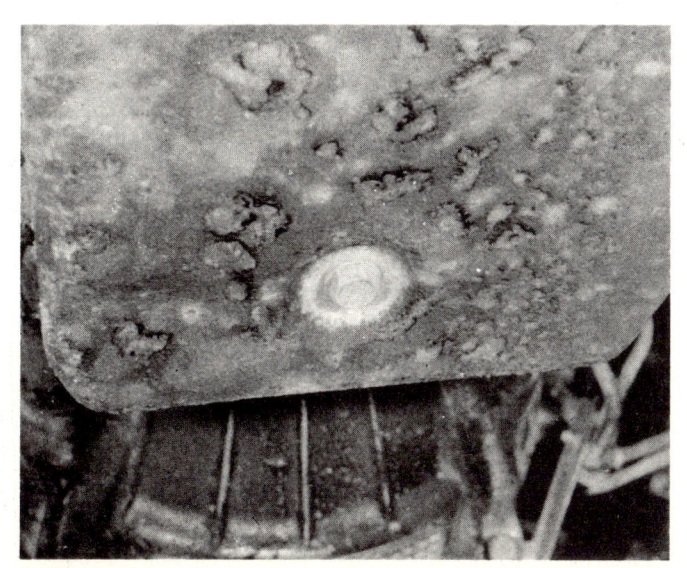

29.2 Transmission drain plug

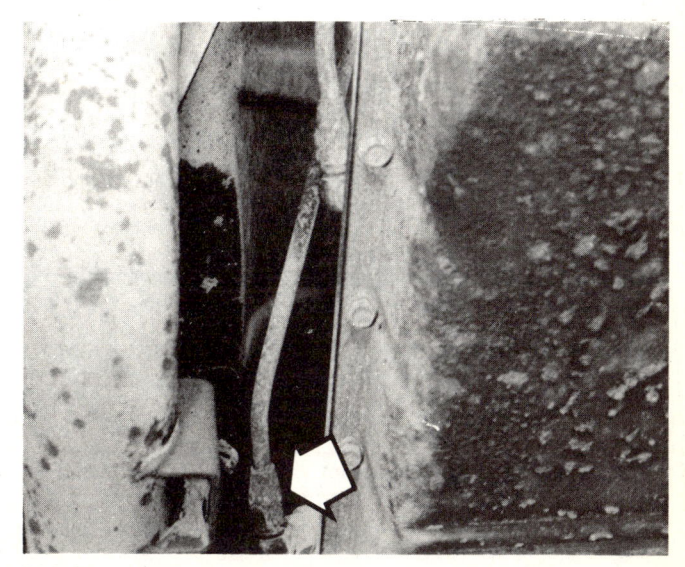

29.3 Oil cooler return union (arrowed)

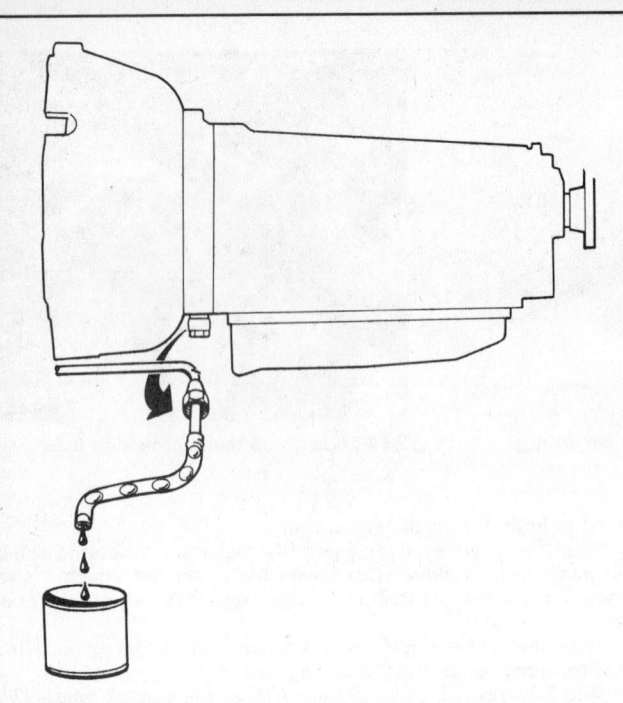

Chapter 6 Manual gearbox, overdrive and automatic transmission

Fig. 6.28 Oil cooler return union disconnected from ZF transmission for fluid changing (Sec 29)

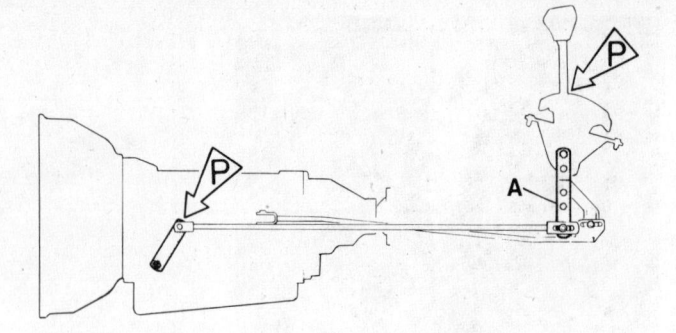

Fig. 6.29 Automatic transmission selector linkage in position P (Sec 30)

A   Actuator lever

30.4 Gear selector reaction lever nut (arrowed, under cable)

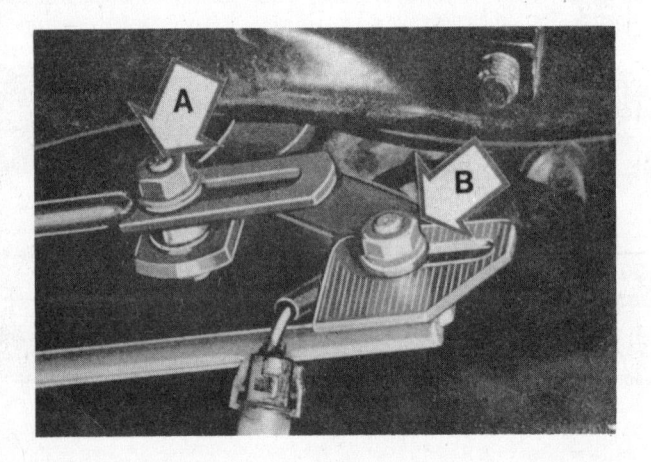

Fig. 6.30 Automatic transmission selector linkage adjustment nuts (Sec 30)

A   Actuator lever   B   Reaction lever

2   Check that the engine will only start in positions P and N, and that the reversing lights only come on in position R (ignition on).
3   Check that the free play from D to N is the same as, or less than, the play from 2 to 1 (AW71) or from 3 to 2 (ZFHP422).
4   If there is insufficient play in D, slacken the reaction lever nut (beneath the vehicle) and move the lever approximately 2 mm (0.08 in) rearwards (photo).
5   If there is insufficient play in 3 or 2, slacken the nut and move the lever approximately 3 mm (0.12 in) forwards.
6   Tighten the nut and recheck the adjustment.

31 Kickdown cable – adjustment

1   With the throttle linkage in the idle position, the distance from the crimped stop on the kickdown cable to the adjuster sleeve should be 0.25 to 1.0 mm (0.01 to 0.04 in). The cable should be taut.

31.2 Checking kickdown cable adjustment at full throttle

2   Have an assistant depress the throttle pedal fully. Measure the distance from the stop to the adjuster again (photo). It should be 50.4 to 52.6 mm (1.99 to 2.07 in). From this position it should be possible to pull the cable out by 2 mm (0.08 in).

Chapter 6 Manual gearbox, overdrive and automatic transmission

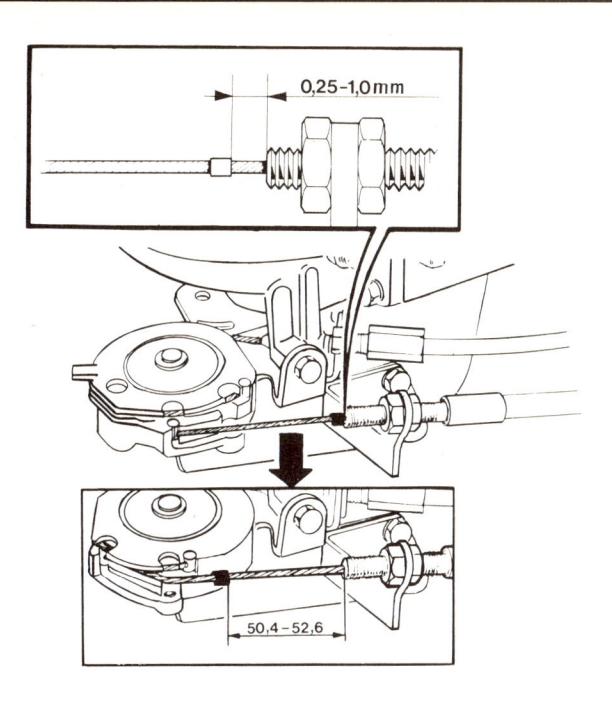

Fig. 6.31 Kickdown cable adjustment at idle (inset, top) and at full throttle (below) (Sec 31)

3 Adjust if necessary by slackening the locknuts, turning the adjuster sleeve and tightening the locknuts.
4 If correct adjustment cannot be achieved, either the throttle linkage adjustment is incorrect (Chapter 3) or the cable crimped stop is incorrectly positioned (Section 32).

32 Kickdown cable – renewal

1 Slacken the cable adjuster at the throttle end. Disconnect the cable inner from the drum and the outer from the bracket (photo).
2 Raise and support the vehicle. Drain the transmission oil pan by removing the drain plug (when fitted) and the dipstick/filler tube nut (photo). **Caution:** *The fluid may be very hot.*
3 Clean the oil pan, then unbolt and remove it. Be prepared for fluid spillage. Recover the gasket.
4 Clean around the cable outer where it enters the transmission. Unhook the cable inner from the cam, using a screwdriver to turn the cam. Cut the cable inner below the stop at the throttle end if there is not enough slack in the inner. Release the cable outer and remove the cable.
5 Fit the new cable to the transmission, attaching the inner to the cam and securing the outer to the transmission case. Use a new O-ring and grease the cable outer where it enters the transmission.
6 Attach the cable outer at the throttle end. Pull the cable inner until light resistance is felt, and in this position crimp the inner stop 0.25 to 1.0 mm (0.01 to 0.04 in) from the adjuster.
7 Reconnect the cable inner to the throttle drum. Adjust the cable as described in Section 31.
8 Clean the inside of the oil pan, including the magnets when fitted.
9 Refit the oil pan, using a new gasket. Reconnect the dipstick/filler tube.
10 Top up the transmission fluid.
11 Lower the vehicle. Road test the transmission, then recheck the fluid level and inspect the oil pan for leaks.

33 Kickdown marker – adjustment

Early models (marker in floor)
1 Release the retainer and screw the marker in towards the floor as far as possible.

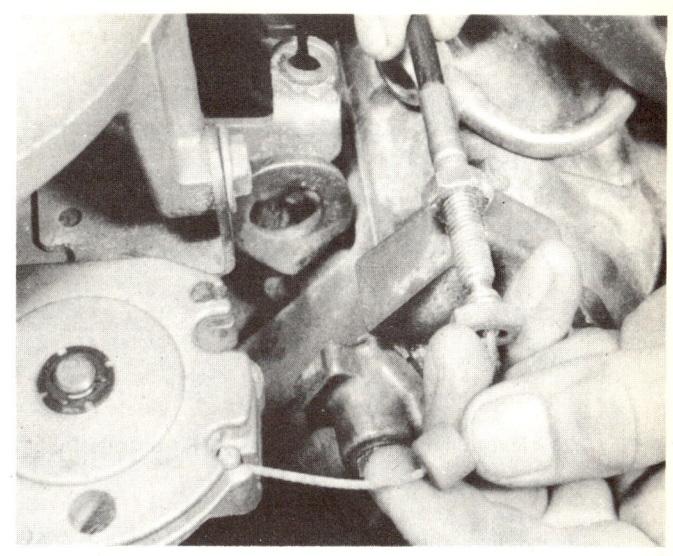

32.1 Disconnecting the kickdown cable at the throttle end

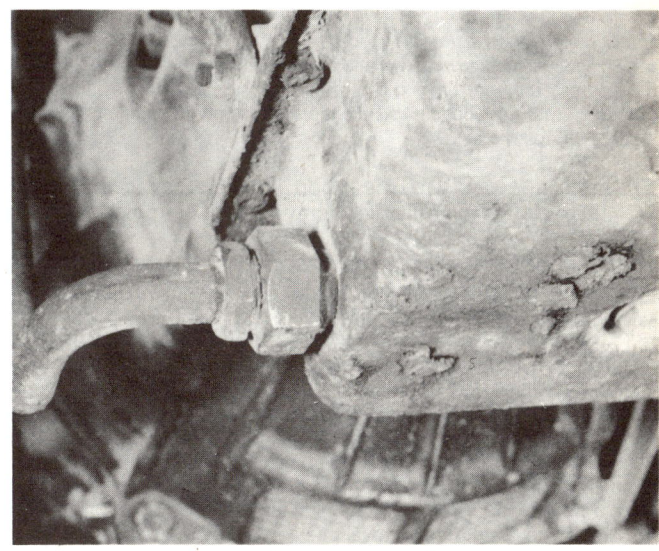

32.2 Transmission dipstick/filler tube nut

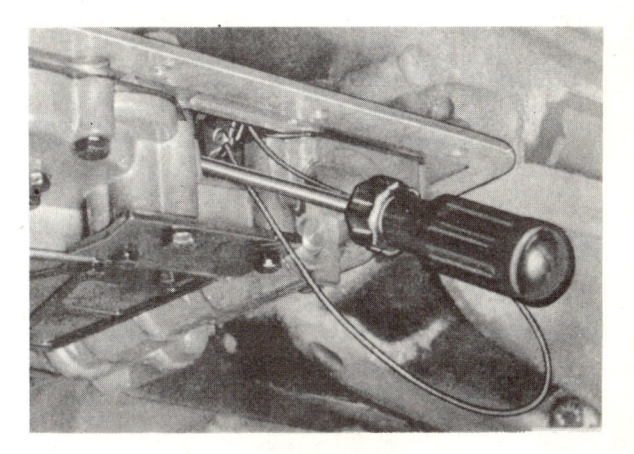

Fig. 6.32 Removing the kickdown cable – AW transmission (Sec 32)

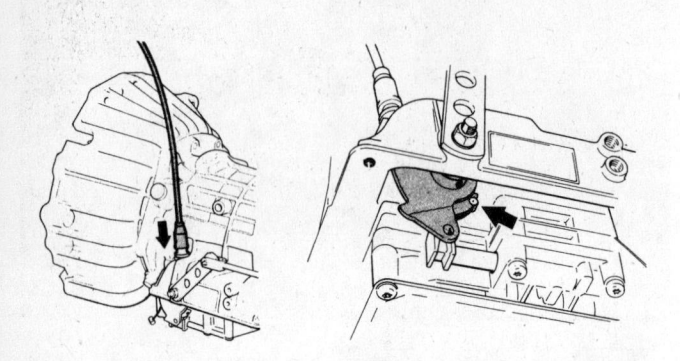

Fig. 6.33 Kickdown cable attachments – ZF transmission (Sec 32)

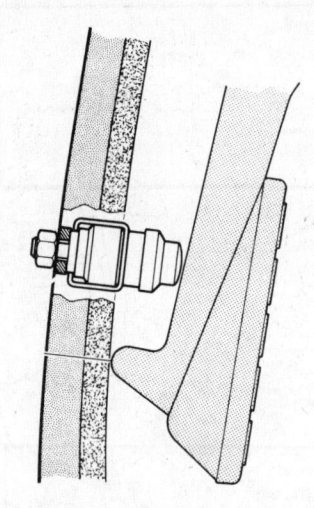

2 Depress the throttle pedal by hand to the start of the kickdown position (beginning of resistance). Hold the pedal in this position and screw the marker upwards to meet it.
3 Refit the retainer.

Later models (marker in pedal)
4 Screw the marker into the pedal as far as possible.
5 Depress the throttle pedal by hand to the start of the kickdown position. Hold the pedal in this position and screw the marker down to meet the floor.

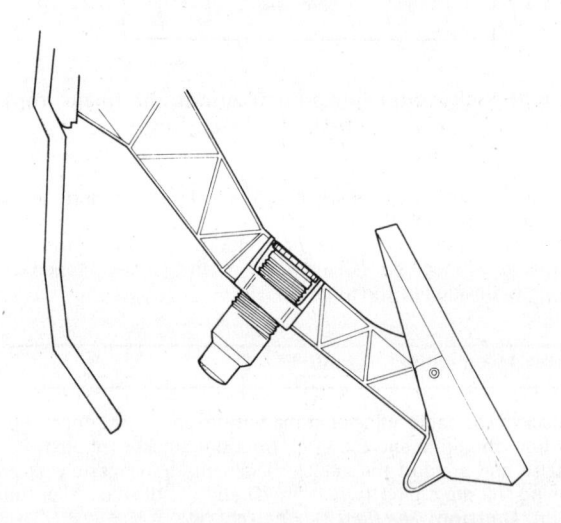

34 Starter inhibitor/reversing light switch – removal and refitting

1 Remove the ashtray and the centre console panel in front of the selector.

Models up to 1984
2 Remove the two screws which secure the left-hand half of the selector cover – one screw at each end of the brush (photo). Lift off the half of the cover.
3 Undo the switch securing screws, disconnect the multi-plug and remove it (photo).
4 If a new switch is being fitted, transfer the lens to it.
5 Refit by reversing the removal operations. Make sure that the stud on the lever enters the slot in the switch.

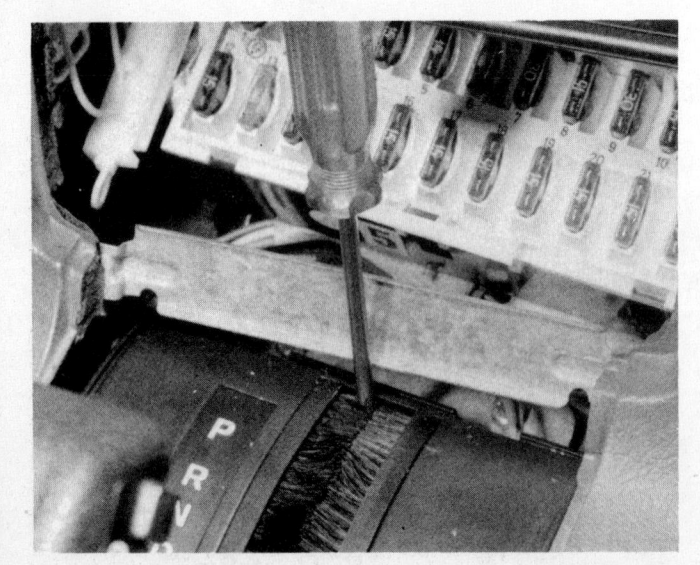

Fig. 6.34 Kickdown marker in floor (Sec 33)

Fig. 6.35 Kickdown marker in throttle pedal (Sec 33)

34.2 Removing a selector cover screw

34.3 Two screws (arrowed) which secure the starter inhibitor switch

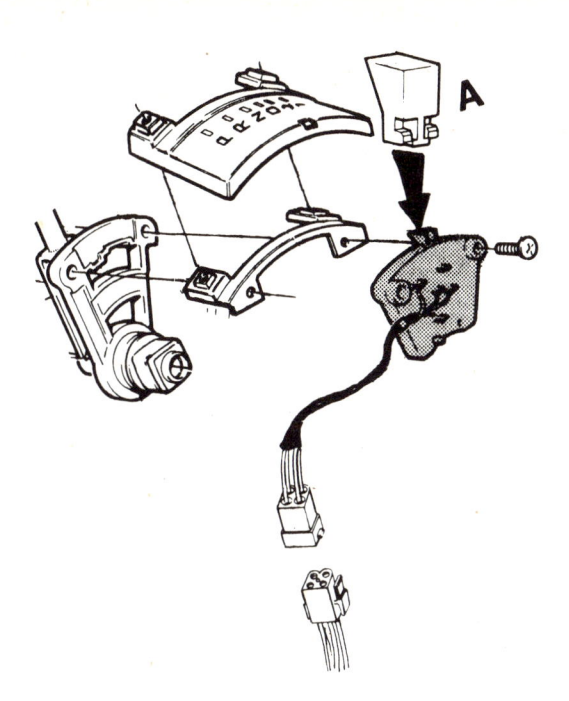

Chapter 6 Manual gearbox, overdrive and automatic transmission

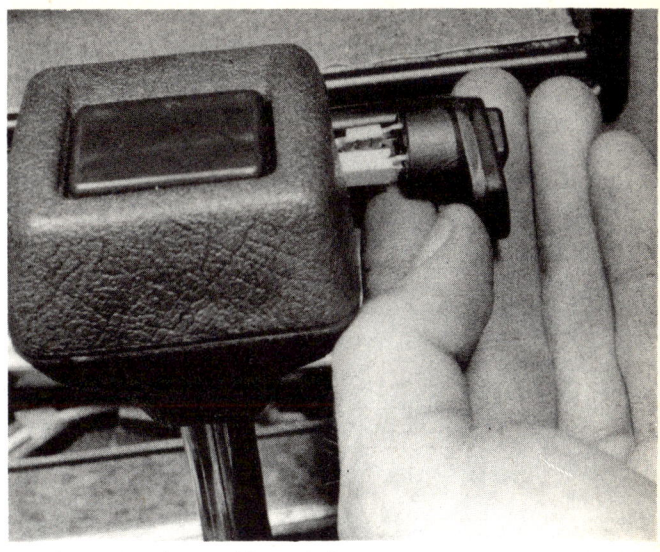

35.1 Removing the overdrive switch

Fig. 6.36 Later type automatic transmission selector cover and associated components (Sec 34)

A Lens

Models from 1985
6 Proceed as above, but note the different selector cover fittings (Fig. 6.36).

35 Overdrive switch (AW71) – removal and refitting

1 Prise the switch out of the side of the selector lever and disconnect it (photo).
2 Refit by reversing the removal operations.

36 Transmission oil seals – renewal

Note: *It is important not to introduce dirt into the transmission when working on it.*

Drive flange
1 Proceed as described in Section 3. The flange central nut may be secured by a lockwasher.

Input shaft/torque converter
2 Remove the transmission (Section 37).
3 Lift the torque converter out of its housing. Be careful, it is full of fluid.
4 Pull or lever out the old seal. Clean the seat and inspect the seal rubbing surface on the torque converter.
5 Lubricate the new seal with ATF and fit it, lips inwards. Seat it with a piece of tube.
6 Refit the torque converter and the transmission.

Selector shaft
7 Remove the selector arm nut and pull the arm off the shaft (photo).
8 Prise the seal out with a small screwdriver. Clean the seat.
9 Grease the new seal and fit it, lips inwards. Seat the seal with a tube or socket.
10 Refit the selector arm and tighten the nut.

All seals
11 Check the transmission fluid level on completion.

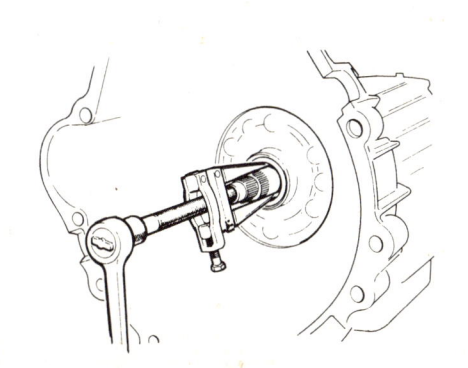

Fig. 6.37 Pulling out the input shaft oil seal (Sec 36)

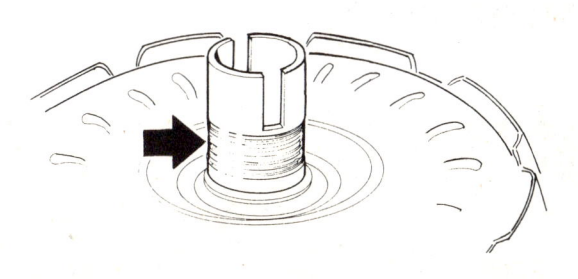

Fig. 6.38 Inspect the seal rubbing surface (arrowed) on the torque converter (Sec 36)

Chapter 6 Manual gearbox, overdrive and automatic transmission

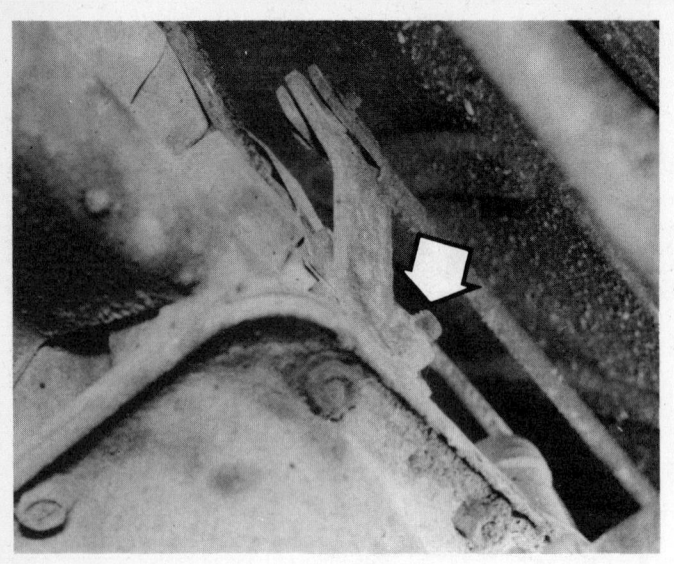

36.7 Remove the selector arm nut (arrowed)

37 Automatic transmission – removal and refitting

Note: *If the transmission is being removed for repair, check first that the repairer does not need to test it in the vehicle.*
1 Select P (AW71) or N (ZF).
2 Disconnect the battery negative lead.
3 Disconnect the kickdown cable at the throttle end.
4 Raise and support the vehicle. Drain the transmission fluid by removing the dipstick/filler tube nut. **Caution:** *The fluid may be very hot.*
5 Disconnect the selector linkage and (when applicable) the overdrive wiring connector from the side of the transmission.
6 Remove the starter motor (Chapter 12). On B28 engines, also remove the other starter motor aperture blanking plate.
7 Remove the dipstick/filler tube.
8 Disconnect the fluid cooler unions at the transmission. Be prepared for spillage. Cap the open unions to keep dirt out.
9 Disconnect the exhaust downpipe and unbolt the exhaust support bracket from the transmission crossmember (photo). Support the exhaust system if necessary.
10 Unbolt the propeller shaft flange.
11 When fitted, remove the cover plate from the bottom of the torque converter housing. Also remove the cooling grilles (photo).
12 Jam the driveplate and remove the bolts which hold the torque converter to the driveplate (photo). Turn the crankshaft to bring the bolts into view. It is possible to work through the starter motor aperture on some models.
13 Support the transmission, preferably with a properly designed cradle. Unbolt and remove the transmission crossmember.
14 Lower the transmission until it takes up a stable position. On the B230 engine, make sure that the distributor is not crushed against the bulkhead.
15 Remove the converter housing-to-engine nuts and bolts.
16 With the aid of an assistant, draw the transmission off the engine, at the same time levering the torque converter away from the driveplate. Keep the transmission tilted rearwards and lower it from the vehicle. It is heavy.
17 Refit by reversing the removal operations, noting the following points:

    (a) Put a smear of grease on the torque converter spigot
    (b) Tighten the torque converter-to-driveplate bolts progressively to the specified torque

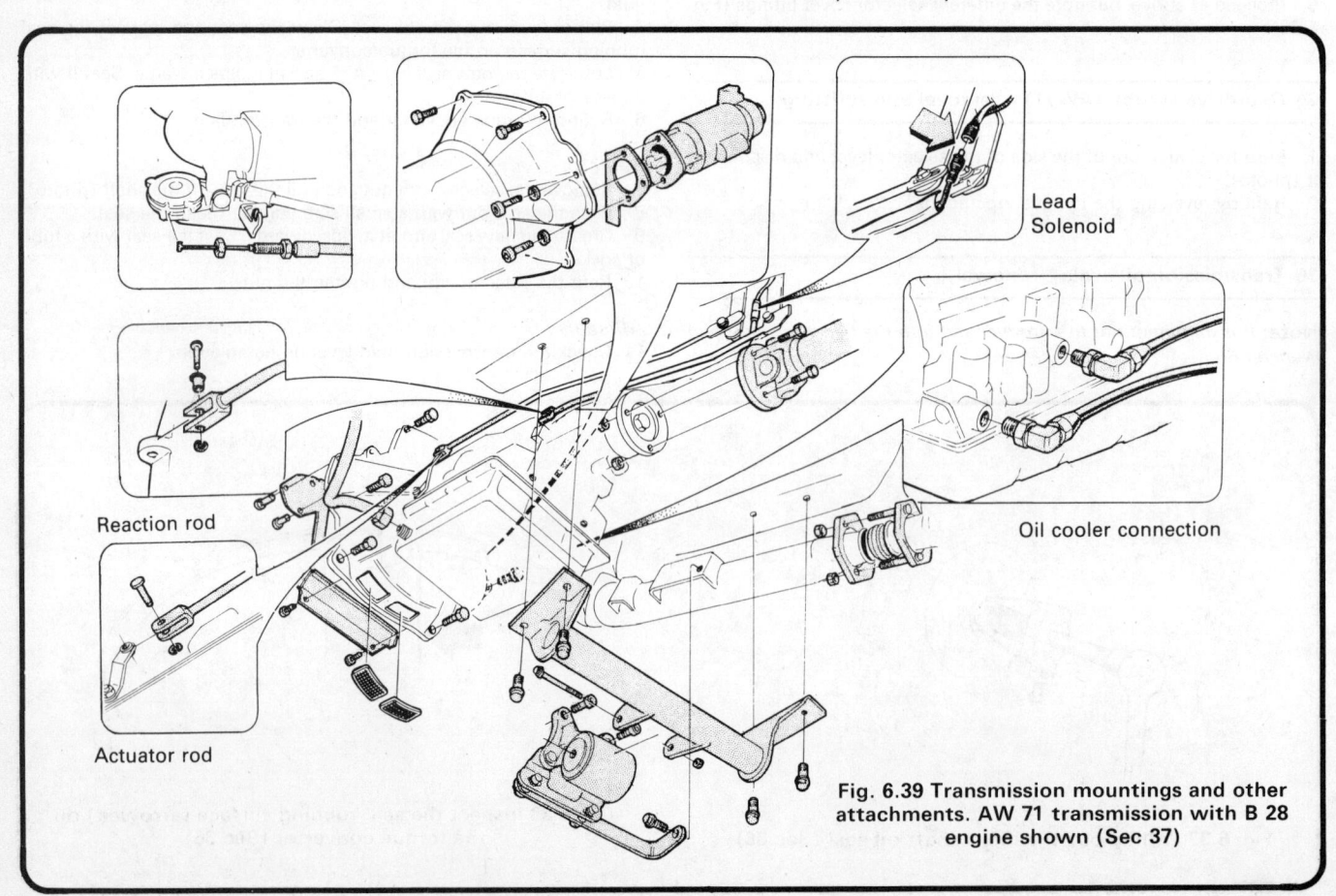

Fig. 6.39 Transmission mountings and other attachments. AW 71 transmission with B 28 engine shown (Sec 37)

Chapter 6 Manual gearbox, overdrive and automatic transmission

37.9 Exhaust bracket and transmission crossmember

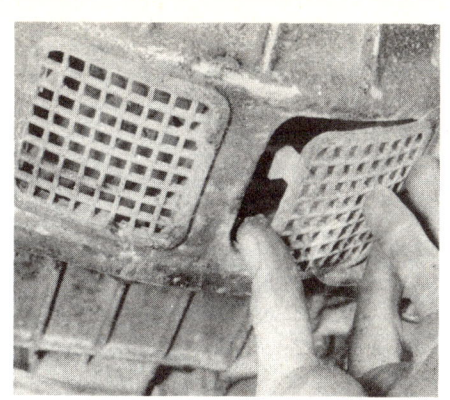

37.11 Removing a torque converter housing grille

37.12 Removing a torque converter-to-driveplate bolt

(c) Do not fully tighten the dipstick tube nut until the tube bracket has been secured
(d) Adjust the selector mechanism (Section 30) and the kickdown cable (Section 31)
(e) Refill the transmission with fluid. If a new transmission has been fitted, flush the oil cooler(s) – see Section 29, paragraph 3 onwards, and Section 27, paragraph 4

38 Fault diagnosis – automatic transmission

In the event of a fault, check the fluid level, the kickdown cable adjustment and the selector mechanism adjustment. Faults which persist when these three items are correct should be referred to a Volvo dealer or transmission specialist.

Chapter 7 Propeller shaft

For modifications, and information applicable to later models, see Supplement at end of manual

Contents

Centre bearing – renewal 6	Propeller shaft – removal and refitting 4
Fault diagnosis – propeller shaft 7	Rubber coupling – removal and refitting 3
General description 1	Universal joints – overhaul 5
Maintenance and inspection 2	

Specifications

General
Shaft type .... Tubular, two-section, with centre support bearing
Number of universal joints .... Two or three (plus rubber coupling on some models)

Torque wrench settings
	Nm	lbf ft
Rubber coupling nuts and bolts	80	60
Plain flange coupling nuts and bolts:		
M10	50	37
M8	35	26

1 General description

A two-section tubular propeller shaft is fitted. Two or three universal joints are used, and on some models a rubber coupling is fitted between the gearbox output flange and the propeller shaft flange. A centre bearing supports the shaft at the junction of the two sections.

The universal joints are secured with circlips instead of by staking, which makes them relatively easy to overhaul.

2 Maintenance and inspection

1 Every 12 000 miles or annually, or if unusual noises or vibrations are noticed, inspect the propeller shaft as follows.
2 Raise and securely support the vehicle, or drive it over a pit.
3 Check the flange bolts and the centre bearing bracket bolts for tightness.
4 Visually check the condition of the rubber coupling, if fitted. Renew it if it is damaged (Section 3).
5 Check for play in the universal joints and centre bearing by attempting to lift, shake and twist the sections of the shaft relative to each other and to the flanges. Repair or renew as necessary (Section 5 or 6).
6 Check that the centre bearing rubber boot is intact. To renew it, the two halves of the shaft must be separated (Section 6).
7 No routine lubrication is required. Occasional oiling of the universal joint circlips is recommended to stop them rusting solid.
8 None of the above checks will detect a seized universal joint, which can cause heavy vibration – the shaft must be removed, or the flanges separated, so that the freedom of movement of the joint may be tested.
9 An out-of-balance or out-of-true shaft can also cause vibration, but these conditions are unlikely to arise spontaneously. Consult a Volvo dealer or a propeller shaft specialist.

Chapter 7 Propeller shaft

Fig. 7.1 Propeller shaft rubber coupling (Sec 3)

A Centre sleeve
B Locating plate

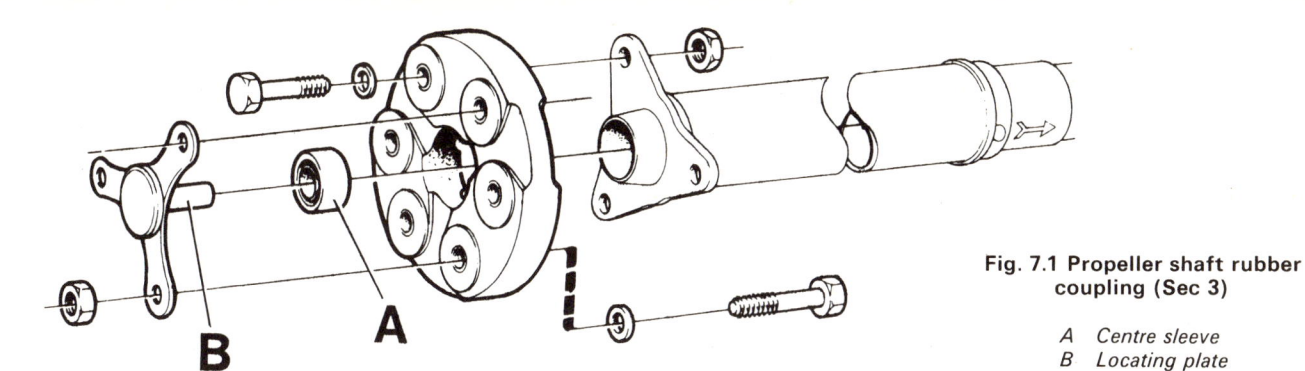

3 Rubber coupling – removal and refitting

1 Raise and securely support the vehicle.
2 Make alignment marks between the shaft and the gearbox output flange.
3 Remove the six nuts and bolts which hold the flanges to the coupling (photo). (It may not be possible actually to remove the forward-facing bolts, which will stay on the flange).
4 Pull the shaft rearwards and lower the front section. Remove the rubber coupling, the centre sleeve and the locating plate (photo).
5 Refit by reversing the removal operations, observing the alignment marks. Apply a little anti-seize compound to the locating plate pin.

4 Propeller shaft – removal and refitting

1 Raise the vehicle on ramps or drive it over a pit.
2 Make alignment marks between the shaft flanges and the gearbox and axle flanges, and between the two sections of the shaft.
3 Remove all the flange nuts and bolts except one at each end (photo). Leave these last ones loose.
4 Have an assistant support the shaft. Remove the bolts which secure the centre bearing carrier (photo).
5 With the aid of the assistant, remove the remaining flange bolts. Remove the shaft and bearing from under the vehicle. Recover the rubber coupling (when fitted).
6 In the absence of an assistant, the shaft may be removed in two sections (rear section first). The two sections simply pull apart. Release the rubber boot from the rear of the centre bearing carrier as this is done (photo).
7 Refit by reversing the removal operations, observing the flange alignment marks. Do not tighten the bearing carrier bolts until the flange bolts have been tightened; the carrier fixings are slotted to allow the bearing to take up an unstrained position.

5 Universal joints – overhaul

1 The joints may need to be overhauled because of excess play; a joint which is stiff will cause vibration and must also be overhauled.

3.3 Rubber coupling nuts and bolts

3.4 Rubber coupling locating plate

4.3 Propeller shaft rear flange – note alignment marks (arrowed)

4.4 Propeller shaft centre bearing bolts

4.6 Withdrawing the rear section from the centre bearing

Chapter 7 Propeller shaft

5.2 Universal joint repair kit

5.5 Raising a bearing cup

2   Obtain an overhaul kit (spider, bearing cups and circlips) for each joint (photo).
3   Clean the joint and apply penetrating oil or releasing fluid to the circlips.
4   Remove the circlips. If they are stuck, tap them with a punch.
5   Rest the yoke of the joint on the open jaws of a vice. Tap the flange with a plastic or copper hammer, or place a piece of tube over the bearing cup and strike that, until the cup protrudes a little way (photo). Do not hit too hard, or clamp the shaft too firmly in the vice – if it is distorted it will be scrap.
6   Grasp the bearing cup with self-locking pliers and withdraw it (photo). Recover any loose rollers.
7   Repeat this process until the spider can be removed from the yoke and all the bearing cups have been removed.
8   Clean the cup seats in the shaft and flange.
9   Carefully remove the cups from the new spider. Check that each cup contains its complement of rollers and that the seals are securely attached. The rollers should already be packed with grease.
10 Offer the spider to the yoke. Fit a cup to the spider, making sure that the rollers are not displaced (photo).
11 Tap the cup lightly to seat it, then press it in using the vice and a tube or socket (photo). The cup should be recessed by 3 to 4 mm (0.12 to 0.16 in).

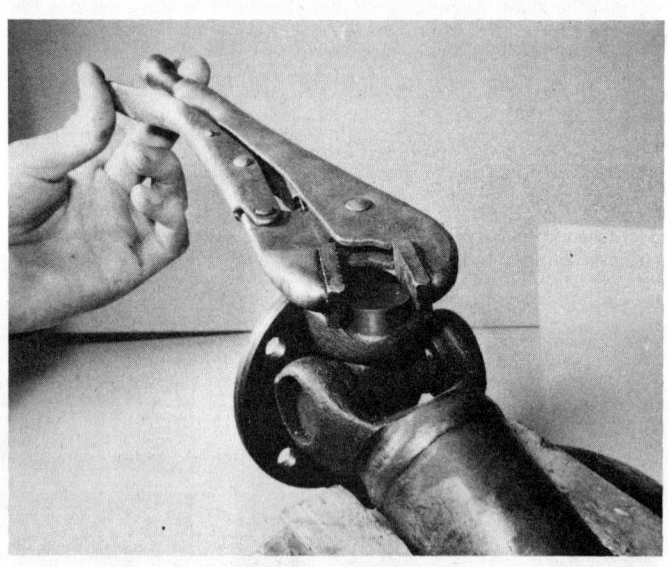

5.6 Removing a bearing cup

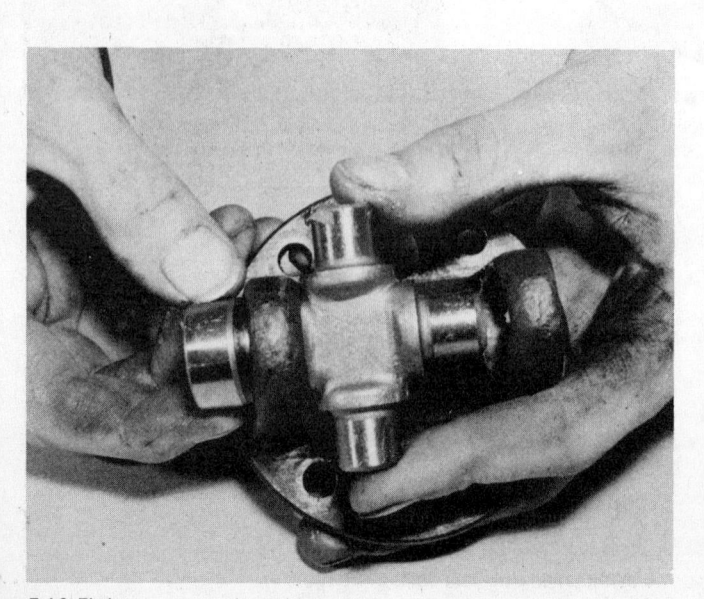

5.10 Fitting a cup to the spider

5.11 Pressing in the cup

Chapter 7 Propeller shaft

5.12 Fitting a circlip

6.2 Driving the shaft out of the centre bearing

12 Fit the circlip to secure the cup (photo).
13 Similarly fit and secure the opposite cup, then assemble the rest of the joint in the same way.
14 Check the joint for freedom of movement. If it is stiff, tap it lightly with a plastic or copper hammer.
15 If vibration persists after overhauling the joints, it may be that the shaft needs to be balanced. This must be done by a specialist.

6 Centre bearing – renewal

Note: *Several different patterns of centre bearing and rubber boot have been fitted (Figs. 7.2 and 7.3). If buying new components in advance, be careful to obtain the correct ones.*

1 Remove the propeller shaft (Section 4) and separate the two sections.
2 Support the front of the bearing and cage on V-blocks or with a piece of split tubing. Press or drive the shaft out of the bearing (photo). Recover the protective rings from both sides of the bearing.
3 If the bearing cage is undamaged, the old bearing can be driven out and a new one pressed in. Otherwise, renew the bearing and cage complete.
4 Fit a new front protective ring to the shaft and tap it home with a wooden or plastic mallet.
5 Fit the new bearing and cage. Seat them with a piece of tube pressing on the bearing inner race.
6 Fit the rear protective ring, keeping it square as it is tapped home.
7 Check that the bearing spins freely, then reassemble the two sections of the shaft, observing the previously made alignment marks. Use a new rubber boot and/or retaining rings if necessary.
8 Refit the shaft to the vehicle.

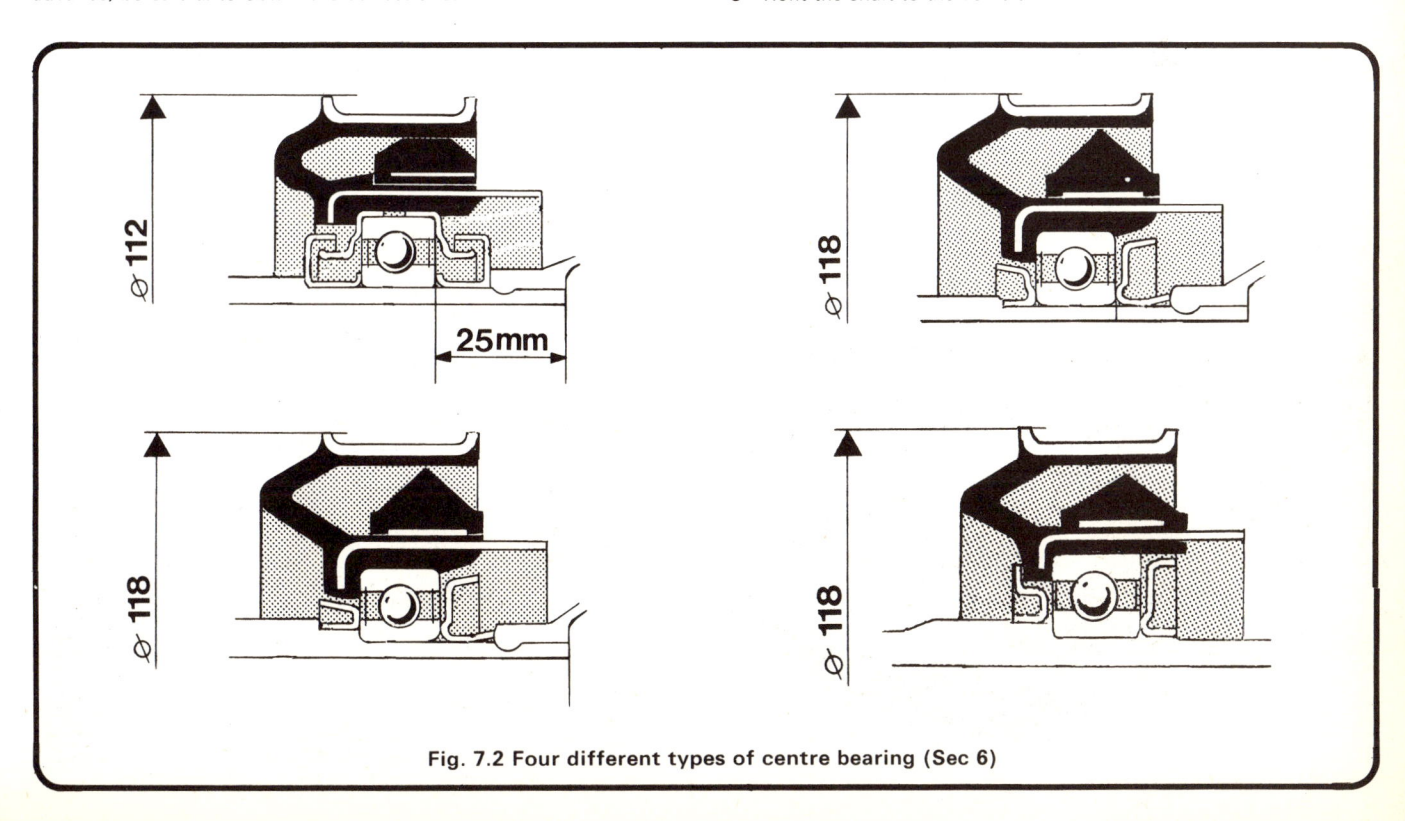

Fig. 7.2 Four different types of centre bearing (Sec 6)

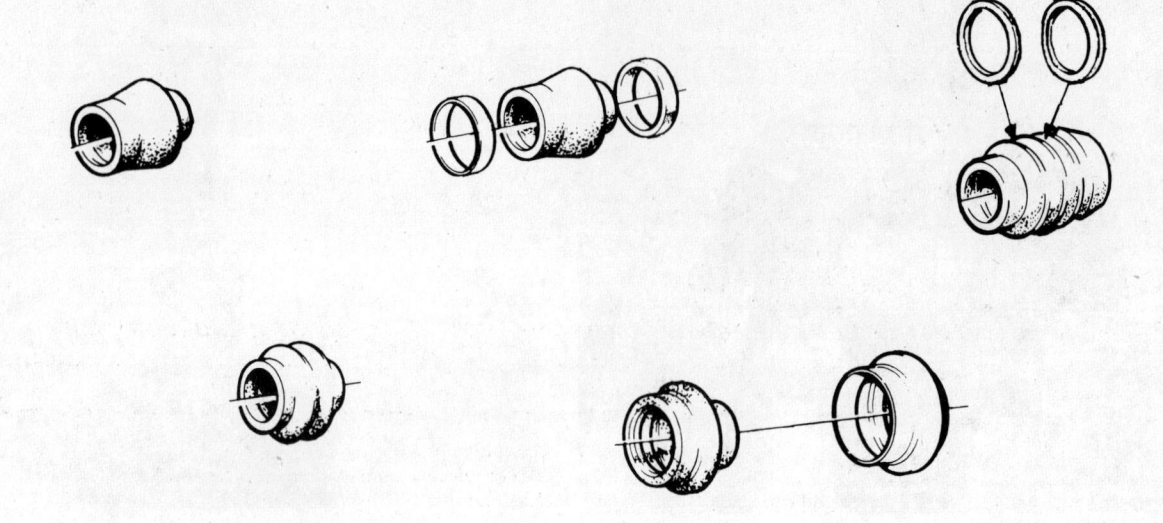

Fig. 7.3 Five different types of rubber boot (Sec 6)

7 Fault diagnosis – propeller shaft

Symptom	Reason(s)
Noise when moving off	Flange bolts loose Splined coupling worn Rubber coupling worn
Noise or vibration when running	Universal joint worn or seized Centre bearing worn Shaft distorted or out of balance

Chapter 8 Rear axle

For modifications, and information applicable to later models, see Supplement at end of manual

Contents

Fault diagnosis – rear axle	7	Oil level – checking	3
General description	1	Pinion oil seal – renewal	4
Halfshaft, bearing and seals – removal and refitting	5	Rear axle – removal and refitting	6
Maintenance and inspection	2		

Specifications

General
Axle type .... Hypoid final drive gears; limited slip differential on some models
Ratio (depending on model and year) .... 3.31, 3.54, 3.73 or 3.91:1

Lubrication
Lubricant type/specification:
  Except limited slip differential .... Hypoid gear oil, viscosity SAE 90EP, to API GL 5 or 6 (Duckhams Hypoid 80W/90S
  Limited slip differential .... Special Volvo oil (No 1 161 276-9), or gear oil as above with Volvo additive (No 1 161 129-0) (Duckhams Hypoid 90DL)
Lubricant capacity:
  Type 1030 .... 1.3 litres (2.3 pints)
  Type 1031 .... 1.6 litres (2.8 pints)

Pinion bearing preload
Turning torque at pinion (wheels free):
  New bearing .... 2.5 to 3.5 Nm (1.8 to 2.6 lbf ft)
  Used bearing .... 1.8 to 3.4 Nm (1.3 to 2.5 lbf ft)

Torque wrench settings

	Nm	lbf ft
Pinion flange nut:		
Solid spacer	200 to 250	148 to 185
Collapsible spacer (see text)	180 to 280	133 to 207
Speedometer sensor locknut	25 to 40	18 to 30
Halfshaft retaining plate bolts	40	30
Trailing arm bracket bolts	45	33
Trailing arm bracket nuts	85	63
Trailing arm to axle	45	33
Panhard rod	85	63
Shock absorber lower mountings	85	63
Torque rods	140	103

1 General description

The rear axle is conventional in design. A rigid casing encloses the final drive unit and two halfshafts. The casing is located by two torque rods bolted to a central subframe, by the two trailing arms and the Panhard rod.

The final drive unit is mounted centrally in the casing. It consists of the differential unit, the crownwheel and pinion. Drive from the propeller shaft is transmitted to the crownwheel by the pinion. The differential unit is bolted to the crownwheel and transmits the drive to the halfshafts. The differential gears and pinions allow the halfshafts to turn at different speeds when necessary, for example when cornering.

On some models the differential is of the 'limited slip' type. Here the difference in speed between the two halfshafts is limited by means of friction clutches. This improves traction on slippery surfaces.

Work on the rear axle should be limited to the operations described in this Chapter. If overhaul of the final drive unit is necessary, consult a Volvo dealer or other specialist.

2 Maintenance and inspection

1 Every 12 000 miles or annually, or if noise or leakage is evident, check the oil level as described in Section 3. Rectify any leakage.
2 Routine oil changing is not specified by the makers, although a drain plug is provided.

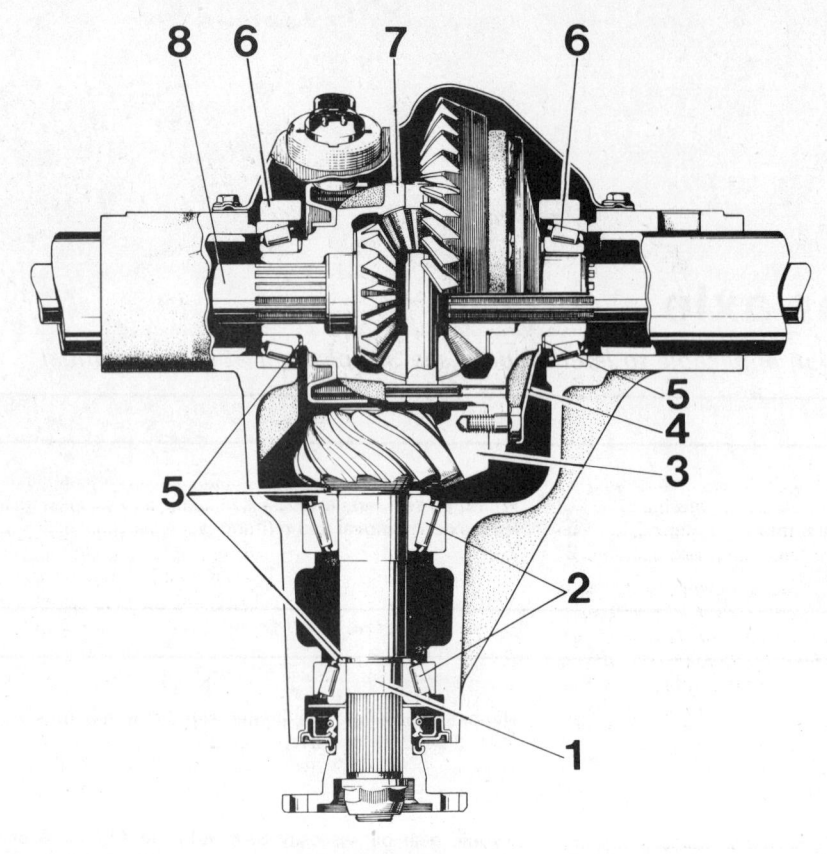

Fig. 8.1 Cutaway view of final drive (Sec 1)

1 Pinion shaft
2 Pinion bearing
3 Crownwheel
4 Lockplate
5 Shims
6 Differential carrier bearings
7 Differential carrier
8 Halfshaft

3 Oil level – checking

1 Park the vehicle on level ground. Raise the vehicle if wished to improve access to the rear axle, but keep it level.
2 Wipe clean around the filler/level plug and unscrew it (photo).
3 Oil should be up to the bottom of the plug hole. Check if necessary by inserting an angled piece of wire to serve as a dipstick.
4 Top up if necessary, using fresh gear oil of the specified type, via the plug hole (photo). Do not overfill: allow any excess to drip out of the hole.
5 Refit and tighten the plug.

4 Pinion oil seal – renewal

Caution: *If the axle has a collapsible spacer in front of the pinion bearing (denoted by the letter 'S' preceding the serial number), care must be taken not to overtighten the pinion flange nut. If the nut is overtightened, it may be necessary to take the axle to a Volvo dealer for a new spacer to be fitted.*

1 Raise and support the rear of the vehicle on ramps, or drive it over a pit.
2 Unbolt the propeller shaft flange from the pinion flange. Make alignment marks between the flanges.
3 Paint or scribe alignment marks between the pinion flange and the flange nut.
4 Restrain the pinion flange with a bar and a couple of bolts. Unscrew the flange nut, counting the number of turns needed to remove it (photo).
5 Pull off the pinion flange (photo). If it is tight, strike it from behind with a copper mallet. Be prepared for oil spillage.
6 Lever out the old oil seal. Clean the seal seat and tap in the new seal, lips inwards.
7 Inspect the seal rubbing surface of the pinion flange. Clean it, or renew the flange, as necessary.
8 Oil the seal lips, then refit the flange.

3.2 Removing the oil filler/level plug

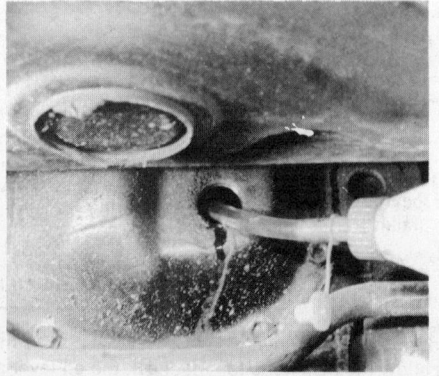

3.4 Topping-up the axle oil

4.4 Pinion flange nut partly unscrewed

Chapter 8 Rear axle

4.5 The oil seal exposed by removal of the flange

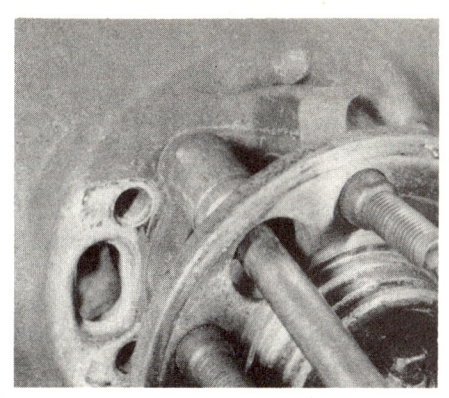

5.2 Undoing a halfshaft retaining plate bolt

5.4 The inner oil seal in the axle tube

9 Refit the flange nut. If the original flange and nut are being used, tighten the nut through the number of turns noted and align the marks made before dismantling. With new components, proceed as follows:

   (a) Axle with solid spacer – tighten the nut to the specified torque
   (b) Axle with collapsible spacer – tighten the nut to the lowest specified torque, then use a spring balance to determine the pinion bearing preload (wheels free, handbrake off). If the preload is lower than specified, carry on tightening the nut until it is correct. Do not exceed the maximum specified preload or tightening torque

10 Refit the propeller shaft and lower the vehicle.
11 Check the axle oil level and top up if necessary.

5 Halfshaft, bearing and seals – removal and refitting

1 Remove the handbrake shoes (Chapter 9, Section 26).
2 Remove the four bolts which secure the halfshaft retaining plate (photo). Recover the handbrake shoe clips.
3 Refit the brake disc the wrong way round (drum facing outwards) and secure it with the wheel nuts, flat faces inwards. Pull on the brake disc to withdraw the halfshaft. Be prepared for oil spillage.

4 With the halfshaft removed, the inner (axle oil) seal may be removed by prising it out of the axle tube (photo). Clean the seal seat and tap the new seal into position using a mallet and a piece of tube.
5 Renewal of the outer (grease) seal and bearing should be left to a Volvo dealer or other specialist, as press tools are required (photo).
6 Before refitting, make sure that the bearing and seal lips are packed with grease.
7 Clean the axle tube and retaining plate mating faces and apply sealant to them.
8 Fit the halfshaft into the axle tube, being careful not to damage the inner seal. Secure it with the retaining plate and the four bolts, tightened to the specified torque. Remember to fit the handbrake shoe clips.
9 Remove the brake disc (if not already done) and refit the handbrake shoes.
10 Check the rear axle oil level and top up if necessary.

6 Rear axle – removal and refitting

1 Slacken the rear wheel nuts. Raise and support the vehicle with the rear wheels free. **Caution:** *If raising the front of the vehicle as well, place the supports under the front control arm brackets, **not** under the*

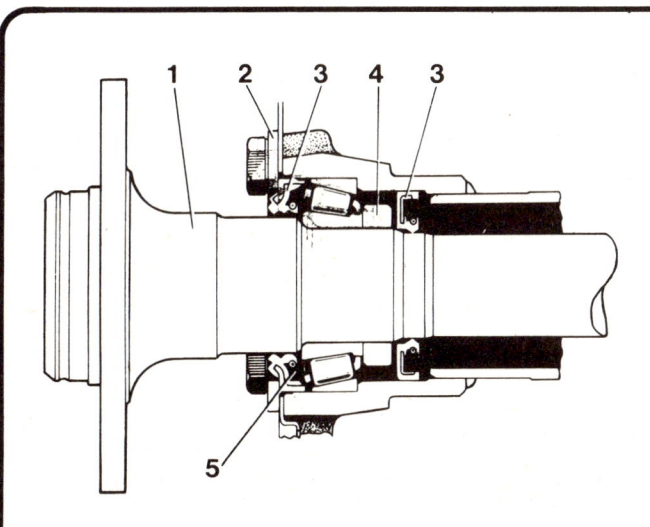

Fig. 8.2 Sectional view of halfshaft, bearing and seals (Sec 5)

1 Halfshaft
2 Lockplate
3 Seals
4 Bearing retainer
5 Grease space

5.5 The outer oil seal is below the bearing

Chapter 8 Rear axle

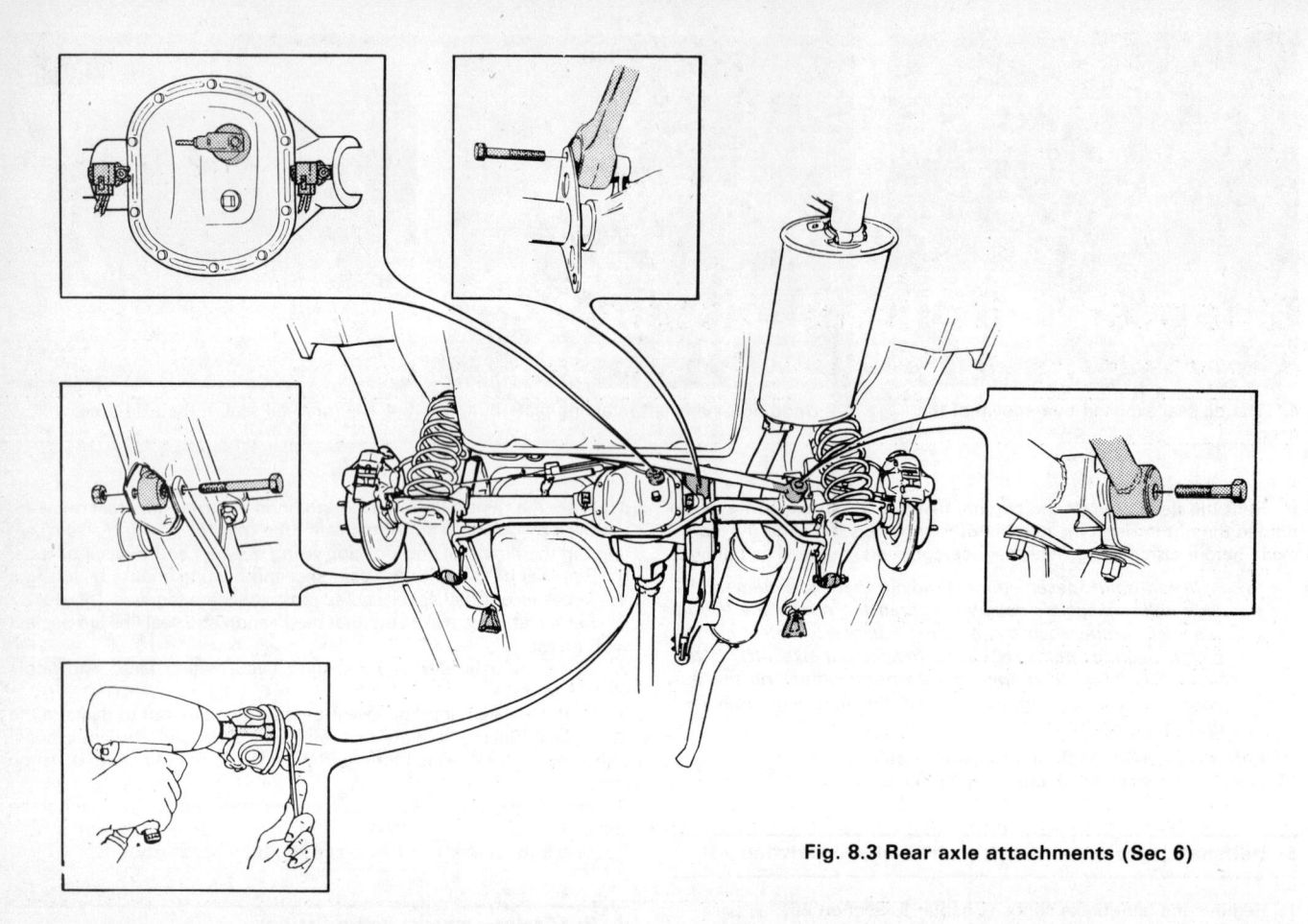

Fig. 8.3 Rear axle attachments (Sec 6)

jacking points. If the front jacking points are used, the vehicle may become nose-heavy.
2 Remove the rear wheels.
3 Remove the rear brake calipers (without disconnecting them), the brake discs and the handbrake shoes. Refer to Chapter 9 for full details.
4 Disconnect the handbrake cables from the brake backplates and from the brackets on the axle.
5 Unbolt the axle torque rods from the subframe, and the lower torque rod from the axle.
6 Support the axle with a cradle and a jack. Take the weight of the axle on the jack.
7 If the exhaust system runs below the axle, remove it.
8 Remove the Panhard rod.
9 Disconnect the speedometer sender/ETC multi-plug(s) (as applicable). The speedometer sender multi-plug may be secured by a locking wire and seal, which must be broken.
10 Unbolt the propeller shaft/axle flanged joint.
11 Unbolt the upper torque rod from the axle.
12 Unbolt the rear shock absorber lower mountings.
13 Remove the trailing arm front mounting bracket nuts and bolts.
14 Lower the axle, at the same time freeing the trailing arm front mountings, and remove it from under the vehicle.
15 The anti-roll bar (if fitted) and the trailing arms may now be removed if wished. The trailing arms are handed: do not mix them up.
16 Refit by reversing the removal operations, noting the following points:

(a) When refitting the trailing arms to the axle, tighten the nuts progressively and in diagonal sequence to the specified torque
(b) Do not finally tighten the torque rods until the weight of the vehicle is back on the wheels (or jack the axle up to simulate this condition)
(c) Check the axle oil level on completion

7 Fault diagnosis – rear axle

Symptom	Reason(s)
Noise on drive or overrun	Oil level low or incorrect grade Bearings worn or damaged Gear teeth worn or damaged
Noise when turning (either way)	Differential worn or damged
Noise when turning left only	Right-hand halfshaft bearing worn (and *vice versa*)
Knock when taking up drive	Propeller shaft flange bolts loose Wheel nuts loose Pinion flange loose Halfshaft splines worn Final drive worn or damaged

Chapter 9 Braking system

For modifications, and information applicable to later models, see Supplement at end of manual

Contents

ABS – fault tracing	32	Handbrake – adjustment	7
ABS components – removal and refitting	31	Handbrake cables – removal and refitting	27
Anti-lock braking system (ABS) – description	30	Handbrake shoes – removal, inspection and refitting	26
Brake discs – inspection	14	Handbrake warning switch – removal and refitting	29
Brake fluid level – checking	3	Hydraulic pipes and hoses – inspection and renewal	8
Brake master cylinder – overhaul	18	Hydraulic system – bleeding	9
Brake master cylinder – removal and refitting	17	Maintenance and inspection	2
Brake pedal – removal and refitting	21	Pressure differential warning valve – overhaul	20
Brake servo – checking operation	6	Pressure differential warning valve – removal and refitting	19
Brake servo – overhaul	23	Rear brake caliper – overhaul	13
Brake servo – removal and refitting	22	Rear brake caliper – removal and refitting	12
Fault diagnosis – braking system	33	Rear brake disc – removal and refitting	16
Front brake caliper – overhaul	11	Rear brake pads – inspection and renewal	5
Front brake caliper – removal and refitting	10	Stop-light switch – removal and refitting	28
Front brake disc – removal and refitting	15	Vacuum pump – overhaul	25
Front brake pads – inspection and renewal	4	Vacuum pump – removal and refitting	24
General description	1		

Specifications

General
System type:
    Footbrake .... Discs all round. Hydraulic operation with servo assistance. Anti-lock braking (ABS) on some models
    Handbrake .... Mechanical to drums on rear wheels
Hydraulic system split:
    Without ABS .... Double triangular
    With ABS .... Front-rear
Hydraulic fluid type/specification .... Hydraulic fluid to DOT 4 (Duckhams Universal Brake and Clutch Fluid)

Brake pads
Lining minimum thickness:
    Front pads .... 3.0 mm (0.12 in)
    Rear pads .... 2.0 mm (0.08 in)

Front brake discs
Diameter:
    Solid .... 280 mm (11.02 in)
    Ventilated .... 262 or 287 mm (10.32 or 11.30 in)
Thickness – solid:
    New .... 14.0 mm (0.55 in)
    Wear limit .... 11.0 mm (0.43 in)
Thickness – ventilated:
    New .... 22.0 mm (0.87 in)
    Wear limit .... 20.0 mm (0.79 in)
Run-out .... 0.08 mm (0.003 in) maximum

Rear brake discs
Diameter .... 281 mm (11.06 in)
Thickness:
    New .... 9.6 mm (0.38 in)
    Wear limit .... 8.4 mm (0.33 in)
Run-out .... 0.10 mm (0.004 in) maximum

Handbrake
Lever travel:
- After adjustment .... 3 to 5 clicks
- In service .... 11 clicks maximum
- Drum diameter .... 160.45 mm (6.317 in) maximum
- Drum run-out .... 0.15 mm (0.006 in) maximum
- Drum out-of-round .... 0.20 mm (0.008 in) maximum

Torque wrench settings

	Nm	lbf ft
Front caliper bracket screws	100	74
Rear caliper mounting bolts*	58	43
Upper guide pin to caliper bracket	25	18
Caliper guide pin bolts	34	25
Front dust shield	24	18
Rear dust shield	40	30
Master cylinder nuts	30	22
Rigid pipe unions	14	10
Flexible hose unions	17	13

*Use new bolts every time

1 General description

The brake pedal operates disc brakes on all four wheels by means of a dual circuit hydraulic system with servo assistance. The handbrake operates separate drum brakes on the rear wheels only by means of cables. An anti-lock braking system (ABS) is fitted to some models, and is described in detail in Section 30.

The hydraulic system is split into two circuits, so that in the event of failure of one circuit, the other will still provide adequate braking power (although pedal travel and effort may increase). Except on models with ABS, the split is 'triangular', ie each circuit serves one rear caliper and half of both front ones.

The brake servo is of the direct-acting type, being interposed between the brake pedal and the master cylinder. The servo magnifies the effort applied by the driver. It is vacuum-operated, the vacuum being derived from the inlet manifold or (on some models) a mechanical vacuum pump.

Instrument panel warning lights alert the driver to hydraulic circuit failure (by means of a pressure differential valve) and on some models

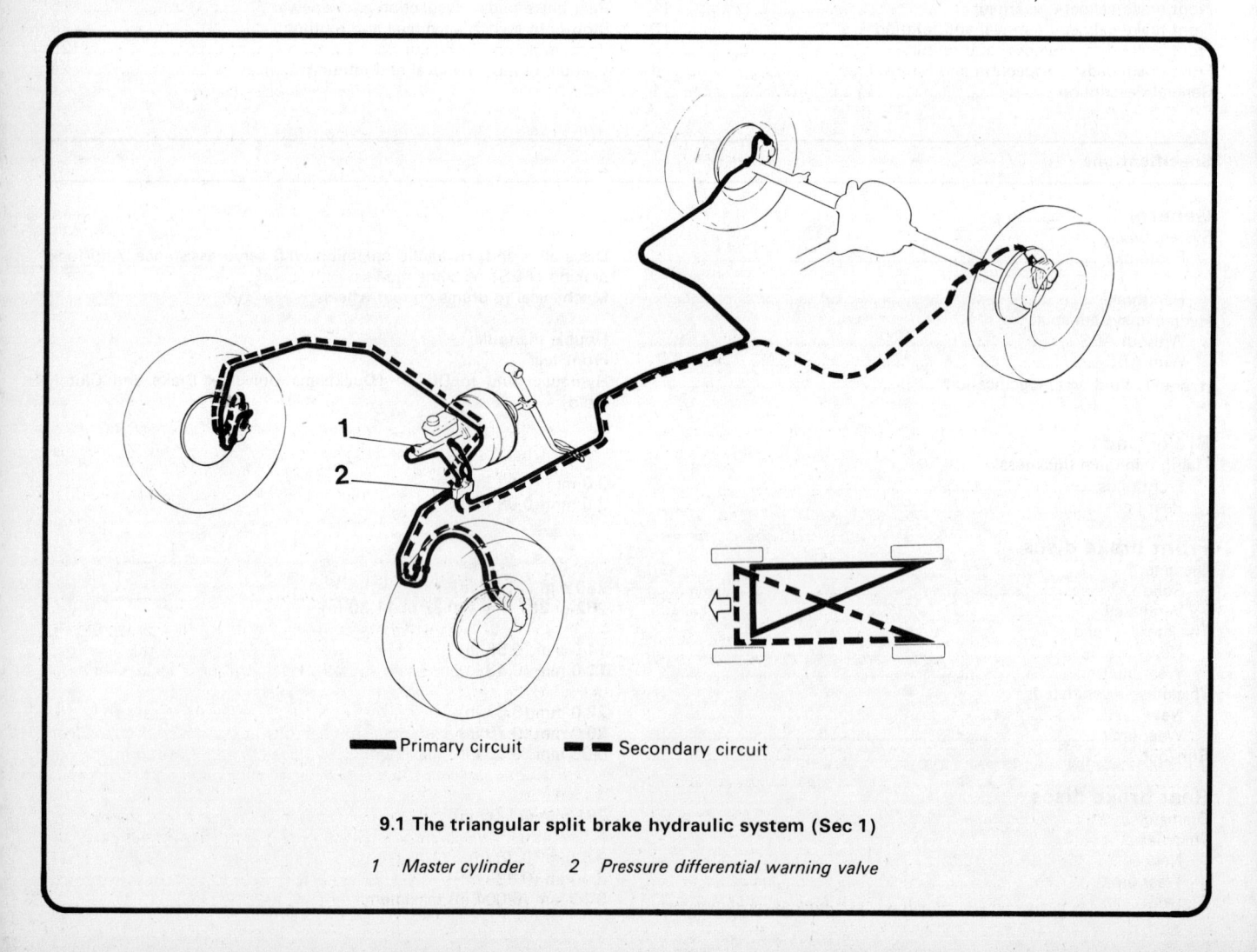

9.1 The triangular split brake hydraulic system (Sec 1)

1 Master cylinder
2 Pressure differential warning valve

Chapter 9 Braking system

Fig. 9.2 Handbrake components (Sec 1)

1 Lever
2 Pivot
3 Adjuster
4 Left-hand (long) cable
5 Cable clamp
6 Axle mounting
7 Right-hand (short) cable
8 Guides
9 Shoe

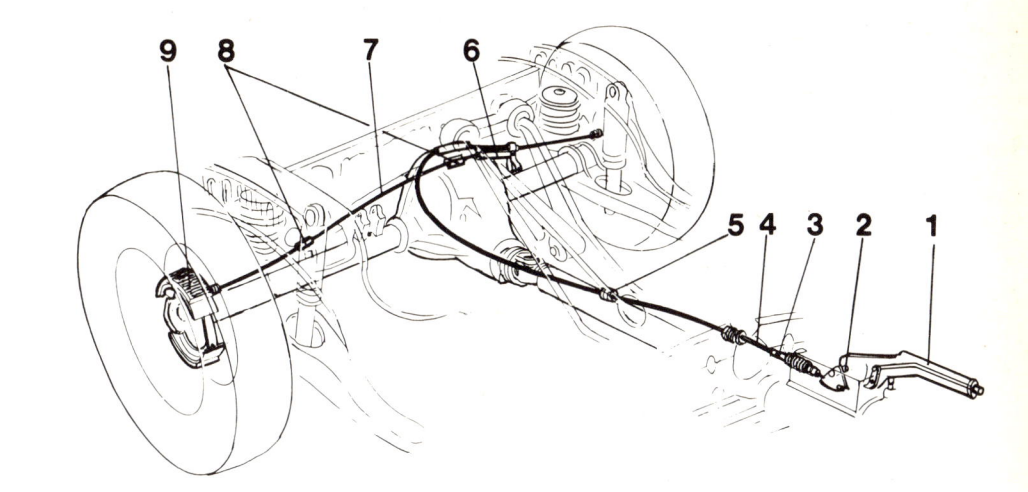

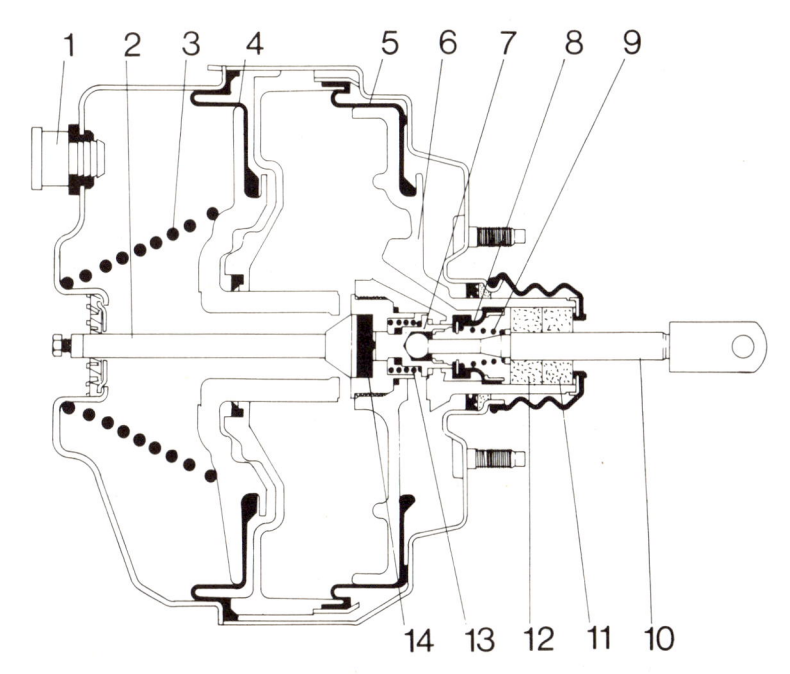

Fig. 9.3 Sectional view through the brake servo (Sec 7)

1 Check valve
2 Front pushrod
3 Spring for diaphragms
4 Front diaphragm
5 Rear diaphragm
6 Housing
7 Valve seat
8 Seal
9 Pushrod spring
10 Rear pushrod
11 Air filter
12 Air filter
13 Spring
14 Reaction disc

to low fluid level. Another warning light reminds when the handbrake is applied. The stop-lights are covered by the bulb failure warning system.

Work on the braking system should be careful and methodical. Scrupulous cleanliness must be observed when working on the hydraulic system. Replacement parts should preferably be the maker's own, or at least of known manufacture and quality.

Warning: Braking system hazards. *Some brake friction materials still contain asbestos. Assume that all brake dust is potentially hazardous for this reason; avoid inhaling it and do not dispense it into the air. Note also that brake fluid is toxic and attacks paintwork. Do not syphon it by mouth, wash thoroughly after skin contact and wash spillage off paintwork immediately.*

2 Maintenance and inspection

1 Weekly, every 250 miles or before a long journey, check the brake fluid level (Secion 3).
2 Every 6000 miles or six months, check that the spare wheel well drain is clear (only on models with ABS).
3 At the same interval, check the brake pad wear (Sections 4 and 5).
4 Every 12 000 miles or annually, check the operation of the servo (Section 6), the adjustment of the handbrake (Section 7) and the condition of the hydraulic pipes and hoses (Section 8).
5 Every 24 000 miles or two years, renew the brake fluid by bleeding (Section 9).

3 Brake fluid level – checking

1 The level of fluid in the brake master cylinder reservoir should be between the 'MAX' and 'MIN' marks on the outside of the reservoir. The reservoir is translucent, so the level can be checked without removing the cap.
2 If topping-up is necessary, wipe clean around the reservoir cap. Remove the cap and top up with fresh hydraulic fluid of the specified type. Only top up to the 'MAX' mark; do not overfill (photo).
3 Refit the reservoir cap. Wash any spilt fluid off paintwork immediately.
4 Regular need for topping-up shows that there is a leak somewhere in the system which should be found and rectified without delay. A **slow** fall in fluid level, as the pads wear, is normal; provided the level stays above the 'MIN' mark, there is no need to top up for this reason. The level will rise again when new pads are fitted.

3.2 Topping-up the brake fluid

4.2 Removing a front caliper lower guide pin bolt

4.3 Pivot the caliper upwards

4 Front brake pads – inspection and renewal

1 Remove a front wheel. The brake pad lining thickness can now be seen through the inspection hole in the caliper. However, for a thorough inspection it is preferable to remove the pads as follows.
2 Remove the caliper lower guide pin bolt, if necessary counterholding the guide pin with an open-ended spanner (photo). On Girling calipers, also slacken the upper guide pin bolt.
3 Pivot the caliper upwards, free the bellows and slide it off the guide pin (photo). Support the caliper so that the hose(s) are not strained. Do not press the brake pedal whilst the caliper is removed.
4 Recover the pads from the caliper bracket, noting their positions if they are to be re-used (photo). Recover the anti-squeal shims (if fitted) from the backs of the pads.
5 Measure the thickness of the pad friction linings. If any one pad lining has worn down to the specified minimum, all four front pads must be renewed. Do not interchange pads in an attempt to even out wear. (Uneven pad wear may be due to the caliper sticking on the guide pins).
6 Clean the caliper and bracket with a damp rag or an old paintbrush.
Caution: *Do not disperse brake dust into the air, and avoid inhaling it, as it may contain asbestos.* Inspect the caliper piston and dust boots for signs of fluid leakage. Also inspect the rubber bellows which cover the guide pins. Repair or renew as necessary (Section 11).
7 Remove any scale or rust from the outer rim of the brake disc with a wire brush or file. Inspect the disc visually; if brake judder has been a problem, carry out a more thorough inspection (Section 14).
8 If new pads are to be fitted, press the caliper pistons back into their bores with a pair of pliers, being careful not to damage the dust boots. As the pistons are pressed back, the fluid level in the master cylinder reservoir wil rise. Remove some fluid if necessary, using a syringe or an old poultry baster. **Do not** *syphon the fluid by mouth: it is poisonous.*
9 Apply anti-seize compound or disc brake lubricant to the backs of the pads and to the caliper guide pins. Do not allow lubricant to contact the disc or pad friction surfaces. Also lubricate both sides of the anti-squeal shims (when fitted) and fit them to the pads.
10 Slide the caliper onto the upper guide pin and engage the bellows. Position the pads with the friction surfaces towards the disc and swing the caliper down over the pads. Make sure that the anti-rattle spring in the caliper is sitting correctly on the pads.
11 Apply thread locking compound to the guide pin bolt, insert it and tighten it to the specified torque (photo). On Girling calipers, also tighten the upper guide pin bolt.
12 Press the brake pedal several times to bring the pads up to the disc.
13 Repeat the operations on the other front brake.
14 Refit the roadwheels, lower the vehicle and tighten the wheel nuts.
15 Check the brake fluid level and top up if necessary.
16 If new pads have been fitted, avoid hard braking as far as possible for the first few hundred miles to allow the linings to bed in.

4.4 One brake pad partly removed

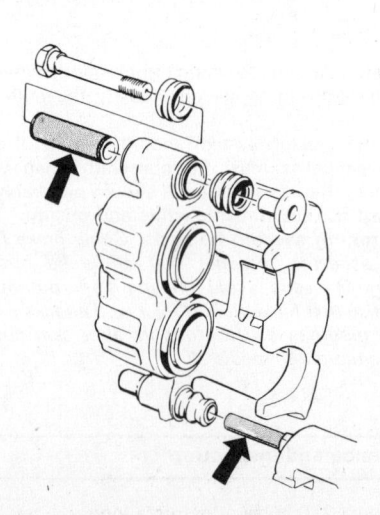

Fig. 9.4 Apply anti-seize compound to the caliper guide pins (arrowed) (Sec 4)

Chapter 9 Braking system

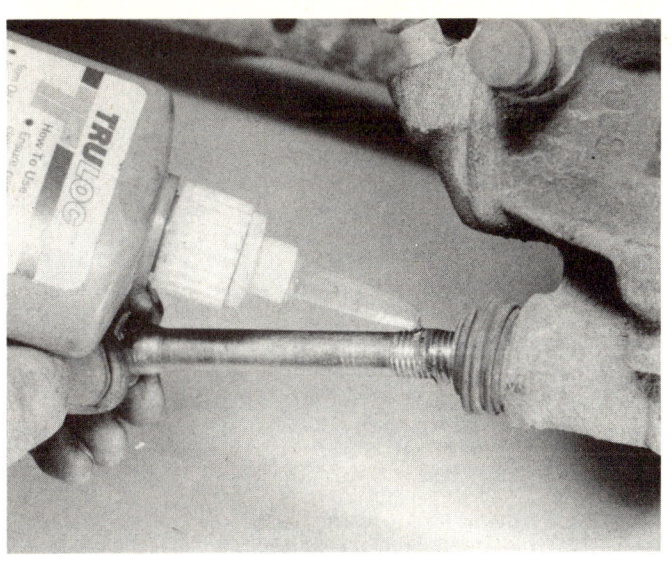

4.11 Use thread locking compound on the guide pin bolt

5 Rear brake pads – inspection and renewal

1 Slacken the rear wheel nuts, raise and support the rear of the vehicle and remove the rear wheel.
2 The pad lining thickness can now be seen through the jaws of the caliper. For a more thorough inspection, or to renew the pads, proceed as follows.
3 Drive the two retaining pins out of the caliper using a hammer and punch (photo). Recover the anti-rattle spring. Obtain a new spring for reassembly.
4 Press each pad away from the disc, using pliers. Do not lever between the pads and the disc.
5 Pull the pads out of the caliper, along with the anti-squeal shims (if fitted). Identify their position if they are to be re-used. Do not press the brake pedal with the pads removed.
6 Measure the thickness of the pad friction linings. If any one pad lining has worn down to the specified minimum, all four rear pads must be renewed. Do not interchange pads in an attempt to even out wear.
7 Clean the caliper with a damp rag or an old paintbrush. **Caution:** *Do not disperse brake dust into the air, and avoid inhaling it, as it may contain asbestos.* Inspect the caliper pistons and dust boots for signs of fluid leakage. Repair or renew as necessary (Section 13).
8 Inspect the visible surface of the brake disc. If deep scoring, cracks or grooves are evident, or if brake judder or snatch has been a problem, carry out a more thorough inspection (Section 14). Remove the caliper if necessary for access to the inboard face of the disc.

9 Anti-squeal shims may be fitted if wished, even if none were fitted before.
10 If new pads are to be fitted, press the caliper pistons back into their bores. Remove a little fluid from the master cylinder reservoir if necessary to prevent overflow.
11 Smear the backs of the pads and both sides of the anti-squeal shims (if used) with anti-seize compound or disc brake lubricant. Keep this off the friction surfaces of the pads.
12 Fit the pads and shims into the jaws of the caliper with the friction surfaces towards the disc (photo).
13 Insert one of the pad retaining pins and tap it home. Fit a new anti-rattle spring and the other pad retaining pin, making sure that the pins pass over the tongues of the spring (photo).
14 Repeat the operations on the other rear brake.
15 Refit the roadwheels, lower the vehicle and tighten the wheel nuts.
16 Pump the brake pedal several times to being the new pads up to the discs.
17 Check the brake fluid level and top up if necessary.
18 If new pads have been fitted, avoid harsh braking as far as possible for the first few hundred miles to allow the linings to bed in.

6 Brake servo – checking operation

1 Normally servo malfunction will be evident to the driver by the increased effort needed at the pedal. A quick check may be made as follows.
2 With the engine stopped, apply the footbrake several times to destroy any residual vacuum.
3 Hold the brake pedal depressed and start the engine. If the servo is working, the pedal will be felt to move downwards slightly. If not, there is a fault in the servo, the non-return valve or hose, or (when fitted) the vacuum pump.

7 Handbrake – adjustment

1 The handbrake should be fully applied within the specified number of clicks of the lever ratchet. Adjustment will be necessary periodically to compensate for lining wear and cable stretch.
2 Remove the rear ashtray and the cigarette lighter/seat belt warning light panel for access to the cable adjuster (photo).
3 Release the locking sleeve from the front of the adjuster, either by driving the sleeve forwards or by pulling the adjuster back. Turn the adjuster nut until handbrake operation occurs within the specified number of clicks. Check that the brake is not binding when the lever is released.
4 Re-engage the locking sleeve and refit the panel and rear ashtray.

8 Hydraulic pipes and hoses – inspection and renewal

1 Inspect the rigid pipes for security in their mountings. The pipes must be free from rust or impact damage.

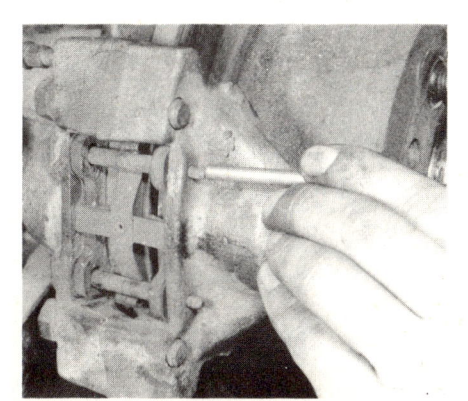

5.3 Driving out a pad retaining pin

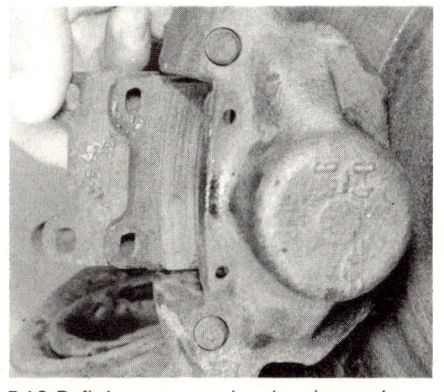

5.12 Refitting a rear pad and anti-squeal shim

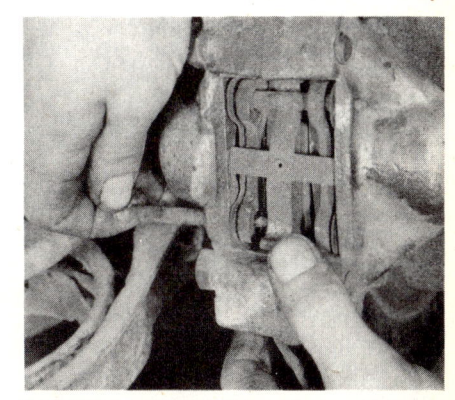

5.13 Fitting a pad retaining pin over the spring tongue

Chapter 9 Braking system

7.2 Handbrake cable adjuster (arrowed) with surrounding trim removed

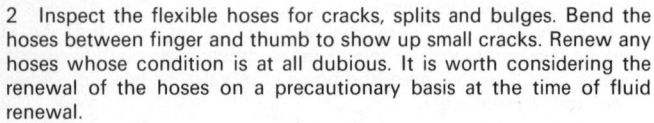

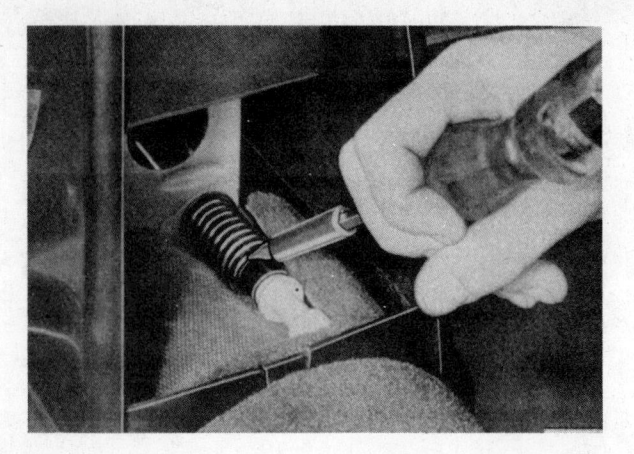

Fig. 9.5 Driving forwards the handbrake cable adjuster locking sleeve (Sec 7)

2 Inspect the flexible hoses for cracks, splits and bulges. Bend the hoses between finger and thumb to show up small cracks. Renew any hoses whose condition is at all dubious. It is worth considering the renewal of the hoses on a precautionary basis at the time of fluid renewal.
3 Details of pipe and hose renewal will vary according to the location of the item in question, but the basic steps are the same.
4 Minimise hydraulic fluid loss by removing the master cylinder reservoir cap, placing a piece of plastic film over the reservoir and tightening the cap over it.
5 Clean around the unions which are to be disconnected. Undo the unions – with a flexible hose, release it at the rigid pipe first, then from the caliper. Free the pipe or hose from any mounting clips and remove (photos).
6 Before refitting, blow through the new pipe or hose with dry compressed air. Any bending needed for a rigid pipe should take place before the unions are connected. If genuine Volvo parts are used, the pipes should fit without bending.
7 When satisfied that the pipe or hose is correctly routed and will not foul adjacent components, refit and tighten the unions.
8 Bleed the hydraulic system as described in Section 9.

9 Hydraulic system – bleeding

1 Whenever the hydraulic system has been overhauled, a part renewed or the level in the reservoir has become too low, air will have entered the system. This will cause some or all of the pedal travel to be used up in compressing air rather than pushing fluid against brake pistons. If only a little air is present, the pedal will have a 'spongy' feel, but if an appreciable amount has entered, the pedal will not offer any appreciable resistance to the foot and the brakes will hardly work at all.
2 To overcome this, brake fluid must be pumped through the hydraulic system until all the air has been passed out in the form of bubbles in the fluid.
3 If only one hydraulic circuit has been disconnected, only that circuit need be bled. If both circuits have been disconnected, or at time of fluid renewal, the whole system must be bled.
4 Bleed the system in the following order:

Without ABS	With ABS
RH rear	LH front
LH rear	RH front
RH front	LH rear
LH front	RH rear

5 There are two bleed screws on each front caliper, and one on each rear caliper (photo).

Bleeding with an assistant

6 Gather together two clear plastic tubes to fit over the bleed screws, a glass jar and a supply of fresh brake fluid.
7 Top up the master cylinder reservoir. Keep it topped up throughout the operation.
8 Attach the tube(s) to the bleed screw(s) of the first caliper to be bled (paragraph 4). Pour a little brake fluid into the jar and place the open ends of the tube(s) in the jar, dipping into the fluid.
9 Slacken the bleed screw(s). Have the assistant depress and release

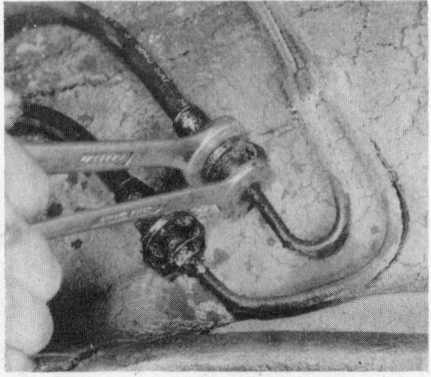

8.5A Releasing a hose-to-pipe union ...

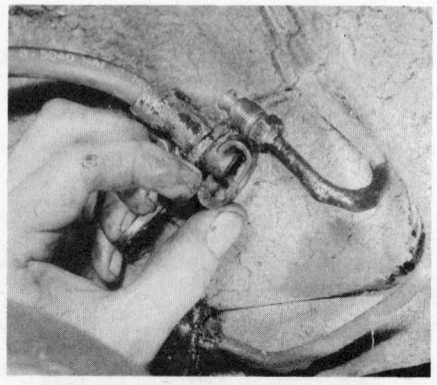

8.5B ... and removing the securing clip

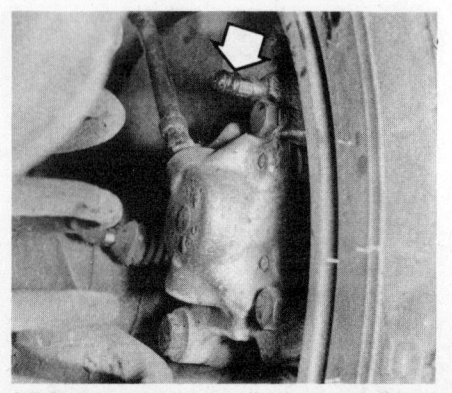

9.5 Bleed screw (arrowed) on a rear caliper

Chapter 9 Braking system

the brake pedal five times, stopping on the fifth downstroke. Tighten the bleed screw(s) and have the assistant release the pedal.
10 Top up the master cylinder reservoir.
11 Repeat paragraphs 9 and 10 until clean fluid, free from air bubbles, emerges from the bleed screw(s).
12 Repeat the process on the remaining calipers in the order given.
13 On completion, check that the brake pedal feels hard. Top up the master cylinder reservoir and refit the cap.
14 Discard the fluid bled from the system as it is not fit for re-use. Dispose of it in a sealed container.

Bleeding using a one-way valve kit
15 There are a number of one-man brake bleeding kits curently available from motor accessory shops. These devices simplify the bleeding process and reduce the risk of expelled air or fluid being drawn back again into the system.
16 To use this type of kit, connect the outlet tube to the bleed screw and then open the screw half a turn. If possible, position the tube so that it can be viewed from inside the car. Depress the brake pedal as far as possible and slowly release it. The one-way valve in the bleed kit will prevent expelled air or fluid from returning to the system at the end of each pedal return stroke (photo). Repeat this operation until clean hydraulic fluid, free from air bubbles, can be seen coming through the bleed tube. Tighten the bleed screw and remove the tube.
17 Repeat the operations on the remaining bleed screws in the correct sequence. Make sure that throughout the process the fluid reservoir level never falls so low that air can be drawn into the master cylinder, otherwise the work up to this point will have been wasted.

Bleeding using a pressure bleeding kit
18 These, too, are available from motor accessory shops and are usually operated by air pressure from the spare tyre.
19 By connecting a pressurised container to the master cylinder fluid reservoir, bleeding is then carried out by simply opening each bleed screw in turn and allowing the fluid to run out, rather like turning on a tap, until air bubbles are no longer visible in the fluid being expelled.
20 Using this system, the large reserve of hydraulic fluid provides a safeguard against air being drawn into the master cylinder during the bleeding process.

10 Front brake caliper – removal and refitting

Conventional system
1 Proceed as for brake pad removal (Section 4), but additionally disconnect the caliper hoses from the hydraulic pipes at the bracket on the inner wing. Identify the hoses so that they can be refitted to the same pipes; be prepared for hydraulic fluid spillage. Keep dirt out of the open unions.
2 If it is wished to remove the caliper bracket, undo the two Allen screws which secure it to the steering knuckle. Obtain new screws for reassembly.
3 When refitting the caliper bracket, apply thread locking compound to the Allen screws and tighten them to the specified torque.
4 Refit the caliper as described in Section 4 and reconnect the hydraulic hoses.
5 Bleed the hydraulic system on completion (Section 9).

ABS
6 Proceed as above, but note that as there is only one hydraulic hose per caliper, the hose-to-pipe union can be left undisturbed. Slacken the hose union at the caliper, remove the caliper from the guide pins and unscrew it from the hose. Check the 'set' of the hose when refitting.

11 Front brake caliper – overhaul

1 With the brake caliper removed (Section 10), clean it externally with methylated spirit and a soft brush.
2 Remove the hydraulic hose(s) and the bleed screws. Empty any remaining hydraulic fluid out of the caliper.
3 Remove the anti-rattle spring (photo).
4 Remove one of the piston dust boots and pull the piston out of its bore (photos). If it is reluctant to move, refit the bleed screws and

9.16 One-way valve bleeder connected to a front bleed screw

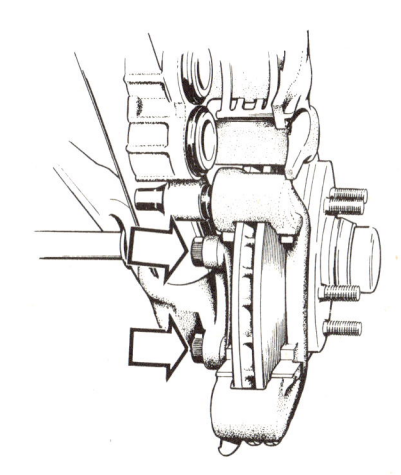

Fig. 9.6 Two Allen screws (arrowed) which secure the front caliper bracket (Sec 10)

11.3 Removing the caliper anti-rattle spring

216   Chapter 9 Braking system

11.4A Remove a piston dust boot  ...   11.4B ... and the piston itself    11.10 Fit a new piston seal into the groove

apply **low** air pressure (eg from a foot pump) to the fluid inlet. **Caution:** *The piston may be ejected with some force.*
5 Hook out the piston seal from the bore using a blunt instrument.
6 Repeat the above operations on the other piston. Identify the pistons if they are to be re-used.
7 Clean the pistons and bores with a lint-free rag and some clean brake fluid or methylated spirit. Slight imperfections may be polished out with steel wool. Pitting, scoring or wear ridging of bores or pistons mean that the whole caliper must be renewed.
8 Renew all rubber components (seals, dust boots and bellows) as a matter of course. Blow through the fluid inlet and bleed screw holes with compressed air.
9 Check that the guide pins slide easily in their housings. Clean or renew them as necessary, and lubricate them with a copper-based anti-seize compound.
10 Lubricate a new piston seal with clean brake fluid. Insert the seal into the groove in the bore, using the fingers only to seat it (photo).
11 Fit a new dust boot to the piston at the end furthest from the piston groove. Extend the dust boot ready for fitting.
12 Lubricate the piston and bore with clean brake fluid, or with assembly lubricant if this is supplied with the repair kit.
13 Offer the piston and dust boot to the caliper. Engage the dust boot with the groove in the piston housing, then push the piston through the dust boot into the caliper bore. Engage the dust boot with the groove on the piston.
14 Repeat the above operations on the other piston and bore.
15 Refit the bleed screws, hydraulic hoses and other disturbed components.
16 Refit the caliper to the vehicle.

12 Rear brake caliper – removal and refitting

1 Remove the rear brake pads (Section 5).
2 Clean around the hydraulic union on the caliper. Slacken the union half a turn.
3 Remove the two bolts which secure the caliper. Of the four bolts on the caliper, these are the two nearest the hub. **Do not** remove the other two bolts, which hold the caliper halves together. Obtain new bolts for refitting.

Fig. 9.7 Caliper dust boot and piston. Fit the boot to the end furthest from the groove (arrowed) (Sec 11)

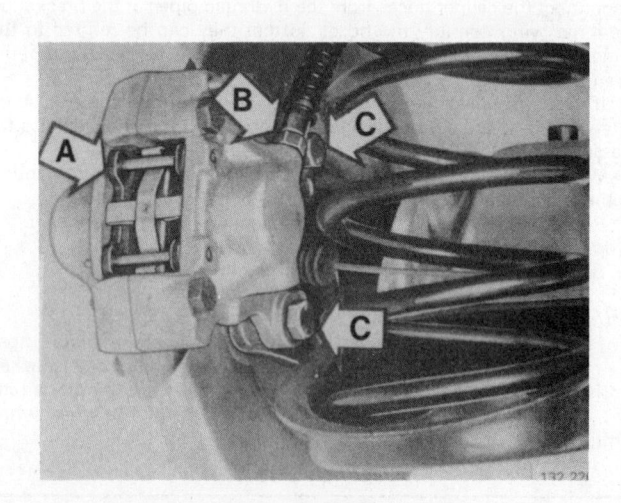

Fig. 9.8 Rear brake caliper removal (Sec 12)

A Rear pads   C Caliper securing bolts
B Hydraulic union

Chapter 9 Braking system

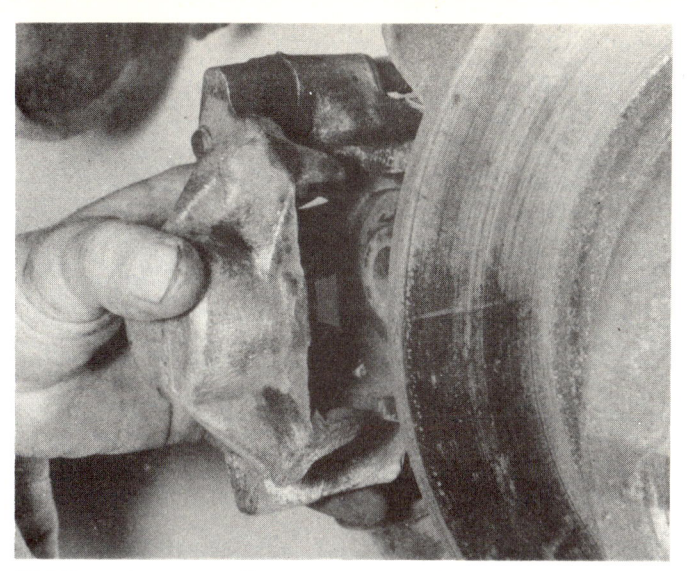

12.4 Removing a rear caliper

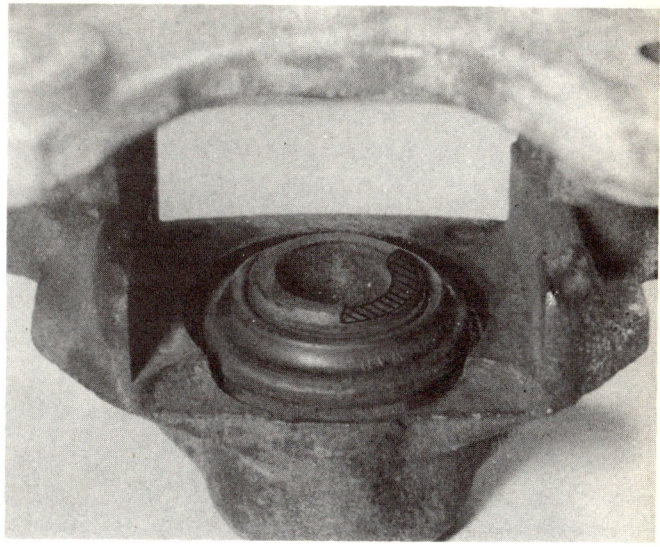

13.3 Pay attention to the position of the piston step (shaded)

4 Remove the caliper from the disc and unscrew it from the hydraulic hose (photo). Be prepared for fluid spillage. Plug or cap open unions.
5 Commence refitting by screwing the caliper onto the flexible hose. Do not tighten the union fully yet.
6 Fit the caliper over the disc and secure it to the axle bracket with two new bolts. Tighten the bolts to the specified torque.
7 Tighten the flexible hose union at the caliper. Check that the routing and 'set' of the hose are such that it does not contact adjacent components. Correct if necessary by releasing the hose union at the brake pipe bracket, repositioning the hose and tightening the union.
8 Refit the brake pads (Section 5).
9 Bleed the appropriate hydraulic circuit (Section 9).

13 Rear brake caliper – overhaul

1 This is essentially the same procedure as that described for the front caliper (Section 11). In addition, note the following points.
2 Do not attempt to separate the caliper halves.
3 Pay attention to the position of the step on the piston (photo). It should be at a 20° angle to the lower surface of the caliper (see Fig. 9.9).

14 Brake discs – inspection

1 Whenever new pads are fitted, or if brake judder or snatch is noticed, inspect the brake discs as follows.
2 Inspect the friction surfaces for cracks or deep scoring (light grooving is normal and may be ignored). A cracked disc must be renewed; a scored disc can be reclaimed by machining provided that the thickness is not reduced below the specified minimum.
3 Check the disc run-out by positioning a fixed pointer near the outer edge, in contact with the friction surface. Rotate the disc and measure the maximum displacement of the pointer with feeler blades. A dial test indicator, if available, will give a more accurate result. Maximum run-out is given in the Specifications. Remember that front wheel bearing wear or maladjustment can also cause disc run-out.
4 Disc thickness variation in excess of 0.015 mm (0.0006 in) can also cause judder. Check this using a micrometer.
5 Whenever the rear discs are removed, check the condition of the handbrake drums. Refinishing, run-out and out-of-round limits are given in the Specifications. The drums are unlikely to wear unless the handbrake is habitually used to stop the vehicle.

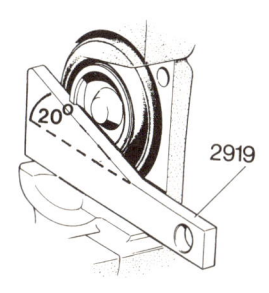

Fig. 9.9 Using a template (Volvo tool 2919) to determine the correct position of the rear caliper piston stop (Sec 13)

15 Front brake disc – removal and refitting

1 Remove the brake caliper and bracket (Section 10), but do not disconnect the hydraulic hoses. Tie the caliper up so that the hoses are not strained.
2 Prise or tap off the hub nut grease cap. Obtain a new cap if the old one is damaged during removal.
3 Remove the split pin from the hub unit. Unscrew and remove the nut.
4 Pull the brake disc outwards to displace the bearing outboard race. Recover the race, then pull the disc off the stub axle.
5 If the bearing inboard race has stayed on the stub axle, pull or lever it off.
6 Clean the stub axle and the oil seal mating face.
7 If a new disc is to be fitted, transfer the bearing tracks to it if they are in good condition, or fit new ones. Renew the oil seal in any case (Chapter 10, Section 15).
8 Clean rustproofing compound off a new disc with methylated spirit and a rag.
9 Grease the stub axle and the oil seal lips. Place the disc and bearing assembly onto the stub axle and push it home. Fit the bearing outboard race.
10 Fit the castellated nut. Adjust the bearings (Chapter 10, Section 4) and secure the nut with a new split pin.
11 Half fill the grease cap with grease, then fit it to the disc and tap it home.
12 Refit the brake caliper and bracket (Section 10).

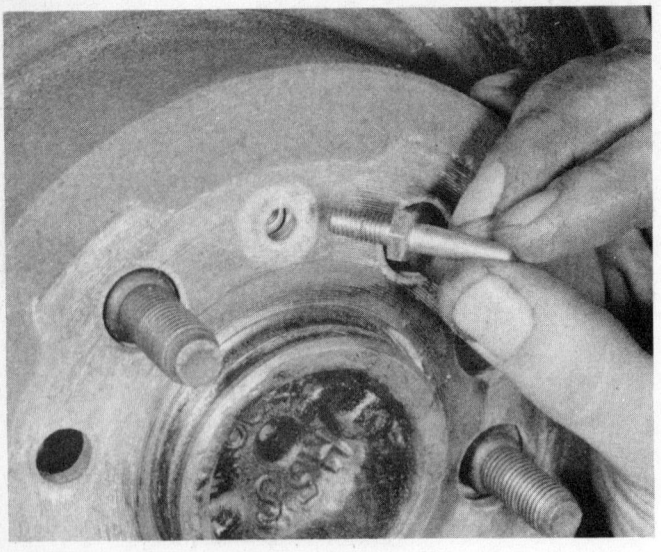

16.2 Removing a wheel locating spigot

Fig. 9.10 Removing a rear brake disc (Sec 16)

16 Rear brake disc – removal and refitting

1 Remove the rear brake caliper without disconnecting the hydraulic hose (Section 12). Tie the caliper up out of the way.
2 If a wheel locating spigot is fitted, unscrew it from the disc (photo).
3 Make sure that the handbrake is released, then pull off the disc. Tap it with a soft-faced mallet if necessary to free it.
4 Refit by reversing the removal procedure. If a new disc is being fitted, remove the traces of rustproofing compound from it.

17 Brake master cylinder – removal and refitting

1 Syphon as much fluid as possible from the master cylinder reservoir, using a hydrometer or old poultry baster. *Do not syphon the fluid by mouth, it is poisonous.*
2 Unbolt the heat shield (when fitted) from around the master cylinder.
3 Disconnect the clutch master cylinder feed pipe from the side of the reservoir (when applicable). Be prepard for fluid spillage. Plug the open end of the pipe (photo).
4 Disconnect the hydraulic unions from the master cylinder. Be prepared for fluid spillage. Cap the open unions to keep dirt out (photo).
5 Remove the nuts which secure the master cylinder to the servo. Pull the master cylinder off the servo studs and remove it (photo). Be careful not to spill hydraulic fluid on the paintwork.
6 Refit by reversing the removal operations. Bleed the complete brake hydraulic system, and if necessary the clutch hydraulic system, on completion.

18 Brake master cylinder – overhaul

1 Empty the fluid out of the master cylinder by pumping the pistons with a screwdriver. Clean the cylinder externally.
2 Pull the reservoir off the master cylinder and recover the seals (photo).

Non-ABS models
3 Depress the pistons and extract the circlip from the mouth of the cylinder (photo).
4 Shake the pistons, spring seat and spring out of the cylinder.
5 Inspect the master cylinder bore. If it is badly corroded or scratched, renew the cylinder complete. Light scoring or surface rust may be removed with steel wool and methylated spirit.
6 Obtain a repair kit, which will contain new pistons with seals already fitted.
7 Clean all parts not being renewed with methylated spirit. Blow through fluid passages with an air line or foot pump.
8 Lubricate the cylinder bore with clean hydraulic fluid. Apply more fluid to the pistons and seals, or smear them with assembly lubricant if this is supplied in the kit.

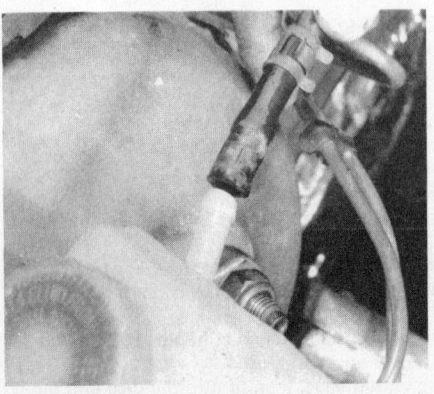

17.3 Disconnecting the clutch feed pipe

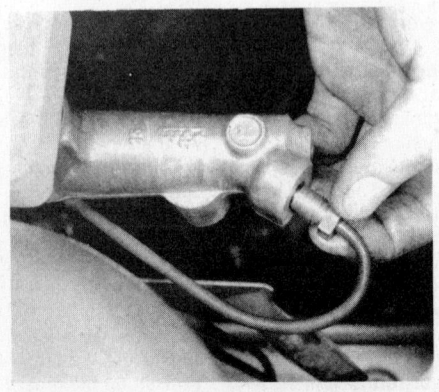

17.4 A master cylinder hydraulic union

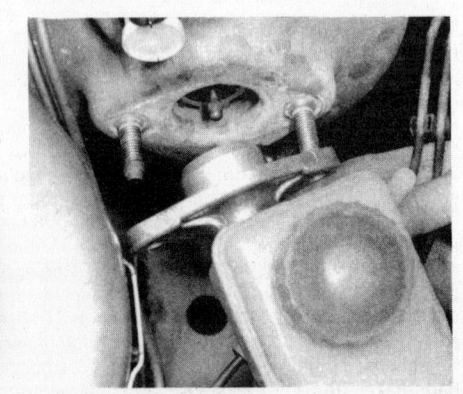

17.5 Removing the master cylinder

Chapter 9 Braking system

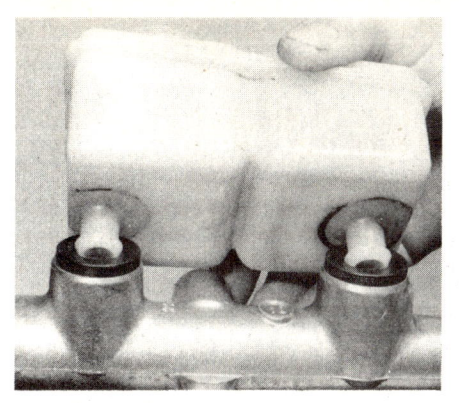

18.2 Removing the reservoir from the master cylinder

18.3 Remove the circlip to release the pistons

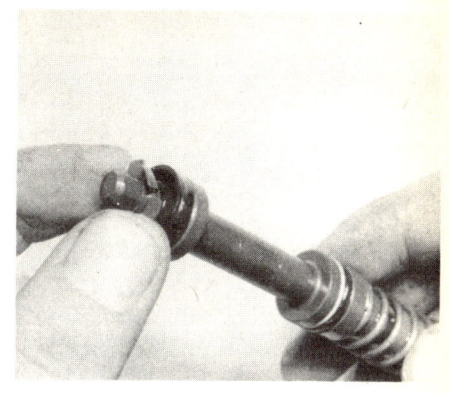

18.9A Fitting the spring seat ...

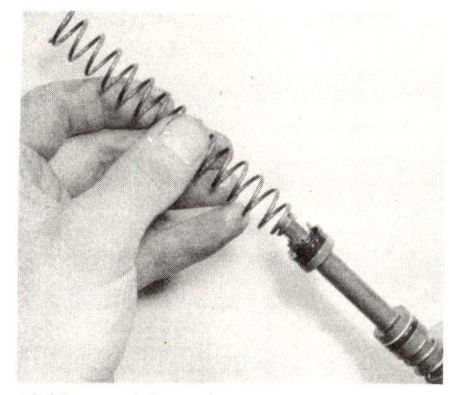

18.9B ... and the spring

18.9C Fitting the pistons into the master cylinder

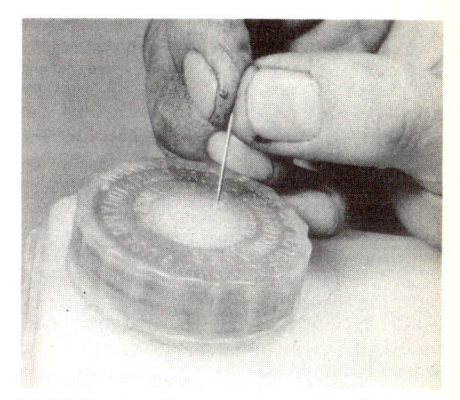

18.10 Clearing the reservoir cap breather hole

9 Assemble the spring, spring seat and pistons. Make sure that all components are perfectly clean, then insert the spring and pistons into the master cylinder. Depress the pistons and insert the circlip (photos).
10 Refit the reservoir and seals; renew the seals if necessary. Make sure that the reservoir cap breather hole is clear (photo).

ABS models
11 The procedure is similar to that just described, but the pistons are retained by a roll pin as well as by a circlip.

19 Pressure differential warning valve – removal and refitting

1 Seal the master cylinder reservoir by blocking the cap vent, or by tightening the cap over a piece of thin plastic film.
2 Clean the valve and its unions. It is located on the left-hand inner wing; access is not good (photo).
3 Disconnect the eight hydraulic unions for the valve, making notes for refitting if there is any possibility of confusion. Be prepared for fluid spillage; cap open unions.
4 Disconnect the electrical lead from the valve.
5 Remove the single securing bolt and remove the valve. Do not drip fluid on the bodywork.
6 Refit by reversing the removal operations. Bleed the complete hydraulic system on completion and check that the brake failure warning light operates correctly.

20 Pressure differential warning valve – overhaul

It is possible to carry out this procedure on the vehicle, but great care must be taken to keep dirt out of the hydraulic system.
1 Thoroughly clean the outside of the valve.
2 Unscrew the switch from the top of the valve. Recover the spring and contact pin (photos).

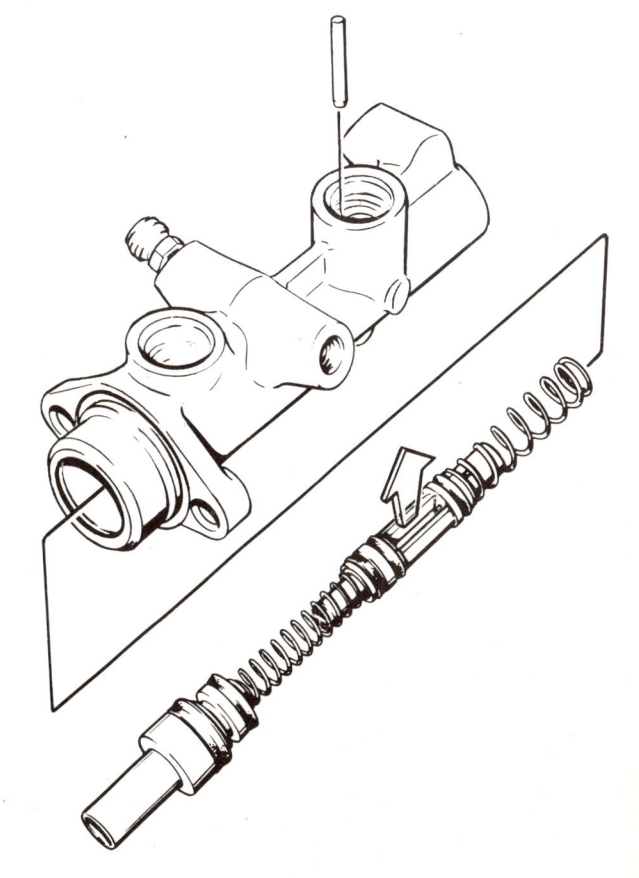

Fig. 9.11 Brake master cylinder components – ABS type. Roll pin fits into piston groove (arrowed) (Sec 18)

Chapter 9 Braking system

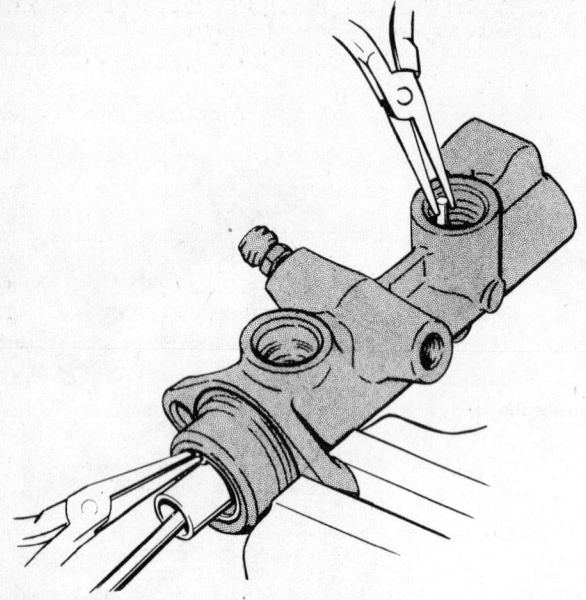

Fig. 9.12 Removing the roll pin and circlip from the ABS type master cylinder (Sec 18)

19.2 The pressure differential warning valve

3 Unscrew the two end plugs from the valve. Recover the O-rings.
4 Remove the piston, spring, plungers and O-rings. Note their fitted order.
5 Clean all parts with methylated spirit and inspect them. If the valve bore is badly worn, rusty or scored, renew the valve complete. Otherwise, obtain a repair kit containing new O-rings (photo).
6 Fit one end plug, using a new O-ring, and tighten it.
7 Assemble the piston, plunges and spring. Lubricate the piston and the new O-rings with clean hydraulic fluid. Fit one of the O-rings to a plunger and insert the assembly into the valve bore, O-ring first.
8 Fit the other O-ring into the bore and press it home with a small tube or other suitable tool.
9 Fit the other end plug and O-ring.
10 If the valve was removed for overhaul, refit it.
11 Bleed the complete hydraulic system (Section 9), then check the success of the overhaul as follows.
12 Have an assistant depress the brake pedal hard for one minute. Watch the valve: if fluid emerges from the switch hole, renew the valve. If no fluid energes, refit the contact pin, spring and switch. Reconnect the switch.

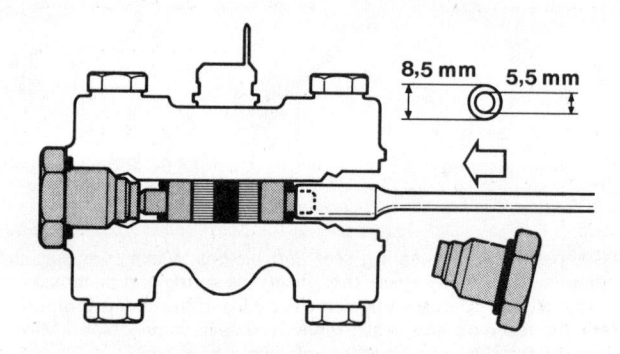

Fig. 9.13 Fitting the second plunger O-ring to the pressure differential warning valve. Tool dimensions as shown (Sec 20)

21 Brake pedal – removal and refitting

If preferred, the pedal box can be removed complete (Chapter 5, Section 4) and the brake pedal removed on the bench. Otherwise, proceed as follows.

1 Remove the steering column/pedal trim.
2 Disconnect the brake pedal from the servo by removing the clevis pin.
3 Remove the pedal pivot bolt and nut. The return spring will force the pedal downwards. Remove the pedal, spring and pivot bolt.
4 The pedal sleeve and bushes may be renewed if required (photo). Grease the new bushes and sleeve before fitting them.
5 Refit by reversing the removal operations.

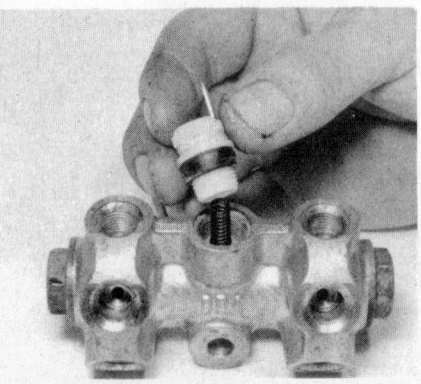

20.2A Unscrew the switch ...

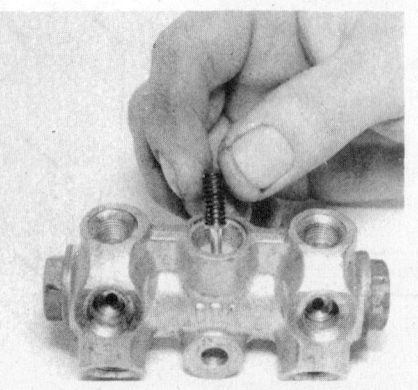

20.2B ... and recover the spring and contact pin

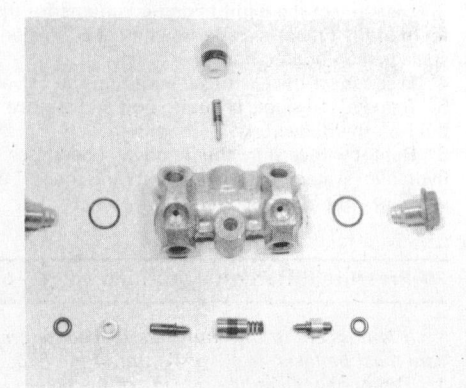

20.5 Valve components laid out for inspection

Chapter 9 Braking system

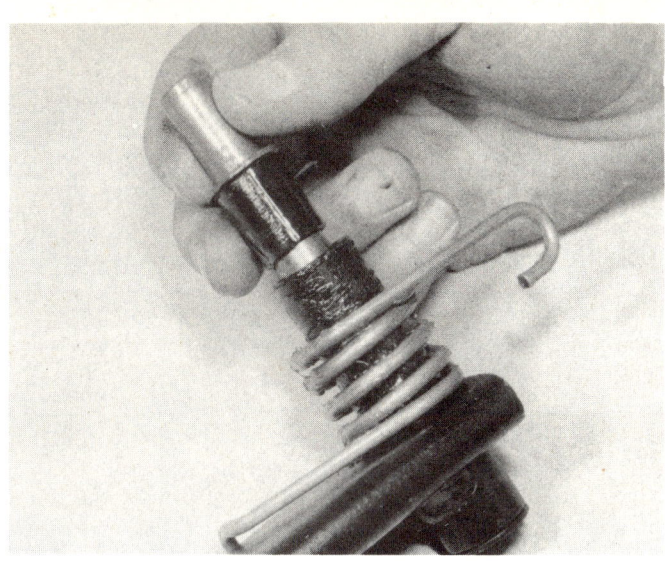

21.4 Removing the brake pedal sleeve and a bush

22 Brake servo – removal and refitting

1 Remove the brake master cylinder (Section 17). If care is taken, the master cylinder can be moved away from the servo without disconnecting the hydraulic unions. It will be necessary to disconnect the clutch master cylinder feed pipe, however.
2 Disconnect the servo vacuum feed, either by disconnecting the hose or by levering out the check valve.
3 Inside the vehicle, remove the steering column/pedal trim. Disconnect the servo clevis from the brake pedal.
4 Remove the four nuts which secure the servo.
5 Withdraw the servo from the engine bay.
6 Refit by reversing the removal operations. If a new servo is being fitted, adjust the pushrod if necessary to give small clearance between the servo pushrod and the master cylinder piston in the resting position.
7 Bleed the hydraulic system on completion if necessary.

23 Brake servo – overhaul

Even if parts are available, overhaul of the servo is not recommended. The peripheral components (check valve, seal and bellows) may be renewed if wished.

24 Vacuum pump – removal and refitting

1 Disconnect the hose from the vacuum pump.
2 Undo the four securing nuts and lift off the pump (photo). Recover the gasket.
3 Refit by reversing the removal operations, using a new gasket if necessary.

25 Vacuum pump – overhaul

1 Remove the pump top cover, which is secured by two screws. Recover the gasket, valve springs, valves and seals. Note which way round the valves are fitted.
2 Make alignment marks between the two halves of the pump. Remove the eight screws and separate the pump halves.
3 Remove the central screw which secures the diaphragm. Remove the diaphragm, washers and spring. Note the fitted sequence of the washers.
4 Undo the four securing screws and remove the bottom cover.
5 Insert feeler blades between the operating lever and the lever bearing arm to support the bearing arm. Drive out the lever pivot pin, then withdraw the feeler blades.
6 Remove the operating rod and lever.
7 Clean all parts and renew as necessary. New gaskets and seals should be used as a matter of course.
8 Commence reassembly by fitting the operating rod and lever to the pump body. Support the bearing arm with feeler blades and drive in the pivot pin; apply a drop of locking fluid to the exposed ends of the pin. Withdraw the feeler blades (photos).
9 Refit the bottom cover, using a new gasket, and secure with the four screws (photo).
10 Assemble the diaphragm, spring, washers and screw. Hold the pump upside down and fit the diaphragm assembly. Secure it with the screw (photo).
11 Refit the top half of the pump, observing the alignment marks made when dismantling. Secure it with the eight screws.
12 Refit the valves, using new seals. Make sure that the valves are fitted the right way round. Refit the springs and the top cover, using a new gasket, and secure it with the two screws (photos).

26 Handbrake shoes – removal, inspection and refitting

1 Slacken the handbrake cable adjuster (Section 7).
2 Remove the rear brake disc (Section 16).
3 Prise the shoes apart and displace the operating mechanism from them at the rear end, and the strut from them at the front (photos).
4 Unhook one of the return springs from one of the shoes, working through the hole in the halfshaft flange.
5 Free the shoes from the U-clips on the backplate and remove them and the springs (photo).

24.2 Vacuum pump securing nuts

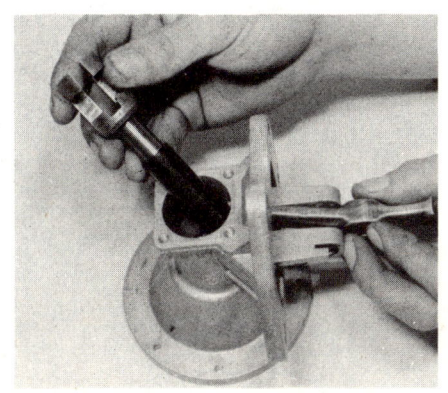

25.8A Fit the operating rod and lever ...

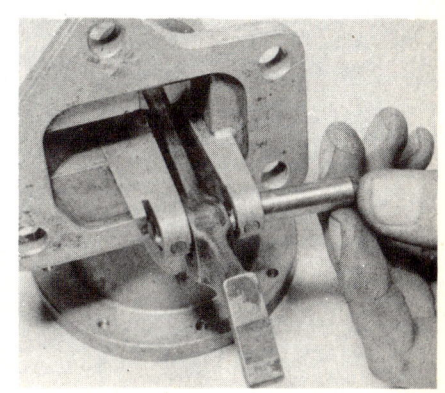

25.8B ... and the lever pivot pin

25.8C Support the arm with feeler blades and tap home the pin

25.9 Fitting the pump bottom cover

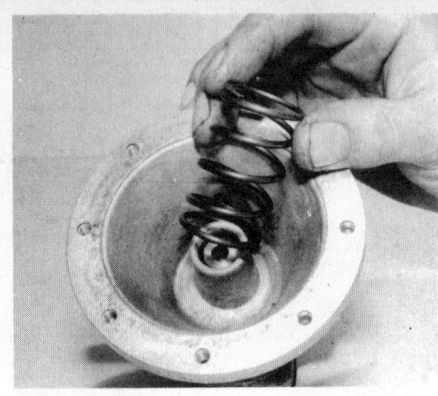

25.10A Fitting the diaphragm spring

25.10B Tightening the diaphragm screw. Circlip pliers are used to counterhold the washer

25.12A Fitting the valves (the right way up) ...

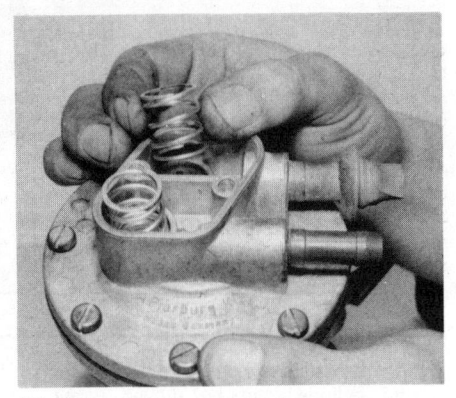

25.12B ... and the valve springs

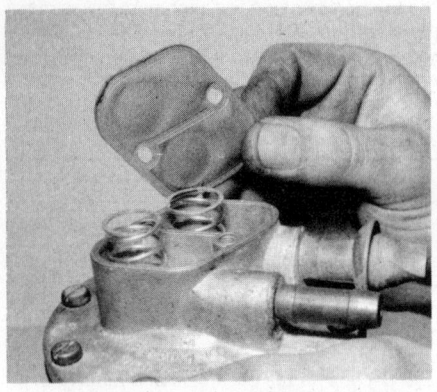

25.12C Fit a new top cover gasket ...

25.12D ... and the top cover itself

26.3A Front of handbrake shoes, showing strut

26.3B Disengaging the rear of the shoes from the operating mechanism

26.5 Handbrake shoe engaged in U-clip

Chapter 9 Braking system

6 Inspect the shoes for wear, damage or oil contamination. Renew them if necessary and rectify the source of any contamination. As with the brake pads, the shoes must be renewed in axle sets.
7 Refit by reversing the removal operations. Adjust the handbrake on completion.

27 Handbrake cables – removal and refitting

Short (right-hand) cable
1 Remove the handbrake shoes on the right-hand side (Section 26).
2 Free the cable from the operating mechanism by pushing out the clevis pin.
3 Remove the clevis pin from the other end of the cable. Free the cable from the guides or retaining clips and remove it.
4 Check the condition of the rubber gaiter and renew it if necessary (photo).

Long (left-hand) cable
5 Inside the vehicle, slacken off the handbrake adjustment as far as possible (Section 7). Release the cable from the lever.
6 Remove the handbrake shoes on the left-hand side (Section 26).
7 Free the cable from the operating mechanism by pressing out the clevis pin.
8 Release the cable from the brake backplate and from the rear axle.
9 Release the cable from the under-floor clamps and grommets and remove it. Transfer the grommets etc to the new cable. Renew the rubber gaiter if necessary.

Both cables
10 Refit by reversing the removal procedure, noting the following points:
  (a) Apply brake anti-seize compound to the operating mechanism and backplate rubbing surfaces. Keep the compound off brake friction surfaces
  (b) Fit the operating mechanism with the arrow visible and pointing upwards (photo).
  (c) Adjust the handbrake on completion (Section 7)

28 Stop-light switch – removal and refitting

1 Remove the steering column/pedal trim.
2 Disconnect the wiring from the switch. Undo the locknut and unscrew the switch (photo).
3 When refitting, screw the switch in so that it operates after 8 to 14 mm (0.32 to 0.55 in) movement of the brake pedal. Reconnect the wires and tighten the locknut.
4 Check for correct operation, then refit the disturbed trim.

29 Handbrake warning switch – removal and refitting

1 Remove the rear console (Chapter 11, Section 36).
2 Remove the switch securing screw (photo). Lift out the switch, disconnect the lead from it and remove it.

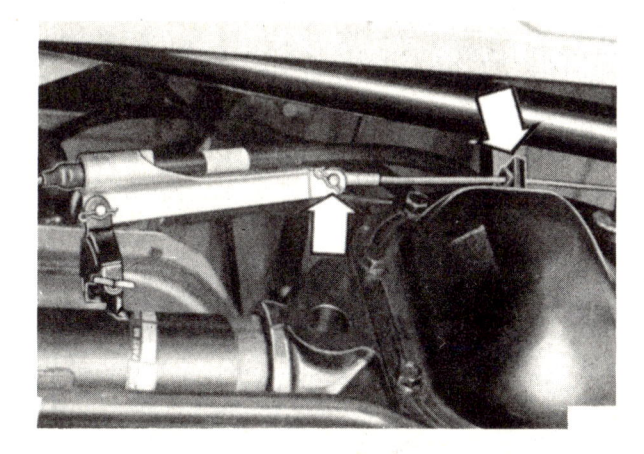

Fig. 9.14 Short handbrake cable attachments (arrowed) on rear axle (Sec 27)

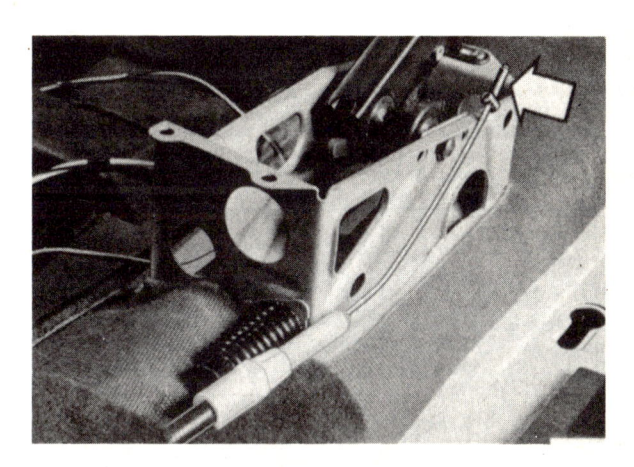

Fig. 9.15 Long handbrake cable end (arrowed) disengaged from the lever (Sec 27)

27.4 Handbrake cable showing rubber gaiter

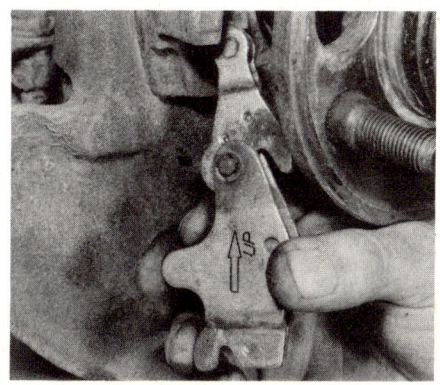

27.10 Refitting the handbrake mechanism – note arrow and 'UP' marking

28.2 Stop-light switch seen through a hole in the brake pedal bracket

224   Chapter 9 Braking system

29.2 Handbrake warning switch – securing screw arrowed

3 Refit by reversing the removal operations. Check for correct operation of the switch before refitting the rear console.

30 Anti-lock braking system (ABS) – description

When fitted, the anti-lock braking system monitors the rotational speed of the wheels under braking. Sudden deceleration of one wheel, indicating that lock-up is occurring, causes the hydraulic pressure to that wheel's brake to be reduced or interrupted momentarily. Monitoring and correction take place several times per second, giving rise to a 'pulsing' effect at the brake pedal when correction is taking place. The system gives even inexperienced drivers a good chance of retaining control when braking hard on slippery surfaces.

The main components of the system are the sensors, the control unit and the hydraulic modulator. (Some of the ABS components are shared with the ETC system, when fitted. See Chapter 4, Section 14).

One sensor is fitted to each front wheel, picking up speed information from a pulse wheel carried on the brake disc. Rear wheel speed information is picked up from the speedometer sensor in the differential housing. For ABS purposes the rear wheels are treated as one unit.

Information from the sensors is fed to the control unit in the boot. The control unit operates solenoid valves in the hydraulic modulator, also in the boot, to restrict if necessary the supply to either front caliper or both rear calipers. The control unit also illuminates a warning light in the event of system malfunction.

The hydraulic modulator contains a pump as well as solenoid valves. It is a semi-active device, increasing the effort applied at the brake pedal. If the modulator fails, adequate braking effort will still be available from the master cylinder and servo, though the anti-lock function will be lost.

On models with ABS the hydraulic circuits are split front-rear instead of triangularly.

To avoid damage to the ABS control unit, do not subject it to voltage surges in excess of 16V, nor to temperatures in excess of 80°C (176°F).

31 ABS components – removal and refitting

Front wheel sensor
1 Follow the sensor wiring back to the suspension turret. Separate the connector, push the wires out of it and feed them back into the wheel arch.

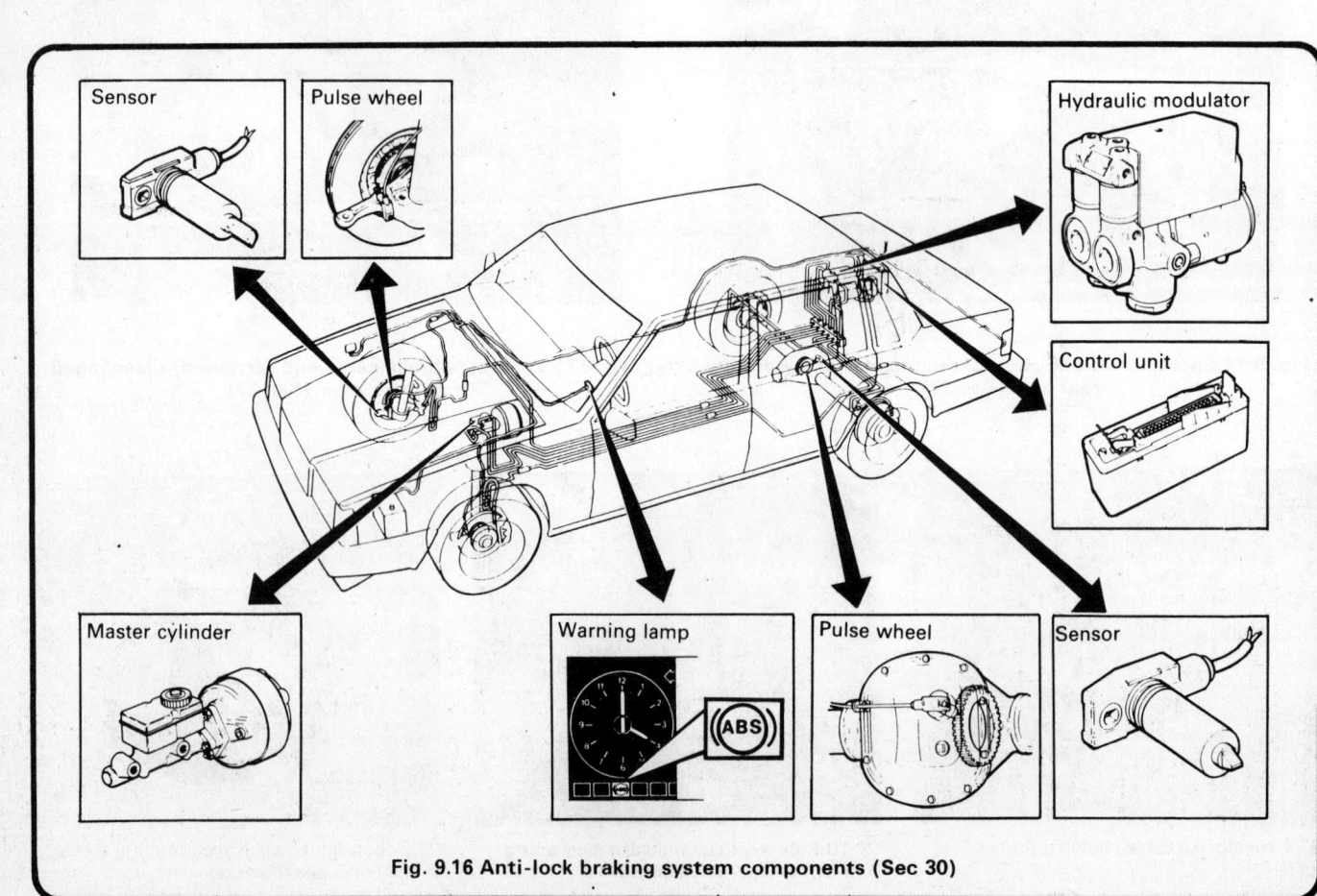

Fig. 9.16 Anti-lock braking system components (Sec 30)

Chapter 9 Braking system

Fig. 9.17 ABS front wheel sensor secured by an Allen screw (arrowed) (Sec 31)

Fig. 9.18 Pulling the pulse wheel off the brake disc (Sec 31)

2 Remove the Allen screw which secures the sensor to the steering knuckle. Withdraw the sensor and its wiring.
3 Refit by reversing the removal operation. Apply a little grease (Volvo No 1 161 037-5, or equivalent) to the body of the sensor.

Front disc pulse wheel
4 Remove the front brake disc (Section 15).
5 Remove the pulse wheel from the disc with a two-legged puller. Be careful not to damage the hub oil seal.
6 Fit the new pulse wheel and seat it with a piece of tube.
7 Refit the brake disc.

Rear wheel sensor
8 This is the same as the speedometer sender (Chapter 12, Section 22), but may be secured by an Allen screw instead of a ring nut. The running clearance for this sender is 0.35 to 0.75 mm (0.014 to 0.030 in) with a target value of 0.60 mm (0.024 in).

Control unit
9 Make sure that the ignition is switched off.
10 Remove the cover from the right-hand well in the boot.
11 Lift the control unit out of its bracket, disconnect the multi-plug and remove it.
12 Refit by reversing the removal operations.

Fig. 9.19 ABS rear wheel sensor secured by an Allen screw (arrowed) (Sec 31)

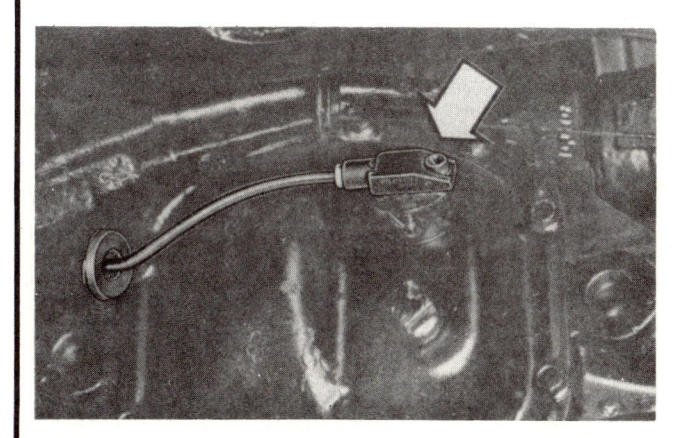

Fig. 9.20 ABS control unit – multi-plug arrowed (Sec 31)

Fig. 9.21 ABS hydraulic modulator securing nuts (arrowed) (Sec 31)

Hydraulic modulator
13 Remove the cover from the right-hand well in the boot.
14 Remove the modulator cover. Unplug the two relays and the electrical connector from the modulator. Release the earth wire.
15 Clean around the modulator hydraulic unions. Make identifying marks or notes.
16 Remove the three nuts from the modulator mountings.
17 Place some rags under the unit. Disconnect the hydraulic unions, being prepared for fluid spillage. Lift out the modulator. Plug or cap open unions.
18 If a new modulator is being fitted, transfer the pipe connectors and the rubber mountings to it.
19 Refit by reversing the removal operations. Bleed the complete hydraulic system on completion.

32 ABS – fault tracing

1 In the event of an ABS fault, first check all fuses and wiring connectors. Besides the fuses in the central electrical unit (Nos 2 and 10 or 12), there is an 80A fuse on the front right-hand inner wing, and a 10A fuse on the transient surge protector next to the control unit.
2 Switch off the ignition and disconnect the multi-plug from the control unit. Remove the cover from the multi-plug so that subsequent tests can be carried out through the side access holes. Do not insert meter probes into the front of the connectors.
3 Measure the resistance between earth and terminals 10, 20, 32 and 34 (Fig. 9.24). It should be zero in each case. If not, check the earthing point at the right-hand tail light. If terminal 32 is not earthed, try fitting a new valve relay to the control unit.
4 Switch on the ignition. Measure the voltage at terminals 2, 3 and 4 of the transient surge protector (Fig. 9.25). It should be 12V. No voltage at terminal 3 may be due to a faulty surge protector; otherwise, look for wiring damage.
5 Still with the ignition on, measure the voltage at the converter terminals (next to the control unit) (Fig. 9.26). If the result is not 12V at all terminals, renew the converter.
6 Returning to the control unit multi-plug, check the voltage between earth and terminals 1, 7, 9, 27 and 28. 12V should be obtained in each case. If not, there is a defect in the transient protector (terminal 1), the converter (7 and 9), the valve relay (27) or the pump relay (28).
7 Check the voltage between terminal 25 and earth whilst an assistant depresses the brake pedal. 12V should be obtained; if not, check the stop-light switch and bulbs.
8 Check the voltage between terminal 29 and earth. It should be 0.5 to 1.0V. If not, the solenoid valve relay is defective.

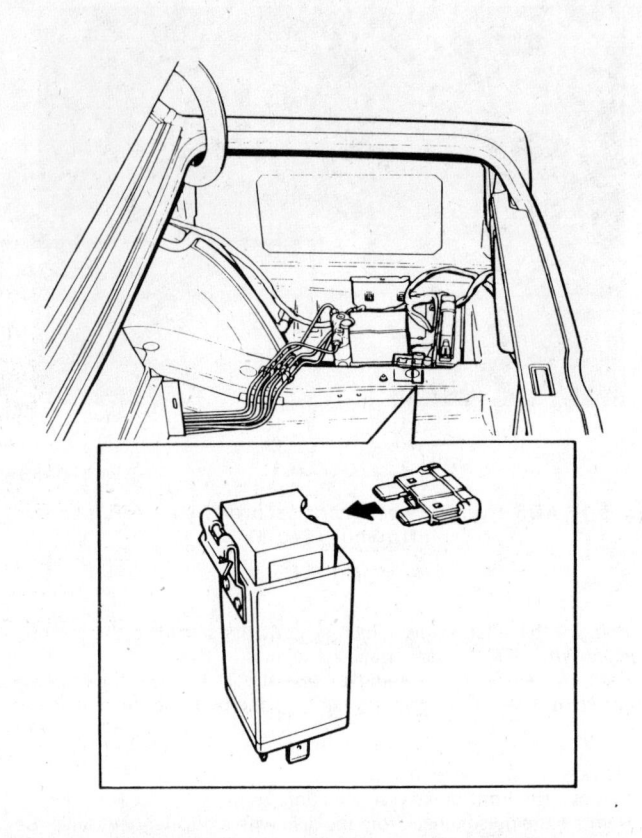

Fig. 9.22 Transient surge protector fuse (Sec 32)

9 Switch off the ignition. Disconnect the hydraulic modulator multi-plug, then switch on the ignition again.
10 Check for 12V between earth and terminals 6, 7, 10 and 12 of the modulator plug. Absence of voltage suggests a blown No 2 fuse (terminal 6), blown ABS warning light bulb (No 7), faulty transient protector (No 10) or blown 80A fuse (No 12).
11 Switch off the ignition and reconnect the modulator multi-plug.
12 Measure the front sensor resistance across terminals 4 and 6, then 21 and 23, of the control unit multi-plug. Resistance should be 0.9 to 2.2 kΩ. If not, check the resistance again at the connectors near the front suspension turrets. Renew the sensor(s) if the resistance is still incorrect.
13 Measure the rear sensor resistance between terminals 7 and 9. It should be 0.6 to 1.6 kΩ. Recheck if necessary at the connector next to the fuel filter pipe.
14 Note that if the rear sensor is too far from the pulse wheel, the ABS will not operate and the warning light will come on. See Chapter 12, Section 22.
15 As a further sensor check, verify that the measured resistance fluctuates if the wheel in question is rotated.
16 Check the resistance of the valve solenoids by measuring between terminal 32 and 2, 18 and 35. In each case the reading should be 0.7 to 1.7 Ω. If not, recheck at the hydraulic modulator (between terminals 4 and 1, 5 and 3). Repair the wiring or renew the modulator as necessary.
17 Switch on the ignition. Briefly earth terminal 28 of the control unit mulit-plug and check that the modulator pump runs. If not, try a new pump relay before condemning the pump.
18 With the ignition still on, connect the voltmeter between terminal 32 and earth. Connect pin 27 to earth: the voltmeter should read 12V. If not, renew the valve relay.
19 If no faults have been revealed, but the ABS is still not working, the fault must be in the control unit.
20 Switch off the ignition, disconnect the test gear and remake the original wiring connections.

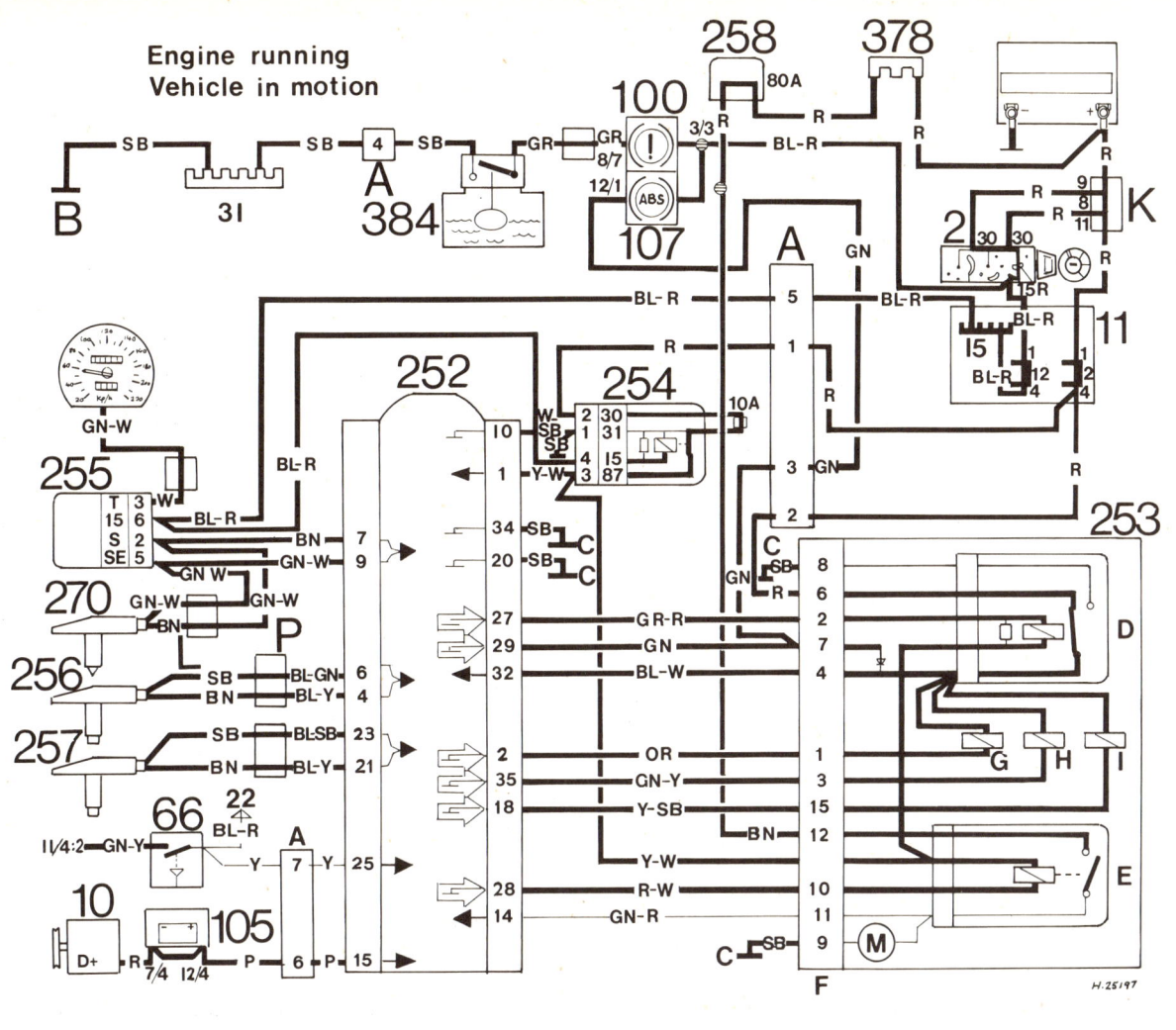

Fig. 9.23 ABS circuit diagram (Sec 32)

2 Ignition/starter switch	105 Ignition (no-charge) warning light	270 Rear wheel/speedometer sensor	F Hydraulic modulator multi-plug
10 Alternator	107 ABS warning light	378 Positive terminal (in engine bay)	G LH front solenoid valve
11 Fusebox	252 Control unit	284 Brake fluid level sensor	H RH front solenoid valve
15 Connector (in central electrical unit)	253 Hydraulic modulator	A Connector (RH A-pillar)	J Rear solenoid valve
22 Stop-light feed	254 Transient surge protector	B Earth point (RH A-pillar)	K Connector (LH A-pillar)
31 Earth point (in central electrical unit)	255 Converter	C Earth point (RH tail light)	L Connector (in boot)
66 Stop-light switch	256 LH front wheel sensor	D Solenoid valve relay	M Pump motor
85 Speedometer	257 RH front wheel sensor	E Pump motor relay	N Connector (RH front inner wing)
100 Brake fluid warning light	258 Fuse (80A)		P Connector (LH front inner wing)

Colour code

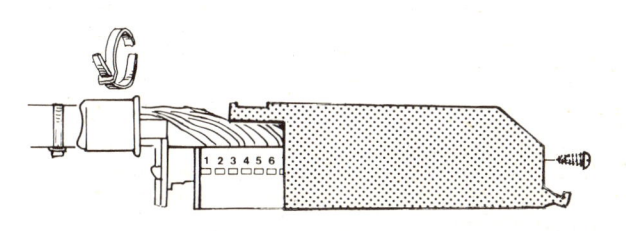

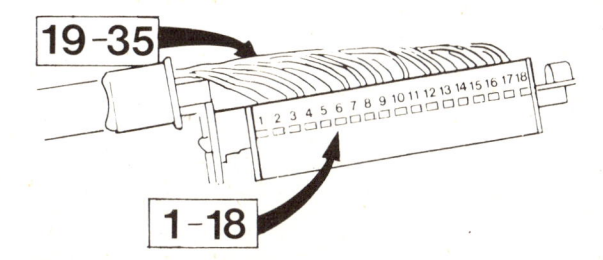

B	Blue	GR	Grey	R	Red	W	White
BN	Brown	OR	Orange	SB	Black	Y	Yellow
GN	Green	P	Pink	VO	Violet		

Fig. 9.24 Remove the multi-plug cover (left) for access to the numbered connectors (Sec 32)

Chapter 9 Braking system

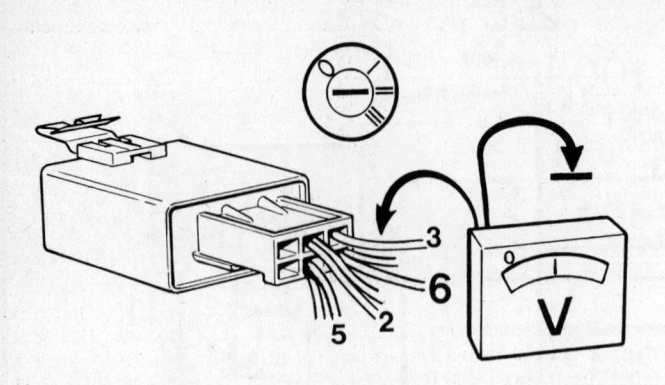

Fig. 9.25 Checking the voltage at the transient surge protector (Sec 32)

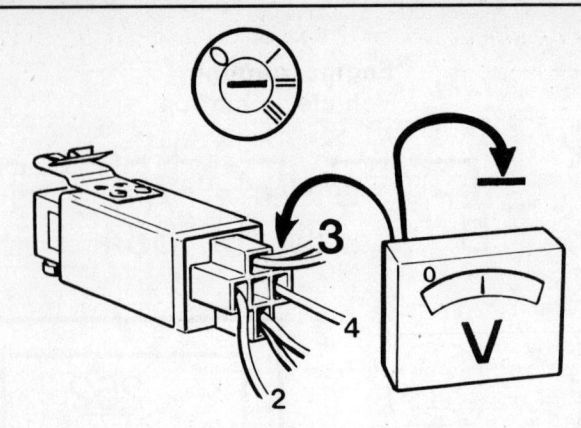

Fig. 9.26 Checking the voltage at the converter (Sec 32)

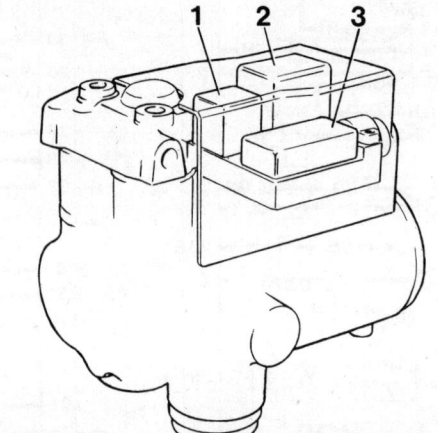

Fig. 9.27 Hydraulic modulator relays and multi-plug (Sec 32)

1 Solenoid valve relay
2 Pump relay
3 Multi-plug

33 Fault diagnosis – braking system

Symptom	Reason(s)
Pedal travel excessive	Air in system Leak in one hydraulic circuit Disc run-out excessive
Pedal spongy	Air in system Rubber hose(s) perished
Pedal creep during sustained application	Fluid leak Master cylinder seals leaking
Excessive pedal effort required	Servo defective or disconnected Linings contaminated or incorrect grade New linings not yet bedded-in
Brakes pull to one side	Linings contaminated on one side Caliper piston seized Type pressures incorrect Steering or suspension fault
Brakes snatch	Linings wet (apply brakes to dry them) Linings worn out Disc worn or cracked Caliper loose Wheel bearing play excessive
Brakes bind	Servo pushrod incorrectly adjusted (too long) Master cylinder breather blocked Brake pipe crushed or blocked Caliper piston seized Air in system Handbrake over-adjusted, or cable seized

Chapter 10 Steering and suspension

For modifications, and information applicable to later models, see Supplement at end of manual

Contents

Fault diagnosis – steering and suspension	33
Front anti-roll bar – removal and refitting	19
Front control arm – removal and refitting	16
Front control arm balljoint – removal and refitting	17
Front radius rod – removal and refitting	18
Front suspension strut – dismantling and reassembly	21
Front suspension strut – removal and refitting	20
Front wheel alignment – checking and adjustment	5
Front wheel bearings – checking and adjustment	4
Front wheel bearings – removal, inspection and refitting	15
General description	1
Maintenance and inspection	2
Panhard rod – removal and refitting	29
Power steering fluid – level checking and bleeding	3
Rear anti-roll bar – removal and refitting	28
Rear shock absorber – removal and refitting	27
Rear spring – removal and refitting	26
Rear subframe and mountings – removal and refitting	24
Rear torque rods – removal and refitting	23
Rear trailing arm – removal and refitting	25
Self-levelling rear suspension – general	30
Steering column – removal and refitting	13
Steering gear – overhaul	9
Steering gear – removal and refitting	8
Steering lock – removal and refitting	14
Steering pump – overhaul	11
Steering pump – removal and refitting	10
Steering rack bellows – renewal	7
Steering wheel – removal and refitting	12
Suspension rubber bushes – renewal	22
Track rod end – removal and refitting	6
Wheels and tyres – general care and maintenance	32
Wheel studs – renewal	31

Specifications

General
Front suspension type	Independent, MacPherson struts, with anti-roll bar
Rear suspension type	Solid axle, coil springs and telescopic shock absorbers. Load levelling facility on some models
Steering type	Rack and pinion, power-assisted

Front wheel alignment
Castor	+4°30' to +5°30'
Camber	−0°12' to +0°48'
Variation between sides (castor or camber)	0°42' maximum
Toe (measured between inner rims)	2.0 ± 0.5 mm (0.08 ± 0.02 in) toe-in

Steering gear
Fluid type/specification	ATF type A, F or G (Duckhams Q-Matic)
Ratio	16.9 : 1
Turns lock to lock	3.5

Wheel rims
Lateral run-out:	
Aluminium	0.8 mm (0.032 in) maximum
Steel	1.0 mm (0.039 in) maximum
Radial run-out:	
Aluminium	0.6 mm (0.024 in) maximum
Steel	0.8 mm (0.032 in) maximum

Tyre pressures (cold)*

Recommended pressures in bar (lbf/in²):	Front	Rear
Saloon, up to 3 occupants	1.9 (28)	1.9 (28)
Saloon, fully laden	2.1 (31)	2.3 (33)
Estate, up to 3 occupants	1.9 (28)	2.1 (31)
Estate, fully laden	2.1 (31)	2.8 (41)
For sustained high speeds (over 72 mph) add	0.3 (4)	0.3 (4)
'Space saver' spare (see text):		
155/R15	3.5 (51)	3.5 (51)
165/14	2.8 (41)	2.8 (41)

Refer to sticker on driver's door for confirmation

Torque wrench settings

	Nm	lbf ft
Front suspension		
Control arm ballpin nut	60	44
Balljoint to strut*:		
Stage 1	30	22
Stage 2	Tighten 90° further	Tighten 90° further
Track rod end ballpin nut	60	44
Control arm to crossmember*	85	63
Radius rod to control arm*	95	70
Radius rod to subframe*		
M12	85	63
M14	140	103

Chapter 10 Steering and suspension

	Nm	lbf ft
Strut top mounting (to body)	40	30
Strut piston rod nut (up to 1984)	150	111
Crossmember to body	95	70

*Use new fastenings every time

Rear suspension	Nm	lbf ft
Trailing arm to axle	45	33
Trailing arm bracket bolts	45	33
Trailing arm bracket nuts	85	63
Rear spring upper mounting	48	35
Shock absorber mountings	85	63
Panhard rod bolts	85	63
Torque rods	140	103
Subframe front mounting	85	63
Subframe rear bush bracket	48	35

Steering	Nm	lbf ft
Steering wheel bolt	32	24
Steering column universal joints	21	16
Steering gear to crossmember	44	33
Track rod end balljoint nut	60	44
Track rod end locknut	70	52
Hydraulic union banjo bolts	42	31

Wheels		
Wheel nuts	85	63

1 General description

Steering is by power-assisted rack and pinion. Power assistance is derived from a hydraulic pump, belt-driven from the crankshaft pulley.

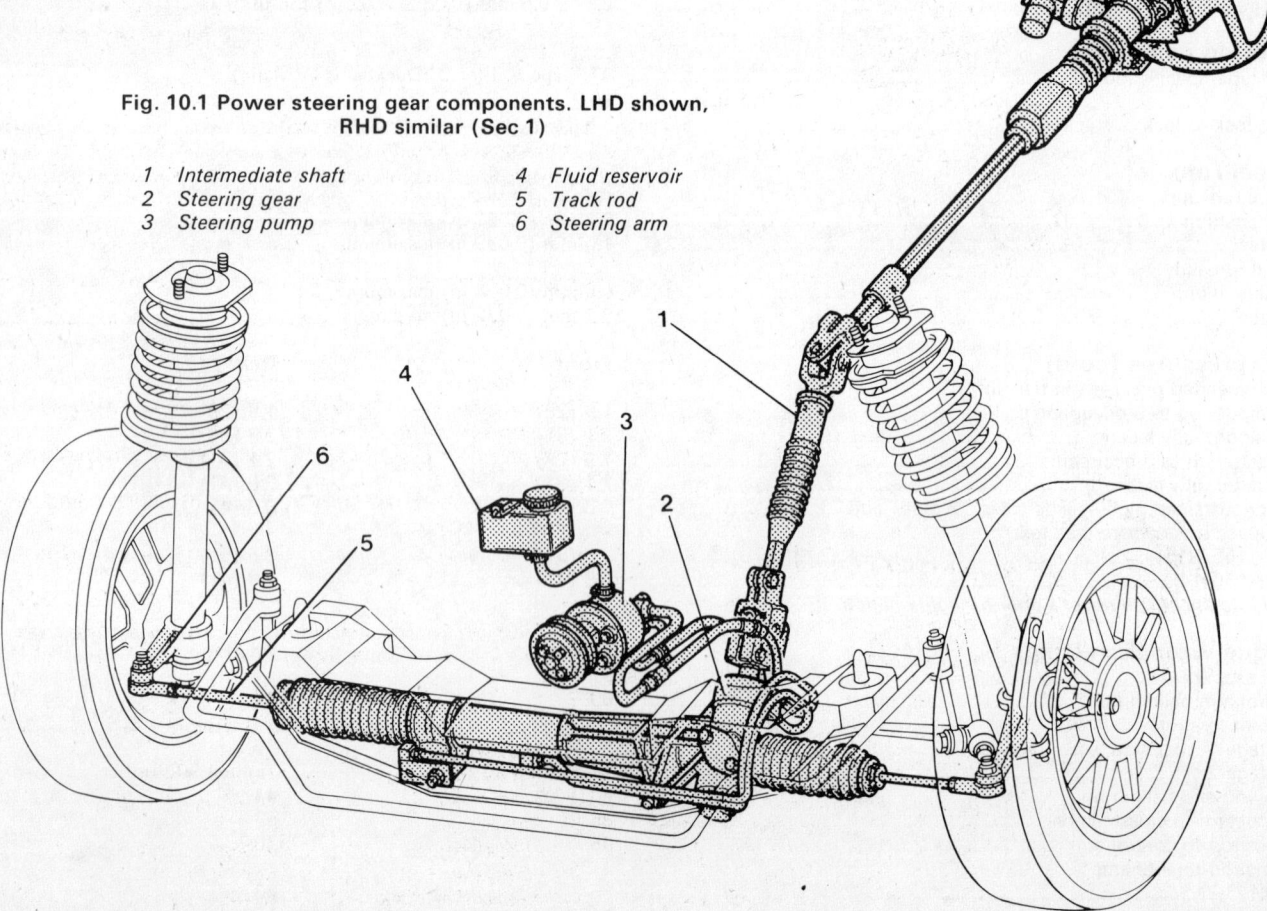

Front suspension is independent. MacPherson struts are used, located at their lower ends by control arms each carrying a balljoint. The control arms are attached to the front crossmember, to a radius rod each and to the front anti-roll bar.

Fig. 10.1 Power steering gear components. LHD shown, RHD similar (Sec 1)

1 Intermediate shaft
2 Steering gear
3 Steering pump
4 Fluid reservoir
5 Track rod
6 Steering arm

Chapter 10 Steering and suspension

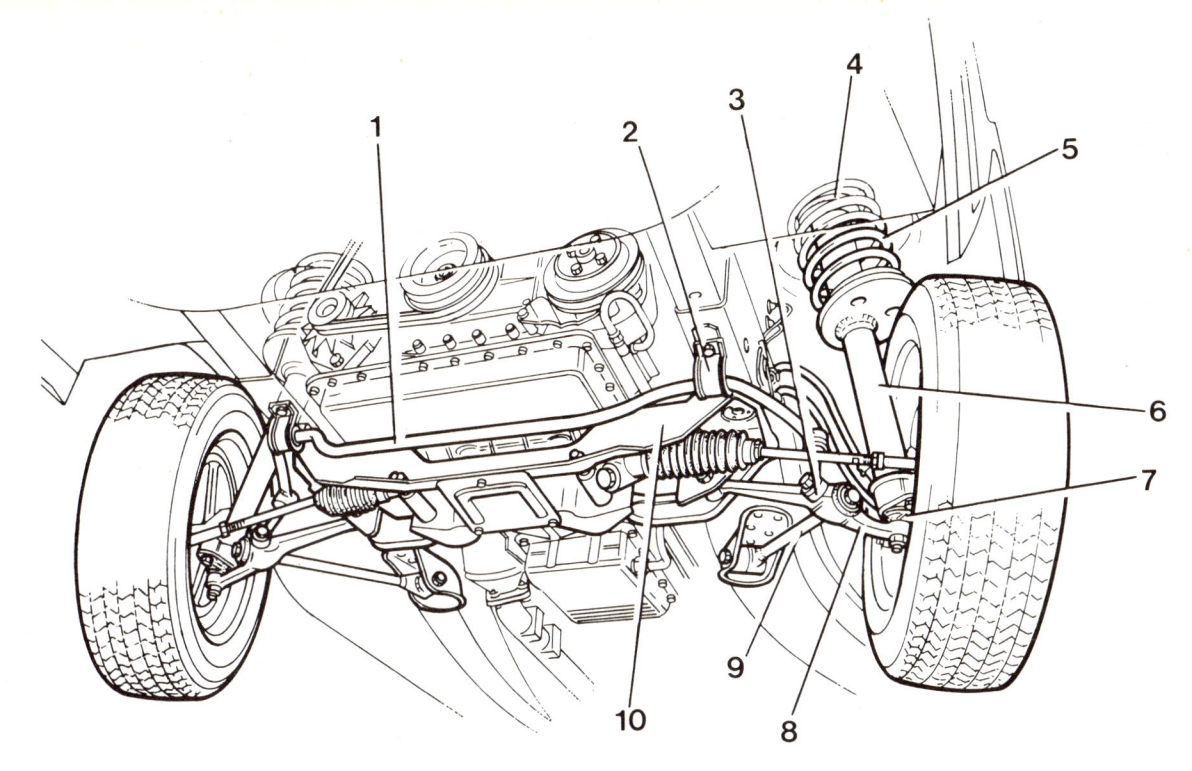

Fig. 10.2 Front suspension components (Sec 1)

1. Anti-roll bar
2. Anti-roll bar bracket
3. Anti-roll bar link
4. Strut upper mounting
5. Spring
6. Strut tube
7. Control arm balljoint
8. Control arm
9. Radius rod
10. Front crossmember

Rear suspension is of the live rear axle type. The axle is supported by two trailing arms, two torque rods and a Panhard rod. The torque rods are attached to a central subframe. A coil spring and a telescopic shock absorber are attached to each trailing arm. An anti-roll bar is fitted to some models.

2 Maintenance and inspection

1 Check the tyre pressures frequently. See Section 32.
2 Every 6000 miles or six months, inspect the tyres more thoroughly, again as described in Section 32. If the wheels are not removed for this check, at least check the tightness of the wheel nuts.
3 At the same intervals check the power-assisted steering fluid level (Section 3).
4 Every 12 000 miles or annually, check the front wheel bearing adjustment (Section 4).

Fig. 10.3 Sectional view through one side of the front suspension (Sec 1)

1. Control arm balljoint
2. Shock absorber seat
3. Control arm
4. Anti-roll bar link
5. Radius rod
6. Anti-roll bar
7. Stub axle
8. Strut tube
9. Spring lower seat
10. Spring
11. Shock absorber
12. Shock absorber retaining nut
13. Piston rod
14. Upper mounting
15. Cover
16. Spring upper seat
17. Bump stop
18. Bellows

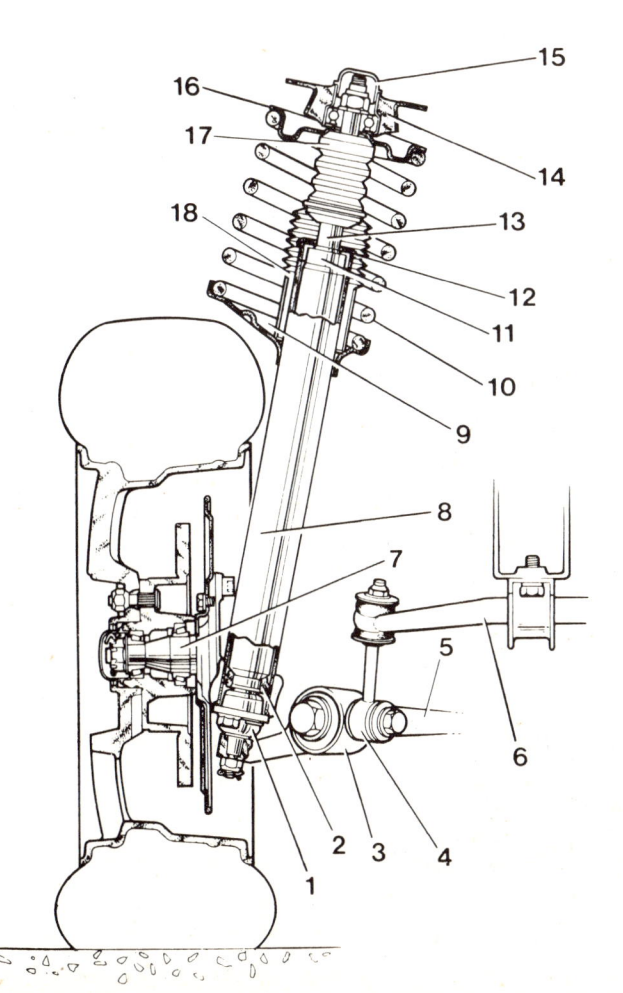

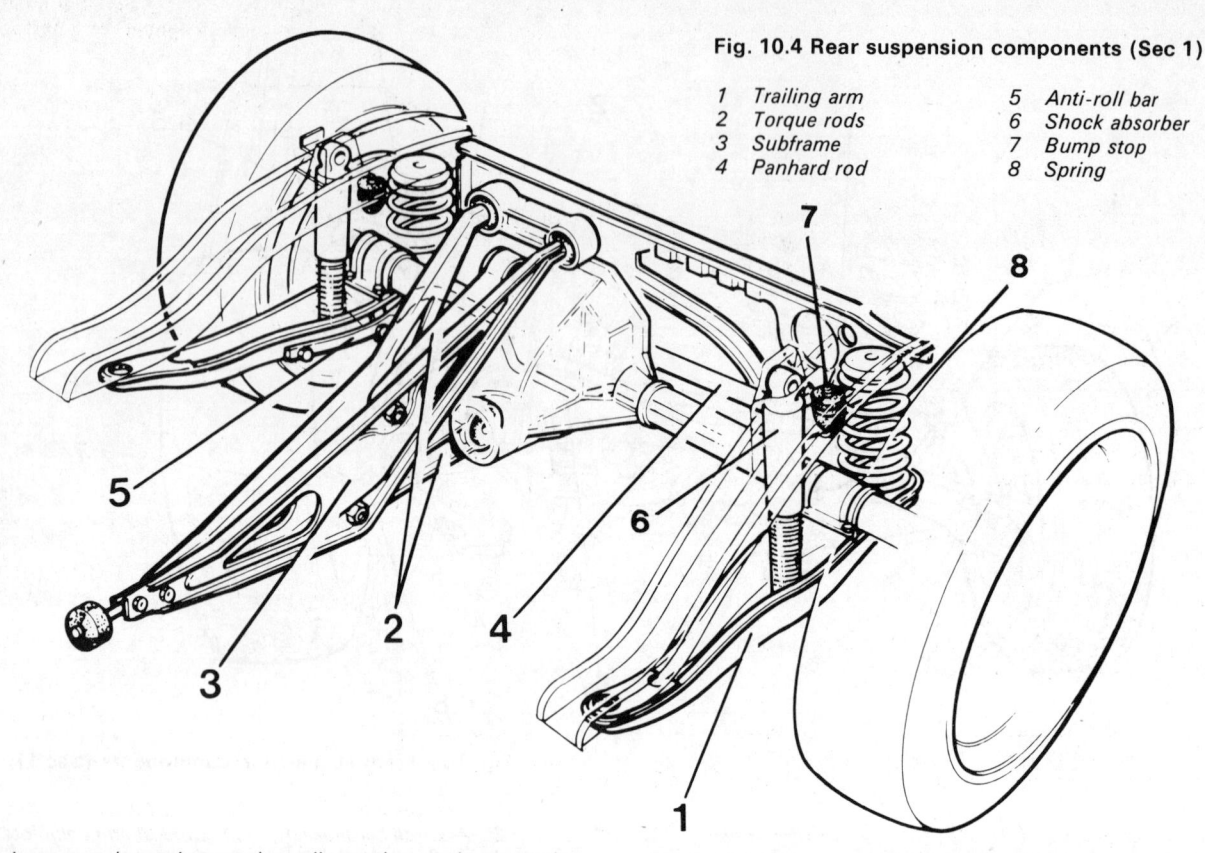

Chapter 10 Steering and suspension

Fig. 10.4 Rear suspension components (Sec 1)

1. Trailing arm
2. Torque rods
3. Subframe
4. Panhard rod
5. Anti-roll bar
6. Shock absorber
7. Bump stop
8. Spring

5 At the same intervals examine all steering and suspension components for wear and damage. Pay particular attention to rubber gaiters, bellows etc (photo). Repair or renew as necessary.
6 To inspect the steering, have an assistant move the steering wheel back and forth with the wheels on the ground. Look for play in the track rod end balljoints and in the rack itself. Radial play in a balljoint should not exceed 0.5 mm (0.020 in); play in the rack may be up to 2.0 mm (0.079 in).
7 Check for excessive play in the front control arm lower balljoints by levering them up and down, and by pushing and pulling the front wheel. Radial play (side-to-side) must not exceed 0.5 mm (0.020 in). Axial play (up-and-down) of up to 3 mm (0.118 in) is permitted. Do not confuse balljoint play with play in the strut top mounting or the control arm bush.

8 Check for wear in front and rear suspension bushes by levering between the component and its attachment.
9 Inspect the shock absorbers visually for leakage or damage. Check their function by bouncing the vehicle at each corner in turn: it should come to rest within one complete oscillation. Continued movement, or squeaking and groaning from the shock absorbers, suggest that renewal is required (Section 21 or 27).
10 Check the security of attachment of the steering gear. Inspect the pipes and unions for leaks. Also check the tension and condition of the pump drivebelt (Chapter 2, Section 7).
11 At the first 12 000 mile service only, check-tighten (slacken and retighten to the specified torque) the following fastenings:

Rear axle-to-trailing arm nuts
Control arm-to-crossmember bolts
Radius rod-to-control arm bolts
Radius rod-to-subframe bolts
Steering gear-to-crossmember bolts
Front crossmember-to-body bolts

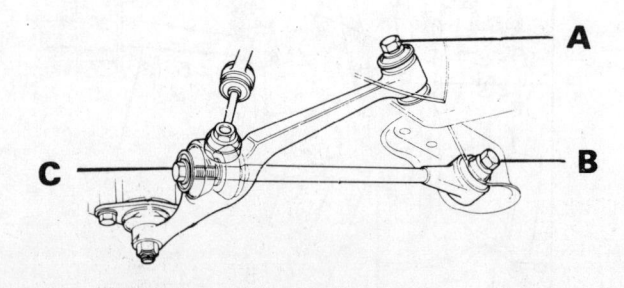

2.5 A split in a shock absorber bellows

Fig. 10.5 Some of the front suspension bolts which must be check-tightened (Sec 2)

A Control arm-to-crossmember
B Radius rod-to-subframe
C Radius rod-to-control arm

Chapter 10 Steering and suspension

12 Front wheel alignment checking is not specified as a routine operation, but it should be carried out whenever abnormal tyre wear is noticed, or after front wheel impact (eg hitting a kerb or pothole at speed).

3 Power steering fluid – level checking and bleeding

1 The steering fluid reservoir may be mounted on the pump, or remotely mounted on the radiator or inner wing. It may have a dipstick, or there may simply be level markings on a translucent container (photos).
2 Fluid level should not exceed the 'MAX' mark, nor drop below the 'LOW' or 'ADD' mark. Some dipsticks are calibrated both for hot and for cold fluid: use the correct markings.
3 If topping-up is necessary, use clean fluid of the specified type. Check for leaks if frequent topping-up is required. Do not run the pump without fluid in it – remove the drivebelt if necessary.
4 After component renewal, or if the fluid level has been allowed to fall so low that air has entered the hydraulic system, bleeding must be carried out as follows.
5 Fill the reservoir to the 'MAX' mark. Start the engine and allow it to idle.
6 Turn the steering wheel from lock to lock a couple of times. Do not hold it on full lock.
7 Top up the fluid if necessary.
8 Repeat paragraphs 6 and 7 until the fluid level ceases to fall. Stop the engine and refit the reservoir cap.

4 Front wheel bearings – checking and adjustment

Checking
1 Raise and support the front of the vehicle with the wheels free.
2 Hold the wheel at top and bottom and try to rock it. Spin the wheel and listen for rumbling or grinding noises. Play should be barely perceptible and noise should be absent.
3 If play or noise is evident, adjust the bearings as follows. If adjustment does not improve matters, remove the bearings for examination (Section 15).

Adjustment
4 Slacken the front wheel nuts. Raise and support the front of the vehicle and remove the front wheel.
5 Prise or tap off the hub nut grease cap. Obtain a new cap if the old one is damaged (photo).
6 Straighten the legs of the hub nut split pin. Remove the split pin; obtain a new one for reassembly.
7 Slacken the hub nut slightly, then tighten it to 57 Nm (42 lbf ft), at the same time rotating the brake disc.
8 Slacken the nut half a turn, then retighten it using the fingers only (nominal torque 1.5 Nm/1.1 lbf/ft).
9 Insert a new split pin to secure the hub nut. Tighten the nut if necessary to align the next split pin hole. Spread the legs of the split pin to secure it.

10 Half fill the grease cap with grease. Refit it and tap it home.
11 Refit the wheel, lower the vehicle and tighten the wheel nuts.

5 Front wheel alignment – checking and adjustment

1 Front wheel alignment is defined by camber, castor, steering axis inclination and toe setting. The first three factors are determined in production; only toe can be adjusted in service. Incorrect toe will cause rapid tyre wear.
2 Toe is defined as the amount by which the distance between the front wheels, measured at hub height, differs from the front edges to the rear edges. If the distance between the front edges is less than that at the rear, the wheels are said to toe-in; the opposite case is known as toe-out.
3 To measure toe, it will be necessary to obtain or make a tracking gauge. These are available in motor accessory shops, or one can be made from a length of rigid pipe or bar with some kind of threaded adjustment facility at one end. Many tyre specialists will also check toe free, or for a nominal sum.
4 Before measuring toe, check that all steering and suspension components are undamaged and that tyre pressures are correct. The vehicle must be at approximately kerb weight, with the spare wheel and jack in their normal position and any abnormal loads removed.
5 Park the vehicle on level ground and bounce it a few times to settle the suspension.
6 Use the tracking gauge to measure the distance between the inside faces of the front wheel rims, at hub height, at the rear of the front wheels. Record this distance; call it measurement 'A'.
7 Push the vehicle forwards or backwards so that the wheels rotate exactly 180° (half a turn). Measure the distance between the front wheel rims again, this time at the front of the wheels. Record this distance; call it measurement 'B'.
8 Subtract measurement 'B' from measurement 'A'. If the answer is positive it is the amount of toe-in; if negative it is the amount of toe-out. Permissible values are given in the Specifications.
9 If adjustment is necessary loosen the track rod end locknuts and the outer bellows clips, then rotate each track rod by equal amounts until the setting is correct. Hold the track rod ends in their horizontal position with a spanner while making the adjustment.

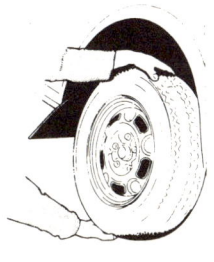

Fig. 10.6 Checking a front wheel bearing for wear (Sec 4)

3.1A Power steering fluid reservoir – remote type with level marking

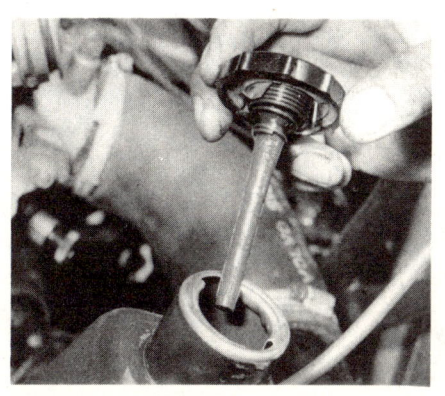

3.1B Removing the dipstick from a pump-mounted reservoir

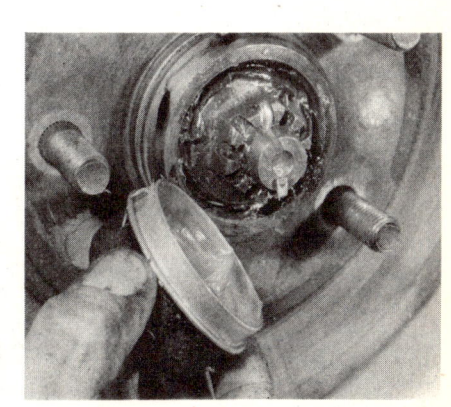

4.5 Removing the front hub grease cap

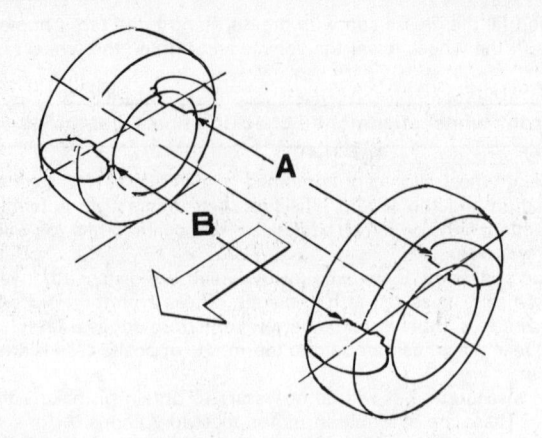

Fig. 10.7 Toe measurements. Arrow points forwards; for A and B see text (Sec 5)

10 Tighten the locknuts and outer bellows clips.
11 Provided the track rods have been adjusted by equal amounts the steering wheel should be central when moving straight-ahead. The amount of visible thread on each track rod should also be equal to within 2 mm (0.08 in). If wheel alignment and track rod length are both correct but steering wheel position is wrong, remove the steering wheel and reposition it (Section 12).

6 Track rod end – removal and refitting

1 Remove the front wheel on the side concerned.
2 Counterhold the track rod and slacken the rod end locknut by half a turn.
3 Unscrew the rod end ballpin nut to the end of its threads. Separate the ballpin from the steering arm with a proprietary separator, then remove the nut and disengage the ballpin from the arm (photo).
4 Unscrew the track rod end from the track rod, counting the number of turns needed to remove it. Record this number.
5 When refitting, screw the track rod end on by the same number of turns noted during removal.
6 Engage the ballpin in the steering arm. Fit the nut and tighten it to the specified torque.
7 Counterhold the track rod and tighten the locknut.
8 Refit the front wheel, lower the vehicle and tighten the wheel nuts.
9 Have the front wheel alignment checked at the first opportunity (Section 5), especially if new components have been fitted.

7 Steering rack bellows – renewal

1 Remove the track rod end on the side concerned (Section 6). Also remove the rod end locknut.
2 Release the two clips which secure the bellows. Peel off the bellows (photos).
3 Clean out any dirt and grit from the inner end of the track rod and (when accessible) the rack. Apply fresh grease to these components.
4 Fit and secure the new bellows, then refit the track rod end.

8 Steering gear – removal and refitting

1 Raise and support the front of the vehicle. Remove the engine undertray.
2 Remove the cover panel from the middle of the front crossmember.
3 Remove the spring clips and slacken the pinch-bolts and nuts on the lower universal joint (photo). Slide the universal joint up the intermediate shaft to free it from the pinion.
4 Disconnect the track rod ends from the steering arms. See Section 6.
5 Clean around the fluid supply and return unions, then disconnect them (photo). Be prepared for fluid spillage. Plug or cap open unions to keep dirt out.

6.3 Using a balljoint separator on the track rod end

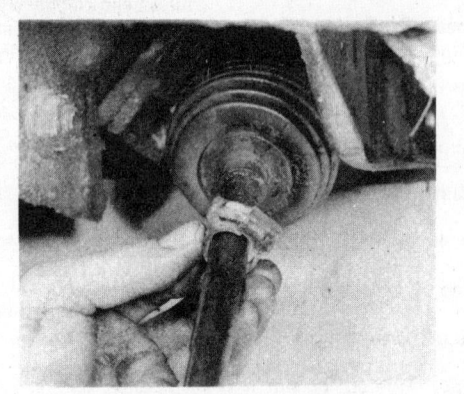

7.2A Removing a steering rack bellows clip ...

7.2B ... and the bellows

8.3 Intermediate shaft lower universal joint

8.5 Steering gear fluid supply and return unions (arrowed)

Chapter 10 Steering and suspension

6 Remove the two mounting bolts and nuts. Remove the steering gear from the crossmember. It may be necessary to displace the front anti-roll bar.
7 Refit by reversing the removal operations, noting the following points:

    (a) Tighten all fastenings to the specified torque
    (b) Use new copper washers on the fluid unions
    (c) Bleed the steering fluid (Section 3)
    (d) Check the front wheel alignment (Section 5)

9 Steering gear – overhaul

Overhaul of the steering gear is not recommended, even if parts are available. Obtain a new or reconditioned unit from a Volvo dealer or other specialist.

10 Steering pump – removal and refitting

1 Slacken the pump pivot and mounting strap nuts and bolts. Push the pump towards the engine and slip the drivebelt off the pulley (photos).
2 Disconnect the pump hydraulic pipes, either from below (remove the undertray) or from the back of the pump. Be prepared for fluid spillage.
3 Remove the pivot and strap nuts and bolts (photo).
4 Lift away the pump. On versions with a remote reservoir, either remove it with the pump or disconnect the hose from the pump.
5 If a new pump is to be fitted, transfer the pulley and mounting brackets to it.
6 Refit by reversing the removal operations, using new copper washers on disturbed banjo unions.
7 Tension the drivebelt (Chapter 2, Section 7).
8 Refill the pump reservoir and bleed the system (Section 3).

11 Steering pump – overhaul

As with the steering gear, overhaul of the pump is not recommended for the home mechanic.

12 Steering wheel – removal and refitting

Note: *If an airbag is fitted, DO NOT attempt to remove the steering wheel.*

1 Disconnect the battery negative lead.
2 Bring the steering wheel to the straight-ahead position.
3 Prise off the steering wheel centre pad (photo).
4 Undo the steering wheel centre bolt (photo).
5 If the steering column or steering gear may be disturbed whilst the wheel is removed, make alignment marks between the steering wheel and column.

Fig. 10.8 Steering gear mounting bolts (arrowed) (Sec 8)

Fig. 10.9 Steering pump hydraulic pipes, seen from below (V6 engine, undertray removed) (Sec 10)

    1 Supply   2 Return

6 Pull the steering wheel off its splines. If it is stuck, refit the bolt by a few turns only and thump the wheel from behind with the hands. Do not use excessive force as the column may be damaged.
7 Refit by reversing the removal operations, observing the alignment marks or the straight-ahead position of the wheel.

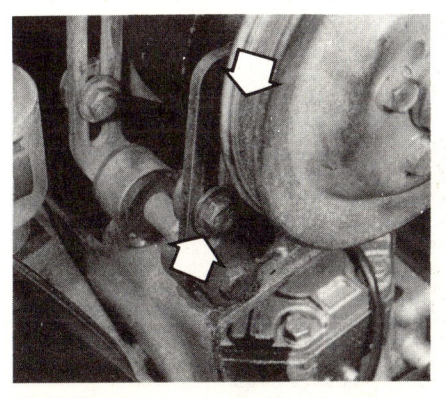

10.1A Steering pump mounting strap – in-line engine

10.1B Removing the steering pump drivebelt

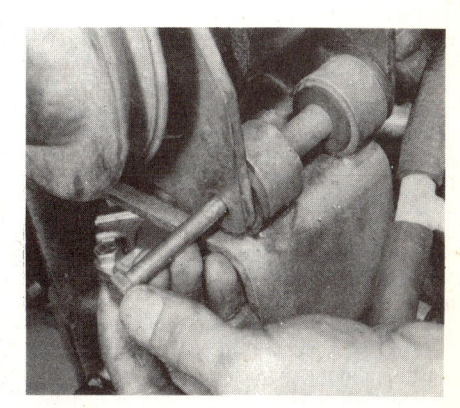

10.3 Removing the pump pivot bolt

236 Chapter 10 Steering and suspension

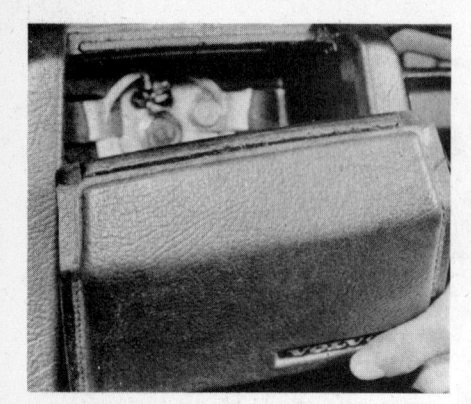

12.3 Removing the steering wheel centre pad

12.4 Undoing the steering wheel centre bolt

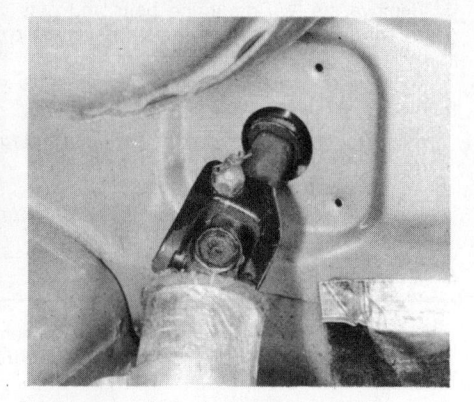

13.2 Intermediate shaft upper universal joint

13.3 Removing the horn contact ring

13.7 Steering column bottom bearing plate

13.8 The three bolts which secure the column top bearing to the crossmember

13 Steering column – removal and refitting

1 Disconnect the battery negative lead.
2 Still under the bonnet, remove the clamp nut and bolt fom the top universal joint of the intermediate steering shaft. The nut is secured by a spring clip (photo).
3 Remove the steering wheel (Section 12) and the steering column switches, complete with baseplate and horn contact ring (photo).
4 Remove the trim panel from below the steering column. It is secured by two screws and two clips. Disconnect the heater duct as the panel is withdrawn.
5 Remove the switch panel to the right of the steering lock.
6 Disconnect the multi-plug from the ignition/starter switch.
7 Remove the three screws which secure the column bottom bearing plate to the bulkhead (photo).
8 Remove the two bolts which secure the column top bearing to the support crossmember (photo). In some markets shear-head bolts will be found here: remove them by drilling and inserting a stud extractor, or by driving their heads round with a punch.
9 Remove the third bolt securing the top bearing. Recover the spacer tube (photo).
10 Remove the three bolts which secure the column support crossmember. To gain access to the right-hand bolts it will be necessary to remove the right-hand lower trim panel, disconnect the Motronic ECU (where applicable) and move the wiring harnesses aside.
11 Remove the steering lock and ignition/starter switch (Section 14).
12 Free the column and withdraw it into the vehicle. Recover the washer from the top bearing spigot (photo).
13 The column bearings may now be removed if necessary. Be careful not to collapse the coupling in the upper section. The overall length of the column must be 727.2 ± 1 mm (28.63 ± 0.04 in).

14 Refit by reversing the removal operations, noting the following points:

 (a) Tighten nuts and bolts to the specified torque (when known)
 (b) When shear-head bolts are used, only tighten them lightly at first. When satisfied that installation is correct, tighten the bolts until their heads break off

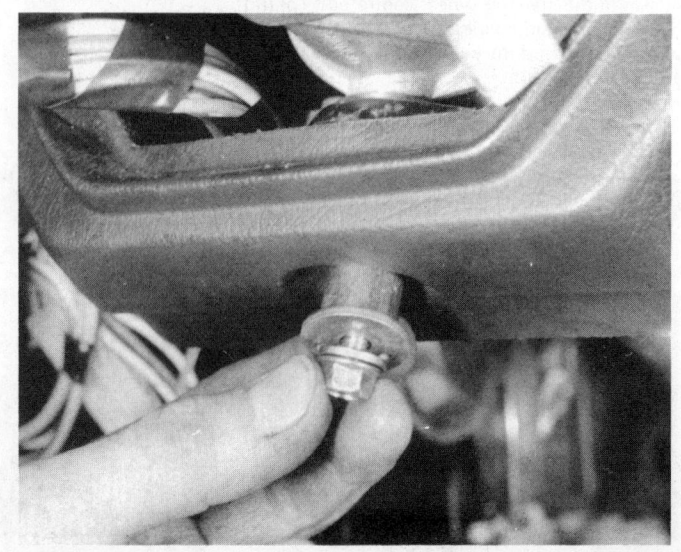

13.9 Removing the third bolt and spacer tube

Chapter 10 Steering and suspension 237

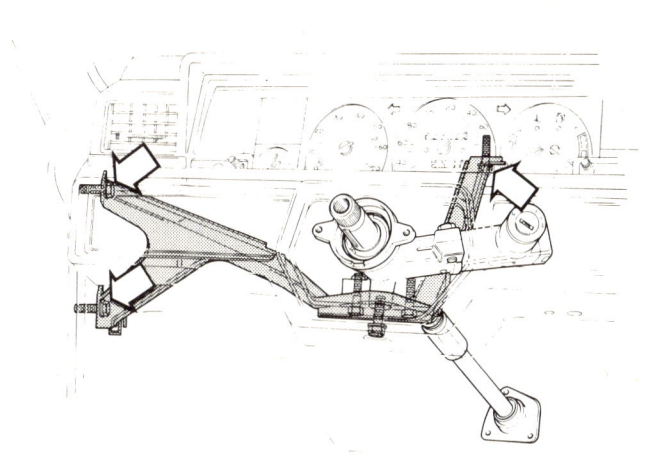

Fig. 10.10 Steering column crossmember support bolts (arrowed). LHD shown, RHD is mirror image (Sec 13)

13.12 Freeing the top bearing spigot

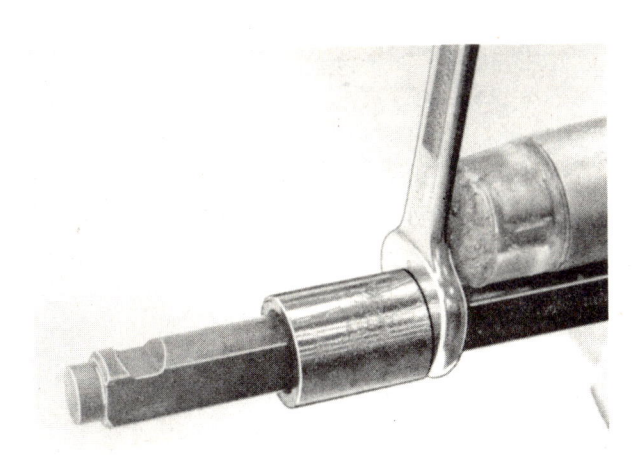

Fig. 10.11 Driving off the column lower bearing (Sec 13)

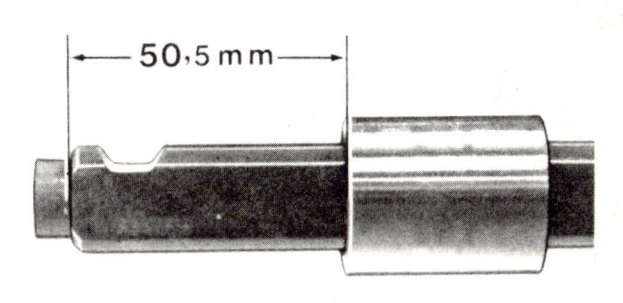

Fig. 10.12 Lower bearing fitted dimension (Sec 13)

Fig. 10.13 Removing the circlip to release the upper bearing components (Sec 13)

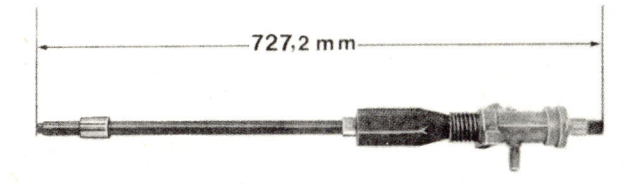

Fig. 10.14 Column overall length when undamaged (Sec 13)

14 Steering lock – removal and refitting

1 Proceed as if to remove the steering column (Section 13, paragraphs 1 and 3 to 10).
2 Remove the lock pinch-bolt from the top bearing housing (photo).
3 Insert the ignition key and turn it to position II. Depress the locking button and begin to withdraw the lock from the bearing housing (photo).
4 The ignition key and lock barrel will obstruct removal by fouling the surrounding trim. Therefore remove the key and free the top bearing housing from the crossmember; do not lose the washer from the spigot.

238 Chapter 10 Steering and suspension

14.2 Steering lock pinch-bolt (arrowed)

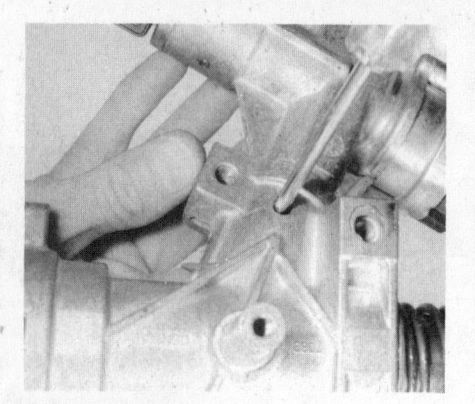

14.3 Depressing the locking button (column removed)

14.4 Withdrawing the steering lock – column fitted. Locking button (arrowed) has just emerged

By moving the bearing housing enough clearance can be gained to withdraw the steering lock complete with ignition/starter switch (photo).
5   Remove the switch from the lock by undoing the two screws.
6   Refit by reversing the removal operations.

15 Front wheel bearings – removal, inspection and refitting

1   Remove the front brake disc (Chapter 9, Section 15). Recover the bearing races and discard the oil seal.
2   Clean the bearing races and the tracks in the hub with paraffin. Inspect them for roughness, blueing or other signs of damage.
3   When renewing bearings, note that the outboard bearing may be made by SKF or by Koyo. Races and tracks of different manufacture must not be mixed.
4   To remove the bearing tracks, tap them out of the disc hub using a hammer and a brass or copper drift (photo).
5   Clean the bearing track seats in the hub.
6   Tap the tracks into position in the hub, being careful to keep them square (photos). Use a socket or tube, or the old tracks, to drive them in.
7   Pack the bearing races with grease, working it well into the rollers by hand. Also put a few fingerfuls of grease into the space between the bearing tracks (photo).
8   Fit the inboard race to the hub. Grease the lips of a new oil seal and fit it so that it is flush with the hub (photo).
9   Fit the hub/disc assembly to the stub axle, which should be well greased. Push the assembly home, then fit the outboard race and the castellated nut (photo).
10 Adjust the bearings (Section 4).
11 Refit the brake caliper and bracket (Chapter 9, Section 10).

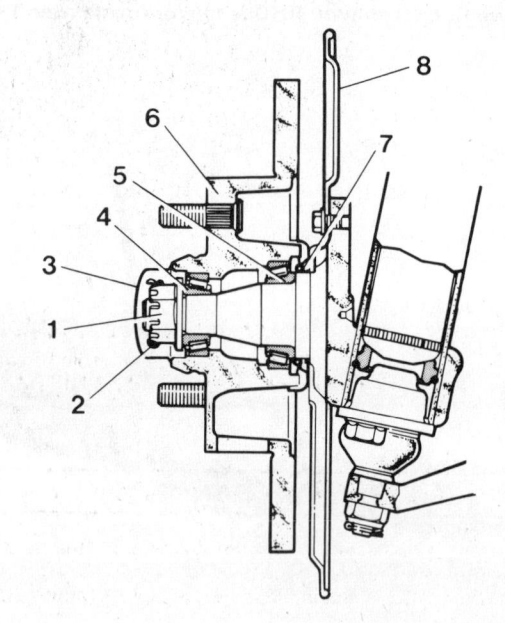

Fig. 10.15 Sectional view through the front hub (Sec 15)

1 Castellated nut	5 Inner bearing
2 Split pin	6 Brake disc
3 Grease cap	7 Oil seal
4 Outer bearing	8 Backplate

15.4 Driving out a bearing track

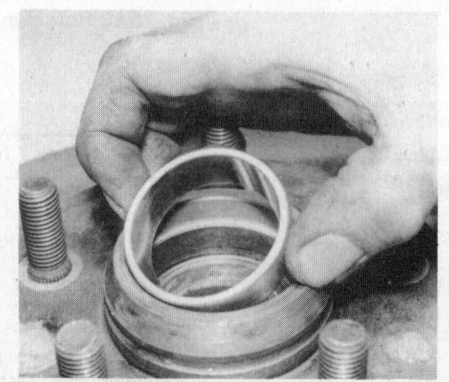

15.6A Fitting a bearing track ...

15.6B ... and driving it home

Chapter 10 Steering and suspension

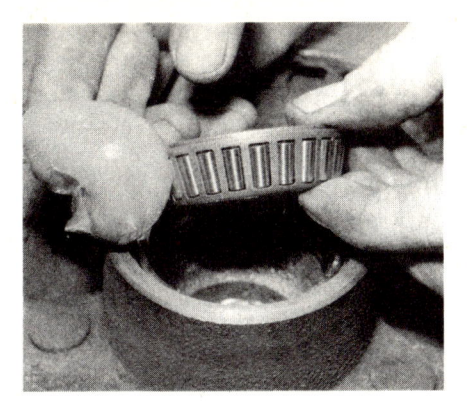

15.7 Greasing a bearing race

15.8 Fitting the oil seal

15.9 Fitting the castellated nut

16.2 Control arm balljoint – split pin partly withdrawn

16.5 Control arm-to-crossmember bolt (arrowed)

16 Front control arm – removal and refitting

1 Slacken the front wheel nuts, raise and support the front of the vehicle and remove the front wheel.
2 Remove the split pin and nut from the ballpin nut (photo). Obtain a new split pin for reassembly.
3 Unbolt the anti-roll bar link and the radius rod from the control arm. Obtain a new bolt for reassembly.
4 Separate the control arm from the balljoint, using a proprietary separator if necessary. Be careful not to damage the balljoint.
5 Remove the control arm-to-crossmember nut and bolt (photo). Remove the control arm from the crossmember. Obtain a new nut and bolt for reassembly.
6 Refit by reversing the removal operations, but do not fully tighten the control arm-to-crossmember nut and bolt until the weight of the vehicle is back on its wheels. Rock the vehicle to settle the suspension, then tighten the nut and bolt to the specified torque.

17 Front control arm balljoint – removal and refitting

1 Proceed as for control arm removal (Section 16, paragraphs 1 to 4) but without unbolting the radius rod.
2 Remove the two bolts which secure the balljoint to the strut (photo). Remove the balljoint.
3 When refitting, use new bolts to secure the balljoint and apply thread locking compound to them. Tighten the bolts in the specified stages, making sure that the balljoint is properly seated.
4 The remainder of refitting is a reversal of the removal procedure.

18 Front radius rod – removal and refitting

1 Slacken the front wheel nuts, raise and support the vehicle and remove the front wheel.
2 Unbolt the radius rod from the control arm and from the subframe. Remove the radius rod.

17.2 Two bolts (arrowed) secure the balljoint to the strut (strut removed)

Chapter 10 Steering and suspension

3 Refit by reversing the removal operations, using new nuts and bolts to secure the radius rod. Do not fully tighten the radius rod-to-subframe nut and bolt until the weight of the vehicle is back on its wheels and it has been rocked a few times.

19 Front anti-roll bar – removal and refitting

1 Raise the front of the vehicle on ramps, or drive it over a pit.
2 Unbolt the two saddle brackets which secure the anti-roll bar (photo).
3 Unbolt the anti-roll bar from its end links, or unbolt the end links from the control arms, as preferred (photo).
4 Refit by reversing the removal operations. Renew the mounting rubbers as necessary. (The saddle bracket rubbers are split and may be renewed without removing the anti-roll bar.)
5 Tighten the link upper nuts to achieve the dimension shown between the washers (Fig. 10.16).

20 Front suspension strut – removal and refitting

1 Slacken the front wheel nuts, raise and support the front of the vehicle and remove the front wheel.
2 Remove the brake caliper (Chapter 9, Section 10), but do not disconnect the hydraulic hoses. Tie the caliper up so that the hoses are not strained.
3 On models with ABS, disconnect or remove the wheel sensor.
4 If the strut is to be renewed, remove the brake disc (Chapter 9, Section 15) and backplate (photo).
5 Remove the split pin from the suspension bottom balljoint nut. Unscrew the nut to the end of the threads. Free the ballpin from the control arm using a proprietary balljoint separator, then remove the nut.
6 Similarly separate the track rod end balljoint from the steering arm.
7 Lever the control arm downwards and free it from the bottom balljoint. If there is not enough movement to allow this, unbolt the anti-roll bar link.
8 Remove the cover from the strut top mounting. Note which way round the mounting is fitted: it is not symmetrical.
9 If the strut is to be dismantled, slacken the piston rod nut, at the same time counterholding the piston rod (photo). **Do not** remove the nut, just slacken it a turn or two.
10 Have an assistant support the strut. Check that all attachments have been removed, then remove the two top mounting nuts. Remove the strut through the wheel arch.
11 Refit by reversing the removal operations, noting the following points:

(a) Observe the correct fitted direction of the top mounting (Fig. 10.17)
(b) Tighten all fastenings to their specified torques

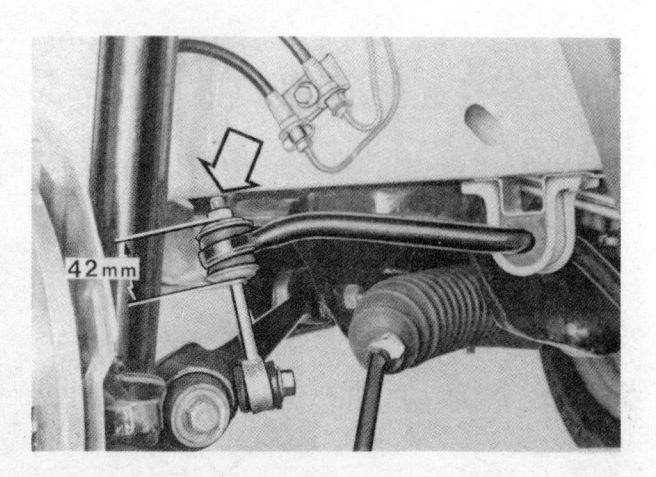

19.2 An anti-roll bar saddle bracket

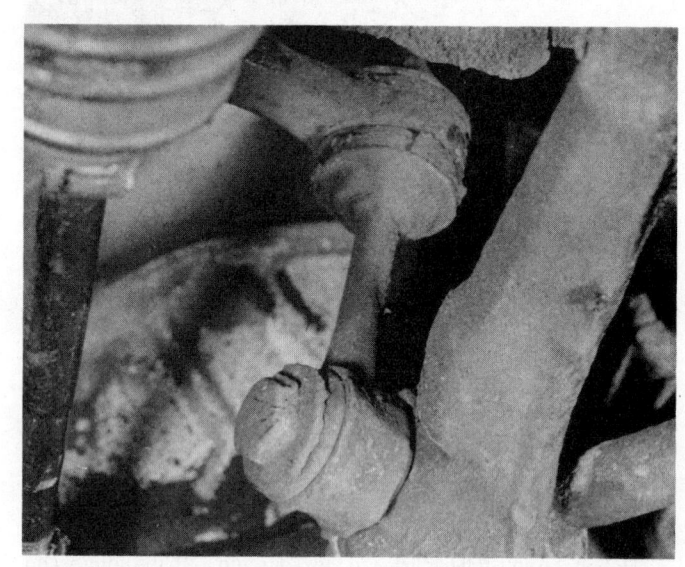

19.3 An anti-roll bar link

Fig. 10.16 Tighten the anti-roll bar link nut (arrowed) to achieve the dimension shown (Sec 19)

20.4 Unbolting the brake disc backplate

Chapter 10 Steering and suspension

20.9 Releasing the piston rod nut (strut removed)

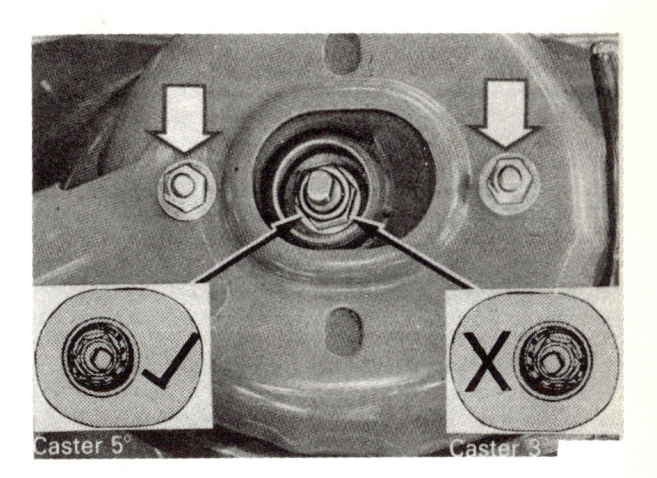

Fig. 10.17 Correct fitting of strut gives correct castor (inset, left). Top mounting nuts arrowed; right-hand strut shown (Sec 20)

21 Front suspension strut – dismantling and reassembly

Warning: *Spring compressors of sound design and construction must be used during this procedure. Uncontrolled release of the spring may result in damage and injury.*

1 Remove the strut from the vehicle (Section 20). Alternatively, the work can be carried out on the vehicle after the following preliminaries:

   (a) Remove the roadwheel and separate the track rod end balljoint
   (b) Unbolt the anti-roll bar link
   (c) Unbolt the brake pipe bracket from the inner wing
   (d) Slacken the piston rod nut, then release the top mounting. Press the control arm down and swing the strut outwards, tying or wiring it to limit movement and to avoid strain on the brake hose

2 Fit spring compressors to catch at least three coils of the spring. Tighten the compressors until the load is taken off the spring seats. Make sure that the compressors are secure.

3 Remove the piston rod nut (which should already have been slackened) and the strut top mounting. Note the position of any washers (photos).
4 Remove the spring upper seat, the spring itself, the washer, bump stop and bellows. (With gas-filled shock absorbers there is no bump stop.) Do not drop or jar the compressed spring (photo).
5 Recover the rubber ring (when fitted) from the spring lower seat (photo).
6 Using a C-spanner or similar tool, unscrew the shock absorber retaining nut (photo).
7 Pull the shock absorber out of its tube (photo).
8 Dismantling of the strut is now complete. Renew components as necessary, remembering that it is good practice to renew springs and shock absorbers in pairs.
9 If the spring is to be renewed, carefully remove the compressors from the old spring and fit them to the new one.
10 Reassemble by reversing the dismantling operations. Note the relationship of the disc to the bellows on models with gas-filled shock absorbers (Fig. 10.18). Do not fully tighten the piston rod nut until the top mounting has been secured to the vehicle.

21.3A Removing the piston rod nut ...

21.3B ... and the strut top mounting

Chapter 10 Steering and suspension

21.4 Removing the bump stop and bellows

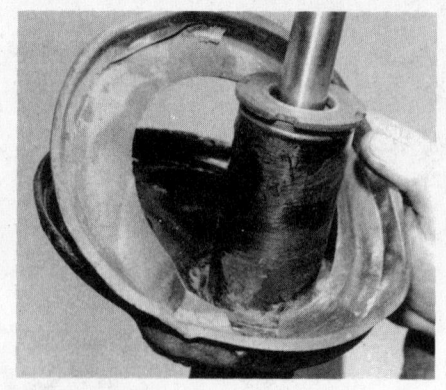

21.5 Removing the rubber ring from the spring lower seat

21.6 Unscrewing the shock absorber nut with a C-spanner

21.7 Removing the shock absorber

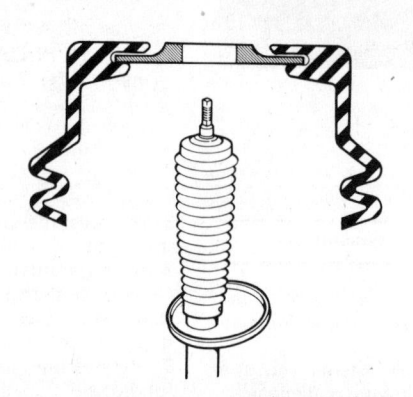

Fig. 10.18 Bellows and disc fitted with gas-filled shock absorbers (Sec 21)

23.2 Torque rod front mounting

22 Suspension rubber bushes – renewal

1 The principle of bush renewal is simple enough: the old bush is pressed out and the new one is pressed in. The reality is slightly more difficult.
2 Various special tools are specified by the makers for bush renewal. They are basically mandrels and tubes of different sizes which are used with a suitable press and sometimes with V-blocks. The amateur may experiment with a bench vice and socket spanners or pieces of tubing, using liquid soap or petroleum jelly as a lubricant. If this is unsuccessful it will be necessary to have the bush renewed by a workshop having press facilities.
3 Bush renewal *in situ* is not recommended.

23 Rear torque rods – removal and refitting

1 Raise and support the rear of the vehicle.
2 Remove the front mounting bolts from both torque rods, even if only one is to be removed (photo).
3 Unbolt and remove the torque rods. Recover the X-link.
4 When refitting, unbolt the subframe front mounting to allow

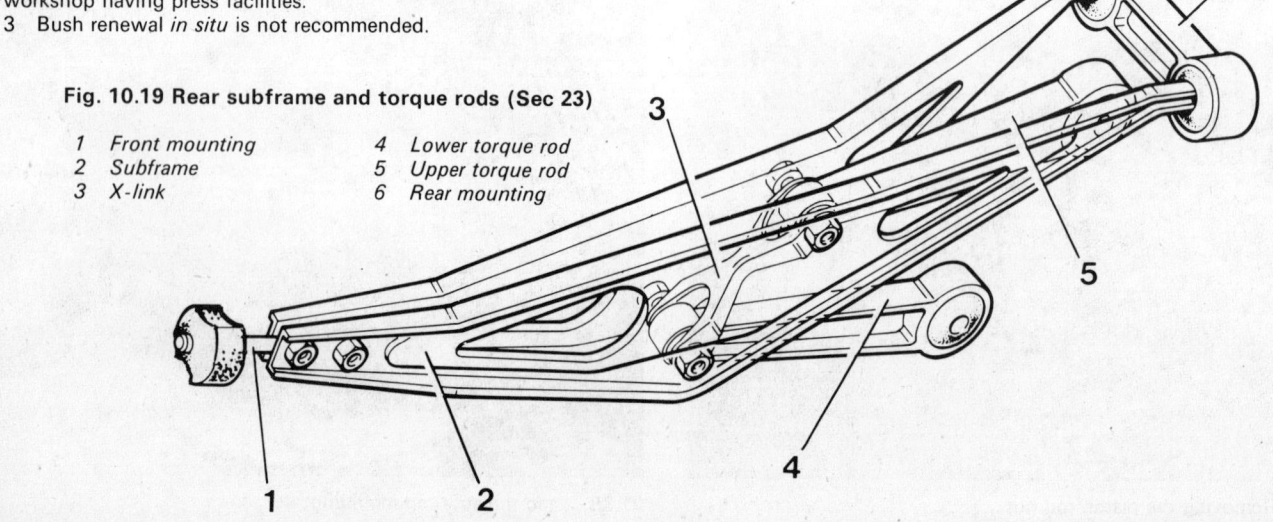

Fig. 10.19 Rear subframe and torque rods (Sec 23)

1 Front mounting
2 Subframe
3 X-link
4 Lower torque rod
5 Upper torque rod
6 Rear mounting

Chapter 10 Steering and suspension

movement of the subframe. Fit the torque rods to the rear axle first without tightening the mountings, then attach them and the X-link to the subframe.
5 Tighten the torque rod-to-subframe mountings to the specified torque.
6 Tighten the subframe front mounting to the specified torque.
7 Get the weight of the vehicle back on the rear wheels, then tighten the torque rod-to-axle mountings to the specified torque.

24 Rear subframe and mountings – removal and refitting

1 Raise and support the rear of the vehicle.
2 Remove the subframe front mounting nuts and bolts.
3 The front mounting rubber and mounting bracket may now be removed if wished, using a chisel and some lubricant around the rubber. Note the orientation of the rubber.
4 To remove the subframe completely, unbolt the torque rods and X-link from it. Also release the handbrake cable from the subframe bracket.
5 Refit one of the front mounting bolts. Hook a G-clamp behind the bolt and use the clamp to pull the subframe out of the rear mountings (Fig. 10.20). The rear mounting bracket can now be unbolted if required.
6
7 Refit by reversing the removal operations, noting the following points:

   (a) Use petroleum jelly as a lubricant for the mounting rubbers
   (b) Carry out the final tightening of the front mounting before that of the torque rods

Fig. 10.20 Using a G-clamp to pull out the rear subframe (Sec 24)

25 Rear trailing arm – removal and refitting

1 Proceed as if for rear spring removal (Section 26, paragraphs 1 to 3).
2 Disconnect the propeller shaft from the rear axle flange, making alignment marks for reference when refitting.
3 Support the trailing arm below the spring pan with a jack.
4 Unbolt the anti-roll bar (when fitted) from both trailing arms. If no anti-roll bar is fitted, remove the shock absorber lower mounting bolt on the side concerned. Slacken the lower mounting on the other side.
5 Lower the jack to release the spring tension.
6 Slacken the trailing arm-to-axle nuts crosswise. Remove the nuts, axle clamp and mounting rubbers. Recover the anti-roll bar bracket, if fitted.
7 Remove the trailing arm bracket nuts and bolts. Prise the front mounting out of the body and remove the trailing arm.
8 Refit by reversing the removal operations, tightening the various fastenings to their specified torques.

26 Rear spring – removal and refitting

1 Slacken the rear wheel nuts on the side concerned. Raise and support the rear of the vehicle so that both rear wheels hang free. Remove the rear wheel.
2 Remove the two bolts which secure the rear brake caliper. Slide the caliper off the disc and tie it up so that the flexible hose is not strained. Obtain new bolts for reassembly.
3 If the exhaust system will be in the way, unhook it from its mountings and lower it or move it aside.
4 Jack up the trailing arm slightly to take the load off the shock absorber. Remove the shock absorber lower mounting nut and bolt. Lower the jack.
5 Remove the nut which secures the spring upper seat (photo).
6 Pull the trailing arm downwards as far as possible. Pull the top of the spring downwards until the upper seat is clear of the mounting stud, then remove the spring and seat rearwards. If difficulty is experienced, either use spring compressors to unload the spring, or disconnect the rear anti-roll bar to allow the trailing arm more downward movement (photo).
7 Inspect the spring seat rubbers and renew them if necessary (photos).
8 Refit by reversing the removal operations, tightening the fastenings to their specified torques. Use new bolts to secure the brake caliper.
9 Refit the roadwheel, lower the vehicle and tighten the wheel nuts.

27 Rear shock absorber – removal and refitting

1 Slacken the rear wheel nuts on the side concerned. Raise and support the rear of the vehicle and remove the rear wheel.
2 Jack up the trailing arm slightly to take the load off the shock absorber. Remove the shock absorber lower mounting nut and bolt (photo). Lower the jack.

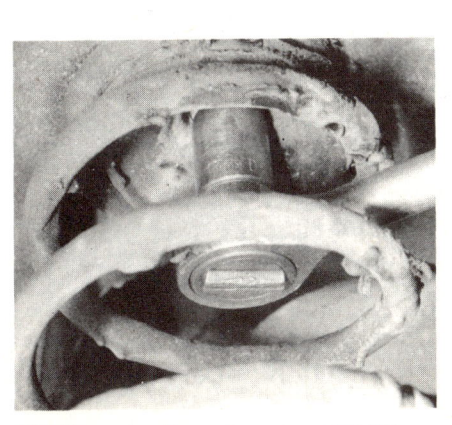

26.5 Removing the spring upper seat nut

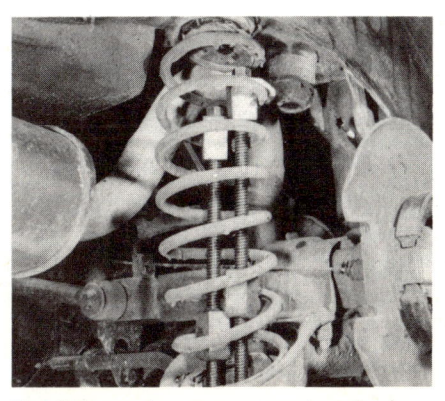

26.6 Using spring compressors to unload the rear spring

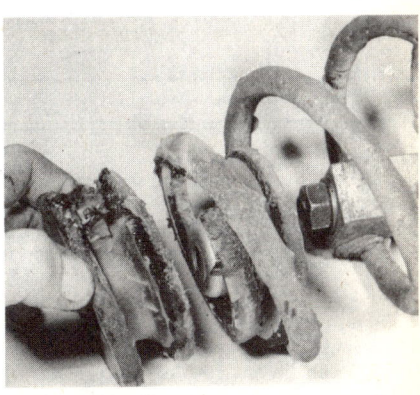

26.7A Removing the spring upper seat rubber from the spring ...

244 Chapter 10 Steering and suspension

26.7B ... and the lower seat rubber from the trailing arm

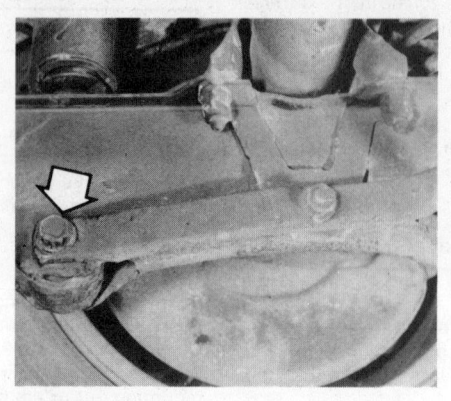

27.2 Rear shock absorber lower mounting bolt (arrowed)

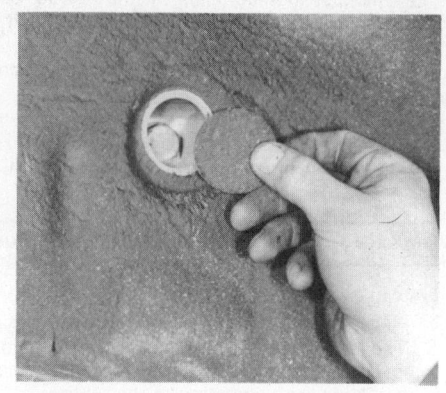

27.3 Exposing the shock absorber upper mouning bolt

3 Remove the rubber bung in the wheel arch which covers the shock absorber upper mounting bolt (photo). Remove the bolt.
4 Pull the shock absorber downwards and remove it.
5 Refit by reversing the removal operations. Tighten the shock absorber mounting and the wheel nuts to the specified torque.

28 Rear anti-roll bar – removal and refitting

1 Raise the rear of the vehicle on ramps or drive it over a pit.
2 Remove the two nuts and bolts on each side which secure the anti-roll bar. The forward bolts also secure the rear shock absorber lower mountings: it may be necessary to jack up under the trailing arms to take the load off these bolts.
3 Remove the anti-roll bar.
4 Refit by reversing the removal operations.

29 Panhard rod – removal and refitting

1 Raise and support the vehicle with the rear wheels free.
2 Unbolt the Panhard rod from the body, then from the rear axle (photo). Remove the rod.
3 If the rod bushes need renewing, have the old ones pressed out and new ones pressed in by a Volvo dealer or other specialist.
4 Refit the rod and tighten the bolts to the specified torque, axle end first. Lower the vehicle.

30 Self-levelling rear suspension – general

Self-levelling rear suspension is fitted to some top range Estate models. No information on this system was available at the time of writing.

31 Wheel studs – renewal

Front

1 Remove the front brake disc (Chapter 9, Section 15).
2 Support the hub area of the disc and press or drive out the old stud. Invert the disc and press the new stud into place.
3 Refit the disc and adjust the hub bearings (Section 4).

Rear

4 Remove the handbrake shoes (Chapter 9, Section 26).
5 If a pusher tool similar to that shown in Fig. 10.21 is available, stud renewal can be carried out *in situ*. Otherwise, remove the halfshaft (Chapter 8, Section 5) and proceed as for the front studs.
6 Refit the halfshafts and/or handbrake shoes and other disturbed components.

32 Wheels and tyres – general care and maintenance

Wheels and tyres should give no real problems in use provided that a close eye is kept on them with regard to excessive wear or damage. To this end, the following points should be noted.

29.2 Panhard rod attachment to the rear axle

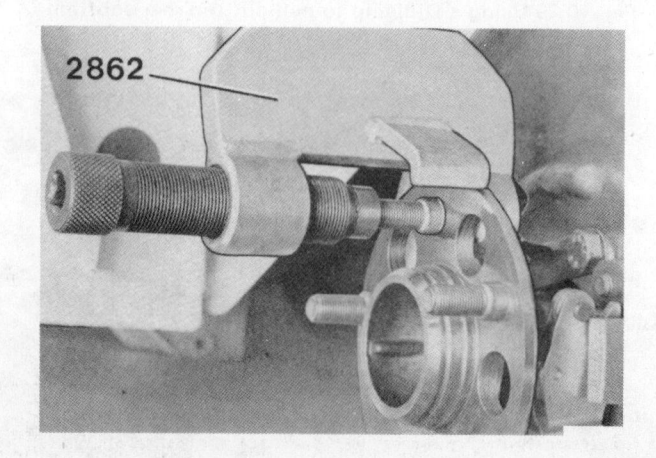

Fig. 10.21 Tool 2862 used to remove and refit wheel studs *in situ* (Sec 31)

Chapter 10 Steering and suspension

Ensure that tyre pressures are checked regularly and maintained correctly. Checking should be carried out with the tyres cold and not immediately after the vehicle has been in use. If the pressures are checked with the tyres hot, an apparently high reading will be obtained owing to heat expansion. Under no circumstances should an attempt be made to reduce the pressures to the quoted cold reading in this instance, or effective underinflation will result.

Underinflation will cause overheating of the tyre owing to excessive flexing of the casing, and the tread will not sit correctly on the road surface. This will cause a consequent loss of adhesion and excessive wear, not to mention the danger of sudden tyre failure due to heat build-up.

Overinflation will cause rapid wear of the centre part of the tyre tread coupled with reduced adhesion, harsher ride, and the danger of shock damage occurring in the tyre casing.

Regularly check the tyres for damage in the form of cuts or bulges, especially in the sidewalls. Remove any nails or stones embedded in the tread before they penetrate the tyre to cause deflation. If removal of a nail *does* reveal that the tyre has been punctured, refit the nail so that its point of penetration is marked. Then immediately change the wheel and have the tyre repaired by a tyre dealer. Do *not* drive on a tyre in such a condition. In many cases a puncture can be simply repaired by the use of an inner tube of the correct size and type. If in any doubt as to the possible consequences of any damage found, consult your local tyre dealer for advice.

Periodically remove the wheels and clean any dirt or mud from the inside and outside surfaces. Examine the wheel rims for signs of rusting, corrosion or other damage. Light alloy wheels are easily damaged by 'kerbing' whilst parking, and similarly steel wheels may become dented or buckled. Renewal of the wheel is very often the only course of remedial action possible.

The balance of each wheel and tyre assembly should be maintained to avoid excessive wear, not only to the tyres but also to the steering and suspension components. Wheel imbalance is normally signified by vibration through the vehicle's bodyshell, although in many cases it is particularly noticeable through the steering wheel. Conversely, it should be noted that wear or damage in suspension or steering components may cause excessive tyre wear. Out-of-round or out-of-true tyres, damaged wheels and wheel bearing wear/maladjustment also fall into this category. Balancing will not usually cure vibration caused by such wear.

Wheel balancing may be carried out with the wheel either on or off the vehicle. If balanced on the vehicle, ensure that the wheel-to-hub relationship is marked in some way prior to subsequent wheel removal so that it may be refitted in its original position.

General tyre wear is influenced to a large degree by driving style – harsh braking and acceleration or fast cornering will all produce more rapid tyre wear. Interchanging of tyres may result in more even wear, but this should only be carried out where there is no mix of tyre types on the vehicle. However, it is worth bearing in mind that if this is completely effective, the added expense of replacing a complete set of tyres simultaneously is incurred, which may prove financially restrictive for many owners.

Front tyres may wear unevenly as a result of wheel misalignment. The front wheels should always be correctly aligned according to the settings specified by the vehicle manufacturer.

Legal restrictions apply to the mixing of tyre types on a vehicle. Basically this means that a vehicle must not have tyres of differing construction on the same axle. Although it is not recommended to mix tyre types between front axle and rear axle, the only legally permissible combination is crossply at the front and radial at the rear. When mixing radial ply tyres, textile braced radials must always go on the front axle, with steel braced radials at the rear. An obvious disadvantage of such mixing is the necessity to carry two spare tyres to avoid contravening the law in the event of a puncture.

In the UK, the Motor Vehicles Construction and Use Regulations apply to many aspects of tyre fitting and usage. It is suggested that a copy of these regulations is obtained from your local police if in doubt as to the current legal requirements with regard to tyre condition, minimum tread depth, etc.

Some models are equipped with a 'space saver' spare wheel and tyre, of narrower section to the regular wheels and tyres. Observe the correct inflation pressure for such spares – see Specifications. The makers recommend that speed be restricted to 60 mph (100 km/h) when the space saver spare is in use, since the handling of the vehicle will be affected. Use of the space saver may be illegal in some countries.

Fig. 10.22 Torque tread wear indicator bar (arrowed) appears when wear is approaching legal limit (Sec 32)

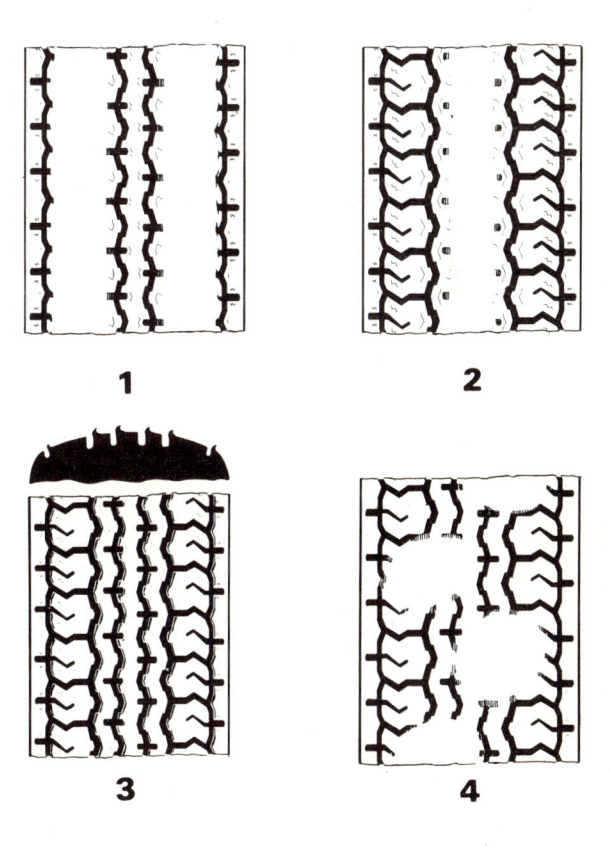

Fig. 10.23 Examples of abnormal tyre wear (Sec 32)

1 Underinflation
2 Overinflation
3 Incorrect toe
4 Imbalance

Chapter 10 Steering and suspension

33 Fault diagnosis – steering and suspension

Symptom	Reason(s)
Excessive play at steering wheel	Worn track rod end balljoints Worn control arm balljoints Worn intermediate shaft coupling Worn steering gear
Vehicle wanders or pulls to one side	Uneven tyre pressures Incorrect wheel alignment Worn track rod end balljoints Worn control arm balljoints Faulty shock absorber Accident damage
Steering heavy or stiff	Low tyre pressures Seized balljoint Seized strut top bearing Incorrect wheel alignment Steering gear damaged or lacking lubricant Power steering fault (see below)
Lack of power assistance	Fluid level low Pump drivebelt slack or broken Pump or steering gear defective
Wheel wobble and vibration	Wheel nuts loose Wheels out of balance or damaged Wheel bearings worn Worn track rod end balljoints Worn control arm balljoints Faulty shock absorber
Excessive tyre wear	Incorrect tyre pressures Wheels out of balance Incorrect wheel alignment Faulty shock absorbers Unsympathetic driving style

Chapter 11 Bodywork and fittings

For modifications, and information applicable to later models, see Supplement at end of manual

Contents

Bonnet – removal and refitting	6
Bonnet release cable – removal and refitting	10
Boot lid – removal and refitting	8
Boot lock – removal and refitting	19
Bumper buffers – removal and refitting	41
Central locking components – removal and refitting	21
Centre console – removal and refitting	35
Door handles, locks and latches – removal and refitting	18
Door mirror – removal and refitting	22
Door mirror glass and motor – removal and refitting	23
Doors – removal and refitting	7
Door surround weatherstrip – removal and refitting	24
Engine undertray – removal and refitting	43
Facia – removal and refitting	37
Front bumper – removal and refitting	39
Front door interior trim – removal and refitting	11
Front door window – removal and refitting	15
Front grille panel – removal and refitting	42
Front seat – removal and refitting	25
Front seat belts – removal and refitting	31
Front seat position adjusters – removal and refitting	28
Front spoiler – removal and refitting	38
Front wing – removal and refitting	44
General description	1
Glovebox – removal and refitting	34
Head restraints – removal and refitting	27
Maintenance – bodywork and underframe	2
Maintenance – upholstery and carpets	3
Major body damage – repair	5
Minor body damage – repair	4
Rear bumper – removal and refitting	40
Rear console – removal and refitting	36
Rear door interior trim – removal and refitting	12
Rear door windows – removal and refitting	16
Rear seat – removal and refitting	26
Rear seat bolts – removal and refitting	32
Seat belts – care and maintenance	30
Seat heating elements – removal and refitting	29
Steering column/pedal trim panel – removal and refitting	33
Sunroof – removal and refitting	46
Sunroof frame and cables – removal and refitting	47
Sunroof motor – removal and refitting	45
Tailgate – removal and refitting	9
Tailgate interior trim – removal and refitting	13
Tailgate lock – removal and refitting	20
Window lift mechanism – removal and refitting	17
Windscreen and other fixed glass – removal and refitting	14

1 General description

Body styles available are 4-door Saloon and 5-door Estate. The body and floorpan are of welded steel construction and form a very strong unit, with crumple zones at front and rear which will deform progressively in case of accident. The doors are also reinforced against side impacts. The tailgate on Estate models is made of aluminium.

Stout bumpers are fitted front and rear, with energy-absorbing buffers to protect against damage in low-speed collisions.

The front wings bolt on for easy renewal. The bonnet has two opening positions: partly open for normal work, and fully open for major work.

Interior trim and fittings are of the high standard expected in a vehicle of this class.

2 Maintenance – bodywork and underframe

The general condition of a vehicle's bodywork is the one thing that significantly affects its value. Maintenance is easy but needs to be regular. Neglect, particularly after minor damage, can lead quickly to further deterioration and costly repair bills. It is important also to keep watch on those parts of the vehicle not immediately visible, for instance the underside, inside all the wheel arches and the lower part of the engine compartment.

The basic maintenance routine for the bodywork is washing – preferably with a lot of water, from a hose. This will remove all the loose solids which may have stuck to the vehicle. It is important to flush these off in such a way as to prevent grit from scratching the finish. The wheel arches and underframe need washing in the same way to remove any accumulated mud which will retain moisture and tend to encourage rust. Paradoxically enough, the best time to clean the underframe and wheel arches is in wet weather when the mud is thoroughly wet and soft. In very wet weather the underframe is usually cleaned of large accumulations automatically and this is a good time for inspection.

Periodically, except on vehicles with a wax-based underbody protective coating, it is a good idea to have the whole of the underframe of the vehicle steam cleaned, engine compartment included, so that a thorough inspection can be carried out to see what minor repairs and renovations are necessary. Steam cleaning is available at many garages and is necessary for removal of the accumulation of oily grime which sometimes is allowed to become thick in certain areas. If steam cleaning facilities are not available, there are one or two excellent grease solvents available, such as Holts Engine Degreasant, which can be brush applied. The dirt can then be simply hosed off. Note that these methods should not be used on vehicles with wax-based underbody protective coating or the coating will be removed. Such vehicles should be inspected annually, preferably just prior to winter, when the underbody should be washed down and any damage to the wax coating repaired using Holts Undershield. Ideally, a completely fresh coat should be applied. It would also be worth considering the use of such wax-based protection for injection into door panels, sills, box sections, etc, as an additional safeguard against rust damage where such protection is not provided by the vehicle manufacturer.

After washing paintwork, wipe off with a chamois leather to give an unspotted clear finish. A coat of clear protective wax polish, like the many excellent Turtle Wax polishes, will give added protection against chemical pollutants in the air. If the paintwork sheen has dulled or oxidised, use a cleaner/polisher combination such as Turtle Wax Hard Shell to restore the brilliance of the shine. This requires a little effort, but such dulling is usually caused because regular washing has been neglected. Care needs to be taken with metallic paintwork, as special non-abrasive cleaner/polisher is required to avoid damage to the finish.

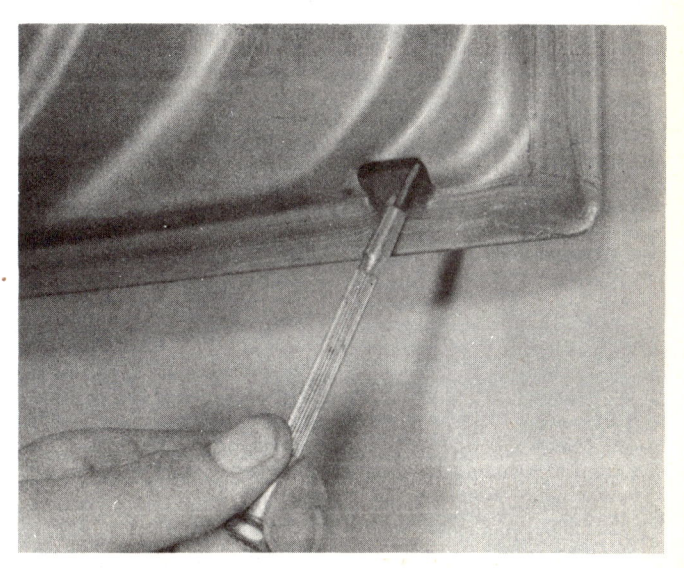

2.0 Clearing a door drain hole

Fig. 11.1 Two opening positions of the bonnet (Sec 1)

Always check that the door and ventilator opening drain holes and pipes are completely clear so that water can be drained out. Bright work should be treated in the same way as paint work. Windscreens and windows can be kept clear of the smeary film which often appears by the use of a proprietary glass cleaner like Holts Mixra. Never use any form of wax or other body or chromium polish on glass.

3 Maintenance – upholstery and carpets

Mats and carpets should be brushed or vacuum cleaned regularly to keep them free of grit. If they are badly stained remove them from the vehicle for scrubbing or sponging and make quite sure they are dry before refitting. Seats and interior trim panels can be kept clean by wiping with a damp cloth and Turtle Wax Carisma. If they do become stained (which can be more apparent on light coloured upholstery) use a little liquid detergent and a soft nail brush to scour the grime out of the grain of the material. Do not forget to keep the headlining clean in the same way as the upholstery. When using liquid cleaners inside the vehicle do not over-wet the surfaces being cleaned. Excessive damp could get into the seams and padded interior causing stains, offensive odours or even rot. If the inside of the vehicle gets wet accidentally it is worthwhile taking some trouble to dry it out properly, particularly where carpets are involved. *Do not leave oil or electric heaters inside the vehicle for this purpose.*

4 Minor body damage – repair

The colour bodywork repair photographic sequences between pages 32 and 33 illustrate the operations detailed in the following sub-sections.
Note: *For more detailed information about bodywork repair, Haynes Publishing produce a book by Lindsay Porter called The Car Bodywork Repair Manual. This incorporates information on such aspects as rust treatment, painting and glass fibre repairs, as well as details on more ambitious repairs involving welding and panel beating.*

Repair of minor scratches in bodywork

If the scratch is very superficial, and does not penetrate to the metal of the bodywork, repair is very simple. Lightly rub the area of the scratch with a paintwork renovator like Turtle Wax Color Back, or a very fine cutting paste like Holts Body + Plus Rubbing Compound to remove loose paint from the scratch and to clear the surrounding bodywork of wax polish. Rinse the area with clean water.

Apply touch-up paint to the scratch using a fine paint brush; continue to apply fine layers of paint until the surface of the paint in the scratch is level with the surrounding paintwork. Allow the new paint at least two weeks to harden: then blend it into the surrounding paintwork by rubbing the scratch area with a paintwork renovator or a very fine cutting paste, such as Holts Body + Plus Rubbing Compound or Turtle Wax Color Back. Finally, apply wax polish from one of the Turtle Wax range of wax polishes.

Where the scratch has penetrated right through to the metal of the bodywork, causing the metal to rust, a different repair technique is required. Remove any loose rust from the bottom of the scratch with a penknife, then apply rust inhibiting paint, such as Turtle Wax Rust Master, to prevent the formation of rust in the future. Using a rubber or nylon applicator fill the scratch with bodystopper paste like Holts Body + Plus Knifing Putty. If required, this paste can be mixed with cellulose thinners, such as Holts Body + Plus Cellulose Thinners, to provide a very thin paste which is ideal for filling narrow scratches. Before the stopper-paste in the scratch hardens, wrap a piece of smooth cotton rag around the top of a finger. Dip the finger in cellulose thinners, such as Holts Body + Plus Cellulose Thinners, and then quickly sweep it across the surface of the stopper-paste in the scratch; this will ensure that the surface of the stopper-paste is slightly hollowed. The scratch can now be painted over as described earlier in this Section.

Repair of dents in bodywork

When deep denting of the vehicle's bodywork has taken place, the first task is to pull the dent out, until the affected bodywork almost attains its original shape. There is little point in trying to restore the original shape completely, as the metal in the damaged area will have stretched on impact and cannot be reshaped fully to its original contour. It is better to bring the level of the dent up to a point which is about $\frac{1}{8}$ in (3 mm) below the level of the surrounding bodywork. In cases where the dent is very shallow anyway, it is not worth trying to pull it out at all. If the underside of the dent is accessible, it can be hammered out gently from behind, using a mallet with a wooden or plastic head. Whilst doing this, hold a suitable block of wood firmly against the outside of the panel to absorb the impact from the hammer blows and thus prevent a large area of the bodywork from being 'belled-out'.

Should the dent be in a section of the bodywork which has a double skin or some other factor making it inaccessible from behind, a different technique is called for. Drill several small holes through the metal inside the area – particulary in the deeper section. Then screw long self-tapping screws into the holes just sufficiently for them to gain a good purchase in the metal. Now the dent can be pulled out by pulling on the protruding heads of the screws with a pair of pliers.

The next stage of the repair is the removal of the paint from the damaged area, and from an inch or so of the surrounding 'sound' bodywork. This is accomplished most easily by using a wire brush or abrasive pad on a power drill, although it can be done just as effectively by hand using sheets of abrasive paper. To complete the preparation for filling, score the surface of the bare metal with a screwdriver or the tang of a file, or alternatively, drill small holes in the affected area. This will provide a really good 'key' for the filler paste.

To complete the repair see the Section on filling and re-spraying.

Repair of rust holes or gashes in bodywork

Remove all paint from the affected area and from an inch or so of the surrounding 'sound' bodywork, using an abrasive pad or a wire brush on a power drill. If these are not available a few sheets of abrasive paper will do the job just as effectively. With the paint removed you will be able to gauge the severity of the corrosion and therefore decide whether to renew the whole panel (if this is possible) or to repair the affected area. New body panels are not as expensive as most people think and it is often quicker and more satisfactory to fit a new panel than to attempt to repair large areas of corrosion.

Remove all fittings from the affected area except those which will act as a guide to the original shape of the damaged bodywork (eg headlamp shells etc). Then, using tin snips or a hacksaw blade, remove all loose metal and any other metal badly affected by corrosion. Hammer the edges of the hole inwards in order to create a slight depression for the filler paste.

Wire brush the affected area to remove the powdery rust from the surface of the remaining metal. Paint the affected area with rust inhibiting paint like Turtle Rust Master; if the back of the rusted area is accessible treat this also.

Before filling can take place it will be necessary to block the hole in some way. This can be achieved by the use of aluminium or plastic mesh, or aluminium tape.

Aluminium or plastic mesh or glass fibre matting is probably the best material to use for a large hole. Cut a piece to the approximate size and shape of the hole to be filled, then position it in the hole so that its edges are below the level of the surrounding bodywork. It can be retained in position by several blobs of filler paste around its periphery.

Aluminium tape should be used for small or very narrow holes. Pull a piece off the roll and trim it to the approximate size and shape required, then pull off the backing paper (if used) and stick the tape over the hole; it can be overlapped if the thickness of one piece is insufficient. Burnish down the edges of the tape with the handle of a screwdriver or similar, to ensure that the tape is securely attached to the metal underneath.

Bodywork repairs – filling and re-spraying

Before using this Section, see the Sections on dent, deep scratch, rust holes and gash repairs.

Many types of bodyfiller are available, but generally speaking those proprietary kits which contain a tin of filler paste and a tube of resin hardener are best for this type of repair, like Holts Body + Plus or Holts No Mix which can be used directly from the tube. A wide, flexible plastic or nylon applicator will be found invaluable for imparting a smooth and well contoured finish to the surface of the filler.

Mix up a little filler on a clean piece of card or board – measure the hardener carefully (follow the maker's instructions on the pack) otherwise the filler will set too rapidly or too slowly. Alternatively, Holts No Mix can be used straight from the tube without mixing, but daylight is required to cure it. Using the applicator apply the filler paste to the prepared area; draw the applicator across the surface of the filler to achieve the correct contour and to level the filler surface. As soon as a contour that approximates to the correct one is achieved, stop working the paste – if you carry on too long the paste will become sticky and begin to 'pick up' on the applicator. Continue to add thin layers of filler paste at twenty-minute intervals until the level of the filler is just proud of the surrounding bodywork.

Once the filler has hardened, excess can be removed using a metal plane or file. From then on, progressively finer grades of abrasive paper should be used, starting with a 40 grade production paper and finishing with 400 grade wet-and-dry paper. Always wrap the abrasive paper around a flat rubber, cork, or wooden block – otherwise the surface of the filler will not be completely flat. During the smoothing of the filler surface the wet-and-dry paper should be periodically rinsed in water. This will ensure that a very smooth finish is imparted to the filler at the final stage.

At this stage the 'dent' should be surrounded by a ring of bare metal, which in turn should be encircled by the finely 'feathered' edge of the good paintwork. Rinse the repair area with clean water, until all of the dust produced by the rubbing-down operation has gone.

Spray the whole repair area with a light coat of primer, either Holts Body + Plus Grey or Red Oxide Primer – this will show up any imperfections in the surface of the filler. Repair these imperfections with fresh filler paste or bodystopper, and once more smooth the surface with abrasive paper. If bodystopper is used, it can be mixed with cellulose thinners to form a really thin paste which is ideal for filling small holes. Repeat this spray and repair procedure until you are satisfied that the surface of the filler, and the feathered edge of the paintwork are perfect. Clean the repair area with clean water and allow to dry fully.

The repair area is now ready for final spraying. Paint spraying must be carried out in a warm, dry, windless and dust free atmosphere. This condition can be created artificially if you have access to a large indoor working area, but if you are forced to work in the open, you will have to pick your day very carefully. If you are working indoors, dousing the floor in the work area with water will help to settle the dust which would otherwise be in the atmosphere. If the repair area is confined to one body panel, mask off the surrounding panels; this will help to minimise the effects of a slight mis-match in paint colours. Bodywork fittings (eg chrome strips, door handles etc) will also need to be masked off. Use genuine masking tape and several thicknesses of newspaper for the masking operations.

Before commencing to spray, agitate the aerosol can thoroughly, then spray a test area (an old tin, or similar) until the technique is mastered. Cover the repair area with a thick coat of primer; the thickness should be built up using several thin layers of paint rather than one thick one. Using 400 grade wet-and-dry paper, rub down the surface of the primer until it is really smooth. While doing this, the work area should be thoroughly doused with water, and the wet-and-dry paper periodically rinsed in water. Allow to dry before spraying on more paint.

Spray on the top coat using Holts Dupli-Color Autospray, again building up the thickness by using several thin layers of paint. Start spraying in the centre of the repair area and then, with a single side-to-side motion, work outwards until the whole repair area and about 2 inches of the surrounding original paintwork is covered. Remove all masking material 10 to 15 minutes after spraying on the final coat of paint.

Allow the new paint at least two weeks to harden, then, using a paintwork renovator or a very fine cutting paste such as Turtle Wax Color Back or Holts Body + Plus Rubbing Compound, blend the edges of the paint into the existing paintwork. Finally, apply wax polish.

Plastic components

With the use of more and more plastic body components by the vehicle manufacturers (eg bumpers, spoilers, and in some cases major body panels), rectification of more serious damage to such items has become a matter of either entrusting repair work to a specialist in this field, or renewing complete components. Repair of such damage by the DIY owner is not really feasible owing to the cost of the equipment and materials required for effecting such repairs. The basic technique involves making a groove along the line of the crack in the plastic using a rotary burr in a power drill. The damaged part is then welded back together by using a hot air gun to heat up and fuse a plastic filler rod into the groove. Any excess plastic is then removed and the area rubbed down to a smooth finish. It is important that a filler rod of the correct plastic is used, as body components can be made of a variety of different types (eg polycarbonate, ABS, polypropylene).

Damage of a less serious nature (abrasions, minor cracks etc) can be repaired by the DIY owner using a two-part epoxy filler repair material like Holts Body + Plus or Holts No Mix which can be used directly from the tube. Once mixed in equal proportions (or applied direct from the tube in the case of Holts No Mix), this is used in similar fashion to the bodywork filler used on metal panels. The filler is usually cured in twenty to thirty minutes, ready for sanding and painting.

If the owner is renewing a complete component himself, or if he has repaired it with epoxy filler, he will be left with the problem of finding a suitable paint for finishing which is compatible with the type of plastic used. At one time the use of a universal paint was not possible owing to the complex range of plastics encountered in body component applications. Standard paints, generally speaking, will not bond to plastic or rubber satisfactorily, but Holts Professional Spraymatch paints to match any plastic or rubber finish can be obtained from dealers. However, it is now possible to obtain a plastic body parts finishing kit which consists of a pre-primer treatment, a primer and coloured top coat. Full instructions are normally supplied with a kit, but basically the method of use is to first apply the pre-primer to the component concerned and allow it to dry for up to 30 minutes. Then the primer is applied and left to dry for about an hour before finally applying the special coloured top coat. The result is a correctly coloured component where the paint will flex with the plastic or rubber, a property that standard paint does not normally possess.

Aluminium components

The tailgate on Estate models (and the bonnet on later 760 models) is made of aluminium. Be careful when hammering out dents in an aluminium panel as the material is easily work-hardened and may crack. Abrasives should be used with great caution on aluminium as it is much softer than steel.

5 Major body damage – repair

Where serious damage has occurred or large areas need renewal due to neglect, completely new sections or panels will need welding in – this is best left to professionals. If the damage is due to impact, it will also be necessary to check completely the alignment of the body shell structure. Due to the principle of construction, the strength and shape of the whole can be affected by damage to a part. In such instances, the services of a Volvo agent with specialist checking jigs are essential. If a body is left misaligned, it is first of all dangerous as the car will not handle properly and secondly uneven stresses will be imposed on the steering, engine and transmission, causing abnormal wear or complete failure. Tyre wear may also be excessive.

6 Bonnet – removal and refitting

1 Disconnect the battery negative lead.
2 Disconnect the washer tube from the bonnet at the T-piece. Unclip the tube from the bulkhead.
3 Remove the under-bonnet light (when fitted) and disconnect the wire from it. Tie a piece of string to the wire, draw the wire through the bonnet cavity into the engine bay, then untie the string and leave it in the bonnet. This will make refitting easier.
4 Mark around the hinge bolts with a soft lead pencil for reference when refitting.
5 With the aid of an assistant, support the bonnet and remove the hinge bolts (photo). Lift off the bonnet.
6 When refitting, place pads of rags under the corners of the bonnet near the hinges to protect the paintwork from damage.
7 Fit the bonnet and insert the hinge bolts. Just nip the bolts up in their previously marked positions.
8 Draw the light lead through with the string. Reconnect and refit the light.
9 Reconnect the washer tube and clip it to the bulkhead.
10 Shut the bonnet and check its fit. The hinge-to-bonnet bolt holes control the fore-and-aft and left-right adjustment. Front height is adjusted by screwing the rubber buffers in or out. Rear height is adjusted at the hinge mounting bolts near the wheel arch.
11 Tighten the hinge bolts when adjustment is correct, and reconnect the battery.

7 Doors – removal and refitting

1 Open the door. Support it with a jack or axle stand, using rags to protect the paintwork.
2 Disconnect the door electrical wiring, either by removing the door trim panel or the adjacent pillar trim. Feed the wiring through so that it hangs free.
3 Mark around the hinge bolts for reference when refitting. With the aid of an assistant, remove the hinge bolts and lift away the door. Recover any hinge shims.
4 Refit by reversing the removal operations. Adjust the fit of the door if necessary, using shims and/or the slotted hinge bolt holes. Do not try to adjust the position of the door lock striker for a good fit.

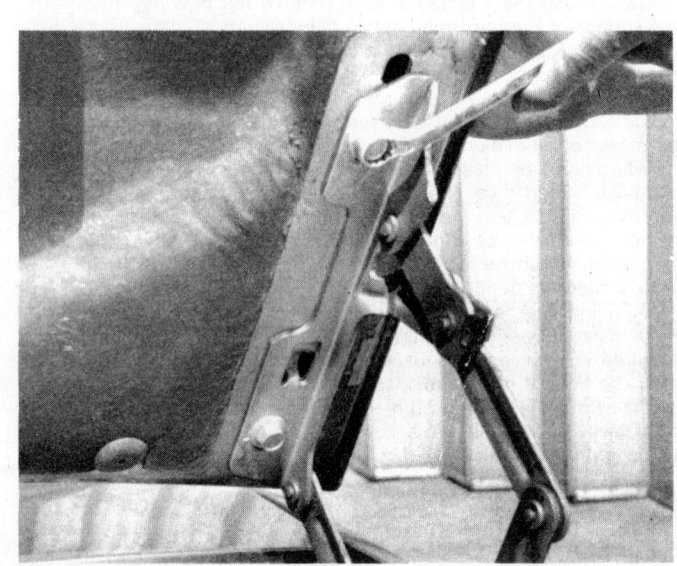

6.5A Undoing a bonnet hinge bolt

6.5B Removing the bonnet

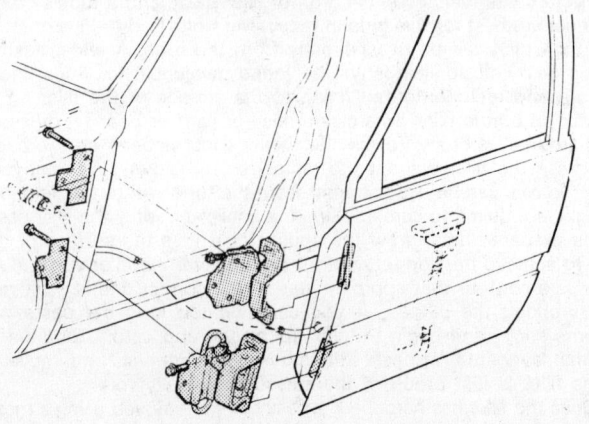

Fig. 11.2 Door hinge details (Sec 7)

Chapter 11 Bodywork and fittings 251

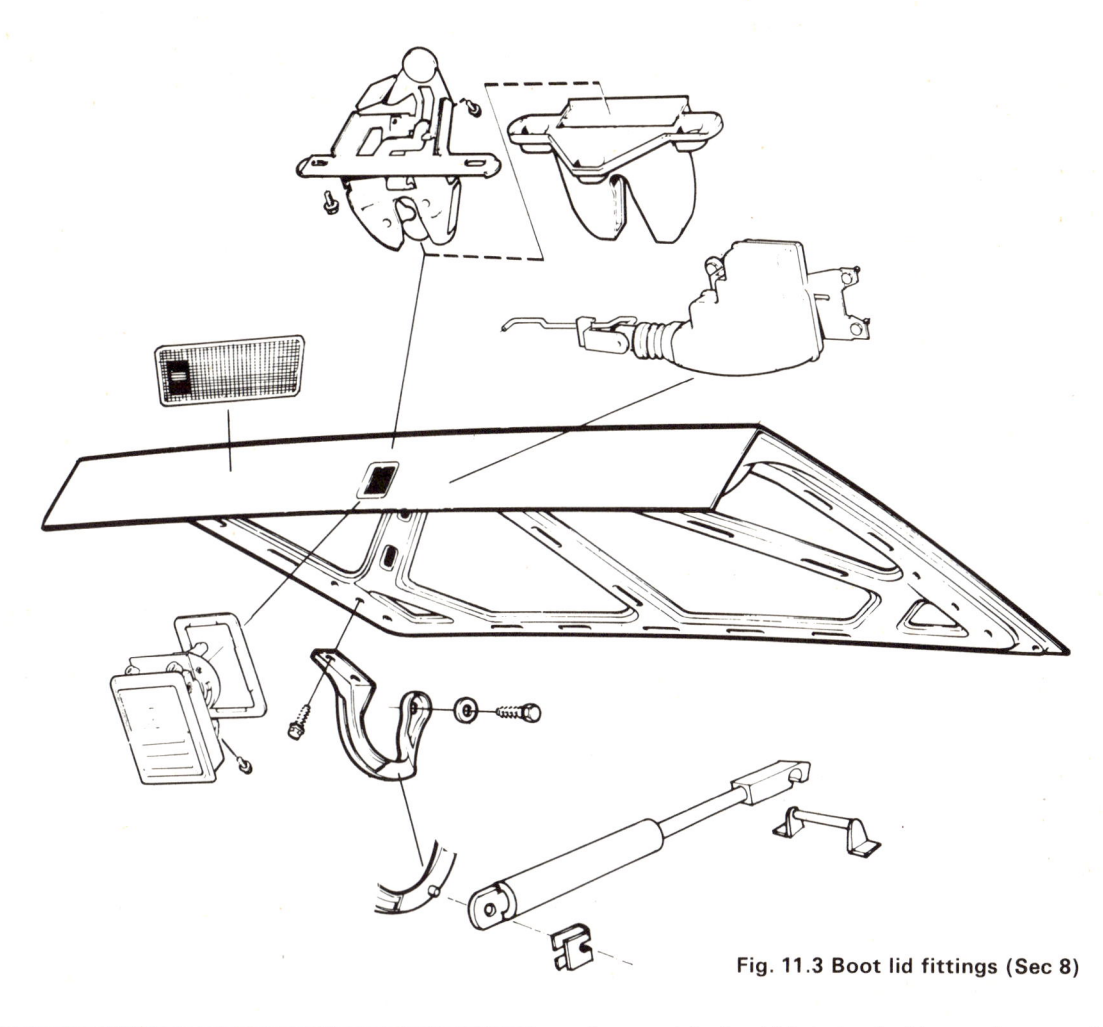

Fig. 11.3 Boot lid fittings (Sec 8)

8 Boot lid – removal and refitting

1 Open the boot. Disconnect the central locking system and/or boot light wiring so that the boot lid is free to be removed.
2 Mark around the hinge bolts. With the aid of an assistant, disconnect the boot lid strut at the hinge end, remove the hinge bolts and lift away the lid (photo).
3 Refit by reversing the removal operations. If height adjustment is necessary, this is carried out at the rear by adjusting the lock bracket, and at the front by adjusting the hinges. Access to the hinge front bolts is via the covers in the rear window pillar trim.

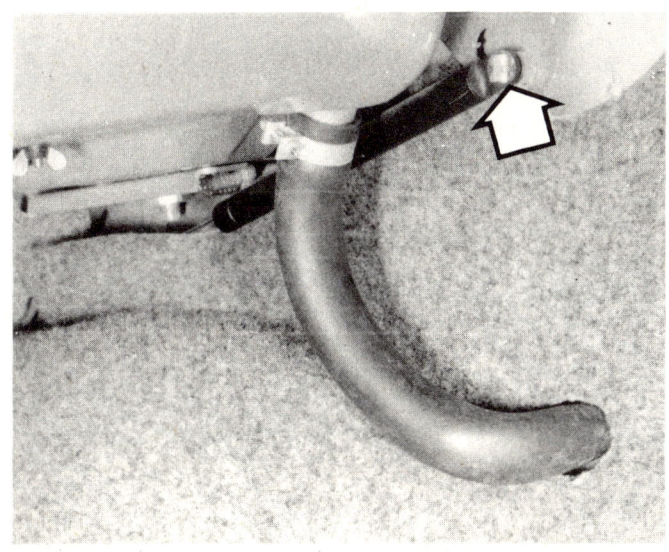

8.2A Boot lid strut – remove spring clip (arrowed) at hinge end

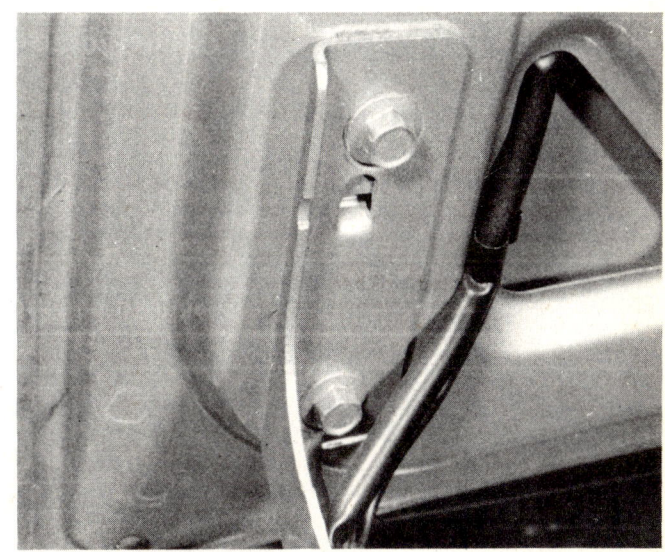

8.2B Boot hinge bolts

9 Tailgate – removal and refitting

1 Disconnect the battery negative lead.
2 Open the tailgate. Disconnect the washer tube at the junction next to the right-hand hinge.
3 Prise out the bungs which conceal two hinge bolts. Slacken the hinge bolts but do not remove them yet (photo).
4 Remove the trim panel from around the load area light. Besides the visible fasteners, there is one concealed behind the light itself (photo).
5 Separate the wiring connectors exposed by removal of the trim panel, making notes for refitting if necessary. Feed the wiring through to the tailgate.
6 Have an assistant support the tailgate. Disconnect the gas struts by removing the wire clips and separating the balljoints (photo).
7 Remove the hinge bolts and lift away the tailgate.
8 Refit by reversing the removal operations. Only provisionally tighten the hinge bolts until satisfied with the fit of the tailgate. Adjust the lock striker and side guide pieces if necessary for a good fit (photo).

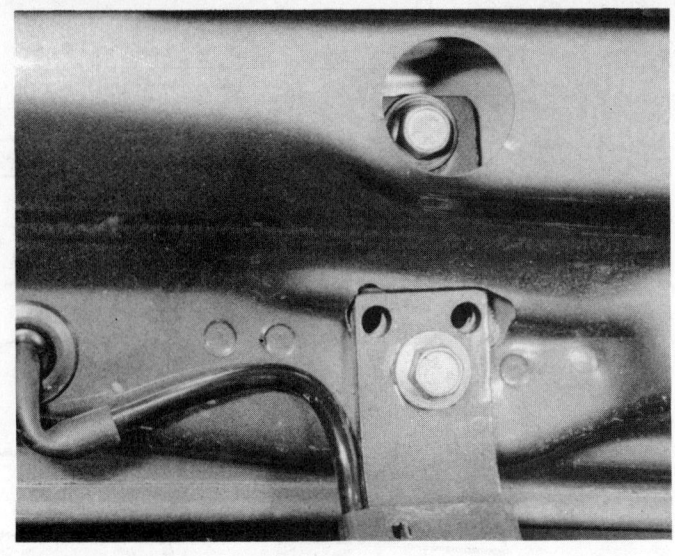

9.3 Tailgate hinge bolts

9.4 Trim panel fastener (arrowed) behind load area light

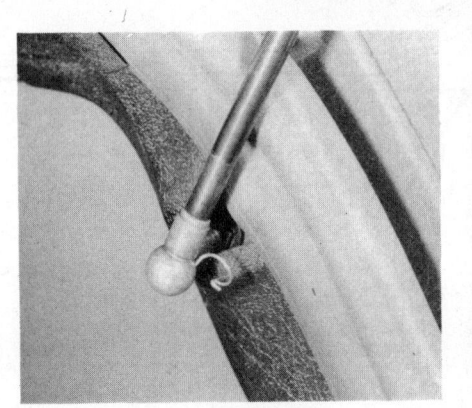

9.6 Disconnecting a tailgate strut

9.8 Tailgate side guide piece

10 Bonnet release cable – removal and refitting

1 Open the bonnet. If the cable is broken, the catches must be released from below, or access can be gained by the destructive removal of the headlights.
2 Unbolt the release catch which is furthest from the release handle. Disconnect the cable inner from it (photos).
3 Release the cable outer from the other catch. Pull the cable free of the catches.

10.2A Unbolt the bonnet release catch ...

10.2B ... and disconnect the cable inner

Chapter 11 Bodywork and fittings

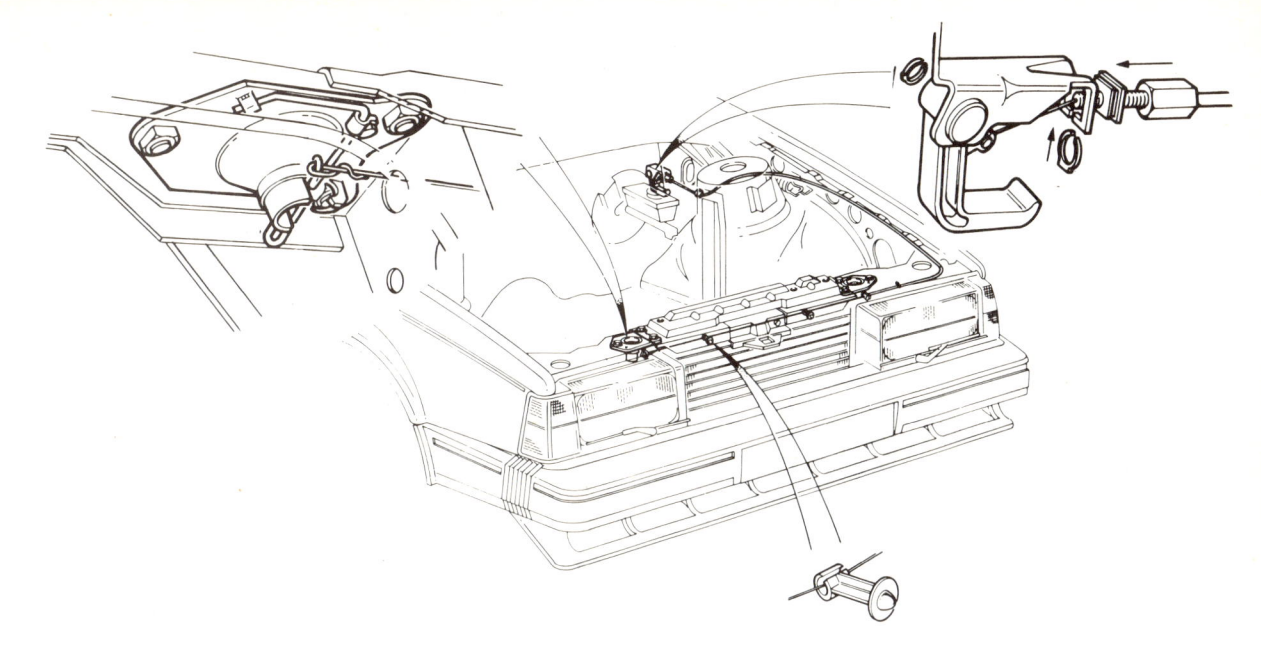

Fig. 11.4 Bonnet release cable fittings. LHD shown, RHD is mirror image (Sec 10)

4 Inside the vehicle, release the cable from the lever by unhooking the inner and removing the slide clip from the outer.
5 Feed the cable into the engine bay and remove it.
6 Refit by reversing the removal operations. Adjust the threaded section of the cable at the release lever end to take most of the slack out of the inner in the resting position.

11 Front door interior trim – removal and refitting

Driver's door – pre-1985 models
1 Disconnect the battery negative lead.
2 Prise out the three screw plugs from the armrest. Remove the three screws (photo).
3 Unclip the door edge marker light lens, noting that the arrow points outwards.
4 Unclip the loudspeaker grille. Remove the four screws which secure the speaker, pull it out of its cavity and disconnect the wires from it. Also remove the fourth armrest screw now exposed.
5 Free the armrest from its clips by tugging firmly. Remove the switch panel from the armrest. Disconnect the edge marker light multi-plug and remove the armrest (photo).
6 Unscrew the interior lock button.
7 Prise out the two clips at the base of the main trim panel (photo). Free the panel from the door clips by tugging or prising and remove it.

8 Remove the large and small water deflectors (photo).
9 Refit by reversing the removal operations.

Passenger's door – pre-1985 models
10 Proceed as above, but note that the door pull must be removed (two screws revealed by prising out the facing) at an early stage. There are also fewer switches to deal with.

1985 and later models
11 Disconnect the battery negative lead.
12 Carefully prise off the speaker grille. On the passenger side, carefully prise off the door pull trim then remove the door pull (two screws).
13 Remove the four screws which secure the speaker, and the single door panel retaining screw which is now exposed. As the speaker is removed, detach the two leads.
14 Unscrew the interior lock button, then remove the interior release handle (one retaining screw after prising out the rubber blanking plug).
15 On models with manually-operated windows, remove the winder handle.
16 On all models, rotate the blanking plug in the door panel well (adjacent to the interior release handle) through 90°, and prise the panel well assembly upwards and away from the door panel. Where appropriate, lift up the electrically-operated window switch assembly.
17 Carefully prise off the door edge marker light lens, then pull out the door panel retaining clip which is now visible.
18 Carefully prise the two plastic clips at the base of the door panel

11.2 Removing an armrest screw plug

11.5 Removing the armrest

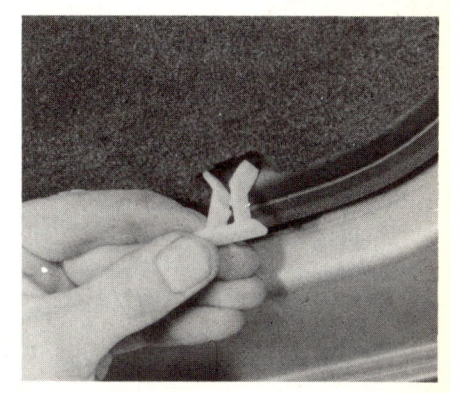

11.7 Door trim panel base clip

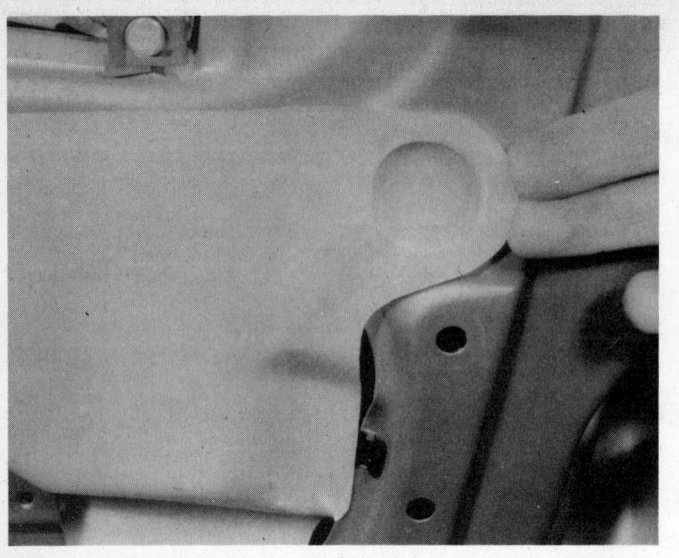

11.8A Removing the large water deflector ...

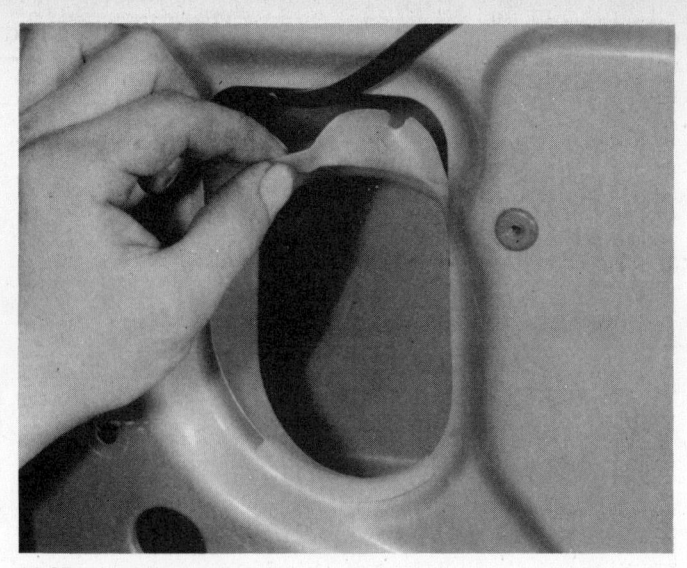

11.8B ... and the small one

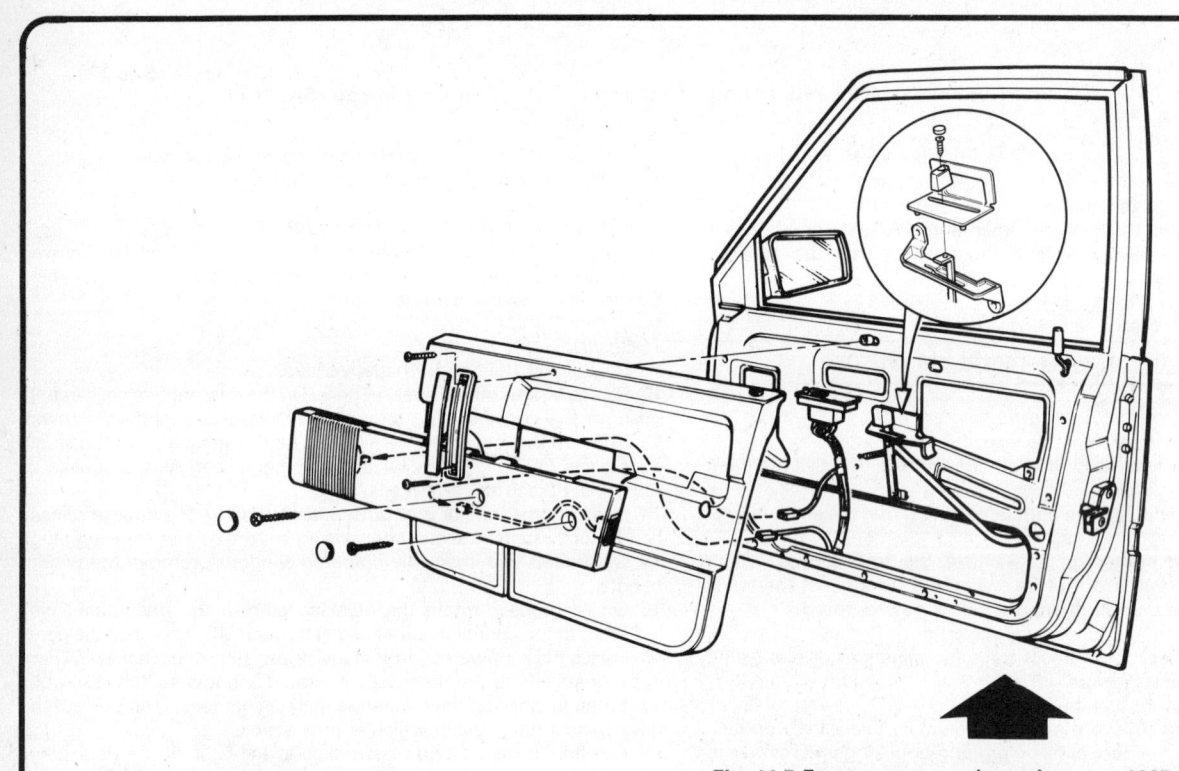

Fig. 11.5 Front passenger door trim – pre-1985 LHD models (Sec 11)

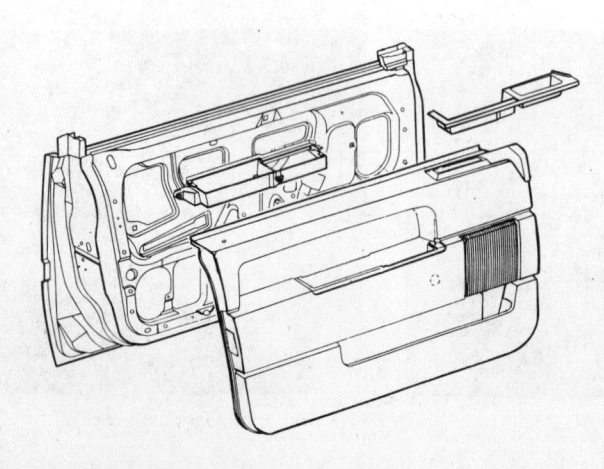

Fig. 11.6 Front driver's door trim – 1985 on LHD models (Sec 11)

Chapter 11 Bodywork and fittings

downwards, then pull the door panel away from the door at the two sides and the bottom edge.
19 Lift the panel upwards and away from the door, detaching the electrical connector from the door edge marker light.
20 Refitting is the reverse of the removal procedure.

12 Rear door interior trim – removal and refitting

1 Disconnect the battery negative lead.

Pre-1985 models
2 Prise out the two screw plugs from the armrest. Remove the screws.
3 Carefully prise the facing out of the pull handle to reveal two screws. Remove these screws and the pull handle (photo).
4 Unclip the loudspeaker grille. Remove the speaker (if fitted) and two more armrest screws now exposed.
5 Pull the armrest off the door. Disconnect the window switch and the edge marker light multi-plug (as applicable) and remove the armrest.
6 Unscrew the interior lock button.
7 Release the trim panel clips by tugging firmly, or by prising with a palette knife or scraper. Lift and remove the trim panel.
8 The waterproof sheet may now be peeled off the door in the area to which access is desired.
9 Refit by reversing the removal operations.

1985 and later models
10 Removal and refitting procedures are virtually identical to those for the front door (Sec 11, paragraph 11 onwards) except that both doors have a pull handle, and there are no clips securing the door panel base.

Fig. 11.7 Rear door trim – pre-1985 models (Sec 12)

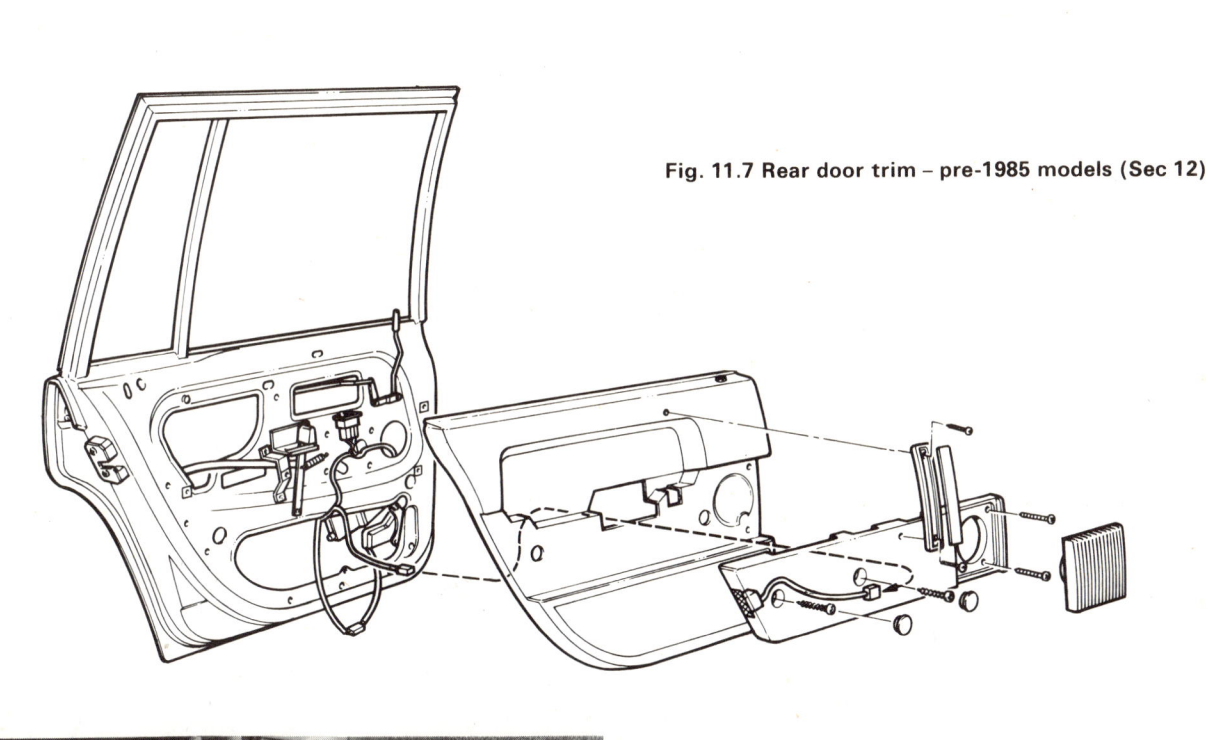

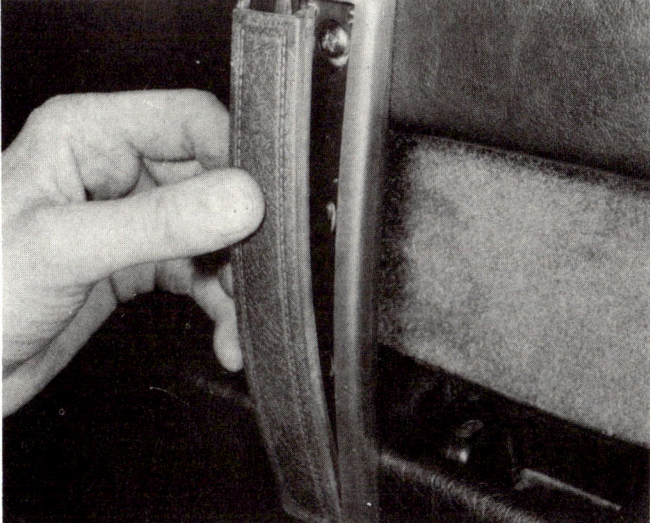

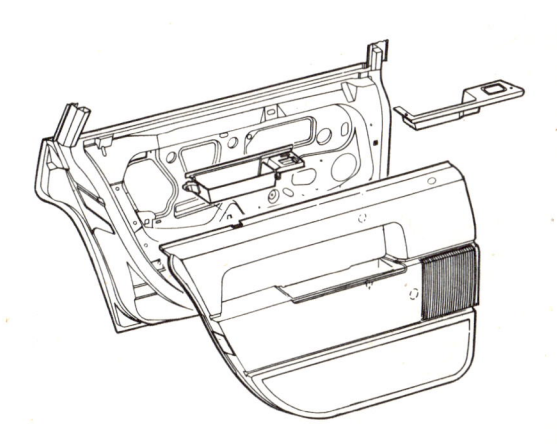

12.3 Removing the pull handle facing

Fig. 11.8 Rear door trim – 1985 on (Sec 12)

13 Tailgate interior trim – removal and refitting

1 Open the tailgate. From the bottom of the trim panel remove the four fasteners by turning them through 90° (photo).
2 Unclip the plastic surround from the interior handle. Remove the two screws now exposed, and the handle trim piece which they also secure (photo).
3 Slide the trim upwards (relative to the closed position of the tailgate) to free the 'keyhole' fasteners along the top edge. Remove the trim panel.
4 Refit by reversing the removal operations.

14 Windscreen and other fixed glass – removal and refitting

Special equipment and techniques are needed for successful removal and refitting of the windscreen, rear window and rear quarter windows. Have the work carried out by a Volvo dealer or a windscreen specialist.

15 Front door window – removal and refitting

1 Remove the door interior trim (Section 11).
2 Raise or lower the window so that the lift arms are accessible. Remove the clip which secures each lift arm to the lift channel (photo).
3 Have an assistant support the window, or wedge or tape it in position. Disengage the lift arms from the channel and lift the glass out of the door.
4 If new glass is being fitted, check whether or not it is supplied with the lift channel attached. If not, it will be necessary to transfer the old channel. Where possible, note the fitted position of the channel relative to the trailing edge of the glass. The channel is removed by judicious use of a rubber mallet. Refer to Fig. 11.9 if the fitted position of the channel is not known.
5 Refit by reversing the removal operations.

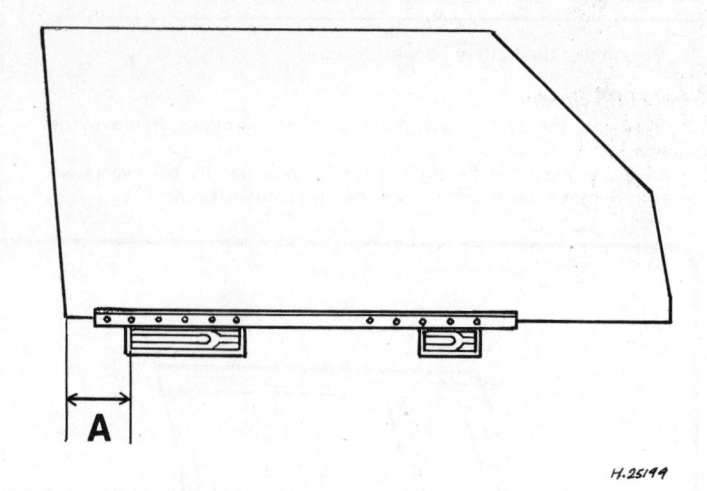

Fig. 11.9 Position of front window channel (Sec 15)

A = 70 mm (2.75 in) approx

13.1 Removing a tailgate trim fastener

13.2 Removing the tailgate interior handle trim

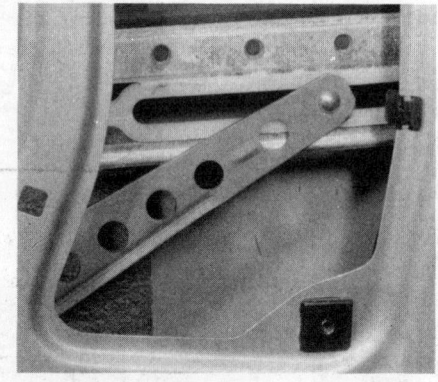

15.2 Window lift arm in the channel. Clip is behind the end of the arm

16 Rear door windows – removal and refitting

1 Remove the door interior trim (Section 12).
2 Remove the drop glass as described for the front window (Section 15).
3 The fixed glass may now be removed after drilling out the blind rivets which secure the guide channel. Remove the guide channel and slide out the glass.
4 When refitting the fixed glass, lubricate the surround with liquid soap.
5 Press the glass home and refit the guide channel, securing it with new blind rivets.
6 If fitting a lift channel to the drop glass, refer to Fig. 11.11 for the correct fitted position.
7 Refit the drop glass and the interior trim.

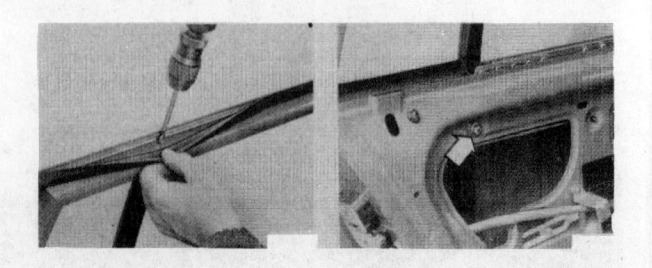

Fig. 11.10 Drilling out the guide channel top rivet (left). Lower rivet is arrowed, right (Sec 16)

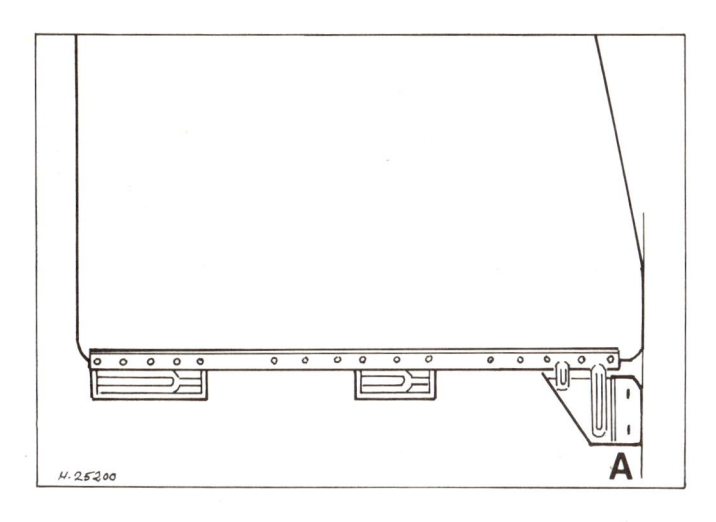

Chapter 11 Bodywork and fittings   257

Fig. 11.11 Correct position of rear window channel (Sec 16)

A = 0 to 1 mm (0 to 0.04 in)

17 Window lift mechanism – removal and refitting

1 Proceed as for door window glass removal (Section 15 or 16), but do not remove the glass completely. Tape or wedge it in the fully raised position.
2 Remove the clip which secures the slide arm in its channel (photo).
3 In the case of electrically-operated windows, remove the motor connectors from the multi-plug, prising them out with a small screwdriver or scriber (photo). It is difficult to do this without damaging the connectors, but if a new motor is to be fitted this does not matter.
4 Remove the nuts which secure the mechanism to the door skin.
5 Push the mechanism into the door cavity and remove it through the large hole at the bottom. It may be necessary to alter the position of the mechanism to allow it to pass through the hole; with electrically-operated windows, do this by carefully connecting a battery to the connectors using jump leads. Do not allow the connectors or the jump lead clips to touch (photo).
6 The motor may be unbolted from the mechanism if wished. Take great care that as the motor is removed, spring pressure does not cause a sudden movement of the toothed quadrant, which could result in injury.
7 Refit by reversing the removal operations. Before refitting the door trim, adjust the stop screw as follows (photo).
8 Slacken the stop screw and press it forwards. Wind the window fully up, press the stop screw rearwards and tighten it.

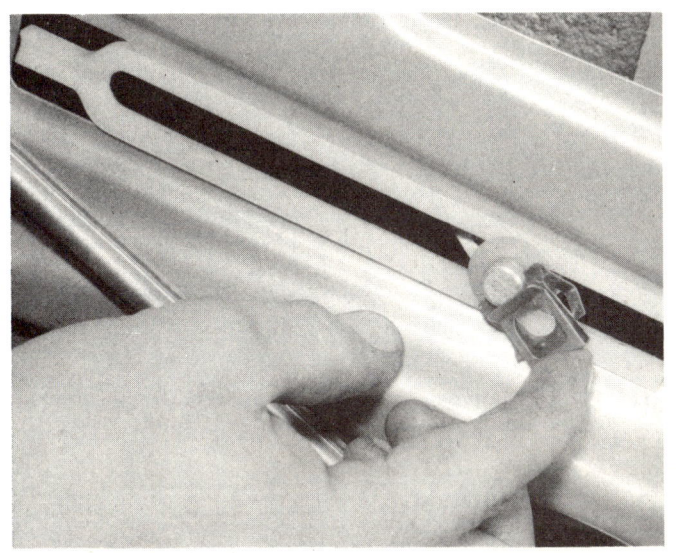

17.2 The clip which secures the slide arm in the channel

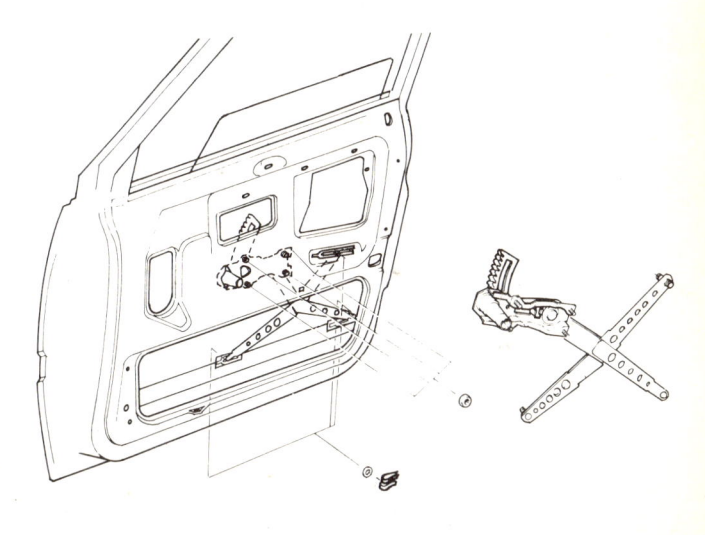

Fig. 11.12 Window lift mechanism fittings – manual type shown (Sec 17)

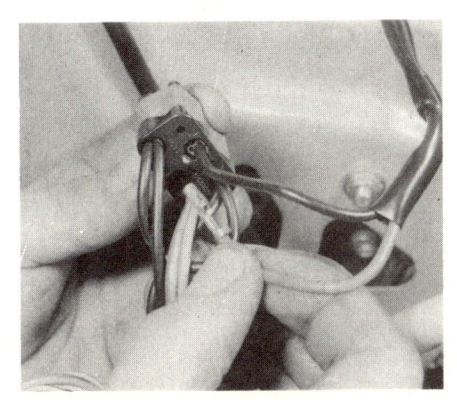

17.3 Removing the window motor connectors from the multi-plug

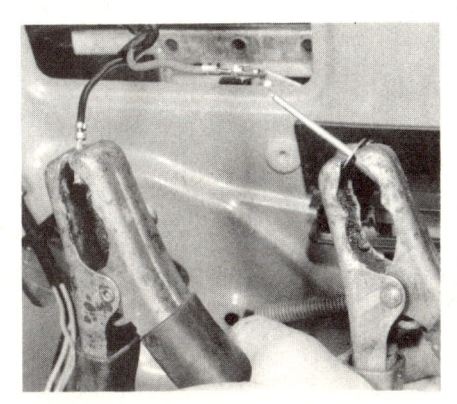

17.5 Using jump leads to apply power to the window motor

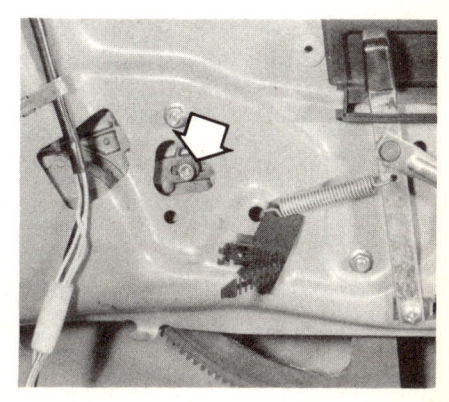

17.7 Window mechanism stop screw (arrowed)

Chapter 11 Bodywork and fittings

18 Door handles, locks and latches – removal and refitting

1 Remove the door interior trim (Section 11 or 12).

Lock barrel

2 On early models, release the central locking switch from the barrel by undoing the retaining clip.
3 Remove the two screws in the door shut face which secure the lock barrel clip (photo).
4 Unhook the lock-to-latch rod, noting which way round it is fitted. Slide the clip off the lock and remove the lock and clip (photo).
5 When refitting, make sure (when applicable) that the central locking switch groove engages with the lug on the lock.

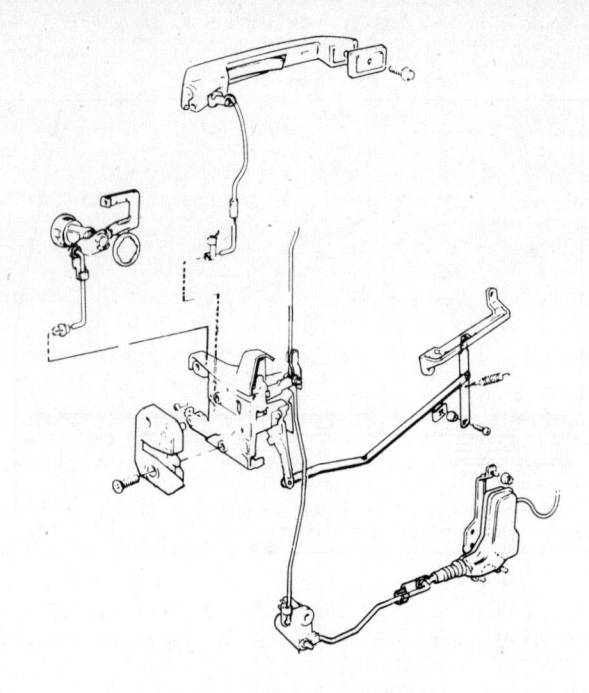

Fig. 11.13 Front door handle, lock and latch (Sec 18)

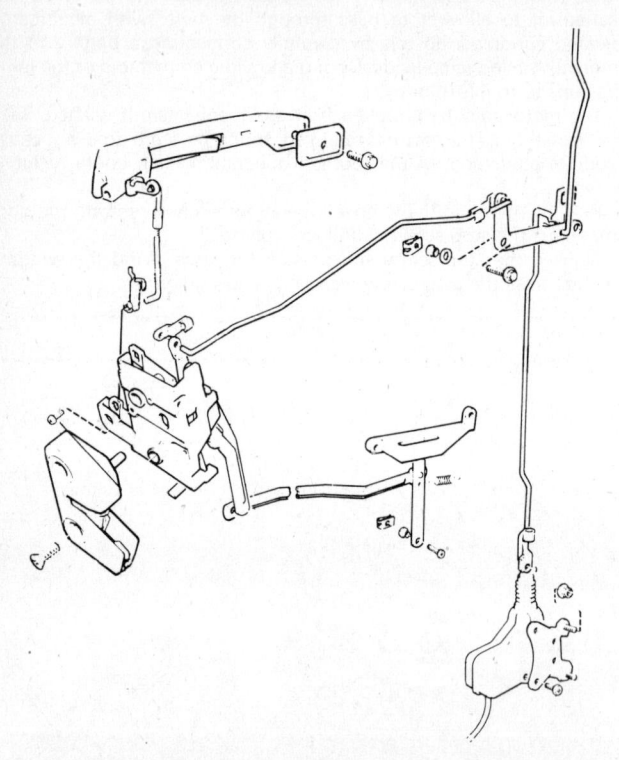

Fig. 11.14 Rear door handle, lock and latch (Sec 18)

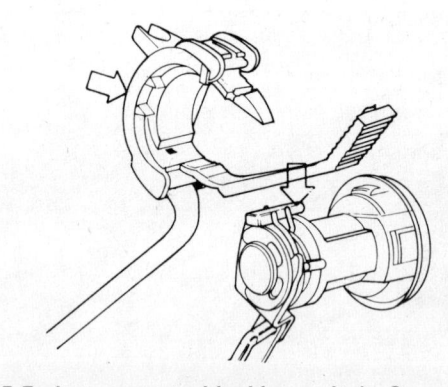

Fig. 11.15 Early type central locking switch. Groove and lug (both arrowed) must engage (Sec 18)

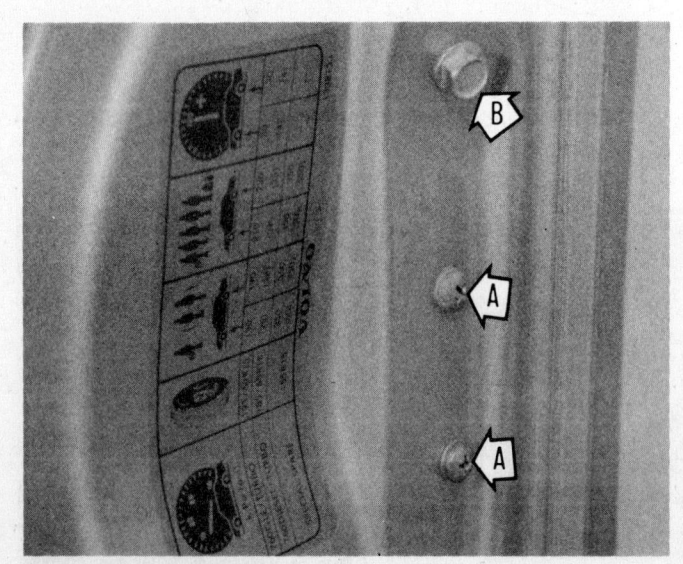

18.3 Lock barrel clip screws (A). Screw B secures one end of the exterior handle

18.4 Door lock barrel removed

Chapter 11 Bodywork and fittings

Exterior handle
6 Remove the two screws which secure the handle. Unhook the link rod and remove the handle (photos).

Latch mechanism
7 Disconnect the lock barrel and exterior handle link rods from the latch.
8 Remove the catch from the door. It is secured by two Allen screws (photo).
9 Remove the single securing screw exposed by removal of the catch (photo).
10 Unclip the interior handle link and remove the latch mechanism (photo).

All items
11 Refit by reversing the removal operations. Check for correct operation before refitting the door trim.
12 Note that the exterior handle link rod contains an adjustable section. The length of the rod should be set so that the latch stop contacts its base, and the handle tongue protrudes at least 22 mm (0.87 in) (Fig. 11.16).

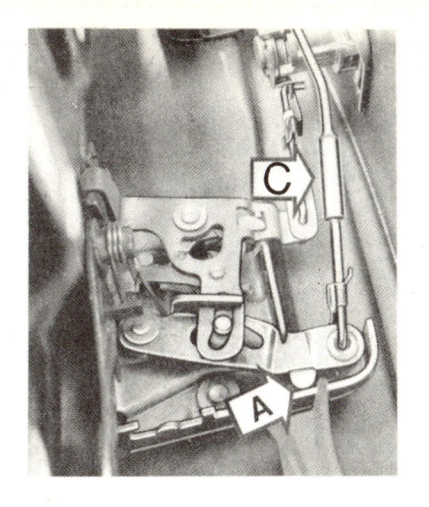

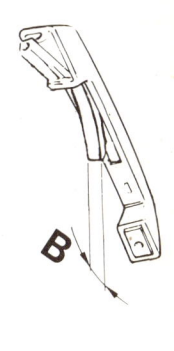

Fig. 11.16 Exterior handle link adjustment (Sec 18)

A Latch stop
B Tongue protrusion
C Adjuster

19 Boot lock – removal and refitting

1 Remove the latch, and unhook the lock motor link rod from the latch driver (photo).
2 Remove the shear-head bolt which secures the lock, using a stud extractor or a hammer and punch.
3 Remove the lock from the boot lid.
4 When refitting, use a new shear-head bolt. Only tighten the bolt lightly until satisfied with the operation of the lock, then tighten the bolt until its head shears off.

20 Tailgate lock – removal and refitting

1 Remove the tailgate interior trim panel (Section 13).
2 Disconnect the link rods from the exterior handle, the lock cylinder and (when applicable) the lock motor.
3 Remove the exterior handle/number plate light assembly, which is secured by two screws and two nuts. Disconnect the wiring.

18.6A This screw secures the other end of the exterior handle

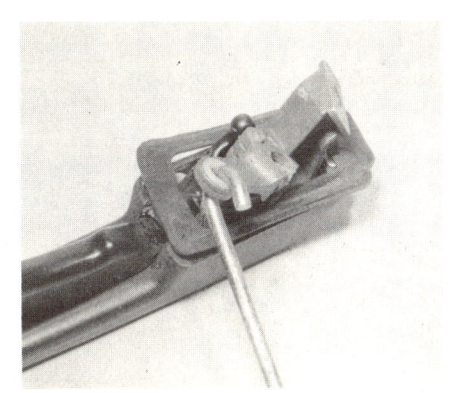

18.6B The link rod engaged with the handle

18.8 Remove the catch ...

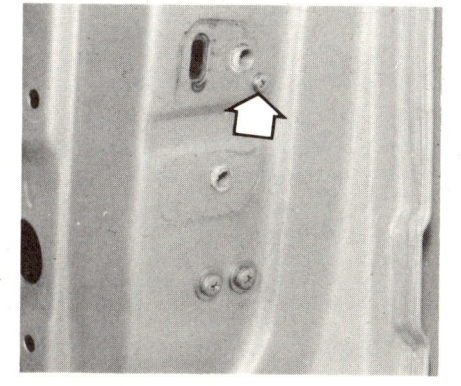

18.9 ... to expose the latch securing screw (arrowed)

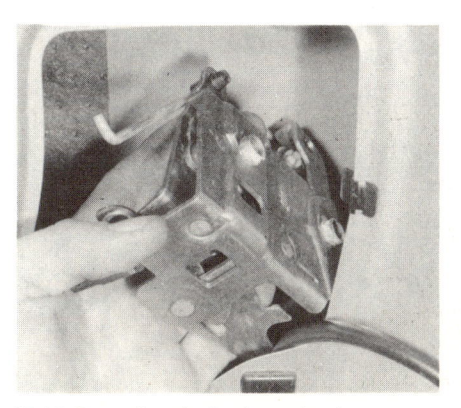

18.10 Removing the latch mechanism

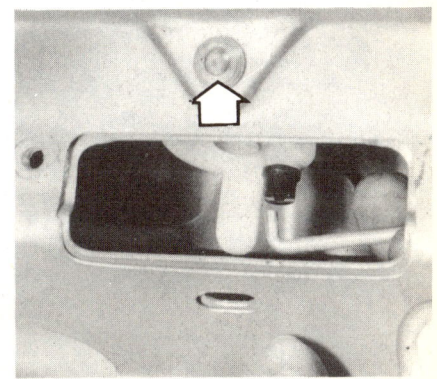

19.1 Unhooking the boot lock motor link rod. Shear-head bolt is arrowed

Fig. 11.17 Removal of the exterior handle/number plate light assembly (Sec 20)

Fig. 11.18 Tailgate lock cylinder E-clip (A) and locking plate (B) (Sec 20)

4 The lock cyclinder and levers can now be removed after releasing the E-clip and locking plate.
5 Refit by reversing the removal operation. Adjust the exterior handle link rod if necessary to give the handle approximately 3 mm (0.12 in) free play.

21 Central locking components – removal and refitting

Driver's door switch
1 Remove the interior trim from the door (Section 11).
2 On early models, unclip the switch from around the lock barrel and remove it. Note how the switch groove engages with the lug on the lock.
3 On later models the switch is located next to the latch mechanism. Remove the single securing screw and lift out the switch. Note how the switch tongue engages with the latch rod (photos).
4 Disconnect the switch multi-plug. If other services share the same plug, prise out the appropriate connectors.
5 Refit by reversing the removal operations.

Door lock motors
6 Remove the interior trim from the door (Section 11 or 12).
7 Unclip the motor link rod from the bellcrank (front door) or lock button link rod (rear door) (photo).
8 Undo the motor securing nuts. Remove the motor and link rod from the door. On some models the link rod is enclosed in a plastic tube (photos):

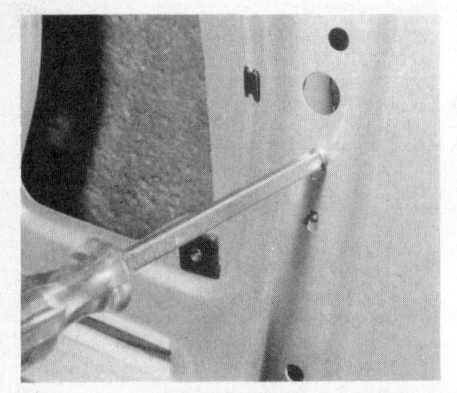

21.3A Remove the securing screw ...

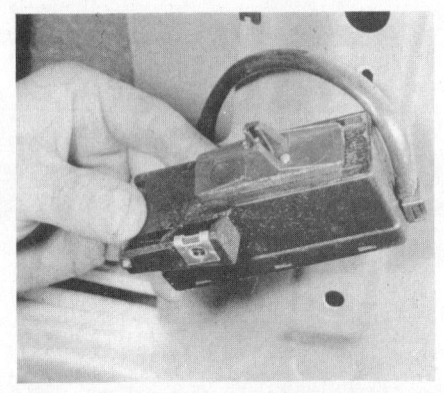

21.3B ... and lift out the switch

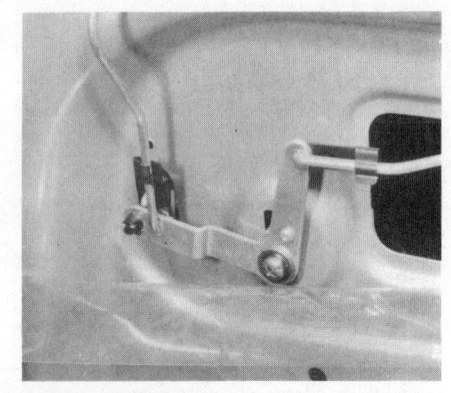

21.7 The lock motor link bellcrank

21.8A Lock motor securing nuts (arrowed)

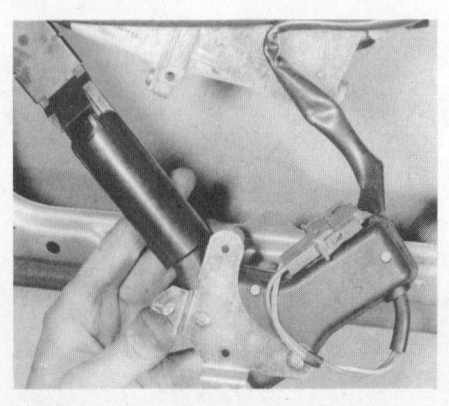

21.8B Removing the lock motor, link rod and tube

Chapter 11 Bodywork and fittings   261

9 Disconnect the multi-plug – see paragraph 4.
10 If a new motor is being fitted, transfer the link rod, mounting plate and any other components to it.
11 Refit by reversing the removal operations.

Boot lock motor
12 Remove the latch cover. Unbolt and remove the latch (photos).
13 Unhook the lock motor link rod from the latch driver.
14 Remove the three securing nuts, disconnect the motor wiring and remove it (photo).
15 Refit by reversing the removal operations. Adjust the position of the latch within the limits of the slotted holes to achieve satisfactory opening and closing.

Tailgate lock motor
16 Remove the tailgate interior trim (Section 13).
17 The lock motor may now be removed in a similar way to the boot lock motor.

Relays
18 On 1982 and 1983 models, two relays in the central electrical unit control the locking and unlocking functions. See Chapter 12, Section 24.
19 On later models no relays are used in the central locking system.

22 Door mirror – removal and refitting

Electrically-operated
1 Remove the door interior trim (Section 11).
2 Prise free the trim plate which covers the mirror mounting (photo).
3 Disconnect the mirror wiring multi-plug. Free the wiring harness from the door (photo).
4 Support the mirror and remove the mounting screw. Lift the mirror off its mountings.
5 Refit by reversing the removal operations.

Manually-operated
6 The operations are similar to those just described, but since there are no wires to disconnect there should be no need to remove the door interior trim panel. No specific information is available.

23 Door mirror glass and motor – removal and refitting

There is no need to remove the mirror from the door for these operations.
1 Press the mirror glass inwards at the bottom until the retaining ring teeth are visible through the access hole.
2 Prise the teeth with a screwdriver to move the ring in an anti-clockwise direction (looking at the glass). This will release the retaining ring from the mounting plate (photo). Remove the glass and retaining ring. When applicable, disconnect the heating element wires.
3 The motor may now be removed after undoing the four retaining screws and disconnecting the wires from it. If the wires cannot be separated from the motor, remove the door interior trim and disconnect the mirror multi-plug (photo).
4 Refit by reversing the removal operations. Observe the 'TOP' marking on the motor, and the 'UNTEN' (bottom) marking on the mirror glass (photo).

24 Door surround weatherstrip – removal and refitting

1 Remove the kick panel from the door sill (photo).
2 Prise the weatherstrip free, starting at the bottom. Use a wide-bladed screwdriver and protect the paintwork by prising against a piece of wood.
3 Refit the weatherstrip starting at the uppermost corner. Tap it home with a rubber mallet.
4 Refit the kick panel.

21.12A Removing the boot latch cover...

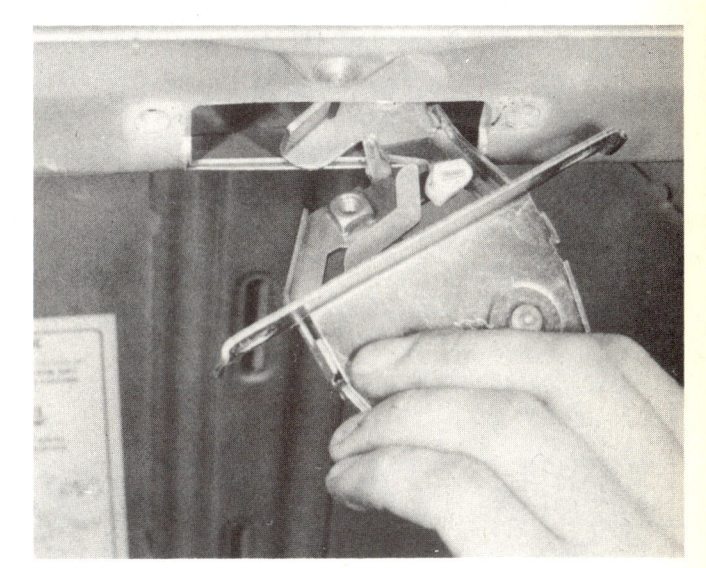

21.12B ... and the latch

21.14 Removing the boot lock motor

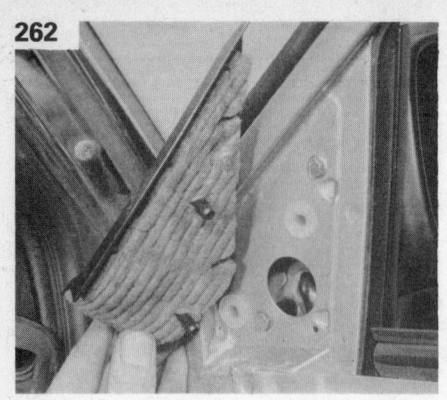

22.2 Removing the door mirror mounting trim plate

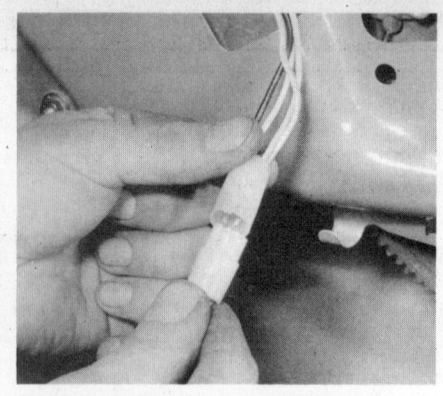

22.3 Disconnecting the mirror multi-plug

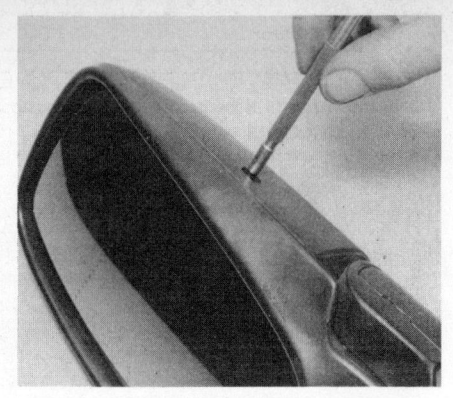

25.2 Releasing the mirror glass retaining ring

23.3 Mirror motor, showing the retaining screws

23.4A Mirror motor is marked 'TOP' ...

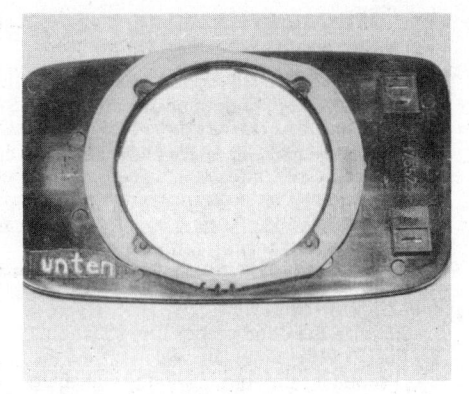

23.4B ... and bottom of glass is marked 'unten'

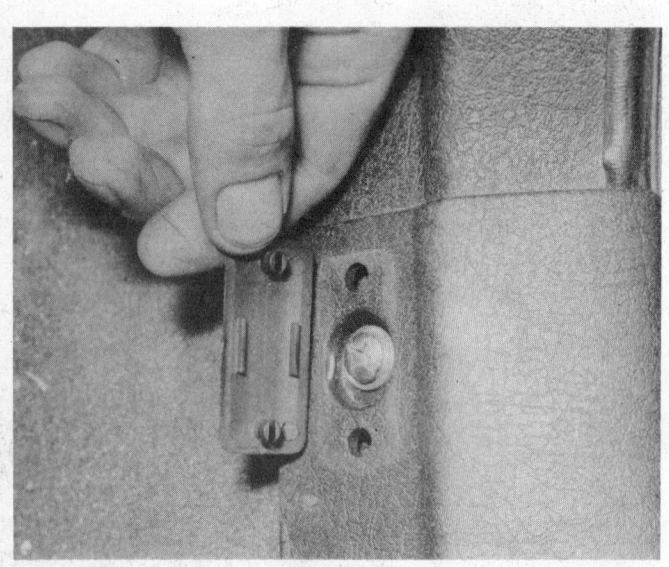

24.1 Sill kick panel screw and cover

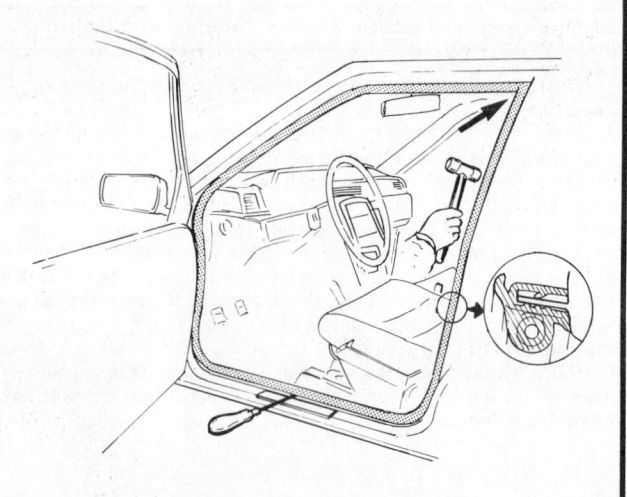

Fig. 11.19 Door surround weatherstrip removal and refitting. Commence refitting at the top corner (arrowed) (Sec 24)

Chapter 11 Bodywork and fittings

25 Front seat – removal and refitting

1 Remove the trim or storage pocket from the outboard side of the seat base. Unbolt the seat belt anchorage thus exposed (photo).
2 Move the seat forwards. Remove the single screw from the rear of each track – these may be concealed by trim covers (photo).
3 Move the seat rearwards. Remove any trim covers, then remove the single screw from the front of each track (photo).
4 Disconnect the seat heater, seal belt switch and adjustment motor multi-plugs (as applicable).
5 Lift the front of the seat, pushing it rearwards at the same time, to free the tracks from their 'keyhole' fixings in the floor. Remove the seat and tracks together.
6 Refit by reversing the removal operations.

26 Rear seat – removal and refitting

Saloon
1 Free the seat cushion from its retaining clips by pushing the front edge down and pulling it rearwards (photo). (On early models the clips are slightly different – Fig. 11.20.) Lift out the cushion.
2 Straighten the tongues of the clips which secure the base of the seat back (photo). Thump the seat back upwards to free it from the top clips and remove it.
3 The armrest may now be unbolted and removed if wished.
4 Refit by reversing the removal operations.

Estate
5 Fold the seat cushions forwards. Remove the hinge retaining nuts and lift out the cushions (photo).
6 Fold down the seat backs. Pull the pins out of the centre mounting (photo) and release the side mounting pins by turning them with pliers. Lift out the seat backs.
7 Refit by reversing the removal operations.

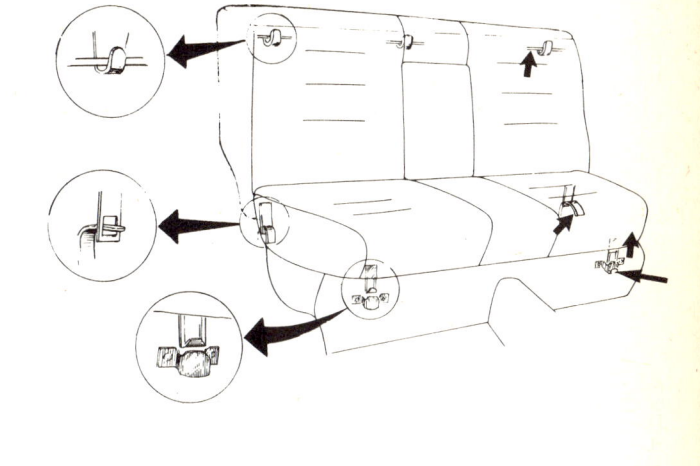

Fig. 11.20 Rear seat fastenings – models up to 1983 (Sec 26)

27 Head restraints – removal and refitting

1 Press the front of the seat backrest about 90 mm (3.5 in) below the top edge, at the same time pulling the head restraint upwards to free it.
2 Pull the head restraint out of the guides and remove it.
3 When refitting, push the head restraint firmly into place until it latches.

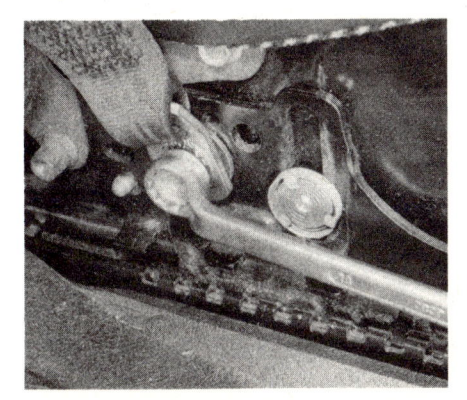

25.1 Unbolting the belt anchorage from the seat

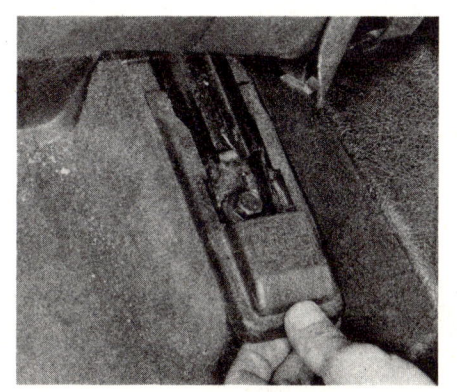

23.2 A rear track bolt ...

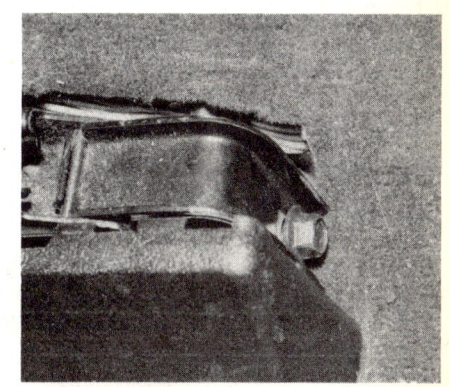

25.3 ... and a front track bolt

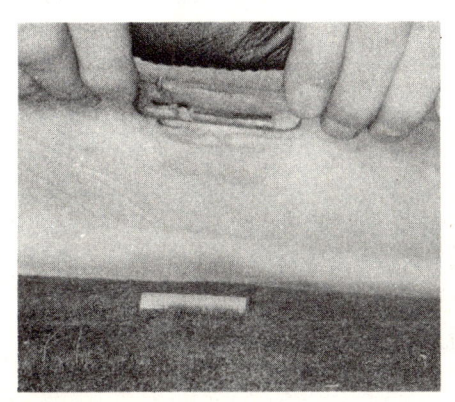

26.1 Unclipping the seat cushion front edge

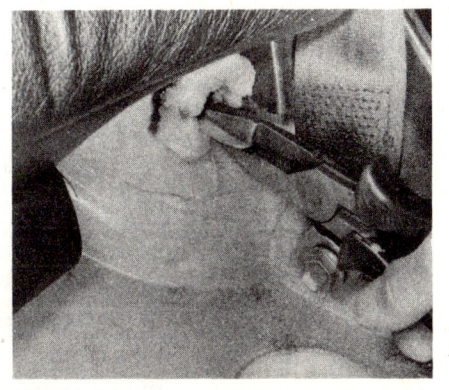

26.2 Straightening a seat back clip tongue

26.5 Two seat cushion hinges (Estate)

Fig. 11.21 Rear seat fastenings – 1984 on (Sec 26)

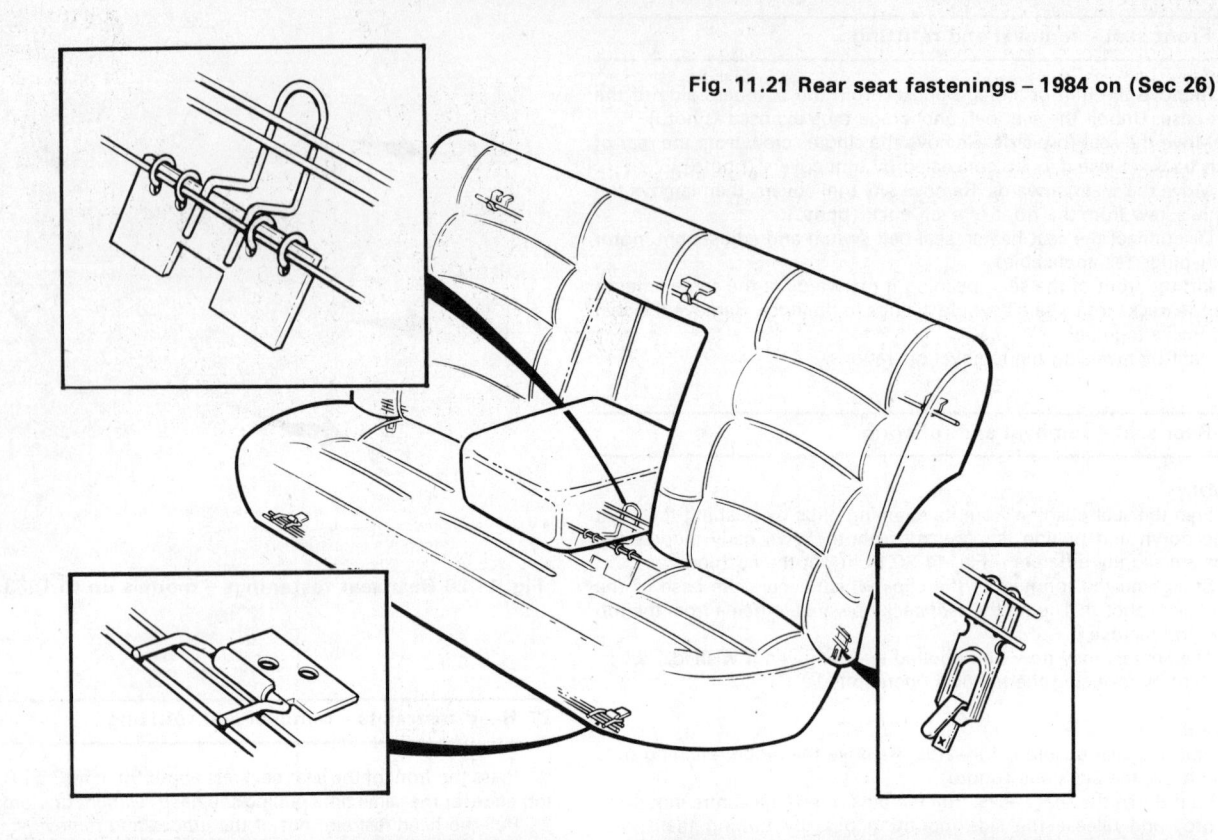

26.6 Seat back centre mounting (Estate) – pins arrowed

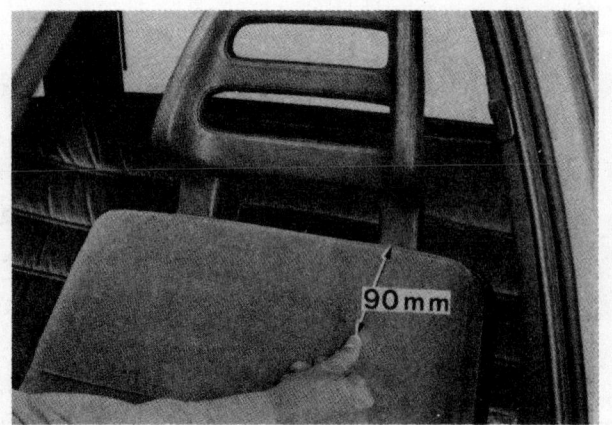

Fig. 11.22 Press where shown to release the head restraint (Sec 27)

Fig. 11.23 Two Allen screws (arrowed) which secure one side of the height adjusters (Sec 28)

28 Front seat position adjusters – removal and refitting

1 Remove the seat and cushion (Sections 25 and 29).

Mechanical height adjusters

2 Raise the adjuster to its highest position. Remove the Allen screws which secure it to the seat.
3 Press the height adjuster lever towards the front of the seat and push downwards on the seat. Separate the seat from the rods.
4 The height adjuster components may now be renewed as necessary.
5 Refit by reversing the removal operations. Insert the rods into the highest holes of the adjuster plates.

Chapter 11 Bodywork and fittings   265

Mechanical reclining adjuster
6 Proceed as for backrest heater renewal (Section 29), but also unbolt the reclining mechanism from the seat base frame. The complete backrest frame and reclining mechanism must be renewed together.

Electrical adjusters
7 No specific information was available at the time of writing. The locations of the motors, and a circuit diagram, are given in Fig. 11.25.

Fig. 11.24 Two bolts (arrowed) securing the reclining adjuster (Sec 28)

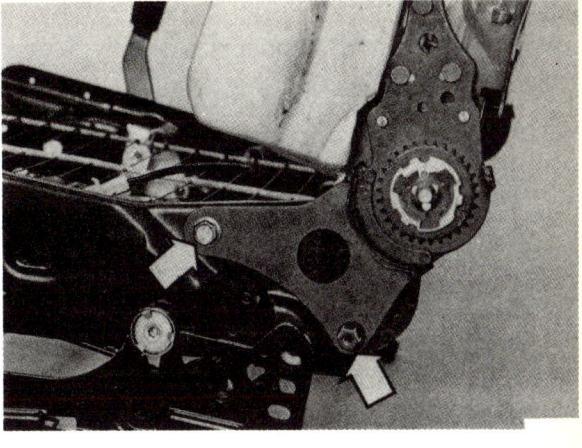

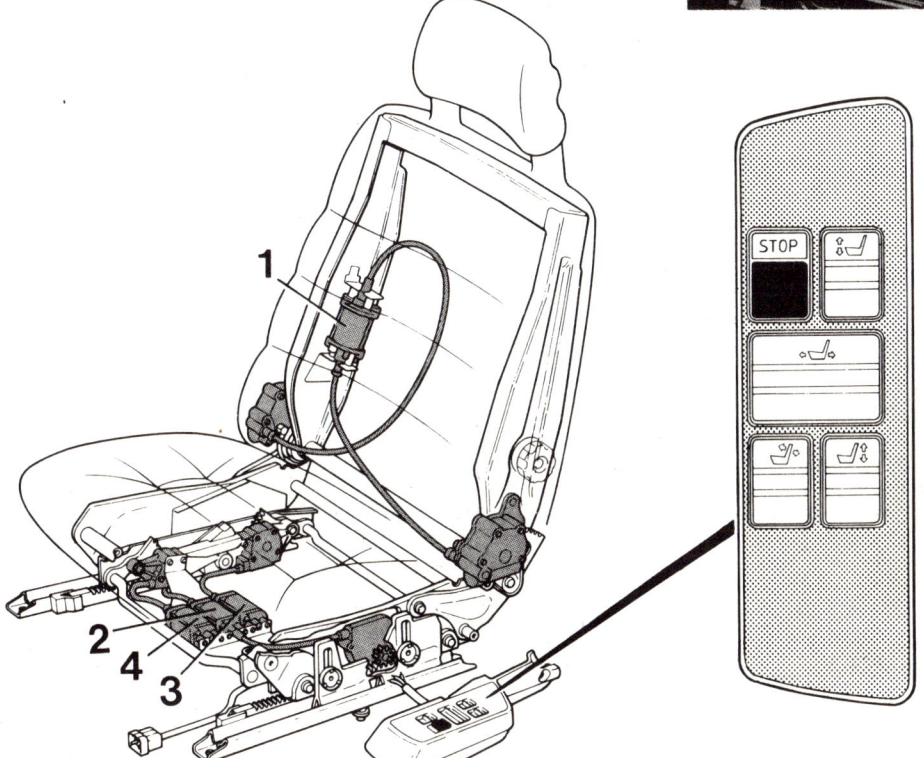

Fig. 11.25 Seat adjustment motor positions and circuit diagram (Sec 28)

1 Reclining motor
2 Fore-aft motor
3 Rear height motor
4 Front height motor
R1 Control relay
R2 Interlock relay
For colour code see main wiring diagrams

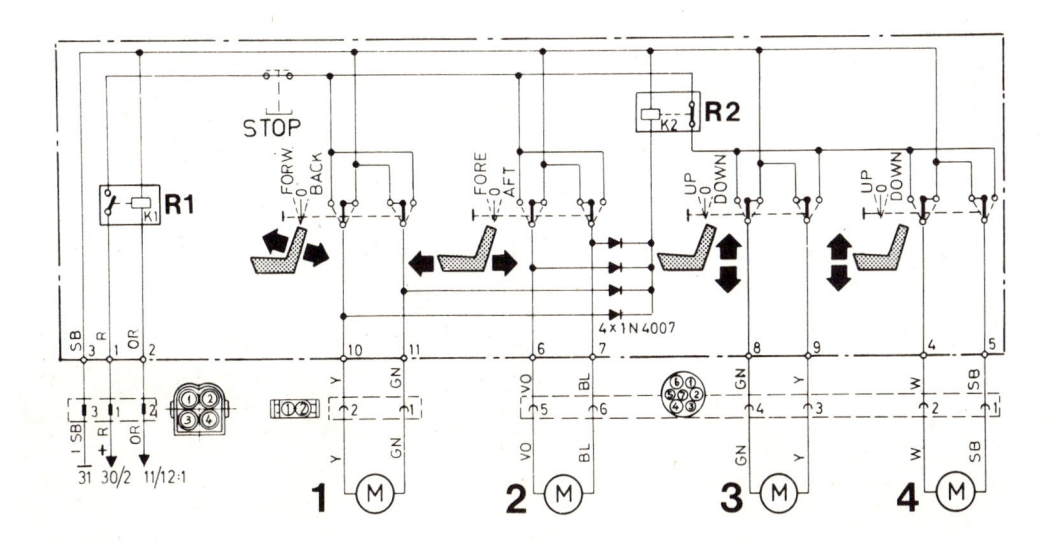

29 Seat heating elements – removal and refitting

1   Remove the front seat (Section 25).
2   Recline the backrest as far as it will go. Insert the seat and free the wiring from the cable ties.

Backrest heater
3   Remove the head restraint (Section 27).
4   Remove the backrest adjuster knob, and the lumbar support adjuster knob and guide.
5   Remove the upholstery retaining rod. Cut the clamp rings which secure the bottom edge of the upholstery. Peel off the upholstery, freeing the centre attachment clamps.
6   The heating element can now be removed.

Cushion heater
7   Remove the upholstery retaining rods. Unhook the side springs and remove the cushion (photo).
8   Cut the clamp rings which secure the upholstery. Peel off the upholstery, unhooking the centre attachments (photo).
9   The heating element and thermostat can now be removed.

All heaters
10   Refit by reversing the removal operations, using new clamp rings when necessary.

Fig. 11.26 Unscrew the special nut (arrowed) to remove the backrest adjuster knob (Sec 29)

Fig. 11.27 Backrest upholstery retaining rod (arrowed) (Sec 29)

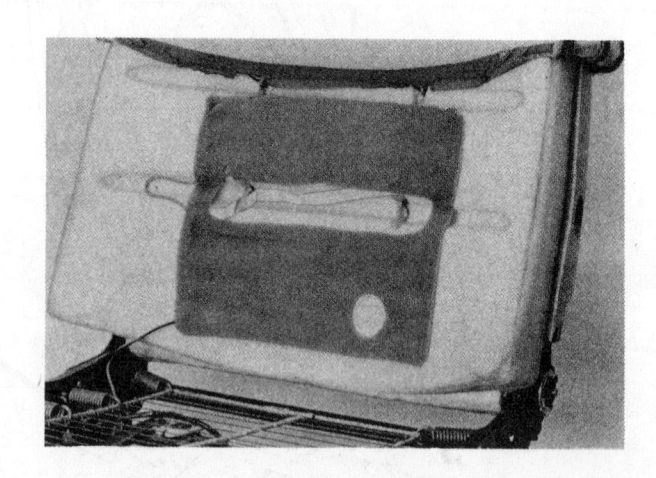

Fig. 11.28 Backrest heating element exposed (Sec 29)

29.7 Unhooking an upholstery retaining rod

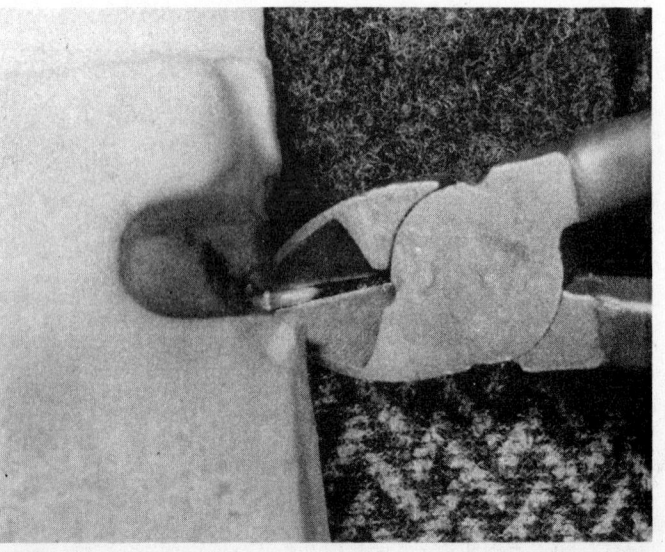

29.8 Cutting a clamp ring

Chapter 11 Bodywork and fittings

30 Seat belts – care and maintenance

1 Maintenance is limited to periodic inspection of the belts for fraying or other damage. Also check the operation of the buckles and retractor mechanisms. In case of damage or malfunction the belt must be renewed.
2 If it is wished to clean the belts, use only an approved upholstery cleaner or a weak solution of detergent, followed by rinsing with water. Do not use solvents, strong detergents, dyes or bleaches. Keep the belt extended until it is dry.
3 Belts which have been subjected to impact loads must be renewed.

31 Front seat belts – removal and refitting

1 Move the front seat forwards. Remove the trim or storage pocket from the outboard side of the seat base. Unbolt the seat belt anchorage.
2 Remove the B-pillar trim panel, which is secured by two screws concealed by plugs. Free the belt guide from the slot in the trim.
3 Remove the reel cover/sill trim panel, which is secured by seven concealed screws.
4 Unbolt the belt upper guide and the inertia reel unit, noting the location of any washers and spacers. Remove the belt and reel (photos).
5 To remove the buckle, it is first necessary to remove the seat (Section 25).
6 Refit by reversing the removal operations.

32 Rear seat belts – removal and refitting

Saloon
1 Remove the rear seat (Section 26).
2 The buckles and floor anchorages can now be unbolted from the seat pan (photo).
3 To gain access to the inertia reels it will first be necessary to remove the parcel shelf loudspeakers (when fitted). Access to their connectors and fastenings is from inside the boot.
4 Remove the parcel shelf securing clips and the parcel shelf itself (photo).
5 Remove the reel cover clips and the reel covers (photo).
6 Unbolt and remove the inertia reels, noting the position of any spacers (photo).
7 Refit by reversing the removal operations.

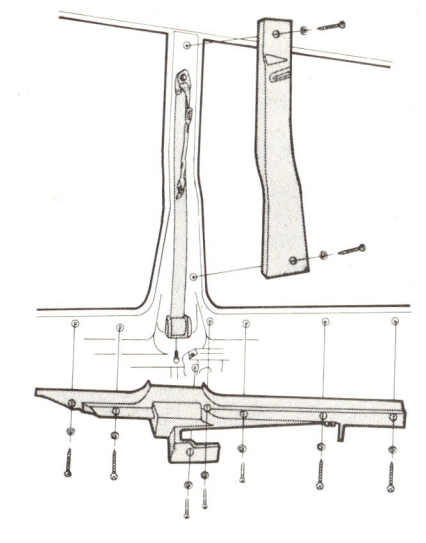

Fig. 11.29 Remove the trim panels for access to the front seat belt (Sec 31)

31.4A Front belt upper guide

31.4B Unbolting an inertia reel unit

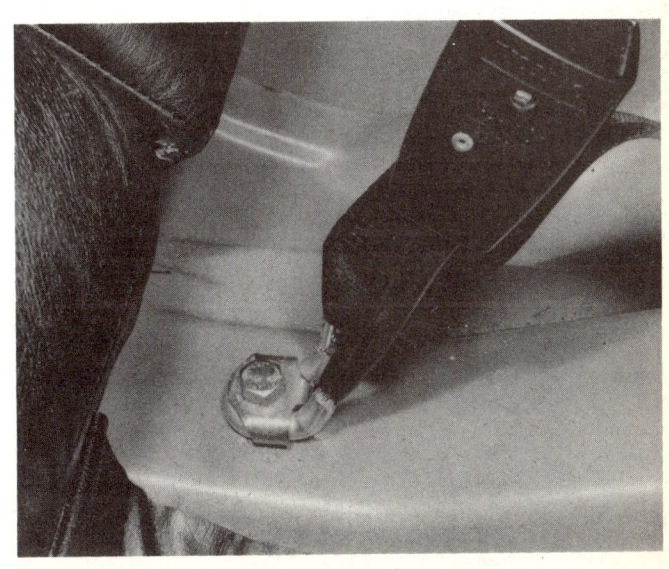

32.2 Rear seat belt buckle bolted to the floorpan

268 Chapter 11 Bodywork and fittings

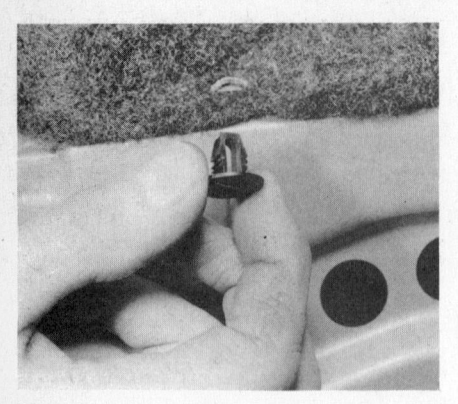

32.4 Removing a parcel shelf clip

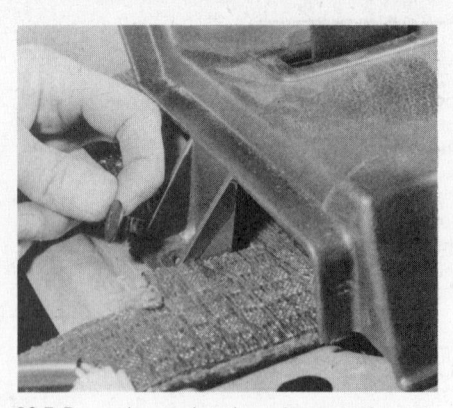

32.5 Removing an inertia reel cover clip

32.6 Rear inertia reel unit with cover removed

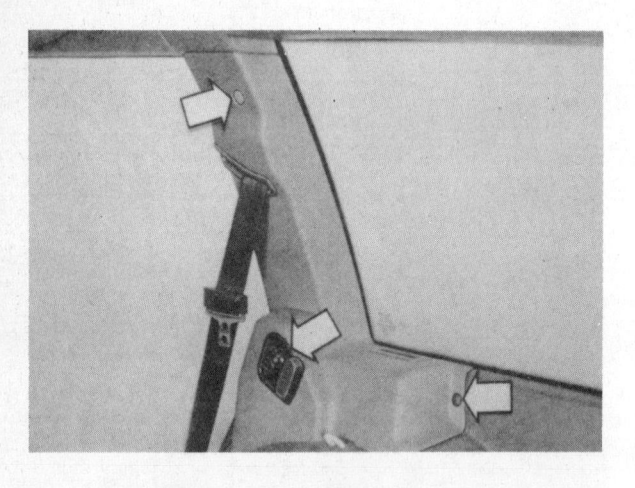

Fig. 11.30 C-pillar trim panel fixing points (arrowed) – Estate (Sec 32)

Estate

8 Access to the buckles and floor anchorages is gained by tipping the seat cushion forwards.
9 The inertia reels are accessible after removing the C-pillar trim. This is attached by a screw at the top, by the seat back stop at the bottom and by a clip at the rear.

33 Steering column/pedal trim panel – removal and refitting

1 The large trim panel below the steering column is secured by two screws and two clips (photo). Remove the screws and turn the clips 90° to release them.
2 Lower the trim panel and disconnect the heater duct from it. Remove the panel.
3 Refit by reversing the removal operations.

34 Glovebox – removal and refitting

1 Open the glovebox. Prise out the two trim pads from the edges of the glovebox and remove the two screws so exposed (photo).
2 Remove the trim panel from below the glovebox. This is secured by three clips which must be turned 90°.
3 Remove the nut at the base of the glovebox (towards the centre of the vehicle) (photo).
4 Lower the glovebox, disconnect the wires from it and remove it.
5 Refit by reversing the removal operations.

35 Centre console – removal and refitting

1 Disconnect the battery negative lead.
2 Remove the steering column/pedal trim (Section 33) and the glovebox (Section 34). This may not be essential but will improve access.
3 Remove the radio (Chapter 12, Section 35).
4 Remove the ashtray and its carrier (photo).

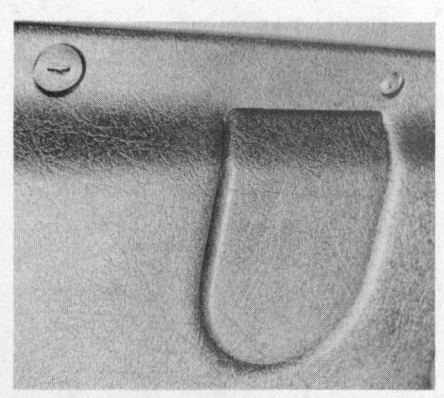

33.1 One of the clips and one of the screws which secure the steering column/pedal trim panel. D-shaped insert is of no significance

34.1 Removing a glovebox screw trim pad

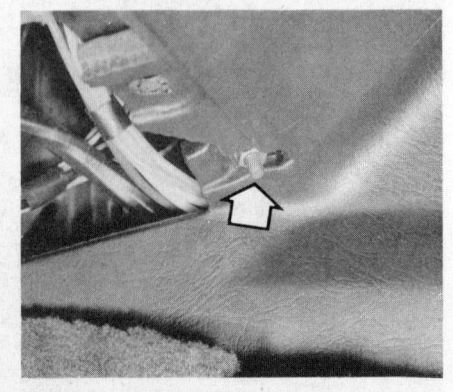

34.3 Glovebox base nut (arrowed)

Chapter 11 Bodywork and fittings

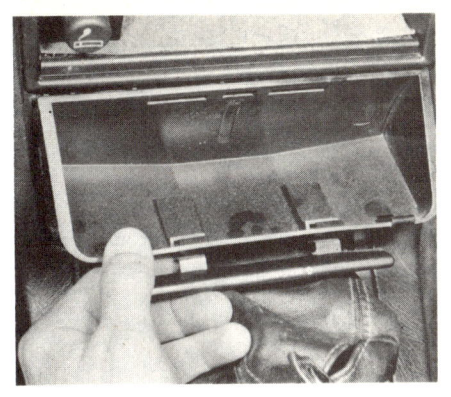

269

35.4 Removing the ashtray carrier

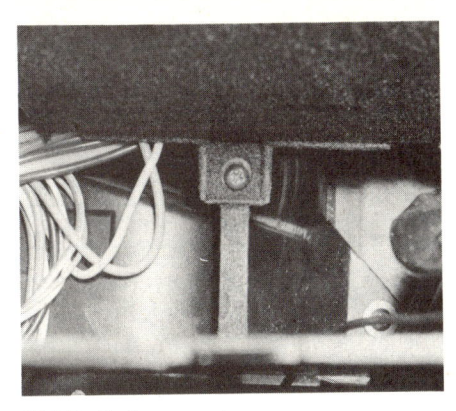

35.6 Radio tray securing screw

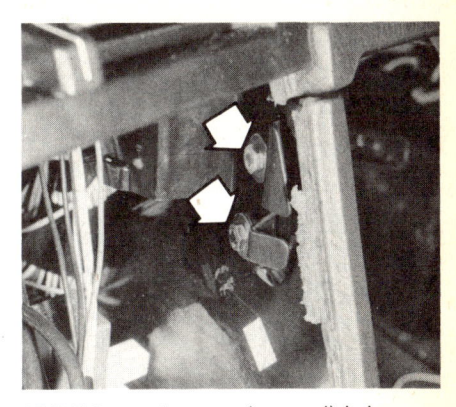

35.7 Side panel screws (arrowed) below the heater controls

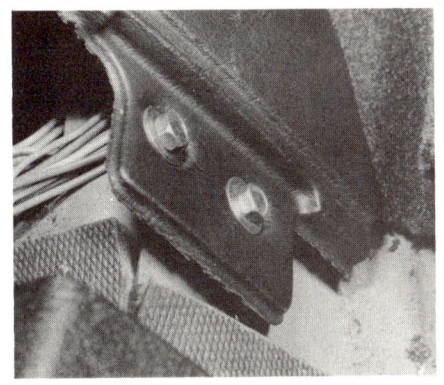

35.8 Side panel rear edge screws

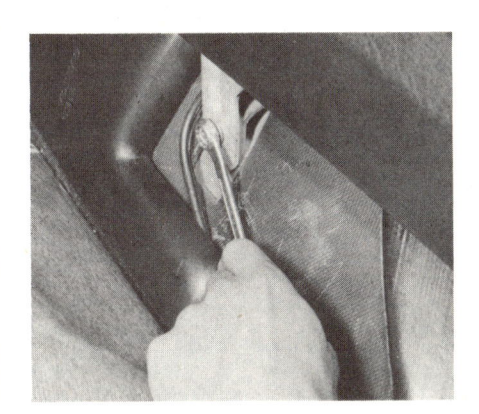

35.9 Removing a side panel forward end screw

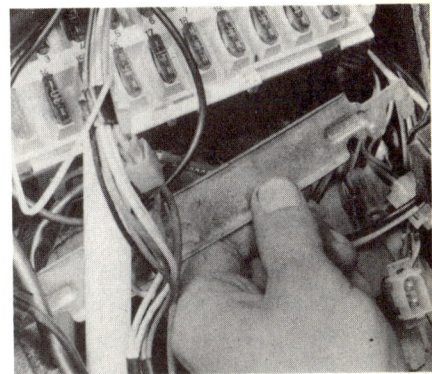

35.10 Removing the side panel connecting strut

5 Remove the cigarette lighter tray. This is secured by one or two screws, exposed by removing the lighter element and cover plate. Disconnect the lighter feed and withdraw the bulb holder as the tray is withdrawn. (On some models the tray may contain audio equipment.)
6 Remove the radio tray, which is secured by a single screw at the back (photo).
7 Remove the side panel screws from below the heater controls (photo).
8 Remove the screws (two on each side) which secure the rear edge of the centre console side panels. It will be necessary to release the rear console and move it rearwards to gain access to these screws (photo).
9 Peel back the carpet from the forward end of the transmission tunnel and remove the two screws (one each side) which secure the forward ends of the side panels (photo).
10 Slacken the screws which secure the side panel connecting strut. Release the strut, which has slotted fixing holes (photo).
11 Remove the centre console side panels.
12 Refit by reversing the removal operations.

36 Rear console – removal and refitting

1 Lift the armrest, empty the storage box and prise out the cover plate from the bottom of the box. Remove the two screws thus exposed (photo).
2 On manual gearbox models, remove the gear lever/handbrake trim. This is retained by two screws. Disconnect any switches.

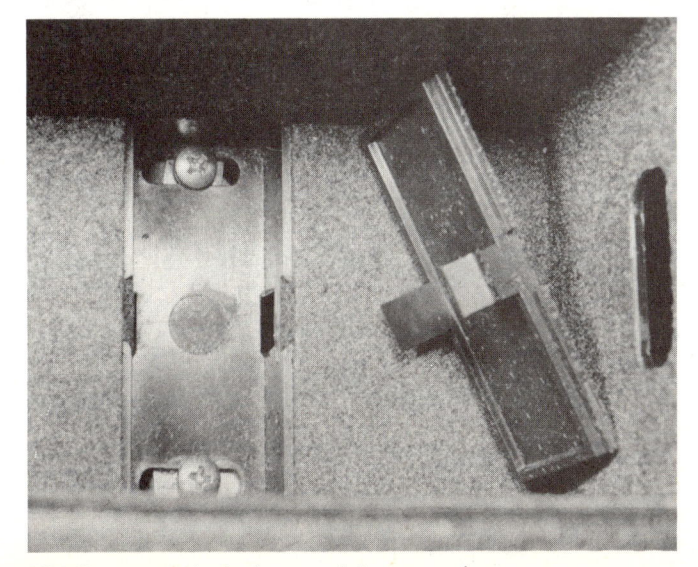

36.1 Two screws in the bottom of the storage box

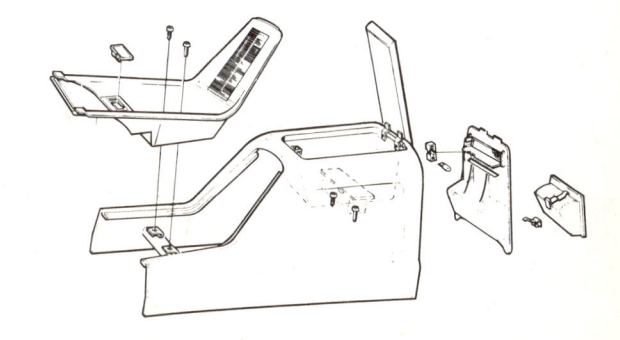

Fig. 11.31 Rear console fittings (Sec 36)

270　　　　　　　　　　　　　　Chapter 11 Bodywork and fittings

3　On automatic transmission models, remove the selector lever trim. See Chapter 6, Section 34.
4　Lift the rear console. Separate the rear ashtray/cigarette lighter/seat belt warning light panel from the console. Remove the console, leaving the panel behind.
5　Refit by reversing the removal operations.

37　Facia – removal and refitting

1　Disconnect the battery negative lead.
2　Remove the steering wheel, the steering column switches and the instrument panel. See Chapters 10 and 12.
3　Remove the steering column/pedal trim, the glovebox, the centre console and the rear console. See Sections 33 to 36.
4　Remove the footwell side trim panels (photo). Also remove the A-pillar trim.
5　Disconnect the switch and lighting multi-plugs.
6　Remove the screw which secures the steering column top bearing to the facia. Recover the spacer tube.
7　Remove the central air vents (photo). Remove the air mix box retaining screw and disconnect the air ducts.
8　Unclip the wiring harness from the facia.
9　On vehicles with automatic climate control, remove the inner temperature sensor.
10　Remove the facia panel complete with switchgear, demister vents etc. Transfer components as necessary if a new panel is to be fitted.
11　Refit by reversing the removal operations.

38　Front spoiler – removal and refitting

1　Have an assistant support the spoiler. Remove the six bolts and D-shaped washers which secure it to the bumper (photo).
2　Free the spoiler from the bumper side section and remove it.
3　Refit by reversing the removal operations.

39　Front bumper – removal and refitting

1　Remove the three nuts which secure each side section (photo). (Depending on equipment and model, it may be necessary to remove the battery, washer reservoir and/or air cleaner for access.)

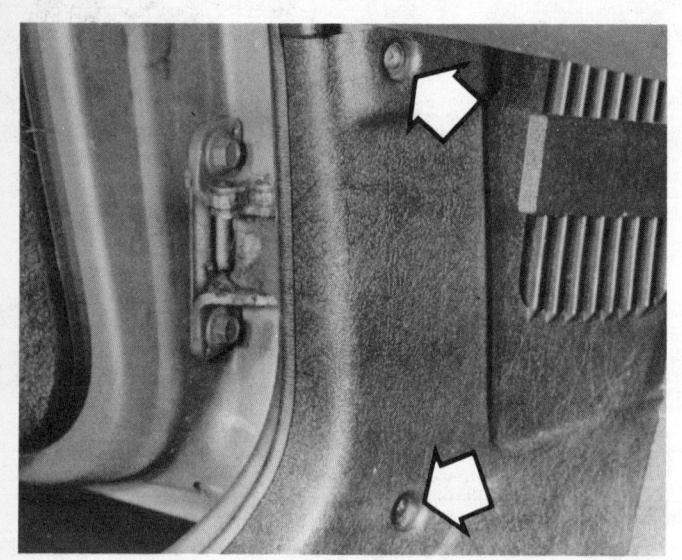

37.4 Footwell side trim panel screws (arrowed)

37.7 Removing the central air vent

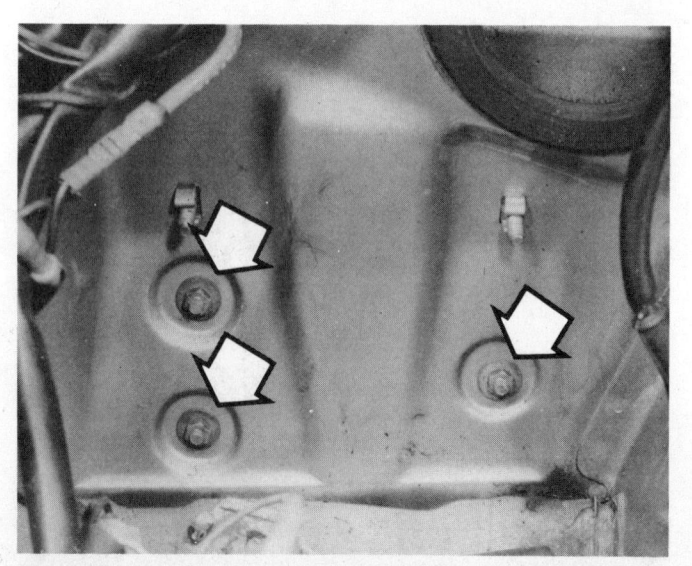

38.1 A front spoiler bolt

39.1 Three nuts (arrowed) which secure the bumper side section. Air cleaner has been removed for access

Chapter 11 Bodywork and fittings

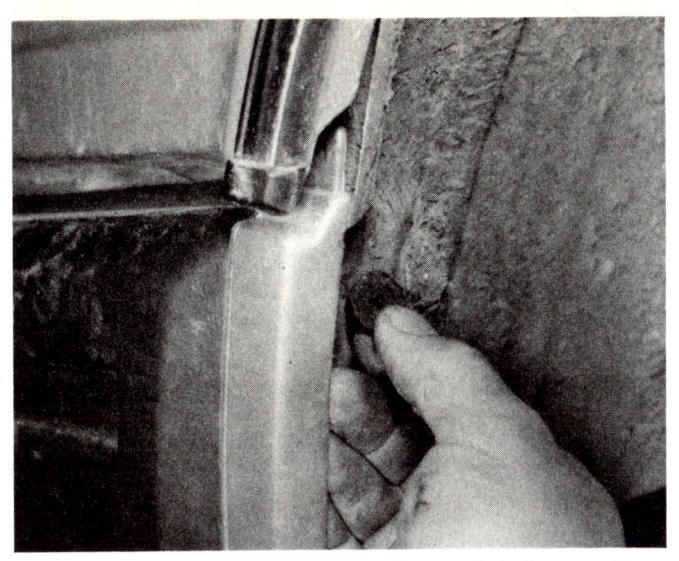

39.2 Removing the plug which secures the rear of the bumper side section

2 Remove the single plug securing the rear edge of each side section (photo).
3 Remove the front spoiler (Section 38).
4 Disconnect or remove the auxiliary lights (when fitted).
5 From inside the bumper remove the four nuts which secure it to the buffers. Remove the bumper.
6 Refit by reversing the removal operations.

40 Rear bumper – removal and refitting

1 This is removed in a similar way to the front bumper, but access to the side section nuts is gained from within the boot or load area wells.

41 Bumper buffers – removal and refitting

1 Remove the front or rear bumper, as appropriate.
2 Remove the two nuts and bolts and the single nut securing the buffer. The single nut is reached from the engine bay (front buffers) or from inside the boot or load area (photo).

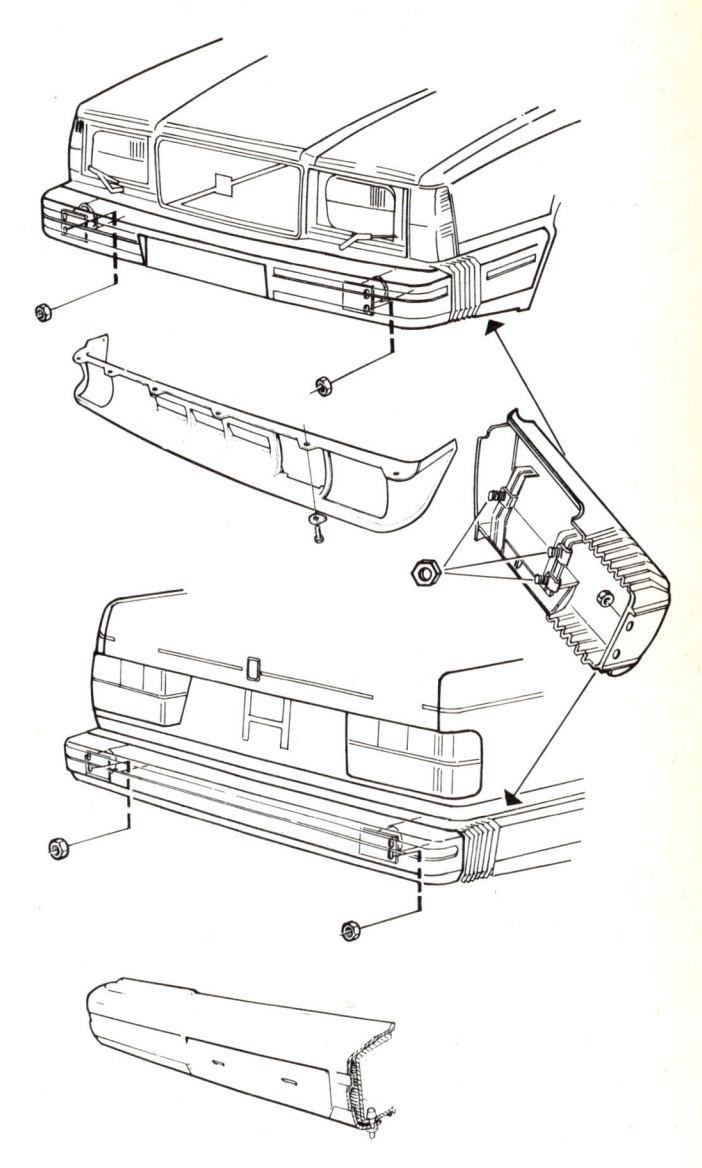

Fig. 11.32 Bumper fitting details (Secs 39 and 40)

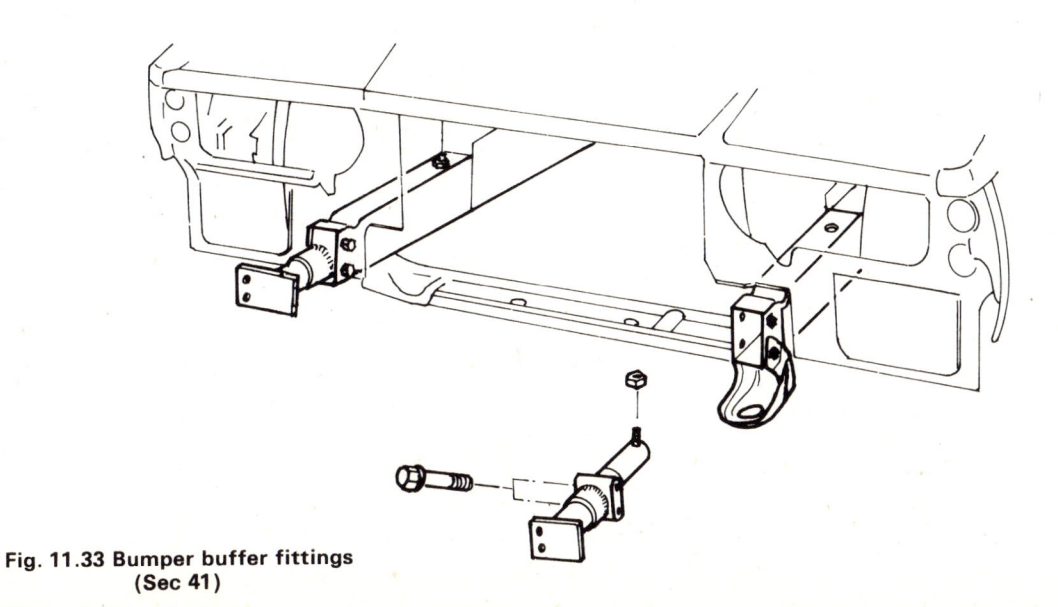

Fig. 11.33 Bumper buffer fittings (Sec 41)

Chapter 11 Bodywork and fittings

41.2A Front bumper buffer securing bolts (arrowed) ...

41.2B ... and single nut in the engine bay

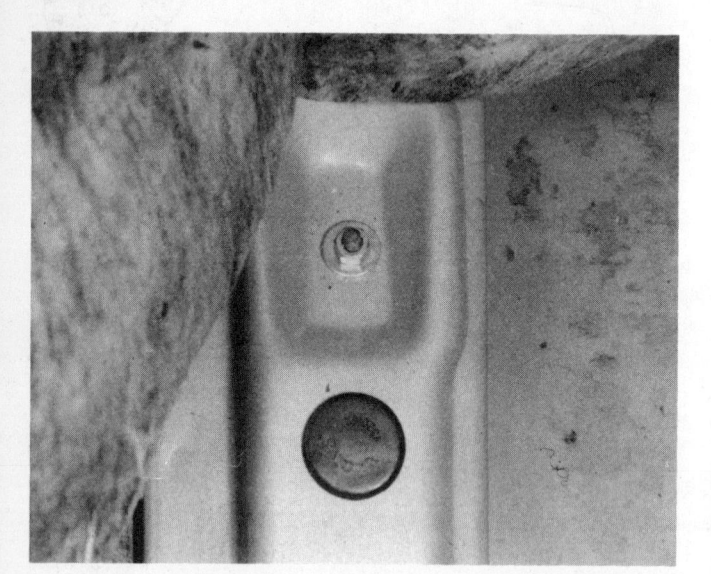

3 Pull the buffer out of its bracket.
4 Do not puncture the buffers, or perform welding on or near them. They contain gas under pressure which could cause injury if suddenly released.
5 Refit by reversing the removal operations.

42 Front grille panel – removal and refitting

1 Open the bonnet. Squeeze the grille panel top retaining clips and remove them (photo).
2 Release the panel from its bottom mountings and remove it (photo).
3 Refit by reversing the removal operations.

43 Engine undertray – removal and refitting

1 Raise and support the front of the vehicle.
2 Remove the undertray securing screws and free it from the lugs

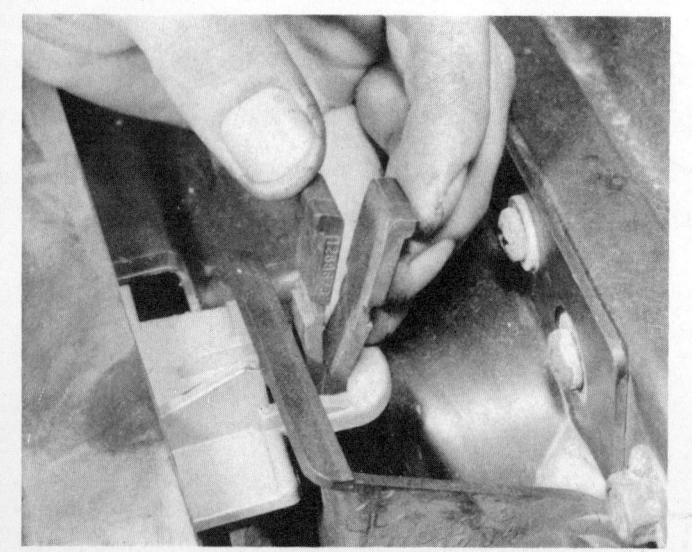

41.2C Rear bumper buffer nut in the boot floor

42.1 Removing a grille panel clip

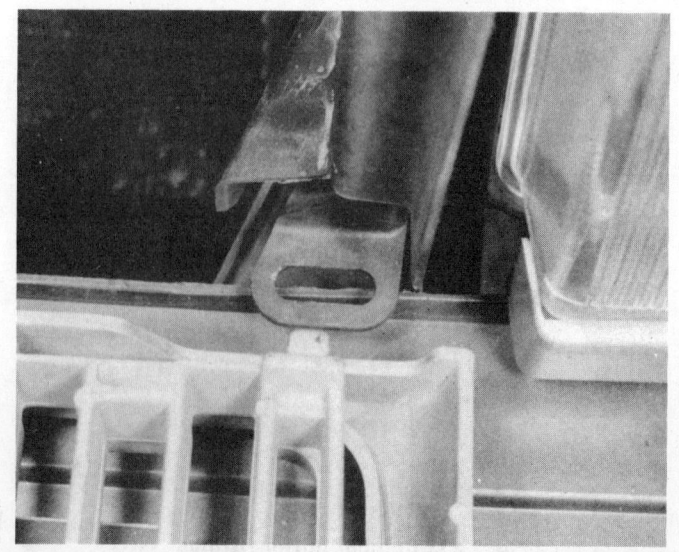

42.2 Grille panel bottom mounting

Chapter 11 Bodywork and fittings

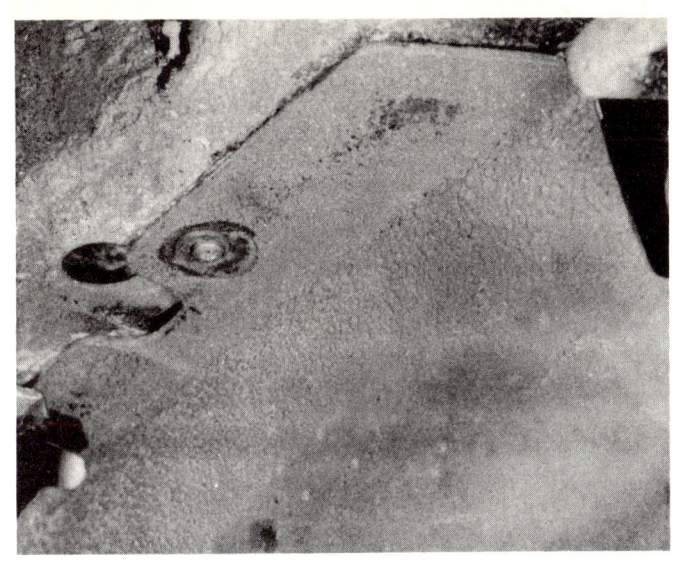

43.2A An undertray side securing screw

43.2B Undertray lug and slot

(photos). When a vacuum tank is fitted in this area, it may share some of the undertray screws.
3   Refit by reversing the removal operations.

44 Front wing – removal and refitting

1   Remove the windscreen wiper arms and the wiper spindle seals.
2   Open the bonnet to the fully raised position. Remove the scuttle panel, which is secured by three screws and some clips.
3   Remove the bumper end section and the direction indicator/parking light unit on the side concerned.
4   Remove the screws which secure the wing. Pull the wing off the wheel arch and remove it.
5   Refit by reversing the removal operations. Apply a suitable sealer (Volvo No 591 278-7, or equivalent) to the wing-to-wheel arch joint.

45 Sunroof motor – removal and refitting

1   Remove the motor cover securing screws and unclip the cover.
2   If a fabric headlining material is covering the motor, carefully cut it away, keeping inside the outline of the motor cover.
3   Remove the two screws and one nut which secure the motor.
4   Lower the motor off the drivegear and disconnect the wiring multi-plug (photo). If the multi-plug is deep in the headlining, either peel back the headlining for access, or cut the wires and fit new connectors when refitting.
5   Before refitting the motor to the drivegear, run it backwards (to the 'fully open' position), then forwards (to the 'fully closed' position) until it stops.
6   Refit and secure the motor and cover.

46 Sunroof – removal and refitting

1   Open the sunroof to the ventilation position (rear edge raised). From outside, release the clips which secure the headlining to the rear edge.

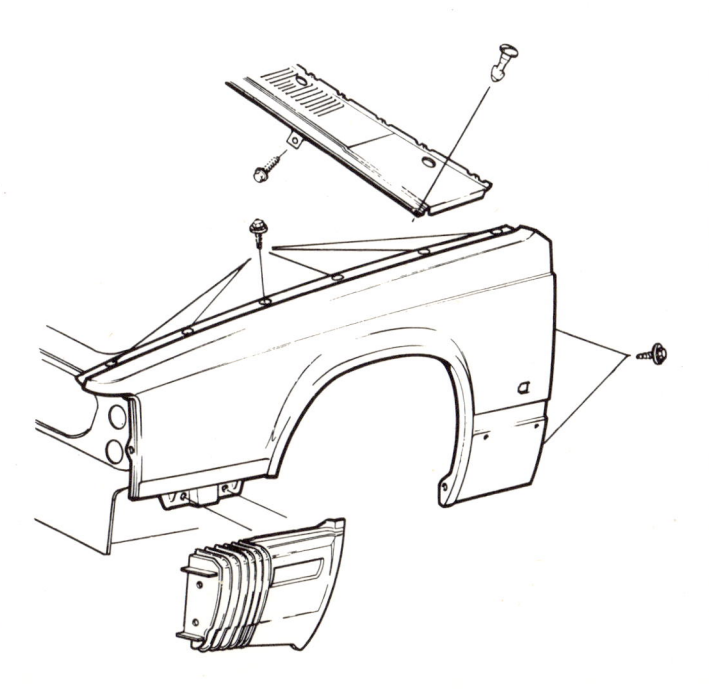

Fig. 11.34 Front wing fitting details (Sec 44)

45.4 Removing the sunroof motor

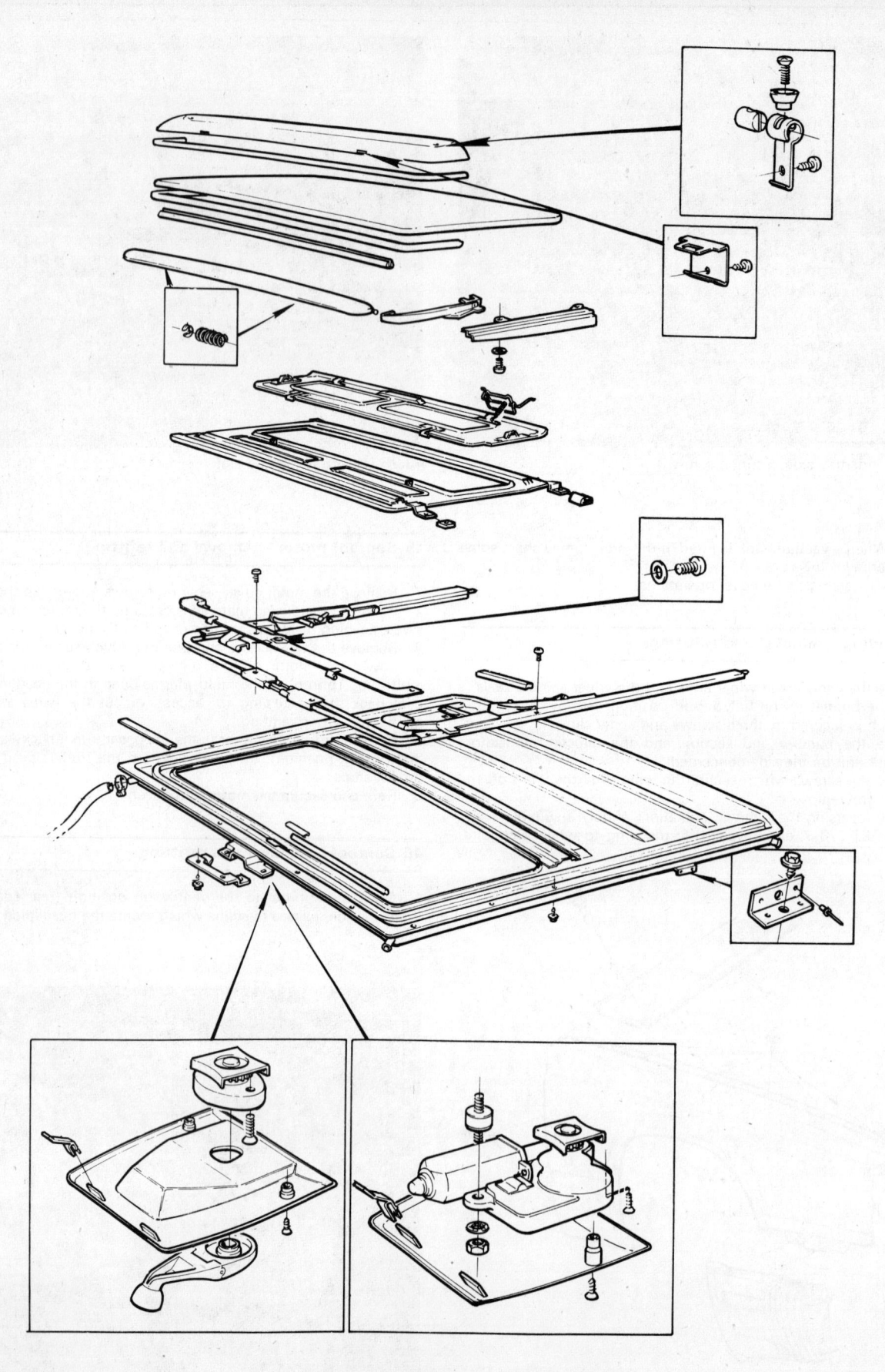

Fig. 11.35 Sunroof components (Secs 45 to 47)

Chapter 11 Bodywork and fittings

2 Back inside, close the sunroof. Free the headlining from the front edge by jerking it rearwards, at the same time pulling down on the main headlining.
3 Release the sunroof spring retainers. Undo the side and front retaining screws and lift out the sunroof.
4 The sealing strip can be renewed after removing the side rails. The join in the strip should be located approximately 100 mm (4 in) from the rear and on the left-hand side.
5 Refit the sunroof and make sure that the sealing strip is a good fit. Fit the retaining screws, positioning the front outside edge of the sunroof flush with the roof or recessed at most by 1.5 mm (0.06 in). The rear edge should be 0.5 to 1.5 mm (0.02 to 0.06 in) proud of the roof. Tighten the screws when the position is correct.
6 Fore-and-aft adjustment, if required, is carried out after slackening the upper adjustment screws (Fig. 11.36). Retighten the screws when adjustment is correct.
7 Test the operation of the sunroof, but do not run it too far back with the headlining detached as it may jam.
8 When satisfied with the adjustment, reattach the headlining and check the full range of movement of the sunroof.

47 Sunroof frame and cables – removal and refitting

1 Remove the sunroof (Section 46) and the crank or motor (Section 45).
2 Protect the paintwork forward of the sunroof aperture with a blanket or similar item.
3 Remove the front mounting, the eight frame retaining screws and the wind deflector.
4 Grasp the control mechanism on both sides. Press it down and pull it forwards until the drain channels emerge.
5 Remove the frame, and if wished the headlining.
6 To remove the cables, free the cable pins from the control mechanism. Remove the guide studs at the rear of the frame and pull out the control mechanism. The cables can then be removed.
7 Refit by reversing the removal operations. Set the control mechanism and the motor or crank to the 'sunroof closed' position when refitting.

Fig. 11.36 One of the sunroof upper adjustment screws (arrowed) (Sec 46)

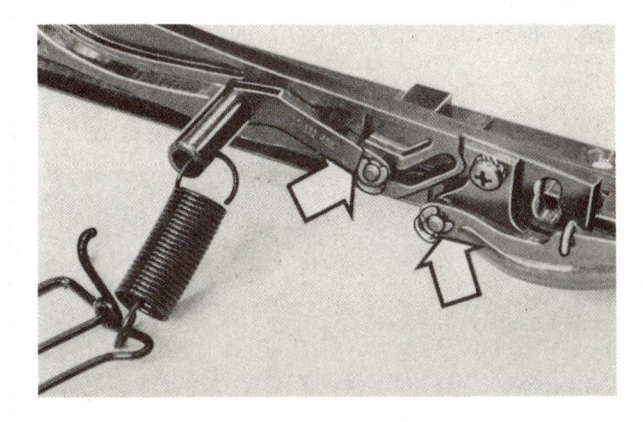

Fig. 11.37 Two circlips (arrowed) which secure the control mechanism (Sec 47)

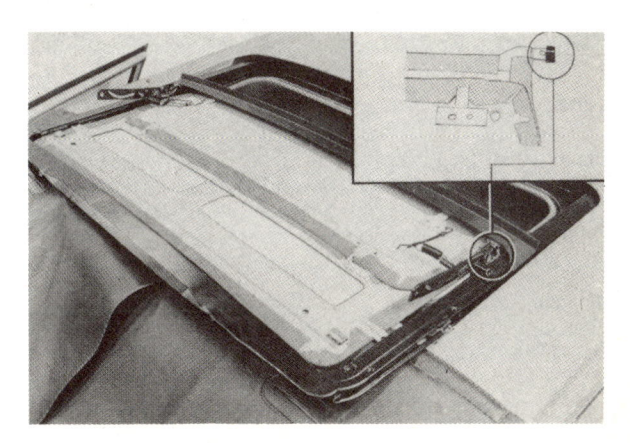

Fig. 11.38 Refitting the sunroof frames. Inset shows correct fitting of headlining clips (Sec 47)

Chapter 12 Electrical system

For modifications, and information applicable to later models, see Supplement at end of manual

Contents

Alternator – brush renewal	9
Alternator – precautions	6
Alternator – removal and refitting	8
Alternator – testing on the vehicle	7
Battery – charging	4
Battery – maintenance	3
Battery – removal and refitting	5
Bulb failure warning system – description	25
Cigarette lighter – removal and refitting	33
Direction indicator system – fault tracing	26
Exterior light bulbs – renewal	13
Exterior light units (except headlights) – removal and refitting	14
Fault diagnosis – electrical system	38
Fuses – general	23
General description	1
Headlight beam alignment	16
Headlight unit – removal and refitting	15
Headlight wiper motor – removal and refitting	31
Heated rear window – general	34
Horn – removal and refitting	27
Ignition/starter switch – removal and refitting	18
Instrument cluster – dismantling and reassembly	21
Instrument cluster – removal and refitting	20
Interior light bulbs – renewal	17
Maintenance and inspection	2
Mobile radio equipment – interference-free installation	37
Radio aerial (original equipment) – removal and refitting	36
Radio/cassette player (original equipment) – removal and refitting	35
Rear wiper motor – removal and refitting	32
Relays – general	24
Speedometer sender unit – removal and refitting	22
Starter motor – overhaul	12
Starter motor – removal and refitting	11
Starter motor – testing on the vehicle	10
Switches – removal and refitting	19
Windscreen/headlight/tailgate washers – general	28
Windscreeen wiper motor and linkage – removal and refitting	30
Wiper blades and arms – removal and refitting	29

Specifications

General
System type	12 volt, negative earth
Battery type	Lead acid
Battery capacity	55 or 66 Ah

Alternator
Make and type	Bosch K1 or N1
Maximum output (at 14V):	
K1	55A
N1	70 or 90A
Brush minimum length	5.0 mm (0.2 in)
Slip ring diameter:	
New	28.0 mm (1.102 in)
Minimum – K2	26.8 mm (1.055 in)
Minimum – N1	27.0 mm (1.063 in)
Rotor resistance:	
K1 and N1 70A	3.4 to 3.7 Ω
N1 90A	2.8 to 3.1 Ω
Voltage regulator type	Integral with brushgear
Regulated voltage at 3000 (engine) rpm:	
5A load	13.8 to 14.9V
30 to 50A load	13.2 to 14.8V

Wiper blades
740 (1982-on)	Champion X-5103
760 (1982 to 1987)	Champion X-5103
760 (1987-on)	Champion X-5303

Starter motor
Make and type	Bosch GF or DW, pre-engaged
Nominal output:	
GF	1.1 kW (1.5 hp)
DW	1.4 kW (1.9 hp)
Commutator minimum diameter:	
GF	33.5 mm (1.319 in)
DW	31.2 mm (1.228 in)
Brush minimum length:	
GF	13.0 mm (0.512 in)
DW	8.0 mm (0.315 in)
Armature endfloat	0.1 to 0.3 mm (0.004 to 0.012 in)

Chapter 12 Electrical system

Light bulbs (typical)	Wattage	Pattern
Headlights	60/55	P45t-38 (H4)
Day running/parking lights	21/5	BA7 15d
Direction indicators	21	BA 15s
Direction indicator side repeaters	5	W 2.1x9.5d
Front foglights/spotlights	55	PK 22s (H3)
Tail lights	5	BA 15s
Stop-lights	21	BA 15s
Combined stop and tail	21/5	BA7 15d
Rear foglights/reversing lights	21	BA 15s
Number plate light	5	BA 9s
Interior (courtesy) light	10	SV 8.5
Reading lights	5	W 2.1x9.5d
Engine bay/load area lights	10	SV 8.5
Glovebox light	2	BA 9s
Vanity mirror light	3	SV 7
Door edge marker light	3	W 2.1x9.5d
Indicator and warning lights	1.2	Integral holder
Instrument illumination	3	W 2.1x9.5d
Control illumination	1.2	W 2x4.6d

Fuses – 1982/83 models

Fuse No	Circuits protected	Rating (A)
1 (or 2)	Fuel pump (main)	25
2 (or 1)	Hazard warning, headlight flasher, central locking	25
3	Front foglights/spotlights	15
4	Stop-lights	15
5	Clock, interior lighting, power aerial, radio (full-time), door edge marker lights	15
6	Cooling fan	25
7	Window motors	30
8	Direction indicators, constant idle system, overdrive relay	15
9	Heated rear window, sunroof motor	30
10	Instruments, reversing lights, seat heaters, seat belt reminder, fuel pump relay, window motor relay, cooling fan relay, air conditioning delay valve, oil level sensor, bulb failure warning	15
11	Day running lights, cruise control, heater blower (low speed), automatic climate control	25
12	Cigarette lighter, radio (ignition-controlled), mirror motors, seat motors	25
13	Horn, windscreen wash/wipe, headlight wash/wipe	25
14	Heater blower (high speeds)	30
15	Fuel pump (auxiliary)	15
16	Rear foglights and relay	15
17	Main beam (LH) and main beam pilot light	15
18	Main beam (RH), front spotlight relay	15
19	Dipped beam (LH)	15
20	Dipped beam (RH)	15
21	Instrument and control lighting, tail/parking lights (LH), number plate light, warning buzzer, front ashtray	15
22	Rear ashtray light, transmission tunnel switch lighting, tail/parking lights (RH), front foglight relay	15

Fuses – 1984 models

Fuse No	Circuits protected	Rating (A)
1	Fuel pump (main), fuel injection system	25
2	Hazard warning, headlight flasher, ABS, central locking	25
3	Front foglights/spotlights	15
4	Stop-lights	15
5	Clock, interior lighting, power aerial, radio (full-time), door edge marker lights	15
6	Cooling fan, seat belt reminder, seat heaters	25
7	Window motors	30
8	Day running lights, bulb failure warning, window motor relay, cooling fan relay	15
9	Heated rear window, sunroof motor, air conditioning	25
10	Instruments, reversing lights, ignition system, cruise control, oil level sensor, ABS	15 or 25
11	Direction indicators, overdrive relay, constant idle system, inlet heater relay	15 or 25

278　　Chapter 12 Electrical system

Fuse No	Circuits protected	Rating (A)
12	Cigarette lighter, radio (ignition-controlled), mirror motors, seat motors	15
13	Horn, windscreen wash/wipe, headlight wash/wipe	25
14	Heater blower, air conditioning	30
15	Fuel pump (auxiliary)	15
16	Rear foglights and relay	15
17	Main beam (LH) and main beam pilot light	15
18	Main beam (RH), front spotlight relay	15
19	Dipped beam (LH)	15
20	Dipped beam (RH)	15
21	Instrument and control lighting, tail/parking lights (LH), number plate light	15
22	Rear ashtray light, transmission tunnel switch lighting, tail/parking lights (RH), front foglight relay	15

Fuses – 1985/86 models

Fuse No	Circuits protected	Rating (A)
1	Fuel pump (main), fuel injection system	25
2	Hazard warning, headlight flasher, ABS, central locking	25
3	Front foglights/spotlight relay, rear foglight relay	15
4	Stop-lights	15
5	Clock, interior lighting, power aerial, radio (full-time), door edge marker lights	15
6	Seat heaters	15
7	Cooling fan	25
8	Window motors	30
9	Direction indicators, seat belt reminder, seat heater relay, window motor relay, air conditioning relay, cooling fan relay	15
10	Heated rear window, sunroof motor, heated mirrors	30
11	Fuel pump (auxiliary)	15
12	Reversing lights, oil level sensor, overdrive, ignition system, cruise control, ABS	15
13	Fuel injection system	15
14	Mirror motors, cigarette lighter, radio (ignition-controlled), rear window wiper	15
15	Horn, windscreeen wash/wipe, headlight wash/wipe	25
16	Heater blower, air conditioning	30
17	Main beam (LH) and main beam pilot light	15
18	Main beam (RH), front spotlights	15
19	Dipped beam (LH)	15
20	Dipped beam (RH)	15
21	Instrument lighting, tail/parking lights (LH), number plate lights	15
22	Rear ashtray light, transmission tunnel switch light, tail/parking lights (RH)	15
23	Seat motor relay	15
24	Spare	–
25	Day running lights	15
26	Seat motors	30

1 General description

The electrical system is of the 12 volt, negative earth type. Electricity is generated by an alternator, belt-driven from the crankshaft pulley. A lead-acid storage battery provides a reserve of power for use when the demands of the system temporarily exceed the alternator output, and for starting.

The battery negative terminal is connected to 'earth' – vehicle metal – and most electrical system components are wired so that they only receive a positive feed, the current returning via vehicle metal. This means that the component mounting forms part of the circuit. Loose or corroded mountings can therefore cause apparent electrical faults.

Many semiconductor devices are used in the electrical system, both in the 'black boxes' which control vehicle functions and in other components. Semiconductors are very sensitive to excessive (or wrong polarity) voltage, and to extremes of heat. Observe the appropriate precautions to avoid damage.

Although some repair procedures are given in this Chapter, sometimes renewal of a well-used item will prove most satisfactory. The reader whose interests extend beyond component renewal should obtain a copy of the *'Automobile Electrical Manual'*, available from the publishers of this book.

Before starting work on the electrical system, read the precautions listed in *'Safety first!'* at the beginning of the manual.

2 Maintenance and inspection

1 Weekly, before a long journey, or when prompted by the appropriate warning light, check that all exterior lights are working. Renew blown bulbs as necessary (Section 13). Top up the washer reservoirs.
2 Every 6000 miles or six months, inspect the battery and check the electrolyte level as described in Section 3.
3 Every 12 000 miles or twelve months, or whenever poor output from the alternator is suspected, check the tension and condition of the alternator drivebelt(s). See Chapter 2, Section 7.
4 Periodically check the condition of the windscreen, rear window

Chapter 12 Electrical system

and headlight wiper blades and arms. Renew as necessary (Section 29).
5 When a power-operated radio aerial is fitted, lubricate it regularly with a proprietary water repellant/lubricant spray, and wipe it clean with a soft cloth. If the aerial is allowed to become dirty or corroded it will jam.

3 Battery – maintenance

1 Periodically inspect the battery terminals for corrosion (photo). If any is found, disconnect the terminals (negative first) and clean them and the terminal posts with abrasive paper or by scraping.
2 Bad corrosion or 'fungus' can be removed by brushing with a solution of sodium bicarbonate in warm water. Do not allow this solution to enter the battery cells.
3 Smear the terminals and posts with petroleum jelly or a proprietary anti-corrosive compound, then refit and tighten the terminals (negative last). Do not overtighten.
4 At the same intervals, check the battery electrolyte level. The electrolyte (liquid) in the cells should cover the plate by about 6 mm (0.25 in), or as marked by level lines on the battery case. If the case is translucent the level may be checked from the outside; otherwise, remove the cell caps and shine a torch or inspection light into the cells to check the level. Remember not to smoke or use naked lights.
5 If topping-up of the electrolyte is necessary, use distilled or de-ionized water, or melted frosting from the ice compartment of a refrigerator. Do not overfill. Run the engine after topping-up to mix the electrolyte.
6 Frequent need for topping-up suggests either a charging system fault (voltage too high) or a battery which is nearing the end of its life.
7 Replacement batteries may be of the 'maintenance-free' type. It is not always possible to inspect or top up the electrolyte on such batteries. Follow the maker's instructions.
8 Never add acid to a battery unless electrolyte has been lost through spillage. Consult an auto-electrician or battery specialist for the procedure to be followed in this case.
9 Whenever the the battery is removed, inspect its tray and the surrounding area for corrosion. Neutralise and repaint as necessary.
10 Keep the top of the battery clean and dry.

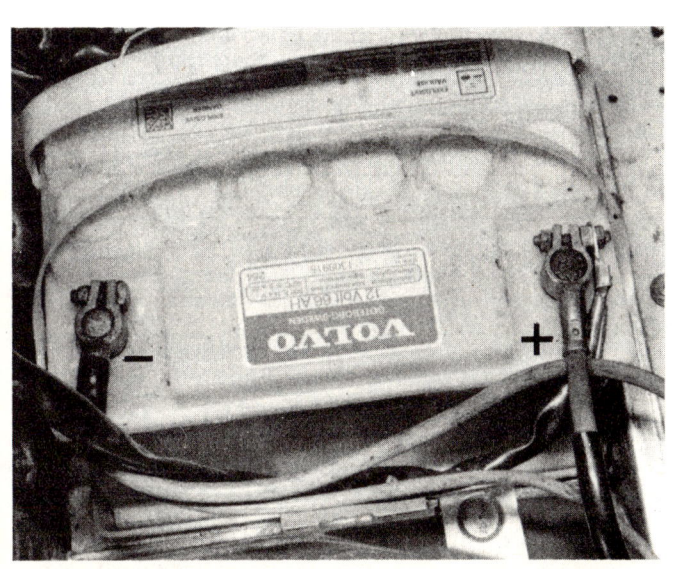

3.1 Battery, showing positive (+) and negative (−) terminals

4 Battery – charging

1 In normal use the battery should not require charging from an external source, unless the vehicle is laid up for long periods, when it should be recharged every six weeks or so. If vehicle use consists entirely of short runs in darkness it is also possible for the battery to become discharged. Otherwise, a regular need for recharging points to a fault in the battery or elsewhere in the charging system.
2 There is no need to disconnect the battery from the vehicle wiring when using a battery charger, but switch off the ignition and if possible leave the bonnet open.
3 Domestic battery chargers (up to about 6 amps output) may safely be used overnight without special precautions. Make sure that the charger is set to deliver 12 volts before connecting it. Connect the leads (red or positive to positive terminal, black or negative to the negative terminal) **before** switching the charger on at the mains.
4 When charging is complete, switch off at the mains **before** disconnecting the charger from the battery. Remember that the battery will be giving off hydrogen gas, which is potentially explosive.
5 Charging at a higher rate should only be carried out under carefully controlled conditions. Very rapid or 'boost' charging should be avoided if possible, as it is liable to cause permanent damage to the battery through overheating.
6 During any sort of charging, battery electrolyte temperature should never exceed 38°C (100°F). If the battery becomes hot, or the electrolyte is effervescing vigorously, charging should be stopped.

5 Battery – removal and refitting

1 Disconnect the battery negative (earth) lead.
2 Disconnect the battery positive leads. These may be protected by a plastic cover. Do not allow the spanner to bridge the positive and negative terminals.
3 Release the battery hold-down clamps. Lift out the battery. Keep it upright and be careful not to drop it – it is heavy.
4 Commence refitting by placing the battery in its tray, making sure it is the right way round. Secure it with the hold-down clamp.
5 Clean the battery terminals if necessary (Section 3), then reconnect them. Connect the positive lead first, then the negative lead.

6 Alternator – precautions

1 To avoid damage to the alternator semiconductors, and indeed to many other components, the following precautions should be observed:

   (a) Do not disconnect the battery or the alternator whilst the engine is running
   (b) Do not allow the engine to turn the alternator when the latter is not connected
   (c) Do not test for output from the alternator by 'flashing' the output lead to earth
   (d) Do not use a battery charger of more than 12 volts output, even as a starting aid
   (e) Disconnect the battery and the alternator before carrying out electric arc welding on the vehicle
   (f) Always observe the correct battery polarity

7 Alternator – testing on the vehicle

1 Should it appear that the alternator is not charging the battery, check first that the drivebelt is intact and in good condition and that its tension is correct (Chapter 2). Always check the condition and security of the alternator electrical connections and the battery leads.
2 Accurate assessment of alternator output requires special equipment and a degree of skill. A rough idea of whether output is adequate can be gained by using a voltmeter (range 0 to 15 or 0 to 20 volts) as follows.
3 Connect the voltmeter across the battery terminals. Switch on the headlights (ignition on) and note the voltage reading: it should be between 12 and 13 volts.
4 Start the engine and run it at a fast idle (approx 2000 rpm). Read the voltmeter: it should indicate 13 to 14 volts.
5 With the engine still running at a fast idle, switch on as many electrical consumers as possible (heated rear window, heater blower etc). The voltage at the battery should be maintained at 13 to 14 volts. Increase the engine speed slightly if necessary to keep the voltage up.

Chapter 12 Electrical system

6 If alternator output is low or zero, check the brushes, as described in Section 9. If the brushes are OK, seek expert advice.
7 Occasionally the condition may arise where the alternator output is excessive. Clues to this condition are constantly blowing bulbs; brightness of lights varying considerably with engine speed; overheating of alternator and battery, possibly with steam or fuses coming from the battery. This condition is almost certainly due to a defective voltage regulator, but expert advice should be sought.
8 Voltage regulator renewal is included in brush renewal (Section 9).
9 Later models vary the charge rate according to battery temperature (Fig. 12.1). In the event of failure or disconnection of the battery temperature sensor, the alternator will vary the rate according to its own internal temperature.

8 Alternator – removal and refitting

On some models, access to the alternator is easiest from below. Remove the undertray if necessary.

1 Disconnect the battery negative lead.
2 Slacken the alternator drivebelt(s) and slip them off the pulley (Chapter 2, Section 7).
3 Disconnect the electrical wiring from the rear of the alternator – this may be a multi-plug or separate screw terminals. Make notes for reconnection if necessary (photo).
4 Support the alternator. Remove the pivot and adjusting strap nuts, bolts and washers, noting the fitted positions of the washers. Lift out the alternator. Do not drop it, it is fragile.
5 Refit by reversing the removal operations. Tension the drivebelt(s) (Chapter 2, Section 7) before reconnecting the battery.

9 Alternator – brush renewal

Depending on model, it may be possible to renew the brushes without removing the alternator from the vehicle, but disconnect the battery negative lead first.

1 From the rear of the alternator remove the two screws which secure the voltage regulator/brush carrier assembly. Withdraw the assembly (photos).
2 Measure the length of each brush protruding from the carrier. If they are worn down to, or below, the minimum specified, the old brushes will have to be unsoldered and new ones soldered into place. Some skill with a soldering iron will be required; excess heat from the soldering iron could damage the voltage regulator. When fitted, the new brushes must move freely in their holders.
3 Clean the slip rings with a cloth moistened with methylated spirit. If they are badly burnt or damaged, seek expert advice.
4 Refit the assembled brush carrier/voltage regulator and secure it with the two screws. If the alternator is on the vehicle, reconnect the battery negative lead.

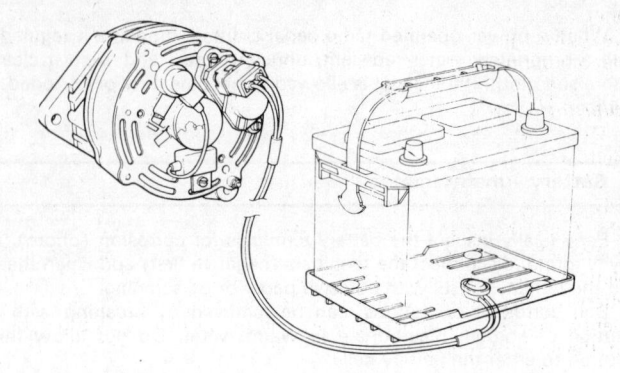

Fig. 12.1 Alternator with battery temperature sensor (Sec 7)

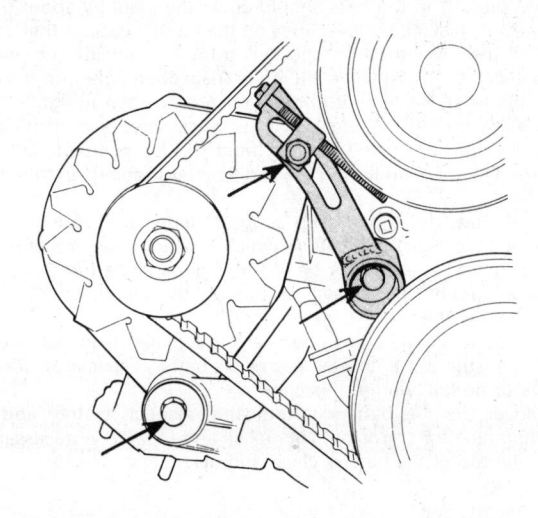

Fig. 12.2 Alternator pivot and adjusting strap nuts and bolts (arrowed) – Turbo model shown, others similar (Sec 8)

10 Starter motor – testing on the vehicle

1 If the starter motor fails to operate, first check that the battery is charged by switching on the headlights (ignition on). If the headlights do not come on, or rapidly become dim, the battery or its connections are at fault.

8.3 Alternator output terminal (arrowed) seen from below

9.1A Undoing the voltage regulator/brush carrier screws

9.1B Removing the voltage regulator/brush carrier

Chapter 12 Electrical system

2 Check the security and condition of the battery and starter solenoid connections. Remember that the heavy lead to the solenoid is always 'live' – disconnect the battery negative lead before using tools on the solenoid connections.

Solenoid check
3 Disconnect the battery negative lead, and all leads from the solenoid.
4 Connect a battery and a 3 watt test lamp between the solenoid body and the solenoid motor terminal (Fig. 12.3). The test lamp should light, if not, the solenoid windings are open-circuit.
5 Connect a battery and an 18 to 21 watt test lamp across the solenoid motor and battery terminals. Connect a further lead from the battery positive terminal to the solenoid spade terminal (Fig. 12.4). The solenoid should be heard to operate and the test lamp should light: if not, the solenoid contacts are defective.

On load voltage check
6 Remake the original connections to the solenoid and reconnect the battery negative lead. Connect a voltmeter across the battery terminals, then disconnect the low tension lead from the coil negative terminal and operate the starter by turning the ignition switch. Note the reading on the voltmeter which should not be less than 9.5 volts.
7 Now connect the voltmeter between the starter motor terminal on the solenoid and the starter motor body. With the coil low tension lead still disconnected, operate the starter and check that the recorded voltage is not more than 1 volt lower than that noted in paragraph 6. If the voltage drop is more than 1 volt a fault exists in the wiring from the battery to the starter.
8 Connect the voltmeter between the battery positive terminal and the motor terminal on the solenoid. With the coil low tension lead disconnected operate the starter for two or three seconds. Battery voltage should be indicated initially, then dropping to less than 1 volt. If the reading is more than 1 volt there is a high resistance in the wiring from the battery to the starter and the check in paragraph 9 should be made. If the reading is less than 1 volt proceed to paragraph 10.
9 Connect the voltmeter between the two main solenoid terminals and operate the starter for two or three seconds. Battery voltage should be indicated initially, then dropping to less than 0.5 volt. If the reading is more than 0.5 volt, the solenoid and connections may be faulty.
10 Connect the voltmeter between the battery negative terminal and the starter motor body, and operate the starter for two or three seconds. A reading of less than 0.5 volt should be recorded; however, if the reading is more, the earth circuit is faulty and the earth connections to the battery and body should be checked.

11 Starter motor – removal and refitting

1 On some models, access to the starter motor is easier from below. Raise the front of the vehicle on ramps if necessary, and remove the undertray.
2 Disconnect the battery negative lead.
3 On the V6 engine, remove the oil filter (Chapter 1, Section 49).

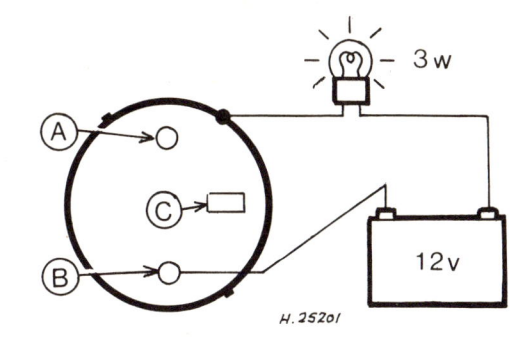

Fig. 12.3 Solenoid winding check (Sec 10)

A Battery terminal
B Motor terminal
C Spade terminal (normally connected to ignition/starter switch)

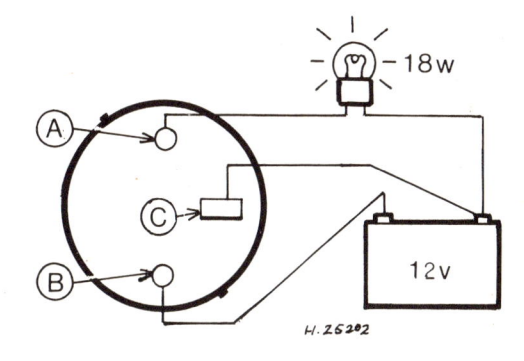

Fig. 12.4 Solenoid winding check. For key see Fig. 12.3 (Sec 10)

4 Disconnect the wires from the starter motor solenoid. Make notes or identifying marks if necessary (photo).
5 Support the starter motor and remove its securing bolts. If a tail bracket is fitted, unbolt it first (photo).
6 Remove the starter motor. When fitted, recover the adaptor plate (photo).
7 Refit by reversing the removal operations.

12 Starter motor – overhaul

1 Before embarking on the overhaul of a starter motor, check the cost and availability of spare parts. Renewal of a well worn motor may be more economical and more satisfactory than repair.

11.4 Starter solenoid connections

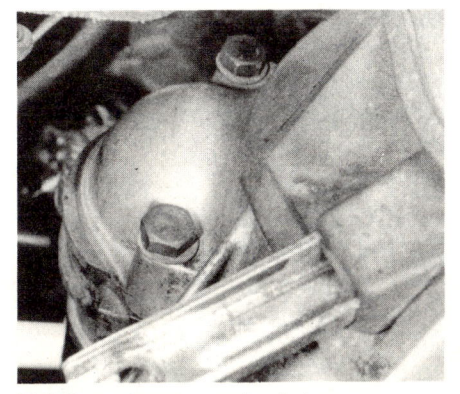

11.5 Starter motor securing bolts

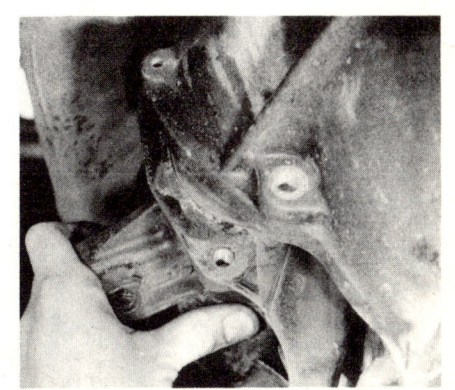

11.6 Removing a starter motor and adaptor plate

Chapter 12 Electrical system

12.2A Removing the armature end cap to expose the E-clip and washer

12.2B Removing the armature washer

12.3 Removing the commutator cover screws

Bosch GF (direct drive)

2 Remove the armature and cap. Remove the E-clip and washer(s) from the end of the armature shaft (photos).
3 Remove the two through-bolts (or studs) and the two screws from the commutator cover. Pull off the cover (photo).
4 Disconnect the motor lead from the solenoid.
5 Hook up the brush springs and remove the brushes from their holders. Remove the brush carrier plate with the negative brushes, leaving the positive brushes soldered to the field coils.
6 Remove the yoke, with field coils and brushes, from the armature.
7 Unbolt and remove the solenoid, unhooking it from the operating arm. Note the arrangement of any spring removed with the solenoid.
8 Remove the rubber plug from behind the operating arm pivot. Also remove the operating arm pivot nut and bolt (if fitted).
9 Remove the armature, operating arm and drive and components together. Unhook the operating arm from the drive.
10 To remove the pinion and clutch, clamp the armature lightly in a vice with padded jaws. Using a piece of tube, drive the stop ring down the shaft to expose the circlip (photo). Remove the circlip and stop ring, followed by the pinion and the one-way clutch.
11 Inspect all components for wear or damage and renew as necessary. The armature shaft brushes can be renewed: soak the new bushes in engine oil for at least half an hour before fitting.
12 Simple continuity checks can be made on the armature and field windings using a multi-meter or a battery and test lamp. Special test equipment is required for thorough checking.
13 A burnt or damaged commutator can sometimes be reclaimed by machining, provided the refinishing limit is not exceeded. This is specialist work. Be wary of using abrasives to clean the commutator, as particles may become embedded in the copper.
14 Renew the brushes if they are worn below the minimum specified. The new brush leads must be soldered to the tails of the old ones. Do not allow solder to run too far up the leads, or flexibility will be lost. It is good practice to renew the springs at the same time as the brushes. Make sure the brushes slide freely in their holders.
15 Renewal of the field coils and pole pieces must be left to a specialist.
16 Before commencing reassembly, apply grease or oil sparingly to the points shown in Fig. 12.5. Be careful not to get lubricant onto the commutator or brushgear. Clean the commutator with methylated spirit.
17 Commence reassembly by fitting the clutch, pinion and stop ring to the armature shaft. Fit a new circlip into the shaft groove, then lever or pull the stop ring over it.
18 Engage the operating arm behind the clutch and fit the assembly into the drive end housing. Fit the pivot nut and bolt (if applicable) (photos).

12.10 Driving the stop ring down the shaft

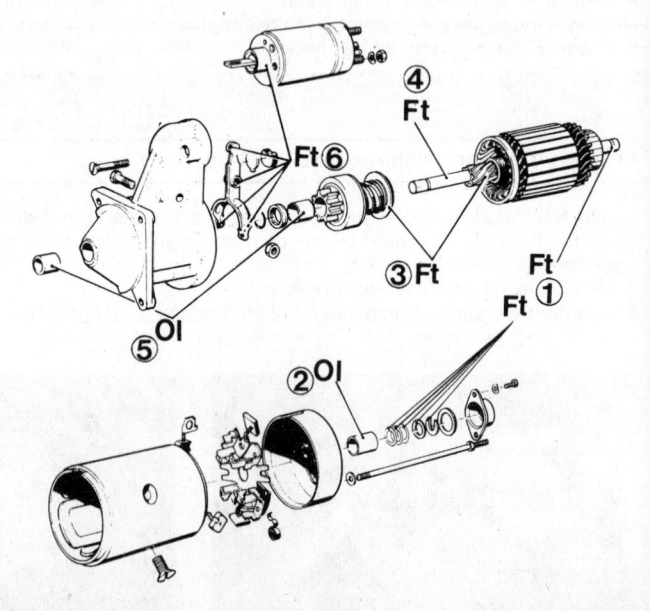

Fig. 12.5 Starter motor lubrication points (Sec 12)

1 Armature shaft and washers
2 Bush
3 Spiral threads
4 Armature shaft
5 Bushes
6 Operating arm and solenoid plunger
Ft Grease
Ol Oil

Chapter 12 Electrical system

19 Refit the rubber plug behind the operating arm, then fit the yoke over the armature (photos).
20 Hook the solenoid onto the operating arm, making sure that the spring (when applicable) is correctly engaged. Fit and tighten the solenoid securing bolts (photo).
21 Reconnect the motor lead to the solenoid (photo).
22 Refit the brush carrier. Insert the brushes into their holders and hook the brush springs on top of them (photos).
23 Refit the commutator cover and secure it with the through-bolts or nuts. Fit the two screws also.
24 Refit the washer(s) and E-clip to the armature shaft. Armature endfloat is controlled by the thickness of the washer(s) – see Specifications for the desired value.
25 Refit the armature end cap.

Bosch DW (reduction gear)

26 Overhaul procedures are similar to that for the Bosch GF starter, except that the armature can be withdrawn from the yoke leaving the reduction gears in place.
27 Permanent magnets may be used instead of field coils on the motor. These magnets are fragile: do not drop the yoke or clamp it in a vice.
28 Remove the cover plate and apply a little silicone grease to the reduction gears before reassembly (photos).

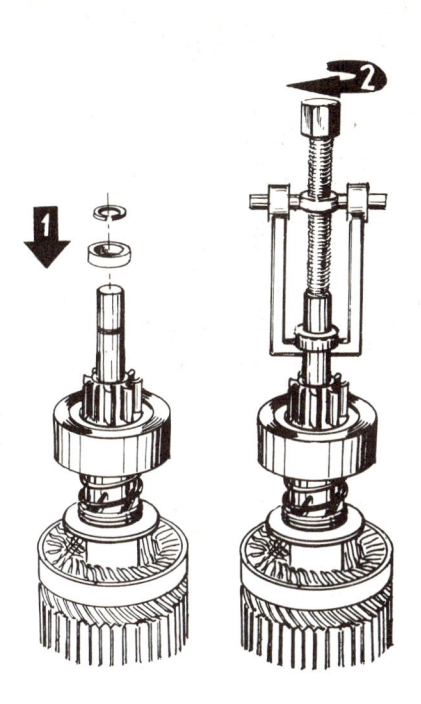

Fig. 12.6 Fit the stop ring and circlip (1), then pull the stop ring over the circlip to keep it in the groove (2) (Sec 12)

13 Exterior light bulbs – renewal

1 With all light bulbs, remember that if they have just been in use, they may be very hot. Switch off the power before renewing a bulb.
2 With quartz halogen bulbs (headlights and similar applications), do not touch the bulb glass with the fingers. Even small quantities of grease from the fingers will cause blackening and premature failure. If a

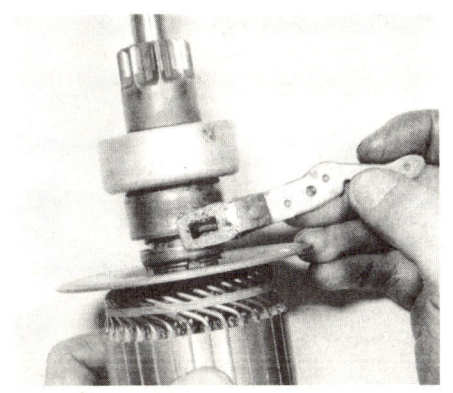

12.18A Engage the operating arm ...

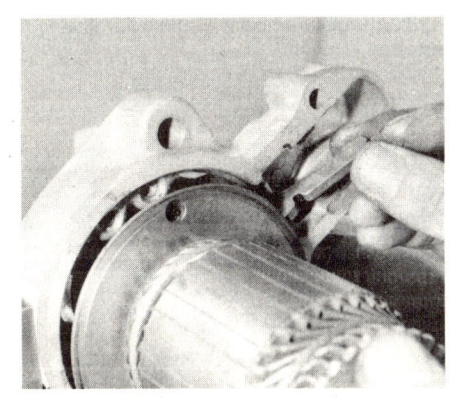

12.18B ... and fit the assembly to the drive end housing

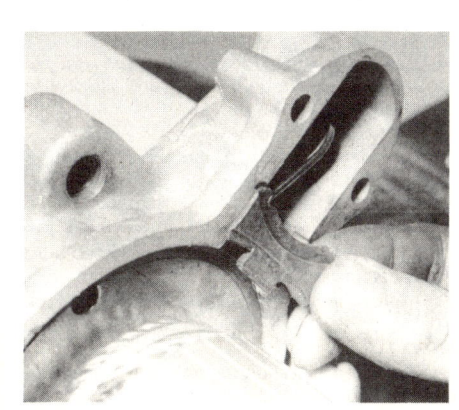

12.19A Fitting the rubber plug

12.19B Fitting the yoke over the armature

12.20 Refitting the solenoid

Chapter 12 Electrical system

12.21 Reconnecting the motor lead

12.22A Refitting the brush carrier – the negative brushes are already in their holders

12.22B Hooking a brush spring onto a positive brush

12.28A Reduction gears with cover plate fitted

12.28B Reduction gears exposed

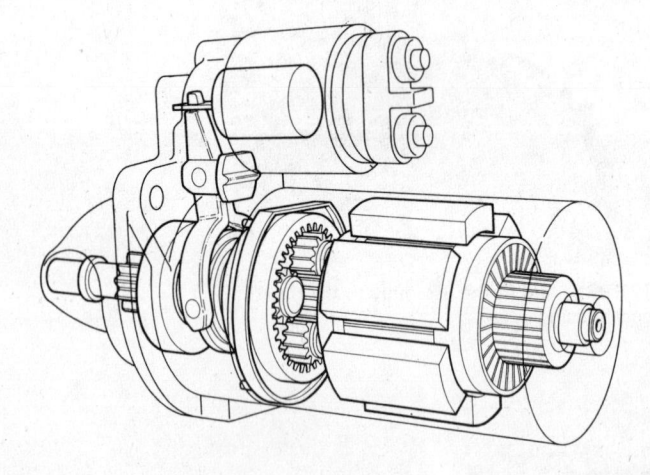

Fig. 12.7 Cutaway view of the reduction gear starter motor (Sec 12)

bulb is accidentally touched, clean it with methylated spirit and a clean rag.
3 Unless otherwise stated, fit the new bulb by reversing the removal operations.

Headlight
4 Open the bonnet. Unclip the plastic cover from the rear of the headlight unit (photo). There is no need to disconnect the multi-plug.
5 Unplug the connector from the bulb. Release the retainer by pushing it and twisting it anti-clockwise. Remove the retainer, spring and bulb (photos).
6 When fitting the new bulb, do not touch the glass (paragraph 2). Make sure that the lugs on the bulb flange engage with the slots in the holder.
7 Observe the 'OBEN/TOP' marking when refitting the plastic cover.

Auxiliary front light
8 Remove the lens/reflector unit, which is secured by two screws and retaining strips (photo).
9 Unplug the bulb wiring connector, release the spring clip and withdraw the bulb (photos).
10 Fit the new bulb, being careful not to touch it with the fingers (paragraph 2). Reconnect the wiring.
11 Refit the lens/reflector unit, observing the 'TOP' marking.

Front direction indicator/day running/parking lights
12 Open the bonnet. Turn the appropriate bulb holder anti-clockwise (without disconnecting it) and withdraw it (photo).
13 Remove the bulb from the holder (photo).
14 When fitting the combined day running/parking light bulb, note that the pins are offset so it will only fit one way round.

Direction indicator side repeater
15 Slide the lens forwards and free it from the rear. Withdraw the bulb holder from the lens without disconnecting the wiring.
16 When refitting, make sure that the rubber seal is seated in the hole.

Rear light cluster (Saloon)
17 Open the boot. Undo the knurled screw on the light unit cover and pivot the cover downwards.
18 Remove the appropriate bulb holder from the unit by twisting the holder anti-clockwise and pulling it (photo).
19 Remove the bayonet fitting bulb from the holder (photo).

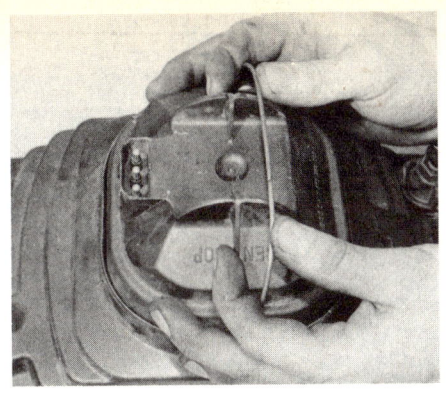

13.4 Unclipping the cover from the headlight unit (unit removed)

13.5A Unplug the connector ...

13.5B ... remove the retainer and spring ...

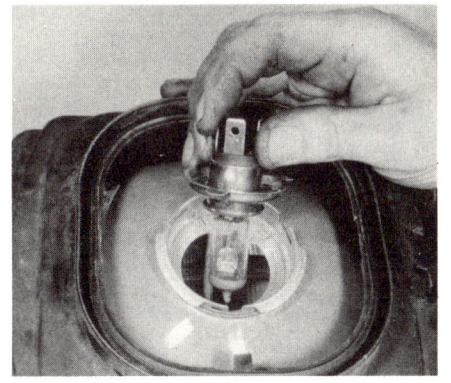

13.5C ... and the bulb. Do not touch the bulb glass

13.8 Auxiliary light lens screw and strip

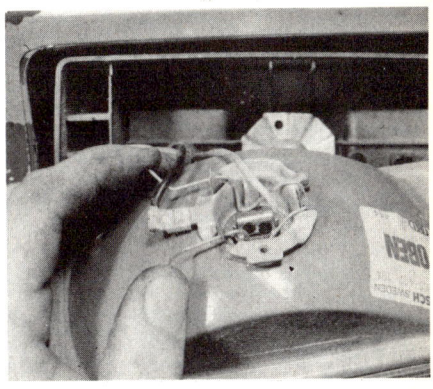

13.9A Release the spring clip ...

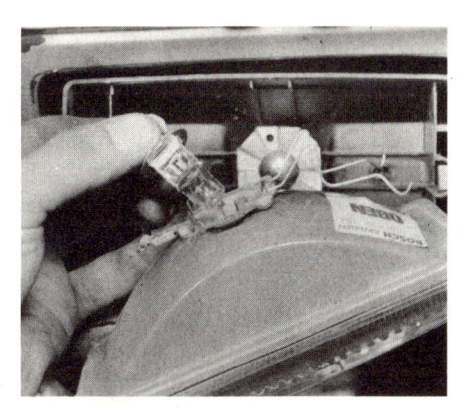

13.9B ... and withdraw the bulb

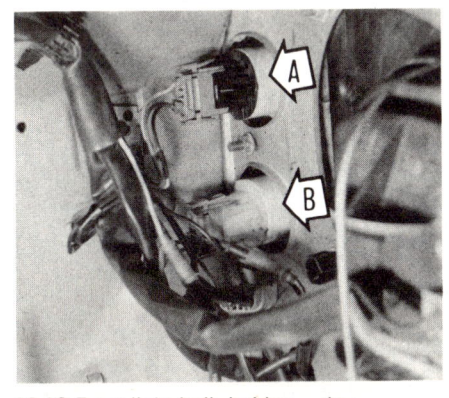

13.12 Front light bulb holders – day running light (A) and direction indicator (B)

13.13 Removing a day running/parking light bulb

13.18 Remove the bulb and holder from the rear light cluster ...

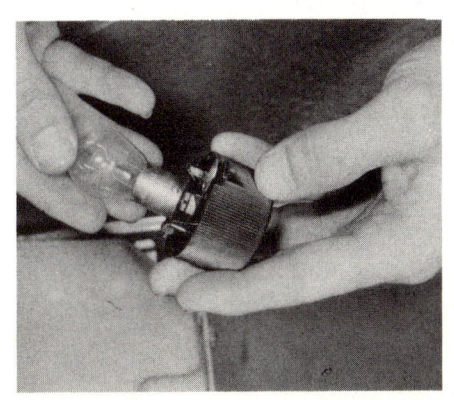

13.19 ... then separate the two

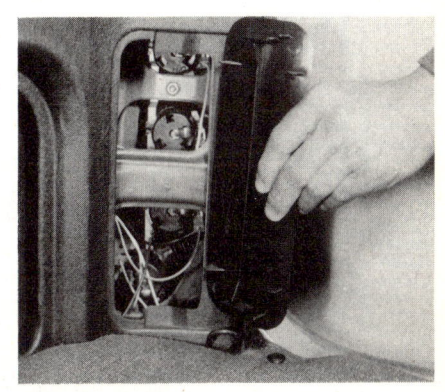

13.20 Rear light unit and cover – Estate

Chapter 12 Electrical system

Rear light cluster (Estate)
20 Open the tailgate. Unclip the light unit cover (photo), then proceed as for Saloon models.

Number plate light
21 Unclip the light unit by sliding it rearwards.
22 Release the bulb by sliding the live contact off its tail. The bulb can then be removed upwards (photo).

14 Exterior light units (except headlights) – removal and refitting

1 Switch off the power before removing a light unit.
2 Except where noted, refit by reversing the removal operations.

Auxiliary front light
3 Follow the wiring back from the light unit and unplug the connector.
4 Remove the nut which secures the light unit to the bracket, or unbolt the bracket complete with the light unit, as wished (photo).

Direction indicator/day running light unit
5 Open the bonnet. Disconnect the two multi-plugs from the bulb holders on the rear of the unit.
6 Remove the single securing nut from the rear of the light unit. Free it from the lugs on the side of the headlight and withdraw it (photo).

Rear light cluster (Saloon)
7 Proceed as for light bulb renewal, but remove the hinged cover completely. It is secured by two nuts.
8 Disconnect the cluster multi-plug (photo).
9 Remove the five flanged nuts and the single screw which secure the unit. Note that the screw also secures earth tags.
10 Remove the cluster from the vehicle (photo).

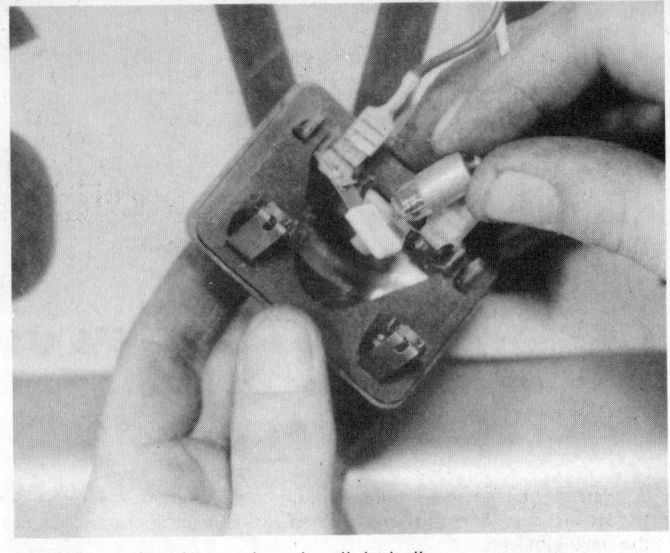

13.22 Removing the number plate light bulb

Rear light cluster (Estate)
11 Proceed as described for Saloon models, making allowances for the detail differences which will be found.

15 Headlight unit – removal and refitting

1 Remove the front grille and the headlight wiper arm.
2 Disconnect the multi-plugs from the headlight and from the day running light/direction indicator light unit (photo). Remove the latter unit, which is secured by one nut.

14.4 Auxiliary light showing securing nut

14.6 Disengaging the front light unit from the headlight

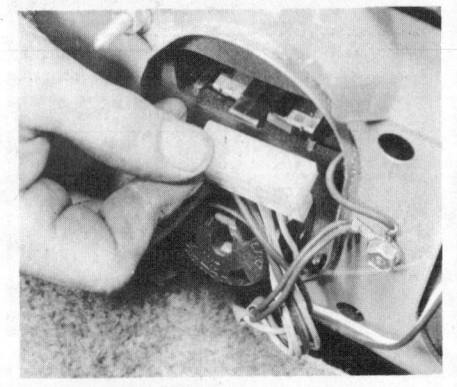

14.8 Disconnecting the rear light cluster multi-plug

14.10 Removing the rear light cluster

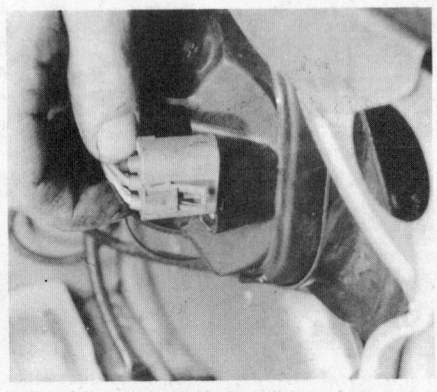

15.2 Disconnecting a headlight multi-plug

Chapter 12 Electrical system

15.4A Removing the trim strip ...

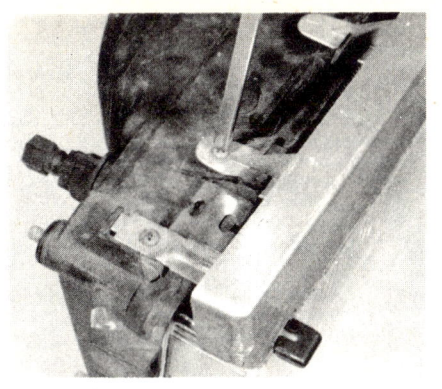

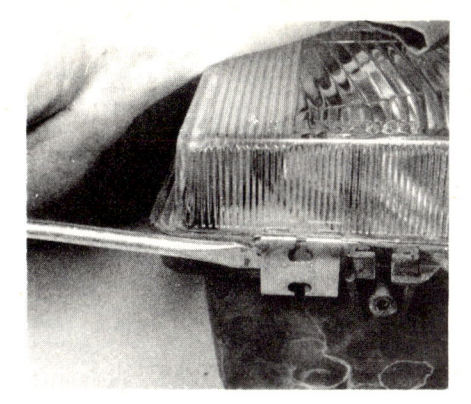

15.4B ... and a retaining clip

15.4C Fitting a new headlight seal

3 Remove the four nuts which secure the headlight unit. Disconnect the washer tube from the wiper arm, if not already done, and remove the headlight unit.
4 The lens and seal may now be renewed after removing the trim strip and the retaining clips (photos).
5 Refit by reversing the removal operations. Remember to feed the washer tube through the trim strip.
6 Have the beam alignment checked on completion (Section 16).

16 Headlight beam alignment

1 Beam alignment should be carried out by a Volvo dealer or other specialist heaving the necessary optical alignment equipment.
2 In an emergency, adjustment may be carried out on a trial and error basis, using the two adjustment screws on the rear of each headlight unit.
3 Holts Amber Lamp is useful for temporarily changing the headlight colour to conform with the normal usage on Continental Europe.

17 Interior light bulbs – renewal

1 See Section 13, paragraphs 1 and 3.
2 Some switch illumination/pilot bulbs are integral with their switches and cannot be renewed separately.

Courtesy/load area lights
3 Pull or prise the light unit from its mountings.
4 Renew the bulb(s), which may be bayonet or end clip fitting (photo).

Glovebox lights
5 Unclip the combined bulb holder/switch unit from the top of the glovebox for access to the bulb (photo).
6 When vanity mirror bulbs are fitted, these are accessible after prising out the light diffuser strip (photo).

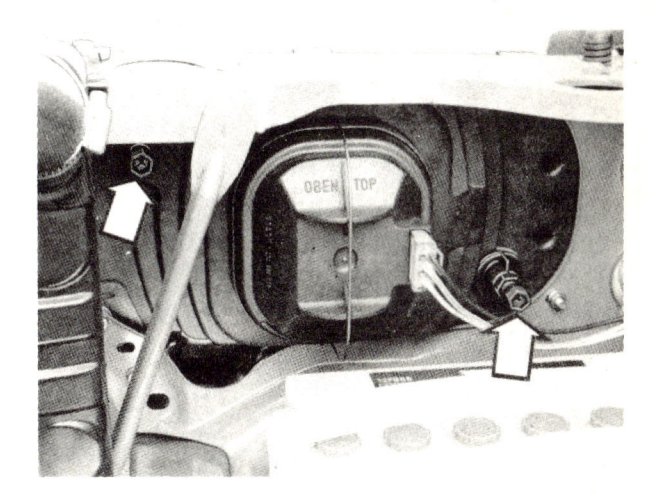

Fig. 12.8 Headlight beam alignment screws (arrowed) (Sec 16)

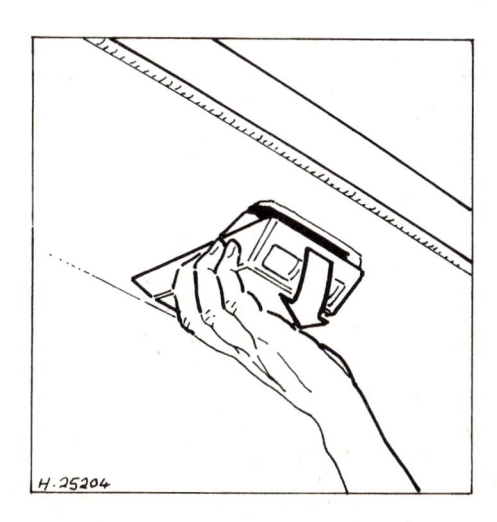

Fig. 12.9 Removing the front courtesy/map reading light (Sec 17)

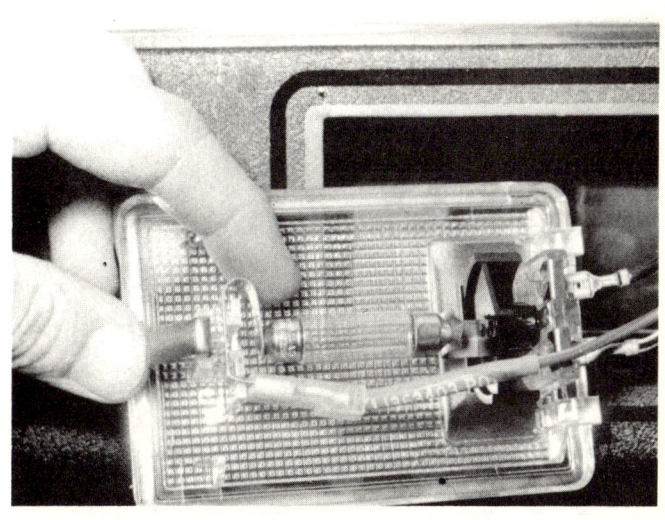

17.4 A load area light with an end clip fitting bulb

288　　　　　　　　　　　Chapter 12 Electrical system

17.5 Glovebox light and switch (seen in a mirror)

17.6 Vanity mirror light bulbs exposed

17.7 Fitting a door edge marker bulb

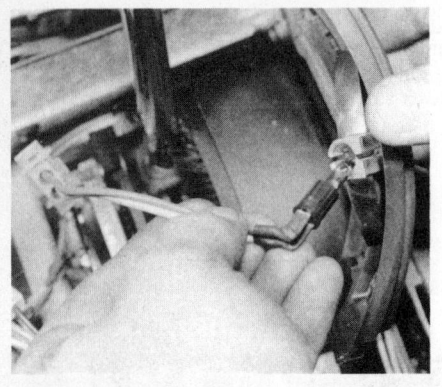

17.8 Extracting the automatic transmission selector light

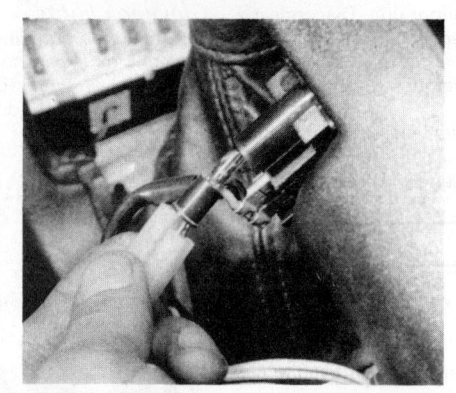

17.10 Removing a rear console switch light

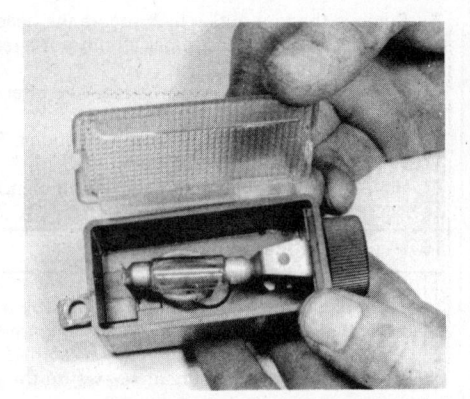

17.12 Under-bonnet light unit (removed)

Door edge marker lights
7　Prise off the lens for access to the bulb. The bulb is of the capless type, so it is a push fit (photo).

Automatic transmission selector light
8　Remove the selector quadrant as if for access to the starter inhibitor switch (Chapter 6, Section 34). The bulb and holder can then be pulled out (photo).

Seat belt buckle light
9　Unclip the bulb holder from the buckle for access to the bulb.

Switch illumination bulbs
10　When these are separable from the switch, they simply pull out (photo).

Instrument panel bulbs
11　See Sections 20 and 21. (The reader with small hands and deft fingers may manage to renew bulbs *in situ* after removing the steering column/pedal trim.)

Under-bonnet light
12　Prise off the lens with a screwdriver. The bulb is of the end clip fitting type (photo).

Cigarette lighter/ashtray light
13　See Section 33.

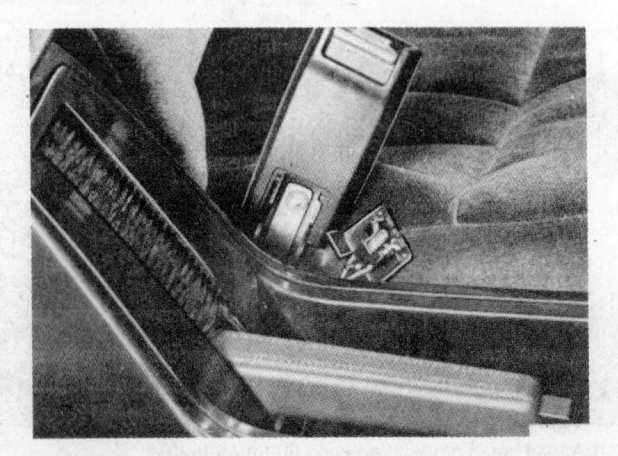

Fig.12.10 The seat belt buckle light (Sec 17)

18 Ignition/starter switch – removal and refitting

1　Disconnect the battery negative lead.
2　Remove the trim panel from below the steering column.
3　Disconnect the multi-plug from the switch (photo).
4　Remove the two screws which secure the switch to the steering lock. Withdraw the switch.
5　Refit by reversing the removal operations. Note that the hole in the centre of the switch is shaped so that it will only engage with the driving spindle in one position (photo).

Chapter 12 Electrical system

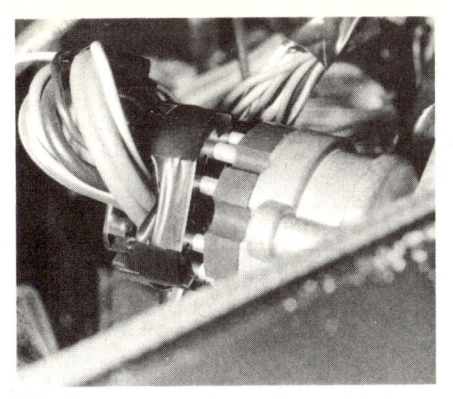

18.3 Disconnecting the ignition switch multi-plug

18.5 Ignition switch and steering lock showing driving hole and spindle

19.4 Removing a column shroud screw

19 Switches – removal and refitting

1 Disconnect the battery negative lead, or satisfy yourself that there is no risk of short-circuit, before removing any switch.
2 Except where noted, a switch is refitted by reversing the removal operations.

Steering column switches
3 Remove the steering wheel (Chapter 10, Section 12).
4 Remove the column shrouds, which are secured by two screws each (photo).
5 Remove the switch in question. Each switch is secured by two screws. Remove the screws, pull the switch out and disconnect the multi-plug (photos).

Horn push switches
6 These are removed by prising them out of the steering wheel. They are difficult to remove without damage (photo).

Facia panel switches
7 Unclip the switch panel (and its surround, if applicable) and withdraw it from the facia (photo).
8 Disconnect the switch multi-plugs, making identifying marks or notes if necessary (photo).
9 Remove the switch concerned by depressing its retaining lugs (photo).

Rear console switches
10 See Chapter 11, Section 36.

19.5A Removing a column switch screw

19.5B Disconnecting a column switch multi-plug

19.6 Removing a horn push switch

19.7 Unclipping a switch panel from the facia

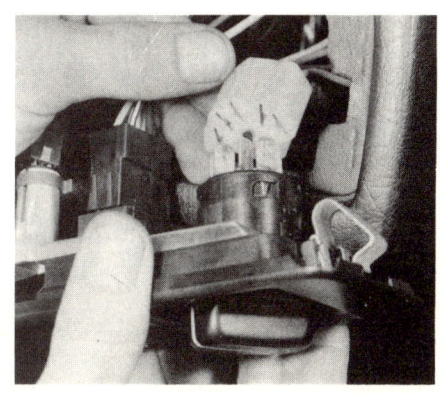

19.8 Disconnecting a switch multi-plug

19.9 Removing a switch from the panel

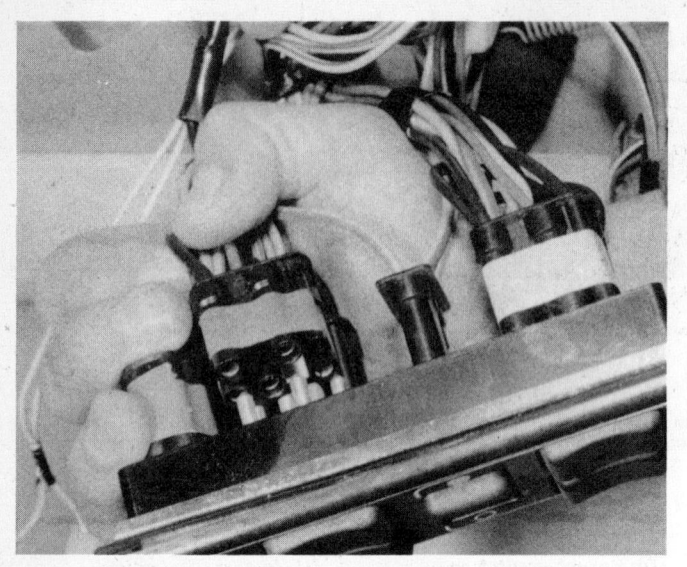

290　　　　　　　　　　　　　　　　Chapter 12 Electrical system

15 Secure the wires with a clothes peg before disconnecting them so that they are not lost in the door pillar.

Other switches

16 Some switches will be found in the Chapter dealing with their system or equipment – for example, temperature-operated switches in Chapter 2, and transmission-operated switches in Chapter 6.

20 Instrument cluster – removal and refitting

1 Disconnect the battery negative lead.
2 Remove the two screws at the bottom corners of the instrument cluster. These may be concealed by plastic covers, which will have to be pulled off first (photo).
3 Pull the cluster towards the steering wheel. (If difficulty is experienced, remove the steering column/pedal lower trim to gain access to the rear of the panel.)
4 Disconnect the multi-plugs and (when applicable) the boot gauge pipe from the rear of the cluster (photo).
5 Lift out the cluster. Do not drop or jar it.
6 Refit by reversing the removal operations.

21 Instrument cluster – dismantling and reassembly

1 Remove the screws which secure the instrument cluster to the transparent panel and surround. Carefully remove the cluster.
2 Individual instruments can now be removed after undoing their securing nuts or screws (photo). Note that the screws are not identical: those which secure conductors are plated.
3 Bulb holders are removed by twisting them 90° and pulling. Some bulbs can be separated from their holders for renewal; others must be renewed complete with holder (photos).
4 The printed circuit can be renewed after removing all the instruments, bulbs and connectors. Be careful when handling the printed circuit, it is fragile.

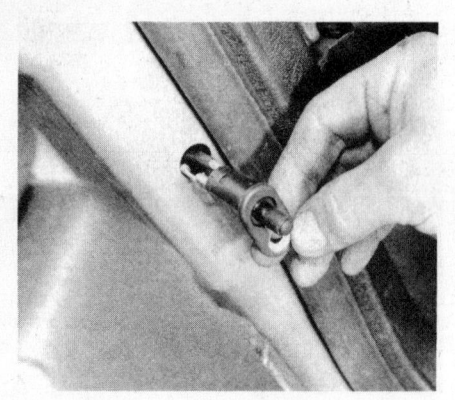

19.12 Disconnecting a window control switch

Window/mirror control switches

11 Remove the door armrest and separate the switch panel from it. See Chapter 11, Section 11.
12 Disconnect the multi-plug from the switch in question (photo).
13 Carefully prise free the retaining lugs and remove the switch from the underside of the switch plate.

Door/tailgate switches

14 Open the door or tailgate. Remove the securing screw and withdraw the switch (photo).

19.14 Removing a door switch

20.2 Exposing an instrument cluster screw

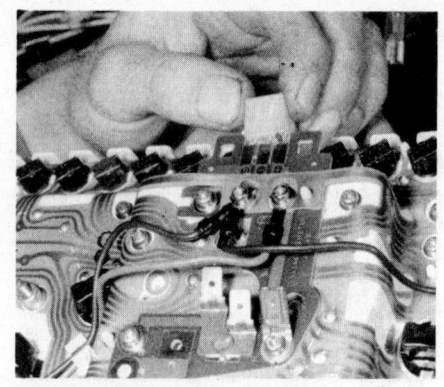

20.4 Disconnecting an instrument cluster multi-plug

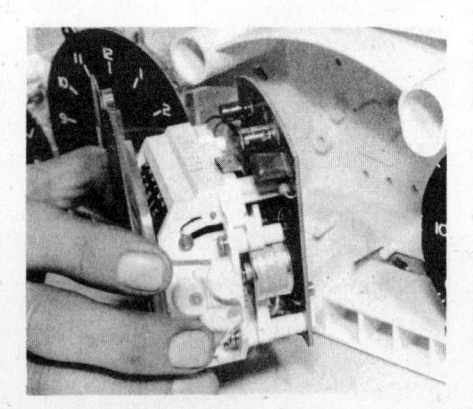

21.2 Removing the speedometer

21.3A Removing a bulb and holder from the printed circuit

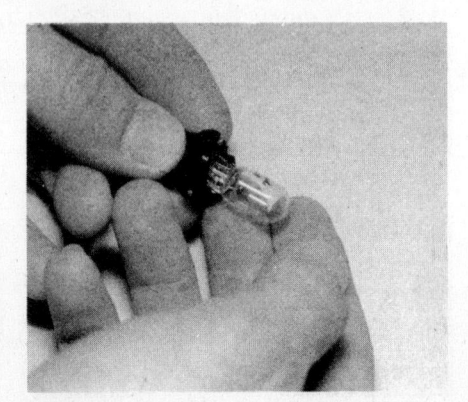

21.3B Separating the capless bulb and holder

Chapter 12 Electrical system   291

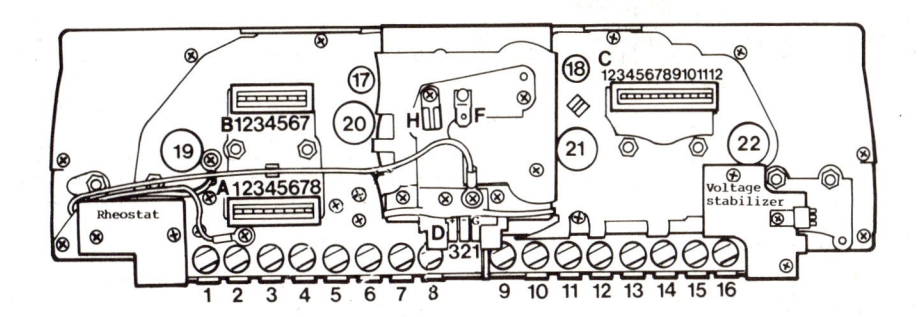

Fig. 12.11 Identification of instrument cluster bulbs – 1984 model shown. A to H are connectors (Sec 21)

1 Overdrive 'off' light (AW 71)
2 Seat belt reminder
3 Spare
4 Low washer fluid warning
5 Bulb failure warning
6 Handbrake warning
7 Brake failure warning
8 High beam pilot light
9 Overdrive 'on' light (M 46)
10 Battery charge failure warning
11 Engine oil pressure warning
12 Choke warning/Turbo boost warning
13 Headlight pilot light
14 Spare
15 Engine oil level warning
16 Spare
17 Direction indicator (RH)
18 Direction indicator (LH)
19 Instrument lighting
20 Instrument lighting
21 Instrument lighting
22 Instrument lighting

22 Speedometer sender unit – removal and refitting

1 Raise the rear of the vehicle on ramps.
2 Although not essential, it will improve access if the rear axle oil filler plug and the Panhard rod are removed. See Chapter 10, Section 29.
3 If there is a sealing wire on the sender connector, cut the wire and remove it.
4 Unplug the sender connector (photo).

Models without ABS

5 Unscrew the ring nut which secures the sender, using a self-locking wrench. This nut may be very tight: be careful not to crush it or it will be impossible to remove. As a last resort, the axle oil may be drained, the differential cover plate removed complete with sender unit, and the assembly dealt with on the bench.
6 Remove the sender unit and shim(s).
7 Refit by reversing the removal operations. If a new sender unit is being fitted, or if other related components have been disturbed, check the clearance between the sender unit and the toothed wheel as follows.
8 Working through the oil filler hole, introduce feeler blades between the sender unit and the toothed wheel and determine the clearance. The desired value is 0.85 mm ± 0.35 mm (0.033 ± 0.014 in). Adjust by adding or removing shims between the sender unit and the differential cover.
9 Refit the oil filler plug and reconnect the sender unit connector. Check the sender for correct operation before fitting a new seal.

Models with ABS

10 Proceed as above, but note the following:
   (a) The sender may be secured with an Allen screw, not a ring nut
   (b) The desired clearance is 0.60 mm (0.024 in), with a tolerated range from 0.35 to 0.75 mm (0.014 to 0.030 in)

23 Fuses – general

1 The fuses are located on the sloping face of the central electric unit, behind the front ashtray. Access is gained by removing the ashtray, then unclipping the ashtray carrier by pressing up the section marked 'electrical fuses – press'.
2 If a fuse blows, the electrical circuit(s) protected by that fuse will cease to operate. Lists of the circuits protected are given in the Specifications; a sticker behind the ashtray gives details for the particular vehicle.
3 To check for a blown fuse, either remove the fuse and inspect its wire link, or (with the power on) connect a 12 volt test light between

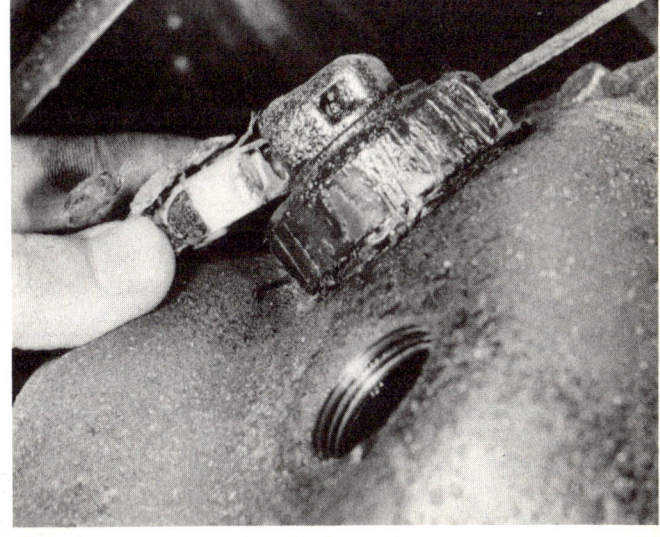

22.4 Disconnecting the speedometer sender

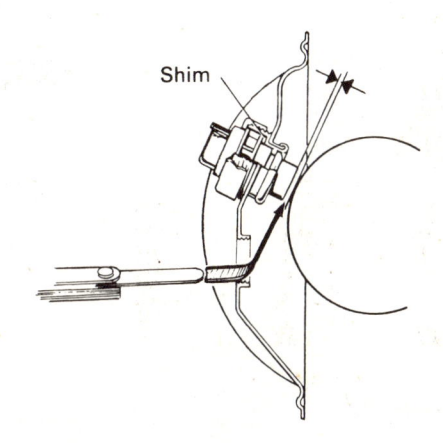

Fig. 12.12 Checking the speedometer sender unit clearance (Sec 22)

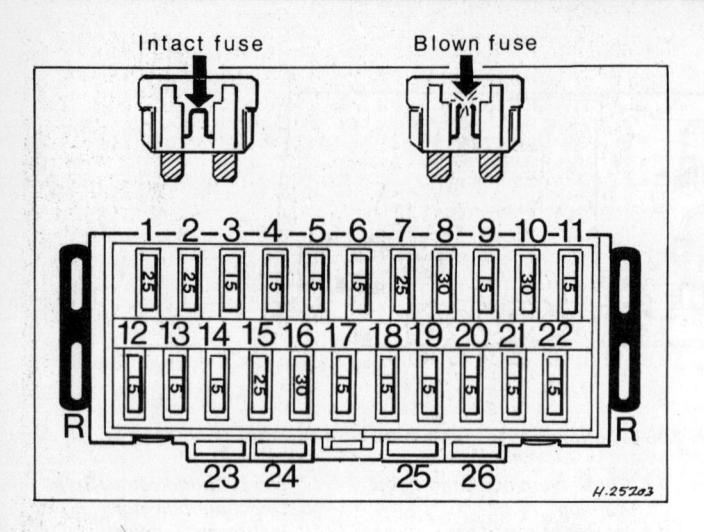

Fig. 12.13 Fuse numbering (1985 model shown) and identification of a blown fuse (Sec 23)

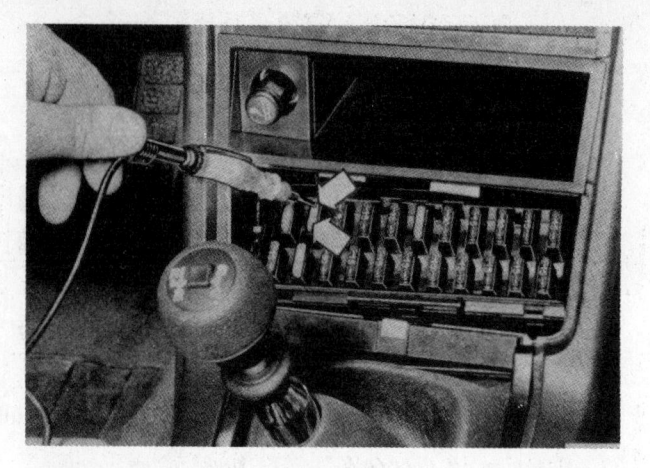

Fig. 12.14 Checking for a blown fuse using a test light (Sec 23)

earth and each of the fuse pegs. If the test light comes on at both pegs, the fuse is OK; if it comes on at one peg only, the fuse is blown.
4 To renew a blown fuse, pull out the old fuse either with the fingers or with the special tool provided. Press in a new fuse of the correct rating (indicated by colour and by a number on the fuse). Spare fuses are provided at each side of the central electrical unit.
5 Never fit a fuse of a higher rating than that specified, nor bypass a blown fuse with wire or metal foil. Serious damage or fire could result.
6 Persistent blowing of a particular fuse indicates a fault in the circuit(s) protected. Where more than one circuit is involved, switch on one item at a time until the fuse blows, so showing in which circuit the fault lies.
7 Besides a fault in the electrical component concerned, a blown fuse can also be caused by a short-circuit in the wiring to the component. Look for trapped or frayed wires allowing a live wire to touch vehicle metal, and for loose or damaged connectors.

24 Relays – general

1 Relays are electrically-operated switches. They are used for two main reasons:

(a) A relay can switch a heavy current at a distance, so allowing the use of lighter gauge control switches and wiring
(b) A relay can receive more than one control input, and can in some circumstances perform logic functions

2 In addition, some relays have a 'timer' function – for instance the intermittent wiper relay.
3 If a circuit which includes a relay develops a fault, remember that the relay itself could be faulty. Testing is by substitution of a known good relay. Do not assume that relays which look similar are necessarily identical for purposes of substitution.
4 Most relays are located on the central electrical unit, in front of the fuses. For identification see Fig. 12.15. For access, remove the ashtray

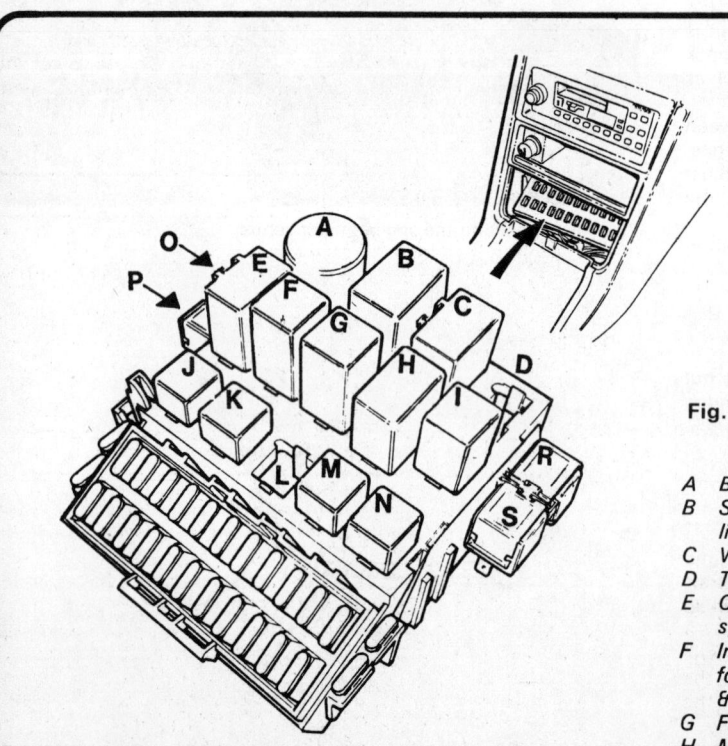

Fig. 12.15 Relay identification in central electrical unit. Not all relays are fitted to all models (Sec 24)

A   Bulb failure warning sensor
B   Seat belt reminder or torque limiter (B 23 ET)
C   Windscreen wiper delay
D   Tailgate wiper delay
E   Continuous injection system/Motronic system
F   Impulse relay (1982)/front foglights/parking lights/main & dipped beams
G   Flasher unit
H   Main lighting relay (1983)
I   Overdrive relay
J   Window winders/electric fan
K   Central locking (unlocking)/main lighting relay (1984 on)
L   Central locking (locking)
M   Auxiliary lights
N   Rear foglights
O   Vacant
P   Vacant
R   Vacant/ignition advance
S   Oil level sensor

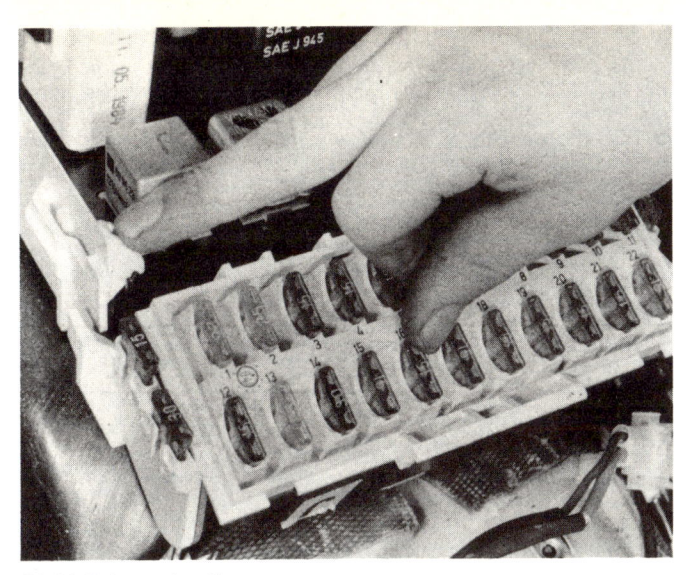

Chapter 12 Electrical system   293

24.4A Release the clips ...

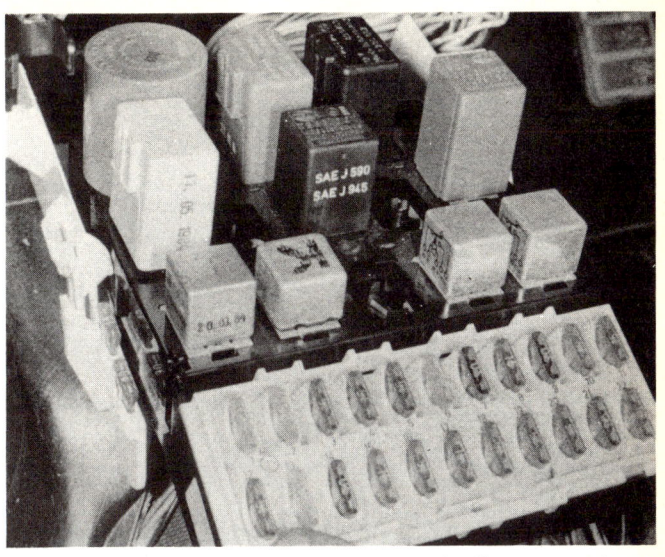

24.4B ... and withdraw the central electrical unit. (Centre console has been removed)

and ashtray holder, then release the clips and draw the unit into the vehicle (photos).
5 Other relays relating to the fuel and ignition systems are found under the bonnet. These are considered in Chapters 3 and 4. Two air conditioning relays are located behind the facia panel.

Wiring for trailer lighting must be connected upstream of the bulb failure warning sensor, otherwise it may be damaged by excessive current flow. Consult a Volvo dealer or an auto-electrician.

25 Bulb failure warning system – description

The bulb failure warning sensor is a special kind of relay. It is mounted on the central electrical unit.
The sensor contains a number of reed switches surrounded by coils of wire. Current to each bulb covered by the system travels through one coil. The coils are arranged in pairs, one pair carrying the current for one pair of bulbs.
When both bulbs of a pair are lit, the magnetic fields produced by the two coils cancel each other out. If one bulb fails, the coil remaining in circuit will produce an uncancelled magnetic field. The magnetic field operates the reed switch, which illuminates the warning light.
From the above it will be realised that no warning will be given if a pair of bulbs fails simultaneously. False alarms may result if bulbs of different wattage, or even of different make, are fitted.

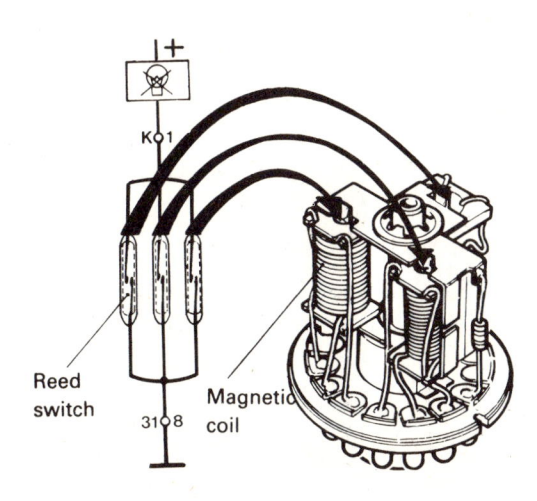

26 Direction indicator system – fault tracing

1 The direction indicator system consists of the flasher unit, control switch, external and repeater lights and the associated wiring. The hazard warning system shares the same flasher unit.
2 If the direction indicators operate abnormally fast or slowly on one side only, check the bulbs and wiring on that side. Incorrect wattage bulbs, dirty bulb holders and poor earth connections can all cause changes in the flashing rate.
3 If the directon indicators do not work at all, check the fuse before suspecting a flasher unit. If the hazard warning system works but the indicators do not, the flasher unit is almost certainly not faulty.
4 The flasher unit occupies position 'G' among the relays on the central electrical unit. Testing is by substitution of a known good unit.

27 Horn – removal and refitting

1 Raise the front of the vehicle on ramps.
2 Working under the front bumper, disconnect the wires from the horn (photo).
3 Unbolt the horn from its bracket and remove it.
4 Refit by reversing the removal operations.

28 Windscreen/headlight/tailgate washers – general

1 The washer systems share a common reservoir, located under the bonnet. For access to the pump(s) and level indicator it will be necessary to remove the air cleaner unit.
2 The level indicator is a float-operated switch which can be removed after unscrewing its retaining ring (photo).
3 To remove a washer pump, unplug its electrical connector, disconnect the hose from it and pull the pump out of its locating spigot (photo). Be prepared for fluid spillage.
4 If a pump malfunctions it must be renewed.
5 Only use clean water, and an approved screen wash additive if wished, in the washer reservoir. Use an additive with antifreeze properties (**not** engine antifreeze) in freezing conditions.

Fig. 12.16 Construction of the bulb failure warning sensor (Sec 22)

294　Chapter 12 Electrical system

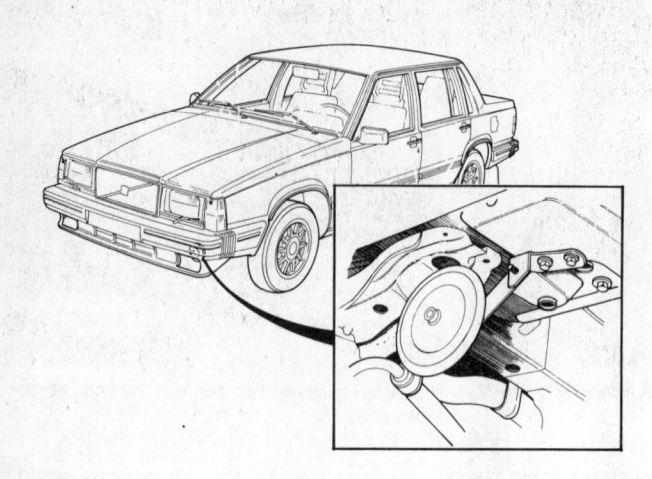

Fig. 12.17 Horn location under the front bumper (Sec 27)

29 Wiper blades and arms – removal and refitting

1 To remove a blade alone, unhook or unclip it from the arm. In the case of headlight wiper blades, also disconnect the washer tube (photos).
2 Before removing an arm end blade, mark the parked position of the blade on the glass with tape or crayon.
3 Remove the nut at the base of the wiper arm and pull the arm off the splines (photo).
4 Refit by reversing the removal operations. In the case of the headlight wiper blades, note that the longer end goes towards the grille.
5 Bias the headlight wiper arms by refitting them with the blades just below the stops (motors parked). Secure the arms, then lift the blades over the stops.

30 Windscreen wiper motor and linkage – removal and refitting

1 Remove the windscreen wiper arms and spindle seals (photo).
2 Raise the bonnet to its fully open position.
3 Remove the scuttle panel, which is secured by three bolts and some clips (photo). Close the bonnet.
4 Unclip the heater air intake cover and disconnect the wiper motor multi-plug (photo).
5 Remove the two bolts and lift out the wiper motor, linkage and cover (photo).
6 The motor may be removed from the linkage by undoing the spindle nut and the three securing screws (photo). Do not attempt to dismantle the motor except out of curiosity; spares are unlikely to be available.
7 Other components of the linkage, including the cable, may be renewed as necessary. There is a tensioning nut at one end of the cable (photo).
8 Refit by reversing the removal operations. Before refitting the wiper arms, switch the wipers on and off to bring the motor into the parked position.

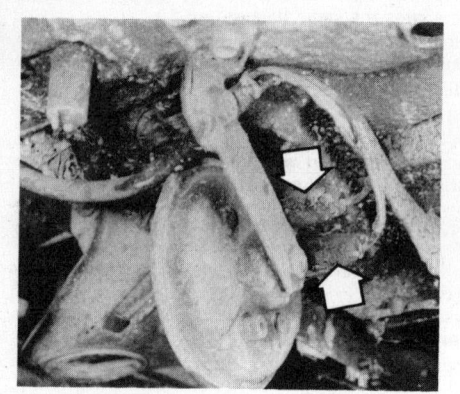

27.2 Horn wiring connectors (arrowed)

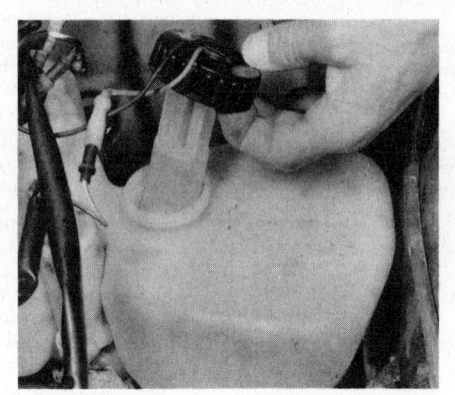

28.2 Washer reservoir level indicator

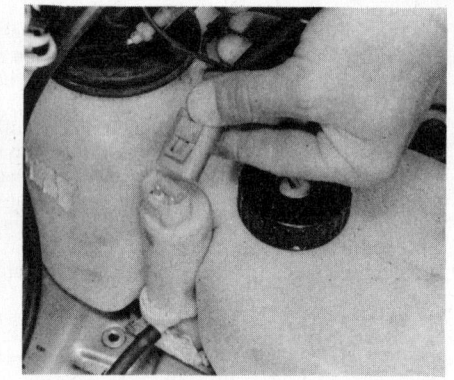

28.3 Disconnecting a washer pump

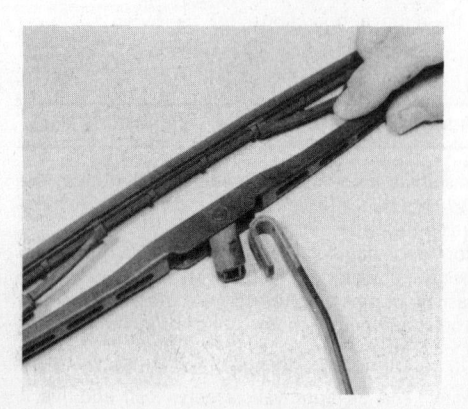

29.1A Unhooking a windscreen wiper blade

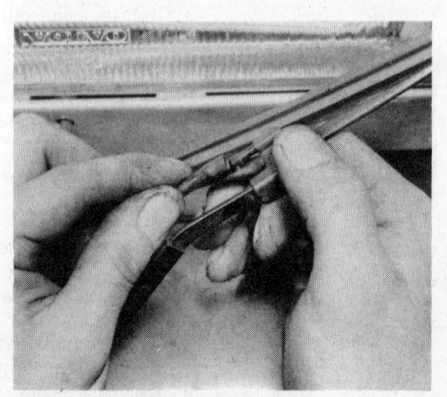

29.1B Disconnecting the washer tube from a headlight wiper blade

29.3 Removing a wiper arm nut

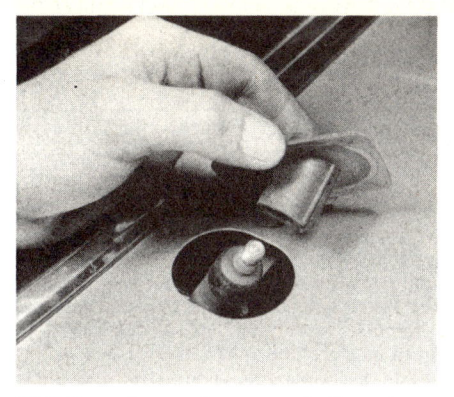

Chapter 12 Electrical system

30.1 Removing a wiper arm spindle seal

30.3 Removing one of the scuttle panel bolts

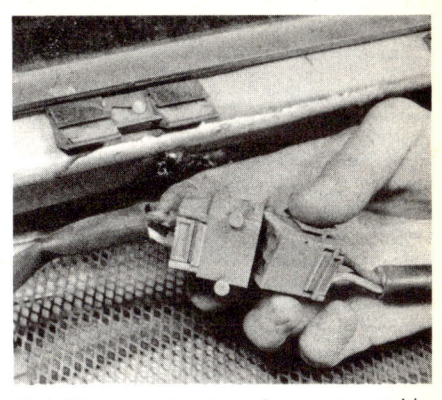

30.4 Disconnecting the wiper motor multi-plug

30.5 Unbolting the wiper assembly

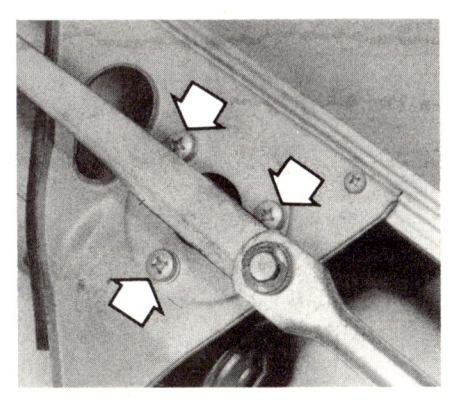

30.6 Wiper motor securing screws (arrowed)

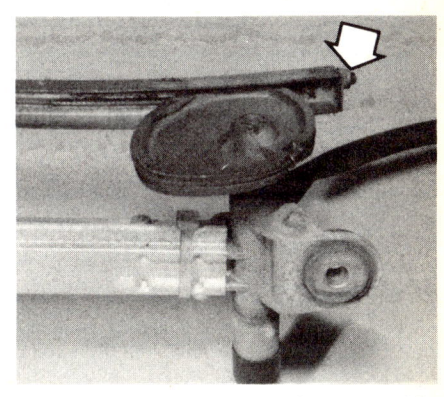

30.7 Part of the wiper linkage – cable tensioning nut is arrowed

31 Headlight wiper motor – removal and refitting

1 Remove the headlight unit on the side concerned (Section 15).
2 Follow the wiper motor wiring back to the multi-plug and disconnect it.
3 Remove the two nuts which secure the motor (photo). Withdraw the motor.
4 Refit by reversing the removal operations. If there is any doubt that the motor is in the 'parked' position, operate the wipers and washers before refitting the wiper arm.

32 Rear wiper motor – removal and refitting

1 Remove the tailgate interior trim (Chapter 11, Section 13).
2 Prise the link balljoint off the wiper motor crank arm (photo).
3 Remove the three bolts which secure the motor. Withdraw the motor (it may be necessary to rotate the crank arm) and disconnect the wiring from it (photo).
4 Refit by reversing the removal operations. Check the operation of the motor before refitting the trim.

31.3 Headlight wiper motor exposed. Securing nuts are behind the two studs (arrowed)

32.3 Rear wiper motor showing crank arm balljoint (arrowed) and securing bolts

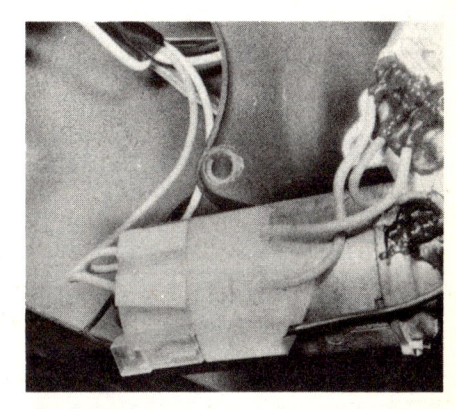

32.3 Removing the rear wiper motor

Chapter 12 Electrical system

33 Cigarette lighter – removal and refitting

1 Make sure that the ignition is switched off.

Front
2 Remove the lighter element and unclip the trim from around the aperture.
3 Remove the screws now exposed (photo).
4 Remove the lighter and tray, disconnecting the wiring feed and bulb holder from the rear (photo). Also disconnect any audio equipment which may be occupying the tray.
5 The lighter may now be removed from the tray if wished.
6 Refit by reversing the removal operations.

Rear
7 This is covered in removal of the rear console (Chapter 11, Section 36).

34 Heated rear window – general

1 All models are equipped with a heated rear window. Heating is achived by passing current through a resistive grid bonded to the inside of the rear window.
2 Do not allow hard or sharp items of luggage to rub against the heating grid. Use a soft cloth or chamois to clean the inside of the window, working along the lines of the grid.
3 Small breaks in the grid can be repaired using special conductive paint, obtainable from motor accessory shops. Use the paint as directed by the manufacturer.
4 The heated rear window draws a high current, so it should not be left switched on longer than necessary. On some models a so-called 'delay relay' is incorporated into the circuit in order to switch the window off after a few minutes.
5 When heated door mirrors are fitted, their heaters are controlled by the heated rear window switch.

35 Radio/cassette player (original equipment) – removal and refitting

1 The radio/cassette player seen in the accompanying photo is removed as follows.
2 Disconnect the battery negative lead.
3 Pull the control knobs off the front of the instrument (photo).
4 Using a hook made of a piece of bent wire, retract the side securing clips, working through the control knob apertures. As the clips are retracted, draw the unit out of its mounting (photos).
5 Withdraw the unit fully and disconnect the wiring.
6 When refitting, reconnect the wiring, then push the unit home until the clips snap into place.
7 Refit the control knobs and reconnect the battery.

36 Radio aerial (original equipment) – removal and refitting

Two types of original equipment aerial are shown here. There may be others.

Saloon (automatic, in boot)
1 Disconnect the battery negative lead.
2 Open the boot and remove the left-hand side trim.
3 Remove the nut and cover which secure the aerial tube to the rear wing (photo).
4 Disconnect the aerial signal and power leads (photos).
5 Remove the tube and drive securing nuts and bolts. Withdraw the aerial into the boot.

33.3 Undoing a cigarette lighter screw

33.4 Removing the lighter bulb and holder. This also serves the ashtray

35.3 Pulling off a radio knob

35.4A Insert a wire hook ...

35.4B ... to retract the side clip (unit removed to show clip)

35.4C Removing the radio/cassette unit

Chapter 12 Electrical system

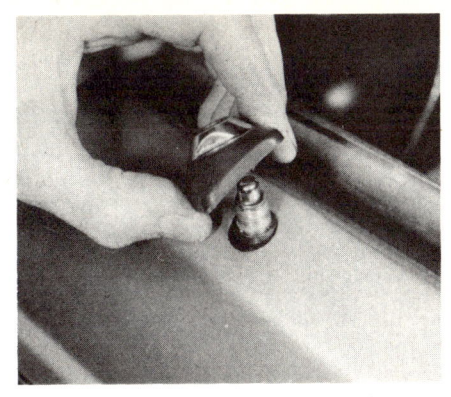

36.3 Removing the aerial tube nut and cover

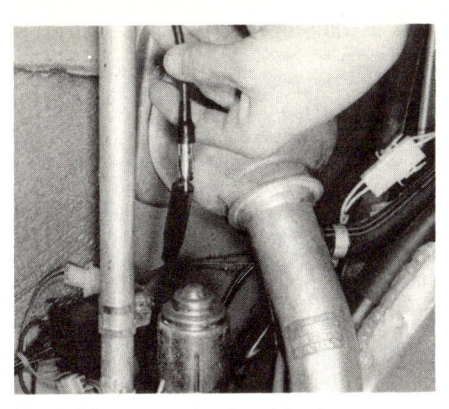

36.4A Disconnect the signal lead ...

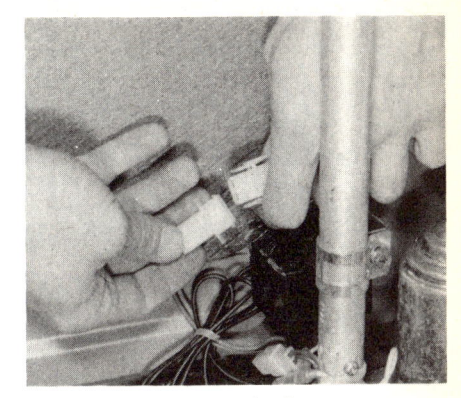

36.4B ... and the power feed

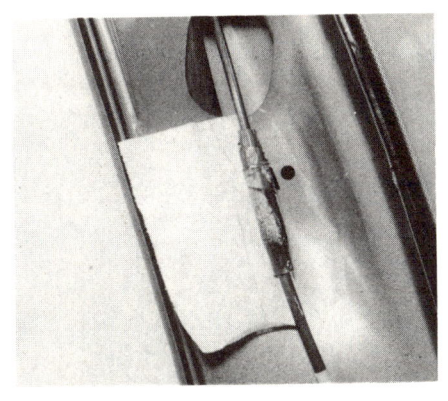

36.7 Signal lead connector (Estate) wrapped in sticky tape

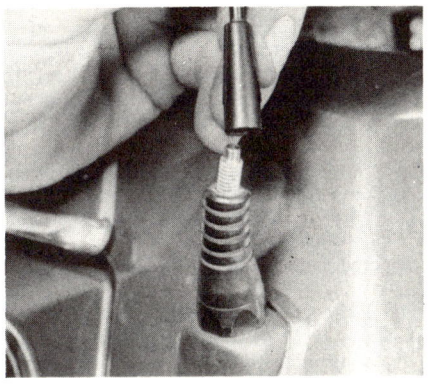

36.8 Unscrewing the aerial rod from the stub

Estate (fixed, on rear pillar)
6 Open the tailgate. Unclip the trim panel which covers the aerial mounting and wiring.
7 Disconnect the aerial signal lead (photo).
8 Outside the vehicle, unscrew the aerial rod from the stub (photo). Recover the spacer.

9 Unscrew the tube securing nut. Recover the lower spacer and seal and withdraw the aerial into the vehicle (photos).

All types
10 Refit by reversing the removal operations, but check for correct operation before refitting disturbed trim.

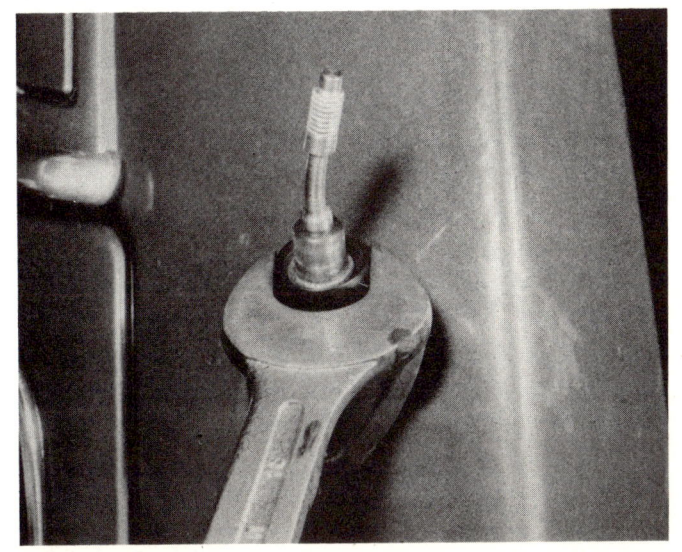

36.9A Remove the aerial tube nut ...

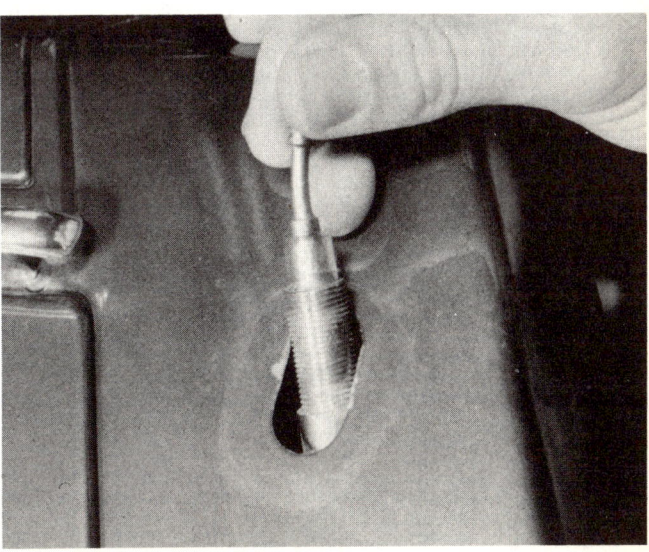

36.9B ... and withdraw the aerial into the vehicle

37 Mobile radio equipment – interference-free installation

Aerials – selection and fitting

The choice of aerials is now very wide. It should be realised that the quality has a profound effect on radio performance, and a poor, inefficient aerial can make suppression difficult.

A wing-mounted aerial is regarded as probably the most efficient for signal collection, but a roof aerial is usually better for suppression purposes because it is away from most interference fields. Stick-on wire aerials are available for attachment to the inside of the windscreen, but are not always free from the interference field of the engine and some accessories.

Motorised automatic aerials rise when the equipment is switched on and retract at switch-off. They require more fitting space and supply leads, and can be a source of trouble.

There is no merit in choosing a very long aerial as, for example, the type about three metres in length which hooks or clips on to the rear of the car, since part of this aerial will inevitably be located in an interference field. For VHF/FM radios the best length of aerial is about one metre. Active aerials have a transistor amplifier mounted at the base and this serves to boost the received signal. The aerial rod is sometimes rather shorter than normal passive types.

A large loss of signal can occur in the aerial feeder cable, especially over the Very High Frequency (VHF) bands. The design of feeder cable is invariably in the co-axial form, ie a centre conductor surrounded by a flexible copper braid forming the outer (earth) conductor. Between the inner and outer conductors is an insulator material which can be in solid or stranded form. Apart from insulation, its purpose is to maintain the correct spacing and concentricity. Loss of signal occurs in this insulator, the loss usually being greater in a poor quality cable. The quality of cable used is reflected in the price of the aerial with the attached feeder cable.

The capacitance of the feeder should be within the range 65 to 75 picofarads (pF) approximately (95 to 100 pF for Japanese and American equipment), otherwise the adjustment of the car radio aerial trimmer may not be possible. An extension cable is necessary for a long run between aerial and receiver. If this adds capacitance in excess of the above limits, a connector containing a series capacitor will be required, or an extension which is labelled as 'capacity-compensated'.

Fitting the aerial will normally involve making a $7/8$ in (22 mm) diameter hole in the bodywork, but read the instructions that come with the aerial kit. Once the hole position has been selected, use a centre punch to guide the drill. Use sticky masking tape around the area for this helps with marking out and drill location, and gives protection to the paintwork should the drill slip. Three methods of making the hole are in use:

(a) Use a hole saw in the electric drill. This is, in effect, a circular hacksaw blade wrapped round a former with a centre pilot drill.
(b) Use a tank cutter which also has cutting teeth, but is made to shear the metal by tightening with an Allen key.
(c) The hard way of drilling out the circle is using a small drill, say $1/8$ in (3 mm), so that the holes overlap. The centre metal drops out and the hole is finished with round and half-round files.

Whichever method is used, the burr is removed from the body metal and paint removed from the underside. The aerial is fitted tightly ensuring that the earth fixing, usually a serrated washer, ring or clamp, is making a solid connection. *This earth connection is important in reducing interference.* Cover any bare metal with primer paint and topcoat, and follow by underseal if desired.

Aerial feeder cable routing should avoid the engine compartment and areas where stress might occur, eg under the carpet where feet will be located. Roof aerials require that the headlining be pulled back and that a path is available down the door pillar. It is wise to check with the vehicle dealer whether roof aerial fitting is recommended.

Loudspeakers

Speakers should be matched to the output stage of the equipment, particularly as regards the recommended impedance. Power transistors used for driving speakers are sensitive to the loading placed on them.

Original equipment speakers will be found in the doors and (when applicable) on the parcel shelf (photos). Access to the parcel shelf speaker fastenings is from inside the boot.

Unit installation

Provision is made for installing the radio in the centre console. Volvo standard radio/cassette players are larger than many other units, so spacers or packing pieces may need to be used with a smaller unit.

Installation of the radio/audio unit is basically the same in all cases, and consists of offering it into the aperture after removal of the knobs (*not* push buttons) and the trim plate. In some cases a special mounting plate is required to which the unit is attached. It is worthwhile supporting the rear end in cases where sag or strain may occur, and it is usually possible to use a length of perforated metal strip attached between the unit and a good support point nearby. In general it is recommended that tape equipment should be installed at or nearly horizontal.

Connections to the aerial socket are simply by the standard plug terminating the aerial downlead or its extension cable. Speakers for a stereo system must be matched and correctly connected, as outlined previously.

Note: *While all work is carried out on the power side, it is wise to disconnect the battery earth lead.* Before connection is made to the vehicle electrical system, check that the polarity of the unit is correct. Volvos use a negative earth system, but radio/audio units often have a reversible plug to convert the set to either + or – earth. *Incorrect connection may cause serious damage.*

37.0A Removing the cover from a front door speaker

37.0B Removing a front door speaker. Note water deflector in aperture

Chapter 12 Electrical system

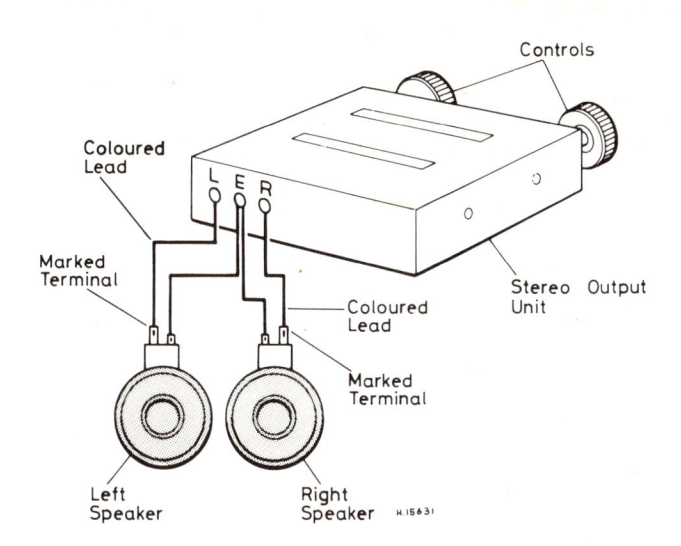

Fig. 12.18 Speaker connections must be made correctly (Sec 37)

it). If the noise disappears it is coming in through the aerial and is *radiation noise*. If the noise persists it is reaching the receiver through the wiring and is said to be *line-borne*.

Interference from wipers, washers, heater blowers, turn-indicators, stop lamps, etc is usually taken to the receiver by wiring, and simple treatment using capacitors and possibly chokes will solve the problem. Switch on each one in turn (wet the screen first for running wipers!) and listen for possible interference with the aerial plug in place and again when removed.

Electric petrol pumps are now finding application again and give rise to an irregular clicking, often giving a burst of clicks when the ignition is on but the engine has not yet been started. It is also possible to receive whining or crackling from the pump.

Note that if most of the vehicle accessories are found to be creating interference all together, the probability is that poor aerial earthing is to blame.

Component terminal markings

Throughout the following sub-sections reference will be found to various terminal markings. These will vary depending on the manufacturer of the relevant component. If terminal markings differ from those mentioned, reference should be made to the following table, where the most commonly encountered variations are listed.

The power lead is often permanently connected inside the unit and terminates with one half of an in-line fuse carrier. The other half is fitted with a suitable fuse (3 or 5 amperes) and a wire which should go to the orange wire which is protected by fuse No 12 (models up to 1984) or No 14 (later models). (When a radio/cassette unit is installed by the makers, this orange wire enters a multi-plug which also carries loudspeaker and aerial motor wires.)

Before switching on for initial test, be sure that the speaker connections have been made, for running without load can damage the output transistors. Switch on next and tune through the bands to ensure that all sections are working, and check the tape unit if applicable. The aerial trimmer should be adjusted to give the strongest reception on a weak signal in the medium wave band, at say 200 metres.

Interference

In general, when electric current changes abruptly, unwanted electrical noise is produced. The motor vehicle is filled with electrical devices which change electric current rapidly.

When the spark plugs operate, the sudden pulse of spark current causes the associated wiring to radiate. Since early radio transmitters used sparks as a basis of operation, it is not surprising that the car radio will pick up ignition spark noise unless steps are taken to reduce it to acceptable levels.

Interference reaches the car radio in two ways:

(a) by conduction through the wiring.
(b) by radiation to the receiving aerial.

Initial checks presuppose that the bonnet is down and fastened, the radio unit has a good earth connection *(not* through the aerial downlead outer), no fluorescent tubes are working near the car, the aerial trimmer has been adjusted, and the vehicle is in a position to receive radio signals, ie not in a metal-clad building.

Switch on the radio and tune to the middle of the medium wave (MW) band off-station with the volume (gain) control set fairly high. Switch on the engine and listen for interference on the MW band. Depending on the type of interference, the indications are as follows.

A harsh crackle that drops out abruptly at low engine speed or when the headlights are switched on is probably due to a voltage regulator.

A whine varying with engine speed is due to the alternator. Try temporarily taking off the drive belt – if the noise goes this is confirmation.

Regular ticking or crackle that varies in rate with the engine speed is due to the ignition system. With this trouble in particular and others in general, check to see if the noise is entering the receiver from the wiring or by radiation. To do this, pull out the aerial plug (preferably shorting out the input socket or connecting a 62 pF capacitor across

Alternator	Alternator terminal (thick lead)	Exciting winding terminal
DIN/Bosch	B+	DF
Delco Remy	+	EXC
Ducellier	+	EXC
Ford (US)	+	DF
Lucas	+	F
Marelli	+B	F

Ignition coil	Ignition switch terminal	'Contact breaker' terminal
DIN/Bosch	15	1
Delco Remy	+	–
Ducellier	BAT	RUP
Ford (US)	B/+	CB/–
Lucas	SW/+	–
Marelli	BAT/+B	D

Voltage regulator	Voltage input terminal	Exciting winding terminal
DIN/Bosch	B+/D+	DF
Delco Remy	BAT/+	EXC
Ducellier	BOB/BAT	EXC
Ford (US)	BAT	DF
Lucas	+/A	F
Marelli		F

Suppression methods – ignition

Suppressed HT cables are supplied as original equipment by manufacturers and will meet regulations as far as interference to neighbouring equipment is concerned. It is illegal to remove such suppression unless an alternative is provided, and this may take the form of resistive spark plug caps in conjunction with plain copper HT cable. For VHF purposes, these and 'in-line' resistors may not be effective, and resistive HT cable is preferred. Check that suppressed cables are actually fitted by observing cable identity lettering, or measuring with an ohmmeter – the value of each plug lead should be 5000 to 10 000 ohms.

If ignition noise persists despite the treatment above, the following sequence should be followed:

(a) Check the earthing of the ignition coil; remove paint from fixing clamp.

(b) If this does not work, lift the bonnet. Should there be no change in interference level, this may indicate that the bonnet is not electrically connected to the car body. Use a proprietary braided strap across a bonnet hinge ensuring a first class electrical connection. If, however, lifting the bonnet increases the interference, then fit resistive HT cables of a higher ohms-per-metre value.

(c) If all these measures fail, it is probable that re-radiation from metallic components is taking place. Using a braided strap between metallic points, go round the vehicle systematically –

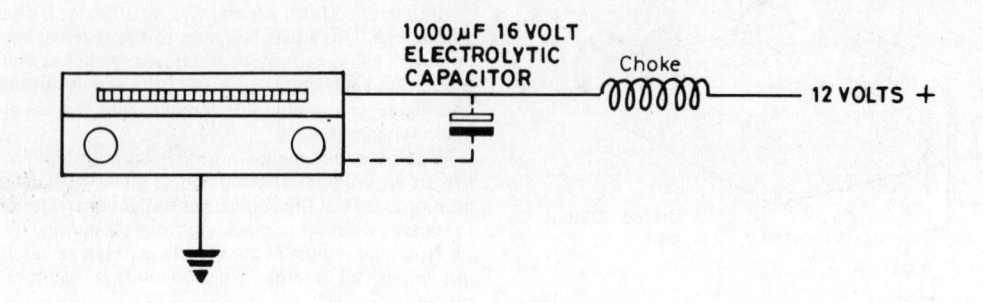

Fig. 12.19 Suppression of line-borne interference (Sec 37)

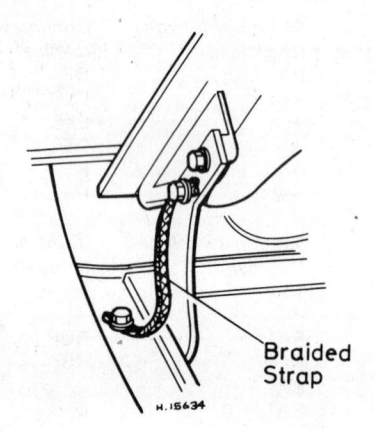

Fig. 12.20 Braided earth strap between bonnet and body (Sec 37)

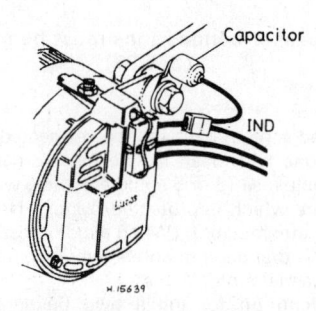

Fig. 12.21 Suppression of interference from electronic voltage regulator integral with alternator (Sec 37)

try the following: engine to body, exhaust system to body, front suspension to engine and to body, steering column to body, gear lever to engine and to body, Bowden cable to body, metal parcel shelf to body. When an offending component is located it should be bonded with the strap permanently.
(d) As a next step, the fitting of distributor suppressors to each lead at the distributor end may help.
(e) Beyond this point is involved the possible screening of the distributor and fitting resistive spark plugs, but such advanced treatment is not usually required for vehicles with entertainment equipment.

Electronic ignition systems have built-in suppression components, but this does not relieve the need for using suppressed HT leads. In some cases it is permitted to connect a capacitor on the low tension supply side of the ignition coil, but not in every case. Makers' instructions should be followed carefully, otherwise damage to the ignition semiconductors may result.

Suppression methods – generators
Alternators should be fitted with a 3 microfarad capacitor from the B+ main output terminal (thick cable) to earth. Additional suppression may be obtained by the use of a filter in the supply line to the radio receiver.
It is most important that alternators are not run without connection to the battery.

Suppression methods – voltage regulators
Integral electronic voltage regulators do not normally generate much interference, but when encountered this is in combination with alternator noise. A 1 microfarad or 2 microfarad capacitor from the warning lamp (IND) terminal to earth for Lucas ACR alternators and Femsa, Delco and Bosch equivalents should cure the problem.

Suppression methods – other equipment
Wiper motors – Connect the wiper body to earth with a bonding strap. For all motors use a 7 ampere choke assembly inserted in the leads to the motor.
Heater motors – Fit 7 ampere line chokes in both leads, assisted if necessary by a 1 microfarad capacitor to earth from both leads.
Electronic tachometer – The tachometer is a possible source of ignition noise – check by disconnecting at the ignition coil CB terminal. It usually feeds from ignition coil LT pulses at the 'contact breaker' terminal. A 3 ampere line choke should be fitted in the tachometer lead at the coil CB terminal.
Horn – A capacitor and choke combination is effective if the horn is directly connected to the 12 volt supply. The use of a relay is an alternative remedy, as this will reduce the length of the interference-carrying leads.
Electrostatic noise – Characteristics are erratic crackling at the receiver, with disappearance of symptoms in wet weather. Often shocks may be given when touching bodywork. Part of the problem is the build-up of static electricity in non-driven wheels and the acquisition of charge on the body shell. It is possible to fit spring-loaded contacts at the wheels to give good conduction between the rotary wheel parts and the vehicle frame. Changing a tyre sometimes helps – because of tyres' varying resistances. In difficult cases a trailing flex which touches the ground will cure the problem. If this is not acceptable it is worth trying conductive paint on the tyre walls.

Chapter 12 Electrical system

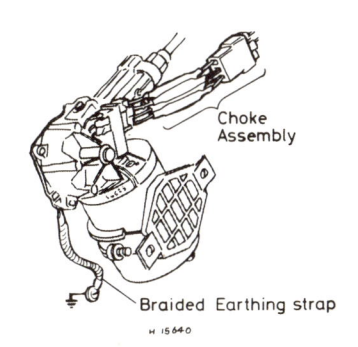

Fig. 12.22 Suppression of interference from the wiper motor (Sec 37)

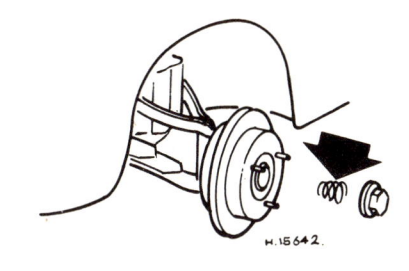

Fig. 12.23 Use of spring contacts at wheels (Sec 37)

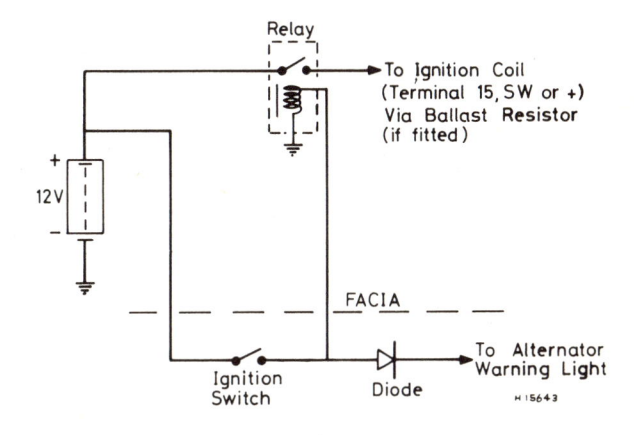

Fig. 12.24 Using an ignition coil relay to suppress case breakthrough (Sec 37)

Fuel pump – Suppression requires a 1 microfarad capacitor between the supply wire to the pump and a nearby earth point. If this is insufficient a 7 ampere line choke connected in the supply wire near the pump is required.

Fluorescent tubes – Vehicles used for camping/caravanning frequently have fluorescent tube lighting. These tubes require a relatively high voltage for operation and this is provided by an inverter (a form of oscillator) which steps up the vehicle supply voltage. This can give rise to serious interference to radio reception, and the tubes themselves can contribute to this interference by the pulsating nature of the lamp discharge. In such situations it is important to mount the aerial as far away from a fluorescent tube as possible. The interference problem may be alleviated by screening the tube with fine wire turns spaced an inch (25 mm) apart and earthed to the chassis. Suitable chokes should be fitted in both supply wires close to the inverter.

Radio/cassette case breakthrough
Magnetic radiation from dashboard wiring may be sufficiently intense to break through the metal case of the radio/cassette player. Often this is due to a particular cable routed too close and shows up as ignition interference on AM and cassette play and/or alternator whine on cassette play.

The first point to check is that the clips and/or screws are fixing all parts of the radio/cassette case together properly. Assuming good earthing of the case, see if it is possible to re-route the offending cable – the chances of this are not good, however, in most cars.

Next release the radio/cassette player and locate it in different positions with temporary leads. If a point of low interference is found, then if possible fix the equipment in that area. This also confirms that local radiation is causing the trouble. If re-location is not feasible, fit the radio/cassette player back in the original position.

Alternator interference on cassette play is now caused by radiation from the main charging cable which goes from the battery to the output terminal of the alternator, usually via the + terminal of the starter motor relay. In some vehicles this cable is routed under the dashboard, so the solution is to provide a direct cable route. Detach the original cable from the alternator output terminal and make up a new cable of at least 6 mm² cross-sectional area to go from alternator to battery with the shortest possible route. *Remember* – *do not run the engine with the alternator disconnected from the battery.*

Ignition breakthrough on AM and/or cassette play can be a difficult problem. It is worth wrapping earthed foil round the offending cable run near the equipment, or making up a deflector plate well screwed down to a good earth. Another possibility is the use of a suitable relay to switch on the ignition coil. The relay should be mounted close to the ignition coil; with this arrangement the ignition coil primary current is not taken into the dashboard area and does not flow through the ignition switch. A suitable diode should be used since it is possible that at ignition switch-off the output from the alternator warning lamp terminal could hold the relay on.

Connectors for suppression components
Capacitors are usually supplied with tags on the end of the lead, while the capacitor body has a flange with a slot or hole to fit under a nut or screw with washer.

Connections to feed wires are best achieved by self-stripping connectors. These connectors employ a blade which, when squeezed down by pliers, cuts through cable insulation and makes connection to the copper conductors beneath.

Chokes sometimes come with bullet snap-in connectors fitted to the wires, and also with just bare copper wire. With connectors, suitable female cable connectors may be purchased from an auto-accessory shop together with any extra connectors required for the cable ends after being cut for the choke insertion. For chokes with bare wires, similar connectors may be employed together with insulation sleeving as required.

VHF/FM broadcasts
Reception of VHF/FM in an automobile is more prone to problems than the medium and long wavebands. Medium/long wave transmitters are capable of covering considerable distances, but VHF transmitters are restricted to line of sight, meaning ranges of 10 to 50 miles, depending upon the terrain, the effects of buildings and the transmitter power.

Because of the limited range it is necessary to retune on a long journey, and it may be better for those habitually travelling long distances or living in areas of poor provision of transmitters to use an AM radio working on medium/long wavebands.

When conditions are poor, interference can arise, and some of the suppression devices described previously fall off in performance at very high frequencies unless specifically designed for the VHF band. Available suppression devices include reactive HT cable, resistive distributor caps, screened plug caps, screened leads and resistive spark plugs.

For VHF/FM receiver installation the following points should be particularly noted:

(a) Earthing of the receiver chassis and the aerial mounting is important. Use a separate earthing wire at the radio, and scrape paint away at the aerial mounting.
(b) If possible, use a good quality roof aerial to obtain maximum height and distance from interference generating devices on the vehicle.
(c) Use of a high quality aerial downlead is important, since losses in cheap cable can be significant.
(d) The polarisation of FM transmissions may be horizontal, vertical, circular or slanted. Because of this the optimum mounting angle is at 45° to the vehicle roof.

Citizens' Band radio (CB)

In the UK, CB transmitter/receivers work within the 27 MHz and 934 MHz bands, using the FM mode. At present interest is concentrated on 27 MHz where the design and manufacture of equipment is less difficult. Maximum transmitted power is 4 watts, and 40 channels spaced 10 kHz apart within the range 27.60125 to 27.99125 MHz are available.

Aerials are the key to effective transmission and reception. Regulations limit the aerial length to 1.65 metres including the loading coil and any associated circuitry, so tuning the aerial is necessary to obtain optimum results. The choice of a CB aerial is dependent on whether it is to be permanently installed or removable, and the performance will hinge on correct tuning and the location point on the vehicle. Common practice is to clip the aerial to the roof gutter or to employ wing mounting where the aerial can be rapidly unscrewed. An alternative is to use the boot rim to render the aerial theftproof, but a popular solution is to use the 'magmount' – a type of mounting having a strong magnetic base clamping to the vehicle at any point, usually the roof.

Aerial location determines the signal distribution for both transmission and reception, but it is wise to choose a point away from the engine compartment to minimise interference from vehicle electrical equipment.

The aerial is subject to considerable wind and acceleration forces. Cheaper units will whip backwards and forwards and in so doing will alter the relationship with the metal surface of the vehicle with which it forms a ground plane aerial system. The radiation pattern will change correspondingly, giving rise to break-up of both incoming and outgoing signals.

Interference problems on the vehicle carrying CB equipment fall into two categories:

(a) Interference to nearby TV and radio receivers when transmitting.
(b) Interference to CB set reception due to electrical equipment on the vehicle.

Problems of break-through to TV and radio are not frequent, but can be difficult to solve. Mostly trouble is not detected or reported because the vehicle is moving and the symptoms rapidly disappear at the TV/radio receiver, but when the CB set is used as a base station any trouble with nearby receivers will soon result in a complaint.

It must not be assumed by the CB operator that his equipment is faultless, for much depends upon the design. Harmonics (that is, multiples) of 27 MHz may be transmitted unknowingly and these can fall into other user's bands. Where trouble of this nature occurs, low pass filters in the aerial or supply leads can help, and should be fitted in base station aerials as a matter of course. In stubborn cases it may be necessary to call for assistance from the licensing authority, or, if possible, to have the equipment checked by the manufacturers.

Interference received on the CB set from the vehicle equipment is, fortunately, not usually a severe problem. The precautions outlined previously for radio/cassette units apply, but there are some extra points worth noting.

It is common practice to use a slide-mount on CB equipment enabling the set to be easily removed for use as a base station, for example. Care must be taken that the slide mount fittings are properly earthed and that first class connection occurs between the set and slide-mount.

Vehicle manufacturers in the UK are required to provide suppression of electrical equipment to cover 40 to 250 MHz to protect TV and VHF radio bands. Such suppression appears to be adequately effective at 27 MHz, but suppression of individual items such as alternators/dynamos, clocks, stabilisers, flashers, wiper motors, etc, may still be necessary. The suppression capacitors and chokes available from auto-electrical suppliers for entertainment receivers will usually give the required results with CB equipment.

Other vehicle radio transmitters

Besides CB radio already mentioned, a considerable increase in the use of transceivers (ie combined transmitter and receiver units) has taken place in the last decade. Previously this type of equipment was fitted mainly to military, fire, ambulance and police vehicles, but a large business radio and radio telephone usage has developed.

Generally the suppression techniques described previously will suffice, with only a few difficult cases arising. Suppression is carried out to satisfy the 'receive mode', but care must be taken to use heavy duty chokes in the equipment supply cables since the loading on 'transmit' is relatively high.

38 Fault diagnosis – electrical system

Symptom	Reason(s)
Starter motor does not turn – no voltage at motor	Battery terminals loose or corroded Battery discharged or defective Starter motor connections loose or broken Starter switch or solenoid faulty Automatic transmission not in P or N Automatic transmission inhibitor switch faulty
Starter motor does not turn – voltage at motor	Starter motor internal defect
Starter motor turns very slowly	Battery nearly discharged or defective Battery terminals loose or corroded Starter motor internal defect
Starter motor noisy or rough	Mounting bolts loose Pinion or flywheel ring gear teeth damaged or worn
Alternator not charging battery	Drivebelt slipping or broken Alternator brushes worn Alternator connections loose or broken Alternator internal defect

Chapter 12 Electrical system

Symptom	Reason(s)
Alternator overcharging battery	Alternator regulator faulty
Battery will not hold charge	Short-circuit (continual drain on battery) Battery defective internally Battery case dirty and damp
Gauge or speedometer gives no reading	Sender unit defective Wire disconnected or broken Fuse blown Gauge or speedometer defective
Fuel or temperature gauge reads too high	Sender unit defective Wire earthed Gauge faulty
Horn operates continuously	Horn push stuck down Cable to horn push earthed
Horn does not operate	Fuse blown Cable or connector broken or loose Switch, slip ring or brush dirty or defective
Lights do not come on	Battery discharged Fuse(s) blown Light switch faulty Bulbs blown Relay defective (when applicable)
Lights give poor illumination	Lenses or reflectors dirty Bulbs blackened Incorrect wattage bulbs fitted
Wiper motor fails to work	Fuse blown Connections loose or broken Relay defective Switch defective Motor defective
Wiper motor works slowly and draws little current	Brushes badly worn Commutator dirty or burnt
Wiper motor works slowly and draws heavy current	Linkage seized or otherwise damaged Motor internal fault
Wiper motor works, but blades do not move	Linkage broken or disconnected Motor gearbox badly worn
Defect in any other components	Fuse blown Relay faulty (when applicable) Supply wire broken or disconnected Switch faulty (when applicable) Earth return faulty (check for loose or corroded mountings) Component itself faulty

Key to Figs. 12.25 to 12.28. Not all items are fitted to all models; some systems do not apply to UK

No	Item	Grid ref*	No	Item	Grid ref*
1	Battery	B2	37	Gear selector light (auto)	H4
2	Ignition switch	D3	38	Instrument illumination	F3, F4
3	Instrument panel connector (3-pole)	N3	39	Engine bay light	B2
4	Ignition coil	B4	40	Boot light	H4
5	Distributor	B3	41	Door edge marker light	L3, L4
6	Spark plugs	C3	42	Heater control illumination	F3, F1
7	Instrument panel connector (7-pole)	N3	43	Vanity mirror light	M5
8	Instrument panel connector (8-pole)	M3	45	Driver's belt lock tight	H3, L3
9	Starter motor	B3	46	Passenger's belt lock light	H3
10	Alternator and voltage regulator	B3	47	Windscreen wiper switch	F2
11	Fusebox	B1, E1, H1	48	Light switch	F3
12	Instrument panel connector (12-pole)	M3	49	Direction indicator/hazard warning/dip switches	F3
13	Headlight main beam	A2, A4, A5	50	Horn switch	B4
14	Headlight dipped beam	A2, A5	51	Heated rear window switch	J3
15	No 15 terminal (in central electrical unit)	K4	52	Rear foglight switch	H5
16	Parking light	A2, A5, J2, J5	53	Passenger seat heater switch	H3
17	Day running light	A2, A5	54	Overdrive switch (auto)	M3
18	Direction indicators	A1, A2, A5, B5, J2, J5	55	Heater blower switch	N1
19	Reversing light	J2, J5	56	Window switch (driver's door)	O3
20	Rear foglight	J2, J5	57	Window switch (front passenger's door)	P4
21	Tail light	J2, J5	58	Window switch (RH rear)	P4
22	Stop-light	J5	59	Window switch (LH rear)	O4
23	Front foglight/spotlight	N5, O5, Q1	60	Door mirror switch (driver's door)	N4
24	Number plate light	J3, J4	61	Door mirror switch (passenger's door)	O4
25	Interior light unit	H3	62	Central locking link rod switch	O4
26	Map reading lights	H3	63	Central locking key switch	O4
27	Courtesy light	H3	64	Sunroof switch	M2
28	Rear reading lights	H3, H4	65	Front foglight/spotlight switch	N5, P5, Q1
29	Positive terminal	C2	66	Stop-light switch	F3
30	No 30 terminal (in central electrical unit)	K4	67	Choke control switch (not UK)	D2
31	Earth connection (in central electrical unit)	K2	68	Handbrake switch	D2
32	Glovebox light	D2, E2, M4	69	Brake failure warning switch	E2
33	Front ashtray light	E2	70	Reversing light switch	H5
34	Rear ashtray light	H4	71	Starter inhibitor switch (auto)	B2
35	Switch illumination	H4	72	Driver's door switch	H5, L4
36	Switch illumination	H4	73	Passenger's door switch	H2, J5, L3
			74	Rear door switches	J2, M3

Key to Figs. 12.25 to 12.28. Not all items are fitted to all models; some systems do not apply to UK (continued)

No	Item	Grid ref*	No	Item	Grid ref*
75	Passenger's seat contact	H3	114	Seat belt reminder (rear)	H3, L3
71	Overdrive switch (manual)	H2	115	Bulb failure sensor	G3
78	CIS microswitch	K1, N3	116	Seat belt buzzer	F3, L2
79	Lambdasond microswitch (not UK)	L2	117	Wiper delay relay	F3
80	Thermal time switch	C5	119	Fuel pump relay	D4
81	Air conditioner pressure sensor	N2, O1	120	Fuel injection impulse relay	D5
82	ACC control panel sensor	O1	121	Flasher unit	F3
83	ACC coolant thermal switch	O2	122	Exhaust temperature sensor (Japan)	N5
85	Speedometer	E3, N3	123	Overdrive relay	H3
86	Tachometer	D3	124	Window winder/cooling fan relay	P3
87	Clock	D3, D4	125	Central locking relay – opening	N3
88	Coolant temperature gauge	E3	126	Central locking relay – closing	O3
89	Fuel gauge	F3	127	Auxiliary light relay	O5, P5, Q2
90	Voltmeter	E4	129	Foglight relay (Sweden)	O5, Q1
91	Oil pressure gauge	E3	130	Glow plug relay (Diesel)	L4
92	Ambient temperature gauge	D3	131	Heater blower relay	N1, P1
93	Instrument voltage regulator	E3	132	Air conditioning delay relay	N1, O1
94	Panel light rheostat	E3	133	Oil level relay	D3
95	Panel illumination	E3	136	Overdrive relay (auto)	M4
96	Oil level warning light	D3	137	Main beam relay	F4
97	Oil pressure warning light	D3	138	Driver's seat heater	H4
98	Choke warning light (not UK)	D3	139	Driver's backrest heater	H4
99	Handbrake warning light	D3	140	Passenger's seat heater	H2
100	Brake failure warning light	E3	141	Passenger's backrest heater	H2
101	Washer fluid level warning light	E3	142	Seat heater thermostats	H2, H4
102	Spare	E3	143	Loudspeaker – LH front	M1
103	Bulb failure warning light	E3	144	Loudspeaker – RH front	M1
104	Preheater pilot light (Diesel)	E3	145	Loudspeaker – LH rear	M1
105	Ignition (no charge) warning light	D4	146	Loudspeaker – RH rear	M1
106	Overdrive pilot light (auto)	D4	147	Aerial (static)	L1
107	Exhaust temperature light (Japan)	D4	148	Aerial (power-operated)	M1
108	Direction indicator pilot light (LH)	D4	149	Radio	L1
109	Main beam pilot light	E4	150	Window motor – driver's door	P3
110	Direction indicator pilot light (RH)	E4	151	Window motor – passenger's door	P4
111	Lambdasond light (not UK)	E4	152	Window motor – RH rear	P4
112	Overdrive pilot light (manual)	E4	153	Window motor – LH rear	O4
113	Seat belt reminder (front)	E4, L3	154	Driver's door mirror	N4

Key to Figs. 12.25 to 12.28. Not all items are fitted to all models; some systems do not apply to UK (continued)

No	Item	Grid ref*	No	Item	Grid ref*
155	Passenger's door mirror	P4	193	Hot start injector	B2
156	Cooling fan motor	K1	194	Idle up solenoid	K4
157	Headlight wiper motor	A3, A4	195	Fuel cut-off solenoid	B2
158	Sunroof motor	M3	196	CIS idle valve	O3
159	Locking motor – passenger's door	N4	197	Oil pressure sensor	D2
160	Locking motor – RH rear	N4	199	Temperature sensor (Diesel)	L5
161	Locking motor – LH rear	N4	200	Air conditioning compressor solenoid	K5
162	Locking motor – boot	N4	201	Overdrive solenoid	H2, N4
163	Windscreen wiper motor	F3	202	Heater control	N2
164	Windscreen washer motor	F3	203	ACC temperature control	O2
165	Heater blower motor	N1, P1	204	ACC ambient temperature sensor	O2
166	Capacitor (interference suppression)	B3	205	ACC cabin temperature sensor	O2
167	Interference suppressors – spark plugs	C3	206	ACC programmer	P2
168	Ignition ballast resistor	B3	207	Horn	A4
169	Heater blower resistor	N1, P1	208	Glow plugs (Diesel)	L4
170	Catalyst element (Japan)	M5	210	Tank pump	H2
171	Lambdasond thermostat (not UK)	L2	211	Main fuel pump	C5
173	Interior light delay	H3	212	Starter cranking contact	C2
174	Oil level control unit	C3	219	Lambdasond test point (not UK)	L2
175	Ignition system control unit	C4	220	CIS test point	N3
176	CIS control unit	O2	221	Heated rear window	J3
177	Lambdasond control unit (not UK)	L2	222	Cruise control speedometer connection	E3
178	Washer fluid level sensor	E2	223	Cigarette lighter	E2
179	Oil level sensor	C4	224	Fan thermoswitch	K1
180	Speedometer sender	E5	225	Cruise control switch	K1
187	Coolant temperature sensor	E5	226	Cruise control control unit	K2
182	Fuel gauge sender	E5	227	Vacuum pump	K2
183	TDC sensor	P4	228	Clutch switch	K2
184	CIS temperature sensor	N3	229	Brake switch	K2
187	Lambda sensor (not UK)	L2	231	Rear foglight relay	O5, Q1
188	Start injector	C5	233	Turbo pressure sensor (Diesel)	D2
189	Control pressure regulator	B5	236	Pressure sensor relay (Diesel)	K1
190	Auxiliary air valve	C5	237	Pressure sensor (Diesel)	K1
191	Frequency valve	L2	250	Auxiliary tank fuel gauge sender	E4
192	Pressure differential switch	M2	251	Connector	D4

'UTOM' means 'except'

*The grid references given here are correct for the 1982 wiring diagram.

Colour code

BL Blue	GR Grey	R Red	W White
BN Brown	OR Orange	SB Black	Y Yellow
GN Green	P Pink	VO Violet	

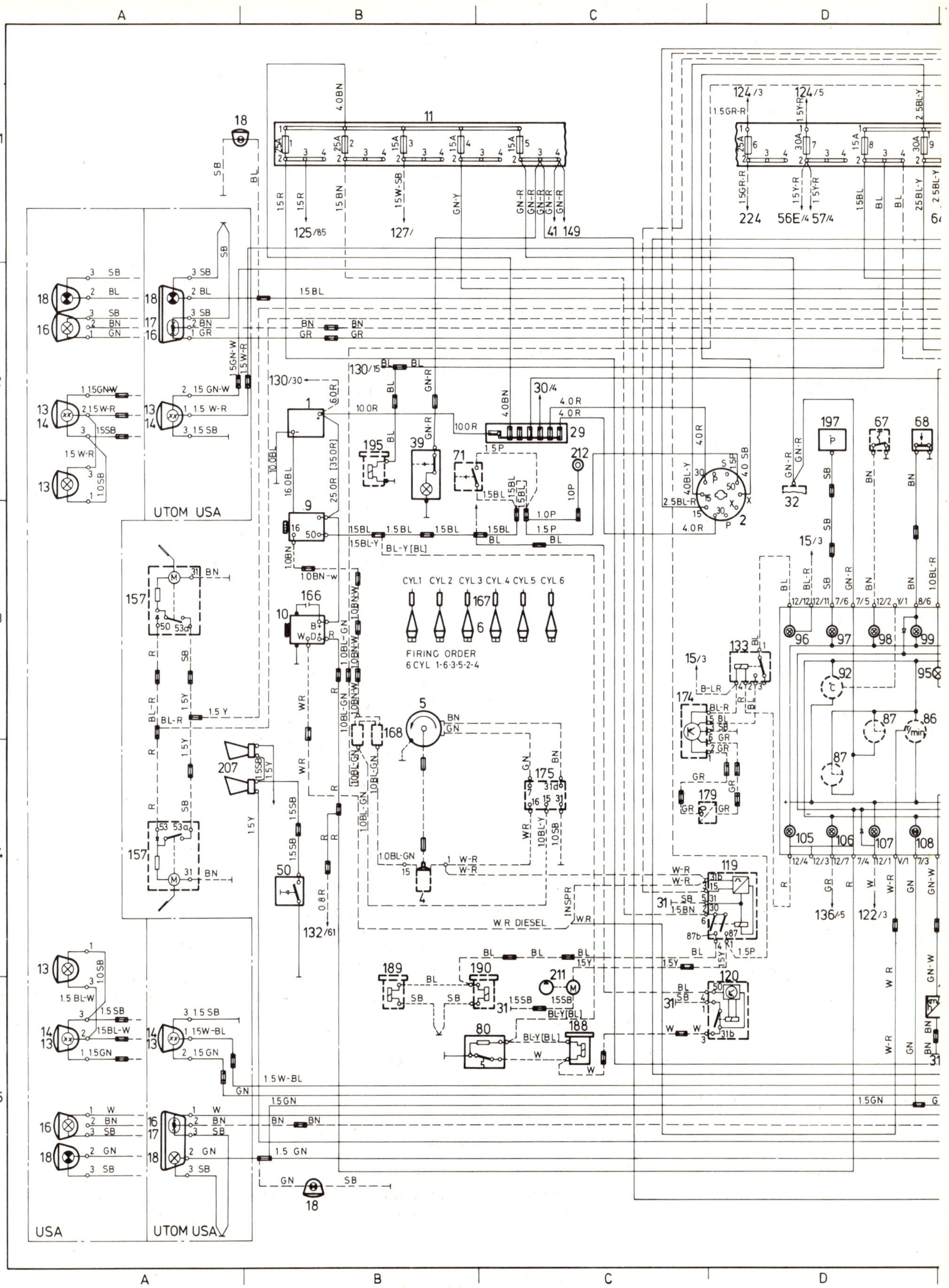

Fig. 12.25 Main wiring diagram for 1982 models.

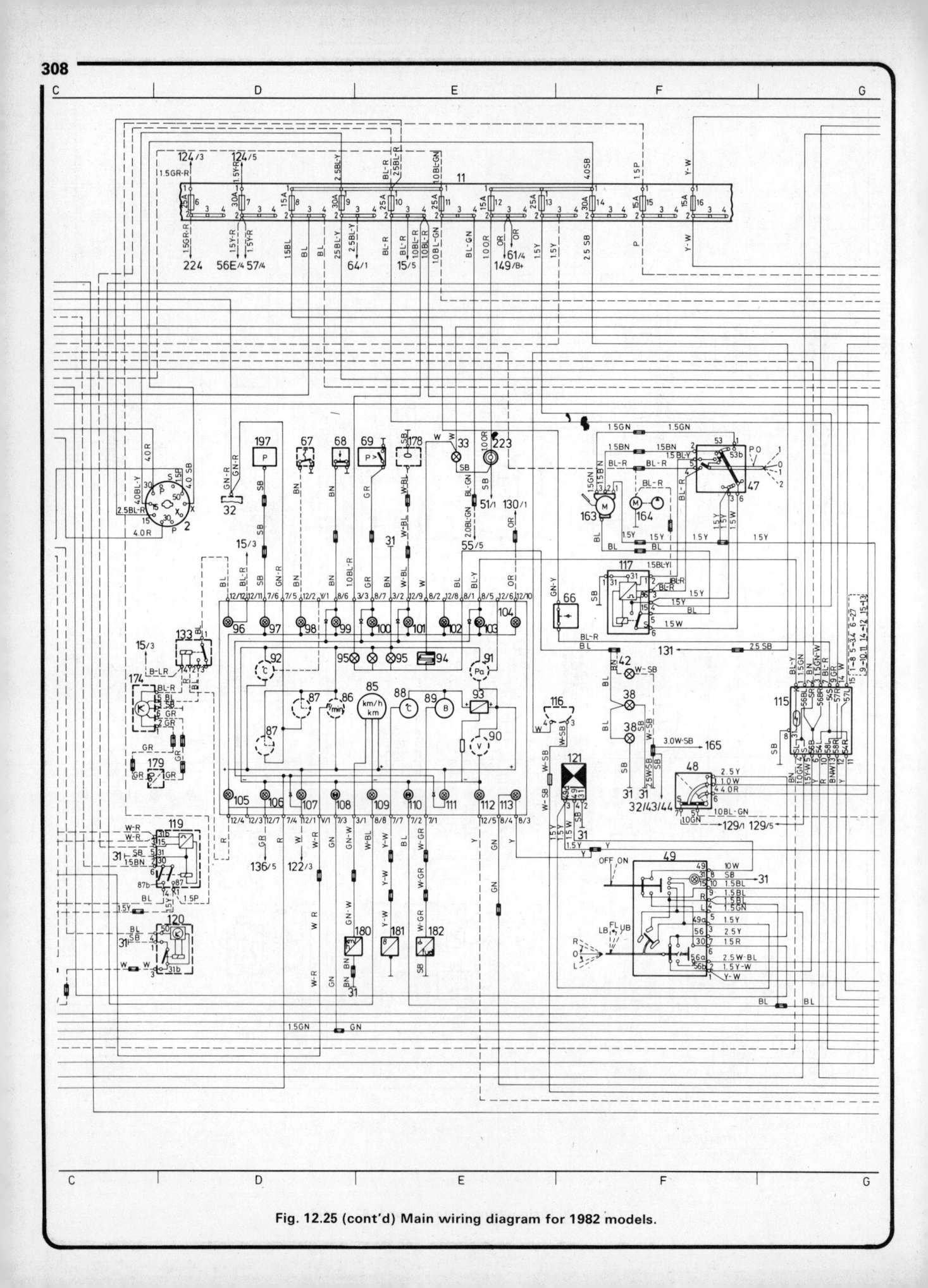

Fig. 12.25 (cont'd) Main wiring diagram for 1982 models.

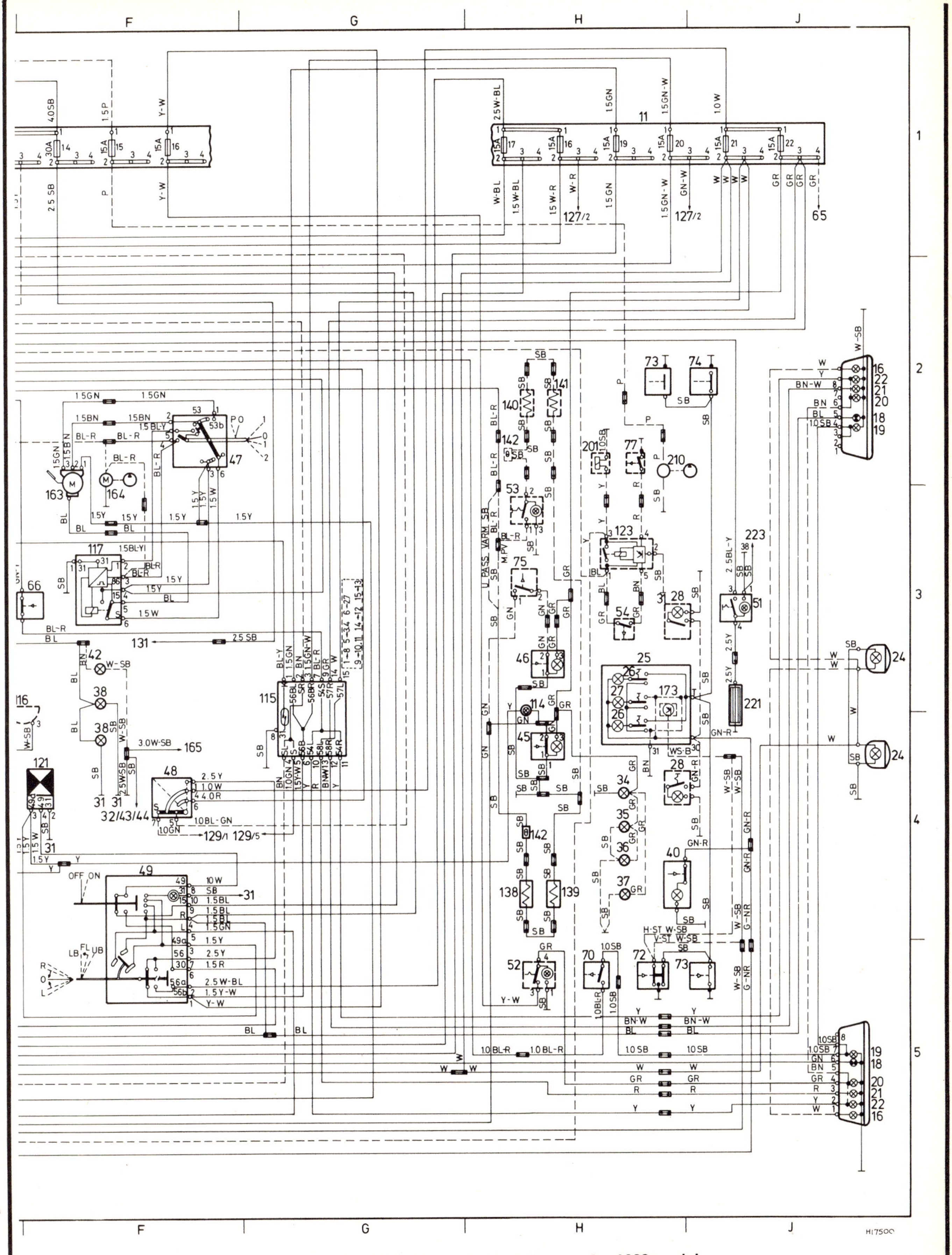

Fig. 12.25 (cont'd) Main wiring diagram for 1982 models.

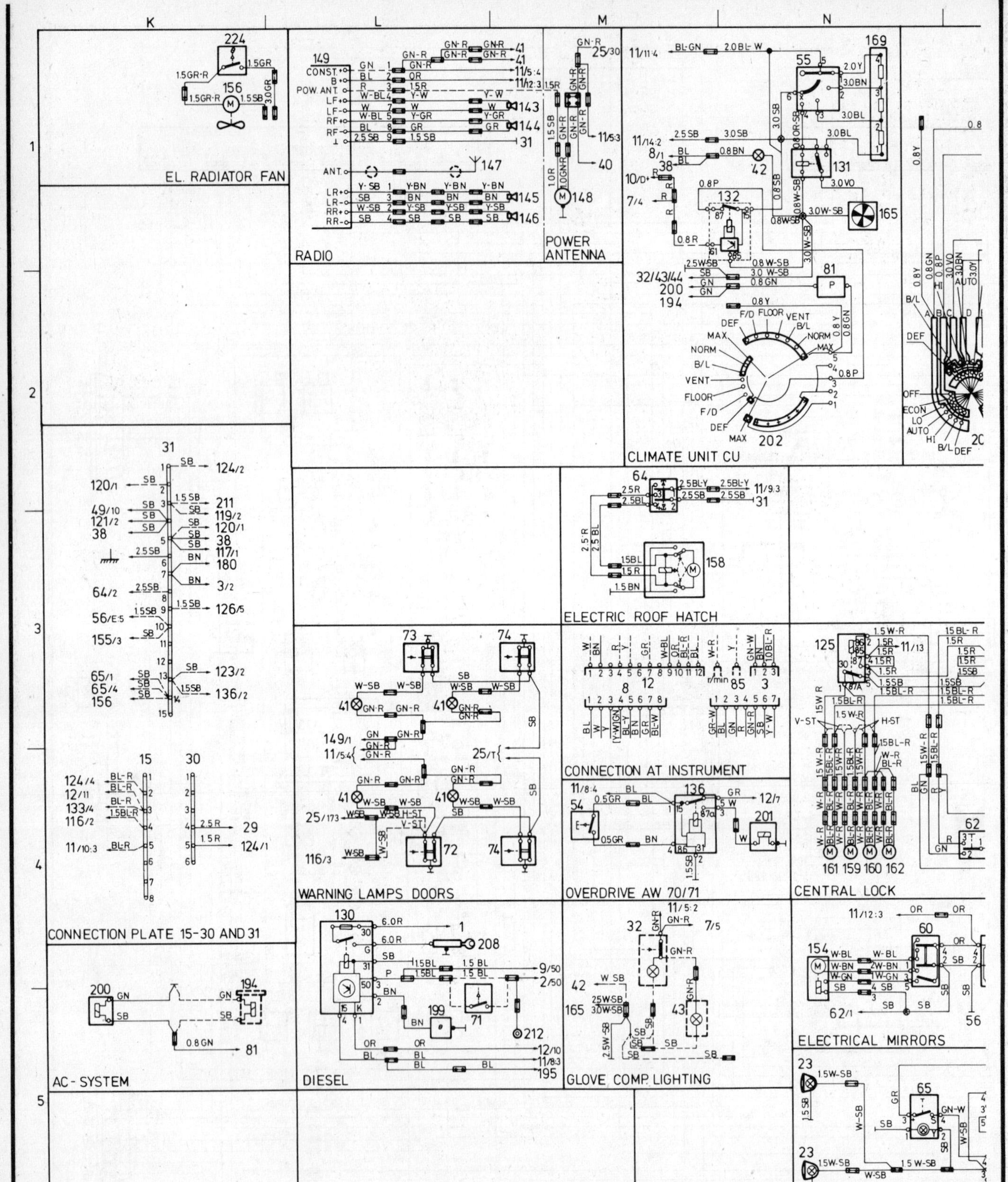

Fig. 12.26 Supplementary wiring diagram for 1982 models.

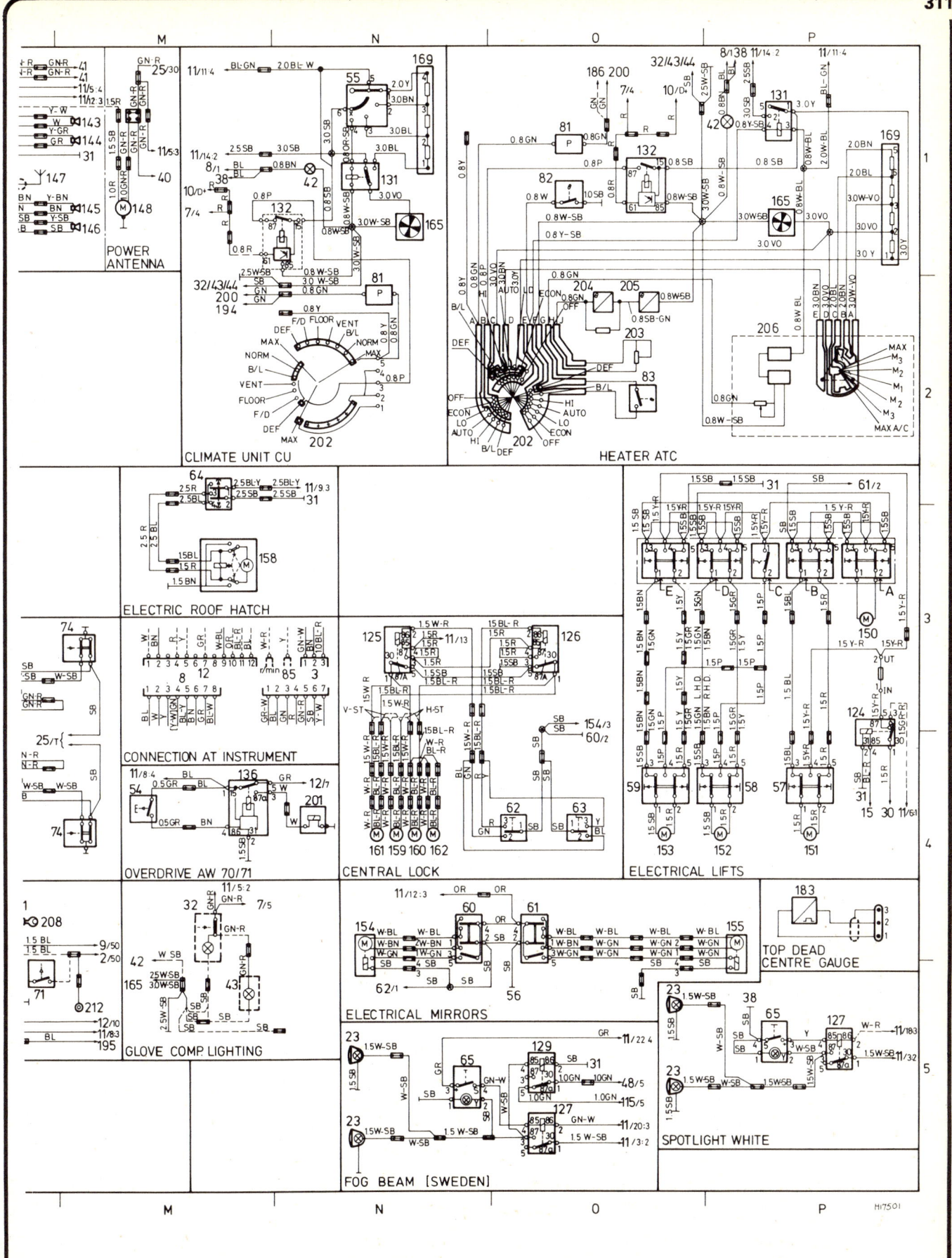

Fig 12.26 (cont'd) Supplementary wiring diagram for 1982 models.

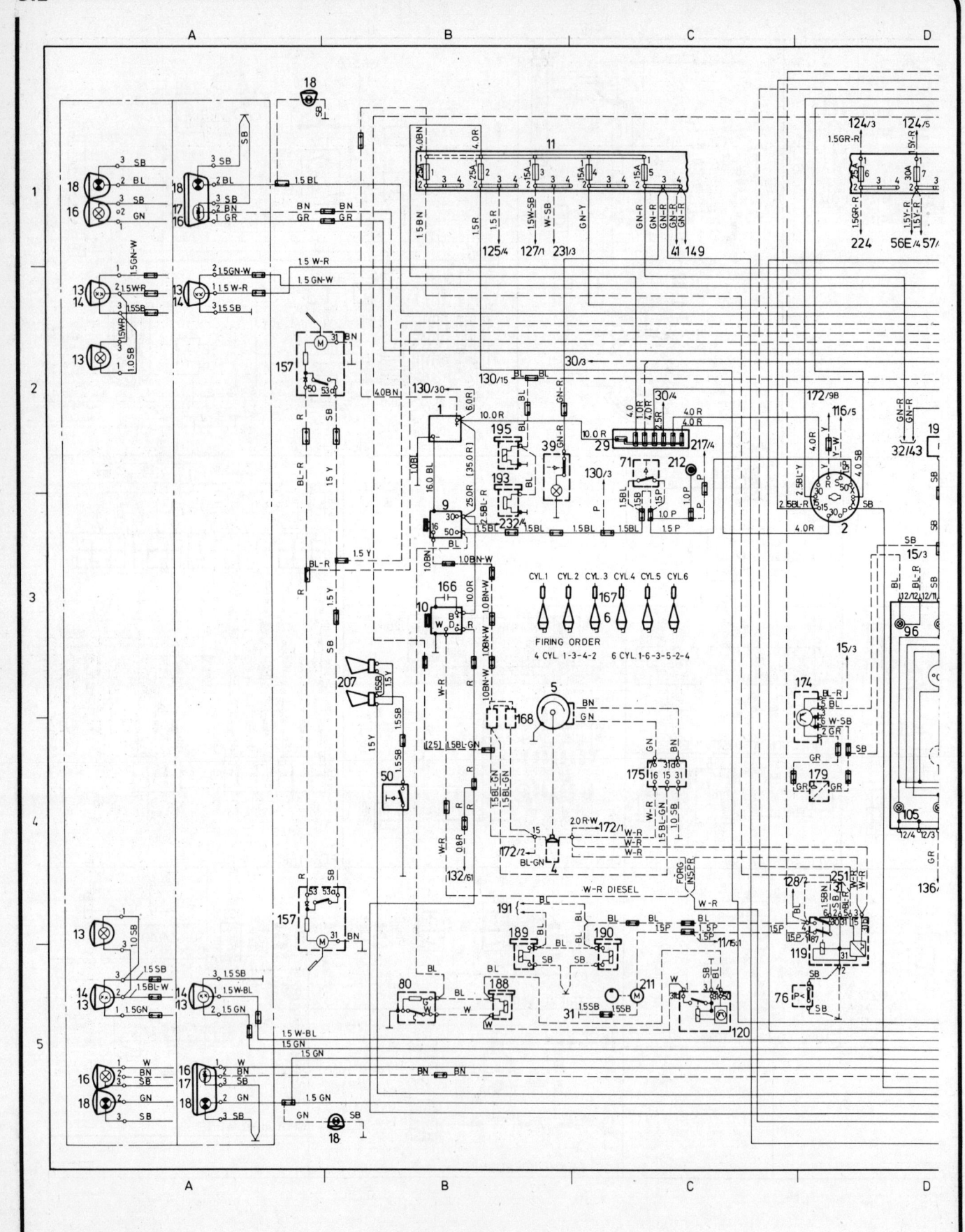

Fig. 12.27 Main wiring diagram for 1983 models.

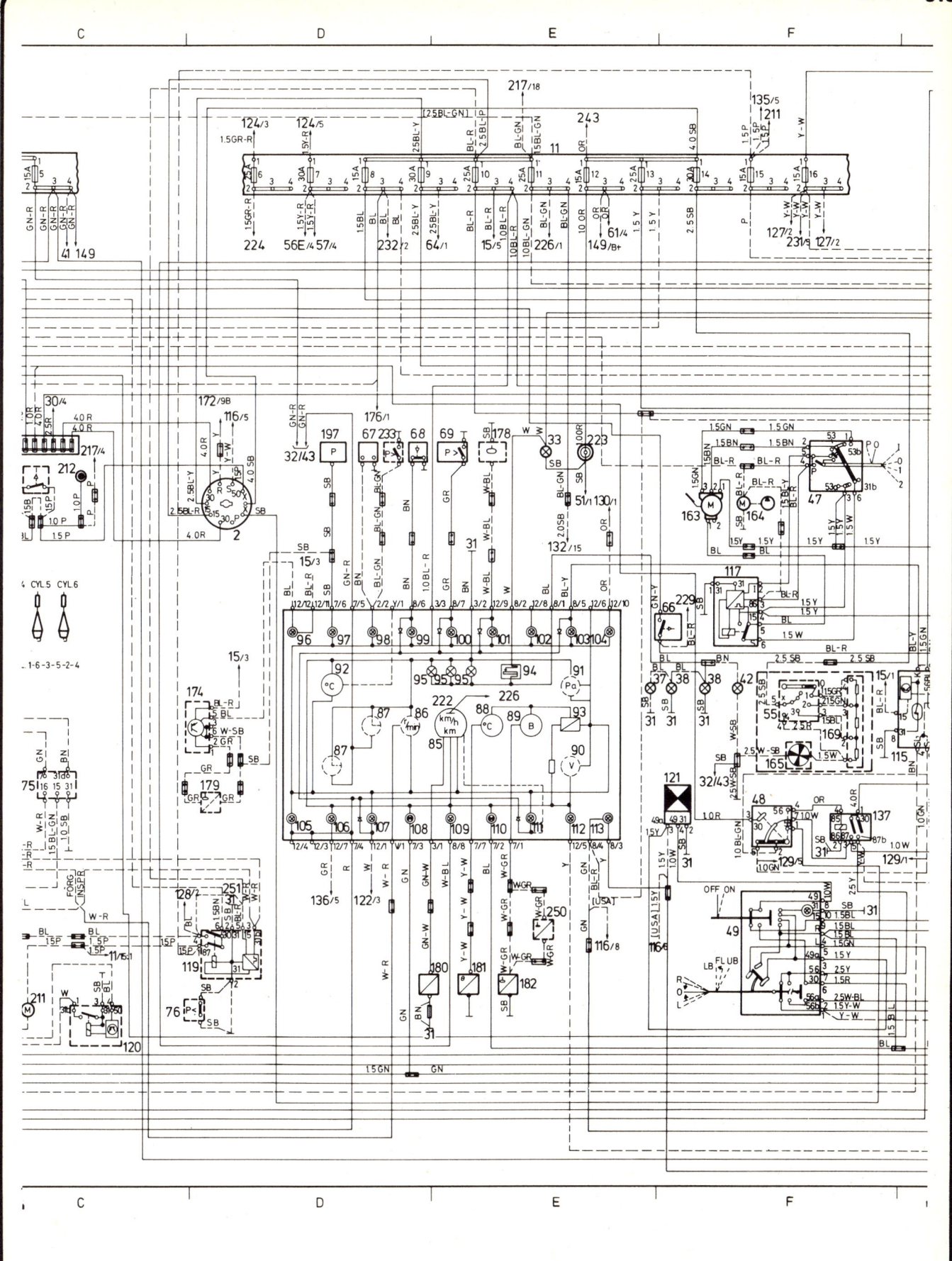

Fig. 12.27 (cont'd) Main wiring diagram for 1983 models.

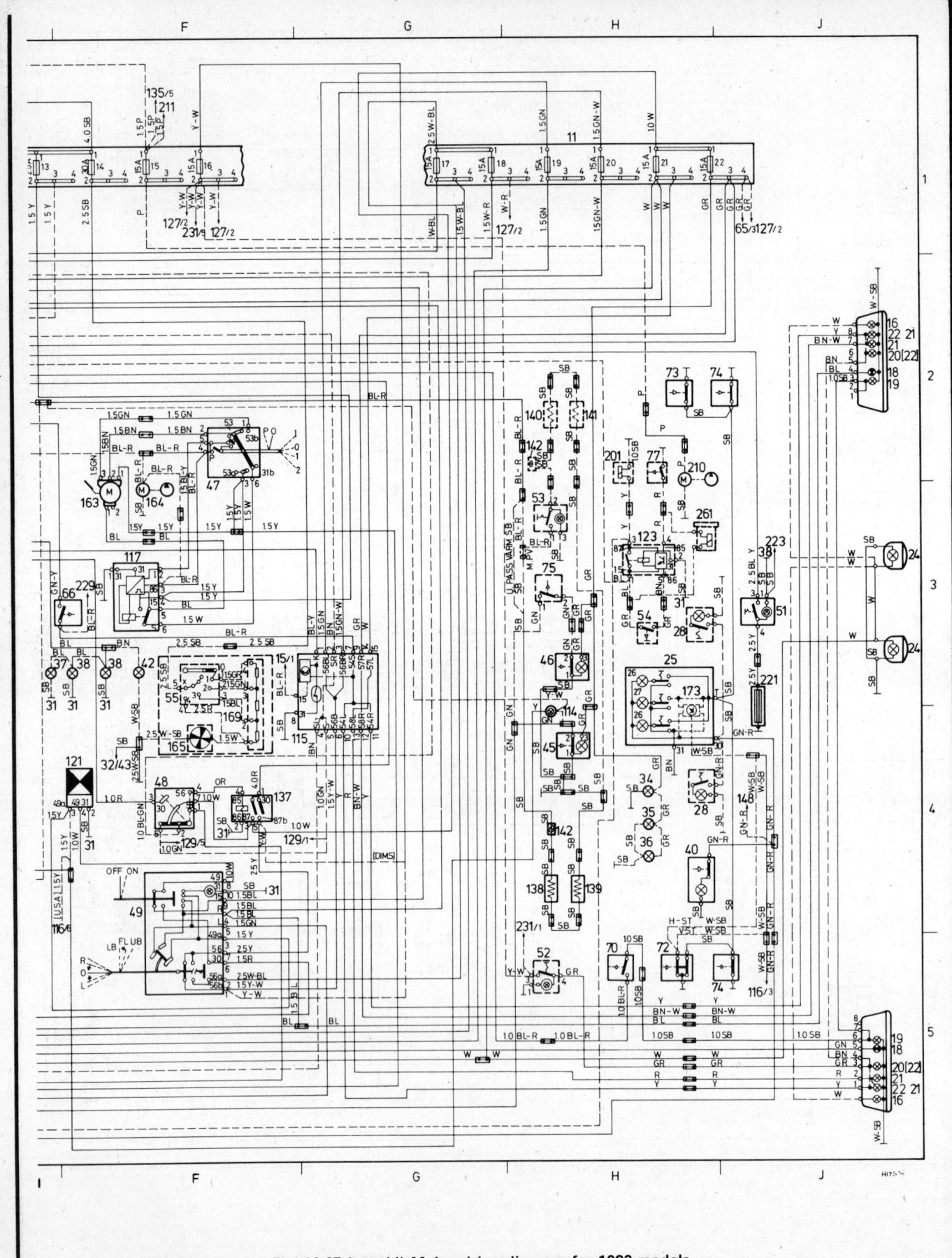

Fig. 12.27 (cont'd) Main wiring diagram for 1983 models.

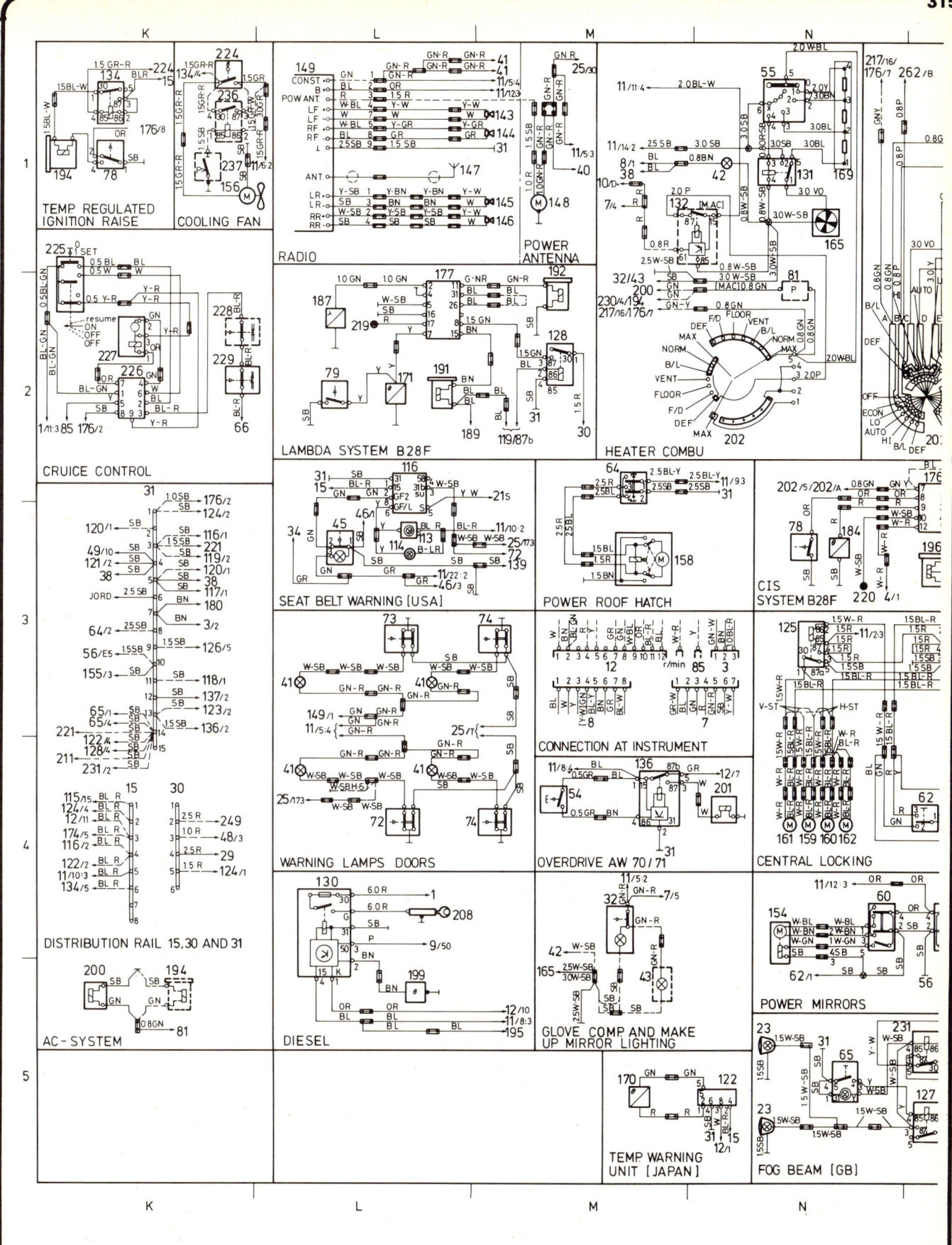

Fig. 12.28 Supplementary wiring diagram for 1983 models.

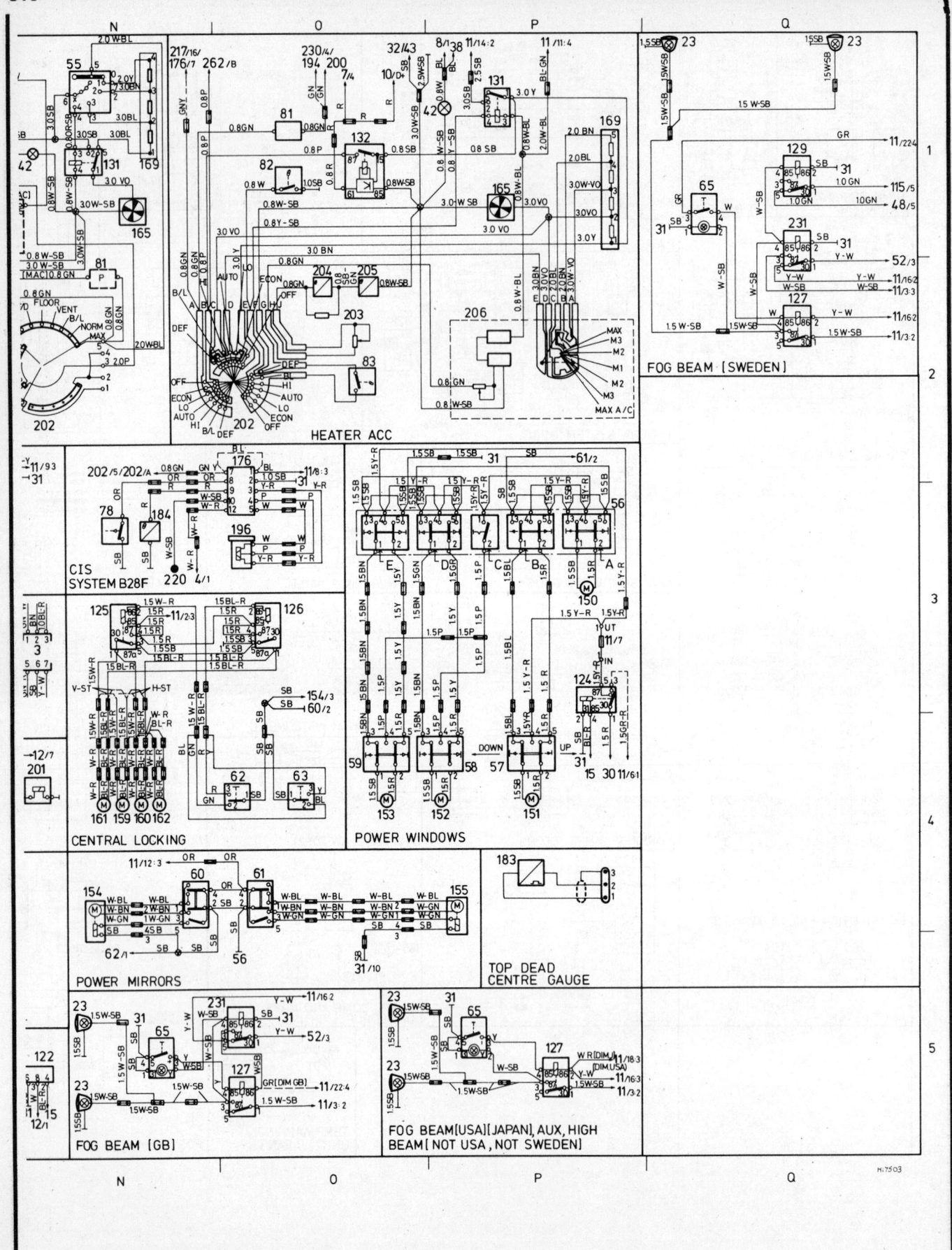

Fig. 12.28 (cont'd) Supplementary wiring diagram for 1983 models.

Key to Figs. 12.29 and 12.30. Not all items are fitted to all models; not all systems apply to UK

No	Item	Grid ref	No	Item	Grid ref
1	Battery	C2	49	Direction indicator/hazard warning/dip switches	G5
2	Ignition/starter switch	E3	50	Horn switch	B4
3	Instrument panel connector (3-pole)	W4	51	Heated rear window switch	L3
4	Ignition coil	C5, N3	52	Rear foglight switch	X6, Z6
5	Distributor	C4, M5, N3, S5	53	Passenger seat heater switch	J3
6	Spark plugs	C4	54	Overdrive switch	V4, W6
7	Instrument panel connector (7-pole)	W4	55	Heater blower switch	G4, W1
8	Instrument panel connector (8-pole)	W4	56	Window switch (driver's door)	Y3
9	Starter motor	C3	57	Window switch (passenger's door)	Z4
10	Alternator and voltage regulator	B3	58	Window switch (LH rear)	Z4
11	Fusebox	C1, E1, H1	59	Window switch (RH rear)	Y4
12	Instrument panel connector (12-pole)	W4	60	Door mirror switch (driver's door)	X4
13	Headlight main beam	A2, A5, A6	61	Door mirror switch (passenger's door)	X4
14	Headlight dipped beam	A2, A6	62	Central locking link rod switch	Y4
15	No 15 terminal (in central electrical unit)	T4	64	Sunroof switch	W3
16	Parking light	A1, A6, L1, M6	65	Front foglight/spotlight switch	Y6, Z6
17	Day running light	A1, A6	66	Stop-light switch	G3
18	Direction indicators	A1, B1, L1, M1, M6	67	Choke control switch (not UK)	E3
19	Reversing light	L1, M1, M6	68	Handbrake switch	F3
20	Rear foglight	L1, M1, M6	69	Brake failure warning switch	F3
21	Tail light	L1, M1, M6	70	Reversing light switch	K5
22	Stop-light	L1, M1, M6	71	Starter inhibitor switch (auto)	D3
23	Front foglight/spotlight	X6, Y6	72	Driver's door switch	K5, V5
24	Number plate light	M3, M4	73	Passenger's door switch	K2, V4
25	Interior light unit	K4	74	Rear door switch	K5, V4, V5
26	Map reading light	K4	75	Passenger's seat contact	K3
27	Courtesy light	K4	76	Turbo overpressure switch	D6, R1
28	Rear reading lights	K4	77	Overdrive switch (manual)	W5
29	Positive terminal	D3	78	CIS microswitch	T1, X3
30	No 30 terminal (in central electrical unit)	T4	79	Lambdasond microswitch (not UK)	U2
31	Earth connection (in central electrical unit)	T3	80	Thermal time switch	B6
32	Glovebox light	K3	81	Air conditioner pressure sensor	X2, V1
33	Front ashtray light	G3	82	ACC control panel sensor	Y1
34	Rear ashtray light	K5	83	ACC coolant thermal switch	Y2
35	Switch illumination	F3	84	Coolant temperature sensor	R2, T3
36	Switch illumination	K5	85	Speedometer	F4, W4
37	Gear selector light (auto)	K5	86	Tachometer	F4
38	Instrument illumination	F3	87	Clock	F4
39	Engine bay light	C2	88	Coolant temperature gauge	F4
40	Boot light	K5	89	Fuel gauge	E4
41	Door edge marker light	U4, V4	90	Voltmeter	E4
42	Heater control illumination	G3, W1, Y1	93	Instrument voltage regulator	G4
43	Vanity mirror light	K3	94	Panel light rheostat	F4
44	Fuel injectors	N2, T2, T3, T4	95	Panel illumination	F4
45	Driver's belt lock light	K4, U3	96	Oil level warning light	E4
46	Passenger's belt lock light	K4	97	Oil pressure warning light	E4
47	Windscreen wiper switch	H3	98	Choke/turbo boost pressure warning light	E4
48	Main light switch	G4			

Key to Figs. 12.29 and 12.30. Not all items are fitted to all models; not all systems apply to UK (continued)

No	Item	Grid ref	No	Item	Grid ref
99	Handbrake warning light	F4	150	Window motor – driver's door	Z4
100	Brake failure warning light	F4	151	Window motor – passenger's door	Z5
101	Washer fluid level warning light	E3	152	Window motor – RH rear	Y5
102	Gearshift indicator	F4	153	Window motor – LH rear	Y5
103	Bulb failure warning light	F4	154	Driver's door mirror	L3, X4
104	Preheater pilot light (Diesel)	G4	155	Passenger's door mirror	L3, Z5
105	Ignition (no charge) warning light	E5	156	Cooling fan motor	U1
106	Overdrive pilot light (auto)	E5	157	Headlight wiper motor	B2, B5
107	Exhaust temperature light (Japan)	E5	158	Sunroof motor	W3
108	Direction indicator pilot light (LH)	F5	159	Locking motor – passenger's door	Y5
109	Main beam pilot light	F5	160	Locking motor – LH rear	W4
110	Direction indicator pilot light (RH)	F5	161	Locking motor – RH rear	W5
111	Lambdasond light (not UK)	F5	162	Locking motor – boot or tailgate	W4
112	Overdrive pilot light (manual)	F5	163	Windscreen wiper motor	G3
113	Seat belt reminder (front)	G5, V3	164	Windscreen washer motor	G3
114	Seat belt reminder (rear)	V3, J4	165	Heater blower motor	G4, X1, Z1
115	Bulb failure sensor	H4	166	Capacitor (interference suppression)	C3
116	Seat belt reminder (not UK)	V3	167	Interference suppressors – spark plugs	C3
117	Windscreen wiper delay relay	G3	168	Ignition ballast resistor	C4
118	Tailgate wiper delay relay	P4	169	Heater blower resistor	X1, Z1
119	Fuel pump relay	E5	170	Catalyst element (not UK)	Z5
120	Fuel injection impulse relay	D6	171	Lambdasond thermostat (not UK)	V2
121	Flasher unit	G4	173	Interior light delay	K4
122	Exhaust temperature sensor (Japan)	Z5	174	Oil level control unit	D4
123	Overdrive relay (manual)	W5	175	Ignition system control unit (Bosch)	D4
124	Window winder/cooling fan relay	Z4	176	CIS control unit	X3
127	Auxiliary light relay	Y6, Z6	177	Lambdasond control unit (not UK)	R5, V2
128	Lambdasond relay (not UK)	V2	178	Washer fluid level sensor	F3
130	Glow plug relay (Diesel)	V5	179	Oil level sensor	E4
131	Heater blower relay	X1, Z2	180	Speedometer sensor	F5
132	Air conditioning delay relay	W1, Y1	181	Coolant temperature sensor	F5
133	Ignition control unit (T2 28)	N3, R6	182	Fuel gauge sender	F5
134	Ignition advance relay (not UK)	T1	184	CIS temperature sensor	X3
135	Motronic/LH Jetronic relay	R2, R5	185	Charge air temperature sensor	52
136	Overdrive relay (auto)	W4	186	Airflow meter	T1
137	Main beam relay	H5	187	Lambda sensor (not UK)	R4, U2
138	Driver's seat heater	J5	188	Start injector	C6
139	Driver's backrest heater	K5	189	Control pressure regulator	C5
140	Passenger's seat heater	J3	190	Auxiliary air valve	C5
141	Passenger's backrest heater	K3	191	Frequency valve	V2
142	Seat heater thermostats	J3, K5	192	Pressure differential switch	V2
143	Loudspeaker – LH front	V1	193	Hot start injector	C3
144	Loudspeaker – RH front	V1	194	Idle up solenoid	T5, T1
145	Loudspeaker – LH rear	V1	195	Fuel cut-off solenoid	C3
146	Loudspeaker – RH rear	V1	196	CIS idle valve	R4, X3
147	Aerial (Static)	V1	197	Oil pressure sensor	E3
148	Aerial (power-operated)	V1	198	Throttle switch (LH Jetronic)	54
149	Radio	U1	199	Temperature sensor (Diesel)	V5

Key to Figs. 12.29 and 12.30. Not all items are fitted to all models; not all systems apply to UK (continued)

No	Item	Grid ref	No	Item	Grid ref
200	Air conditioning compressor solenoid	T6	253	ABS modulator	Q3
201	Overdrive solenoid	V5, W4	254	ABS surge protector	Q2
202	Heater control	W3, X3	255	ABS converter	Q1
203	ACC temperature control	Y2	256	ABS sensor – LH front	P2, Q1
204	ACC ambient temperature sensor	P5, Y2	257	ABS sensor – RH front	P2, Q1
205	ACC cabin temperature sensor	Y2	258	ABS fusebox	Q2
206	ACC programmer	Z2	259	Charge pressure sensor	R4
207	Horn	B4	260	Ignition system control unit (EZ-K)	S6
208	Glow plugs (Diesel)	V5	261	EGR solenoid (not UK)	W5
210	Tank pump	K2	262	ACC vacuum valve	X1
211	Main fuel pump	D6	263	Overdrive oil pressure switch	S3
212	Starter cranking contact	D3	264	Charge pressure switch	S3
213	Throttle switch (Motronic)	S1	265	Torque limiter relay	P2, S3
214	Crankshaft position sensor	R1	266	Heated rear window timer	L3
215	Engine speed sensor	R1	267	EZ-K test point	S6
216	Motronic control unit	S1	268	Gear position sensor	R3
217	LH-Jetronic control unit	S3	270	ABS speedometer sensor	Q1
218	Knock sensor	56	271	Fuel shut-off valve	V6
219	Lambdasond test point (not UK)	U2	272	Throttle pedal switch	S6
220	CIS test point	S4, X3	273	EZ-K temperature sensor	S6
221	Heated rear window	L2	274	EGR relay (not UK)	Q6
223	Cigarette lighter	F3	275	EGR idling switch (not UK)	Q5
224	Fan thermoswitch	U1	277	EGR 3-way valve (not UK)	R5
225	Cruise control switch	T1	278	Aneroid switch (not UK)	P6
226	Cruise control control unit	T2	279	Altitude compensation solenoid valve (not UK)	P6
227	Vacuum pump	T2	280	Driver's seat heater relay	J3
228	Clutch switch	U2	284	Air mass meter	S4
229	Brake switch	G3, U2	286	ETC sensor – LH rear	N1
231	Rear foglight relay	X6	287	ETC sensor – RH rear	N1
232	Hot spot relay	T5	288	ETC pressure sensor	P1
233	Turbo pressure sensor (Diesel)	E3	289	ETC power stage	N1, T3
234	Hot spot thermostat	U5	290	ETC control unit	P1
235	Hot spot PTC resistor	U5	292	Idle advance solenoid valve	T6, W6
236	Pressure sensor relay (Diesel)	U1	293	Idle advance relay	P5, T6, U6, V6, W6
237	Pressure sensor (Diesel)	U1	295	Gearshift indicator relay	Q5
238	Rear washer pump	P4	298	ETC relay	N1
239	Rear wiper connector	P3	349	Instrument connector (6-pole)	P5
240	Rear wash/wipe switch	P3	375	Seat heater switch	J5
241	Rear wiper motor	P3	376	Ballast resistor (Jetronic)	N1, S2
242	Power seat emergency stop	Q4	378	Positive terminal (engine bay)	C2
243	Power seat 'on' switch	Q4	379	Trip computer electronic unit	P5
244	Power seat control unit	Q3	384	Brake fluid level sensor	Q1, F3
245	Power seat motor – fore/aft	R4	395	ETC switch	P1
246	Power seat motor – up/down (front)	R3	403	Battery temperature sensor	C3
247	Power seat motor – up/down (rear)	R3	404	Vacuum switch	R6
248	Power seat motor – backrest inclination	R4	405	Choke heater	C3
250	Auxiliary tank fuel gauge sender	F5	406	Renix ignition unit (not UK)	P5
251	Connector	E5	407	Renix impulse sender (not UK)	P5
252	ABS control unit	R1	408	Gearshift indicator switch	Q5

BL	Blue	GR	Grey	R	Red	W	White
BN	Brown	OR	Orange	SB	Black	Y	Yellow
GN	Green	P	Pink	VO	Violet		

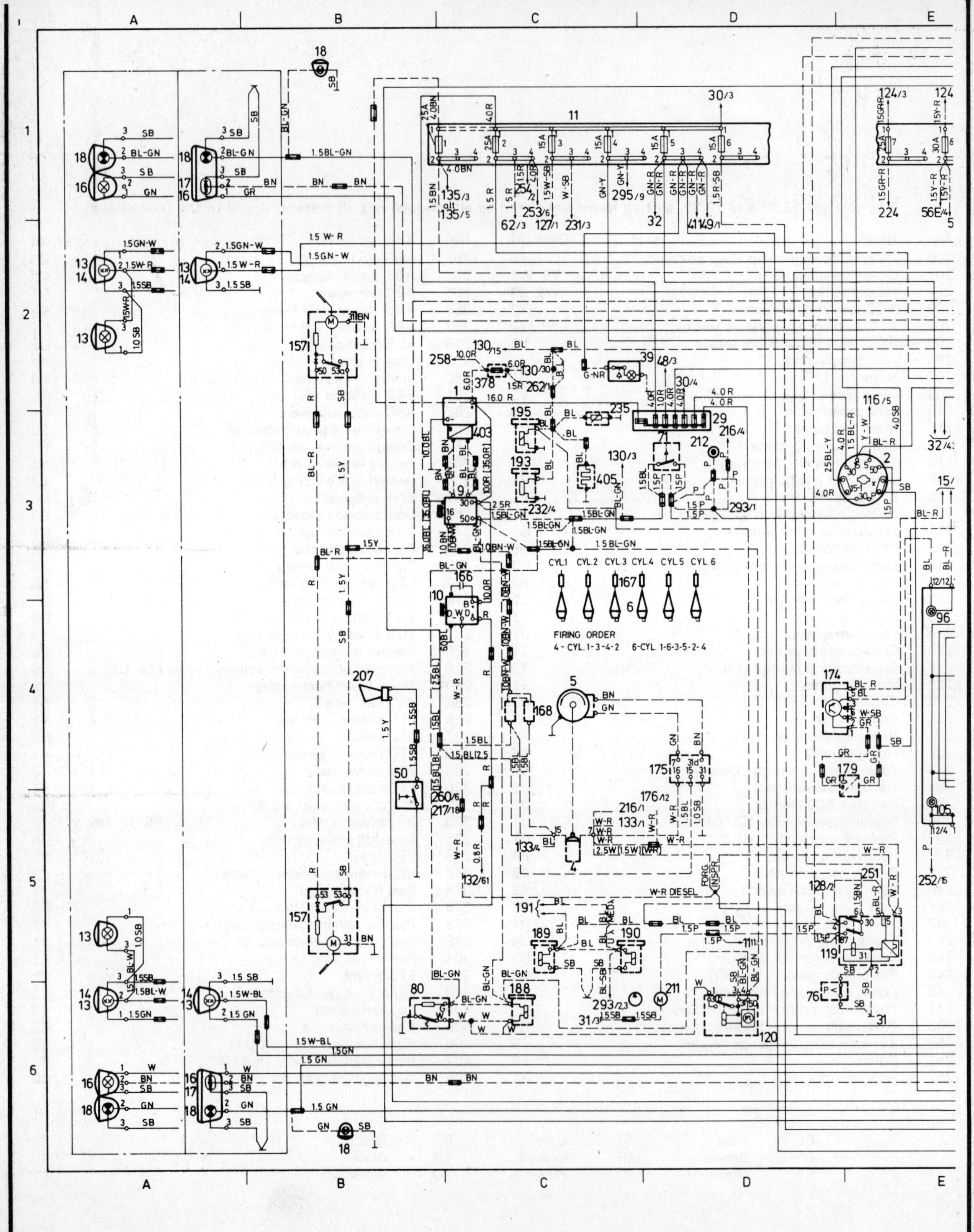

Fig. 12.29 Main wiring diagram for 1984 to 1988 models.

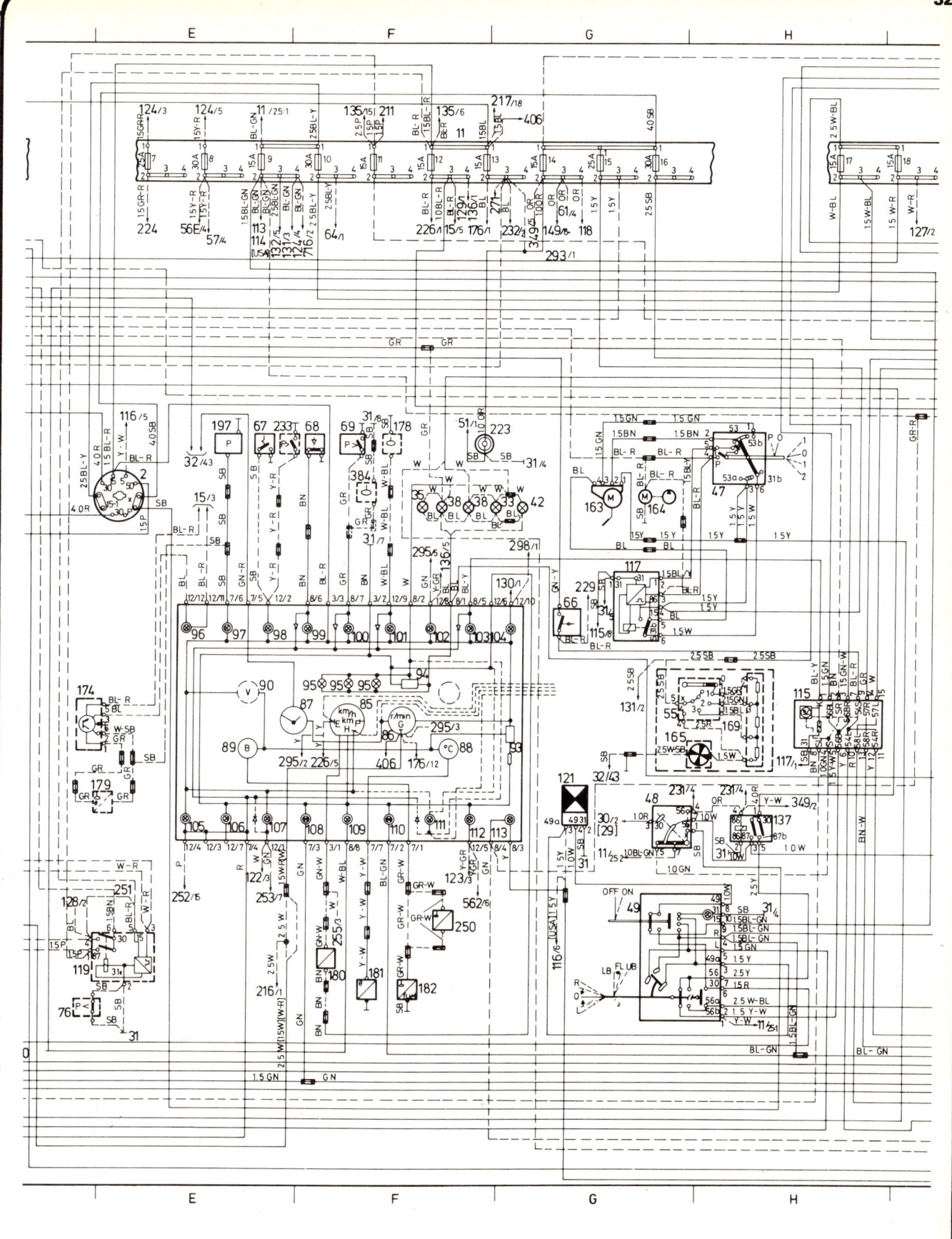

Fig. 12.29 (cont'd) Main wiring diagram for 1984 to 1988 models.

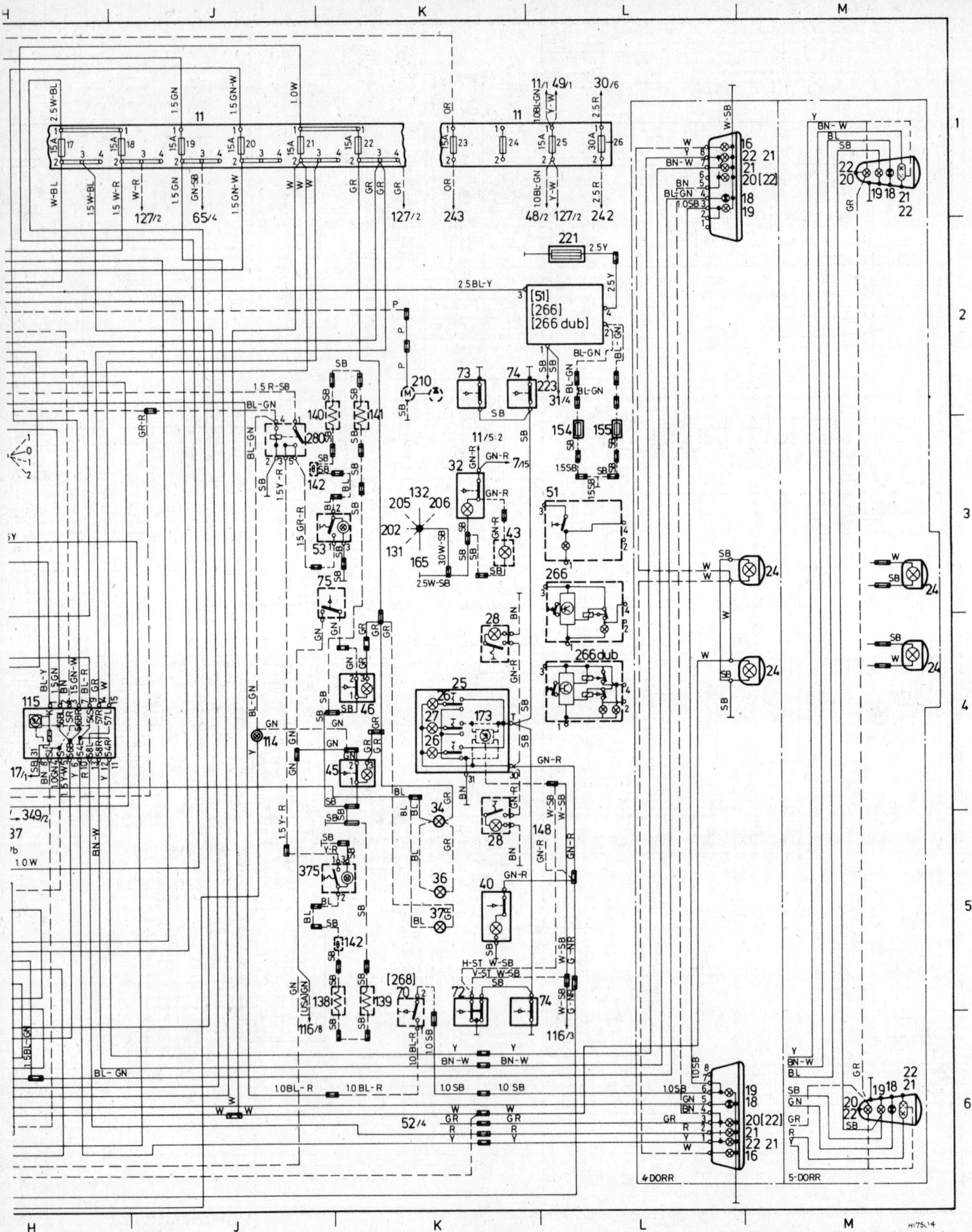

Fig. 12.29 (cont'd) Main wiring diagram for 1984 to 1988 models.

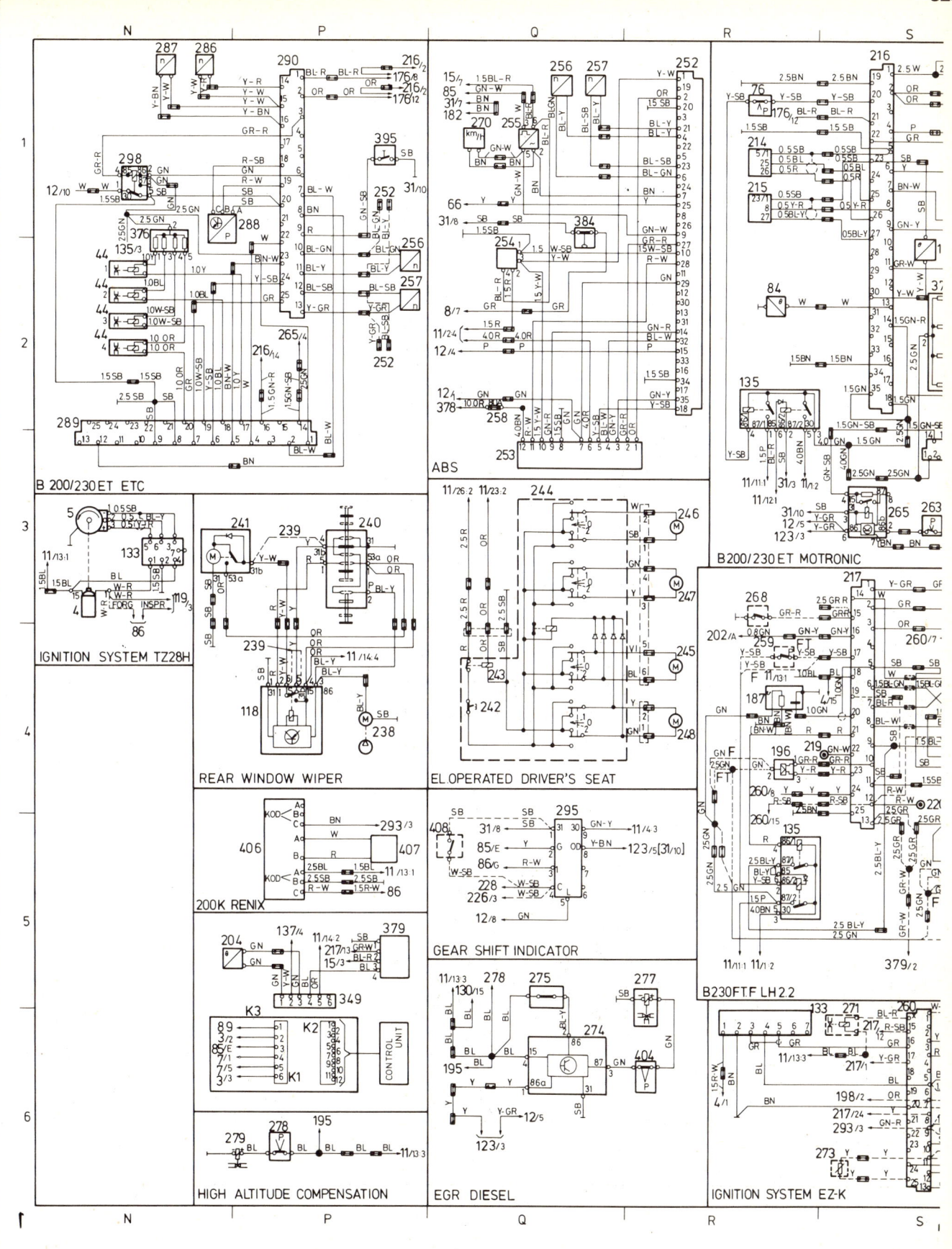

Fig. 12.30 Supplementary wiring diagram for 1984 to 1988 models.

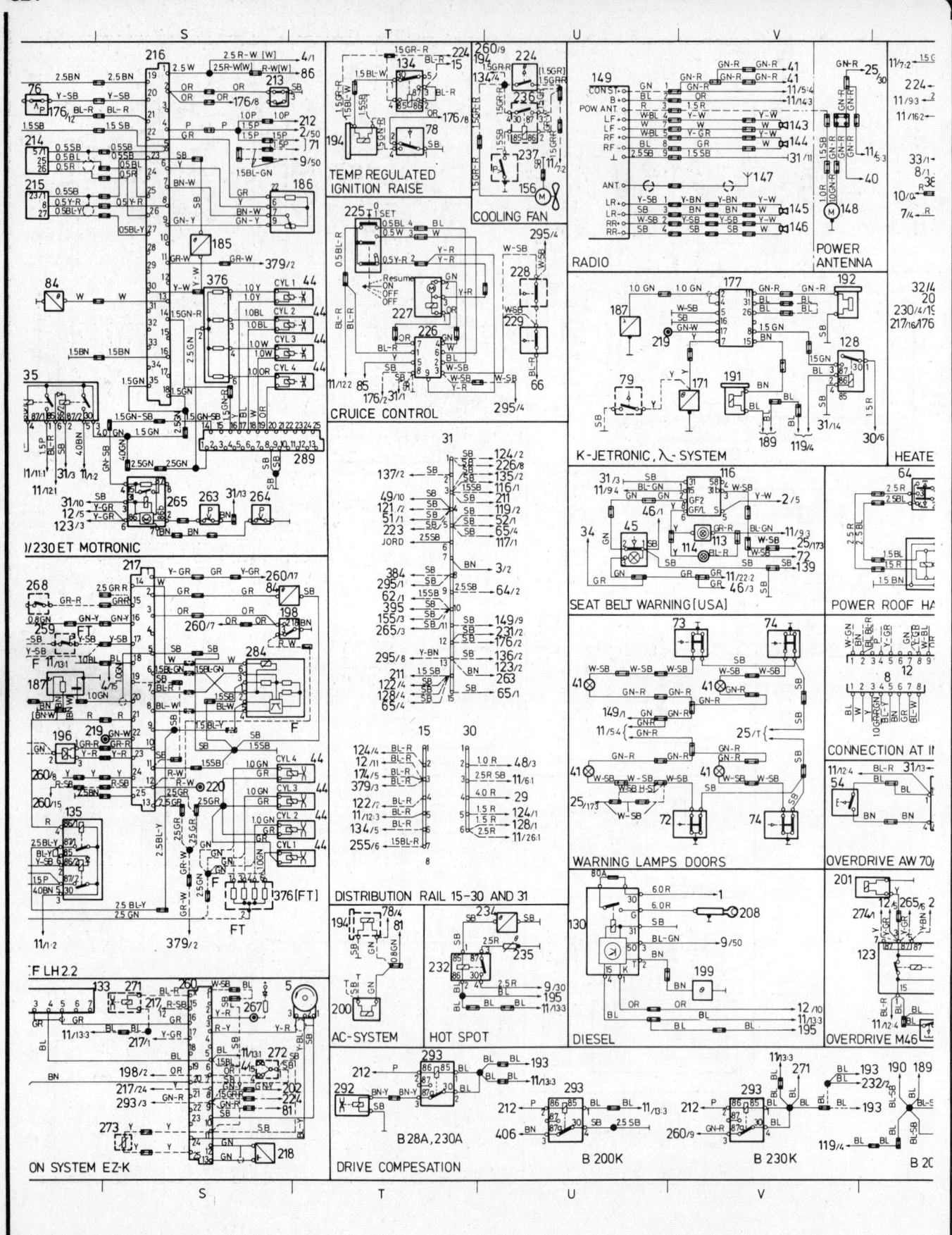

Fig. 12.30 (cont'd) Supplementary wiring diagram for 1984 to 1988 models.

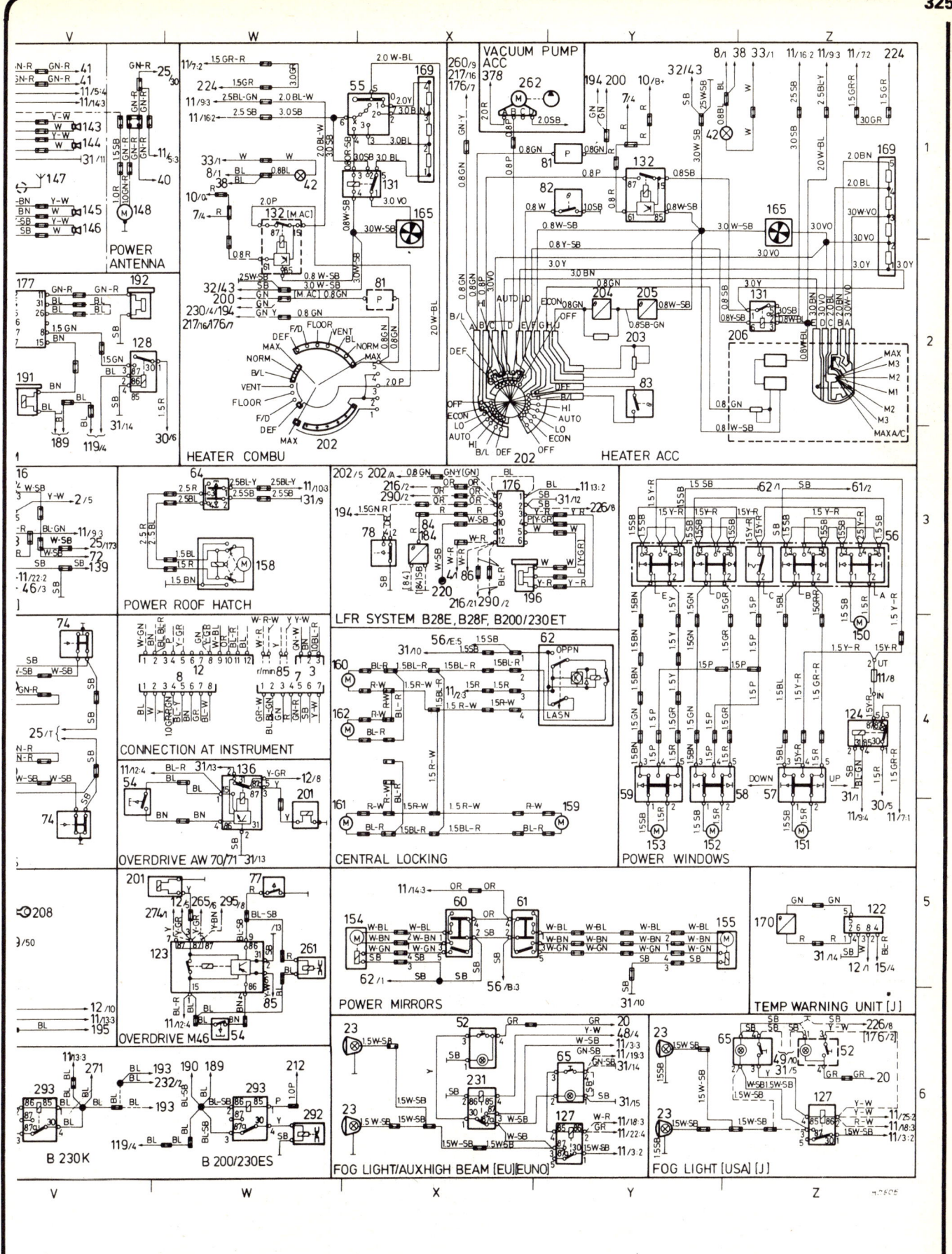

Fig. 12.30 (cont'd) Supplementary wiring diagram for 1984 to 1988 models.

Chapter 13 Supplement:
Revisions and information on later models

Contents

Introduction 1
Specifications 2
Routine maintenance 3
   Spark plug renewal (all models)
   Deleted operations – 1988 and 1989 models
   Revised operations – 1989 models
   Extra operations – all models with EGR/Pulsair
   Extra operation – models with independent rear suspension
In-line engine 4
   B200E engine – description
   Oil pressure switch (all models) – removal and refitting
   Camshaft oil feed all models)
   Camshaft cover gasket (all B200/B230) – fitting
   Crankcase ventilation system (B230K, 1988 on) – general
V6 engine 5
   B280 engine – general description
   Valve clearances (B280) – checking and adjustment
   Oil cooler (B280) – removal and refitting
   Oil pressure switch (all models) – removal and refitting
   Distributor drive oil seal (B280) – renewal
   Crankshaft and main bearings (B280) – removal and refitting
   Pistons and connecting rods (B280) – examination and renovation
   Cylinder liner seal identification – B280
   Engine mountings (B280) – removal and refitting
   Cylinder head bolts
Cooling, heating and air conditioning systems 6
   Cooling system – flushing (all models)
   Cooling system (B280) – draining
   Water pump (B280) – removal and refitting
   Automatic climate control vacuum hose connections – all models with ACC
   Electronic Climate Control (760, 1988 on) – description
   ECC sensors – description
   ECC control panel and control unit – removal and refitting
Fuel and exhaust systems 7
   Unleaded fuel and catalytic converter
   Air cleaner element (B280) – renewal
   Air cleaner unit (B280) – removal and refitting
   Idle mixture adjustment – models with Pulsair system
   Pierburg 2B7 carburettor (B230K, 1987 on) – description
   Control pressure regulator ventilation (all models with continuous injection system)
   Pulsair system (B230K, 1987 on) – description
   Pulsair system – maintenance
   EGR system (B230K, 1987 on) – description
   EGR system – maintenance
   Fuel gauge senders (later models) – testing
   Fuel tank pump/fuel gauge sender (1988 on) – removal and refitting
   Fuel tank (760 Saloon, 1988 on) – removal and refitting
   Inlet manifold (B280) – removal and refitting
   LH-Jetronic fuel injection system – description
   LH-Jetronic system – idle speed and mixture adjustment
   LH-Jetronic system – fuel injector check
   LH-Jetronic system – testing procedures
   LH-Jetronic system – component removal and refitting
   Water-cooled turbocharger (1987 on) – general

Ignition system 8
   Description (B280)
   Distributor (B280) – removal and refitting
   Control unit (B280) – removal and refitting
   Power stage (B280) – removal and refitting
   Knock sensors – general
   Flywheel sensor (B280) – removal and refitting
Manual gearbox and overdrive 9
   Mainshaft gear bearings (manual gearbox, 1988 on)
   5th gear and synchro unit (M57 II) – dismantling and reassembly
   Overdrive (J/P hybrid) – general
Propeller shaft 10
   Removal and refitting (models with independent rear suspension)
   Removal and refitting (models with saddle fuel tank)
Final drive and driveshafts (760 Saloon, 1988 on) 11
   Description
   Oil level – checking
   Pinion oil seal – renewal
   Side oil seals – renewal
   Final drive unit – removal and refitting
   Driveshafts – removal and refitting
Braking system 12
   Brake fluid level warning switch (later models)
   Brake pads (1989 on) – wear measurement
   Rear brake pads (models with independent rear suspension) – inspection and renewal
   Front brake caliper (all Bendix/DBA) – overhaul
   Rear brake caliper (models with independent rear suspension) – removal and refitting
   Rear brake caliper (models with independent rear suspension) – overhaul
   Front brake disc (pre-1988 models) – renewal
   Front brake disc (1988 on) – removal and refitting
   Handbrake (models with independent rear suspension) – adjustment
   Handbrake cables (models with independent rear suspension) – removal and refitting
   Handbrake shoes (models with independent rear suspension) – removal and refitting
   Anti-lock braking system (ABS) – modifications (1988 on)
   ABS hydraulic modulator (1988 on) – removal and refitting
   ABS control unit (1988 on) – removal and refitting
   ABS rear sensor (models with independent rear suspension) – removal and refitting
   Hydraulic system – bleeding (models with ABS, 1988 on)
Steering and suspension 13
   Steering column (760, 1988 on) – removal and refitting
   Front wheel bearing repair kits – all models with adjustable bearings
   Front wheel bearings (1988 on) – description
   Front wheel bearings (ball type) – renewal
   Aluminium front control arm (740, 1987 models)
   Front suspension strut upper mounting (1985 on)
   Independent rear suspension (760 Saloon, 1988 on) – description
   Independent rear suspension – rear wheel alignment

Chapter 13 Supplement: Revisions and information on later models

Rubber bushes (independent rear suspension) – renewal
Lower link (independent rear suspension) – removal and refitting
Track rod (independent rear suspension) – removal and refitting
Support arm (independent rear suspension) – removal and refitting
Rear axle member lower section (independent rear suspension) – removal and refitting
Upper link (independent rear suspension) – removal and refitting
Rear hub carrier (independent rear suspension) – removal and refitting
Rear wheel bearing (independent rear suspension) – renewal
Rear spring and shock absorber (models with independent rear suspension) – removal and refitting
Bodywork and fittings 14
Doors (1988 on) – adjustment
Bonnet (760, 1988 on) – removal and refitting
Front door interior trim (1989 on) – removal and refitting
Door lock motor (1989 on) – removal and refitting
Driver's door central locking switch (1989 on) – removal and refitting
Window lift motor (1988 on) – removal and refitting
Seat adjuster motor control panel – removal and refitting
Seat adjuster motors – removal and refitting
Seat heater control unit (1988 on) – general
Glovebox (760, 1988 on) – removal and refitting
Front bumper (760, 1988 on) – removal and refitting
Front wing liners
Electrical system 15
Battery disconnection – all models with radio
Headlight unit 760, 1988 on) – removal and refitting
Headlight unit (760, 1988 on) – dismantling and reassembly
High level brake light (1986 on) – bulb renewal
Vanity mirror light bulbs (760) – renewal
Steering column switches (760, 1988 on) – removal and refitting
Instrument cluster (760, 1988 on) – removal and refitting
Instrument fuse 1985 on) – renewal
Fuses (760, 1988 on)
Relays (760, 1988 on) – general
Windscreen wiper motor and linkage (760, 1988 on) – removal and refitting
Radio/cassette player (quickly detachable type) – removal and refitting
Wiring diagrams (later models)

1 Introduction

This Supplement contains information which has become available after the publication of the first edition of the manual. Most of the information relates to 1987 and later models, but some is applicable to all models.

The Sections in the Supplement follow the same order as the Chapters in the main part of the book. The Specifications are grouped together at the beginning of the Supplement for convenience, but they too follow Chapter order.

To use the Supplement to its best advantage, it should be consulted before referring to the appropriate main Chapter. In this way note can be taken of any differences which are liable to affect the proposed operations.

2 Specifications

These Specifications are revisions of, or additional to, those given in the previous Chapters. Where no new value is given, the old value may be assumed to apply

In-line engine type B200E

General
Application	740 GL and SE, from August 1987
Bore	88.9 mm (3.500 in) nominal
Stroke	80 mm (3.150 in)
Cubic capacity	1986 cc (121.2 cu in)
Compression ratio	10.0:1
Maximum power	89 kW (119 bhp) @ 5700 rpm
Maximum torque	158 Nm (117 lbf ft) @ 4800 rpm

Cylinder bores
Standard sizes:
C	88.90 to 88.91 mm (3.5000 to 3.5004 in)
D	88.91 to 88.92 mm (3.5004 to 3.5008 in)
E	88.92 to 88.93 mm (3.5008 to 3.5012 in)
G	88.94 to 88.95 mm (3.5016 to 3.5020 in)
Oversize 1	89.29 mm (3.5154 in)
Oversize 2	89.67 mm (3.5303 in)

Pistons
Piston diameter	0.02 mm (0.0008 in) less than bore
Running clearance	0.01 to 0.03 mm (0.0004 to 0.0012 in)

Piston rings
Clearance in groove:
Top compression	0.060 to 0.092 mm (0.0024 to 0.0036 in)
Second compression	0.030 to 0.062 mm (0.0012 to 0.0024 in)
Oil control	0.020 to 0.055 mm (0.0008 to 0.0022 in)

End gap (for 88.90 mm/3.5000 in bore):
Top compression	0.30 to 0.50 mm (0.012 to 0.020 in)
Second compression	0.30 to 0.55 mm (0.012 to 0.022 in)
Oil control	0.25 to 0.50 mm (0.010 to 0.020 in)

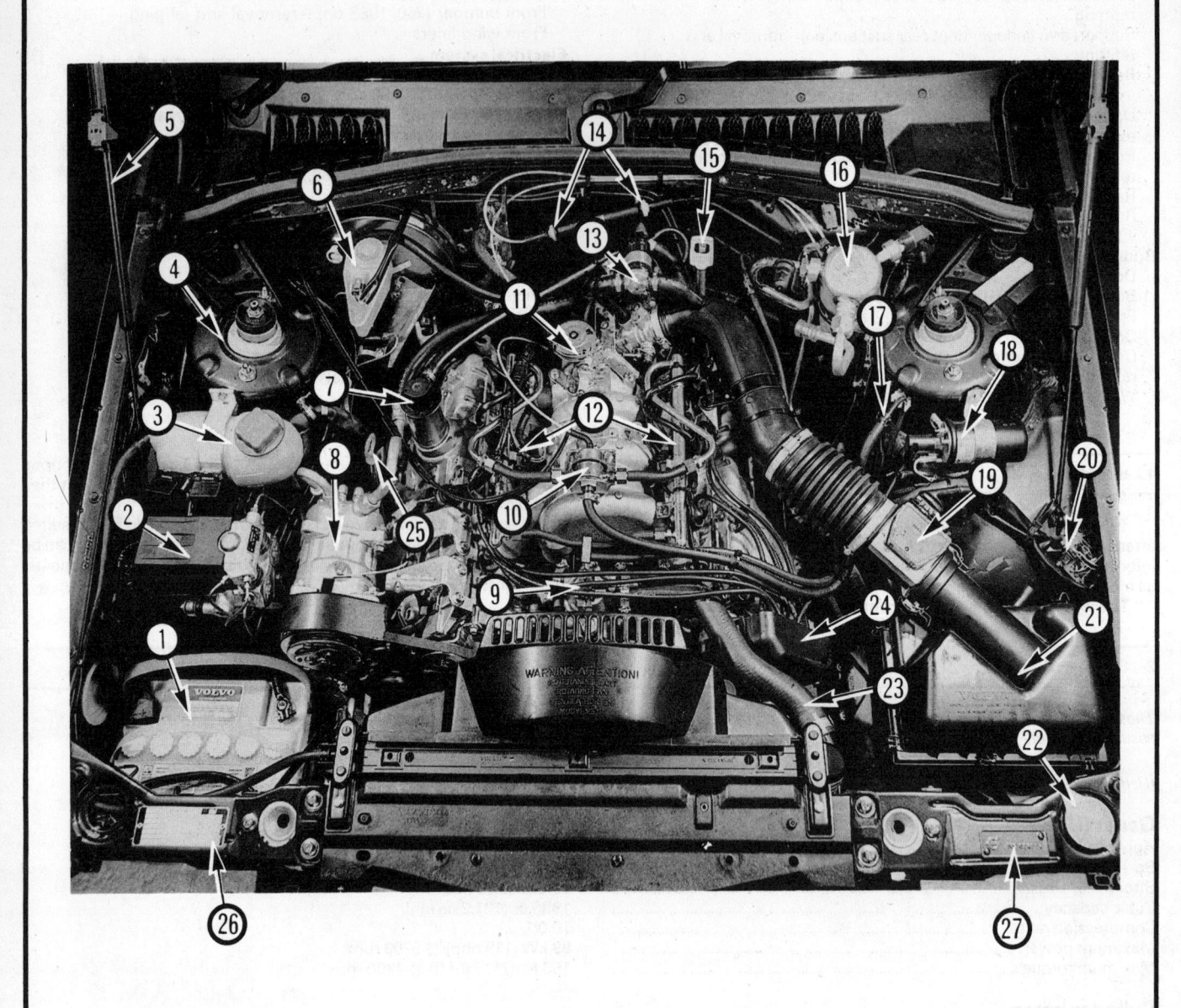

Under-bonnet view of a 1989 760 GLE

1 Battery
2 ABS hydraulic modulator
3 Expansion tank
4 Suspension turret
5 Bonnet strut
6 Brake fluid reservoir
7 Engine oil filler cap
8 Air conditioning compressor
9 Thermostat housing
10 Fuel pressure regulator
11 Throttle cable drum
12 Fuel injector rails
13 Air control valve
14 Vacuum non-return valves
15 Transmission fluid dipstick
16 Air conditioning receiver/drier
17 Fuel return line
18 Ignition coil
19 Air mass meter
20 Electrical connectors
21 Air cleaner
22 Washer reservoir cap
23 Radiator top hose
24 Distributor cover
25 Engine oil dipstick
26 Identification plate
27 Body number

Chapter 13 Supplement: Revisions and information on later models

Camshaft
Identification letter V

Other specifications as given for B230 engine (Chapter 1)

In-line engine type B230K (1987 on)

General
Compression ratio 10.5:1
Maximum power .... 86 kW (112 bhp) @ 5220 rpm
Maximum torque .... 194 Nm (143 lbf ft) @ 2520 rpm

All in-line engines

Torque wrench setting
	Nm	lbf ft
Oil pressure switch	30 to 40	22 to 30

V6 engine type B280E

General
Application .... 760 GLE, 1987 on
Compression ratio 10.0:1
Maximum power .... 125 kW (168 bhp) @ 5400 rpm
Maximum torque .... 240 Nm (177 lbf ft) @ 4500 rpm

Cylinder liners
Protrusion above block (depends on gasket, consult your dealer) ..... 0.14 to 0.21 mm (0.0055 to 0.0083 in)
Liner seal thickness – early type:
   1 tab .... 0.10 ± 0.01 mm (0.0039 ± 0.0004 in)
   2 tabs .... 0.12 ± 0.01 mm (0.0047 ± 0.0004 in)
   3 tabs .... 0.15 ± 0.02 mm (0.0059 ± 0.0008 in)
Liner seal thickness – later type:
   Orange .... 0.116 ± 0.018 mm (0.0046 ± 0.0007 in)
   Clear .... 0.136 ± 0.018 mm (0.0054 ± 0.0007 in)
   Blue .... 0.166 ± 0.028 mm (0.0065 ± 0.0011 in)

Pistons
Diameter (matched to liners):
   Grade A .... 90.920 to 90.930 mm (3.5795 to 3.5799 in)
   Grade B .... 90.930 to 90.940 mm (3.5799 to 3.5803 in)
   Grade C .... 90.940 to 90.950 mm (3.5803 to 3.5807 in)
   Clearance in bore .... 0.070 to 0.090 mm (0.0028 to 0.0035 in)

Gudgeon pins
Number of sizes .... One only
Clearance in piston .... 0.007 to 0.017 mm (0.0003 to 0.0007 in)
Clearance in connecting rod .... Nominal (firm push-fit)
Securing method .... Circlips

Inlet valves
Head diameter .... 45.3 mm (1.784 in)
Stem diameter:
   26.5 mm (1.04 in) from head .... 7.958 to 7.980 mm (0.3133 to 0.3142 in)
   Just below collet groove .... 7.973 to 7.995 mm (0.3139 to 0.3148 in)
Valve head angle .... 44° 30′

Exhaust valves
Head diameter .... 38.5 mm (1.5158 in)

Valve seat inserts
Fit in cylinder head .... 0.136 to 0.200 mm (0.0054 to 0.0079 in) interference
Valve seat angle (inlet and exhaust) .... 45°

Camshafts
Identification letter .... R
Colour code .... White
Identification number:
   Left 957
   Right 959
Maximum lift (at lobe):
   Inlet .... 6.08 mm (0.2394 in)
   Exhaust .... 5.85 mm (0.2303 in)

Chapter 13 Supplement: Revisions and information on later models

Crankshaft
Main bearing journal diameter:
   Standard .... 70.043 to 70.062 mm (2.7576 to 2.7583 in)
   Undersize .... Not permitted
Connecting rod bearing journal diameter:
   Standard .... 59.971 to 59.990 mm (2.3611 to 2.3618 in)
   Undersize .... Not permitted
Connecting rod bearing shell thickness .... 1.838 to 1.848 mm (0.0724 to 0.0728 in)

Lubrication system
Oil capacity (drain and refill):
   Engine only .... 5.5 litres (9.7 pints)
   Engine and oil filter .... 6.0 litres (10.6 pints)
Oil pump gear side clearance .... 0.020 to 0.095 mm (0.0008 to 0.0037 in)

Torque wrench settings
	Nm	lbf ft
Main bearing cap side bolts	20 to 25	15 to 18

Other specifications as given for B28 engine (Chapter 1)

All V6 engines

Torque wrench settings
	Nm	lbf ft
Oil pressure switch	30 to 40	22 to 30
Cylinder head bolts (with fixed washer):		
Stage 1	60	44
Slacken, then Stage 2A	40	30
Stage 2B	Angle-tighten a further 160° to 180°	

Cooling system

General
Coolant capacity:
   B200E .... 8.5 litres (15.0 pints) approx.
   B280E .... 10.0 litres (17.6 pints) approx.

Thermostat (B280E)
Opening commences .... 86 to 88°C (187 to 190°F)
Fully open at .... 97°C (207°F)

Temperature gauge sender (1987 on)
Resistance at temperature of:
   60°C (140°F) .... 560 Ω
   90°C (194°F) .... 206 Ω
   100°C (212°F) .... 153 Ω

Fuel and exhaust systems

General
System type:
   B200E .... Continuous fuel injection, normally aspirated
   B230K (1987 on) .... Pierburg 2B7 carburettor
   B280E .... LH-Jetronic 2.2
Unleaded fuel capability .... See Section 7

Continuous injection system (B200E)
Idle adjustments
Idle speed .... 900 rpm
CO level at idle:
   Setting value 1.0%
   Checking value .... 0.5 to 2.0%

Other specifications as B230E (Chapter 3)

LH-Jetronic system (B280E)
Idle adjustments
Idle speed (in 'P'):
   Regulated .... 750 rpm
   Basic .... 700 rpm
CO level at idle:
   Setting value 1.0%
   Checking value .... 0.5 to 2.0%

Air mass meter
Electrical resistance:
   Filament .... 2.5 to 4.0 Ω
   CO potentiometer .... 0 to 1000 Ω

Pressure regulator
Line pressure .... 2.5 bar (36 lbf/in^2) above inlet manifold pressure

Chapter 13 Supplement: Revisions and information on later models

Fuel injectors
Delivery at specified line pressure:
    Early type (No. 725) .... 170 ml (6 fl oz) per minute
    Later type (No. 734) .... 137 g (4.8 oz) per minute

Coolant temperature sensor
Resistance at:
    −10°C (+14°F) .... 8100 to 10 770 Ω
    +20°C (68°F) .... 2280 to 2720 Ω
    +80°C (176°F) .... 292 to 362 Ω

Air control valve
Resistance (between terminals 3 and 4, and 4 and 5) .... 20 Ω approx

Fuel pumps
Main pump capacity at 3.0 bar (42 lbf/in^2), 20°C (68°F) and input voltage:
    12V .... 140 litres (30.8 gallons) per hour
    11V .... 120 litres (26.4 gallons) per hour
    10V .... 90 litres (19.8 gallons) per hour
Current consumption:
    Main pump 5A
    Tank pump .... 3 to 4A

Pierburg 2B7 carburettor (B230K, 1987 on)

Idle adjustments
Idle speed:
    Manual .... 800 rpm
    Automatic (in 'P') .... 900 rpm
CO level at idle (Pulsair system blocked):
    Setting value 1.0%
    Checking value .... 0.5 to 1.5%

Calibration

Jet sizes:	Primary	Secondary
Idle (air/fuel)	47.5/115	–
Auxiliary (air/fuel)	45/130	–
Main	115	142.5

Overhaul data
Fast idle gap:
    Manual .... 5.0 mm (0.197 in)
    Automatic .... 5.6 mm (0.221 in)

Other data as for 2B5 carburettor (Chapter 3)

Fuel gauge senders (later models)
Sender resistance – tubular type from 1986, lever type from 1987:
    Main tank, empty .... 0 Ω
    Main tank (60 litres/13 gallons), full .... 280 Ω
    Main tank (80 litres/18 gallons), full .... 363 to 370 Ω
    Auxiliary tank, empty .... 0 Ω
    Auxiliary tank, full .... 60 Ω

Ignition system

General
System type – B200E and B280E .... EZ-K

Spark plugs
Make and type:
    B200E .... Champion N7YCC or N7YC
    B230E (1987-on) .... Champion RN7YCC or RN7YC
    B230K (1987) .... Champion N7YCC or N7YC
    B230K (1988-on) .... Champion RN7YCC or RN7YC
    B230ET .... Champion RN7YCC or RN7YC
    B280E .... Champion RS9YCC or RS9YC

Electrode gap:
    N7YCC, RN7YCC and RS9YCC .... 0.8 mm (0.032 in)
    N7YC, RN7YC and RS9YC .... 0.7 mm (0.028 in)

Ignition timing
B200E .... 12° BTDC @ 900 rpm
B280E .... 10° BTDC @ 750 rpm (not adjustable)

Torque wrench settings

	Nm	lbf ft
Knock sensor screw (see text):		
Type I	11	8
Type II and B280E	20 ± 5	15 ± 4
Distributor shaft centre screw (B280E)	70 to 90	52 to 66

Manual gearbox and overdrive

General
Gearbox designation – 5-speed, 1986 on	M47 II
Lubricant capacity – M47 II	1.6 litres (2.8 pints)
Overdrive type – 1987 on	J/P hybrid

Torque wrench settings
	Nm	lbf ft
Drain plug (all models)	27 to 40	20 to 30
5th gear synchro nut – M47 II	120	89

Automatic transmission

General
Designation – with B200E engine	AW70
Stall speed:	
AW70/B200E	2100 rpm
AW71/B280E	2100 rpm

Rear axle (760 Saloon, 1988 on)

General
Type	Discrete final drive unit feeding twin driveshafts
Ratio	3.54, 3.73 or 3.91:1

Lubrication
Lubricant capacity	1.4 litres (2.5 pints)

Pinion bearing preload
Turning torque at pinion (wheels free):
New bearing	1.2 to 2.8 Nm (0.9 to 2.1 lbf ft)
Used bearing	1.0 to 2.5 Nm (0.7 to 1.8 lbf ft)

Torque wrench settings
	Nm	lbf ft
Pinion flange nut (see text)	120 to 280	89 to 207
ABS/speedo sensor bolt	8 to 12	6 to 9
Rear casing bolts	20 to 28	15 to 21
Side bracket bolts	40 to 56	30 to 41
Lockplate bolts	40 to 56	30 to 41
Oil drain and filler plugs	27 to 40	20 to 30
Driveshaft Allen screws (renew every time):		
Stage 1	30	22
Stage 2	Tighten 90° further	Tighten 90° further
Driveshaft nut (renew every time):		
Stage 1	190	140
Stage 2	Tighten 60° further	Tighten 60° further

Braking system

Front brake discs
Run-out (1988 on)	0.06 mm (0.0024 in) maximum
ABS (1991):	
Diameter	280 mm (11.02 in)
Thickness:	
New	26 mm (1.02 in)
Wear limit	23 mm (0.91 in)

Rear brake pads (models with independent rear suspension)
Lining minimum thickness	3.0 mm (0.12 in)

Torque wrench settings
	Nm	lbf ft
Rear caliper (models with independent rear suspension):		
Guide pin bolts	34	25
Bracket bolts*	58	43

*Use new bolts every time

Steering and suspension

General
Rear suspension type (760 Saloon, 1988 on)	'Multi-link' independent

Wheel alignment (760 Saloon, 1988 on)
Front:
Toe (between inner rims)	2.5 ± 0.5 mm (0.10 ± 0.02 in) toe-in

Rear:
Camber	Variable – see text
Toe	0.5 ± 0.8 mm (0.02 ± 0.03 in) toe-in

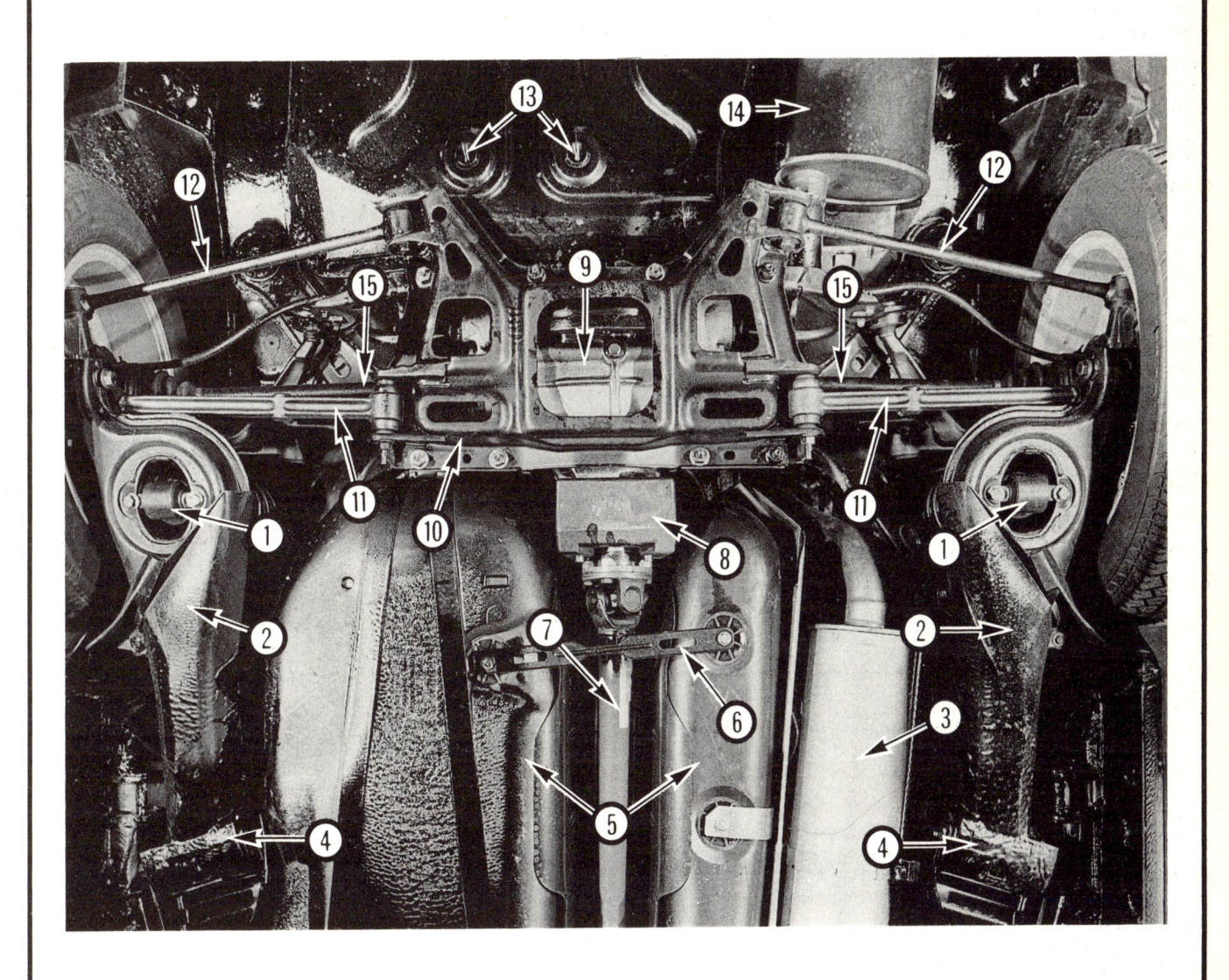

Rear underside view of a 760 GLE with independent rear suspension

1. Shock absorber lower mountings/spring seats
2. Support arms
3. Intermediate silencer
4. Support arm front mountings
5. Fuel tank
6. Bracing strap
7. Propeller shaft
8. Vibration damper
9. Final drive unit
10. Rear axle member lower section
11. Lower links
12. Track rods
13. Boot drain holes
14. Rear silencer
15. Driveshafts

Chapter 13 Supplement: Revisions and information on later models

Tyre pressures (760, 1988 on)

	Front	Rear
Recommended pressures in bar (lbf/in²):		
Up to 3 occupants	2.0 (29)	1.9 (28)
Fully laden	2.1 (31)	2.6 (38)

Torque wrench settings

	Nm	lbf ft
All 760, 1988 on		
Steering column bearing bolts	24	18
Steering intermediate shaft joint clamp bolts	24	18
All later models		
Front suspension piston rod nut – 1985 on	70	52
Front wheel hub nut – with ball-bearings (see text):		
Stage 1	100	74
Stage 2	Tighten 45° further	Tighten 45° further
Independent rear suspension		
Support arm front mounting to body:		
Bolts	48	35
Nut, stage 1	70	52
Nut, stage 2	Tighten 60° further	Tighten 60° further
Upper link to hub carrier	115	85
Upper link to axle member:		
Front, stage 1	70	50
Front, stage 2	Tighten 60° further	Tighten 60° further
Rear	85	63
Track rod to hub carrier	85	63
Track rod to axle member	70	52
Final drive unit to axle member	160	118
Shock absorber to support arm	56	41
Shock absorber to body	85	63
Support arm to hub carrier:		
Stage 1	60	44
Stage 2	Tighten 90° further	Tighten 90° further
Support arm to mounting:		
Stage 1	125	92
Stage 2	Tighten 120° further	Tighten 120° further
Lower link to hub carrier or axle member:		
Stage 1	50	37
Stage 2	Tighten 90° further	Tighten 90° further
Rear axle member to body:		
Stage 1	70	52
Stage 2	Tighten 60° further	Tighten 60° further
Rear axle member upper to lower sections:		
Stage 1	70	52
Stage 2	Tighten 30° further	Tighten 30° further

Nuts and bolts which are angle tightened should be renewed every time

Electrical system

General
Battery rating – later models:
Cold cranking .... 330, 440, 450, 500, 520 or 600A
Reserve capacity .... 70, 85, 90, 100, 110 or 120 minutes

Alternator – additional types
Make and designation .... Bosch N1 31/80 and 31/100
Output at 14 V/1500 rpm 31A
Output at 14 V/6000 rpm .... 80 or 100A
Rotor resistance:
31/80 .... 2.9
31/100 .... 2.6

Other data as for earlier type N1 (Chapter 12)

Alternator voltage regulator (1985 on)
Regulated voltage at 3000 (engine) rpm:
5A load .... 14.1 to 14.9V
30 to 50A load .... 13.8 to 14.6V

Starter motor (later models) – additional type
Make .... Hitachi
Nominal output .... 1.4 kW (1.9 bhp)
Commutator minimum diameter .... 39 mm (1.535 in)
Brush minimum length .... 11 mm (0.43 in)

Chapter 13 Supplement: Revisions and information on later models

Light bulbs (later models) — Wattage — Pattern

High-level brake light:
- 1986 .... 20 — BA 9s
- 1987 on .... 21 — BA 15s

Number plate light .... 4 — BA 9s

Fuses (760, 1988 on)

Fuse No	Circuit protected	Rating (A)
1	Tail/parking lights (LH), number plate lights	10
2	Tail/parking lights (RH)	10
3	Main beam (LH)	15
4	Main beam (RH)	15
5	Not used in UK	–
6	Dipped beam (LH)	15
7	Dipped beam (RH)	15
8	Foglights (front)	15
9	Foglights (rear)	10
10	Instrument and control lighting	5
11	Reversing lights, turn signals, cruise control	15
12	Dim-dip system	15
13	Heated rear window, heated mirrors	25
14	Bulb failure warning, overdrive relay, power window relay, sunroof motor relay, seat belt warning	10
15	Not used	–
16	Spare	–
17	Spare	–
18	Radio	5
19	Electronic climate control, mirror motors, rear window wiper, seat motor relay, cigarette lighter	15
20	Horn, windscreen wipers, headlight wipers	25
21	Electronic traction control, constant idle speed system	5
22	ABS	5
23	Spare	–
24	Spare	–
25	Hazard warning lights, central locking	25
26	Clock, interior lights, door markers	10
27	Stop-lights	15
28	Heater blower, electronic climate control	30
29	Radio aerial, trailer lighting	30
30	Auxiliary fuel pump	10
31	Main fuel pump, fuel injection system	25
32	Radio amplifier	10
33	Radio	10
34*	Window motors, sunroof motor	30
35*	Seat heaters, seat motors	30
–	ABS (on transient protector relay)	10

*Circuit breakers

Dimensions, weights and capacities (1988 on)

Dimensions
- Overall length .... 4.79 m (15 ft 8.6 in)
- Track:
  - Front .... 1.47 m (4 ft 9.9 in)
  - Rear (except 760 Saloon) .... 1.46 m (4 ft 9.5 in)
  - Rear (760 Saloon) .... 1.52 m (4 ft 11.8 in)
- Overall height:
  - All except 760 Estate .... 1.41 m (4 ft 7.5 in)
  - 760 Estate .... 1.44 m (4 ft 8.7 in)

Weights
Kerb weight (depending on equipment):
- 740 .... 1300 to 1500 kg (2867 to 3308 lb)
- 760 .... 1471 to 1540 kg (3244 to 3396 lb)

Capacities (approx)
- Cooling system (B200E) .... 8.5 litres (15 pints)
- Gearbox oil (M47 II) .... 1.6 litres (2.8 pints)
- Final drive oil (760 Saloon) .... 1.4 litres (2.5 pints)

Other values as given at the beginning of the manual

3 Routine maintenance

1 The maintenance schedule at the beginning of the manual is generally applicable to later models, but note the following points.

Spark plug renewal (all models)
2 'Long life' spark plugs (Bosch prefixes WR and HR, or equivalent) need only be renewed every 12 000 miles or 12 months.

Deleted operations – 1988 and 1989 models
3 The spare wheel well drain check (models with ABS) is deleted from 1988.
4 Front wheel bearing adjustment is no longer necessary on models with ball-bearing front hubs (from 1988). See Section 13.
5 The brake servo check is not specified for 1989 models.

Revised operations – 1989 models
6 The engine valve clearances need not be checked at 48 000 miles/4 years. They must still be checked at 24 000 miles/2 years and 72 000 miles/6 years.
7 Idle speed and CO level need now be checked only every 12 000 miles or annually.

Extra operations – all models with EGR/Pulsair
8 Every 12 000 miles or annually, check the operation of the EGR and Pulsair systems (Section 7).
9 Every 24 000 miles or two years, clean the EGR pipe and valve (Section 7).

Extra operation – models with independent rear suspension
10 Every 12 000 miles or annually, inspect the driveshaft rubber bellows for splits or other damage.

4 In-line engine

B200E engine – description
1 The B200E engine is of the same family as the B230 described in Chapter 1. It is a 2-litre unit with continuous fuel injection.
2 Repair and overhaul operations are as described for B230 engines. Refer to the Specifications at the beginning of this Chapter for details specific to the B200.

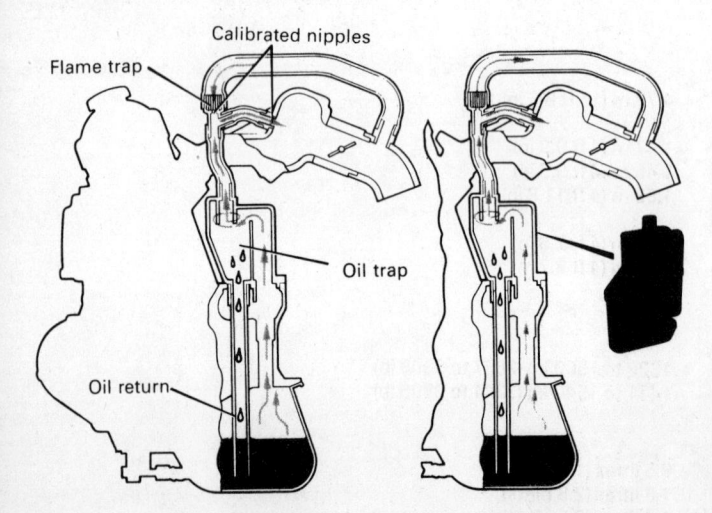

Fig. 13.1 Crankcase ventilation system – later B230K. Arrows show gas flow at low load (right) and full load (left) (Sec 4)

Oil pressure switch (all models) – removal and refitting
3 The oil pressure switch is located on the right-hand side of the engine block, between the oil filter and the alternator. Depending on model and equipment, access may be easier from below.
4 Disconnect the electrical lead from the switch. Wipe clean around the switch, unscrew it from the block and remove it. Recover the sealing washer (if fitted).
5 Refit by reversing the removal operations. Put a smear of sealant on the switch threads and renew the sealing washer if necessary.
6 Run the engine and check for correct operation of the oil pressure warning light. Inspect the switch to see that there are no leaks.

Camshaft oil feed (all models)
7 The camshaft oil feed arrives via an annulus around a cylinder head bolt (No 3 in the tightening sequence – see Fig. 1.11 in Chapter 1). The hole for this bolt is thus an important oilway, and should be thoroughly cleaned whenever the head is removed.

Camshaft cover gasket (all B200/B230) – fitting
8 During 1987 an asbestos-free gasket was introduced for the camshaft cover. When fitting this gasket, silicone sealant must be applied to the camshaft front and rear bearing caps.

Crankcase ventilation system (B230K, 1988 on) – general
9 The crankcase ventilation system has been redesigned to incorporate an improved oil trap, located above the inlet manifold. Details are shown in Fig. 13.1.
10 Maintenance is still as described in Chapter 1, Section 5.

5 V6 engine

B280 engine – general description
1 The B280 engine closely resembles the B28 which it supersedes. Internal differences are confined mainly to the crankshaft, which has been modified to equalise the firing interval between cylinders. Camshaft profiles have also been changed. A welcome modification for the home mechanic is a change to floating gudgeon pins, meaning that pistons can be removed from their connecting rods without the need for a press.
2 Peripheral changes to the engine include the provision of a new inlet manifold, the relocation of the distributor drive to the front of the left-hand camshaft and the provision of an oil-to-coolant oil cooler.
3 The LH-Jetronic fuel injection system fitted to this engine is considered in Section 7.

Valve clearances (B280) – checking and adjustment
4 Disconnect the battery negative lead. Cover the battery with a piece of wood or plastic, or remove it completely, so that its terminal will not be short-circuited by the air conditioning compressor.
5 Remove the oil filler cap and the crankcase ventilation hoses attached to it.
6 Disconnect the air conditioning compressor control lead.
7 Unbolt the compressor mountings from the engine. Free the drivebelts and move the compressor to one side, complete with mounting bracket and belt tensioner pulley. Be careful not to disconnect or strain the refrigerant hoses.
8 Unbolt the compressor drivebelt idler pulley from the right-hand rocker cover. Slacken the bolt which holds the idler to the timing chain cover and move the idler out of the way.
9 Remove the engine oil dipstick.
10 Unbolt any remaining cable clamps from the right-hand rocker cover.
11 Remove the air cleaner and the air mass meter (Section 7). Also remove the air inlet trunking which connects the air mass meter to the throttle body housing.
12 Disconnect the HT leads from the left-hand bank of spark plugs and move them aside. Note the inductive pick-up on No 1 HT lead.
13 Unbolt the HT lead clip from the left-hand rocker cover (photo).
14 Free the remaining wiring harnesses and fuel lines from their brackets, cutting cable ties where necessary. Use new ties on re-assembly.

Chapter 13 Supplement: Revisions and information on later models

15 Remove the rocker cover securing bolts, noting the location of the different lengths of bolt. Noted also the earth strap under one of the left-hand bolts (photo).
16 Remove the rocker covers and recover the gaskets.
17 The procedure is now as described in Chapter 1, Section 50, paragraph 8 onwards. Note however that there is now only one notch on the crankshaft pulley, which relates to No 1 cylinder, and that the timing scale itself is slightly different (photo).

Oil cooler (B280) – removal and refitting

18 The oil cooler is located between the oil filter and the left-hand side of the cylinder block.
19 Disconnect the battery negative lead.
20 Remove the engine undertray.
21 Drain the cooling system, including the left-hand cylinder bank (Section 6).
22 Remove the air cleaner unit (Section 7).
23 Remove the oil filter.
24 Remove the centre bolt from the oil cooler (photo).
25 Disconnect the coolant hoses from the oil cooler.
26 Remove the oil cooler, being prepared for some spillage of oil and coolant.
27 Refit by reversing the removal operations, using a new oil filter.

Oil pressure switch (all models) – removal and refitting

28 The oil pressure switch is located on the right-hand side of the block, at the front of the engine. Access is easiest from below (photo).
29 Proceed as described for in-line engines in Section 4. Note however that several different types and lengths of switch have been fitted over the years. It is important that the correct switch be fitted, otherwise loss of oil pressure may result.

Distributor drive oil seal (B280) – renewal

30 Remove the distributor and its shaft (Section 8).
31 Free the oil seal by tapping one edge inwards, then prise it from its location. Be careful not to damage the housing or the surface of the drive piece (photo).
32 Clean the seal housing and inspect the drive piece. Renew the drive piece if its sealing surface is scored.
33 Grease the lips of a new seal and press it into position, using the distributor shaft Allen screw, some tubing and a washer (photos).
34 Refit the distributor shaft and distributor.

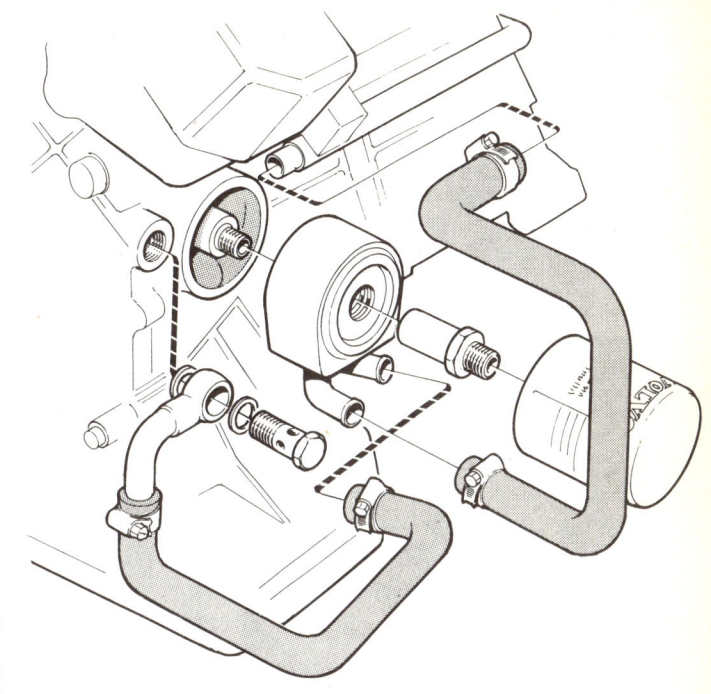

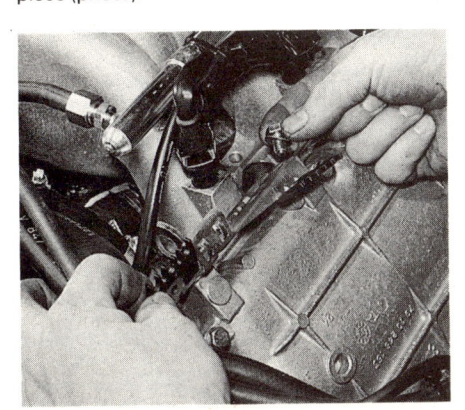

Fig. 13.2 Oil cooler fitted to B280 engine (Sec 5)

5.13 Unbolt the HT lead clip

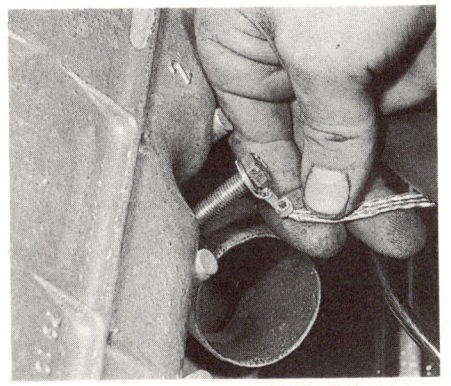

5.15 The earth strap secured by one of the left-hand bolts

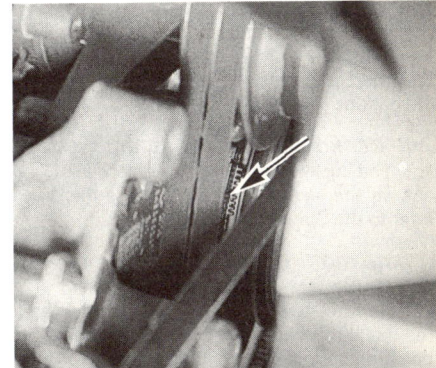

5.17 Timing scale (arrowed) on B280 engine

5.24 Oil cooler centre bolt (arrowed)

5.28 The oil pressure switch

5.31 Removing the distributor drive oil seal

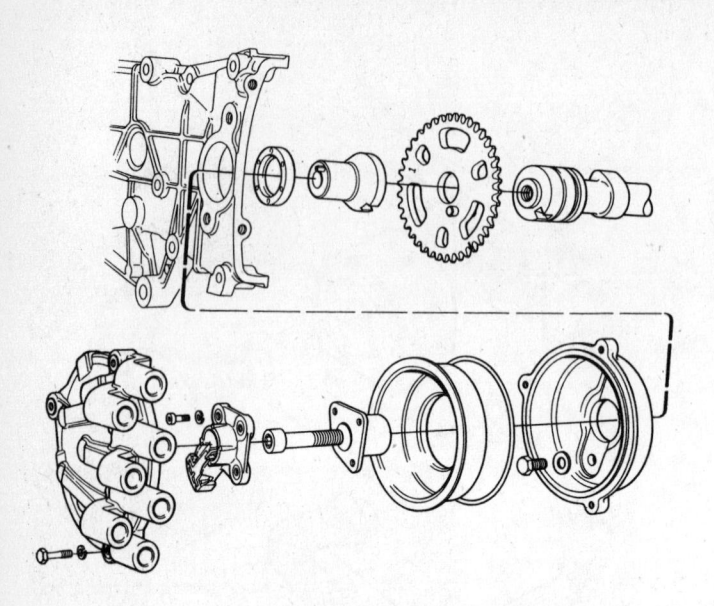

Chapter 13 Supplement: Revisions and information on later models

Fig. 13.3 Distributor drive components – B280 (Sec 5)

Crankshaft and main bearings (B280) – removal and refitting

35 Proceed as in Chapter 1, Sections 72, 86 and 87, but note that main bearing caps Nos 2 and 3 are secured to the block by two side bolts each. Remove these bolts before attempting to remove the bearing caps.

36 When reassembling, only insert the side bolts finger tight until the cap nuts have been fully tightened (Chapter 1, Section 87). When this has been done, tighten the side bolts to the specified torque.

Pistons and connecting rods (B280) – examination and renovation

37 The use of floating gudgeon pins secured by circlips means that the pistons and rods may be separated if necessary. Proceed as described in Chapter 1, Section 31. Note the correct orientation of rods and pistons (Fig. 13.5).

Cylinder liner seal identification – B280

38 The thickness of early type liner seals is indicated by the number of tabs on the seal. On later types the thickness is indicated by colour. Refer to the Specifications for details.

Engine mountings (B280) – removal and refitting

39 Disconnect the battery negative lead.
40 Remove the two securing bolts from the top of the fan shroud and free the shroud from its lower mountings. There is no need to remove the shroud completely.
41 Make arrangements to support and lift the engine, preferably using lifting tackle from above. At a pinch it is possible to use a jack under the sump, with a piece of wood to spread the load, but this is not recommended.
42 Unbolt the exhaust system flanged joint behind the front silencer.
43 Unbolt the engine dampers at one end of their mountings (photo). Compress the dampers slightly and swing them aside.
44 Remove the sump guard (if not already done). If working on the right-hand side, unbolt the transmission fluid line bracket from the bellhousing.
45 Remove the three nuts and two bolts securing each mounting. One of the nuts is not immediately obvious: it is accessible from the underside of the engine support crossmember (photos).
46 Raise the engine slightly and withdraw the mounting (photo). Recover the spacer.
47 Refit by reversing the removal operations.

Cylinder head bolts

48 New bolts have been introduced which do not require retorquing after the engine has been run. Consult your dealer for further information.

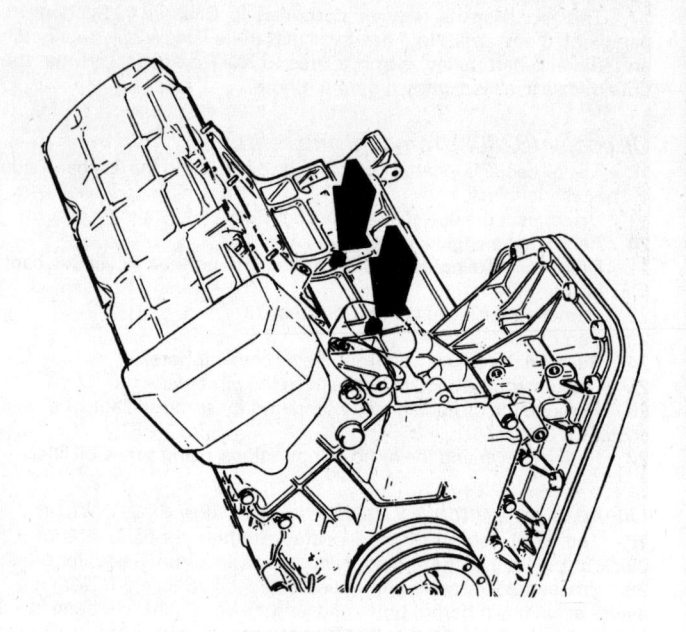

Fig. 13.4 Two of the main bearing cap side bolts (arrowed) – B280 (Sec 5)

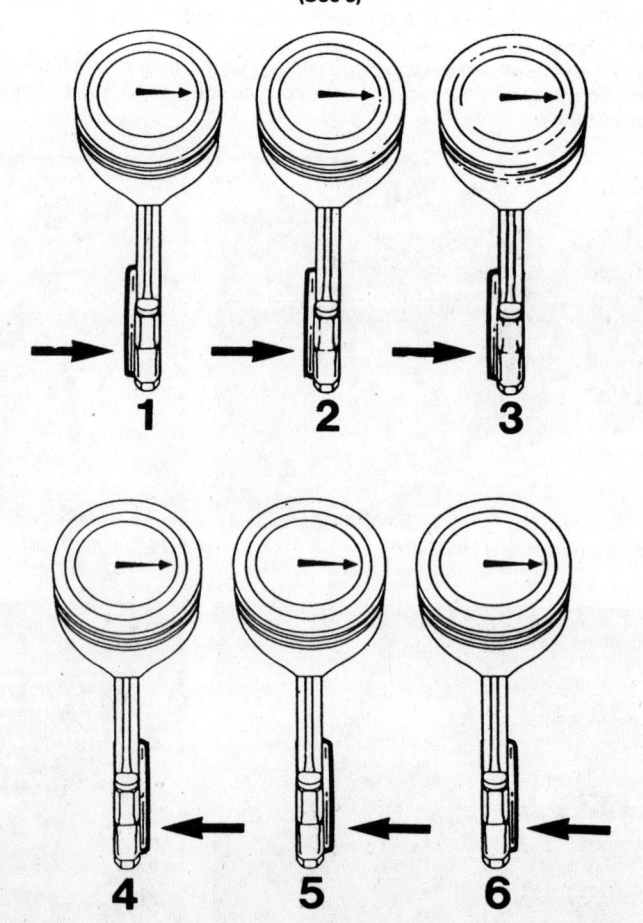

Fig. 13.5 Relationship of connecting rod offsets (large arrows) with arrows on piston crowns (Sec 5)

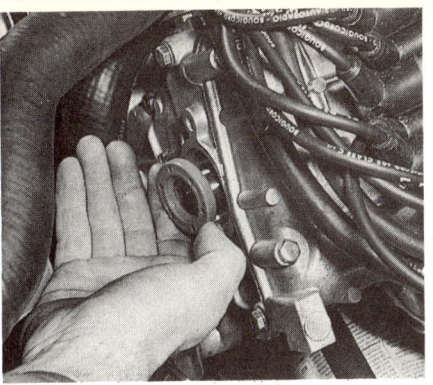

Chapter 13 Supplement: Revisions and information on later models

5.33A Fitting a new seal

5.33B Pressing the new seal home

5.43 Unbolting an engine damper

5.45A One of the engine mounting bolts (arrowed)

5.45B Engine mounting nut accessible through the crossmember

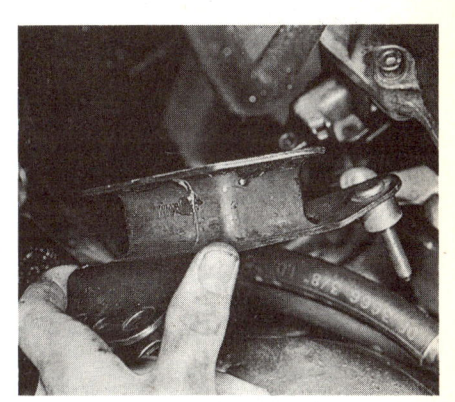

5.46 Removing an engine mounting

6 Cooling, heating and air conditioning systems

Cooling system – flushing (all models)

1 The makers now recommend that the cooling system be flushed at every coolant change. A flushing additive is available from Volvo dealers.

Cooling system (B280) – draining

2 The drain tap on the right-hand side of the engine is unchanged (see Chapter 2, Fig. 2.2).

3 The drain tap on the left-hand side of the engine is no longer fitted; in its place there is a banjo union which is part of the oil cooler coolant circuit. The left-hand cylinder bank can be drained by slackening this union (photo).

Water pump (B280) – removal and refitting

4 In theory there is no need to remove the inlet manifold before removing the water pump, though access is obviously poor with the manifold in place. The procedure is otherwise as described in Chapter 2, Section 12.

Automatic climate control vacuum hose connections – all models with ACC

5 The vacuum line connections to the programmer must not be pushed right home, or an apparent loss of vacuum can result. The connection is correct when the spigot on the programmer enters the connector only as far as the entry of the vacuum line on the side of the connector.

Electronic Climate Control (760, 1988 on) – description

6 Electronic Climate Control (ECC) is a development of the Automatic Climate Control (ACC) described in Chapter 2. The main difference is in the control system, which now incorporates a microprocessor, solenoid valves and a servo motor. Less use is made of vacuum than in the old system, with a consequent increase in reliability.

7 From the driver's point of view the two systems are very similar. When the automatic function is engaged, the selected temperature is maintained in the cabin by mixing of hot and cold air, using the heating and air conditioning systems as necessary.

8 The microprocessor control unit, mounted on the back of the control panel, incorporates a built-in fault diagnosis facility. A fault is signalled to the driver by the flashing of the air conditioner control button. If the fault is serious, the button will flash continuously while the engine is running. If the fault is less serious, the button will flash for about 20 seconds after the engine is started.

9 There is little that the home mechanic can do to the system other

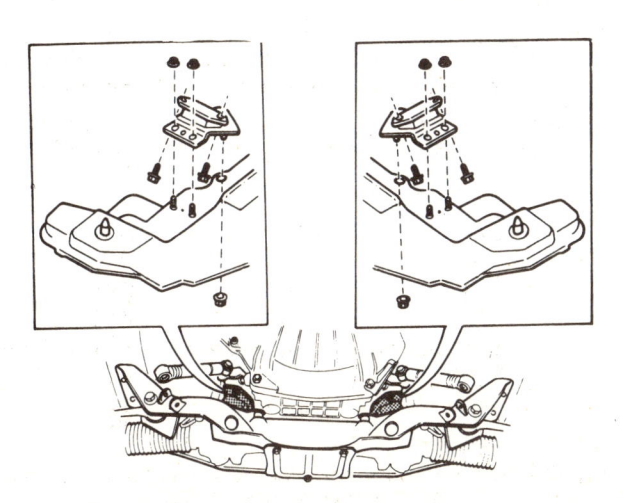

Fig. 13.6 Engine mounting details – B280 (Sec 5)

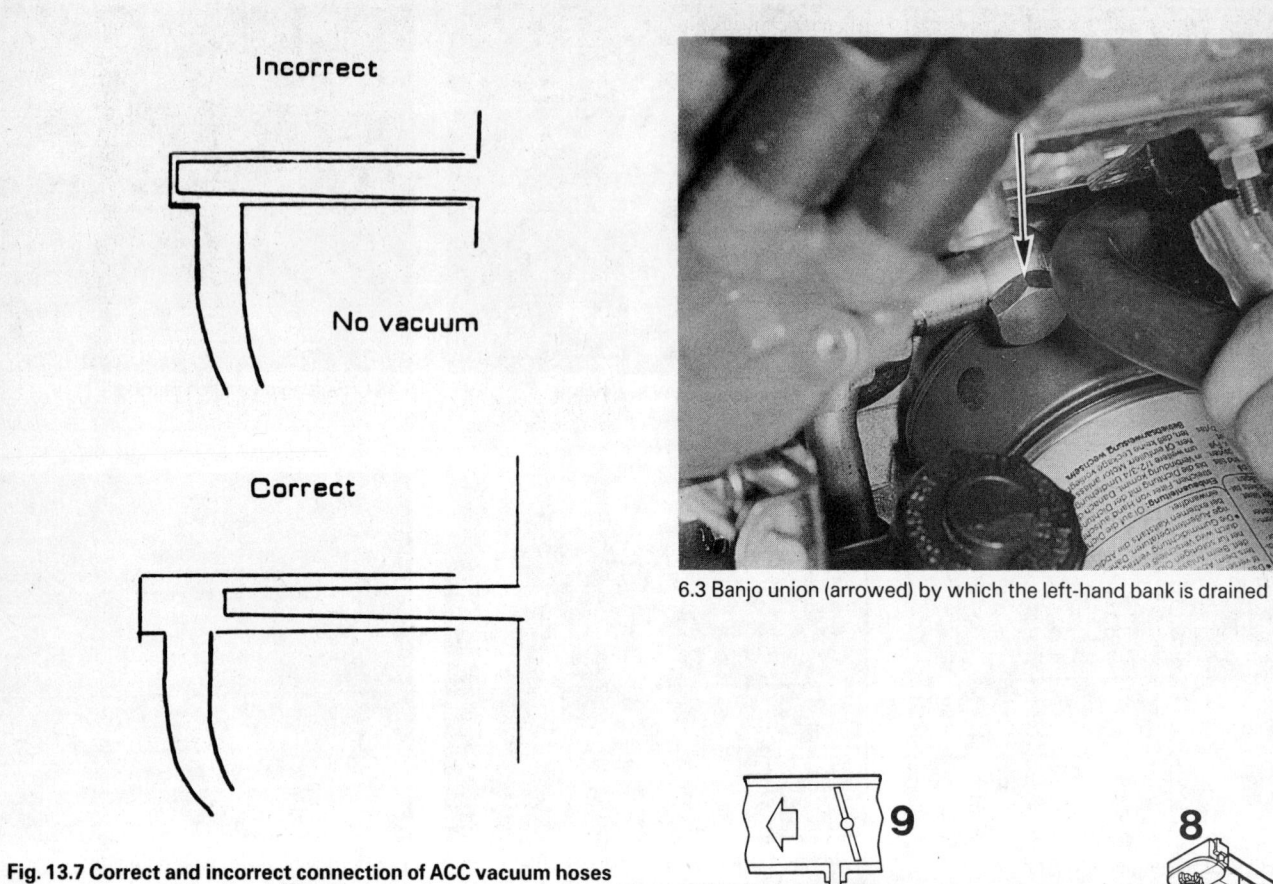

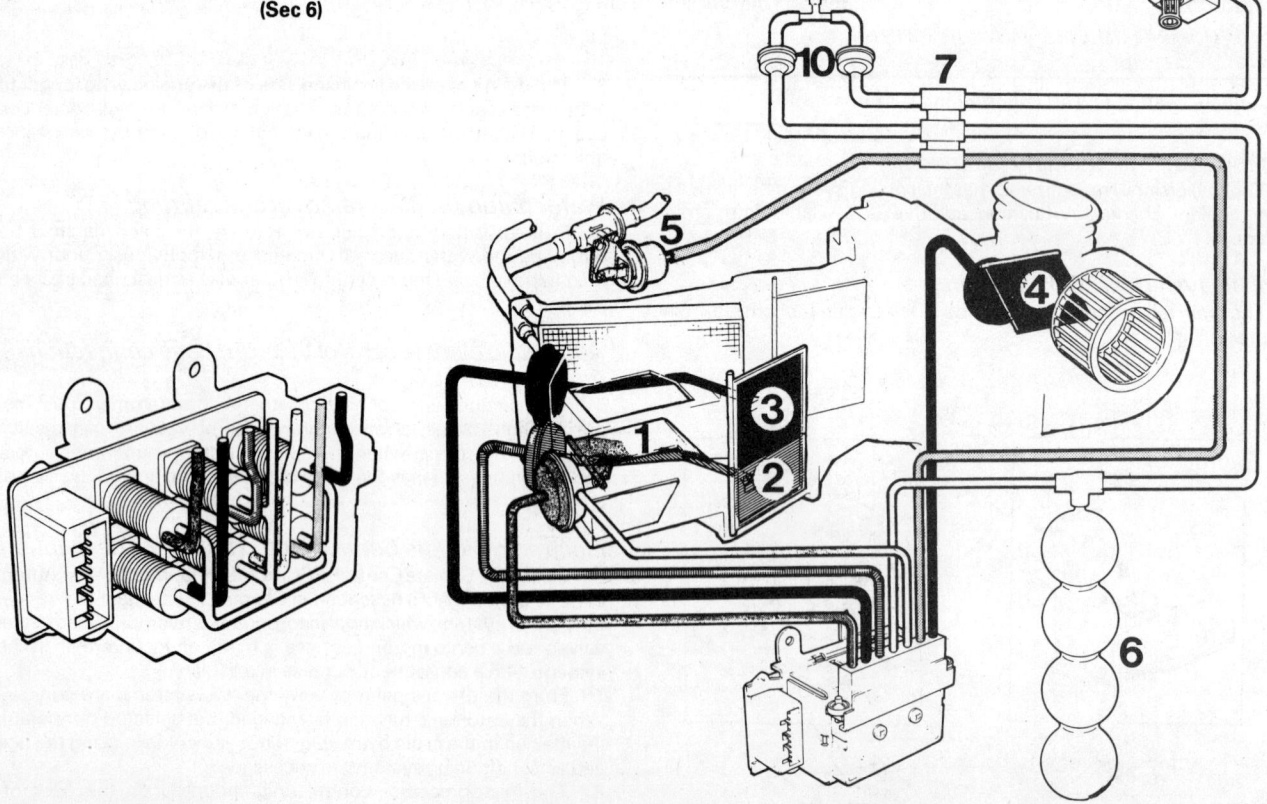

340   Chapter 13 Supplement: Revisions and information on later models

6.3 Banjo union (arrowed) by which the left-hand bank is drained

Fig. 13.7 Correct and incorrect connection of ACC vacuum hoses (Sec 6)

Fig. 13.8 Electronic climate control vacuum connections (Sec 6)

1 Floor/defroster shutter
2 Bi-level shutter
3 Vent shutter
4 Recirculation shutter
5 Water valve
6 Vacuum reservoir
7 Bulkhead connector
8 Interior air temperature sensor
9 Inlet manifold
10 Non-return valves

Chapter 13 Supplement: Revisions and information on later models

than verifying the integrity of electrical and vacuum connections. Fault diagnosis and repair should be entrusted to a Volvo dealer.

ECC sensors – description

10 The four sensors peculiar to the ECC are the solar sensor, the water temperature sensor, the interior air temperature sensor and the exterior air temperature sensor.

11 The solar system is mounted on top of the dashboard, in the left-hand loudspeaker grille. Its function is to reduce cabin air temperature by up to 3°C (5°F) in bright sunlight.

12 The interior air temperature sensor is located inside the interior light (photo). It measures cabin air temperature. A hose running from the sensor to the inlet manifold maintains a flow of air through the sensor when the engine is running.

13 The water temperature sensor is located next to the heater matrix. This sensor in fact measures air temperature adjacent to the matrix. When the automatic function is selected, this sensor prevents the fan running at maximum speed before the matrix has heated up.

14 The exterior air temperature sensor is mounted on the blower motor casing. It measures the temperature of the air passing through the blower; when the recirculation function is selected, this air comes from inside the car, but otherwise it comes from outside.

ECC control pump and control unit – removal and refitting

15 Disconnect the battery negative lead.
16 Remove the switch panel and ECC panel surround.
17 Remove the four screws now exposed and draw the ECC panel and control unit into the car. (This will give sufficient access for bulb renewal if this is the reason for removal.)
18 Disconnect the multi-plugs from the rear of the unit and remove it.
19 Do not attempt to dismantle the control unit, unless out of curiosity. There are no serviceable parts inside.
20 Refit by reversing the removal operations.

7 Fuel and exhaust systems

Unleaded fuel and catalytic converter

1 From 1988 model year, a green fuel filler cap is fitted to those models which may be run on UK premium unleaded fuel (95 RON) without making any adjustment.
2 Use of unleaded fuel is not essential except on models fitted with a catalytic converter.
3 Owners of models not having a green filler cap who wish to use unleaded fuel should consult a Volvo dealer.

Catalytic converter – precautions
(a) DO NOT use leaded petrol
(b) Always keep the ignition and fuel systems well maintained
(c) If the engine develops a misfire, do not drive the car until the fault is cured
(d) DO NOT push- or tow-start the car
(e) DO NOT switch off the ignition at high engine speeds – ie do not "blip" the throttle immediately before switching off
(f) DO NOT use fuel or engine oil additives
(g) DO NOT continue to use the car if the engine burns oil to the extent of leaving a visible trail of blue smoke
(h) The catalytic converter operates at very high temperatures, so DO NOT park the car over inflammable material after a long run
(i) The catalytic converter is FRAGILE – do not strike it with tools during servicing work

Air cleaner element (B280) – renewal

4 Proceed as described in Chapter 3, Section 5, but note that the two clips which hold the air cleaner lid to the air mass meter must also be released (photo).

Air cleaner unit (B280) – removal and refitting

5 Remove the air cleaner element.
6 Disconnect the hot air trunking.
7 Release the air cleaner unit from its rubber mountings by tugging firmly. If the mountings come away with the air cleaner, remove them and refit them to their locations on the inner ring and mounting bracket.
8 Refit by reversing the removal operations, pushing the air cleaner firmly onto the rubber mountings.

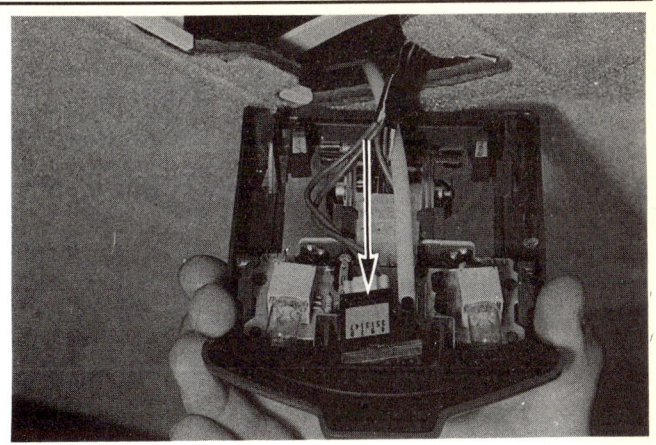

6.12 Interior air temperature sensor (arrowed) inside interior light

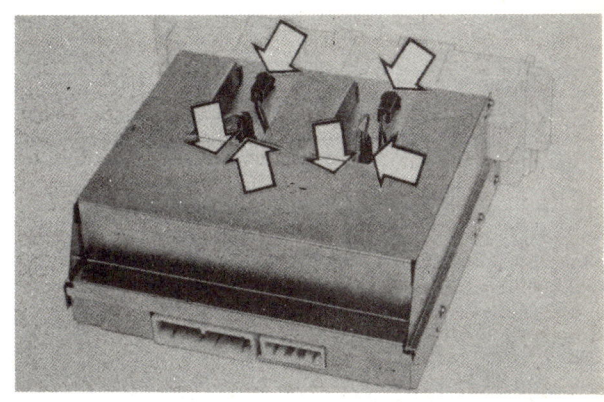

Fig. 13.9 Location of light bulbs (arrowed) in ECC control unit (Sec 6)

Idle mixture adjustment – models with Pulsair system

9 On models so equipped, the Pulsair system must be blocked off when checking or adjusting the CO level, or false readings will result. This is most easily done by clamping the hose between the air cleaner and the non-return valves.

Pierburg 2B7 carburettor (B230K, 1987 on) – description

10 From 1987 the 2B5 carburettor fitted to earlier B230K engines was superseded by the 2B7. As explained in Chapter 3, the carburettor seen in the photos there is in fact a 2B7; the overhaul and adjustment procedures in Chapter 3 also apply. Specifications peculiar to the 2B7 are given at the beginning of this Chapter.
11 The operation of the automatic choke vacuum unit is also slightly different on this carburettor: it takes place in three stages instead of the previous two. For details see Fig. 13.10.

Control pressure regulator ventilation (all models with continuous injection system)

12 The base of the control pressure regulator must be open to atmosphere, otherwise incorrect fuel pressure may result. Problems of difficult starting, high fuel consumption and poor driveability are possible signs of a problem in this area.
13 If corrosion is visible on the control pressure regulator baseplate, proceed as follows.
14 Disconnect the battery negative lead.
15 Remove the control pressure regulator and its bracket (Chapter 3, Section 34).
16 Separate the regulator and bracket, making alignment marks between the two for reference on reassembly.
17 Drill a hole in the middle of the bracket as shown in Fig. 13.11. Deburr or countersink the hole.
18 Refit the bracket to the regulator. The drilled hole must coincide with the ventilation hole on the regulator.
19 Refit the control pressure regulator to the engine.

Pulsair system (B230K, 1987 on) – description

20 The Pulsair system is part of the emission control package fitted to later B230K engines. It uses fluctuations in pressure in the exhaust

342   Chapter 13   Supplement: Revisions and information on later models

7.4 Unclipping the air cleaner lid from the air mass meter

7.21A Pulsair system non-return valves

7.21B Pulsair system shut-off valve

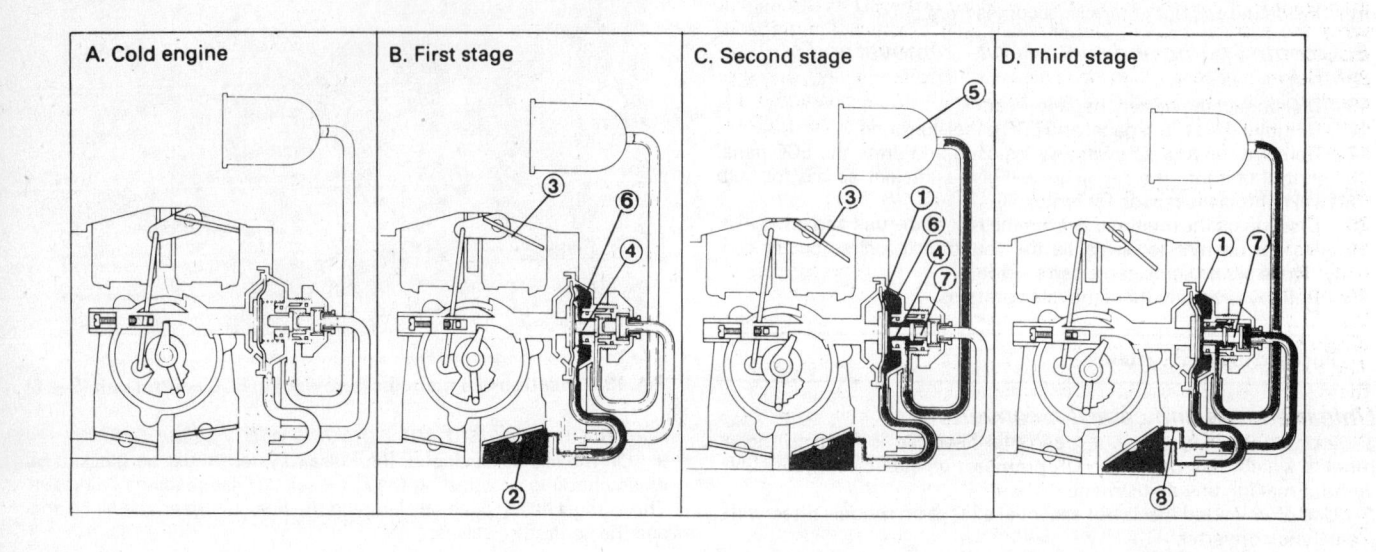

A. Cold engine   B. First stage   C. Second stage   D. Third stage

Fig. 13.10 Operation of automatic choke vacuum unit – 2B7 carburettor (Sec 7)

1 Diaphragm
2 Throttle valve
3 Choke valve
4 Piston
5 Vacuum reservoir
6 Vacuum chamber
7 Valve
8 Vacuum passage

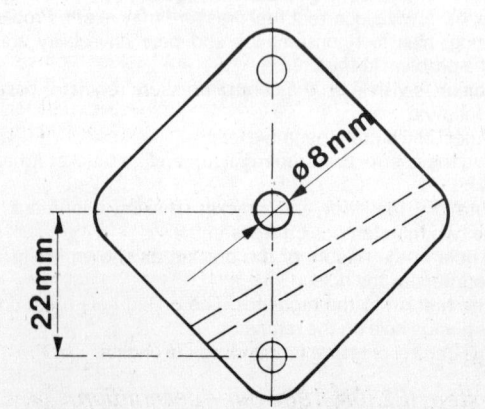

Fig. 13.11 Hole position in control pressure regulator bracket (Sec 7)

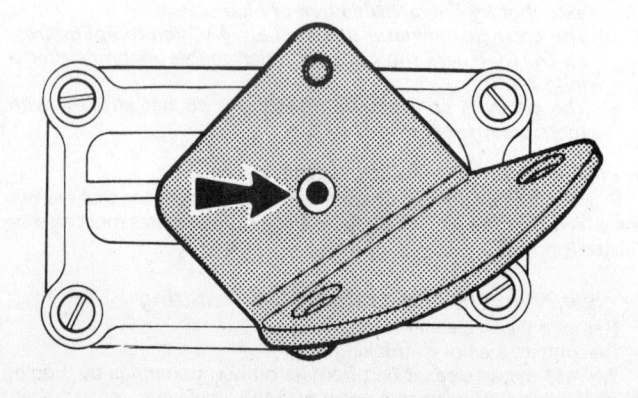

Fig. 13.12 Hole in bracket coincides with hole in regulator (arrowed) (Sec 7)

Chapter 13 Supplement: Revisions and information on later models

manifold to draw in pulses of air from the air cleaner housing. The air provides oxygen which combines with unburnt hydrocarbons to reduce exhaust pollution.

21 The components of the system are an air distribution manifold, two non-return valves, a shut-off valve and associated pipework (photos). The shut-off valve is controlled by inlet manifold vacuum and stops air flowing to the exhaust manifold for a couple of seconds at the beginning of deceleration, so preventing backfiring. The non-return valves prevent exhaust gas from arriving in the air cleaner.

Pulsair system – maintenance

22 At the specified intervals, check that all hoses and connections in the system are in good condition.
23 Disconnect the non-return valves from the air cleaner hose. Run the engine and feel the openings of the valves. Air should be sucked in and exhaust gas must not be emitted. Renew the valves if necessary. Reconnect the hose.
24 If the shut-off valve is not closing as it should, backfiring on the overrun will have been noticed. To verify that the valve is open at idle, measure the exhaust gas CO level first with the system blocked off (paragraph 9), then with the system unblocked. The CO level must fall when the system is unblocked: if not, the shut-off valve is probably defective.
25 Before renewing a shut-off valve which is apparently not opening, check that the ignition timing and the engine valve clearances are correct. Retarded timing or excessive valve clearances can both be causes of excessive inlet manifold vacuum.

EGR system (B230K, 1987 on) – description

26 The EGR (Exhaust Gas Recirculation) system is also part of the emission control package. Its function is to introduce a controlled quantity of exhaust gas into the inlet manifold at part load conditions. This has the effect of reducing the emission of oxides of nitrogen.
27 The components of the system are a control valve, a vacuum amplifier, a thermostatic valve and the associated pipework (photos). The control valve receives a signal from the vacuum amplifier, causing the valve to open or close and so allow more or less exhaust gas to pass. The vacuum amplifier receives signals from the air cleaner-to-carburettor ducting, from the carburettor venturi and from the inlet manifold itself. The manifold vacuum signal arrives via the thermostatic valve, which is closed at low temperatures and so stops the system operating during warm-up.
28 When the system is operating correctly, exhaust gas recirculation does not take place at idle nor at full load. Malfunction at idle will cause rough running and stalling. Malfunction at full load will cause loss of power.

EGR system – maintenance

29 At the specified intervals, check the function of the system as follows.
30 Open the bonnet and locate the EGR valve. Shine a light on it so

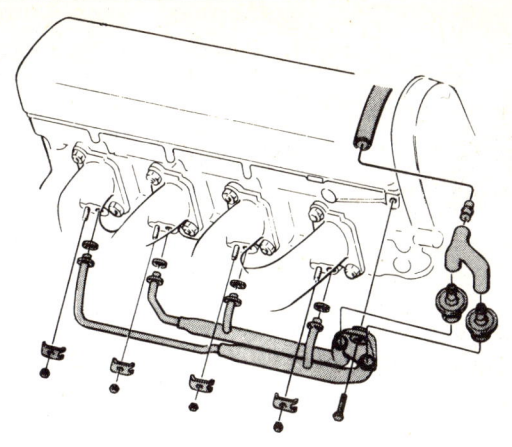

Fig. 13.13 Pulsair system air distribution manifold (Sec 7)

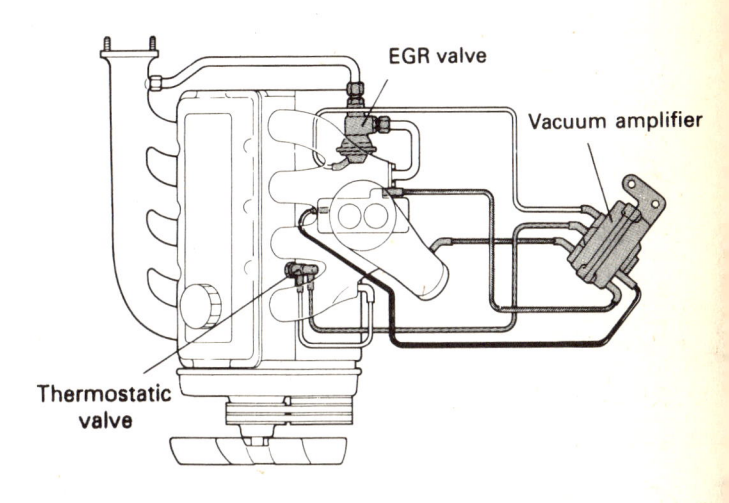

Fig. 13.14 EGR system components (Sec 7)

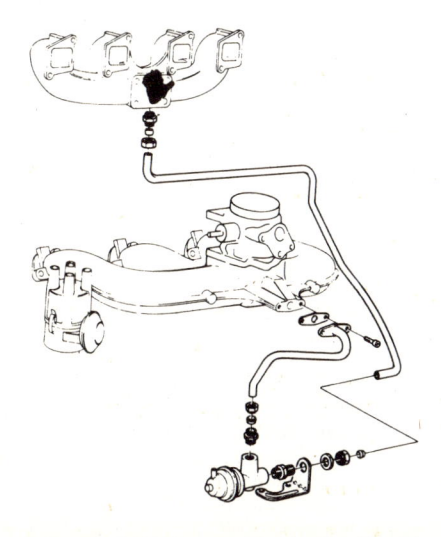

Fig. 13.15 EGR pipework and valve (Sec 7)

that the movement of the valve rod may be observed. Start the engine from cold and warm it up. No movement of the valve rod should take place, regardless of throttle opening, at coolant temperatures below 55°C (131°F). If the valve operates at low coolant temperatures, the thermostatic valve is defective.
31 With the engine warm, verify that the EGR valve rod moves at part throttle and closes again at idle. If not, the valve or the vacuum amplifier (or their connections) must be at fault. The EGR valve and the vacuum amplifier are matched in production and must be renewed as a pair.
32 Periodically clean those parts of the system which are in direct contact with exhaust gas – ie the EGR valve seat and the manifold connecting pipes. Remove these components for cleaning. Tap the pipes with a copper mallet to loosen carbon deposits, then blow through them with compressed air. Clean out unions and calibrated orifices with a screwdriver or scraper.
33 When cleaning the EGR valve seat, hold the valve in the hand, not in a vice. Only clean the valve by scraping: do not use solvents or caustic chemicals, as these may damage it.

Fuel gauge senders (later models) – testing

34 From 1986 (tubular type senders) or 1987 (lever type) the resistance characteristics of the fuel gauge senders were reversed. The senders now register minimum resistance when the tanks are empty. Details are given in the Specifications.

Fuel tank pump/fuel gauge sender (1988 on) – removal and refitting

35 The procedures in Chapter 3, Sections 13 and 15, are basically unchanged. Note however that the pump/sender unit is now secured to the neck of the tank by a large plastic ring nut (photo). Unscrew this nut,

344 Chapter 13 Supplement: Revisions and information on later models

7.27A EGR control valve

7.27B EGR vacuum amplifier

7.35 Tank pump/gauge sender unit secured by a plastic ring nut

using a 'soft' tool such as a strap wrench, to remove the pump/sender unit.
36 When refitting the unit, align the arrows on each side of the unit with the seam of the fuel tank.

Fuel tank (760 Saloon, 1988 on) – removal and refitting
37 Disconnect the battery negative lead. Observe strict fire precautions throughout the operation.
38 Inside the boot, remove the spare wheel. Peel back the carpet around the fuel filler tube. Remove the left-hand drain tube and the cover from around the filler tube.
39 Remove the cover plate to expose the tank pump/fuel gauge sender unit. Disconnect the fuel hoses and the electrical connector from the unit.
40 Syphon or pump the fuel from the tank into suitable sealed containers, working through the hole in the pump/gauge sender.
41 Raise and support the vehicle. Remove the three screws which secure the protective shield at the front of the tank. Remove the shield.
42 Remove the two inner bolts from the tank front mounting bracket. Slacken the outer bolts by about 10 mm (not quite half an inch) but do not remove them yet. Release the straps from the front mounting bracket and let them hang down.
43 Unbolt the bracing strap from the base of the tank. Also remove the nut on the right-hand side of the tank (up near the silencer).
44 Free the tank and lower it so that it is resting on the front mounting bracket and the rear suspension members.
45 Remove the rear half of the propeller shaft, making identification marks for reassembly.
46 Have an assistant support the tank, or make up a cradle to support it on a trolley jack. Be careful not to damage the tank; it is only made of plastic.
47 Remove the remaining bolts from the front mounting bracket and remove the bracket. Lower the tank and draw it forwards, at the same time disconnecting the breather hose on top of the tank. Remove the tank from under the vehicle.
48 If a new tank is to be fitted, transfer the pump/gauge unit, heat shields, bump stops etc from the old tank.
49 Refit by reversing the removal operations.

Inlet manifold (B280) – removal and refitting
50 Disconnect the battery negative lead.
51 Slacken the fuel filler cap to release any pressure in the tank.
52 Cut the cable ties which secure the electrical and vacuum lines to the injector rails. Also cut the tie which secures the fuel supply line to the air inlet trunking. Obtain new ties for reassembly.
53 Disconnect the throttle and kickdown cables from the throttle drum and from their brackets.
54 Remove the air mass meter and the trunking which connects it to the throttle body housing. It will be necessary to disconnect the air control valve and the crankcase ventilation hose from the trunking (photo).
55 Disconnect the vacuum hose and the crankcase ventilation hose from the throttle body housing.
56 Disconnect the vacuum hose and the fuel return line from the fuel pressure regulator (photos). Be prepared for fuel spillage.

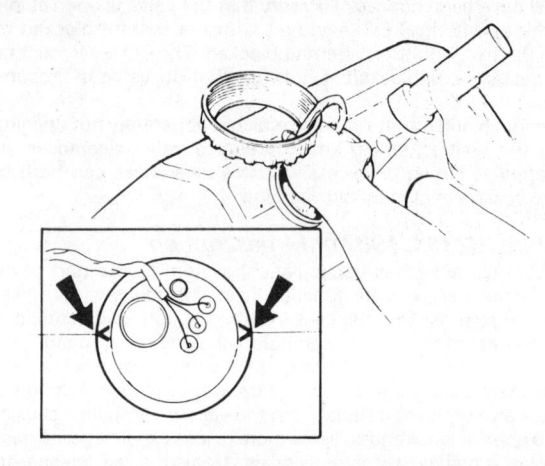

Fig. 13.16 Arrows on pump/gauge sender unit (arrowed) must align with seam on tank (Sec 7)

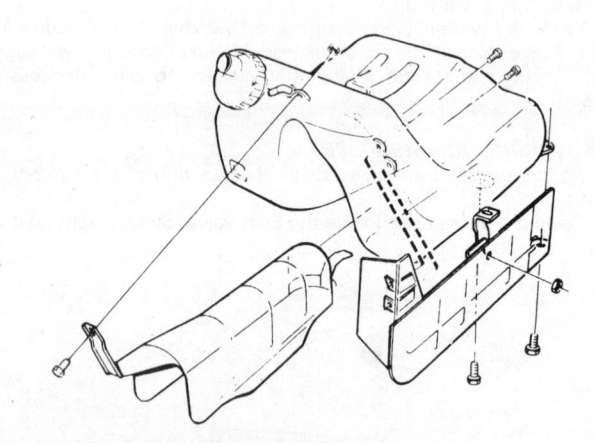

Fig. 13.17 Fuel tank heat shields (Sec 7)

57 Disconnect the fuel pressure regulator from one or other injector rail, counterholding the union (photo). Remove the fuel pressure regulator bracket bolts.
58 Disconnect the wiring plugs from the fuel injectors. Also disconnect the throttle switch plug at the throttle body housing.
59 Disconnect the two knock sensor multi-plugs (photo). Note how the sensor wires pass through the manifold branches: they will have to be fed back through when refitting.
60 Unbolt the earth straps from the manifold (photo).
61 Disconnect the fuel supply lines from the front of the injector rails. Be prepared for fuel spillage.

Chapter 13 Supplement: Revisions and information on later models

7.54 Disconnecting the air control valve hose from the air intake trunking

7.56A Disconnecting the fuel pressure regulator vacuum hose ...

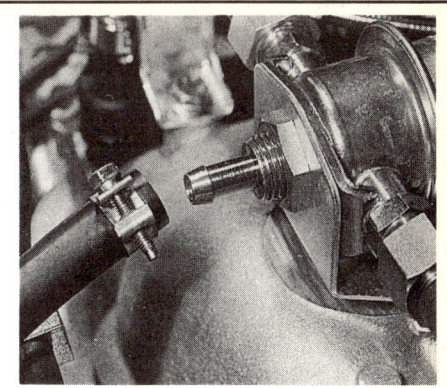

7.56B ... and the fuel return line

7.57 Disconnecting the fuel pressure regulator from an injector rail

7.59 Disconnecting a knock sensor multi-plug

7.60 Manifold earth straps

62 Unbolt the injector rails and remove them, pulling the injectors from their locations. Move the injector wiring harness aside.
63 Disconnect the temperature sensor multi-plug at the water pump (photo).
64 Disconnect the air control valve multi-plug. Also disconnect the valve hose from the throttle body housing.
65 Move aside any remaining wires, pipes or hoses which will obstruct removal. Remove the four bolts which secure the manifold and lift off the manifold complete with throttle body housing and throttle drum (photo).
66 Refit by reversing the removal operations. Use new O-rings between the manifold and the cylinder heads (photo). Remember to feed the knock sensor cables through the manifold as it is being fitted.
67 On completion, check the adjustment of the throttle and kickdown cables, then check the idle speed if necessary.

LH-Jetronic fuel injection system – description

68 The LH-Jetronic system fitted to the B280 engine is similar to the electronic fuel injection system fitted to the four-cylinder turbocharged engines (see Chapter 3, Section 44). The main difference is that with the LH-Jetronic system it is the mass of the inducted air which is measured, not its volume. Air mass is measured simply by observing the cooling effect of the air on a hot wire; this elegant method avoids the need to use cumbersome and potentially unreliable moving flaps to measure airflow.
69 The main components of the system are shown in Fig. 13.19, and their relationship in Fig. 13.20. It will be noticed that the ignition system is controlled by a separate unit, though both fuel injection and ignition control units communicate with each other and are interdependent.
70 The air mass meter is located in the air inlet trunking between the air cleaner and the throttle housing. It incorporates a platinum wire which, when the system is operating, is heated to 100°C (180°F) above the temperature of the incoming air. The current needed to maintain this temperature is used by the control unit to calculate air mass. Each time the engine is switched off, the control unit briefly heats the wire to 1000°C (1832°F) to burn off any deposits.
71 The hot wire in the air mass meter is the most vulnerable component in the system. Should the wire break, the control unit provides a 'limp home' signal to the injectors, enabling the car to be driven at low speed until a new air mass meter can be fitted.
72 Other functions of the control unit include enrichment during starting, warm-up, acceleration and full-load conditions; fuel cut-off during overrun; idle speed control; fuel pump cut-off when the engine is stopped, and injection cut-off at excessive engine speed.

LH-Jetronic system – idle speed and mixture adjustment

73 Refer to Chapter 3, Section 8, paragraphs 1 to 3. Note however that the oil filler cap should be removed while checking the CO level.
74 Locate the air control valve diagnostic connector – a small (2-pole) connector tucked under the cooling system expansion tank. Temporarily disable the valve by connecting the red and white wire in the connector to the battery earth (negative) terminal (photo).
75 Basic idle speed should now be as given in the Specifications. Adjust if necessary by turning the knurled adjuster screw on the throttle housing (photo).
76 Disconnect the earthing wire from the diagnostic connector. Idle speed should now rise to the regulated value and remain steady.

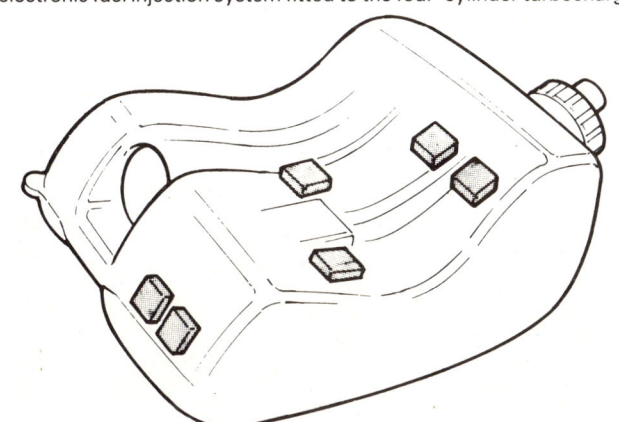

Fig. 13.18 Fuel tank bump stops (Sec 7)

346 Chapter 13 Supplement: Revisions and information on later models

Fig. 13.19 LH-Jetronic system main components (Sec 7)

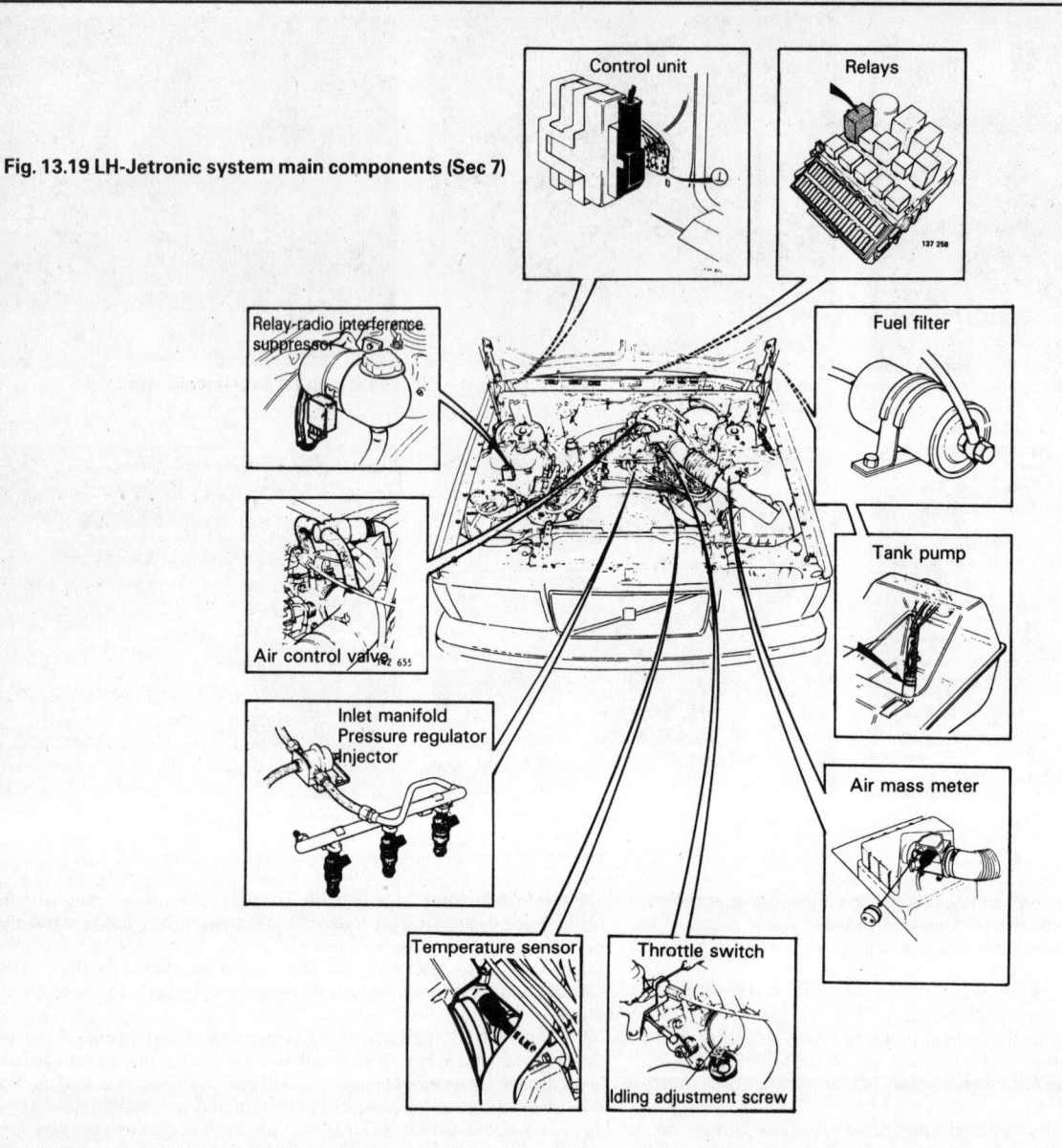

Fluctuations may be caused by leaks around the throttle housing or air control valve.

77 Read the CO level. Adjustment should only be attempted if it is outside the limits specified for checking values. The adjustment screw is located in the air mass meter, beneath a tamperproof cap next to the multi-plug (photo). The cap is removed by drilling two small holes in it and pulling it out with circlip pliers.

78 Turn the CO adjustment screw clockwise to increase CO level, anti-clockwise to reduce it, until the specified setting value is obtained.

79 Disconnect the test gear and when necessary fit a new tamperproof plugs. Refit the oil filler cap.

LH-Jetronic system – fuel injector check

80 All six injectors operate simultaneously. Therefore if an injector is suspected of being faulty (indicated by a regular misfire which is not ignition-related) a quick check may be made as follows.

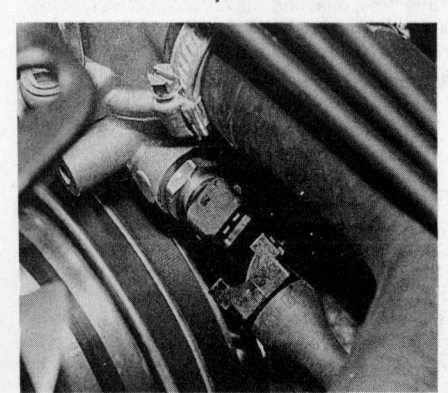

7.63 Disconnecting the temperature sensor multi-plug

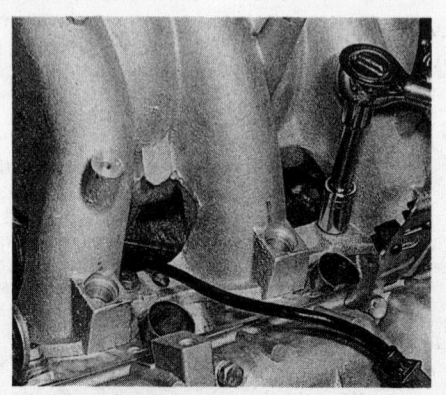

7.65 Removing a manifold securing bolt

7.66 An inlet manifold O-ring

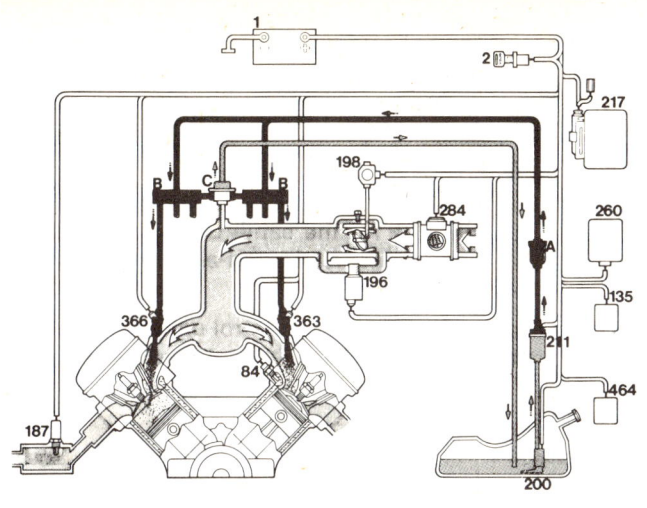

Chapter 13 Supplement: Revisions and information on later models

Fig. 13.20 Relationship of LH-Jetronic system components (Sec 7)

1 Battery	217 Control unit
2 Ignition switch	260 Ignition system control unit
84 Temperature sensor	284 Air mass meter
135 Relay	363 Injector (left-hand bank)
187 Lambda sensor (not UK)	366 Injector (right-hand bank)
196 Air control valve	464 Relay
198 Throttle switch	A Fuel filter
200 Tank pump	B Fuel supply rails
211 Main fuel pump	C Fuel pressure regulator

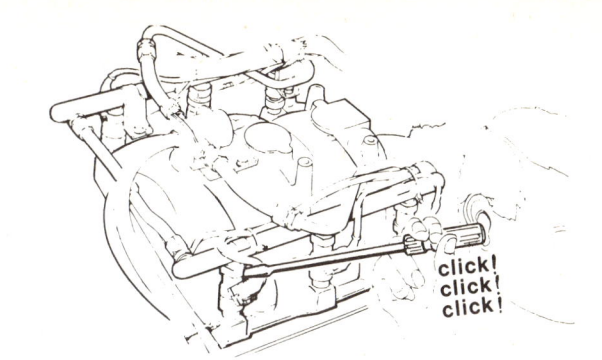

Fig. 13.21 Listening to the fuel injectors (Sec 7)

81 Open the bonnet and run the engine at idle. Using a long screwdriver or a stethoscope, listen to each injector in turn. A correctly functioning injector will make a regular clicking noise.
82 If an injector is found which does not click, swap over connectors with an injector which is functioning correctly. If the fault follows the connectors it is in the connectors or wiring; if the same injector fails to click, the fault is in the injector.
83 As a further check, disconnect one injector at a time. The idle will become rougher (though speed will not drop) when a good injector is disconnected. No change will be noticed when a faulty injector is disconnected.
84 A defective fuel injector must be renewed.

LH-Jetronic system – testing procedures

85 To carry out this procedure in its entirety, a fuel pressure gauge, a vacuum gauge and an electrical multi-meter will be required. The home mechanic may prefer to entrust the procedure to a Volvo dealer or other specialist.

Preliminary checks
86 Check that the earth straps on the inlet manifold are secure and making good contact.
87 Check that the fuel pump fuses are intact.
88 Check the security of the electrical connections on the air mass meter, the air control valve, the knock sensors, the coolant temperature sensor and the injectors.
89 Check that there are no air leaks on the induction side. If the problem relates to idling, check the throttle butterfly setting and the throttle position switch (paragraphs 132 to 138).

Fuel pressure checks
90 Connect a pressure gauge into the fuel rail supply line, using a T-piece so that the supply is uninterrupted.
91 The fuel pumps must now be made to run. On 1987 models this is done by removing the fuel pump relay and bridging terminals '30' and '87/2' in the relay socket. On later models this cannot be done because the relay is permanently fixed to the board (see Section 15) so it would seem that the pumps must be started by earthing terminals 17 and 21 on the control unit connector (paragraph 101). In either case the ignition must be on.
92 With the fuel pumps running, the pressure gauge should show the line pressure given in the Specifications. See Chapter 3, Section 45, paragraphs 9, 10 and 11, for further action.
93 Switch off the ignition, disconnect the pressure gauge and remake the original connections.

Electrical checks
94 Remove the trim from the driver's footwell. Make sure the ignition is off, then disconnect the multi-plug from the Jetronic control unit. Check that the control unit's earth connections (on the multiple tag nearby) are secure.
95 Remove the cover from the multi-plug. Remove the protective side strips from the plug to give access to the test holes in the sides of the connector. Only insert the meter probes through these side holes. Do not insert probes into the mating face.
96 Using a multi-meter set to measure resistance, check for continuity between earth and terminals 5, 11, 19 and 25. If there is no continuity, check the earthing points on the inlet manifold.
97 Temporarily disconnect the ignition control unit multi-plug. Check that there is now no continuity between earth and (Jetronic connector) terminal 12. Reconnect the ignition control unit.

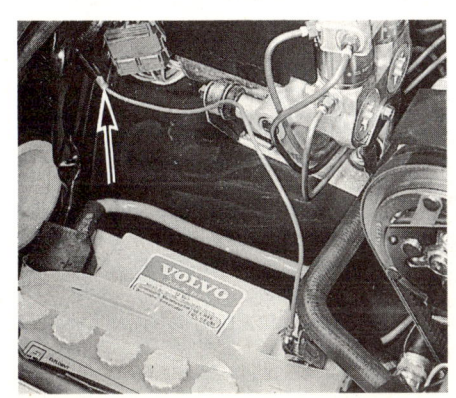

7.74 Air control valve connector (arrowed) connected to the battery earth terminal

7.75 Idle speed adjuster screw (arrowed)

7.77 CO adjustment screw is beneath the tamperproof cap (arrowed)

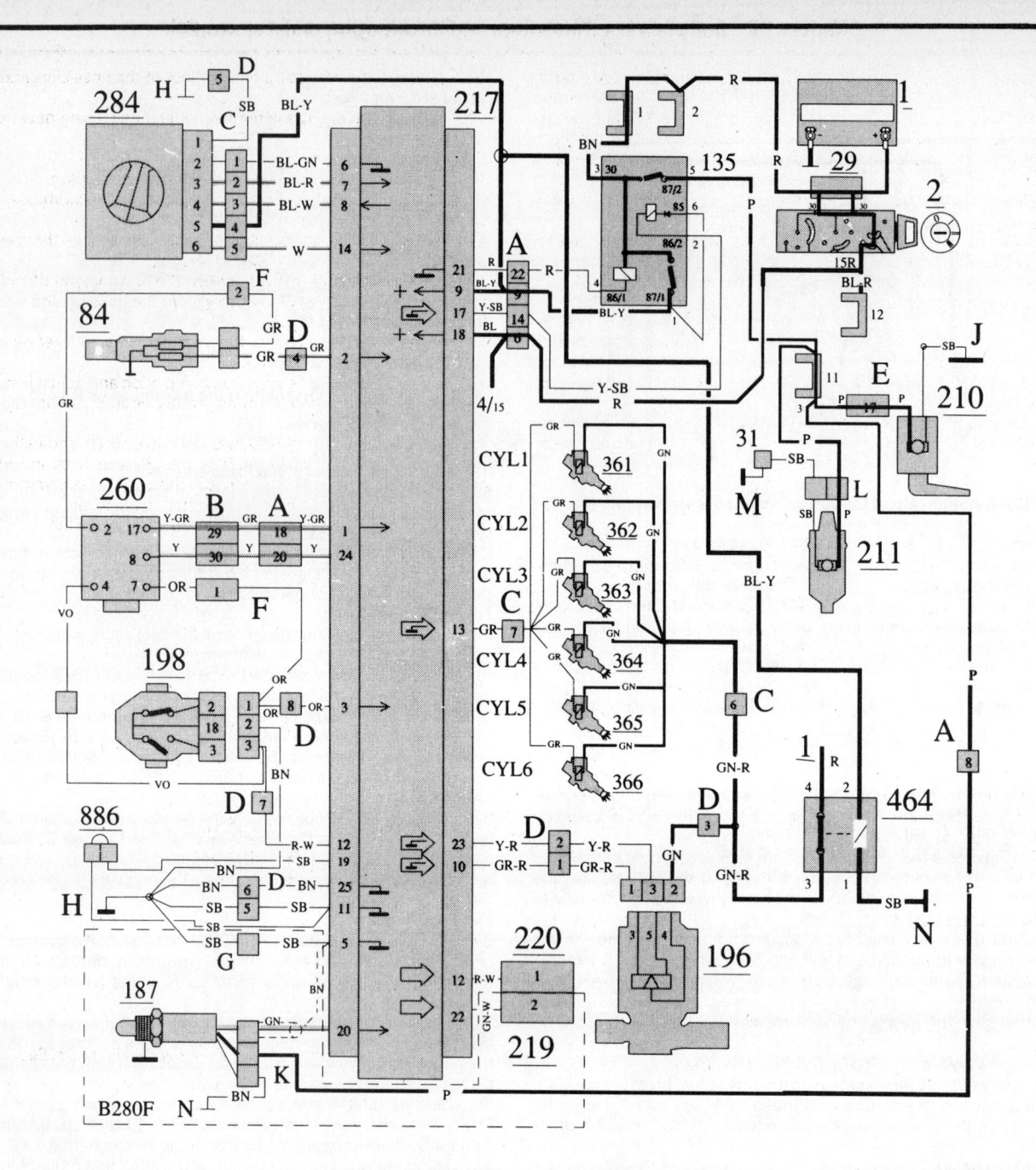

Fig. 13.22 LH-Jetronic system wiring diagram (Sec 7)

1 Battery	210 Tank pump	364 Injector	E Connector (LH A-post)
2 Ignition switch	211 Main fuel pump	365 Injector	F Connector (LH suspension turret)
4 Coil	217 Jetronic control unit	366 Injector	G Connector (RH suspension turret)
29 Positive test point	219 Lambda test point (not UK)	464 Radio interference suppression relay	H Earth point (inlet manifold)
31 Earth rail in central electrical unit	220 Air control valve diagnostic connector	886 Connector	J Earth point (luggage area)
84 Coolant temperature sensor	260 Ignition system control unit	A Connector (RH A-post)	K Connector (bulkhead)
135 Relay	284 Air mass meter	B Connector (LH A-post)	L Connector (central electrical unit)
187 Lambda sensor (not UK)	361 Injector	C Connector (RH suspension turret)	M Earth point (RH A-post)
196 Throttle switch	362 Injector	D Connector (RH suspension turret)	N Earth point (RH front wing)
198 Throttle switch	363 Injector		

Small numbers are terminal identification numbers
For colour code see main wiring diagrams

Chapter 13 Supplement: Revisions and information on later models

98 Read the voltage between terminal 1 and earth while operating the starter motor. A voltage should be registered: if not, the Jetronic control unit is not receiving an engine speed signal from the ignition control unit.
99 Measure the voltage at terminal 18. Battery voltage should be present when the ignition is on and when the starter motor is operating.
100 Connect the voltmeter between earth and terminal 9. Switch on the ignition. Connect terminal 21 to earth: the meter should show battery voltage, indicating that the main relay has closed.
101 With terminal 21 still earthed and the ignition on, connect terminal 17 to earth. Both fuel pumps should start, indicating that the fuel pump relay has closed. Disconnect the earth wire from terminal 17.
102 Under the bonnet, peel back the rubber cover from the air mass meter multi-plug. Measure the voltage there between earth and terminals 1 and 5. With terminal 21 earthed and the ignition on, battery voltage should be shown at both terminals. Switch off the ignition and remove the earth wire from terminal 21. Reinstate the air mass meter plug cover.
103 Back at the Jetronic multi-plug, measure the resistance between terminals 6 and 7. It should be between 2.5 and 4.0Ω (air mass meter filament resistance).
104 Similarly measure the resistance between terminals 6 and 14. It must be between 0 and 1000Ω, depending on the position of the CO adjustment potentiometer.
105 Measure the resistance between earth and terminal 2. The desired value is that given in the Specifications for the coolant temperature sensor.
106 Check the function of the throttle position switch by measuring the resistance between earth and terminal 3. With the throttle closed or wide open, resistance should be zero; at all intermediate positions it should be infinite. Adjustment of the throttle position switch is described in paragraph 137.
107 Reassemble the connector and reconnect it to the control unit (make sure the ignition is off before connecting).
108 If the engine would not run at the beginning of the electrical test procedure and still refuses to run at the end of it, correct values having been obtained in the tests, either the control unit is at fault or the fault lies in some other area. In either event it would be wise to consult a specialist before attempting tests by substitution.
109 Remake all electrical connections and refit any disturbed trim etc on completion.

LH-Jetronic system – component removal and refitting
110 Disconnect the battery negative lead.
Air mass meter
111 Disconnect the multi-plug from the air mass meter (photo).
112 Release the two clips which secure the air mass meter to the air cleaner, and the hose clip which secures it to the trunking. Remove the air mass meter.
113 Refit by reversing the removal operations. Check the CO level on completion if a new meter has been fitted.
Injectors
114 Disconnect the wiring plug from each injector. Free the wiring harness from the fuel rails, cutting cable ties where necessary.
115 Uncouple the fuel supply and return pipes from the rails, counterholding the unions when slackening them. Be prepared for fuel spillage.
116 Remove the two bolts which secure each fuel rail to the inlet manifold. Pull each rail upwards to release the injectors from the manifold and remove the rails complete with injectors (photo).
117 Individual injectors may now be removed from the rails by removing the securing clips and pulling them out (photos).
118 Refit by reversing the removal operations. Check that the injector O-rings are in good condition and renew them if necessary; smear them with petroleum jelly or silicone grease as an assembly lubricant.
Fuel pressure regulator
119 Disconnect the vacuum and fuel hoses from the regulator. Be prepared for fuel spillage. Counterhold the unions when slackening the nuts.
120 Remove the regulator from its bracket, being prepared for further spillage.
121 Refit by reversing the removal operations.
Air control valve
122 Disconnect the multi-plug from the valve.
123 Carefully pull the valve air hoses off the stubs on the inlet trunking and throttle housing.
124 Remove the valve and hoses. Slacken the hose clips and remove the hoses if required.
125 Refit by reversing the removal operations, using new hoses and clips if necessary.
Throttle housing
126 Disconnect the throttle position switch multi-plug.
127 Disconnect the air control valve hose, the vacuum hose and the air inlet trunking from the housing.
128 Prise out the spring clip and disconnect the throttle linkage balljoint from the operating lever (photo).
129 Make alignment marks between the housing and the inlet manifold. Remove the three nuts which secure the housing and withdraw it (photo). Recover the gasket.
130 When refitting, observe the alignment marks if refitting the old components. Use a new gasket.
131 Check the basic idle speed on completion and adjust if necessary. If basic idle speed cannot be brought low enough, adjust the throttle butterfly setting as follows.
132 Slacken the butterfly adjuster locknut. Undo the adjuster until the throttle is completely closed. Screw the adjuster back in until it just contacts the operating lever, then from this position screw it in a further quarter turn. Hold the adjuster and tighten the locknut.
133 Check the operation of the throttle position switch.
Throttle position switch
134 Disconnect the multi-plug from the switch.

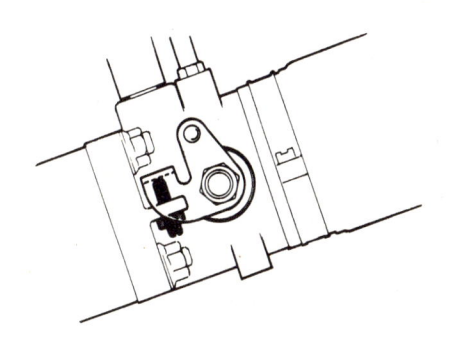

Fig. 13.23 Throttle adjuster screw and locknut (Sec 7)

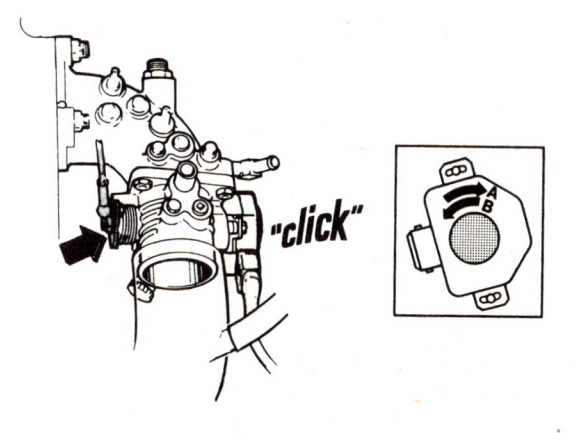

Fig. 13.24 Adjusting the throttle position switch (Sec 7)

A Turn clockwise
B Turn anti-clockwise

Chapter 13 Supplement: Revisions and information on later models

7.111 Disconnecting the air mass meter multi-plug

7.116 Removing a fuel rail with injectors

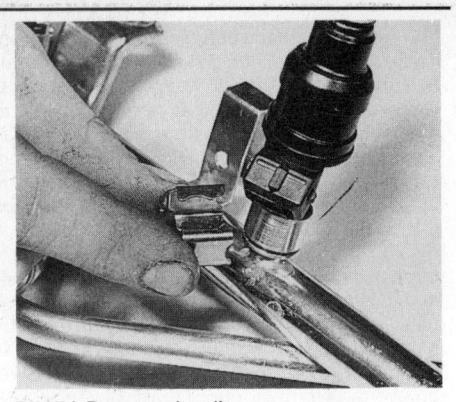

7.117A Remove the clip ...

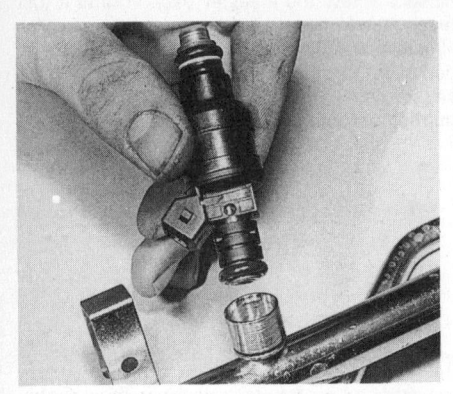

7.117B ... and pull out the injector

7.128A Prise out the spring clip ...

7.128B ... and disconnect the throttle linkage balljoint

7.129 Removing the throttle housing

7.141 Jetronic control unit

135 Make alignment marks between the switch and the throttle housing. Remove the two Allen screws which secure the switch and withdraw the switch.
136 Refit by reversing the removal operations. Check that a click is heard from the switch as soon as the throttle is opened; if not, adjust as follows.
137 Slacken the switch screws. Turn the switch clockwise (viewed from behind) within its limits of travel, then turn it slowly anti-clockwise until a click is heard. Hold the switch in this position and tighten the screws.
138 Check again that the click is heard when the throttle is opened; repeat the adjustment if necessary.

Coolant temperature sensor
139 Proceed as described for the temperature gauge sender in Chapter 2, Section 14.

Control unit
140 Remove the trim from the driver's footwell.
141 Disconnect the multi-plug from the control unit (photo). Unbolt the control unit and remove it from its bracket.
142 Refit by reversing the removal operations.

Water-cooled turbocharger (1987 on) – general
143 From 1987 model year, a water-cooled turbocharger is fitted. Water cooling keeps the operating temperature of the turbo bearings lower than previously. Water continues to circulate by convection after the engine has stopped, so cooling the turbocharger if it is hot after a long run. A service kit is available for installing the water-cooled unit to earlier models.

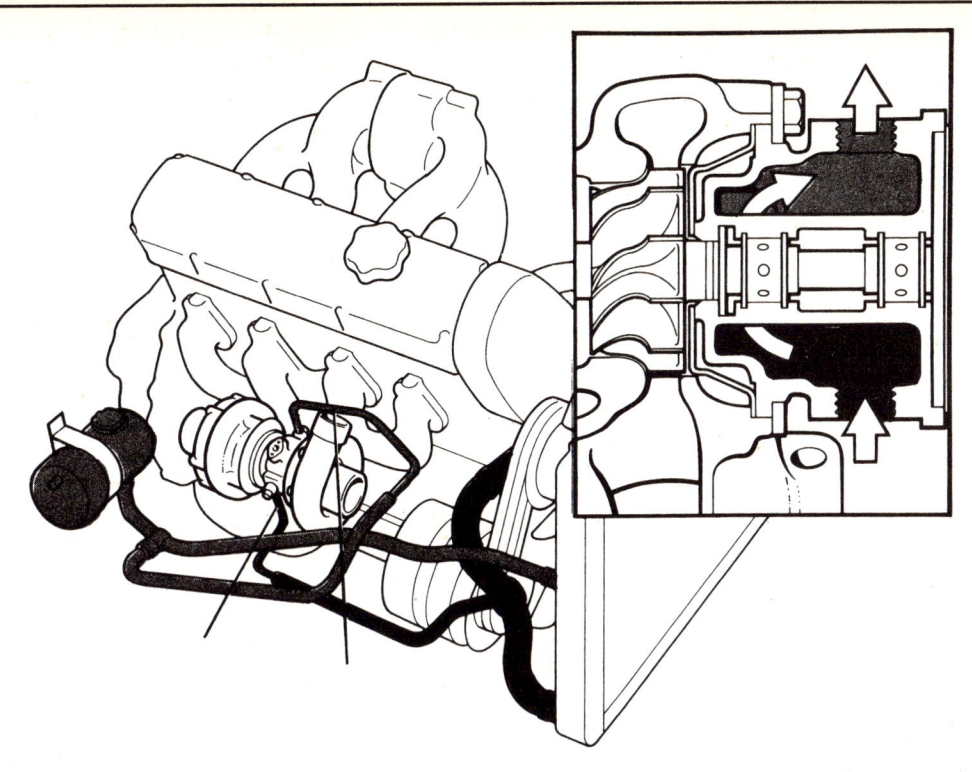

Chapter 13 Supplement: Revisions and information on later models

Fig. 13.25 Water-cooled turbo pipework. Inset shows circulation round bearings (Sec 7)

144 Removal and refitting are basically still as described in Chapter 3, Section 39, but additionally the cooling system must be drained before removal and the water pipes disconnected from the turbo. Reconnect the pipes when refitting and refill the cooling system on completion.

8 Ignition system

Description (B280)

1 The ignition system fitted to B280 engines is designated EZ 115 K. It is a development of the EZ-K system fitted to B230E/K engines and described in Chapter 4. The main difference in the new system is that ignition timing is now determined by a flywheel sensor, and so cannot be adjusted.

2 Minor changes have also been made in order to accommodate features peculiar to the V6. There are two knock sensors, one for each cylinder bank. Since the flywheel sensor cannot indicate which bank is firing when a knock occurs, an inductive transducer is fitted to No 1 spark plug lead to supply this information to the control unit (photo).

3 The temperature sensor and throttle position switch are shared with the LH-Jetronic system. The control units of the two systems also interact to some extent: the anti-knock function operates both by retarding ignition timing and by richening the fuel-air mixture. The Jetronic control unit receives engine speed information from the ignition system control unit and supplies load information to it.

Distributor (B280) – removal and refitting

4 Disconnect the battery negative lead.
5 Unclip the shield from the top of the distributor (photo).
6 Remove the 3 bolts which secure the distributor cover and cap. Remove the cover and withdraw the cap (photo).
7 Undo the 3 Allen screws which secure the rotor arm. Lift off the arm and screws. The screws are captive in the arm (photo).
8 Lift out the flash shield with its O-ring (photo).
9 Remove the 10mm Allen screw which secures the distributor shaft. Remove the shaft if it is loose; if not, leave it for now (photo).
10 Remove the two bolts which secure the distributor body. If the shaft is still in place, pull the body forwards against the shaft and tap the body with a mallet to release the shaft. Remove the body and shaft (photo).

11 Refit by reversing the removal operations, engaging the hole in the distributor shaft with the peg on the drive (photo). Use a new flash shield O-ring if necessary.

Control unit (B280) – removal and refitting

12 Disconnect the battery negative lead.
13 Remove the trim from around the driver's footwell.
14 Identify the control unit. It is distinguished by being nearly square and black, as opposed to the Jetronic control unit which is silver (and labelled 'Jetronic'), or the ABS control unit (1988 on) which is silver and longer than it is deep.
15 Free the control unit from its mountings and disconnect the multi-plug from it (photo).
16 Refit by reversing the removal operations.

Power stage (B280) – removal and refitting

17 Disconnect the battery negative lead.
18 Remove the air cleaner (Section 7).
19 Disconnect the multi-plug. Remove the two screws which secure the power stage to its heat sink and remove it (photo).
20 Refit by reversing the removal operations. Ensure that there is good contact between the power stage and the heat sink; use a thermally conductive paste if available.

Knock sensors – general

21 Three types of knock sensor have been fitted – see Fig. 13.27.
22 Different tightening torques apply to the different sensors – see Specifications.
23 To remove a sensor from the B280 engines, the following preliminary work is necessary:

   (a) Remove the inlet manifold (Section 7)
   (b) Drain the cooling system and remove the Y-pipe (Chapter 2, Section 3, and Section 6 of this Chapter) (photo)

Flywheel sensor (B280) – removal and refitting

24 Disconnect the battery earth lead.
25 Disconnect the sensor wiring plug. Remove the Allen screw and withdraw the sensor from its housing (photo).
26 Refit by reversing the removal operations.

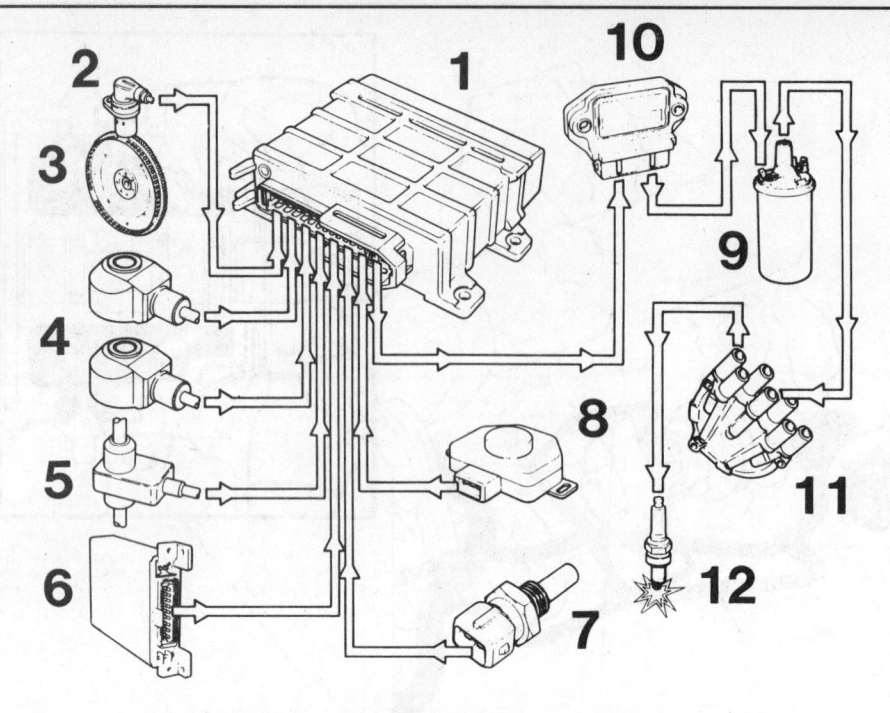

352 Chapter 13 Supplement: Revisions and information on later models

Fig. 13.26 Main components of EZ 115 K ignition system (Sec 8)

1 Control unit
2 Flywheel sensor
3 Flywheel
4 Knock sensor
5 No 1 cylinder inductive transducer
6 Jetronic control unit
7 Temperature sensor
8 Throttle switch
9 Coil
10 Power stage
11 Distributor
12 Spark plug

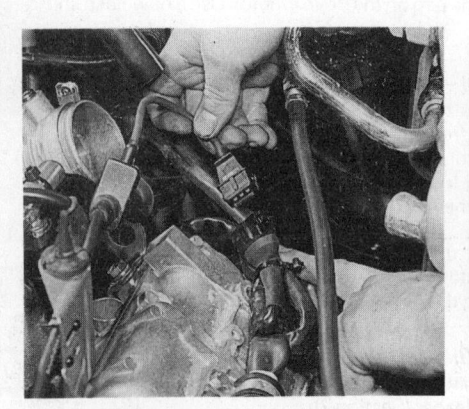

8.2 Inductive transducer connector

8.5 Unclip the shield from the distributor

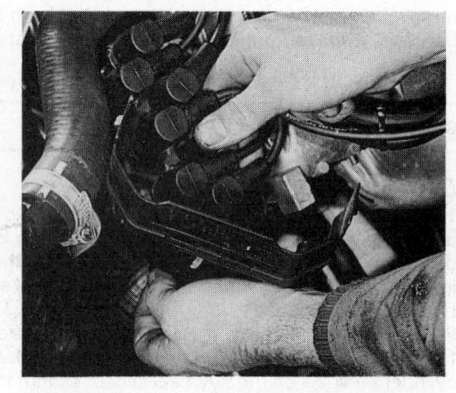

8.6 Removing the distributor cover and cap

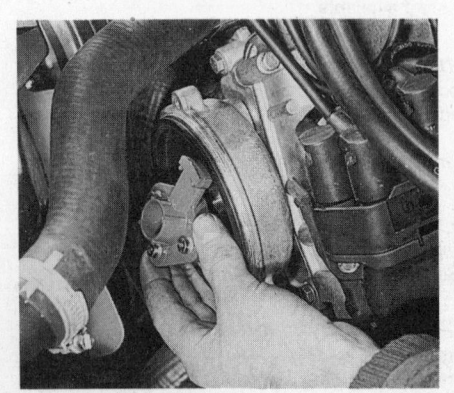

8.7 Removing the rotor arm

8.8 Removing the flash shield

8.9 Undoing the distributor shaft screw

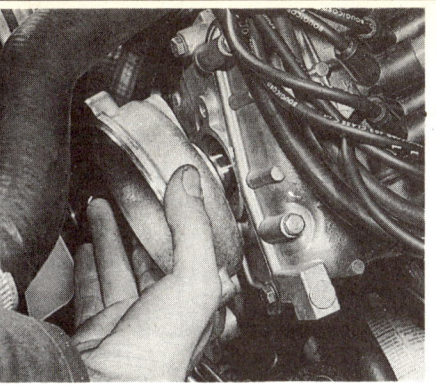

Chapter 13 Supplement: Revisions and information on later models

8.10 Removing the distributor body

8.11 Refitting the distributor shaft. Note peg and hole (arrowed)

8.15 The ignition system control unit

8.19 Ignition system power stage

8.23 Knock sensor on the B280 engine – removal is blocked by the Y-pipe

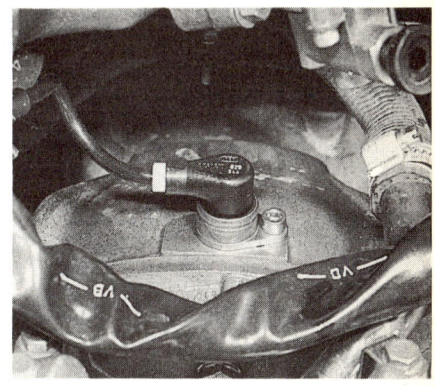

8.25 Flywheel sensor on the B280 engine

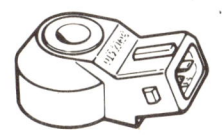

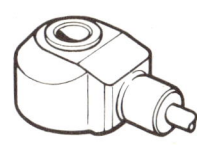

I B 23/200/230   II   B 280

Fig. 13.27 Three types of knock sensor (Sec 8)

9 Manual gearbox and overdrive

Mainshaft gear bearings (manual gearbox, 1988 on)
1 All mainshaft gears are now supported on needle-roller bearings. Some pre-1988 models used plain bushes. Overhaul procedures are not affected.

5th gear and synchro unit (M47 II) – dismantling and reassembly
2 When dismantling the M47 II gearbox, 5th gear and synchro unit are pulled off the layshaft together. The job of separating them, if necessary, should be entrusted to a Volvo dealer or gearbox specialist, since special tools are required to slacken and tighten the 5th gear synchro nut.

Overdrive (J/P hybrid) – general
3 From 1987 the type J overdrive is superseded by a type known as J/P hybrid. This change coincides with the use of a larger diameter mainshaft on the M46 gearbox.
4 The practical consequence of this change is to affect interchangeability of gearboxes and overdrive units. Type J overdrive units can only be fitted to early type M46 gearboxes (with the smaller mainshaft). Hybrid overdrive units can only be fitted to later type gearboxes.

10 Propeller shaft

Removal and refitting (models with independent rear suspension)
1 Proceed as in Chapter 7, Section 4, but note that a new shaft may have a balance mark on it, in the form of a pink dot near the final drive flange end. This dot must be aligned with a similar mark on the final drive flange.

Removal and refitting (models with saddle fuel tank)
2 Remove the bracing strap from the bottom of the tank before attempting to remove the shaft rear section.

11 Final drive and driveshafts (760 Saloon, 1988 on)

Description
1 In conjunction with the independent rear suspension (see Section 13), the old solid rear axle has been replaced by a separate final drive unit and twin open driveshafts.
2 The final drive unit contains a differential unit, a crownwheel and a pinion. Drive from the propeller shaft is transmitted by a flanged coupling, as before. Two side flanges transmit the drive to the driveshafts. A sensor on the rear of the final drive casing provides speed information for the speedometer and for the ABS.
3 The driveshafts each have two constant velocity joints to permit movement with the suspension. They transmit drive to stub shafts in the rear hub carriers by means of splines.

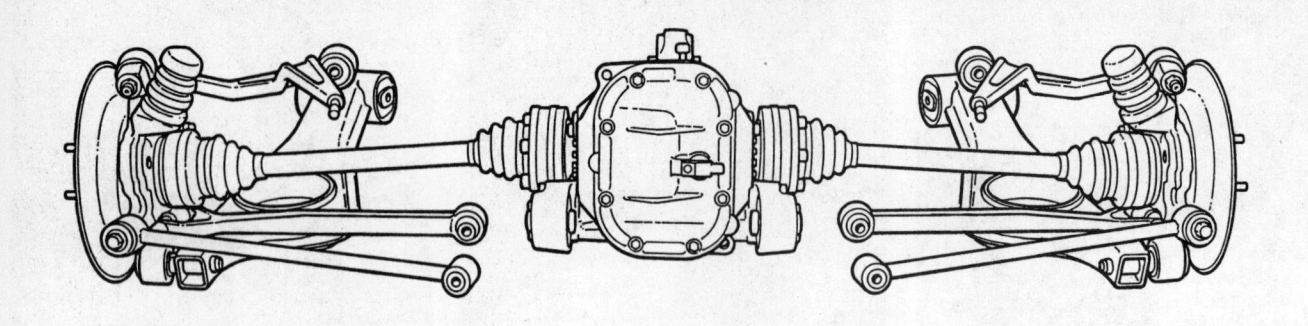

Fig. 13.28 Final drive and driveshafts fitted with independent rear suspension (Sec 11)

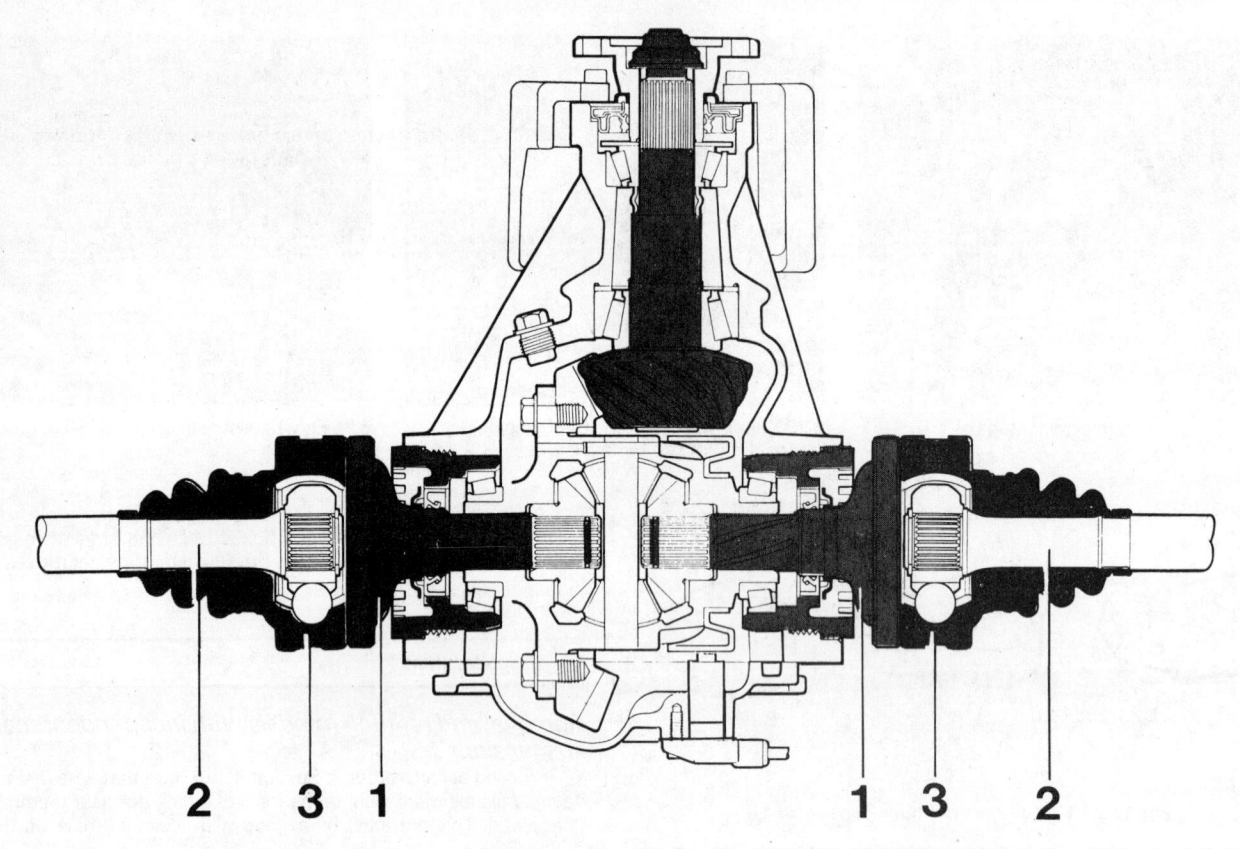

Fig. 13.29 Sectional view of final drive unit and inboard ends of driveshafts (Sec 11)

1 Output shafts   2 Driveshafts   3 Constant velocity joints

Oil level – checking
4 Proceed as in Chapter 8, Section 3, but note that the filler/level plug is on the front of the final drive casing (photo).

Pinion oil seal – renewal
5 Proceed as in Chapter 8, Section 4, following instructions for collapsible spacer type axles. The relevant tightening and turning torques are given in the Specifications at the beginning of this Chapter.

Side oil seals – renewal
6 This is not a DIY job, since it involves disturbing the differential carrier side bearings. Remove the final drive unit and have the seals renewed by a Volvo dealer.

Final drive unit – removal and refitting
7 Remove the rear axle member lower section (Section 13).
8 Disconnect the driveshafts and the propeller shaft from the final drive unit. Mark the propeller shaft alignment if necessary.
9 Take the weight of the final drive unit on a jack. Remove the three bolts which hold the unit to the rear axle member upper section.
10 Lower the final drive unit. Disconnect or remove the speedometer/ABS sender as it becomes accessible.
11 Remove the final drive unit from under the vehicle.
12 Refit by reversing the removal operations. Check the oil level in the final drive on completion.

Driveshaft – removal and refitting
13 Remove the wheel trim from the rear wheel on the side concerned. Apply the handbrake, engage a gear (or 'P') and chock the wheels.
14 Slacken the driveshaft nut. This nut is extremely tight. A ¾-inch drive 36 mm socket will be required. Remove the nut.
15 Slacken the wheel nuts, raise and support the rear of the vehicle and remove the rear wheel. Make sure that the vehicle supports do not

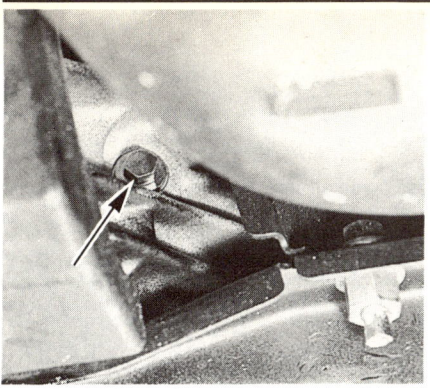

Chapter 13 Supplement: Revisions and information on later models

11.4 Final drive oil filler/level plug (arrowed)

11.17 Removing a driveshaft flange screw

11.20 Angle-tightening the driveshaft nut

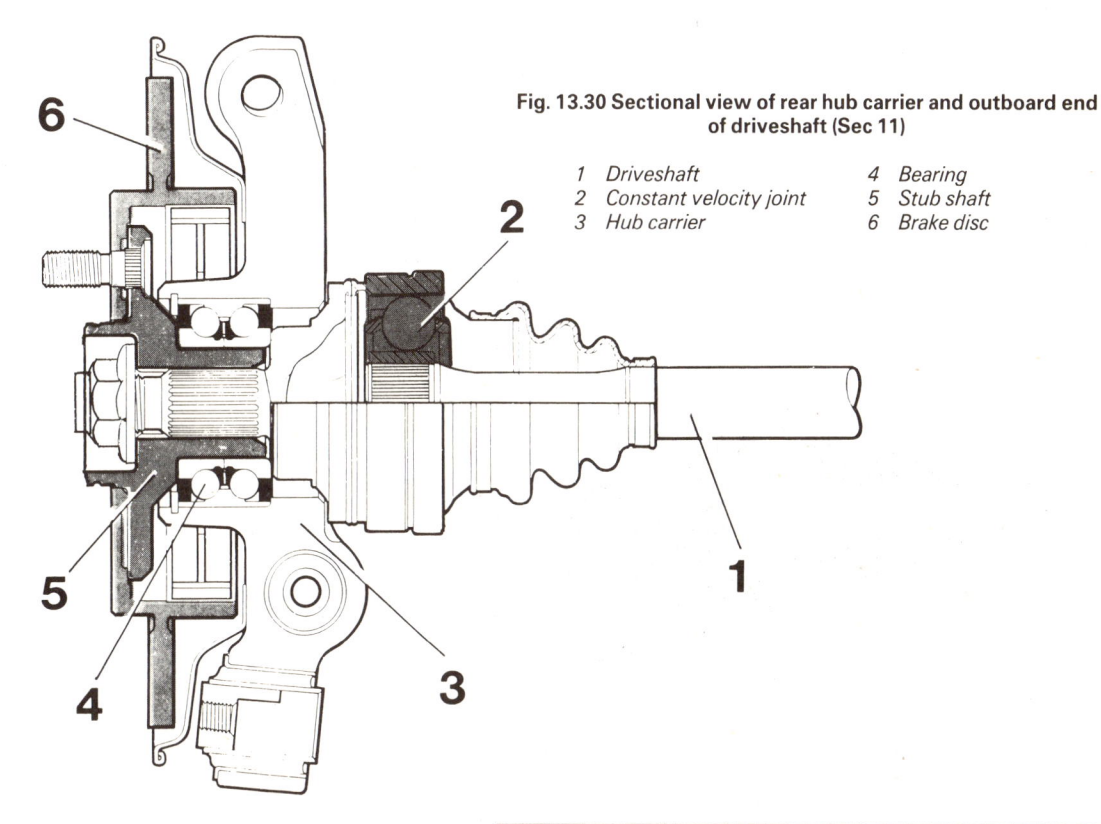

Fig. 13.30 Sectional view of rear hub carrier and outboard end of driveshaft (Sec 11)

1 Driveshaft
2 Constant velocity joint
3 Hub carrier
4 Bearing
5 Stub shaft
6 Brake disc

obstruct the driveshaft or the rear axle member lower section.
16 Remove the eight nuts and bolts which secure the axle member upper and lower sections together. Note the handbrake cable guide under one of the rear bolts. Pull the axle member lower section downwards as far as it will comfortably go without straining the linkages.
17 Undo and remove the six Allen screws which secure the driveshaft inboard flange. Recover the three double washers (photo). An assistant should be available to apply and release the handbrake, disengage gear etc.
18 Lower the driveshaft and withdraw the outboard end from the hub. Tap it out if it is tight. Withdraw the driveshaft from under the vehicle.
19 Overhaul of a worn driveshaft is unlikely to be possible. Bellows kits may be available; if so, follow the instructions in the kit.
20 Refit by reversing the removal operations, noting the following points:

(a) Use new Allen screws and a new driveshaft nut
(b) Align the axle members with a couple of bolts or drifts when reassembling – see Section 13
(c) Do not attempt final tightening of the driveshaft nut until the weight of the car is back on its wheels (photo)

12 Braking system

Brake fluid level warning switch (later models)

1 Some later models are fitted with a brake fluid level warning switch instead of a pressure differential valve. The switch is float-operated and is incorporated in the master cylinder reservoir cap.
2 Operation of the switch may be checked by removing the cap with the ignition on. As the float leaves the reservoir the warning lamp on the instrument cluster should light.

Brake pads (1989 on) – wear measurement

3 It is possible to measure pad wear on the above models without removing the wheels. At the front, a measurement is taken between a flat on the caliper body and another flat on the bracket (photo). If the measurement exceeds 35 mm (1.38 in) the pads are excessively worn and must be renewed.
4 At the rear this method only applies to cars with independent rear suspension. The measurement is taken between the flat on the caliper body and the surface of the guard plate between the hub carrier and the caliper bracket. The measurement must not exceed 25 mm (0.98 in).
5 Whilst this method of assessing pad wear is convenient, it is no

substitute for a thorough inspection of the pads themselves, which also gives the opportunity to examine the discs, calipers and flexible hoses.

Rear brake pads (models with independent rear suspension) – inspection and renewal

6 As with the earlier type of rear caliper, the pads may be inspected through a window in the caliper body after removal of the rear wheel. If any one pad is worn down to the specified minimum, all four must be renewed as follows.

7 With the wheel removed, press the caliper piston back into its bore by levering between the back of the outboard pad and the caliper body. The brake fluid level in the master cylinder will rise when this is done, so remove a little fluid if necessary to prevent overflow.

8 Remove the lower guide pin bolt, counterholding the pin with an open-ended spanner. Slacken, but do not remove, the upper guide pin bolt (photo).

9 Pivot the caliper body upwards and remove the brake pads (photo). Do not press the brake pedal whilst the pads are removed.

10 Inspect the caliper for fluid leaks or other signs of malfunction. Inspect the brake disc (Chapter 9, Section 14). Also check that the rubber bellows on the guide pins are in good condition.

11 Fit the new pads, friction surface towards the disc, and swing the caliper body down over them.

12 Refit the lower guide pin bolt. Tighten both guide pin bolts to the specified torque, counterholding the guide pins if necessary.

13 Depress the brake pedal a few times to bring the new pads up to the disc.

14 Repeat the operations on the other rear wheel.

15 Refit the wheels, lower the vehicle and tighten the wheel nuts.

16 Check the brake fluid level and top up if necessary.

17 Remember that new brake pads need to be bedded in before full braking efficiency can be obtained.

Front brake caliper (all Bendix/DBA) – overhaul

18 If problems are experiences with the guide pins sticking on Bendix/DBA calipers (indicated by excessive wear, brake drag and/or vibrations), a repair kit can be obtained which contains improved components. Commonly it is only the lower pin which sticks and a repair kit can be obtained for this alone.

19 Calipers produced from the end of 1986 already contain the improved components.

Rear brake caliper (models with independent rear suspension) – removal and refitting

20 Slacken the rear wheel nuts, raise and support the vehicle and remove the rear wheel.

21 If the caliper is to be removed completely, slacken the hydraulic hose union half a turn.

22 Remove the two bolts which secure the caliper bracket to the rear hub carrier. Obtain new bolts for refitting.

23 Withdraw the caliper, bracket and pads from the disc. Unscrew the caliper from the hydraulic hose if wished, being prepared for fluid spillage. If the caliper is not to be removed completely, tie it up so that the hose is not strained.

24 Refit by reversing the removal operations, using new bolts to secure the caliper bracket and tightening them to the specified torque.

25 If the hydraulic hose was disconnected, the hydraulic system must be bled on completion.

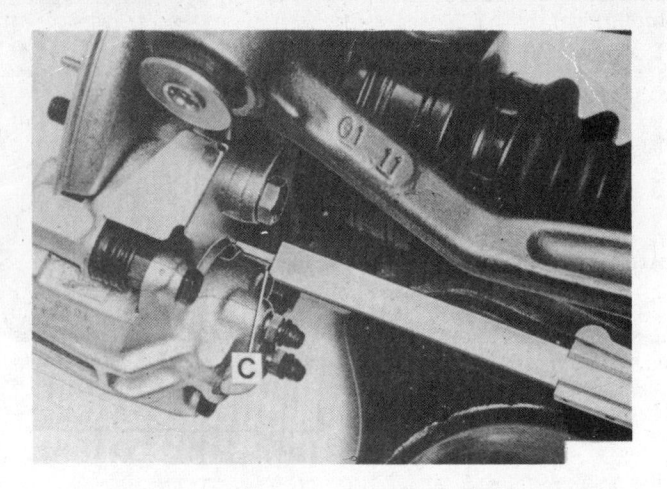

Fig. 13.31 Rear brake pad wear measurement points (Sec 12)

C Flat on caliper body

Rear brake caliper (models with independent rear suspension) – overhaul

26 This procedure is very similar to that described in Chapter 9, Section 11. The main difference is that the rear caliper has only one piston.

Front brake disc (pre-1988 models) – renewal

27 The original type of brake disc with integral hub will eventually be unavailable. A service kit will be available, consisting of separate disc and hub units. After initial fitting and adjustment, the disc can subsequently be removed and refitted independently of the hub.

Front brake disc (1988 on) – removal and refitting

28 This procedure is covered in Section 13, as part of the front hub bearing renewal procedure. There is no need to remove the hub if it is only the disc which requires attention.

29 Note that versions may be encountered with a separate disc and hub, but with adjustable (taper-roller) bearings. The presence of a separate disc is not an indication that non-adjustable bearings are fitted.

12.3 Front brake pad wear measurement points

12.8 Slackening the upper guide pin bolt

12.9 Pivot the caliper body upwards and remove the pads

Chapter 13 Supplement: Revisions and information on later models

Handbrake (models with independent rear suspension) – adjustment

30 The procedure is basically as described in Chapter 9, Section 7. Note however that it is necessary to remove the rear console for access to the cable adjuster.
31 Because effective cable length may vary with suspension movement, final adjustment should be carried out with all four wheels on the ground.

Handbrake cables (models with independent rear suspension) – removal and refitting

32 On models with independent rear suspension, three handbrake cables are used. A long cable runs from the handbrake lever to the rear suspension members. Two short cables, one on each side, transmit the movement of the long cable to the brake shoe assemblies.

Long cable

33 Remove the rear console, slacken the cable adjustment right off and disconnect the cable from the handbrake lever.
34 Remove the rear seat cushion and pull back the carpet. Free the cable grommet from the floor.
35 Raise and support the vehicle. Slacken, but do not remove, the fuel tank retaining bolts (see Section 7) so that the tank is lowered by 10 to 15 mm (say half an inch).
36 Unbolt the clamp which secures the cable to the underside of the floor.
37 Disconnect the long cable from its junctions with the small cables (photo). Remove the long cable, drawing it over the petrol tank.
38 Refit by reversing the removal operations. Note that the correct position of the floor clamp on the new cable is shown by a paint mark.
39 Adjust the cable before refitting the rear console.

Short cable

40 Remove the rear console and slacken the cable adjuster.
41 Remove the handbrake shoes on the side concerned.
42 Disconnect the short cable at its junction with the long cable. (It may be easier if both short cables are disconnected).
43 Withdraw the cable clip from the brake backplate (photo).
44 Withdraw the operating mechanism from the backplate and disconnect the cable inner from it, noting which way the arrow on the mechanism faces (photos). Remove the cable.
45 Refit by reversing the removal operations. Adjust the handbrake before refitting the console.

Handbrake shoes (models with independent rear suspension) – removal and refitting

46 The procedure is basically as described in Chapter 9, Section 26. Note however that a heavy horseshoe spring is now fitted instead of a coil return spring. The open ends of the spring engage in holes in the shoes, and the closed end is secured by a clip (photos). The spring is removed and refitted by levering the open ends out of or into their holes using a large screwdriver.

Anti-lock braking system (ABS) – modifications (1988 on)

47 From 1988 model year, the ABS hydraulic modulator is located under the bonnet, on the right-hand inner wing, instead of in the boot. The control unit is relocated from the boot to the area of the driver's footwell.
48 The introduction of a new front hub bearing (Section 13) means that the ABS front pulse wheels are located on the inside of the hubs rather than on the discs themselves. It does not appear that the hub and pulse wheel can be separated.
49 On models with independent rear suspension, the procedure for gaining access to the rear wheel sensor has changed.
50 For 1991, thicker discs and single-piston calipers are fitted. The procedures in Chapters 9 and 13 are not affected.

ABS hydraulic modulator (1988 on) – removal and refitting

51 Disconnect the battery negative lead. Remove the securing screw and lift the cover off the hydraulic modulator (photo).
52 Remove both relays from the modulator. Undo the cable clamp screw and disconnect the multi-plug. Also disconnect the earth strap (photos).

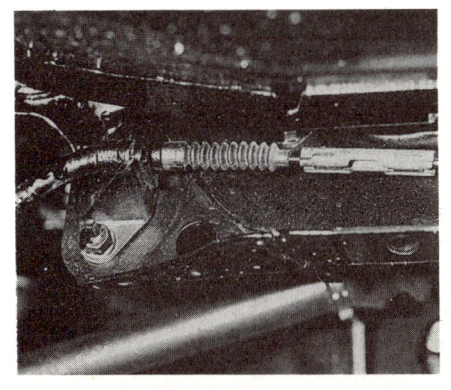

12.37 Handbrake long cable attachment to left-hand short cable

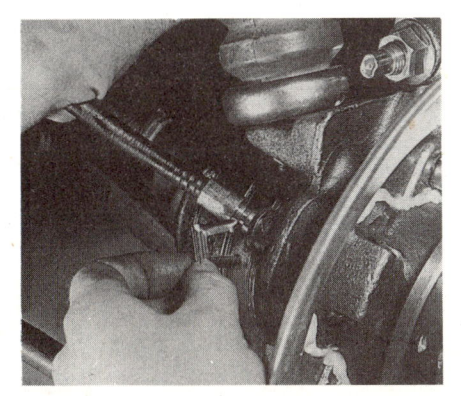

12.43 Removing the cable clip from the backplate

12.44A Operating mechanism on backplate – note arrow

12.44B Operating mechanism freed – cable clevis arrowed

12.46A Handbrake shoe spring being removed

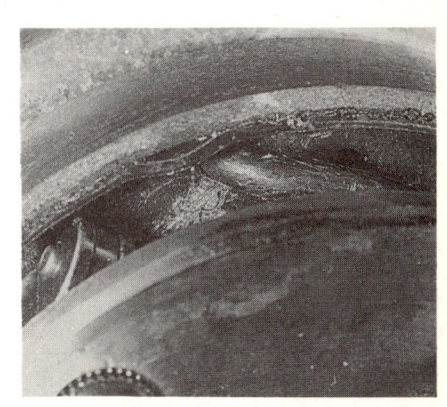

12.46B Detail showing spring engagement in shoe

12.46C Spring clip fits over closed end of spring

12.51 Removing the hydraulic modulator cover

12.52A Removing a relay from the modulator

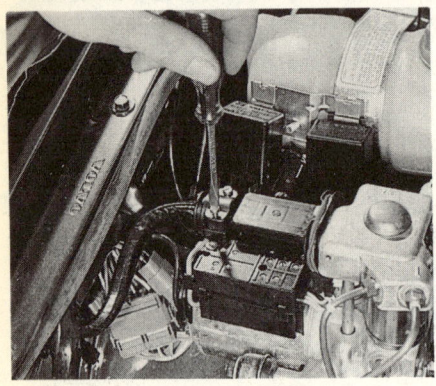

12.52B Undo the cable clamp screws ...

12.52C ... and disconnect the multi-plug

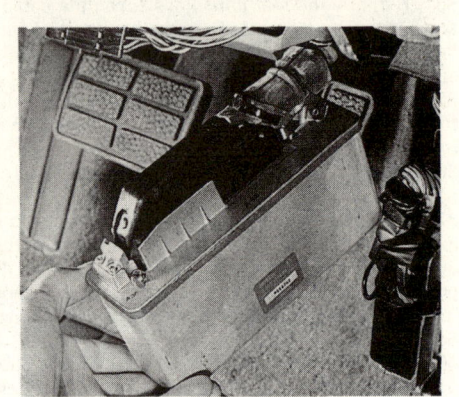

12.59 ABS control unit

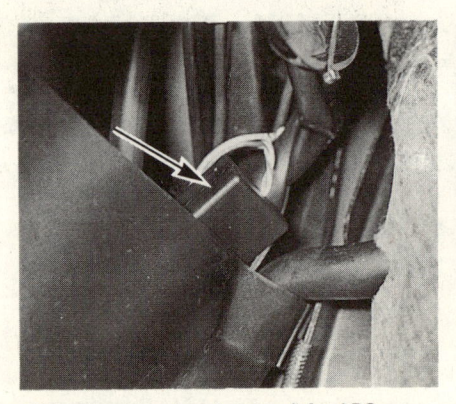

12.62 Connector box (arrowed) for ABS rear sensor wiring

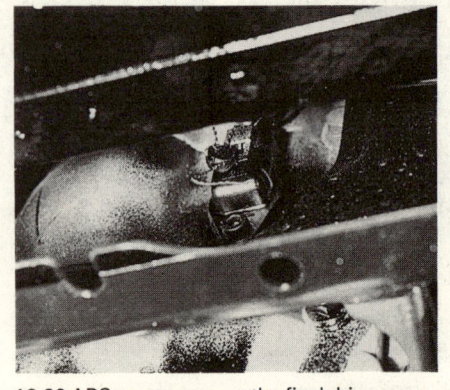

12.66 ABS rear sensor on the final drive rear cover

53   Inspect the hydraulic pipes for identification marks. Mark the pipes if necessary, following the identification letters on the modulator:
V – Inlet, front (vorn) circuit
H – Inlet, rear (hintern) circuit
l – outlet, left front
r – outlet, right front
h – outlet, rear

54   Place rags under the modulator to catch spilt fluid. Disconnect the hydraulic unions.
55   Remove the two nuts and one bolt which secure the modulator. Remove the modulator, being careful not to spill hydraulic fluid on the paintwork.
56   Refit by reversing the removal operation. Bleed the complete hydraulic system on completion.

ABS control unit (1988 on) – removal and refitting

57   Disconnect the battery negative lead.
58   Remove the trim from below the instrument panel and around the right-hand side of the driver's footwell.
59   Identify the control unit, free it from its strap and withdraw it (photo). Disconnect the wiring plug from the unit and remove it.
60   Refit by reversing the removal operations.

ABS rear sensor (models with independent rear suspension) – removal and refitting

61   As on earlier models, the ABS sensor doubles as the speedometer sender. Access is as follows.
62   Inside the boot, remove the spare wheel and fold back the carpet. Remove the protective cover from around the fuel filler pipe to expose the ABS sensor connector box (photo).
63   Open the box, break the seal and disconnect the connector. Free the ABS sensor cable where it passes through a grommet in the boot floor.

Chapter 13 Supplement: Revisions and information on later models

64 Raise and support the rear of the vehicle. Support the rear axle lower member with a jack.
65 Remove the four bolts which secure the rear axle upper member to the body. Carefully lower the axle members/final drive assembly until the ABS sensor is accessible. Be careful that the driveshafts do not foul the fuel tank.
66 Break the seal on the ABS sensor, undo the Allen screw and withdraw the sensor (photo). Unclip the cable, noting how it is routed. Access may be improved by disconnecting the handbrake right-hand short cable.
67 Refit by reversing the removal procedure, using new bolts to secure the axle upper member and tightening them as specified.

Hydraulic system – bleeding (models with ABS, 1988 on)
68 On all 1988 and later models with ABS, the brake hydraulic system should be bled in the following order:

Rear wheels (in either order)
Front wheels (in either order, but commence with the upper bleed screw on each)

69 On models with ABS and independent rear suspension, it will be noticed that each rear caliper has two bleed screws. Only the upper screw should be used for bleeding.

Fig. 13.32 Depressing the lock securing catch with a thin punch (arrowed) (Sec 13)

13 Steering and suspension

Steering column (760, 1988 on) – removal and refitting
1 Disconnect the battery negative lead.
2 Remove the steering wheel, the trim below the steering column and the steering column switches.
3 Remove the horn contact ring. Remove the four screws which secure the column upper mounting plate to the dashboard.
4 Disconnect the ignition/starter switch multi-plug. Insert the ignition key and turn it to position I. Depress the lock securing catch using a thin punch through the hole in the lock housing, and pull the lock barrel out with the key.
5 Under the bonnet, remove the lockpins from the intermediate steering shaft joint bolts. Remove the bolt from the top joint and slacken the bolt on the lower joint.
6 Separate the intermediate shaft from the base of the column by sliding the shaft towards the steering gear.
7 Under the dashboard, remove the four bolts which secure the column bearing. Remove the column.
8 Refit by reversing the removal operations.

Fig. 13.33 Four bolts (arrowed) securing the steering column bearing (Sec 13)

Front wheel bearing repair kits – all models with adjustable bearings
9 The instruction in some front wheel bearing repair kits tell the reader to complete adjustment by slackening the nut if necessary to align a split pin hole. This is incorrect. The instructions in Chapter 10, Section 4, paragraphs 7 to 9, should be followed.

Front wheel bearings (1988 on) – description
10 During 1988 model year, changes were made to the design of the front wheel bearings. The first change was to discard the integral disc/hub assembly in favour of separate disc and hub units. Adjustable taper-roller bearings were still used.
11 The second change was to discard taper-roller bearings in favour of angular contact ball-bearings. The new type of bearing can be recognised by the absence of a split pin to secure the hub nut. The nut is tightened to a much higher torque than previously so the split pin is no longer needed.
12 Periodic adjustment of the new type of bearing is not required. If the bearings are slack or noisy they must be renewed.

Front wheel bearings (ball type) – renewal
13 The hub contains the bearing outer races. If the bearings are worn the complete hub must be renewed.
14 Slacken the wheel nuts. Raise and support the front of the vehicle and remove the wheel.
15 Remove the brake pads, brake caliper (without disconnecting it) and the caliper bracket. See Chapter 9, Section 10.
16 Remove the spigot pin which holds the disc to the hub. Lift off the disc (photos).
17 Prise or chisel off the bearing dust cap. Obtain a new cap for reassembly.
18 Undo the hub nut now exposed. This nut is very tight so make sure the car is well supported. Use a new nut on reassembly.
19 Pull the hub and bearings off the stub axle (photo). The inner race of the inboard bearing may remain on the stub axle; if so pull it off.
20 Clean the stub axle and lightly grease it before fitting the new hub and bearings.
21 Fit the nut and tighten it to the stage one specified torque. Tighten the nut further through the angle specified for stage two (photo).
22 Refit the remaining components in the reverse order to removal.

Aluminium front control arm (740, 1987 models)
23 Some 1987 740 models are fitted with aluminium front control arms instead of steel ones. The two types of arm are not interchangeable. If renewal is necessary, the same type of arm must be fitted as was already present.

Front suspension strut upper mounting (1985 on)
24 Models from 1985 on incorporate an axial ball-bearing in the front suspension strut upper mounting. If this bearing has to be renewed, note that it must be fitted with the yellow side up, and the grey or orange side down. Incorrect fitting will result in the bearing being too tightly clamped, with consequent problems of stiff steering and noise.

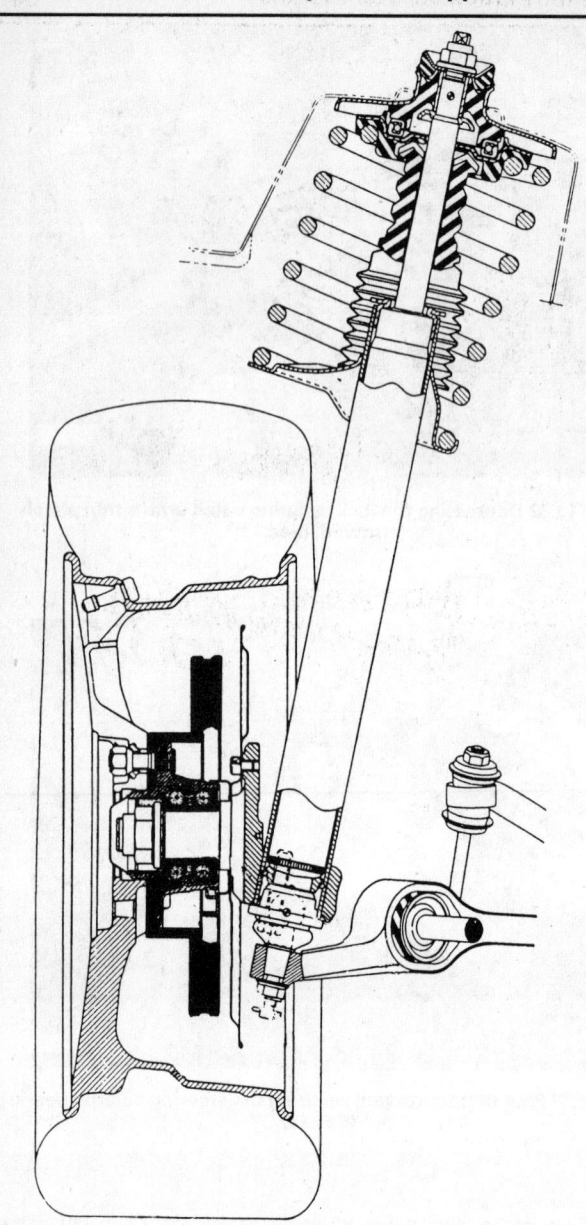

360   Chapter 13 Supplement: Revisions and information on later models

Fig. 13.35 Ball bearing (arrowed) on suspension strut upper mounting (Sec 13)

Independent rear suspension (760 Saloon, 1988 on) – description

25 Independent rear suspension is fitted to 760 Saloon models from 1988 model year. The makers call the system 'Multi-link'. Advantages over the old type of rear suspension include:

Improved roadholding
Reduced risk of wheel spin
Improved ride quality

26 The main components of the independent rear suspension are shown in Fig. 13.36. The final drive unit and driveshafts, which are new components associated with the suspension, are dealt with in Section 11.
27 The design of the suspension means that rear wheel alignment can be adjusted by means of eccentric mounting bolts and spacers.
28 Rear wheel camber is not fixed but varies with load, expressed as the angle of the lower links above or below the horizontal (Fig. 13.38).

Independent rear suspension – rear wheel alignment

29 Checking and adjusting of rear wheel alignment should only be carried out by a Volvo dealer or other specialist having the necessary optical alignment equipment.

Rubber bushes (independent rear suspension – renewal

30 Refer to Chapter 10, Section 22.
31 Rear wheel alignment must be checked after bush renewal.

Lower link (independent rear suspension) – removal and refitting

32 Proceed as for track rod removal and refitting (paragraphs 33 to 36), but remove the lower link instead of the track rod (photos).

Fig. 13.34 Sectional view of front hub with separate brake disc and ball-bearings (Sec 13)

13.16A Remove the spigot pin ...

13.16B ... and lift off the brake disc

13.19 Removing the hub (with ABS pulse wheel) from the stub axle

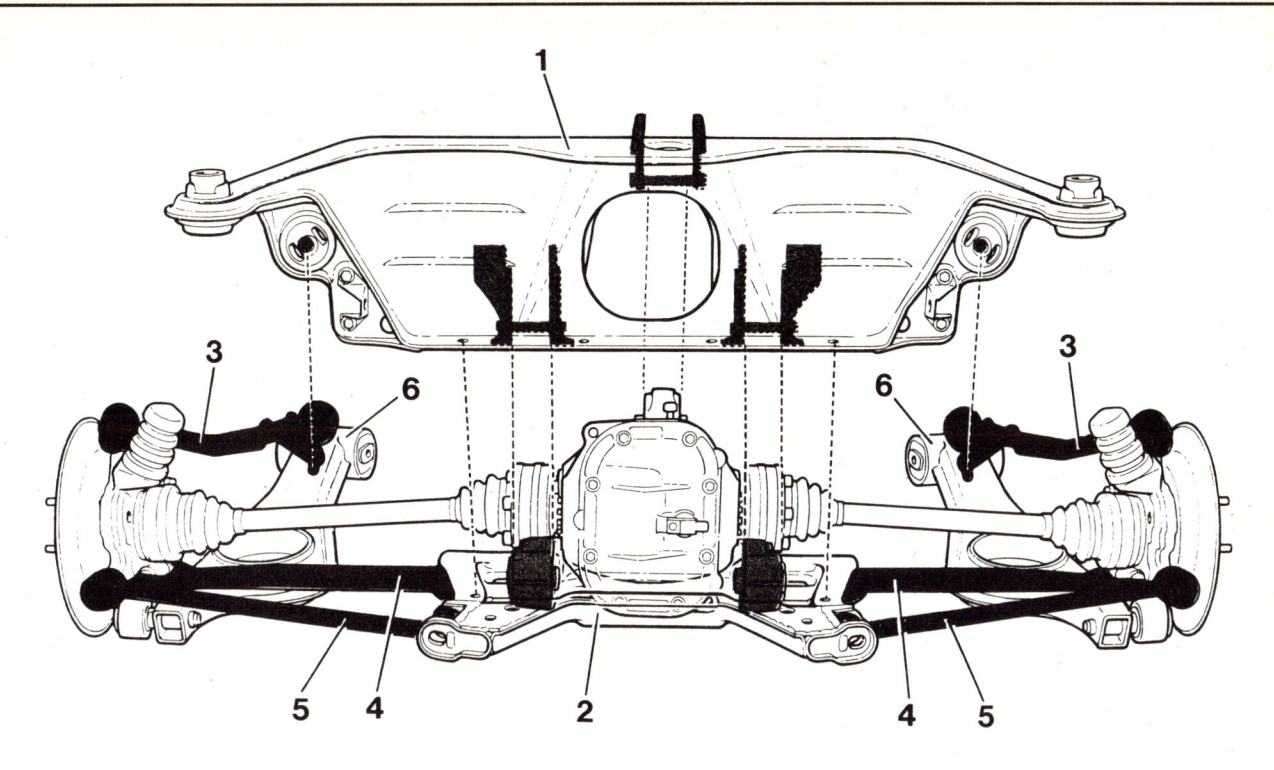

Chapter 13 Supplement: Revisions and information on later models 361

Fig. 13.36 Independent rear suspension components (Sec 13)

1 Rear axle member (upper Section)
2 Rear axle member (lower Section)
3 Upper links
4 Lower links
5 Track rods
6 Support arms

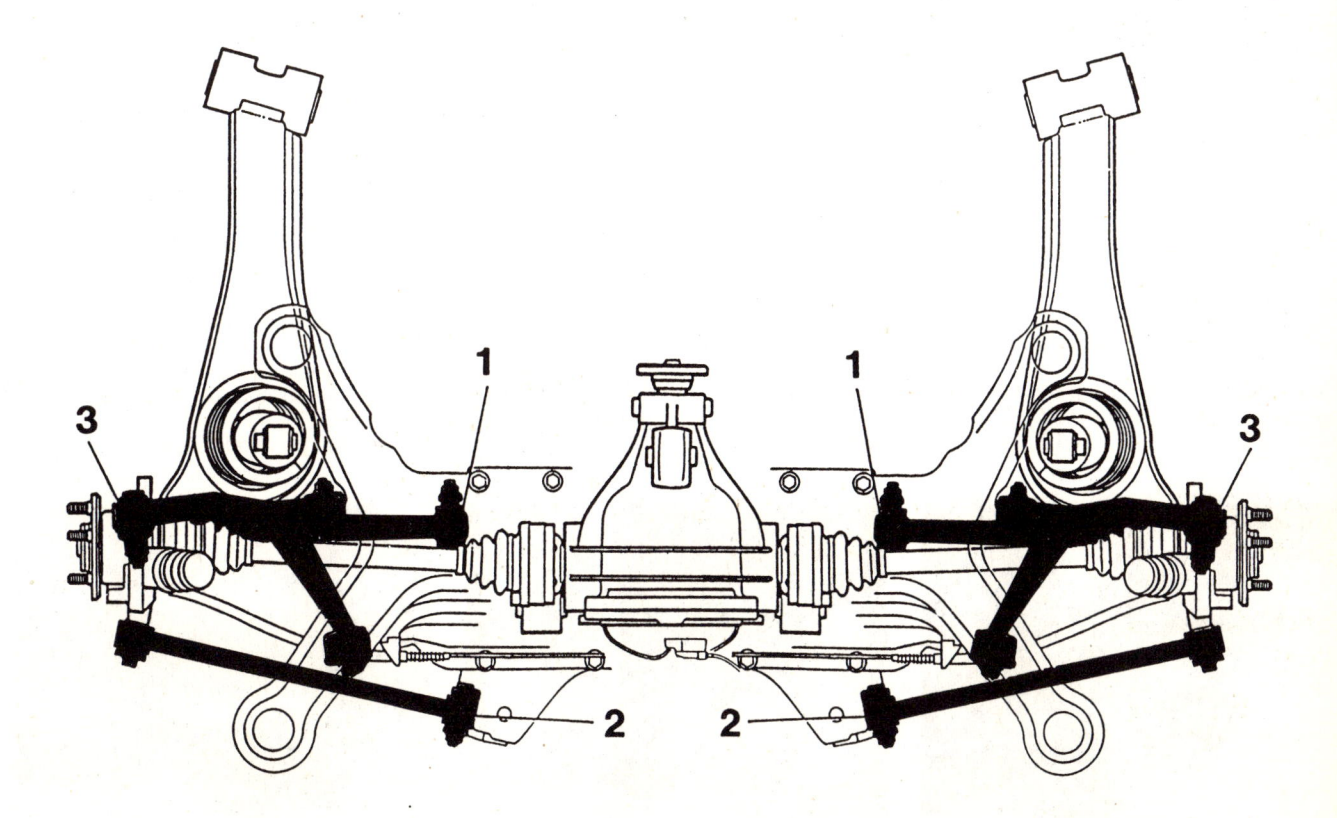

Fig. 13.37 Rear wheel alignment adjustment points (Sec 13)

1 Camber　　　2 Toe　　　3 Toe variation

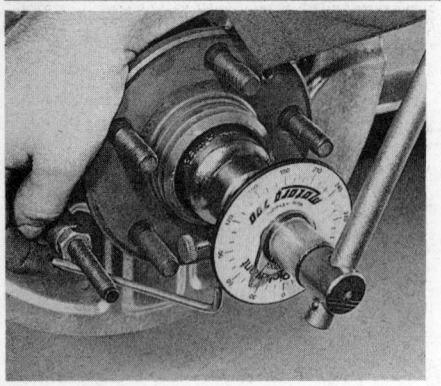

Chapter 13 Supplement: Revisions and information on later models

13.21 Angle-tightening the front hub nut

13.32A Lower link eccentric inner mounting

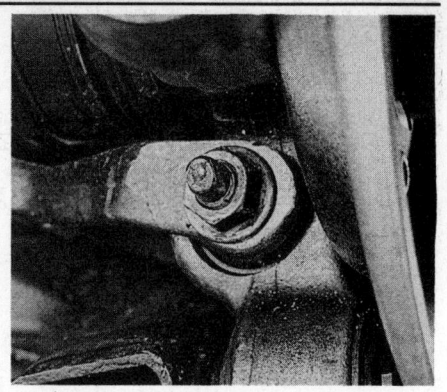

13.32B Lower link attachment to hub carrier

Track rod (independent rear suspension) – removal and refitting

33 Slacken the rear wheel nuts on the side concerned. Raise and support the vehicle with the suspension support arm free. Remove the rear wheel.
34 Mark the position of the track rod eccentric mounting on the axle member to provide an approximately correct setting for reassembly (photo).
35 Remove the nuts and bolts which secure the track rod ends to the axle member and hub carrier (photo). Remove the track rod.
36 Refit by reversing the removal operations. Tighten the track rod mountings to the specified torque and have the rear wheel alignment checked on completion.

Support arm (independent rear suspension) – removal and refitting

37 Slacken the rear wheel nuts on the side concerned. Raise and support the vehicle with the support arm free. Remove the wheel.
38 Remove the bolt which secures the support arm to the hub carrier (photo), and the nuts and bolts which secure the arm front mounting to the body. Also unbolt the guard plate from the arm.
39 Separate the support arm from the hub carrier by levering it off. Note the mating flats on the bush and support arm spigot.
40 Take the weight of the support arm, using a jack and a piece of wood.
41 Unbolt the shock absorber top mounting (photos). Lower the jack and remove the support arm complete with spring and shock absorber.
42 The spring and shock absorber may now be removed if wished. The shock absorber is secured to the support arm by two bolts.
43 Refit by reversing the removal operations, tightening all fastenings to the specified torque and using new nuts and bolts when these are subjected to angle tightening.

Rear axle member lower section (independent rear suspension) – removal and refitting

44 The rear axle member lower section is removed complete with lower links and track rods. In this way the wheel alignment is not disturbed.
45 Slacken the rear wheel nuts on both sides. Raise and support the vehicle with the support arms free. Remove the rear wheels.
46 Unbolt the lower link mountings from the hub carriers.
47 Unbolt the support arms from the hub carriers. Separate the arms from the carriers by levering.
48 Unbolt the track rods from the hub carriers. Pull off the track rods, either by hand or with the aid of a small puller.
49 Remove the eight nuts and bolts which hold the axle member upper and lower sections together. Note the handbrake cable guide under one of the rear bolts.
50 Support the axle member lower section. Swing the hub carriers outwards and free the lower links and track rods from them. Remove the axle member section with links and track rods.
51 Commence refitting by roughly positioning the axle member lower section. Insert a couple of the nuts and bolts to secure it to the upper section, but do not tighten them yet.
52 Insert two bolts, dowels or similar items, 12 mm (0.47 in) in diameter, into the two centering holes on the front edge of the axle members (photo).
53 Fit all the axle member nuts and bolts (not forgetting the handbrake cable guide) and tighten them to the specified torque. Remove the centering bolts.
54 Reconnect the lower links to the hub carriers. Tighten their fastenings to the specified torque, at the same time pulling the hub carrier inwards.
55 Reconnect and tighten the support arms and the track rods.
56 Refit the wheels, lower the vehicle and tighten the wheel nuts.

Upper link (independent rear suspension) – removal and refitting

57 Proceed as if preparing for axle member lower section removal (paragraphs 45 to 48). Also unbolt the brake caliper and support it out of the way.
58 Unbolt the upper link from the hub carrier. Separate the link from

13.34 Rear track rod eccentric mounting

13.35 Rear track rod attachment to hub carrier

13.38 Bolt (arrowed) securing support arm to hub carrier

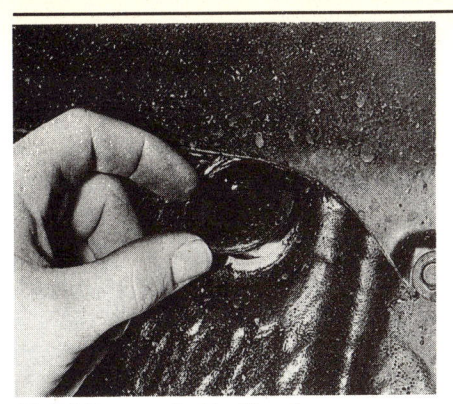

Chapter 13 Supplement: Revisions and information on later models

13.41A Remove the blanking plug ...

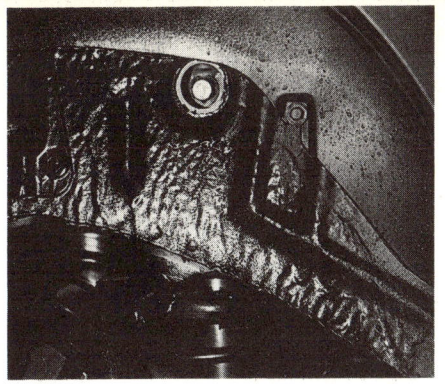

13.41B ... for access to the shock absorber top mounting

13.52 A dowel inserted into an axle member centering hole

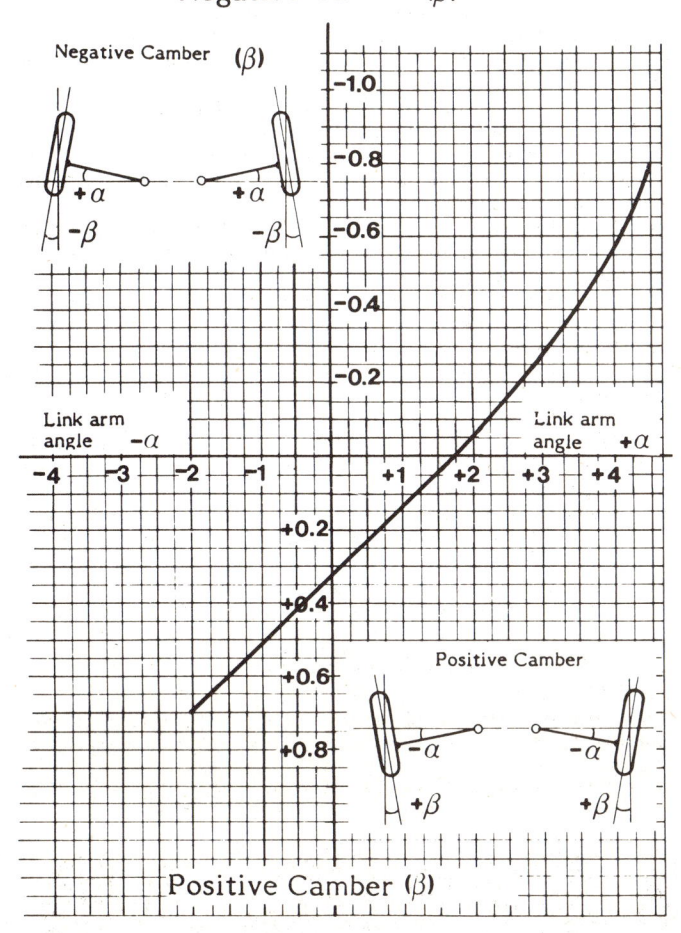

Fig. 13.38 Relationship between link arm angle α and camber angle β (Sec 13)

the carrier, collecting any spacers (photos).
59 Unbolt the upper link from the axle member. Remove the link.
60 Refit by reversing the removal operations, noting the following points:

(a) Pull the top of the hub carrier outwards when tightening the upper link mounting
(b) Pull the bottom of the hub carrier inwards when tightening the lower link mounting

Rear hub carrier (independent rear suspension) – removal and refitting

61 Slacken the driveshaft nut (see Section 11).
62 Raise and support the rear of the vehicle and remove the rear wheel. Make sure the supports do not obstruct the suspension components.
63 Remove the brake caliper (without disconnecting it), the brake disc and the handbrake shoes. Disconnect the handbrake cable. Refer to Section 12 if necessary.
64 Remove the driveshaft nut (photo).
65 Remove the nuts and bolts which secure the upper link, the lower link, the track rod and the support arm to the hub carrier. Separate the carrier from the links and arms, recovering and noting any spacers on the upper link mounting. The support arm mounting may be stiff: lever the carrier off it if necessary.
66 Support the driveshaft and remove the hub carrier, tapping the end of the driveshaft to free it if necessary (photo).
67 Refit by reversing the removal operations, using new nuts and bolts if the specified tightening procedure includes angle tightening.
68 If new bushes, link rods etc have been fitted, have the rear wheel alignment checked on completion.

Rear wheel bearing (independent rear suspension) – renewal

69 Remove the rear hub carrier as described earlier.
70 Support the hub carrier and drive or press the stub shaft out of the bearing. The inner race will probably stay on the shaft; if so, pull it off (photo).
71 Remove the circlip from the outboard side of the carrier.
72 Remove the brake backplate and the bump stop for ease of handling. Press or drive out the bearing from the hub carrier, pressing from the inboard side.
73 Clean and lightly grease the bearing seat in the hub carrier. Also clean and grease the stub shaft.
74 Offer the new bearing to the carrier (it is symmetrical so it can be fitted either way round). Press the bearing in from the outboard side. Use the old bearing outer race to press with: it will not get jammed because the lead-in in the carrier is chamfered.
75 Refit the circlip (photo).
76 Refit the brake backplate, applying silicone sealant to the joint between the hub carrier and the backplate (photo).
77 Support the bearing inner race and press in the stub shaft.
78 Refit the hub carrier.

Rear spring and shock absorber (models with independent rear suspension) – removal and refitting

79 The procedures are basically still as described in Chapter 10, Sections 26 and 27, but note the following points:

(a) The spring can only be removed after the shock absorber
(b) There is no spring upper seat nut
(c) If sufficient clearance cannot be gained by disconnecting the support arm from the hub carrier, remove the support arm completely

Chapter 13 Supplement: Revisions and information on later models

13.58A Upper link attachment to hub carrier

13.58B Spacers between link and carrier

13.64 Removing the driveshaft nut

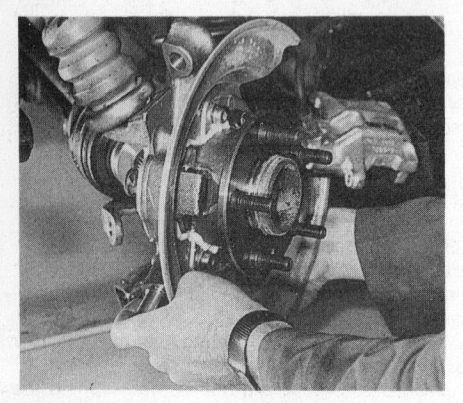

13.66 Removing the rear hub carrier

13.70 Pulling the inner race off the stub shaft

13.75 Fitting the bearing circlip

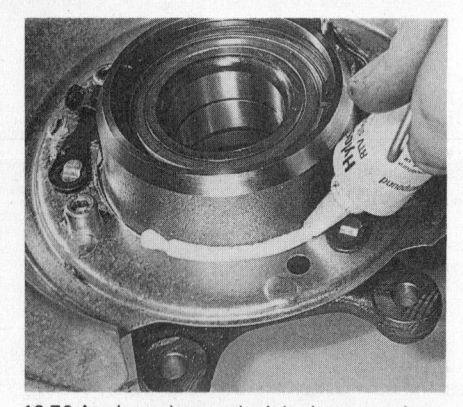

13.76 Apply sealant to the joint between the hub carrier and the backplate

14.1 Guide pin (arrowed) determines hinge position

14.2 Bonnet gas strut upper attachment

14 Bodywork and fittings

Doors (1988 on) – adjustment

1 Slotted hinge bolt holes are no longer the means of adjustment for door fit. A guide pin on each hinge is used to ensure correct fit (photo).

Bonnet (760, 1988 on) – removal and refitting

2 The procedure is similar to that described in Chapter 11, Section 6, but the gas struts which are used instead of springs must be disconnected at their upper ends (photo). Support the bonnet when doing this.

Front door interior trim (1989 on) – removal and refitting

3 Disconnect the battery negative lead.
4 Prise out the control switches from the armrest and disconnect them (photo).
5 Remove the loudspeaker trim by sliding it forwards (photo).
6 Remove the clip from the well behind the interior handle by turning the clip a quarter-turn with a screwdriver (photo).
7 Unclip the edge marker light lens.
8 Prise out the three clips from the bottom edge of the trim panel.
9 Unclip the panel by tugging it firmly. Disconnect the loudspeaker and edge marker light connectors and remove it.
10 To remove the armrest frame, first prise the plug out of the end of the interior handle cap (photo). Remove the exposed screw and remove the cap.
11 Unclip any wiring from the armrest frame. Remove the two securing screws and lift off the frame (photo).
12 Refit by reversing the removal operations.

Chapter 13 Supplement: Revisions and information on later models

14.4 Disconnecting a control switch from the armrest

14.5 Removing the loudspeaker trim

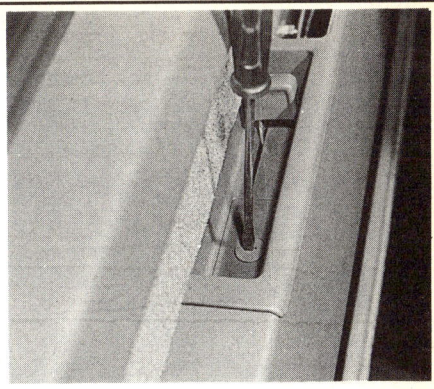

14.6 Turn the clip through 90° to release it

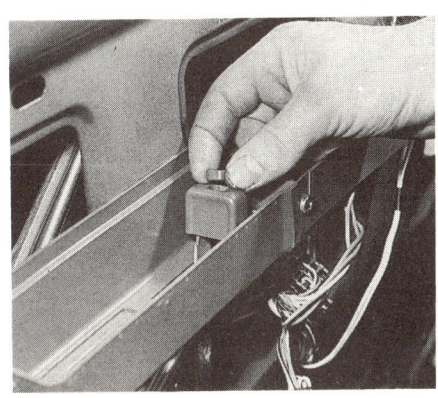

14.10 Remove the plug for access to the screw

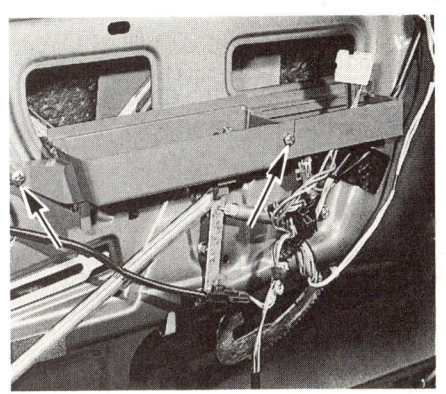

14.11 Two screws (arrowed) securing the armrest

Door lock motor (1989 on) – removal and refitting

13 Remove the door interior trim panel.
14 Remove the two screws from below the catch, and the single screw near the interior lock button, to release the glass rear guide (photos).
15 Remove the two Torx screws which secure the catch. Remove the catch, and the single screw thus exposed (photo).
16 Remove the glass rear guide (photo).
17 Unclip the handle operating levers. Unhook the motor and remove it complete with interior lock button rod (photo).
18 The motor itself can be removed from the locking mechanism by undoing the two securing screws (photo).
19 Refit by reversing the removal operations.

Driver's door central locking switch (1989 on) – removal and refitting

20 Proceed as for door lock motor removal. The driver's door switch looks the same as a door lock motor, but the casing only contains a switch.

Window lift motor (1988 on) – removal and refitting

21 An additional connector has been fitted in the wiring harness so that the motor can be removed without damaging the main wiring connector – see Chapter 11, Section 17.

Seat adjuster motor control panel – removal and refitting

22 Remove the seat (Chapter 11, Section 25).
23 Cut the cable tie which secures the wiring harness to the motor mounting plate.
24 Disconnect the motor multi-plugs.
25 Remove the two screws which secure the control panel.
26 Withdraw the control panel, feeding the cables and connectors through the side of the seat.
27 Refit by reversing the removal operations. Use a new cable tie.

Seat adjuster motors – removal and refitting

28 Remove the seat (Chapter 11, Section 25).

Fore-and-aft and height motors

29 Invert the seat on a clean bench or floor so that the base is accessible (photo).
30 Remove the four small bolts which secure the motor mounting plate.
31 Ease the mounting plate away from the seat. Remove the four bolts which secure the motor in question; note the spacers fitted to the front height adjuster motor (photo).
32 Remove the motor from the bracket and withdraw the cable from it (photo).
33 Disconnect the motor multi-plug (photo). This plug is shared by all three motors; to remove one motor it will be necessary to prise its connectors out of the plug, or to cut the wires and make new connections for a new motor.
34 Refit by reversing the removal operations.

Backrest motor

35 Remove the lumbar support adjustment knob. Also remove the covers at the base of the backrest on both sides.
36 Recline the backrest as far as possible, using a screwdriver or similar tool inserted in the hole in the reclining gear mechanism at the base of the backrest.
37 On the underside of the seat disconnect the multi-plugs which feed the heater pads and the backrest motor. It will also be necessary to free or cut the black wire from the three-pole connector. Cut cable ties as necessary.
38 Remove the four bolts which secure the backrest. Lift the backrest off the seat base.
39 Cut the rings which secure the upholstery to the base of the backrest. Peel back the upholstery and unhook the clips inside to gain access to the motor.
40 Remove the nut which secures the motor bracket to the backrest frame. Withdraw the motor with cables and bracket still attached.

366 Chapter 13 Supplement: Revisions and information on later models

14.14A Removing the two screws below the catch

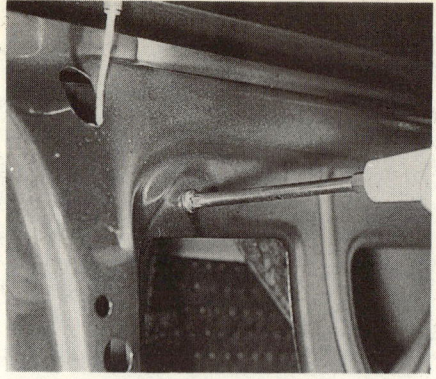

14.14B Glass guide screw near the lock button

14.15 The single screw hidden by the catch

14.16 Removing the glass rear guide

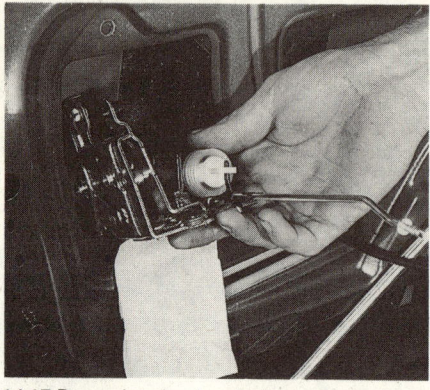

14.17 Removing the door lock motor

14.18 Lock motor showing securing screws

14.29 General view of the base of a motorised seat

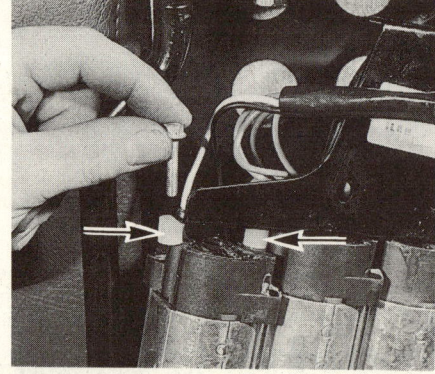

14.31 Removing a front height adjuster motor bolt. Note spacers (arrowed)

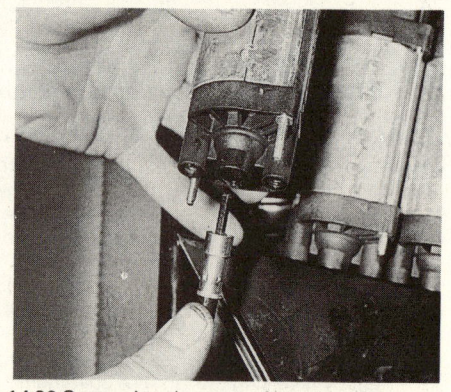

14.32 Separating the motor from the cable

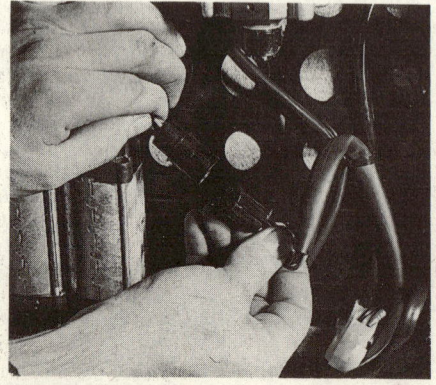

14.33 Seat motor multi-plug

14.44 Seat heater control unit

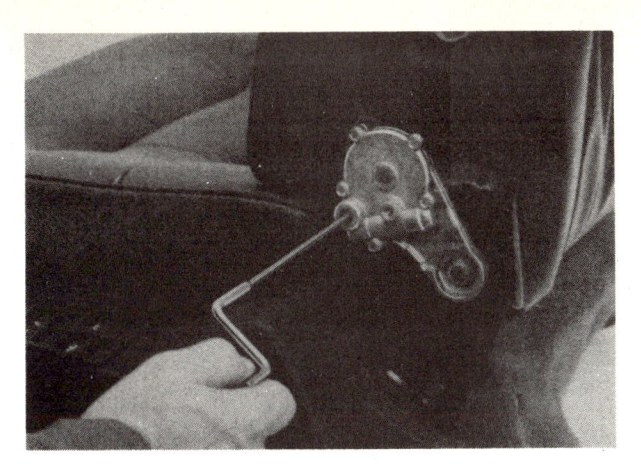

Chapter 13 Supplement: Revisions and information on later models

Fig. 13.39 Reclining the power seat backrest by hand (Sec 14)

41 Unbolt the bracket from the motor and free it from the cables. Disconnect its wiring plug and remove it.
42 Refit by reversing the removal operations, using new cable ties, upholstery rings etc.

Seat heater control unit (1988 on) – general
43 Up to 1988, seat heating was controlled simply by a thermostat and a switch. From 1988 seats have control units (one for each front seat) which vary the seat heater output according to temperature. Maximum output is only provided at under-seat temperatures below 10°C (50°F). Low output is provided at temperatures between 10 and 18°C (50 and 64°F). No output is provided at higher temperatures, apart from a burst of a few seconds when the ignition is first switched on.
44 If the seat heater does not behave as described, the control unit may be at fault. Access is obtained by removing the seat, cutting the mounting strap and unplugging the control unit (photo).
45 Testing of the control unit is by substitution of a known good unit. Assuming that only one control unit will fail at a time, a good unit may be borrowed from the other seat to confirm the fault.

Glovebox (760, 1988 on) – removal and refitting
46 Remove the panel from below the glovebox. This is secured by clips at the top (turn through 90° to release) and screws at the bottom.
47 Remove the two screws at the base of the glovebox lid, one at each end (photo).
48 Open the glovebox. Remove the plastic covers from above the lid straps and remove the screws thus exposed.
49 Free the glovebox from its mounting clips and remove it.
50 Refit by reversing the removal operations.

Front bumper (760, 1988 on) – removal and refitting
51 The procedure is as given in Chapter 11, Section 39, except that the side sections of the bumper are no longer separate from the front section.

Front wing liners
52 Models from 1990 have plastic wing liners fitted, to increase corrosion protection. They are secured by screws.

15 Electrical system

Battery disconnection – all models with radio
1 Make sure that the radio is switched off before disconnecting or reconnecting the battery. Failure to observe this precaution can result in damage to the microprocessor fitted to many modern car radios.

Headlight unit (760, 1988 on) – removal and refitting
2 Remove the front grille panel (Chapter 11, Section 42). On the left-hand side also remove the air cleaner (Section 7).
3 Remove the bulbholders, with bulbs, from the direction indicator/parking light unit. Push and twist the holders to release them (photo).
4 Unclip the direction indicator/parking light unit and remove it (photo).
5 Unplug the headlight and auxiliary front light connector (photo).
6 Unplug the headlight wiper motor connector and disconnect the headlight washer tube (photo).
7 Remove the four screws, two on each side, which secure the headlight unit. Note that the screws on the grille side also secure a baffle plate; on the left-hand side they also secure the air cleaner air intake (photos).

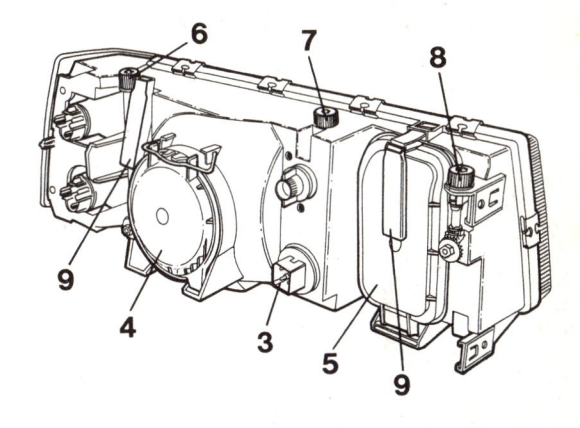

Fig. 13.40 Rear view of later type headlight unit (Sec 15)

3 Connector
4 Main headlight cover
5 Auxiliary light cover
6 Headlight adjuster screw (horizontal)
7 Headlight adjuster screw (vertical)
8 Auxiliary light adjuster screw (vertical)
9 Ventilation holes

14.47 Removing a glovebox securing screw

15.3 Removing a bulb and holder from the direction indicator/parking light unit

15.4 Removing the direction indicator/parking light unit

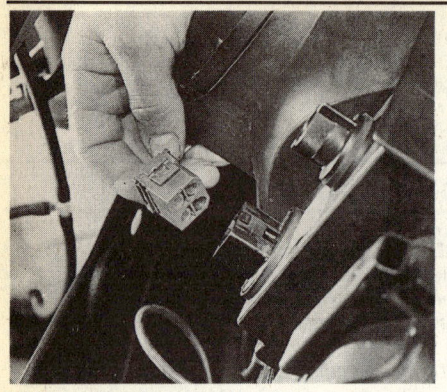

Chapter 13 Supplement: Revisions and information on later models

15.5 Headlight/auxiliary front light connector

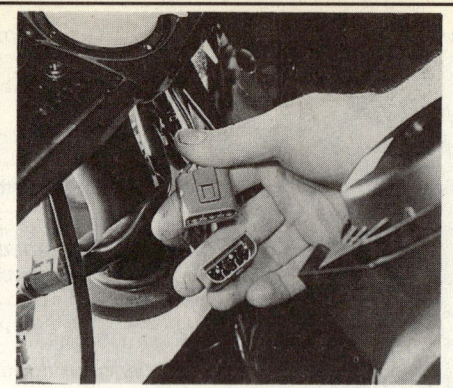

15.6A Headlight wiper motor connector

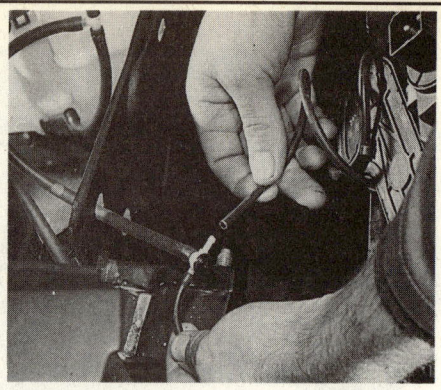

15.6B Headlight washer tube connector

8 Remove the headlight unit complete with wiper motor, arm and blade.
9 Refit by reversing the removal operations. Note that the securing screws allow the headlight unit to be moved to align it relative to the body, then bumper and the other headlight; do not tighten the screws until satisfied with the position of the unit. If the headlight is fitted too high, the bonnet will strike it.
10 Have the headlight beam alignment checked on completion. The adjustment screws are now on top of the unit (photo).

Headlight unit (760, 1988 on) – dismantling and reassembly
11 Remove the headlight unit as just described.
12 Remove the two screws which secure the headlight wiper motor. Unclip the front trim strip and remove the wiper motor complete with arm, blade and trim strip (photos).
13 Remove the four wiper blade stops, which are secured by one screw each (photo).
14 Remove the eight clips which secure the lens to the reflector.
15 Remove the lens and extract the seal.
16 Remove the bulb covers from the rear of the unit. Disconnect and remove the bulbs, being careful not to touch the glass with the fingers.
17 Renew components as necessary and reassemble by reversing the dismantling procedure. Pay particular attention to the condition and fit of the lens-to-reflector seal.

High level brake light (1986 on) – bulb renewal
18 Remove the brake light cover. On Estates it simply pulls off; on Saloons the catch at the base the cover must be depressed with a screwdriver (photo).

15.7A Two screws (arrowed) securing the grille side of the headlight. Note air cleaner intake

15.7B Two screws securing the indicator side of the headlight

15.10 Headlight (A) and auxiliary light (B) alignment screws

15.12A Removing the headlight wiper motor screws

15.12B Unclip the front trim strip

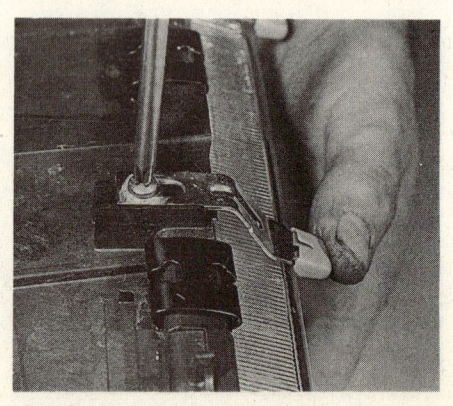

15.13 Removing a wiper blade stop

Chapter 13 Supplement: Revisions and information on later models

19 Squeeze the catches on each side of the reflector/bulbholder and withdraw it (photo).
20 Remove the old bulb and fit a new one.
21 Fit the reflector and press it home until the catches engage. Check for correct operation, then refit the cover.

Vanity mirror light bulbs (760) – renewal

22 Move the light switch to the left ('off' position).
23 Lever the mirror surround away from the sun visor, starting on the right-hand side and working anti-clockwise. It will be necessary to hold the surround away from the sun visor to stop it slipping back into place. When levering at the top of the surround, do so at least 10 mm (0.4 in) below the top edge.
24 When the two clips at the top have been released, pivot the mirror and surround downwards and lift it out (photo).
25 Renew the light bulbs as required (photo).
26 Engage the mirror surround bottom lugs. Press firmly on the top edge of the surround (not on the mirror itself) to engage the clips.

Steering column switches (760, 1988 on) – removal and refitting

27 Disconnect the battery negative lead. Remove the Allen screw which secures the column adjuster control knob (photo). Remove the knob.
28 Remove the column shrouds, which are secured by six screws.
29 Remove the two screws which secure each switch. Withdraw the switches and disconnect their wiring plugs (photo).
30 Refit by reversing the removal operations.

Instrument cluster (760, 1988 on) – removal and refitting

31 Disconnect the battery negative lead.
32 Prise out the bright trim strip to the right of the cluster.
33 Remove the screw which secures the left-hand bright trim strip. This screw is accessible from inside the glovebox (photo). Prise out the trim strip.
34 Unclip the switch panels, disconnect the multi-plugs and remove the panels. (The main lighting switch can be left in position.)
35 Remove the air direction grilles by turning them upwards as far as possible, then unclipping them from their pivots using firm hand pressure.
36 Remove the seven screws which secure the cluster surround. These are located as follows: one in each air grille housing, two (under covers) below the air grille housings, and two (under covers) on the underside of the instrument hood (photos). Remove the surround.
37 Remove the four screws, one in each corner, which secure the instrument cluster itself. Carefully draw the cluster out of its recess and disconnect the multi-plugs from it (photo). Remove the cluster.
38 Refit by reversing the removal operations.

Instrument fuse (1985 on) – renewal

39 Some makes of instrument cluster incorporate a fusible link in the printed circuit (photo). If this link blows, a repair strip or portion of circuit card should be obtained and secured in the same position. The fault which caused the original link to blow must also be rectified.

Fuses (760, 1988 on)

40 The fuses are housed at the right-hand end of the facia. For access, open the driver's door, then unclip the fuse cover panel (photo).
41 Fuse details are given in the Specifications, and on the inside of the cover panel. Spare fuses are housed in a drawer at the base of the fuse board.
42 Fuses Nos 34 and 35 are in fact circuit breakers. If a circuit breaker trips, it should reset itself after about 20 seconds. If it fails to reset, renew it, first rectifying the cause of any overload.
43 A separate fuse protects the ABS circuits. This fuse is mounted on a relay, known as the ABS Transient Surge Protector, located to the right of the instrument panel (photo).

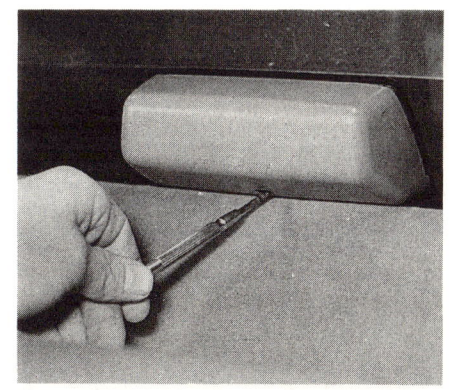

15.18 Releasing the brake light cover – Saloon

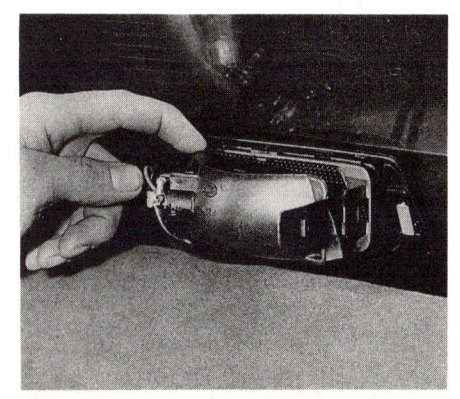

15.19 Removing the brake light reflector/bulbholder

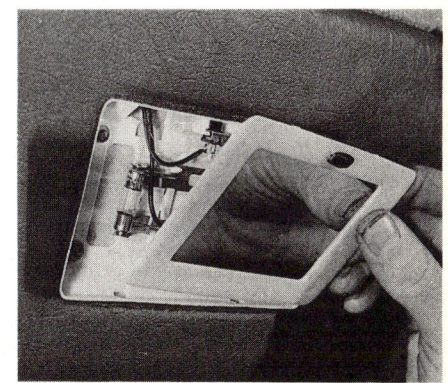

15.24 Removing the illuminated vanity mirror

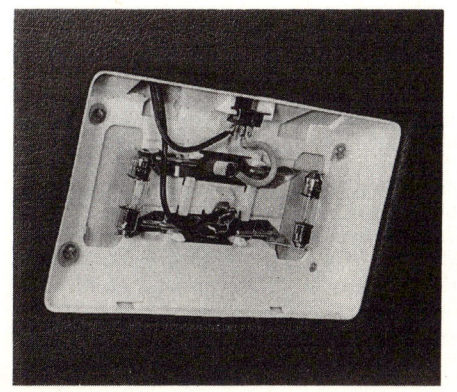

15.25 Vanity mirror light bulbs exposed

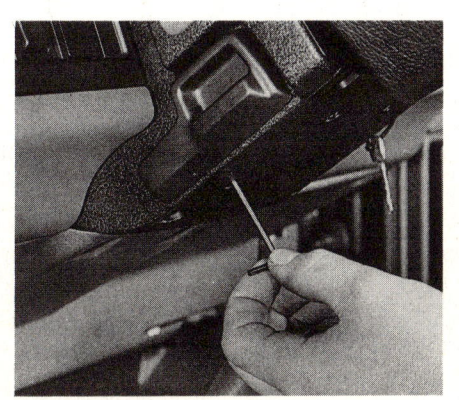

15.27 Removing the column adjuster control knob screw

15.29 Removing a steering column switch

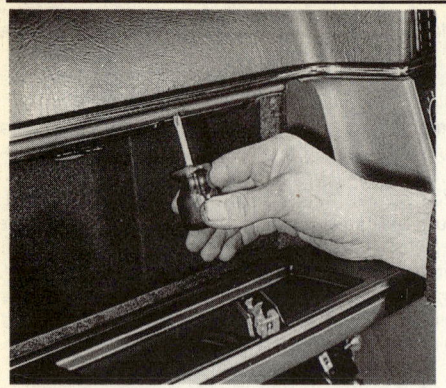

370 Chapter 13 Supplement: Revisions and information on later models

15.33 Removing the trim strip screw

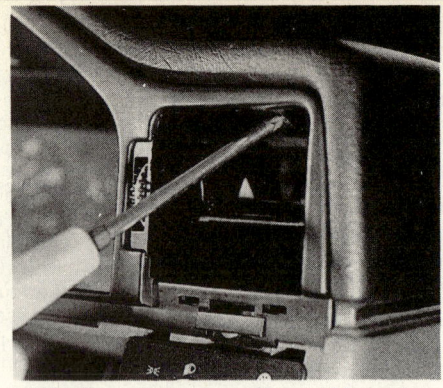

15.36A Removing a screw from within an air grille housing

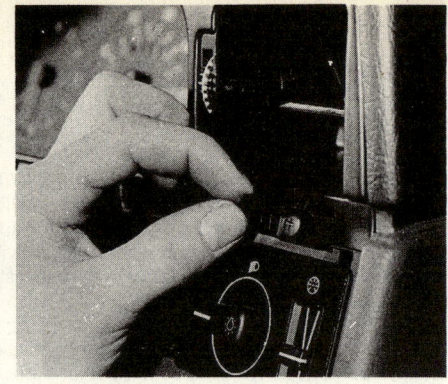

15.36B Removing a screw cover below an air grille housing

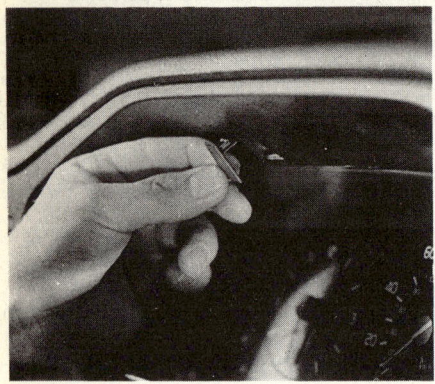

15.36C Removing an instrument hood screw cover

15.37A Removing an instrument cluster screw

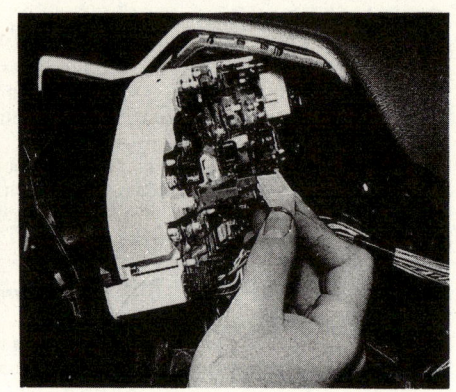

15.37B Removing the instrument cluster

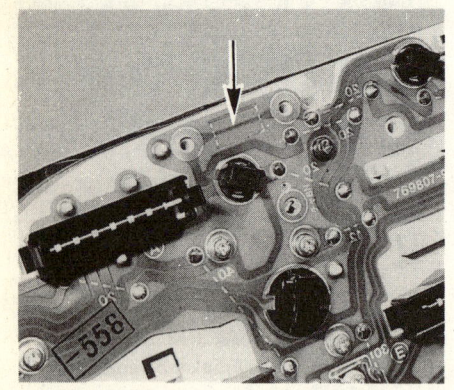

15.39 Fusible link (arrowed) in the printed circuit

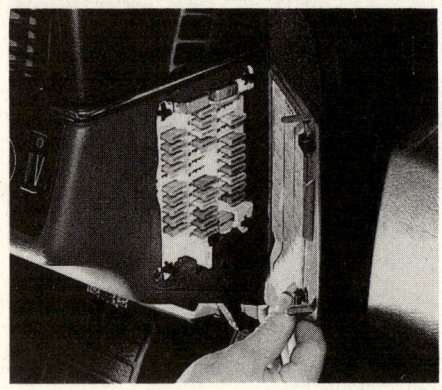

15.40 Fuse board at the right-hand end of the facia

15.43 ABS fuse mounted on the transient surge protector

Relays (760, 1988 on) – general

44 From 1988 model year, the relays may be located on the right-hand side of the centre console. Remove the side trim for access (photo).

45 Relay identification is as follows:

    A Main lighting relay (part)
    B Motronic/Jetronic relay
    C Central locking relay
    D Foglamp relay
    E Main lighting relay (part)
    F Bulb failure warning relay (front)
    G Overdrive relay
    J Power boost relay
    K Rear wiper delay relay
    L Windscreen wiper delay relay
    M Seat belt warning relay

46 Relays in positions A, B, F, J, L and M are permanently attached to the board. Depending on equipment, not all the relays listed are fitted to all models.

47 Bulb failure warning for the rear lights on 760 models is controlled by a relay mounted behind the trim on the left-hand side of the boot or luggage area (photo).

48 The electric cooling fan relay on 760 models is located under the bonnet, in front of one or other suspension strut housing (photo).

49 The heated rear window relay is now incorporated in the heated rear window control switch.

50 The direction indicator/hazard warning flasher unit is incorporated in the hazard warning switch.

51 Access to the switch-incorporated units is by unclipping the switch panel, disconnecting the multi-plugs and unclipping the switch from its location (photo).

Chapter 13 Supplement: Revisions and information on later models

15.44 Main relay board exposed

15.47 The rear bulb failure warning unit

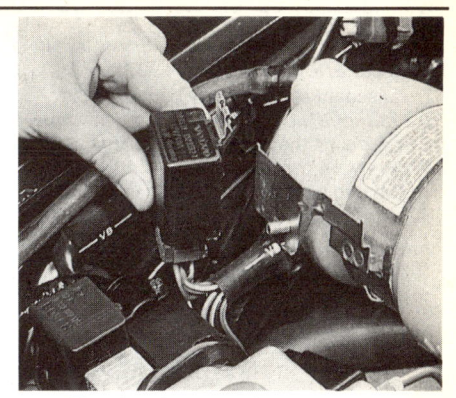

15.48 The electric cooling fan relay

15.51 Removing the heated rear window switch/relay and the hazard warning switch/flasher unit

15.54 Removing an end screw from the wiper water shield

15.57 Wiper motor multi-plug. Bolt to left secures earth tag

15.58A Two screws (arrowed) securing the right-hand end of the linkage ...

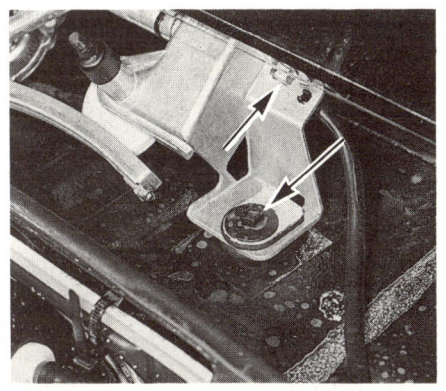

15.58B ... and two screws on the left

15.62 Releasing a radio securing clip

15.63A Pull the radio from its recess ...

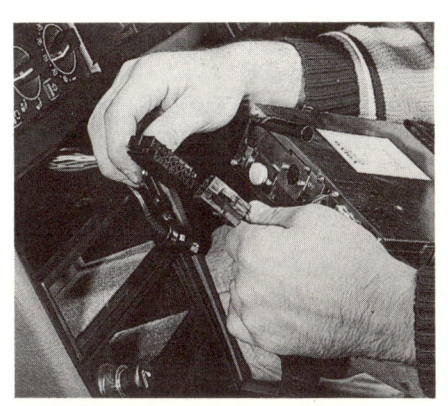

15.63B ... and disconnect the wiring from it

Windscreen wiper motor and linkage (760, 1988 on) – removal and refitting

52 Disconnect the battery negative lead.
53 Remove the nuts which secure the wiper arms. Mark the positions of the arms on the screen if necessary, then pull them off their spindles.
54 Remove the six screws from the front of the water shield. Note that the two end screws are longer (photo).
55 Release the drain hoses from each end of the water shield.
56 Remove the water shield. Note how the clips on its rear edge engage with the bottom of the windscreen.
57 Disconnect the wiper motor multi-plug (photo).
58 Remove the four bolts which secure the motor and linkage to the bulkhead (photos). Also remove the screw which secures the motor wiring earth tag. Lift out the motor and linkage. Note the position of any spacers.
59 Refit by reversing the removal operations. Make sure that the motor is in the parked position before refitting the wiper arms.

Radio/cassette player (quickly detachable type) – removal and refitting

60 On some models a quickly detachable radio is fitted. The unit is removed, as an anti-theft measure or for any other reason, as follows.
61 Make sure that the radio is switched off.
62 Push and release the two securing clips to raise them (photo).
63 Pull the radio from its location by means of the two clips. Disconnect the wiring from the rear of the radio and remove it (photos).
64 Refit by reversing the removal operations, pushing the radio home until it clicks into place.

Wiring diagrams (later models)

65 For reasons of space it has not been possible to include a complete range of wiring diagrams for all later models.

Key to Figs. 13.41 and 13.42. Not all items are fitted to all models; some systems do not apply to the UK

No	Description	Coordinate	No	Description	Coordinate
1	Battery	B2	81	AC pressure sensor	AC1, A2
2	Ignition switch	E3	82	Temperature sensor, heater fan	AB2
3	Instrument connection (4-pole)	Z4	84	Coolant temperature sensor	P5, Q1, U2, V3, AA3, N2, AE2, AF5
4	Ignition coil	C5			
5	Distributor	C4, R4	85	Speedometer	F4
6	Spark plug	C3	86	Tachometer	F4
7	Instrument connection (7-pole)	Z4	87	Clock	F4
8	Instrument connection (8-pole)	Z4	88	Engine temperature gauge	F4
9	Starter motor	B3	89	Fuel gauge	E4
10	Alternator (incl. regulator)	B3	90	Volt meter	E4
11	Fusebox	B-K1	91	Service reminder light	F5
12	Instrument connection (12-pole)	Z4	92	Indicator lamp, diagnosis	E5
13	High beam	A2, A5	94	Panel lighting intensity (rheostat F)	F4
14	Low beam	A2, A5	95	Instrument lighting	F4
15	Relay unit, 15 circuit	W4	96	Indicator light, exhaust gas temperature (Japan)	E4
16	Parking light (USA incl. rear)	A1, A6, L1, L6	97	Indicator lamp, oil pressure	E4
17	Hazard lights	A1, A6	98	Indicator lamp, choke	E4
18	Turn signal bulb	A1, A6, B1, B6, L1, L6	98	(alt) Indicator lamp, boost pressure (turbocharger)	
19	Back-up light bulb	L1, L6	99	Indicator lamp, parking brake	F4
20	Rear fog light bulb	L1, L6	100	Indicator lamp, brake failure	F4
22	Brake light bulb	L1, L4, L6	101	Indicator lamp, washer level	F4
23	Fog light bulb	A2, A5	102	Indicator lamp, AW70/71 auto.trans. overdrive	F4
24	Numberplate lighting	L3, L4	103	Indicator lamp, bulb failure	G3
25	Courtesy light bulb	N5	104	Indicator lamp, glow plugs (diesel)	G3
26	Reading light, front	N5	104	Indicator lamp, (Diesel)	
27	Inside lights	N5	105	Indicator lamp, battery charge	E5
29	Positive terminal	D2	107	Indicator lamp, ABS	E5
30	Relay unit (30 circuit)	W4	108	Indicator lamp, left turn signal	F5
31	Relay unit (31 circuit)	W3	109	Indicator lamp, high beam	F5
32	Glove compartment light	M4	110	Indicator lamp, right turn signal	F5
33	Ashtray light, front	G3	112	Indicator lamp, M46 overdrive	F5
34	Ashtray light, rear	J4	113	Indicator lamp, front seat belts	F5, X3
35	Sunroof switch light	F3	114	Indicator lamp, rear seat belts	J5, X3
36	Seat heater switch light, passenger	J4	115	Bulb failure warning sensor	I4
37	Gear selector light, auto. trans	J3	116	Fasten seat belt light (USA)	X3
38	Instrument and panel lighting	F3, G3	117	Windscreen wipers, intermittent relay	G3
39	Engine compartment light	C2, N3	118	Rear windscreen wiper, intermittent relay	R4
40	Trunk light	M5	119	Relay, fuel pump	E5
41	Door-open warning light	M5, N5	121	Flasher unit, hazard lights	G4
42	Heater controls light	G3, Z1, AB1	122	Relay, exhaust gas temperature (Japan)	Ö5
43	Sunvisor light	N4	123	Relay, M46 overdrive	Z5
45	Seat belt lock, driver	J5, X4	124	Relay, power windows	AC4
46	Seat belt lock, passenger	J4, X4	125	Relay, central lock	AA4
47	Windscreen wiper and washer switch	H3	127	Relay, auxiliary lights	I6, AB6, AC6
48	Headlight switch	H4	130	Relay, glow plugs (diesel)	R5
49	Turn signal/hazard light switch	H5	131	Relay, fan	AA1, AC2
49	High beam/low beam		132	Relay, AC	Z1
50	Horn	A5	135	Relay, Motronic, LH-2.2, LH-2.4	M1, U2, U5, AC4
51	Heated rear window switch	K2, K3	136	Relay, overdrive AW70/71	Z4
52	Rear fog light switch	T5, AB6, AC6	137	Relay, headlights	H4
53	Seat heater switch, passenger	J3, Q3	138	Seat heater pad, driver	K5, R3
54	Overdrive switch	Z4, Z5	138	Seat heater pad, passenger	J2
55	Heater fan switch	J3, Z1	139	Seat heater backrest, driver	K5, P3
56	Power window switch, driver	A3	139	Seat heater backrest, passenger	K2, Q3
57	Power window switch, passenger	AC4	142	Thermostat, driver seat	J5, P3
58	Power window switch, driver side, rear	AB4	142	Thermostat, passenger seat	J2, Q3
59	Power window switch, passenger side, rear	AB4	143	Loudspeaker, passenger door	Y1, Y2
60	Power door mirror switch, driver	AA5	144	Loudspeaker, driver door	Y1, Y2
61	Power door mirror switch, passenger	AA5	145	Loudspeaker, rear left	Y1, Y2
62	Central lock linkage	Z4	146	Loudspeaker, rear right	Y1, Y3
64	Sunroof switch	Z3	147	Aerial	Y1, Y2
65	Fog light switch	T6, AA6, AC6	148	Power aerial	Y1, Y2
66	Brake light contact	G3, T5, W2	149	Radio	X1, X2
67	Choke contact	E3	150	Motor, driver power window	AC3
68	Parking brake contact	F3	151	Motor, passenger power window	AC5
70	Back-up light contact	J6	152	Motor, power window, rear, driver side	AB5
71	Start inhibitor switch, auto.trans	D3	153	Motor, power window, rear, passenger side	AB5
72	Driver door switch	M4	154	Power mirror, driver side	K2, Z5
73	Passenger door switch	N4	155	Power mirror, passenger side	K2, Ä5
74	Door switch, rear	M6, N6	156	Motor, electric cooling fan	W1
75	Passenger seat contact	J4	157	Motor, headlight wiper	B2, B5
76	Pressure sensor, turbocharger	U4	158	Motor, sunroof opening	Z3
77	Overdrive contact (M46)	Z5	159	Motor, central lock, passenger door	AB4
80	Thermal time-switch	B6	160	Motor, central lock, rear door, driver side	Z5

373

Key to Figs. 13.41 and 13.42. Not all items are fitted to all models; some systems do not apply to the UK (continued)

No	Description	Coordinate	No	Description	Coordinate
161	Motor, central lock, rear door, passenger side	AB5	240	Contact, wiper/washer, rear window	S3
162	Motor, central lock, trunk	AA5	241	Motor, wiper, rear window	R3
163	Motor, windscreen wiper	G3	251	Kickdown unit	Z5
164	Motor, windscreen washer	G3	252	Control unit, ABS	T1
165	Motor, heater fan	AA2, AB1	253	Hydraulic unit, ABS	T3
166	Capacitor	C3	254	Transient surge protector, ABS	S2
169	Resistor, heater	AA1, AB1	256	Sensor ABS, front left	T1
170	Catalytic converter, thermo element	AC5	257	Sensor ABS, front right	T1
173	Delay relay, courtesy light	N4	258	Hydraulic pump, ABS	T3
176	Control unit, CIS	AA3	260	Control unit, EX-K ignition system	P1, Q4, AD2
178	Washer level sensor	F3	262	Vacuum pump, ACC	AA1
180	Speedometer sensor	F5, T2	263	Relay, vacuum pump, ACC	AA1
181	Engine temperature sensor	F5	266	Delay relay, heated rear window	K2, K4
182	Fuel level sensor	F5	267	Test outlet, EZ-K	Q5
185	Charge air temperature sensor	AF4	271	Fuel turn-off valve	P4
186	Air flow meter	V1	272	Microswitch, engine	Q5
187	Lambdasond, heated or unheated	AD4	274	Time-relay, EGR	T3
188	Starter valve	AD5, M2	275	Microswitch, idle	T3
189	Control pressure valve	C5	276	Transformer, EGR	Q1
190	Auxiliary air valve	C5	277	Three-way valve, EGR	T3
193	Hot start valve	B5	278	Aneroid switch	T4
194	Solenoid valve, AC compensation/ignition advance	X4	279	Solenoid, altitude compensation	T4
195	Solenoid valve, carburettor	S5	280	Relay, heater, driver seat	P2
195 (alt)	Fuel valve (diesel)		284	Air mass meter	N1, V4
196	Idle valve	AD5	289	Power stage	P3
197	Oil pressure sensor	E3	292	Solenoid, idle increase	Z6
198	Throttle switch, LH-Jetronic	AF4	293	Relay, idle increase	W6, X6, Z6
199	Temperature sensor (diesel)	S5	295	Relay, gear shift indicator	S5
200	Actuator solenoid	X4	346	Roof light, trunk	M6
201	Actuator solenoid, overdrive	Y5, Z4	347	Door contact, rear	N6
202	Climate control	Z3, AA2	361	Injector 1	N3, V2, V6, AF6
203	Temperature control	Ä2, AB2	362	Injector 2	N3, V2, V6, AF6
204	Outside temperature sensor	Ä2, AB2	363	Injector 3	N3, V2, V6, AF6
205	Inside temperature sensor	Ä2, AB2	364	Injector 4	N3, V2, V6, AF6
206	Program controller	Ö2, AC2	375	Seat heater switch, driver	L5, P3
207	Horn	A4	376	Ballast resistor	U6, V2
208	Glow plug (diesel)	S5	377	PTC resistor	C3
210	Tank pump	F5	378	Positive pole, engine compartment, ABS	T2
211	Fuel pump	D5	384	Brake fluid level sensor	F3, T1
212	Service socket	D3, S5, W6, X6, AD5	403	Temperature sensor, battery	B2
213	Throttle switch, Motronic	P1	404	Microswitch, vacuum control valve	T3
214	Crankshaft position indicator	U1	405	Half-automatic heated choke	B5
215	Engine RPM sensor	U1	406	Control unit, Renix	Q3
216	Control unit, Motronic	U1	407	Impulse generator, Renix	R3
217	Control unit, LH-2.2	U3	413	Impulse generator, EZ-K	Q2, AE2
218	Knock sensor	Q2, Q6, AE3	419	Power stage, EZ-K	P1, P4
219	Test point, Lambdasond	U4	420	Power stage, REX-1	AO2
220	Test point, idle	V4, AA3	424	Solenoid valve, charge pressure limiter	R6
221	Heated rear window	K2	438	Seat heater contact	J3
223	Cigarette lighter	G2	440	Pressure sensor, REX-1	AF4
224	Thermostat, electric cooling fan	W1	464	Relay, injectors	N3, U6, V3, AE6
225	Cruise control switch	W1	472	Control unit, LH-2.4	M1, M2
226	Control unit, cruise control	W2	473	Control unit, REX-Regina	AE4
227	Vacuum pump, cruise control	W2	479	Control unit, DIM-DIP	G5
228	Bridge connector, clutch	T5, W2	482	Diagnostic outlet	N2, Q2, P2, AD3
229	Bridge connector, brake	T5, W2	488	Relay, hazard lights	T5
231	Relay, fog lights, rear	T5, AA6	490	Power aerial switch	Y2
232	Relay, hot spot	X5	886	Transfer box 1234705	N1, V4
233	Turbocharger pressure sensor (turbo diesel)	R6, E3	901	Amplifier, radio	X2
234	Thermostat, hot spot	Y5	928	SRS	B4
235	PIC resistor, hot spot	X5	929	Ignition module, SRS	B5
238	Motor, window washer, rear	S4	931	Safety circuit, SRS	G4
239	Cut-over, window washer, rear	R4			

375

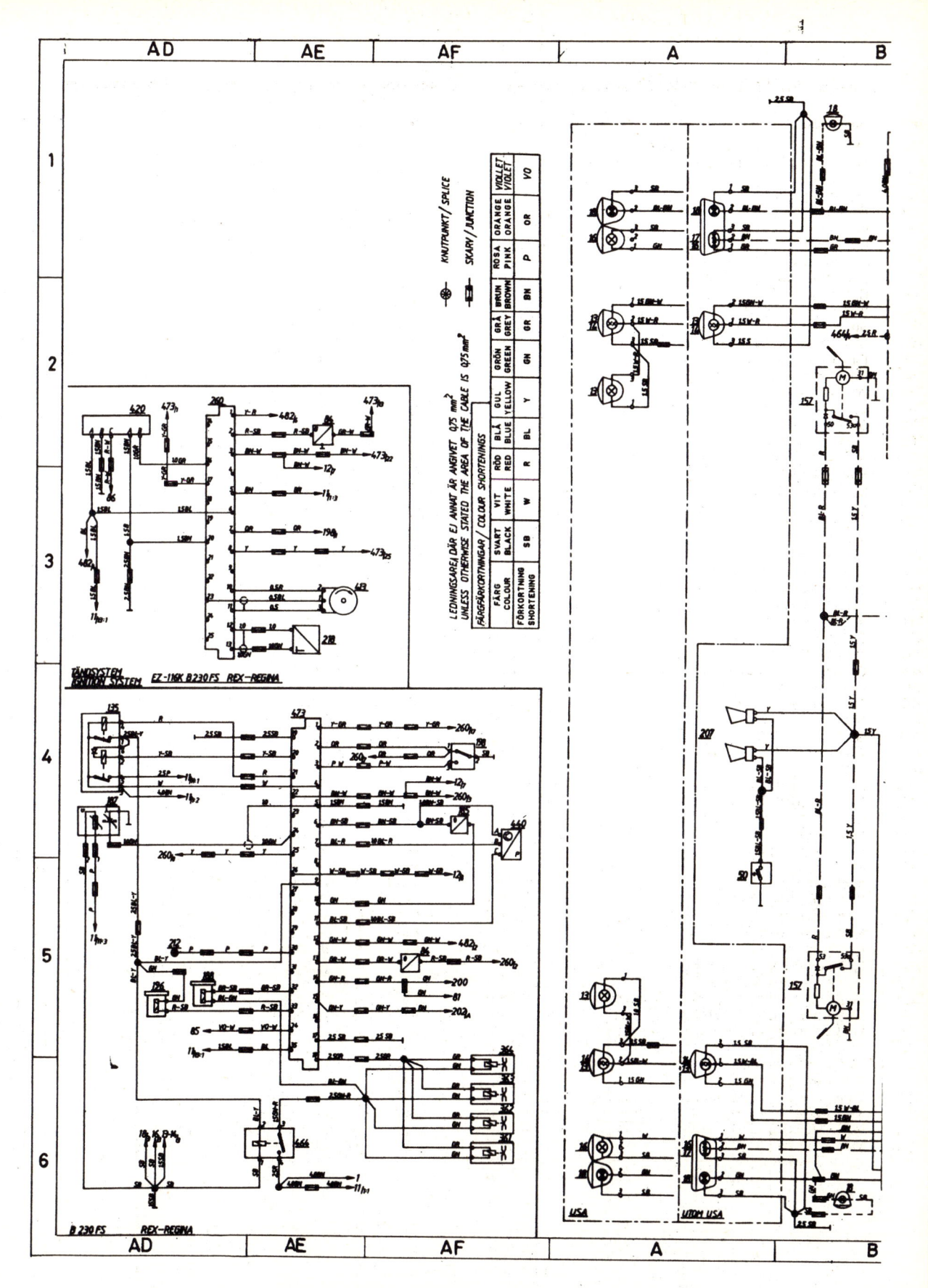

Fig. 13.41 Main wiring diagram for 740 models

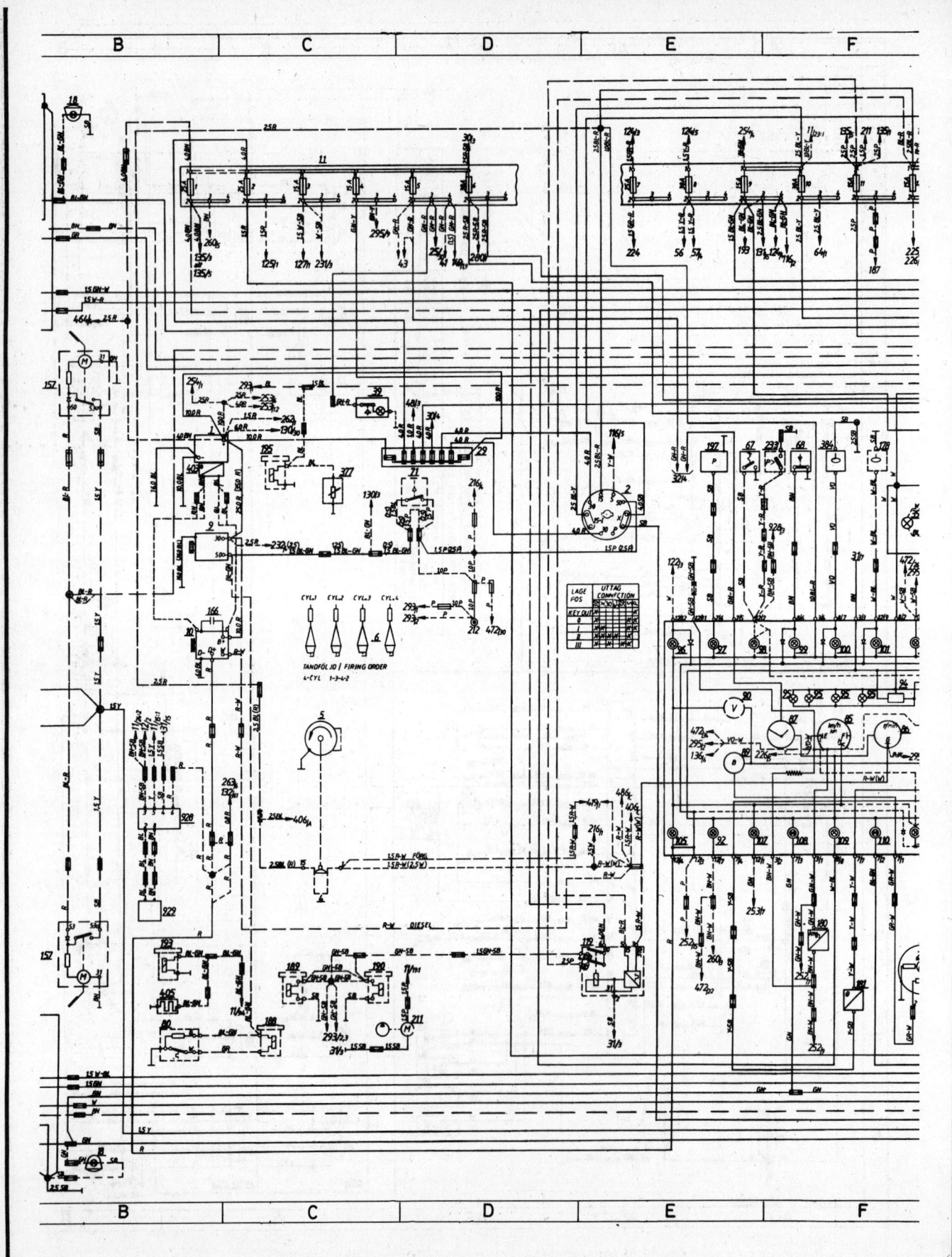

Fig. 13.41 Main wiring diagram for 740 models (continued)

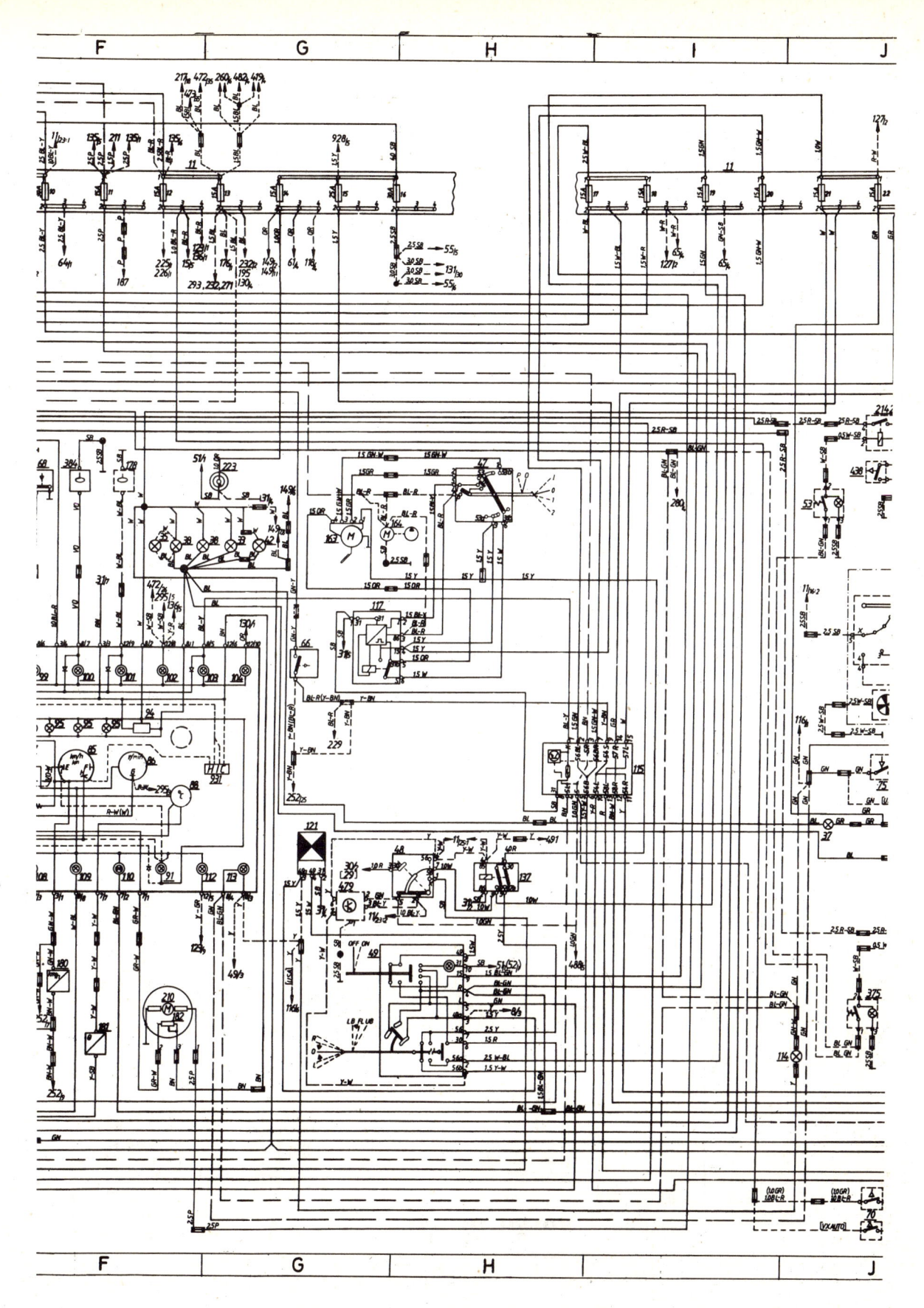

Fig. 13.41 Main wiring diagram for 740 models (continued)

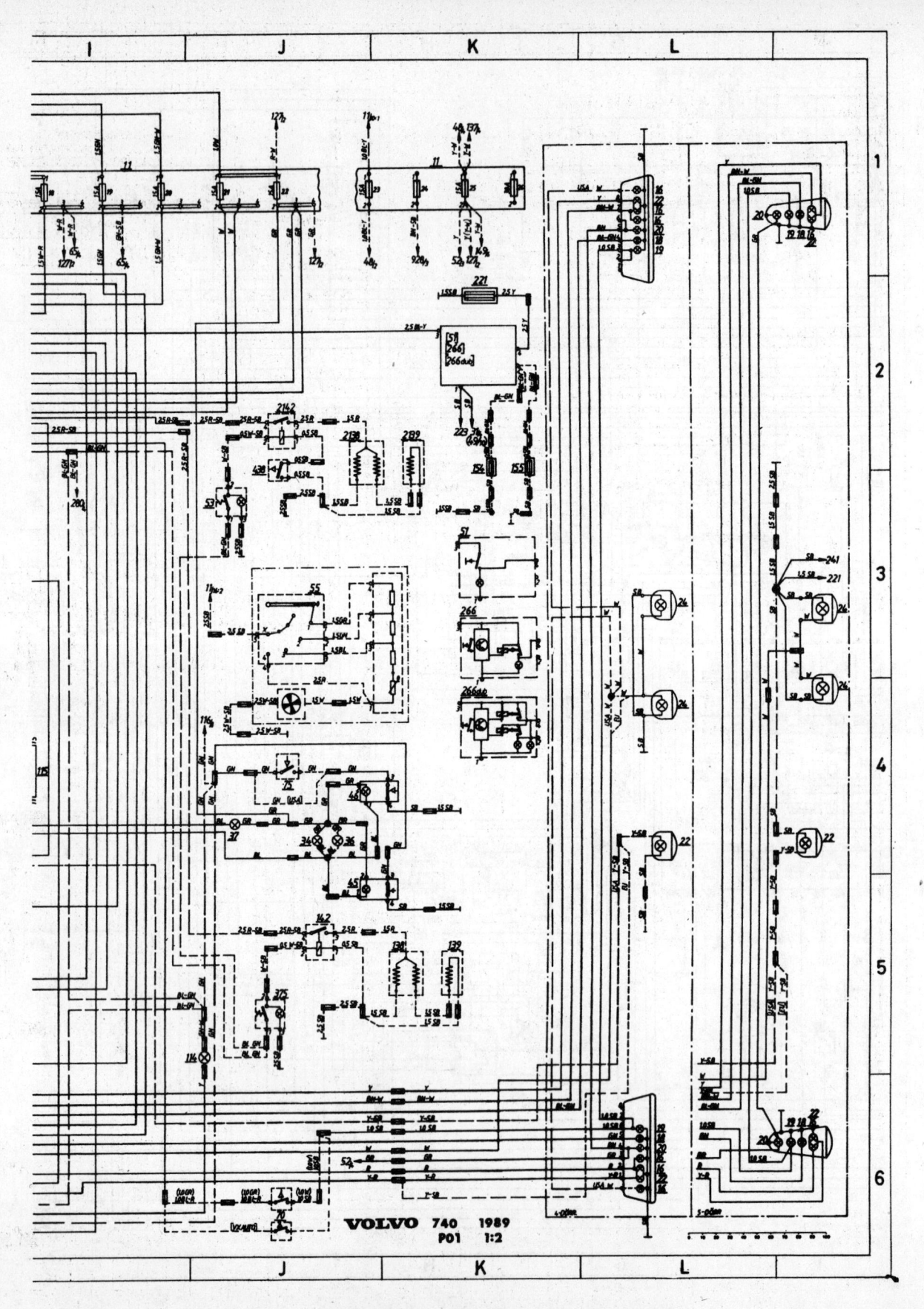

Fig. 13.41 Main wiring diagram for 740 models (continued)

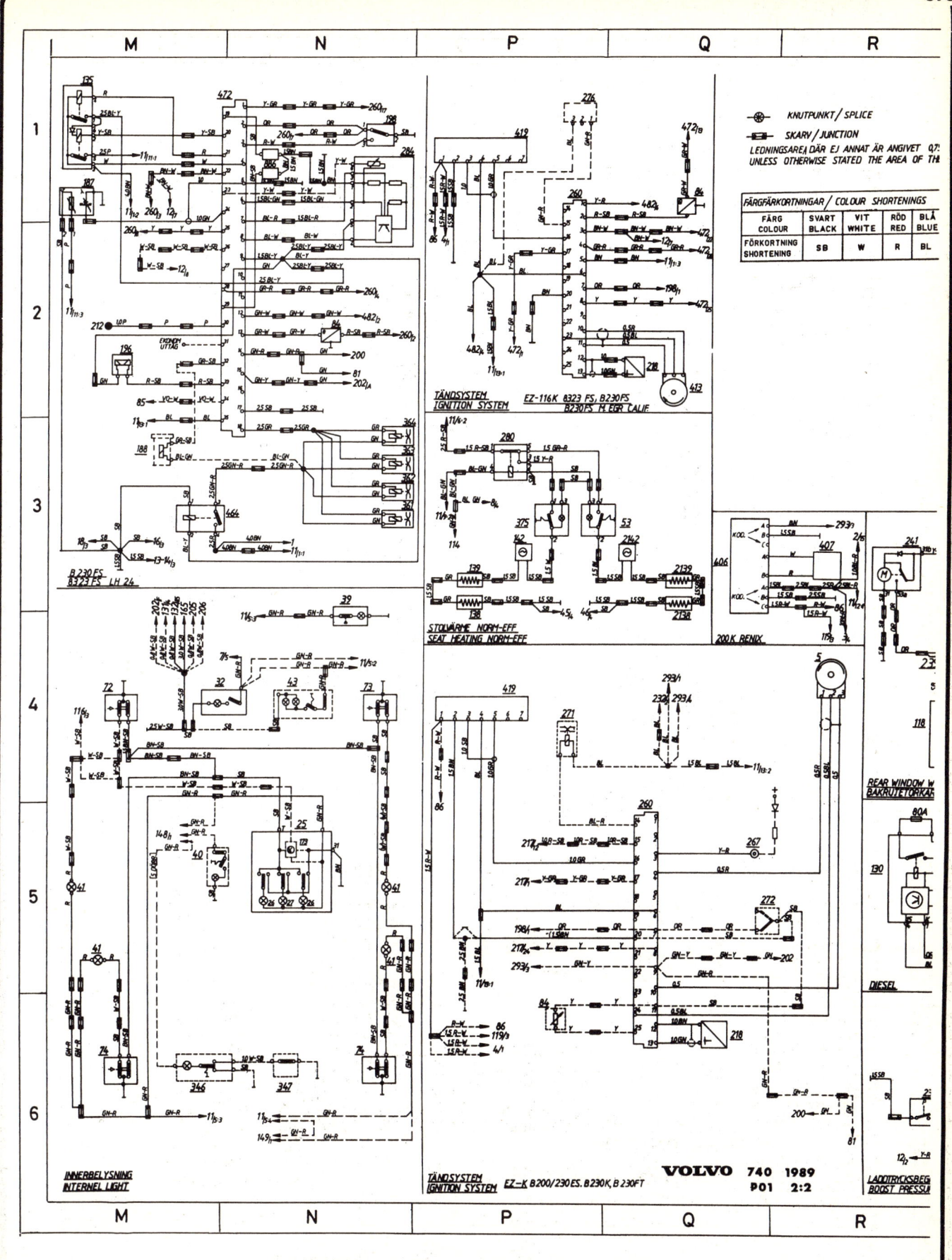

Fig. 13.42 Supplementary wiring diagram for 740 models

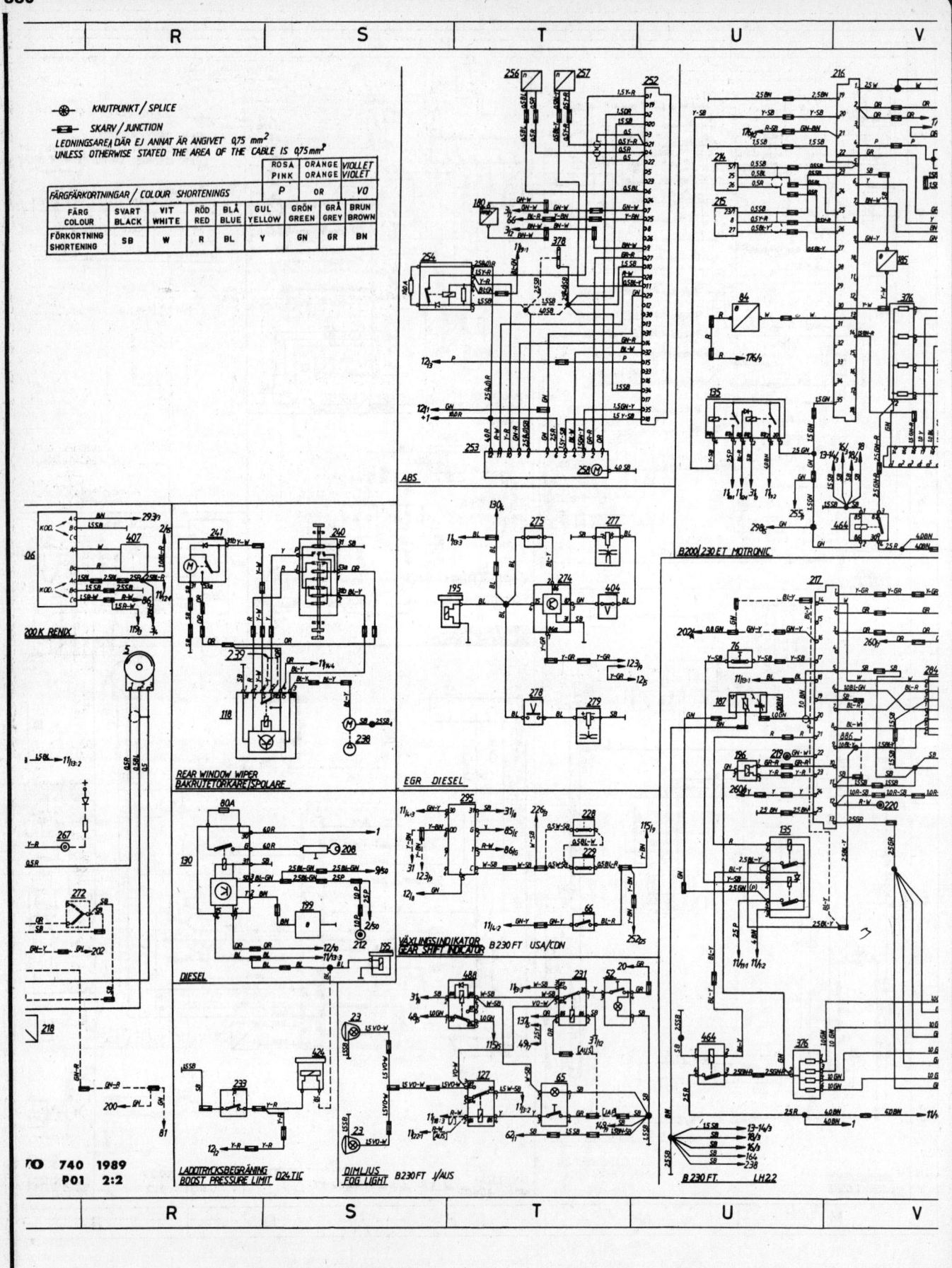

Fig. 13.42 Supplementary wiring diagram for 740 models (continued)

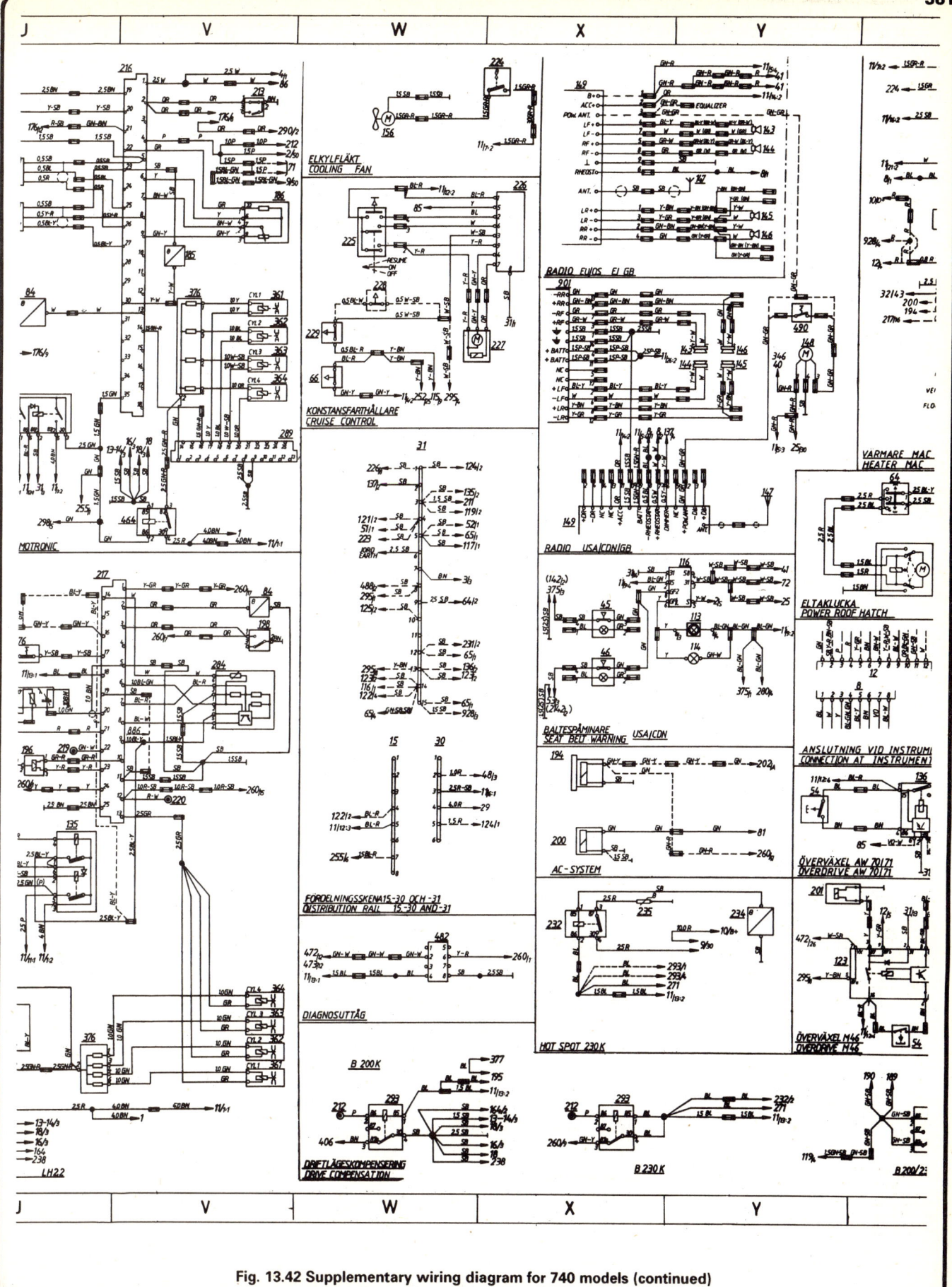

Fig. 13.42 Supplementary wiring diagram for 740 models (continued)

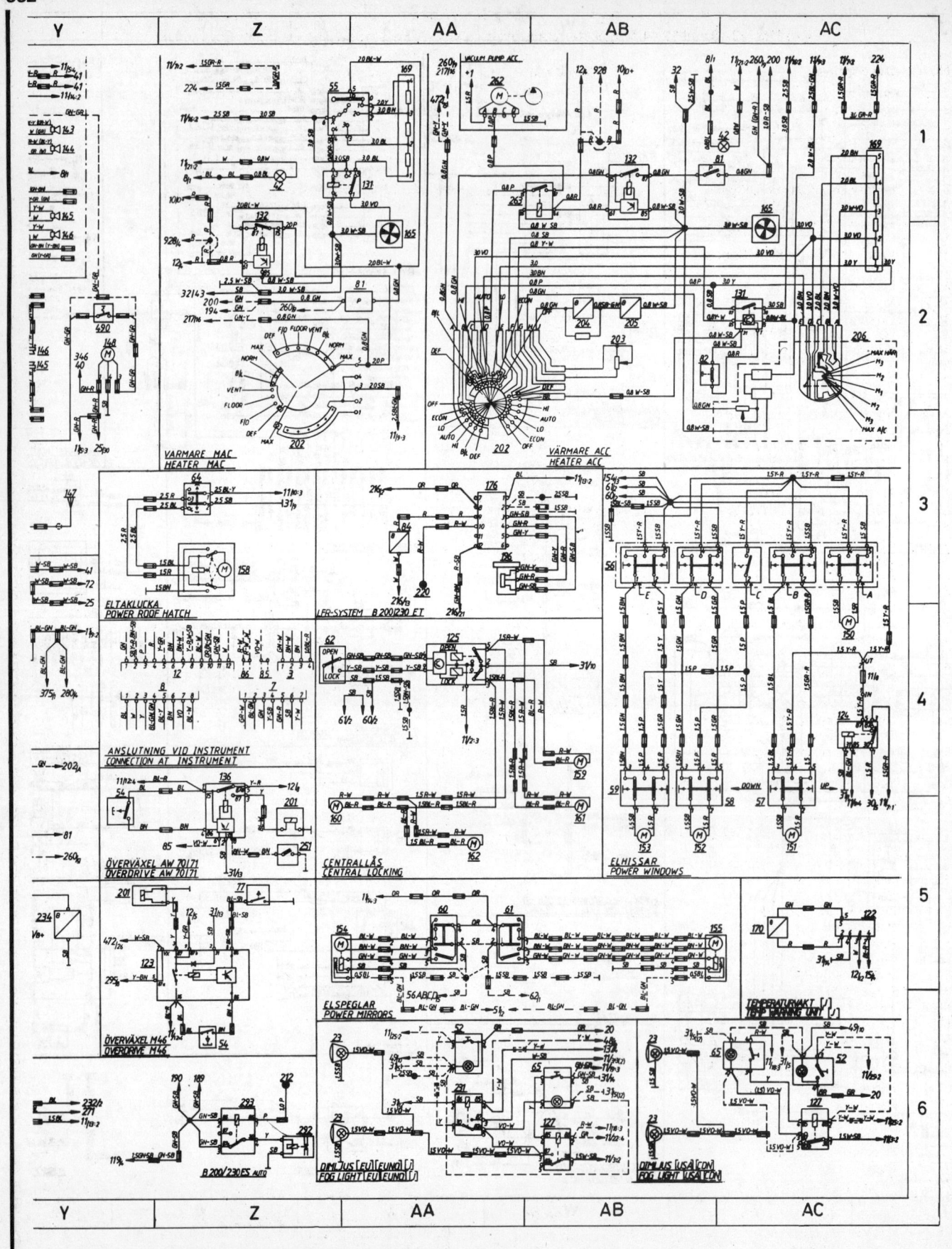

Fig. 13.42 Supplementary wiring diagram for 740 models (continued)

Key to Figs. 13.43 and 13.44. Not all items are fitted to all models; some systems do not apply to UK

No	Description	Coordinate	No	Description	Coordinate
1	Battery 12V	D3	84	Coolant temperature sensor	Q2, P3, P5, R3, S3, V1
2	Ignition switch	E3	85	Speedometer	G4, Z4
3	Instrument connection (4-pole)	T5	86	Tachometer	H4
4	Ignition coil 12V	D6, N5, P5	87	Clock	F4
5	Distributor	D5, R5	88	Engine temperature gauge	H4
6	Spark plug	E5	89	Fuel gauge	F4
8	Instrument connection (8-pole)	T5	90	Indicator light, rear foglight	F5
9	Starter motor 800W	Q3	91	Indicator lamp, service	F5
10	Alternator (incl. regulator)	Q4	92	Indicator lamp, diagnosis	F5
11	Fusebox	B-K1	93	Speed warning chime	G4
12	Instrument connection (12-pole)	T5	94	Panel lighting intensity (rheostat)	H4
13	High beam 60W	A1, A6	95	Instrument lighting	G4
14	Low beam 55W max	A1, A6	96	Indicator lamp, temperature warning	H5
15	Relay unit, 15 circuit		97	Indicator lamp, oil pressure	G5
16	Parking light 4CP/5W (USA: rear also)	A1, A6, L1, L6	98	Indicator lamp, boost pressure (turbocharger)	H5
17	Hazard light 32CP/21W	A1, A6	99	Indicator lamp, parking brake	G5
18	Instrument connection (18-pole)	T6	100	Indicator lamp, brake failure	G5
21	Tail light 4CP/5W	L2, L6	101	Indicator lamp, washer level	F5
22	Brake light 32CP/21W	L1, L5, L6	102	Indicator lamp, W/70/71 auto.trans. overdrive	H5
23	Fog light 55W	A1, A6, L1, L6	103	Indicator lamp, bulb failure	F5
24	Numberplate lighting 4CP/5W	L4	104	Indicator lamp, glow plugs (diesel)	H5
25	Courtesy light	W5	105	Indicator lamp, battery charge	F5
26	Reading light, front 5W	W5	106	Indicator lamp, trailer	F5
27	Reading light, front 5W	W5	107	Indicator lamp, ABS	G5
27	Courtesy light 10W	W5	108	Indicator lamp, left turn signal	G4
28	Reading light, rear 5W	V5, W5	109	Indicator lamp, high beam	G5
29	Turn signal bulb 32 CP/21W	A1, A6, B1, B6, L2, L6	110	Indicator lamp, right turn signal	G4
30	Relay unit (30 circuit)	T4	112	Indicator lamp, M46 overdrive	G5
31	Relay unit (21 circuit)	T4	113	Indicator lamp, front seat belts	H5
32	Glove compartment light 2W	W4	114	Indicator lamp, rear seat belts	I4
33	Ashtray light, front 1.2W	I3	115	Bulb failure warning sensor (14-pole)	K5
34	Ashtray light, rear 1.2W	I3	116	Fasten seat belt light (USA)	I4
35	Sunroof switchlight 1.2W	I3	117	Windscreen wipers, intermittent relay	B5
36	Seat heater switch light, passenger 1.2W		118	Rear windscreen wiper, intermittent relay	U6
37	Gear selector light, auto. trans. 1.2W	I3	120	Bulb failure warning sensor (9-pole)	C2
38	Instrument and panel lighting 1.2W	H3	121	Flasher unit and switch, hazard lights	C4
39	Engine compartment light 15W	W3	122	Relay, temperature warning (Japan)	R5
40	Trunk light	V5	123	Relay, M46 overdrive	N6
41	Door-open warning light	V4, V5, W4, W5	124	Relay, power windows and electric cooling fan	Y5
42	Heater controls light	H3	125	Relay, central lock	X6
43	Sun visor light	W4	127	Relay, auxiliary lights	C3
44	Glove compartment light switch	W4	130	Relay, glow plugs (diesel)	S5
45	Seat belt lock light, driver 1.2W	J4	131	Relay, fan	V2
46	Seat belt lock light, passenger 1.2W	J3	135	Relay, Motronic, LH-2.2	N3, Q2, Q3
47	Windscreen wiper switch	B4	136	Relay, overdrive AW70/71	Q6
48	Headlight switch	B2	137	Relay, headlights	C3
49	Turn signal/hazard lights/headlight switch	C4	138	Seat heater pad, driver 30/130W	K4
50	Horn switch	A4		Seat heater pad, passenger 30/130W	K3
51	Heated rear window switch	K2	139	Seat heater backrest, driver	K4
52	Seat heater switch, driver	J4		Seat heater backrest, passenger	K3
53	Seat heater switch, passenger	J3	140	Loudspeaker, instrument panel, left	U2
54	Overdrive switch M46	P6	141	Loudspeaker, instrument panel, right	U2
56	Power window switch, driver, front	X3	142	Thermostat, driver seat	J4
57	Power window switch, passenger front	Y5		Thermostat, passenger seat	J3
58	Power window switch, driver, rear	Y5	143	Loudspeaker (4 Ω), passenger door	U1
59	Power window switch, passenger, rear	X5	144	Loudspeaker (4 Ω), driver door	U2
60	Power door mirror switch, driver	X5	145	Loudspeaker (4 Ω), rear left	U2
61	Power door mirror switch, passenger	X5	146	Loudspeaker (4 Ω), right rear	U1
62	Central lock linkage switch	W6	147	Aerial	T2
64	Sunroof switch	T6	148	Power aerial 3A	U3
65	Fog light switch, front/rear	B3	149	Radio	T2
66	Brake light contact	J5, R6, S5	150	Motor (5A), driver power window	Y4
68	Parking brake switch	G5	151	Motor (5A), passenger power window	Y5
70	Back-up light contact	J6	152	Motor (5A), driver power window, rear	X5
71	Start inhibitor switch, auto.trans	E3	153	Motor (5A), passenger power window, rear	X5
72	Driver door switch	V3	154	Power mirror, driver side	W5
73	Passenger door switch	X3	155	Power mirror, passenger side	Y5
74	Door contact, rear	V4, X4	156	Motor (13A), electric cooling fan	W2
75	Passenger seat contact	I4	157	Motor (1A), headlight wiper	A2, A5
76	Pressure sensor, turbocharger	N3	158	Motor, sunroof opening	T6
77	Overdrive contact (M46)	P6, R6	159	Motor, central lock, passenger door	Y6
78	Positive terminal	E2	160	Motor, central lock, rear door, driver side	W6
79	Fuse box light	W4	161	Motor, central lock, rear door, passenger side	Y6
80	Thermal time-switch	O4	162	Motor, central lock, trunk	Y6
81	AC pressure sensor	U3	163	Motor (3.5A), windscreen wiper	B4

383

Key to Figs. 13.43 to 13.44. Not all items are fitted to all models; some systems do not apply to the UK (continued)

No	Description	Coordinate	No	Description	Coordinate
164	Motor (2.6A), windscreen washer	A2	253	Hydraulic unit, ABS	S3
166	Capacitor	D4	254	Transient surge protector, ABS	S2
167	Suppressor resistor	E4	256	Sensor ABS, front left	S1
170	Thermo-element, catalytic converter	R5	257	Sensor ABS, front right	S1
176	Control unit, CIS	S3	260	Control unit, EX-K ignition system	N5, Q4
178	Washer level sensor	F5	265	Delay relay, heated rear window	K2
182	Fuel level sensor	R3	266	Delay relay, heated rear window and door mirrors	K3
185	Charge air temperature sensor	Q2	267	Test point, EZ-K	Q5
186	Air flow meter	R1	270	Speedometer sensor	S1
187	Lambdasond	N3, Q4	284	Air mass meter	R4
188	Cold start injector	E4	289	Power stage, DME	R2
195	Solenoid valve, carburettor	S5	295	Relay, gear shift indicator	G5, R5
195	(alt) fuel valve (diesel)	S5	296	Control unit, DIM-DIP	T3
196	Idle valve	N4, Q4, S3	346	Roof light, trunk	W5
197	Oil pressure sensor	G6	347	Door contact, rear	W5
198	Throttle switch	P3, P5, R3, R5	361	Injector 1	P4, R2, R4
199	Temperature sensor (diesel)	S5	362	Injector 2	P4, R2, R4
200	AC compressor (3.9A)	U3	363	Injector 3	P4, R2, R4
201	Actuator solenoid, overdrive	P6, R6	364	Injector 4	P4, R2, R4
207	Horn (5A + 5A)	A3	365	Injector 5	R4
208	Glow plug (diesel)	S5	366	Injector 6	R4
210	Tank pump	F3	376	Series resistance	P4, R2
211	Fuel pump (6.5A)	N4, Q2, Q4	378	Ground, ABS	S1
212	Service socket	E4	384	Brake fluid level sensor	G5, S1
213	Throttle switch, Motronic	R1	413	Impulse generator, EZ-K	P5
214	Crankshaft position sensor	Q1	416	HT lead sensor	N5
215	Engine RPM sensor	Q1	417	Service socket, EZ-K	P5
216	Control unit, Motronic	Q1	419	Power stage, EZ-K	N4, Q4
217	Control unit, LH-2.2	N3, R3	424	Solenoid valve, charge pressure limiter	S6
218	Knock sensor	N4, Q4	425	Temperature sensor, charge pressure limiter	S6
219	Test point, Lambdasond	N4, Q4	438	Seat heater contact	L3
220	Test point, idle system	P4, R4, S3	456	Relay unit (30 circuit)	J4
221	Heated rear window 150W	L2	457	Relay unit, fuse box	W4
222	Cigarette lighter (7A)	I3	458	Relay unit (30 circuit), fuse box	W4
223	Cigarette lighter light	I2	464	Relay, injectors	N3, Q4, R2
224	Thermostat, electric cooling fan	W3	490	Power aerial switch	U2
225	Cruise control switch	S4	491	Dimmer	B3
226	Control unit, cruise control	T4	495	Heater controls, ECC 130	Y1
227	Vacuum pump, cruise control	S5	496	Heater controls, ECC sensor	W1
228	Bridge connector, clutch pedal	R6, S5	497	Solenoid valve, ECC	W1
229	Bridge connector, brake pedal	S5	498	Servo motor, ECC heater	Y3
231	Relay, rear fog lights	C3	499	Power unit, ECC	X2
233	Turbocharger pressure sensor (turbo diesel)	S6	501	Fan motor, ECC	X2
238	Motor, window washer, rear (2.6A)	U6	502	Ambient air temperature, ECC	V1
240	Switch, wiper/washer, rear	U5	503	Sun heat sensor, ECC	W1
241	Motor, wiper, rear	U5	504	Inside air temperature sensor, ECC	V1
242	Power seat emergency stop	V6	870	Clock switch	H3
243	Relay, power seat	V6	886	Transfer box 1234705	P4, R4
244	Control unit, adjustable power seat	V5	900	Accessories	H3
245	Power seat motor, fore-aft	W6	901	Amplifier, radio	T1
246	Power seat motor, up-down, front	W5	928	SRS	C5
247	Power seat motor, up-down, rear	W5	929	Ignition module, SRS	C5
248	Power seat motor, backrest tilt	W6	930	Indicator lamp, SRS	H5
251	Kickdown inhibitor	R6	931	Safety circuit, SRS	H5
252	Control unit, ABS	T1			

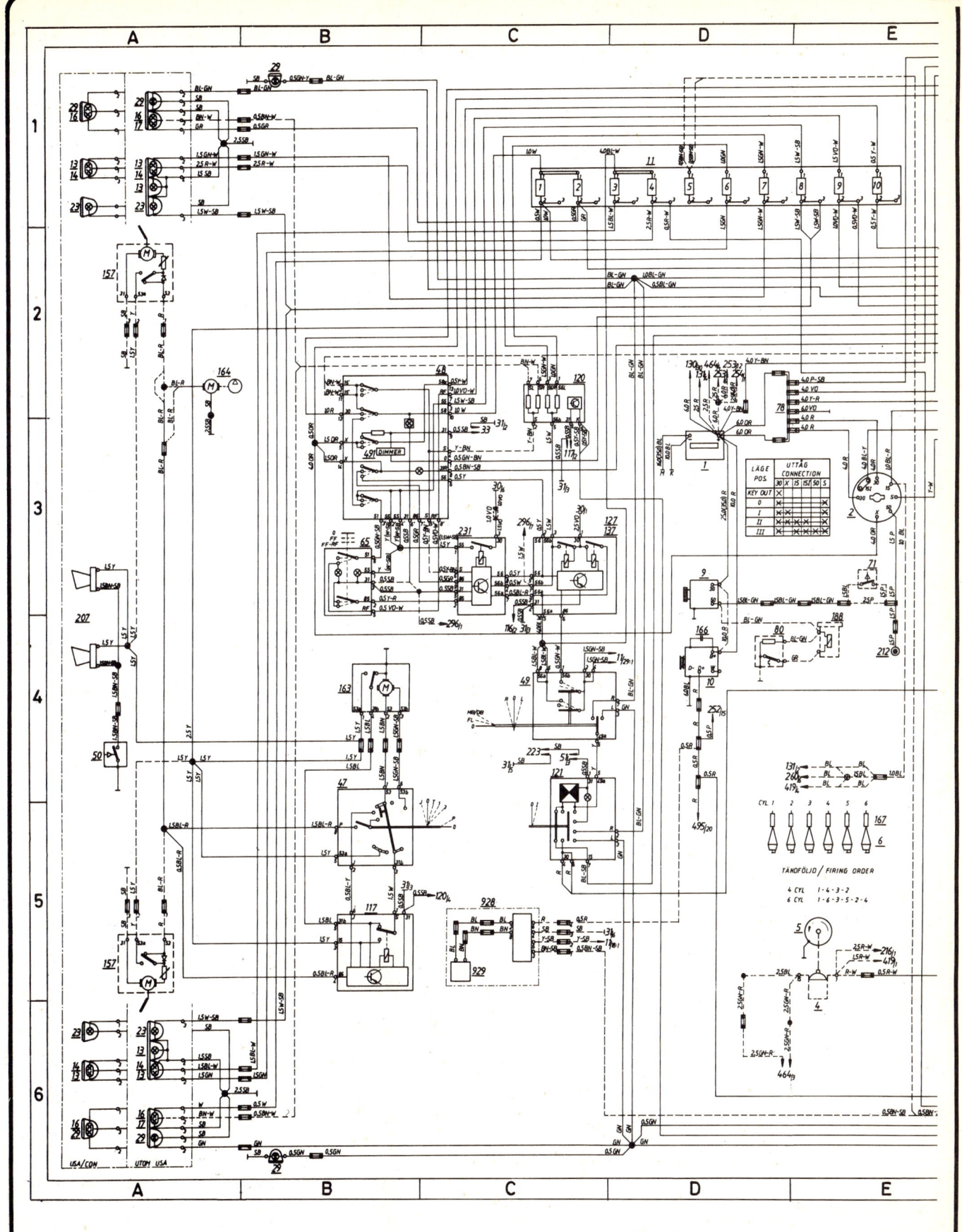

Fig. 13.43 Main wiring diagram for 760 models

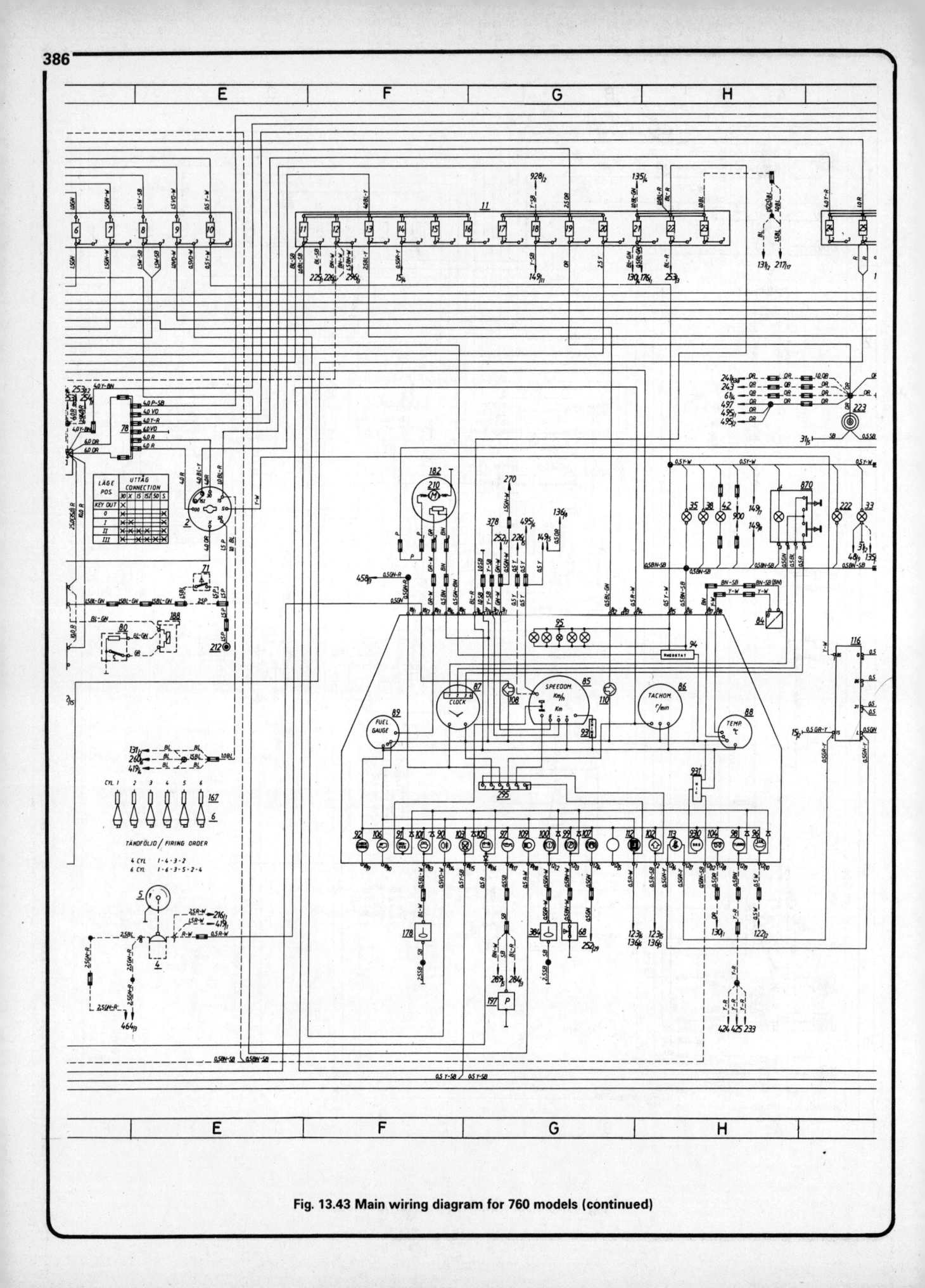

Fig. 13.43 Main wiring diagram for 760 models (continued)

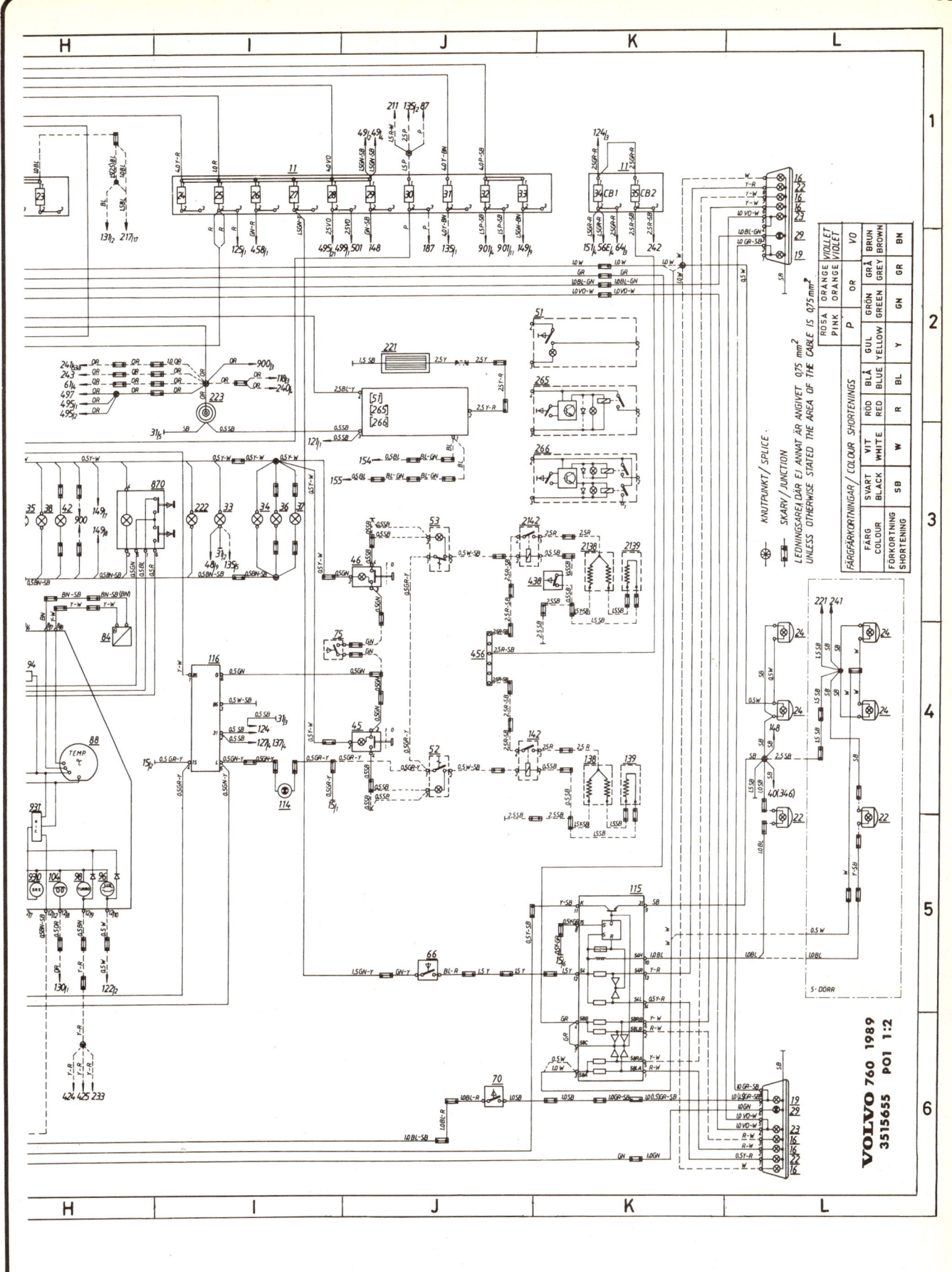

Fig. 13.43 Main wiring diagram for 760 models (continued)

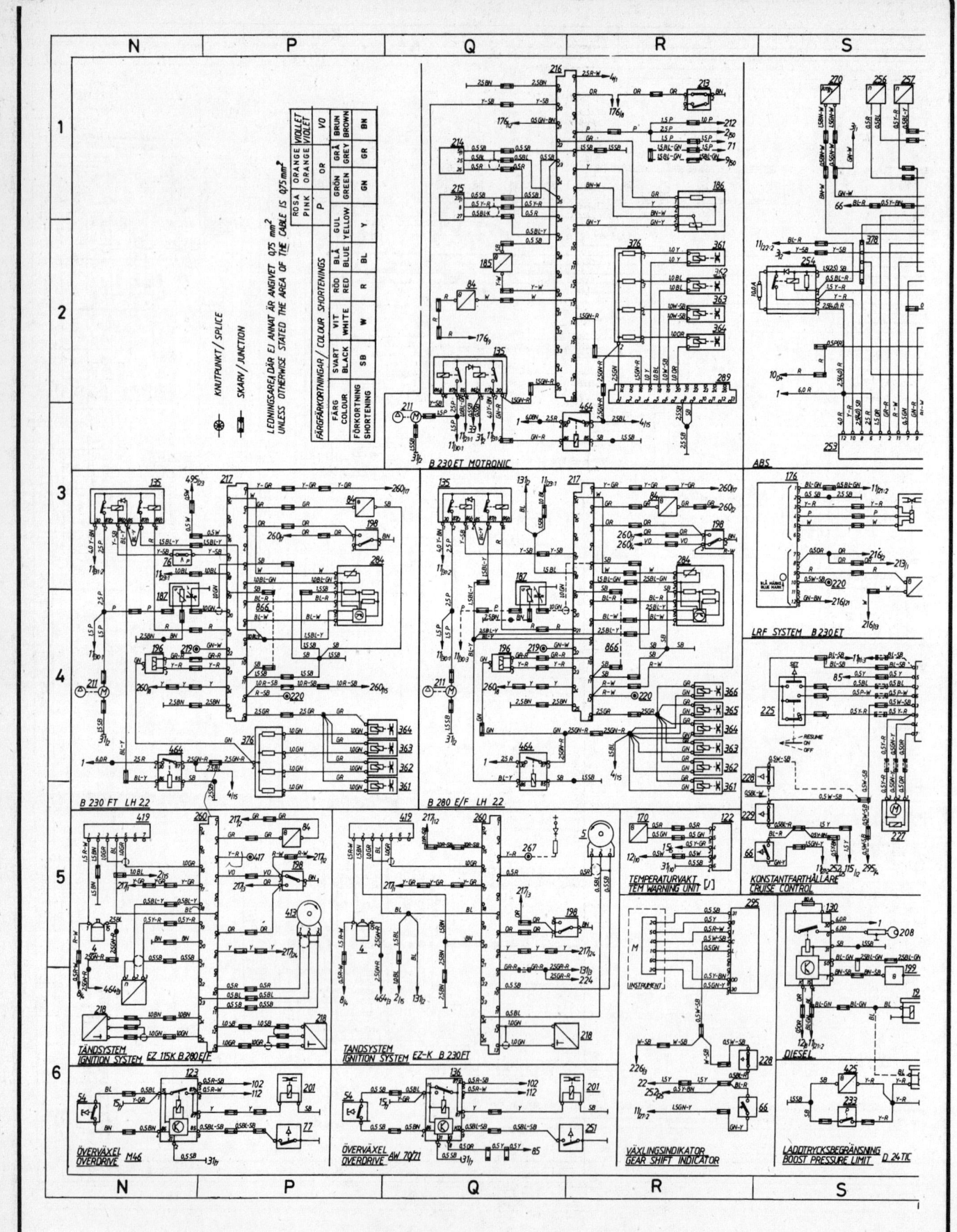

Fig. 13.44 Supplementary wiring diagram for 760 models

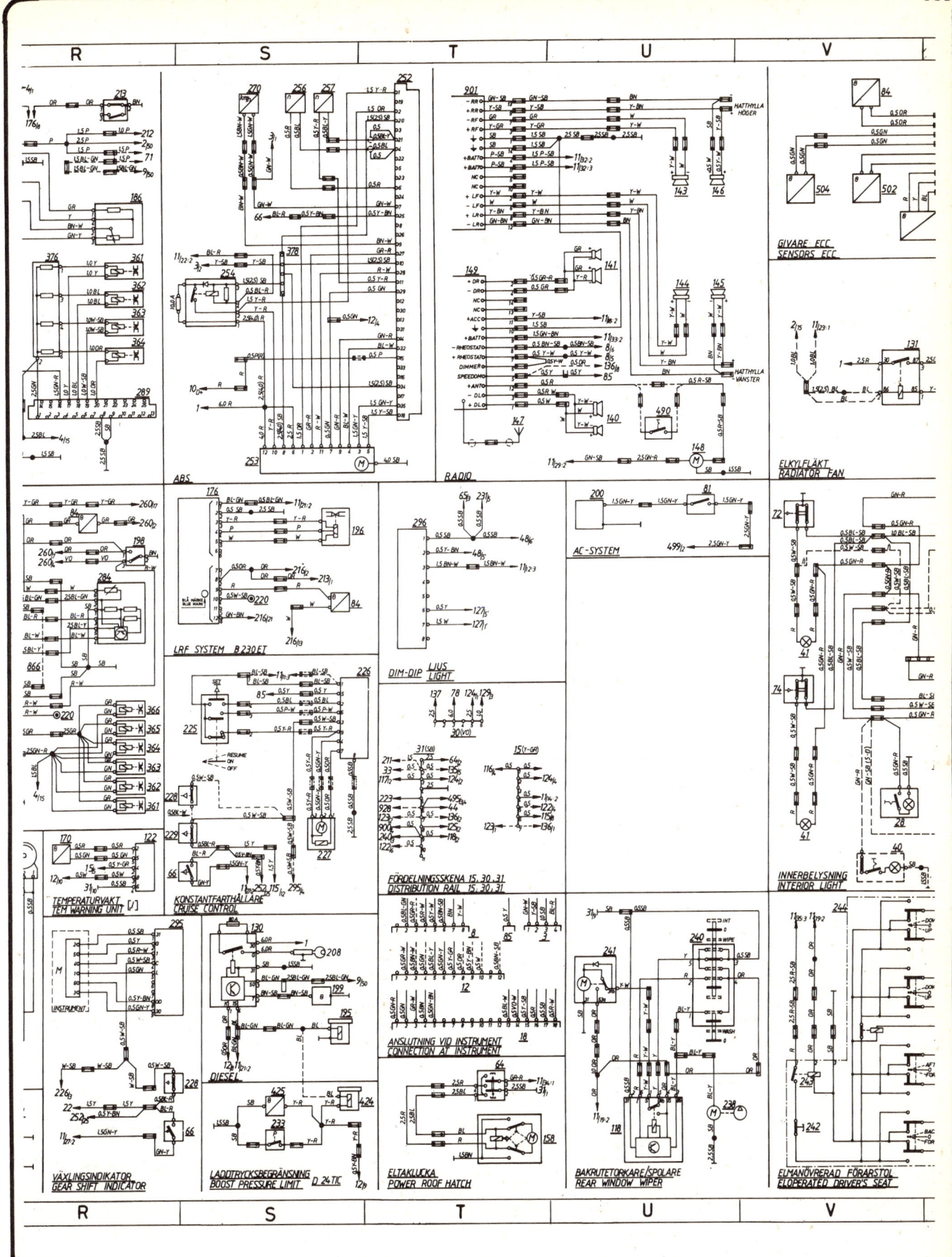

Fig. 13.44 Supplementary wiring diagram for 760 models (continued)

Fig. 13.44 Supplementary wiring diagram for 760 models (continued)

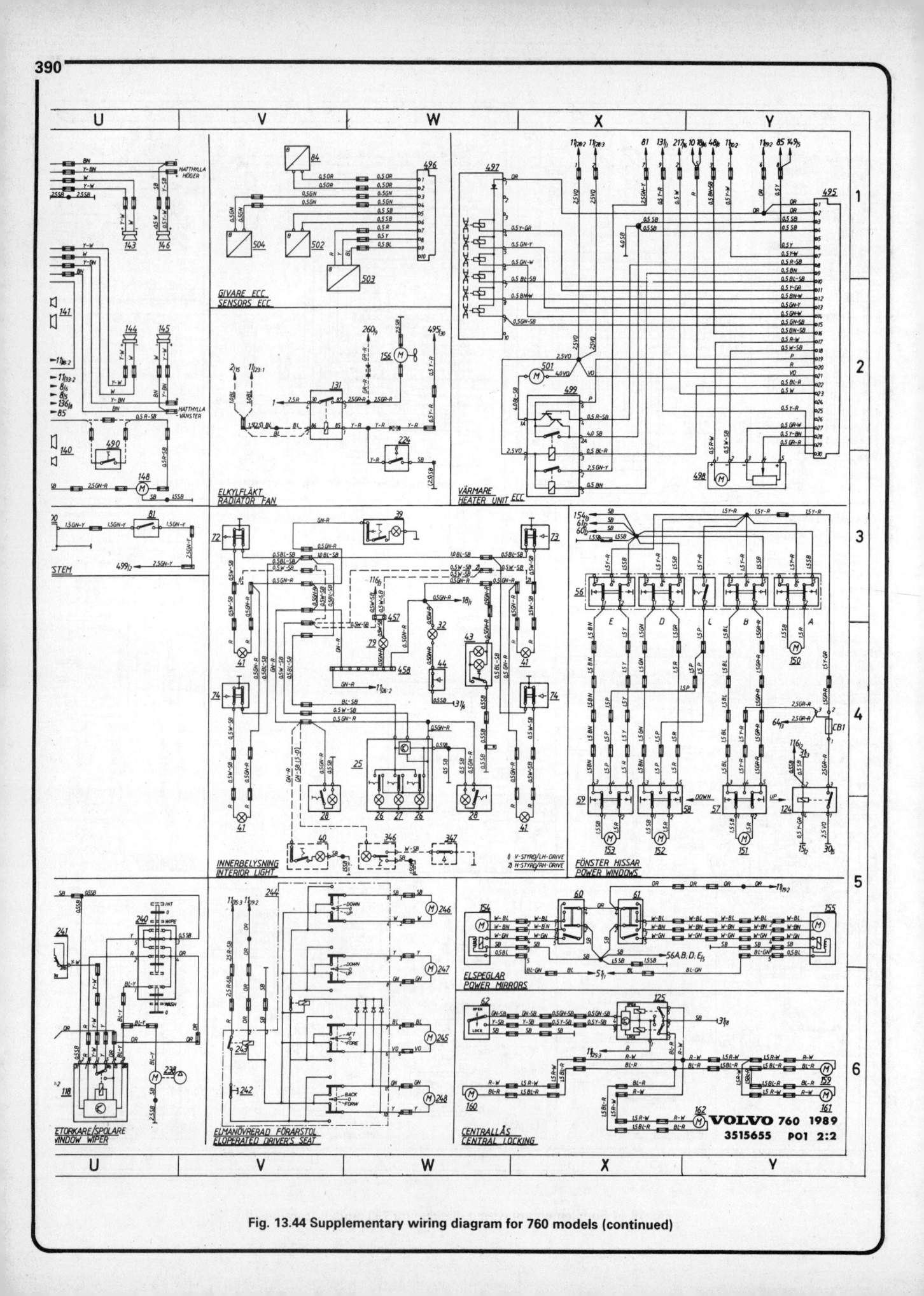

Index

A

About this manual – 4
Accelerator
    cable removal, refitting and adjustment – 108
    pedal removal and refitting – 109
Acknowledgements – 2
Advance systems (ignition)
    description and testing – 150
Aerial removal and refitting – 296
Air cleaner
    element renewal – 99, 341
    removal and refitting – 100, 341
Air conditioning
    compressor drivebelt – 81
    control panel removal and refitting – 88
    fault diagnosis – 95
    general description – 88
    maintenance – 88
    precautions – 88
Alternator
    brush renewal – 280
    drivebelt – 81
    precautions – 279
    removal and refitting – 280
    testing – 279
Antifreeze mixture – 80
Anti-lock braking system (ABS)
    bleeding – 359
    caliper – 215
    component removal and refitting – 223, 357, 358
    control unit – 358
    description – 224
    fault tracing – 226
    master cylinder – 219
    modifications – 357
Anti-roll bar removal and refitting
    front – 240
    rear – 244
Ashtray light bulb renewal – 288
Automatic choke adjustment – 121
Automatic climate control
    programmer removal and refitting – 94
    sensors
        description – 92
        testing, removal and refitting – 92
    vacuum hose connections – 339
Automatic transmission
    fault diagnosis – 199
    fluid
        level checking – 193
        renewal – 193
    gear selector – 193, 288
    general description – 191
    kickdown cable – 194, 195
    maintenance and inspection – 192
    oil seals – 197
    overdrive switch – 197
    removal and refitting – 198
    reversing light switch – 196
    selector – 193, 288
    specifications – 165, 332
    starter inhibitor switch – 196
Auxiliary front light
    bulb renewal – 284
    removal and refitting – 286
Auxiliary shaft
    examination and renovation – 51
    refitting – 54
    removal – 47
Axle see **Rear axle** or **Final drive and driveshafts**

B

Ballast resistor
    testing, removal and refitting – 155
Battery
    charging – 279
    disconnection – 367
    maintenance – 279
    removal and refitting – 279
Bearings
    clutch release – 162
    engine (in-line)
        crankshaft spigot – 43
        main – 47, 48, 53
    engine (V6)
        crankshaft spigot – 67
        main – 71, 74, 338
    halfshaft – 207
    mainshaft (manual gearbox) – 353
    propeller shaft – 203
    roadwheel – 233, 238, 336, 359, 363
Bleeding hydraulic system
    brakes – 214, 359
    clutch – 161
    power steering – 233
Blower motors (heater) – 92
Bodywork and fittings – 247 et seq, 364 et seq
Bodywork and fittings
    bonnet – 250, 252, 364
    boot lid – 251
    bumpers – 270, 271, 367
    central locking – 260, 364
    centre console – 268
    damage repair
        major – 250
        minor – 248
    doors – 250, 253, 255, 256, 258, 259, 261, 364
    engine undertray – 272
    facia – 172
    general description – 247
    glovebox – 268, 367
    grille – 272
    head restraints – 263
    locks – 237, 258, 259, 260, 364
    maintenance
        bodywork and underframe – 247
        upholstery and carpets – 248

Index

mirrors – 261, 369
rear console – 269
seat belts – 267
seats – 263, 264, 267, 365, 367
spoiler – 270
steering column trim – 268
sunroof – 273, 275
tailgate – 252, 256, 259
undertray – 272
weatherstrip – 261
wings – 273
windows – 256, 257, 365
windscreen – 256
Bodywork repair – *see colour pages between pages 32 and 33*
Bonnet
light bulb renewal – 288
release cable removal and refitting – 252
removal and refitting – 250, 364
Boot lid
lock removal and refitting – 259
removal and refitting – 251
Braking system – 209 *et seq*, 355 *et seq*
Braking system
ABS – 224, 226, 357, 358, 359
bleeding – 214, 359
caliper
front – 215, 356
rear – 216, 217, 356
disc
front – 217, 356
rear – 218
fault diagnosis – 226, 228
fluid level – 211, 355
general description – 210
handbrake – 213, 221, 223, 357
high level lights – 368
hoses – 213
maintenance and inspection – 211, 336
master cylinder – 218
pads
front – 212, 355
rear – 213, 355, 356
pedal – 220
pipes – 213
pressure differential warning valve – 219
servo – 213, 221
specifications – 209, 332
stop-light switch – 223
vacuum pump – 221
Bulb failure warning system – 293
Bulb renewal
ashtray light – 288
automatic transmission selector light – 288
auxiliary front light – 284
boot light – 287
brake light (high level) – 368
cigarette lighter illumination – 288
courtesy light – 287
day running light – 284
direction indicator – 284, 286
door edge marker light – 288
headlight – 284
instrument panel light – 288
load area light – 287
number plate light – 286
parking light – 284
rear light cluster – 284, 286
seat belt buckle light – 288
switch illumination – 288
under-bonnet light – 288
vanity mirror light – 369
Bumper removal and refitting
buffers – 271

front – 270, 367
rear – 271
Bushes (independent rear suspension)
renewal – 360

C

Cable
accelerator – 108
clutch – 158
handbrake – 223
heater – 90
kickdown – 194, 195
sunroof – 275
throttle – 108
Caliper (brake)
overhaul
front – 215, 356
rear – 217, 356
removal and refitting
front – 215
rear – 216, 356
Camshaft
in-line engine
cover gasket – 336
drivebelt removal, refitting and tensioning – 38
examination and renovation – 51
oil feed – 336
removal and refitting – 40
V6 engine
examination and renovation – 73
removal and refitting – 66
Carburettor fuel system
adjustments – 100, 116, 121, 341
air cleaner – 99, 100
automatic choke adjustment – 121
carburettor
description – 114, 341
overhaul – 117
removal and refitting – 116
EGR system – 343
fault diagnosis – 114, 122
fuel filter – 100
fuel pump – 105, 106
hot start valve – 122
idle speed adjustment – 100, 336
manifolds – 112, 113
Pulsair system – 341, 343
mixture adjustment – 100, 336
tamperproof screws – 99
Cassette player removal and refitting – 296, 371
Catalytic converter – 341
Central locking
components removal and refitting – 260, 365
Centre bearing (propeller shaft)
renewal – 203
Choke adjustment – 121
Cigarette lighter
bulb renewal – 288
removal and refitting – 296
Clutch – 158 *et seq*
Clutch
bleeding – 161
cable – 158
component inspection – 162
fault diagnosis – 163
general description – 158
maintenance and inspection – 158, 162

Index

master cylinder - 159
pedal - 159
release bearing - 162
removal and refitting - 161
slave cylinder - 160
specifications - 158
Compression test
  in-line engine - 57
  V6 engine - 77
Coil testing, removal and refitting - 155
Column (steering)
  removal and refitting - 236, 359
Connecting rods
  in-line engine
    examination and renovation - 50
    removal and refitting - 44
  V6 engine
    examination and renovation - 72, 338
    refitting - 75
    removal - 70
Console removal and refitting
  centre - 268
  rear - 269
Console switch (rear)
  removal and refitting - 289
Constant idle speed system
  description - 102
  adjustments - 102
Continuous injection system
  components removal and refitting - 128
  control pressure regulator ventilation - 341
  control unit overhaul - 129
  description - 123
  fault diagnosis - 114, 131
  testing procedures - 124
Control arm
  general - 359
  balljoint removal and refitting - 239
Control pressure regulator ventilation
  continuous injection system - 341
Control unit removal and refitting
  ABS - 358
  injection - 152, 351
  Motronic - 153
  seat heater - 367
Conversion factors - 22
Coolant
  draining - 80, 339
  filling - 80
  flushing - 80, 339
  level sensor - 85
Cooling, heating and air conditioning - 78 et seq, 339 et seq
Cooling and heating system
  air conditioning - 88, 92, 94, 95
  antifreeze mixture - 80
  automatic climate control - 92, 94, 339
  control cable - 90
  control panel - 88
  coolant level sensor - 85
  draining - 80, 339
  drivebelts - 81
  electric fan - 83
  electronic climate control - 339, 341
  fan - 82, 83
  fault diagnosis - 94, 95
  filling - 80
  flushing - 80, 339
  general description - 78, 86
  maintenance and inspection - 79
  matrix - 90
  motors - 91, 92
  oil coolers - 86
  radiator - 82
  specifications - 78, 330
  temperature gauge - 84, 85
  thermostat - 84

thermoswitch - 83
viscous coupled fan - 82
water pump - 83, 339
water valve - 92
Courtesy light bulb renewal - 287
Crankcase ventilation system
  in-line engine - 37, 336
  V6 engine - 60
Crankshaft
  in-line engine
    examination and renovation - 48
    refitting - 3
    removal - 47
    spigot bearing removal and refitting - 43
  V6 engine
    examination and renovation - 71
    refitting - 74
    removal and refitting - 70, 338
    spigot bearing removal and refitting - 67
Cruise control - 111
Cylinder block and bores (in-line engine)
  examination and renovation - 50
Cylinder block and liners (V6 engine)
  examination and renovation - 71
Cylinder head
  in-line engine
    decarbonising, valve grinding and renovation - 52
    dismantling - 51
    reassembly - 53
    removal and refitting - 41
  V6 engine
    cylinder head bolts - 338
    decarbonising, valve grinding and renovation - 73
    dismantling - 73
    reassembly - 73
    removal and refitting - 63, 70, 76
Cylinder liners (V6 engine)
  examination and renovation - 71
  identification - 338
  refitting and checking protrusion - 73
  removal - 71

D

Damaged body repair
  major - 250
  minor - 248
Day running lights
  bulb renewal - 284
  removal and refitting - 286
Direction indicators
  bulb renewal - 284
  fault tracing - 293
  removal and refitting - 286
Discs (brake)
  inspection - 217
  removal and refitting
    front - 217, 356
    rear 219
  renewal 359
Distributor
  drive oil seal - 337
  overhaul - 147
  removal and refitting - 145, 351
Door
  adjustment - 364
  edge marker bulb renewal - 288
  handles - 258
  interior trim - 253, 255, 364
  latches - 258
  locks - 258, 364
  mirror - 261
  removal and refitting - 250
  switches - 289, 290
  weatherstrip - 261

Index

window – 256
window lift mechanism – 257, 290, 365
Drivebelts (accessory)
  removal, refitting and tensioning – 81
Driveplate
  in-line engine
    examination and renovation – 51
    removal and refitting – 42
  V6 engine
    examination and renovation – 72
    removal and refitting – 66
Driveshaft removal and refitting – 354

E

EGR system
  description – 343
  maintenance – 343
Electrical system – 276 *et seq*
Electrical system
  aerial – 296
  alternator – 279, 280
  battery – 279, 367
  brake lights (high level) – 368
  bulb failure warning system – 293
  bulbs – 283, 287, 368, 369
  cigarette lighter – 296
  direction indicator fault tracing – 293
  fault diagnosis – 293, 302
  fuses – 291, 369
  general description – 278
  headlights – 286, 287, 293, 295, 367, 368
  heated rear window – 296
  horn – 293
  ignition/starter switch – 288
  instrument cluster – 290, 369
  interference suppression – 298
  lights – 283, 286, 287
  maintenance and inspection – 278
  radio/cassette player – 296, 372
  relays – 292, 370
  specifications – 276, 334
  speedometer – 291
  starter motor – 280, 281
  switches – 151, 190, 196, 197, 223, 288 to 290, 336, 337, 355, 372
  wash/wipe – 293, 294, 295, 372
  wiring diagrams – 304 to 325, 372 to 390
Electronic climate control
  control pump and control unit – 341
  description – 339
  sensors – 341
Electronic traction control – 153
Engine – 27 *et seq*, 336 *et seq*
Engine (in-line)
  auxiliary shaft – 47, 51, 54
  bearings – 43, 47, 48, 53
  bores – 50
  camshaft – 38, 40, 51, 336
  compression test – 57
  connecting rods – 44
  connection to transmission – 55
  crankcase ventilation system – 37, 336
  crankshaft – 43, 47, 48, 53
  cylinder block – 50
  cylinder head – 40, 41, 51, 52, 53
  dismantling – 39, 47
  driveplate – 42, 51
  examination and renovation – 48
  fault diagnosis – 57
  flywheel – 42, 51
  general description – 34, 336
  maintenance and inspection – 35, 336
  management relays – 155
  manifolds – 112, 113

  mountings – 45
  oil filter – 35
  oil pressure switch – 336
  oil pump – 44, 48
  oil seal – 42
  operations possible with engine installed – 39
  pistons – 44
  reassembly – 39, 54
  refitting – 56
  removal – 45, 46
  separation from transmission – 46
  specifications – 28, 327
  start-up after overhaul – 56
  sump – 43
  tappets – 51
  undertray – 272
  valves – 36, 52, 336
Engine (V6)
  bearings – 67, 70, 71, 74, 338
  camshaft – 66, 73
  compression test – 77
  connecting rods – 70, 72, 75, 338
  connection to transmission – 76
  crankcase ventilation system – 60
  crankshaft – 66, 67, 70, 71, 74, 338
  cylinder block – 71
  cylinder head – 63, 70, 73, 76, 338
  cylinder liners – 70, 71, 73, 338
  dismantling – 61, 69
  distributor drive oil seal – 337
  driveplate – 66, 72
  examination and renovation – 71, 338
  fault diagnosis – 77
  flywheel – 66, 72
  general description – 58, 336
  maintenance and inspection – 58, 336
  management relays – 155
  manifolds – 112, 113, 344
  mountings – 67, 338
  oil cooler – 337
  oil filter – 58
  oil level sensor – 77
  oil pressure switch – 337
  oil pump – 63, 71
  oil seal 66, 337
  operations possible with engine installed – 60
  pistons – 70, 72, 75, 338
  reassembly – 61, 76
  refitting – 76, 77
  removal – 68, 69
  rocker gear – 73
  separation from transmission – 69
  specifications – 31, 329
  start-up after overhaul – 77
  sump – 67
  timing chains and sprockets – 61, 72
  timing scale – 76
  undertray – 272
  valves – 59, 336
Exhaust system
  inspection and repair – 109
  manifold – 113
  removal and refitting – 110

F

Facia
  removal and refitting – 270
  switches – 289
Fan removal and refitting
  electric – 83
  thermoswitch – 83
  viscous coupled – 82
Fault diagnosis
  automatic transmission – 199
  braking system – 226, 228

Index

clutch – 163
cooling system – 94
electrical system – 293, 302
engine
    in-line – 57
    V6 – 77
fuel system – 114, 122
general – 23
heating and air conditioning – 95
ignition system – 157
manual gearbox – 191
overdrive – 191
propeller shaft – 204
rear axle – 208
steering and suspension – 246
Final drive and driveshafts – 353 *et seq*
Final drive and driveshafts
    description – 353
    driveshaft removal and refitting – 354
    final drive unit removal and refitting – 354
    lower section removal and refitting – 362
    maintenance and inspection – 336
    oil level checking – 354
    pinion oil seal renewal – 354
    side oil seals renewal – 354
    specifications – 332
Fixed glass removal and refitting – 256
Fluid
    automatic transmission
        level checking – 193
        renewal – 193
    braking system
        level checking – 211
        level warning switch – 355
    power steering
        level checking and bleeding – 233
Flywheel
    in-line engine
        examination and renovation – 51
        ring gear renewal – 42
    V6 engine
        examination and renovation – 72
        removal and refitting – 66
        ring gear renewal – 66
Flywheel sensors
    removal and refitting – 152, 351
Fuel and exhaust systems – 96 *et seq*, 341 *et seq*
Fuel and exhaust systems
    air cleaner – 99, 100, 341
    carburettor – 114 to 122, 341
    catalytic converter – 341
    constant idle speed system – 102, 341
    continuous pressure regulator ventilation – 341
    cruise control – 111
    EGR system – 343
    exhaust – 109, 110
    exhaust manifold – 113
    fault diagnosis – 114, 122, 131, 141
    fuel filter – 100
    fuel gauge – 107, 108, 343
    fuel injection – 123 to 131, 135 to 141, 341, 345 to 350
    fuel pumps – 105, 106, 343
    fuel tanks – 104, 344
    general description – 99, 345
    hot start valve – 122
    idle speed adjustment – 100, 336, 345
    inlet manifold – 112, 344
    maintenance and inspection – 99, 336
    mixture adjustment – 100, 336, 341, 345
    Pulsair system – 336, 341, 343
    specifications – 96, 330
    tamperproof adjustment screws – 99
    throttle cable – 108
    throttle pedal – 109
    turbocharger – 131 to 134, 350
    unleaded fuel – 99, 341

Fuel filter renewal – 100
Fuel gauge
    senders removal, testing and refitting – 107, 343
    testing – 108
Fuel injection systems
    air cleaner – 99, 100
    constant idle speed – 102
    continuous injection – 123 to 131, 341
    control pressure regulator ventilation – 341
    fault diagnosis – 114, 131
    fuel filter – 100
    fuel pump – 105, 106
    idle speed adjustment – 100, 336, 345
    LH-Jetronic – 345 to 350
    mixture adjustment – 100, 336, 345
    Motronic system – 135 to 141
    tamperproof screws – 99
    testing – 347
    turbocharger – 131 to 134, 350
Fuel pump removal and refitting
    main pump – 106
    tank pump – 105, 343
Fuel tank removal and refitting
    auxiliary tank – 104
    main tank – 104, 344
Fuses (general) – 291, 369

G

Gearbox *see* **Manual gearbox**
Gear lever (manual gearbox)
    pullrod renewal – 189
    removal and refitting – 188
Gear selector (automatic transmission)
    checking and adjustment – 193
General dimensions, weights and capacities – 6, 335
General repair procedures – 11
Glass removal and refitting
    door – 256
    fixed – 256
    tailgate – 256
    windscreen – 256
Glovebox removal and refitting – 268, 367
Grille removal and refitting – 272

H

Halfshaft, bearing and seals – 207
Handbrake
    adjustment – 213, 357
    cables – 223, 357
    shoes – 221, 357
    warning switch – 223
Headlights
    beam alignment – 287
    bulb renewal – 284
    dismantling and reassembly – 368
    removal and refitting – 286, 367
    wash/wipe – 293, 294, 295
Head restraints removal and refitting – 263
Heated rear window – 296
Heated seats – 266
Heater *see* **Cooling and heating system**
Horn
    removal and refitting – 293
    push switch removal and refitting – 289
Hot start valve testing – 122
Hub carrier (independent rear suspension)
    removal and refitting – 363
Hydraulic pipes and hoses – 213
Hydraulic system bleeding
    brakes – 214
    clutch – 161

I

Idle speed adjustment – 100, 336, 345
Ignition advance systems
    description and testing – 150
Ignition control unit
    removal and refitting – 152
Ignition/starter switch
    removal and refitting – 288
Ignition system 143 et seq, 351
Ignition system
    advance systems – 150
    ballast resistor – 155
    coil – 155
    control unit – 152, 153, 351
    distributor – 145, 147, 337, 351
    electronic traction control – 153
    engine management relays – 155
    fault diagnosis – 157
    flywheel sensor – 152, 351
    knock sensor – 143, 351
    maintenance and inspection – 144, 336
    Motronic system – 155
    power stage – 351
    spark plugs – 144, 336
    specifications – 142, 331
    switch – 288
    throttle position switch – 151
    timing – 149
    torque limiter system – 153
Ignition timing
    checking and adjustment – 149
Independent rear suspension
    axle member lower section – 362
    bushes renewal – 360
    description – 360
    lower link – 360
    shock absorber – 363
    specifications – 332
    spring – 363
    support arm – 362
    track rod – 362
    upper link – 362
    wheel alignment – 360
    wheel bearing – 363
Indicators see **Direction Indicators**
Inlet manifold removal and refitting – 112, 344
Input shaft (manual gearbox)
    dismantling – 174
    reassembly – 175
Instrument
    cluster
        dismantling and reassembly – 290
        removal and refitting – 290, 369
    fuse renewal – 369
    panel bulb renewal – 288
Intercooler removal and refitting – 134
Interference suppression (radio) – 298
Introduction to the Volvo 740 and 760 – 5

J

Jacking – 7

K

Kickdown (automatic transmission)
    cable
        adjustment – 194
        renewal – 195
    marker adjustment – 195
Knock sensor (EZ-K system)
    general
    removal and refitting – 152

L

Layshaft (manual gearbox)
    dismantling – 174
    reassembly – 176
LH-Jetronic fuel injection system
    component removal and refitting – 349
    description – 345
    injector check – 346
    idle speed adjustment – 345
    mixture adjustment – 345
    testing procedures – 347
Light bulbs see **Bulbs**
Locks
    central locking – 260, 365
    removal and refitting
        boot – 259
        door – 258, 365
        steering – 237
        tailgate – 259
Lower link (independent rear suspension)
    removal and refitting – 360
Lubricants and fluids – 21

M

Main bearings
    in-line engine
        examination and renovation – 48
        refitting – 53
        removal – 47
    V6 engine
        examination and renovation – 71
        refitting – 74, 338
        removal – 70, 338
Mainshaft (manual gearbox)
    bearings – 353
    dismantling – 174
    reassembly – 176
Maintenance and inspection see **Routine maintenance**
Manual gearbox, overdrive and auto transmission – 164 *et seq*, 356
Manual gearbox
    component examination and renovation – 175
    dismantling – 170, 173, 353
    fault diagnosis – 191
    gear lever – 188, 189
    general description – 166
    input shaft – 174, 175
    layshaft – 174, 176
    mainshaft – 174, 176, 353
    maintenance and inspection – 169
    oil seals – 169
    reassembly – 178, 179, 353
    removal and refitting – 169
    reversing light switch – 190
    specifications – 164, 332
Master cylinder
    overhaul
        brakes – 218
        clutch – 159
    removal and refitting
        brakes – 218
        clutch – 159
Mirror (door)
    glass and motor removal and refitting – 261
    removal and refitting – 261
    switch – 290
Mixture adjustment – 100, 336, 331, 345
Motronic system
    component removal and refitting – 139
    control unit removal and refitting – 153

Index

description – 135
fault diagnosis – 114, 141
testing procedures – 137, 155
Mountings removal and refitting
  in-line engine – 45
  V6 engine – 67, 338

N

Number plate light
  bulb renewal – 286

O

Oil coolers removal and refitting – 86, 337
Oil filter renewal
  in-line engine – 35
  V6 engine – 58
Oil level sensor (V6 engine)
  testing, removal and refitting – 77
Oil pump
  in-line engine
    examination and renovation – 48
    removal and refitting – 44
  V6 engine
    examination and renovation – 71
    removal and refitting – 63
Oil pressure switch removal and refitting
  in-line engine – 336
  V6 engine – 337
Oil seal renewal
  automatic transmission – 197
  distributor drive – 337
  engine
    in-line – 42
    V6 – 66, 337
  manual gearbox – 169
  rear axle/final drive – 206, 354
Overdrive
  description – 181, 353
  dismantling – 184
  fault diagnosis – 191
  overhaul – 187
  reassembly – 188
  removal and refitting – 184
  switch removal and refitting – 190, 197

P

Pads (brake) inspection and renewal
  front – 212, 355
  rear – 213, 355, 356
Panhard rod removal and refitting – 244
Parking lights
  bulb renewal – 284
Pedal removal and refitting
  accelerator – 109
  brake – 220
  clutch – 159
Pinion oil seal renewal – 206
Pistons
  in-line engine
    examination and renovation – 50
    removal and refitting – 44
  V6 engine
    examination and renovation – 72, 338
    refitting – 75
    removal – 70

Power stage (ignition)
  removal and refitting – 351
Power steering see **Steering**
Propeller shaft – 200 *et seq*, 353
Propeller shaft
  centre bearing – 203
  fault diagnosis – 204
  general description – 200
  maintenance and inspection – 200
  removal and refitting – 201, 353
  rubber coupling – 201
  specifications – 200
  universal joints – 201
Pressure differential warning valve
  overhaul – 219
  removal and refitting – 201
Pulsair system
  description – 341
  maintenance – 343
Pump (steering)
  overhaul – 235
  removal and refitting – 235

R

Rack bellows (steering) renewal – 234
Radiator
  grille – 272
  removal and refitting – 82
Radio/cassette player
  aerial – 296
  interference suppression – 298
  removal and refitting – 296, 367, 372
Radius rod removal and refitting – 239
Rear axle – 205 *et seq*
Rear axle see also **Final drive and driveshafts**
  bearing – 207
  fault diagnosis – 208
  general description – 205
  halfshaft – 207, 336
  maintenance and inspection – 205
  oil level checking – 206
  oil seal renewal – 206
  removal and refitting – 207
  specifications – 205
Rear light cluster
  bulb renewal – 284, 286
  removal and refitting – 286
Relays
  general – 292, 370
  ignition/engine management – 155
Repair procedures – 11
Reversing light switch removal and refitting
  automatic transmission – 196
  manual gearbox – 190
Roadwheels
  alignment – 233, 360
  bearings
    checking and adjustment – 233, 336
    removal, inspection and refitting – 238, 363
  general care and maintenance – 244, 336
  studs renewal – 244
Rocker gear (V6 engine)
  examination and renovation – 73
Routine maintenance
  automatic transmission – 192
  bodywork and underframe – 247
  braking system – 211, 336
  clutch – 158
  cooling, heating and air conditioning – 79, 88
  electrical system – 278, 279
  engine
    in-line – 35, 336
    V6 – 58, 336

final drive and driveshafts – 336
fuel and exhaust systems – 99, 336
general – 15
ignition system – 144, 336
maintenance intervals – 15, 336
manual gearbox and overdrive – 170
propeller shaft – 200
rear axle – 205
suspension and steering – 231, 336
upholstery and carpets – 248
Rubber coupling (propeller shaft)
removal and refitting – 201

S

Safety first! – 14
Seat belts
buckle light bulb renewal – 288
care and maintenance – 267
removal and refitting – 267
Seats
adjuster control panel removal and refitting – 365
adjusters removal and refitting – 264, 365
heating
control unit – 367
elements removal and refitting – 266
removal and refitting – 263
Self-levelling rear suspension – 244
Servo (brake)
checking operation – 213, 336
overhaul – 221
removal and refitting – 221
vacuum pump – 221
Shock absorber (rear)
removal and refitting – 243, 363
Shoes (handbrake)
removal, inspection and refitting – 221
Slave cylinder
overhaul
clutch – 160
removal and refitting
clutch – 160
Spark plugs
removal, inspection and refitting – 144, 336
Spark plug condition – *see colour pages between pages 32 and 33*
Spare parts – 9
Speedometer sender unit
removal and refitting – 291
Spoiler removal and refitting – 270
Spring (rear suspension)
removal and refitting – 243, 363
Starter inhibitor (automatic transmission)
removal and refitting – 196
Starter motor
overhaul – 281
removal and refitting – 281
testing – 280
Steering and suspension – 229 *et seq*, 359 *et seq*
Steering
bellows – 234
column removal and refitting – 236, 268, 359
column switches – 289, 369
fault diagnosis – 246
fluid – 233
general description – 230
lock – 237
maintenance and inspection – 231
overhaul – 235
pump
drivebelt – 81
overhaul – 235
removal and refitting – 235

rack bellows – 234
removal and refitting – 234
roadwheel
alignment – 233
bearings – 233, 238, 336
care and maintenance – 244, 336
studs – 244
specifications – 229, 332
track rod end – 234
wheel removal and refitting – 235
Stop-light switch – 223
Subframe and mountings (rear)
removal and refitting – 243
Sunroof
frame and cables removal and refitting – 275
motor removal and refitting – 273
removal and refitting – 273
Supplement: Revisions and information on later models – 325 *et seq*
Supplement: Revisions and information on later models
automatic transmission – 332
bodywork and fittings – 364 to 367
braking system – 332, 355 to 359
cooling and air conditioning systems – 330, 339 to 341
dimensions, weights and capacities – 335
electrical system – 334, 367 to 372
engine
in-line engine – 327, 329, 336
V6 engine – 329, 330, 336 to 338
final drive and driveshafts – 353 to 355
fuel and exhaust systems – 330, 341 to 350
ignition system – 331, 351 to 352
manual gearbox – 332, 333
overdrive – 332, 353
propeller shaft – 353
rear axle – 332
routine maintenance – 336
steering and suspension – 332, 359 to 363
Suspension
anti-roll bar – 240, 244
bushes – 242
control arm – 239, 359
fault diagnosis – 246
general description – 230
independent rear suspension – 360 to 363
maintenance and inspection – 231
Panhard rod – 244
radius rod – 239
rear subframe – 243
roadwheel
alignment – 233
bearings – 233, 238, 336
care and maintenance – 244, 336
studs – 244
self-levelling – 244
shock absorber – 243
specifications – 229, 332
spring – 243
strut – 240, 241, 359
subframe – 243
torque rods – 242
trailing arm – 243
Sump removal and refitting
in-line engine – 43
V6 engine – 67
Support arm (independent rear suspension)
removal and refitting – 362
Switches removal and refitting
central locking – 365
console – 289
door – 290
facia panel – 289
handbrake warning – 223
horn push – 289
ignition/starter – 288
mirror control – 290
oil pressure – 336, 337

Index

overdrive - 190, 197
rear console - 289
reversing light - 190, 196
starter inhibitor - 196
steering column - 289, 369
stop-light - 223
tailgate - 290
window - 290

T

Tailgate
  glass removal and refitting - 256
  interior trim removal and refitting - 256
  lock removal and refitting - 259
  removal and refitting - 252
  switches - 290
  wash/wipe - 293, 294, 295
Tamperproof adjustment screws (fuel system) - 99
Tappets examination and renovation - 51
Temperature gauge
  sender removal and refitting - 84
  testing - 85
Thermostat removal, testing and refitting - 84
Thermoswitch (electric fan)
  removal and refitting - 83
Throttle
  cable removal, refitting and adjustment - 108
  pedal removal and refitting - 109
Throttle position switch adjustment - 151
Timing chain and sprockets (V6 engine)
  examination and renovation - 72
  removal and refitting - 62
Timing (ignition)
  checking and adjustment - 149
Timing scale (V6 engine)
  checking and adjusting - 76
Tools and working facilities - 12
Torque rods removal and refitting - 242
Towing - 7
Track rod end
  removal and refitting - 233
Track rod (independent rear suspension)
  removal and refitting - 233
Traction control - 153
Trailing arm removal and refitting - 243
Turbocharger
  boost pressure checking and adjustment - 134
  description - 131, 350
  inspection and repair - 133
  intercooler removal and refitting - 135
  removal and refitting - 132
  precautions - 132
  wastegate actuator inspection and renewal - 134
Tyres
  general care and maintenance - 244
  specifications - 229

U

Undertray (engine) removal and refitting - 272
Universal joints (propeller shaft)
  overhaul - 201
Unleaded fuel - 99, 341
Upper link (independent rear suspension)
  removal and refitting - 362

V

Vacuum pump (braking system)
  overhaul - 221
  removal and refitting - 221
Valve clearances
  in-line engine - 36, 336
  V6 engine - 59, 336
Vanity mirror light bulb renewal - 369
Vehicle identification numbers - 9
Viscous coupled fan removal and refitting - 83

W

Wash/wipe - 293, 294, 295
Wastegate actuator
  inspection and renewal - 134
Water pump
  drivebelt - 81
  removal and refitting - 83, 339
Water valve (heater)
  removal and refitting - 92
Weatherstrip (door)
  removal and refitting - 261
Wheel changing - 7
Wheels see **Roadwheels**
Window lift
  mechanism removal and refitting - 257
  motor removal and refitting - 365
Window removal and refitting
  door - 256
  fixed glass - 256
Windscreen
  removal and refitting - 256
  wash/wipe - 293, 294, 372
Wing removal and refitting - 273
  wing liners - 367
Wipers
  blades and arms - 294
  motor and linkage - 294, 295, 372
Wiring diagrams - 304 to 325, 372 to 390
Working facilities - 12